Applied Probability and Statistics (*Continued*)

CHAKRAVARTI, LAHA and ROY · Handbook of Methods of Applied Statistics, Vol. I

CHAKRAVARTI, LAHA and ROY · Handbook of Methods of Applied Statistics, Vol. II

CHERNOFF and MOSES · Elementary Decision Theory

CHIANG · Introduction to Stochastic Processes in Biostatistics

CLELLAND, deCANI, BROWN, BURSK, and MURRAY · Basic Statistics with Business Applications

COCHRAN · Sampling Techniques, *Second Edition*

COCHRAN and COX · Experimental Designs, *Second Edition*

COX · Planning of Experiments

COX and MILLER · The Theory of Stochastic Processes

DAVID · Order Statistics

DEMING · Sample Design in Business Research

DODGE and ROMIG · Sampling Inspection Tables, *Second Edition*

DRAPER and SMITH · Applied Regression Analysis

GOLDBERGER · Econometric Theory

GUTTMAN and WILKS · Introductory Engineering Statistics

HALD · Statistical Tables and Formulas

HALD · Statistical Theory with Engineering Applications

HOEL · Elementary Statistics, *Second Edition*

HUANG · Regression and Econometric Methods

JOHNSON and LEONE · Statistics and Experimental Design: In Engineering and the Physical Sciences, Volumes I and II

LANCASTER · The Chi Squared Distribution

MILTON · Rank Order Probabilities: Two-Sample Normal Shift Alternatives

PRABHU · Queues and Inventories: A Study of Their Basic Stochastic Processes

SARHAN and GREENBERG · Contributions to Order Statistics

SEAL · Stochastic Theory of a Risk Business

WILLIAMS · Regression Analysis

WOLD and JURÉEN · Demand Analysis

WONNACOTT and WONNACOTT · Introduction to Econometric Methods

YOUDEN · Statistical Methods for Chemists

Tracts on Probability and Statistics

BILLINGSLEY · Ergodic Theory and Information

BILLINGSLEY · Convergence of Probability Measures

CRAMÉR and LEADBETTER · Stationary and Related Stochastic Processes

RIORDAN · Combinatorial Identities

TAKÁCS · Combinatorial Methods in the Theory of Stochastic Processes

Rank Order Probabilities

A WILEY PUBLICATION IN
APPLIED STATISTICS

Rank Order Probabilities

Two-Sample Normal Shift Alternatives

ROY C. MILTON
The University of Wisconsin

JOHN WILEY & SONS, INC.
New York · London · Sydney · Toronto

Library of Congress Catalogue Card Number: 73-100323

SBN 471   60700   2

Printed in the United States of America

Foreword

Substantial interest in the evaluation of probabilities of rank orders began in the early 1950's. The people most directly concerned were Robert Bechhofer, Z. W. Birnbaum, W. J. Dixon, Daniel Teichroew, and myself. In looking through my correspondence file from the National Bureau of Standards at that time, I see that these individuals were very much concerned about the preparation of tables which would facilitate the evaluation of the power functions of nonparametric procedures and which would help in the design of experiments for multiple-decision problems. Chapter 2 of this monograph covers the nonparametric applications that were in mind in the early 1950's, and Chapter 4 describes the applications related to the design of experiments. Thereafter, interest did not lag; see, for example, the work of J. Klotz (1964).

Consider the vector $\mathbf{x} = (x_1, \ldots, x_N)$. Associated with this vector is the vector of ranks, $\mathbf{r} = (r_1, \ldots, r_N)$. Here r_i is equal to the number of the x_j less than or equal to x_i. The symbols just defined will often be thought of as realizations of the random quantities $\mathbf{X} = (X_1, \ldots, X_N)$ and $\mathbf{R} = (R_1, \ldots, R_N)$. For many problems in statistics the basic quantities of interest are the random vectors, $\mathbf{R}$, and their possible values, $\mathbf{r}$. In particular, in the rank order analysis of data these are the natural quantities to be examined. Typically $\mathbf{r}$ is a vector of N integers, each of which is less than or equal to N in value. When the X's come from a distribution function with a density, the values for $\mathbf{r}$ having positive probability are the permutations of the first N integers. The first problem that one faces is, given the probabilistic structure of the vector $\mathbf{X}$, to determine the probabilities of the following form: $P\{\mathbf{R} = \mathbf{r}\}$. In general, the evaluation of these probabilities is extremely difficult, typically involving the evaluation of $(N-1)$-fold integrals.

In a few cases, however, $P\{\mathbf{R} = \mathbf{r}\}$ can be evaluated explicitly. For example, if the X's form an exchangeable process with a density function, each of the probabilities is equal to $1/N!$ If the random variables X_i are mutually independent and have a distribution function of the form F^{θ_i}, where F is continuous, the probabilities can be computed [see Savage (1956)]. A few other special cases can also be handled, such as those involving mutually independent random variables with uniform distributions over intervals of different lengths. It is not possible to evaluate without numerical integration the important situation where the random variables X_i have normal distributions. It would never be feasible to prepare tables of $P\{\mathbf{R} = \mathbf{r}\}$ for the case where the X's have an arbitrary multivariate normal distribution. For all but the smallest values of N the number of parameters involved would be excessive.

Consequently, we have focused our attention on the most important case in applied statistics—two samples differing in mean only. Under these circumstances the only parameters involved are the two sample sizes, m and n, and the ratio of the difference of the means divided by the standard deviation common to the two normal populations, d. Even so, tables of these probabilities are awkward. The difficulty is that the number of rank orders is essentially $\binom{N}{n}$, where $N = m + n$. For this reason, coupled with the large number of rank orders for each combination of sample sizes, as well as the different possible values for the difference in the mean, we obtain very large tables.

Chapter 1 describes in a substantial amount of detail the method of construction of the tables, their accuracy, and related tables. In Chapter 2 the tables are used to ascertain the power function of some standard non-parametric tests, which was the major motivation for the preparation of these tables. Chapter 3 presents tables necessary to facilitate sequential probability ratio procedures based on rank data when sampling from normal populations. The preparation of the material for this chapter was not planned at the outset but proved easy as a result of the preparation of the basic tables in Chapter 1. Chapter 4 goes on to give more details about the problems of multiple selection than were originally discussed in the early 1950's. Again, the basic nature of the tables in Chapter 1 allowed a more extensive development here than had been previously anticipated. Chapter 5 contains some specialized results and uses values for $P\{\mathbf{R} = \mathbf{r}\}$ that were not originally obtained in Chapter 1. The methods of computing developed in Chapter 1, however, made it possible to perform the computations for Chapter 5. In Chapter 5 an asymptotic formula of Hodges and Lehmann (1962) is compared with exact numerical results.

At this time, of course, it is difficult to think of just what other applications these basic tables will have. Some thought has suggested three possibilities. (1) For a given rank order and for varying values of the difference in the two means, one can search for the maximum likelihood estimate of that difference. In this case the maximum likelihood estimate is based on the ranks only. Although the number of different values for the means which were used for tabulation purposes is not very large, a few rough checks suggest that the search for the maximum can be done fairly effectively. (2) In Chapter 1, Theorem 6.1 has a modification of some interest. In particular, it is shown there that under certain circumstances a difference is positive. It has been conjectured by Savage and Saxena that the ratio of the two quantities is actually increasing in the difference in the means, d. The tabled values support the conjecture. (3) An examination of formula (2.1) in Chapter 1 shows that the quantity which has been tabulated is closely related to the moment-generating function of linear functions of order statistics from the normal distribution. Whether or not tabulations are done finely enough so that partial knowledge of the moment-generating function would help extensively in finding out about the distribution function is not at all clear. But the fact that the relationship exists and that partial tabulations have now begun seems encouraging. It is my impression that these linear functions of order statistics do have a use in statistical practice. The unanticipated (at the time these computing projects were begun) results in Chapters 3, 4, and 5, as well as the points just mentioned, constitute a strong argument for the importance of the effort of preparing the basic tables of Chapter 1.

In looking over the correspondence that I carried on in the early 50's regarding this problem, it is clear, even to a nonexpert like myself, that the development of numerical analysis and computing methodology changed what was essentially an impossible computational chore into a task which could be done with a reasonable expenditure of funds. This is one of the most direct observations that I have on the growth of our ability to compute. It is my guess that future generations will not wish to extend the tables as they are presented in this monograph. Rather I predict that, as computing becomes easier and less expensive to perform, people will prepare specialized tables of functions of the kind which have been tackled here. No doubt more difficult problems will eventually be attacked, such as those involving dependent random variables.

I am thankful to Roy Milton for having finished a task that quite a few of us had thought about and had considered almost impossible.

I. RICHARD SAVAGE

Tallahassee, Florida
May 1969

Acknowledgments

This work was begun as a doctoral dissertation at the University of Minnesota in June 1963 at the suggestion and under the guidance of Professor I. Richard Savage. Upon Professor Savage's departure to Florida State University, Professor Charles Kraft joined him as my co-advisor. Chapter 1 and Table A were completed in August 1964 after my departure to the Atomic Bomb Casualty Commission (ABCC), Hiroshima, Japan. The dissertation was completed in June, 1965, and it has also appeared as Technical Report No. 53 (January, 1965) and No. 53a (January, 1966) of the Department of Statistics, University of Minnesota.

The final work was completed under the joint advisorship of Professors Savage, Kraft, and Milton Sobel. All advisors have contributed much toward the completion of this research. To Professor Savage I am deeply indebted for the formulation of the problem and for his continued interest, encouragement, guidance, and generous assistance through comment and criticism. I am very grateful to Professor Kraft for his suggestions and valuable contributions upon accepting co-advisorship during the final stages of preparation of Chapter 1. I am also grateful to Professor Milton Sobel for the experience I acquired, both in related problems and in computer work, while assisting him in earlier research, which has aided in the completion of the present work. Professor Max Engeli's suggestion, to consider the use of "extrapolation to the limit" to improve the accuracy of the computations, was vital to the successful completion of the entire work.

Computations were done on the University of Minnesota's CDC 1604 computer and the Atomic Bomb Casualty Commission's IBM 1440 computer. Both the Numerical Analysis Center (Minnesota) and ABCC were extremely generous in aiding this research through use of their facilities. The research was supported also by Office of Naval Research contract Nonr-710(31), NR 042-003, and by a grant from the Society of the Sigma Xi and RESA Research Fund.

Upon my return from Japan in 1966, I was encouraged by Professor Savage to publish this work. A preliminary version of the material in Chapter 3 was presented at the 36th Session of the International Statistical Institute in Sydney, Australia, in August, 1967, with abstract in the *Bulletin of the I.S.I.*, Volume 42. The material in Chapter 5 appeared in the *Annals of Mathematical Statistics*, Vol. 38, No. 5, October, 1967 (pp. 1401–1403), and is included here by permission of the editor of that journal. In 1968 I began to investigate the possibility of automatic typesetting of mathematical tables from computer tape, and the final results are seen in this volume.

I am indebted to the University of Wisconsin Computing Center for the opportunity to pursue publication and for the facilities supporting this effort. Computations on the CDC 3600 computer, necessary to reorganize and recode the tables into a form suitable for input to the phototypesetting system, were made possible through support, in part, from the National Science Foundation, other United States government agencies, and the Wisconsin Alumni Research Foundation through the University of Wisconsin Research Committee. The facilities of Madison Campus Administrative Data Processing, made generously available to me, were also helpful in this effort.

The final typing of the manuscript was expertly done by Mrs. Diana Webster.

Roy C. Milton

Madison, Wisconsin
August 1969

Contents

CHAPTER 1 Rank Order Probabilities: Two-Sample Normal
           Shift Alternatives   1

 1. Introduction   1
 2. Notation   2
 3. Description of Table A   3
 4. Related tables   5
 5. Method of computation   6
 6. Accuracy of computations   13
 7. Availability of Table A for mechanical or electronic
    processing   19

CHAPTER 2 Small-Sample Power of Two-Sample Nonparametric
           Tests Against the Normal Shift Alternative   21

 1. Summary   21
 2. Introduction   22
 3. Power of the Wilcoxon two-sample rank-sum test   23
 4. Power of the Terry-Hoeffding-Fisher-Yates two-sample
    normal scores (c_1) test   25
 5. Power of the Mood-Brown two-sample median test   26
 6. Power of the Kolmogorov-Smirnov two-sample test   27
 7. Most powerful two-sample test based on ranks   27
       Table 1   Power relationship among MPRT, Wilcoxon,
       and c_1 tests   30
 8. Selected power comparisons   28
       Table 2   Comparison of power of two-sample tests
       (one-sided)   32
       Table 3   Effect of change in sample size on power
       of two-sample tests (one-sided)   34

xi

Table 4   Comparison of the power of one- and
two-sided two-sample tests   36

9. Selected efficiency comparisons   29
Table 5   Hodges-Lehmann efficiency of the Wilcoxon test   37
Table 6   Hodges-Lehmann efficiency of the c_1 test   38
Table 7   Hodges-Lehmann efficiency of the median test   39
Table 8   Hodges-Lehmann efficiency of the
Kolmogorov-Smirnov test   40
Table 9   Hodges-Lehmann efficiency of the Wilcoxon
test: unequal sample sizes   43

CHAPTER 3   Sequential Two-Sample Rank Tests of the Normal
Shift Hypothesis

1. Summary   43
2. Introduction   43
3. The sequential tests   44
4. Computation of r_t 46
5. Properties of the sequential tests   46
Table 1   Values of d' such that $\mathscr{E}_{d'}(\log p_t) = 0$ 47
Table 2   Values of ASN function converted to total
sample sizes for the rank-sum test   48
Table 3   Values of $\mathscr{E}_{d=0}(\log p_t)$, $\mathscr{E}_{d=d'}(\log^2 p_t)$, and
$\mathscr{E}_{d=d_1}(\log p_t)$ for the rank-sum test   52
Table 4   Sample size required for one-sided nonsequential
t test, and optimum ASN for sequential rank-sum test,
for normal shift alternative d_1 56
6. Other sequential rank tests   56

CHAPTER 4   Rank Order Probabilities and Selection Procedures   58

1. Introduction   58
2. Goals and procedures   58
3. A table for use in selection procedures   60

CHAPTER 5   Exact Probabilities of Rank Orders for Two Widely
Separated Normal Distributions   61
Table 1 $P(\mathbf{z}^0 \mid d)$ and $P(\mathbf{z}^1 \mid d)$ for large d 62
References   67
Symbols   71

TABLES
Table A   Rank Order Probabilities $P_{m,n}(\mathbf{z} \mid d)$ 75

Table B-1   Power of one-sided two-sample Wilcoxon test vs.
              normal shift alternative d 219
Table B-2   Power of two-sided two-sample Wilcoxon test vs.
              normal shift alternative d 224
Table B-3   Power of one-sided two-sample normal scores
              test vs. normal shift alternative d 228
Table B-4   Power of two-sided two-sample normal scores
              test vs. normal shift alternative d 233
Table B-5   Power of one-sided two-sample median test vs.
              normal shift alternative d 237
Table B-6   Power of two-sided two-sample median test vs.
              normal shift alternative d 239
Table B-7   Power of one-sided two-sample Kolmogorov-Smirnov
              test vs. normal shift alternative d 241
Table B-8   Power of two-sided two-sample Kolmogorov-Smirnov
              test vs. normal shift alternative d 245
Table B-9   Power of one-sided two-sample Student t test vs.
              normal shift alternative d 248
Table B-10   Power of two-sided two-sample Student t test vs.
              normal shift alternative d 253
Table C-1   Distribution of Wilcoxon two-sample statistic
              under normal shift alternative d 259
Table C-2   Log $R(w \mid d)$, the common logarithm of the rank-sum
              likelihood ratio under normal shift alternative d 275
Table D $P(n'; r, m, n, d)$ 292

Index   301

Rank Order Probabilities

CHAPTER 1

Rank Order Probabilities:
Two-Sample Normal Shift Alternatives

1. INTRODUCTION

Consider the situation where random variables $X_1, \ldots, X_m$ and $Y_1, \ldots, Y_n$ are normally distributed with means μ_X and μ_Y, respectively, and common variance σ^2, all $m + n$ random variables being mutually independent and $d = (\mu_Y - \mu_X)/\sigma$. Let $\mathbf{U} = (U_1, \ldots, U_{m+n})$, $U_1 < \cdots < U_{m+n}$, denote the order statistics of the random variables $(X_1, \ldots, X_m, Y_1, \ldots, Y_n)$, and let $\mathbf{Z} = (Z_1, \ldots, Z_{m+n})$ denote a random vector of zeros and ones, where the ith component Z_i is 0 (or 1) if U_i is an X (or Y). Denote by $f(x - \theta)$ the normal density with mean θ and variance 1, $(2\pi)^{-\frac{1}{2}} \exp\left[-(x - \theta)^2/2\right]$. If $\mathbf{z} = (z_1, \ldots, z_{m+n})$ is a fixed vector of zeros and ones, the probability of the rank order $\mathbf{z}$, $\Pr\{\mathbf{Z} = \mathbf{z}\}$, is given by

$$(1.1) \qquad P_{m,n}(\mathbf{z} \mid d) = m! \, n! \int \cdots \int_R \prod_{i=1}^{m+n} f(t_i - z_i d) \, dt_i,$$

where the region of integration R is $-\infty < t_1 \le t_2 \le \cdots \le t_{m+n} < \infty$. In this chapter a method for computing $P_{m,n}(\mathbf{z} \mid d)$ is described, and values of $P_{m,n}(\mathbf{z} \mid d)$ to 9 decimal places are presented in tabular form for all $\mathbf{z}$ for $1 \le n \le m \le 7$ and $n = 1, m = 8(1)12; d = .2(.2)1.0, 1.5, 2.0, 3.0$.

Tables of $P_{m,n}(\mathbf{z} \mid d)$ are useful in statistical investigations concerning, for example, (1) the small-sample exact power of two-sample nonparametric tests for location against normal alternatives, (2) the construction of the most powerful rank test for location with respect to normal alternatives, (3) sequential two-sample procedures based on ranks and on rank sums with respect to normal alternatives, (4) certain multiple decision or ranking procedures, and (5) the admissibility of two-sample rank order tests [see, for example, Bradley (1967); Bradley, Martin, and Wilcoxon (1965); Bradley,

1

Merchant, and Wilcoxon (1966); Klotz (1964); Savage, Sobel, and Woodworth (1966); Wilcoxon, Rhodes, and Bradley (1963)].

2. NOTATION

The notation presented here follows in part that used by Savage, Sobel, and Woodworth (1964).

According to the particular usage, $P_{m,n}(\mathbf{z} \mid d)$ will be also denoted variously by $P_{m,n}(\mathbf{z})$, $P(\mathbf{z} \mid d)$, and $P(\mathbf{z})$.

The *complement of* $\mathbf{z}$, $\mathbf{z}^c = (z_1^c, \ldots, z_{m+n}^c)$, is defined to be the vector whose ith component is $z_i^c = 1 - z_i$; the *transpose of* $\mathbf{z}$, $\mathbf{z}^t = (z_1^t, \ldots, z_{m+n}^t)$, is defined to be the vector whose ith component is $z_i^t = z_{m+n+1-i}$.

Several useful relationships that follow immediately from the symmetry and translation properties of the normal density, that is, $f(x) = f(-x)$ and $f(x - d) = f[(x - d) - 0]$, are included here without proof in the form given by Savage, Sobel, and Woodworth (1964):

$$\text{(i) } P(\mathbf{z} \mid d) = P(\mathbf{z}^t \mid -d);$$

$$\text{(ii) } P(\mathbf{z} \mid d) = P(\mathbf{z}^c \mid -d);$$

$$\text{(iii) } P(\mathbf{z} \mid d) = P(\mathbf{z}^{tc} \mid d) = P(\mathbf{z}^{ct} \mid d).$$

Note that $\mathbf{z}^{tc} = (\mathbf{z}^t)^c = (\mathbf{z}^c)^t$ and $(\mathbf{z}^{tc})^{tc} = \mathbf{z}$.

By Theorem 8 [Savage and Sobel (1963), p. 24]

$$(2.1) \qquad P(\mathbf{z} \mid d) = \binom{m + n}{n}^{-1} e^{-nd^2/2} \mathscr{E} \exp\left(d \sum_{i=1}^{m+n} z_i V_i\right),$$

where $V_1, \ldots, V_{m+n}$ are the order statistics of a sample of size $m + n$ drawn from a standard normal population and $\mathscr{E}$ denotes expectation. The first terms in the Maclaurin expansion of $P(\mathbf{z} \mid d)$ yield approximations to $P(\mathbf{z} \mid d)$, for small values of d. Differentiating the right-hand side of (2.1) twice with respect to d and evaluating the first and the second derivatives at $d = 0$ gives the two-term approximation

$$(2.2) \qquad P(\mathbf{z} \mid d) \doteq \binom{m + n}{n}^{-1} \left\{1 + d\mathscr{E}\left(\sum_{i=1}^{m+n} z_i V_i\right) + \frac{d^2}{2}\left[\mathscr{E}\left(\sum_{i=1}^{m+n} z_i V_i\right)^2 - n\right]\right\}$$

$$= \binom{m + n}{n}^{-1} + d(c_1) + \frac{d^2}{2}(c_2),$$

where

$$c_1 = \left(\mathscr{E} \sum_{i=1}^{m+n} z_i V_i\right)\binom{m + n}{n}^{-1} \quad \text{and} \quad c_2 = \left[\mathscr{E}\left(\sum_{i=1}^{m+n} z_i V_i\right)^2 - n\right]\binom{m + n}{n}^{-1}.$$

3. DESCRIPTION OF TABLE A

Table A contains 9-decimal-place values of $P_{m,n}(\mathbf{z} \mid d)$ for all possible $\mathbf{z}$ for $1 \leq n \leq m \leq 7$ and $n = 1, m = 8(1)12; d = .2(.2)1, 1.5, 2, 3$. In checking these tables no error larger than 10^{-8} was found. Nine places are given since many values are known to be accurate to 9 places; furthermore, the ninth place serves to facilitate ordering of the values of $P(\mathbf{z})$ when these values have 7 or 8 leading zeros (a situation in which the ninth place was found to be correct wherever a check was made).

The rank order values of $\mathbf{z}$ are ordered according to the values c_1 and c_2 associated with each $\mathbf{z}$, as follows:

(i) The $\mathbf{z}$'s are ordered according to decreasing order of their c_1 values.

(ii) In case of ties in c_1 values, ordering is done according to decreasing order of the respective c_2 values; no ties occurred in c_2 values. The quantities c_1 and c_2 are as in (2.2); if they are multiplied by $\binom{m+n}{n}$ they become the c_1 and c_2 statistics of Terry (1952). The tabled values of c_1 and c_2 were computed from Teichroew's tables (1956) of expected values of order statistics and products of order statistics and are given to 7 decimal places. The column headed w contains the value of the Wilcoxon (1945) two-sample statistic for each $\mathbf{z}$; that is,

$$w = \sum_{i=1}^{m+n} i z_i,$$

where the z_i are the elements of the vector $\mathbf{z} = (z_1, \ldots, z_{m+n})$.

The right-most column contains values of $\mathbf{z}^{tc}$ corresponding to $\mathbf{z}$ appearing in the left-most column. For the case $m = n$, the symbol $\times$ following a value in the $\mathbf{z}^{tc}$ column indicates that $\mathbf{z} = \mathbf{z}^{tc}$ for that particular $\mathbf{z}$ value. Also, in this case, identical rows of the table are eliminated when $\mathbf{z} \neq \mathbf{z}^{tc}$ by omitting from the column labeled $\mathbf{z}$ half of the $\mathbf{z}$ values that also appear as $\mathbf{z}^{tc}$ values. For example, for $m = n = 2$, both $\mathbf{z}' = (0110)$ and $\mathbf{z}'' = (1001)$ would normally appear in column $\mathbf{z}$. However, since $\mathbf{z}' = (\mathbf{z}'')^{tc}$ and $P(\mathbf{z}) = P(\mathbf{z}^{tc})$, $\mathbf{z}''$ appears only in column $\mathbf{z}^{tc}$. The number K of unordered pairs $(\mathbf{z}, \mathbf{z}^{tc})$ for which $\mathbf{z} \neq \mathbf{z}^{tc}$ is seen, using a result of Savage (1964), to be

(3.1)
$$K = \frac{\binom{2Q}{Q}}{2} - 2^{Q-1},$$

where $m = n = Q$. The total number of rank orders, for a given m and n, is denoted in the table as

$$(3.2) \qquad\qquad C(m, n) = \binom{m + n}{n}.$$

Note that $P_{m,n}(\mathbf{z} \mid 0) = 1/C(m, n)$. For $m \neq n$, the number of distinct rank orders (i.e., rank orders whose probabilities are not equal) is $C(m, n)$; for $m = n = Q$, this number is equal to

$$(3.3) \qquad\qquad C(m, n) - K = \frac{\binom{2Q}{Q}}{2} + 2^{Q-1}.$$

Example. For $m = n = 7$, we have $C(7, 7) = 3432$, $K = 1652$, and the number of distinct rank orders is 1780.

The sections of Table A are arranged according to increasing values of $m + n$, from $2 \leq m + n \leq 14$. The values of $P_{m,n}(\mathbf{z})$ for a given small value of $m + n$ normally appear on one page. Values for large $m + n$ are listed on successive pages, and the heading of each page indicates, by a phrase such as "Rank Orders 101 thru 150 of 210" (an example from $m = 6, n = 4$), what portion of the table appears on that page.

Finally, a bullet ($\bullet$) notation is employed in the body of $P(\mathbf{z})$ values to mark "crossovers" to assist in analysis of the ordering of probabilities of rank orders [see Savage, Sobel, and Woodworth (1966)]. When, for two successive adjacent $\mathbf{z}$ values, say $\mathbf{z}$ and $\mathbf{z}'$, it is observed that $P(\mathbf{z} \mid d_1) \geq P(\mathbf{z}' \mid d_1)$ but $P(\mathbf{z} \mid d_2) < P(\mathbf{z}' \mid d_2)$, where $0 \leq d_1 < d_2$ and

(i) $P(\mathbf{z} \mid 0) = 1/C(m, n)$ and
(ii) d_1 and d_2 are consecutive tabular values of d, $d_1 = 0$ permitted,

then $P(\mathbf{z}' \mid d_2)$ is followed by a bullet.

Example (from $m = 5, n = 2$).

z	$d = 1.0$	$d = 1.5$	$d = 2.0$ $\cdots$
.	.	.	.
.	.	.	.
.	.	.	.
0010010 $\cdots$	.036963626	.019928955	.007763526 $\cdots$
0001100 $\cdots$	.036581623	.020281040$\bullet$	.008326564 $\cdots$
.	.	.	.
.	.	.	.
.	.	.	.

No examples of the following form were found: $P(\mathbf{z} \mid d_1) > P(\mathbf{z}' \mid d_1)$, $P(\mathbf{z} \mid d_2) < P(\mathbf{z}' \mid d_2)$, and $P(\mathbf{z} \mid d_3) > P(\mathbf{z}' \mid d_3)$ for $0 \leq d_1 < d_2 < d_3$.

4. RELATED TABLES

a. Klotz (1961) has computed $P_{m,n}(\mathbf{z} \mid d)$ for $1 \leq m$, $n \leq 9$, such that $m + n \leq 10$, and $d = .25, .5(.5)2.0(1.0)6.0$, with results reported to 8 decimals. Comparison of his results with Table A herein and routine inspection of certain values in Klotz's table indicate that his table varies in accuracy from 8 decimals to 4 decimals (the latter for higher-dimensional integrals). In particular, Klotz gives $P(\mathbf{z})$ for $\mathbf{z} = (1, \ldots, 1)$, whose correct value is 1 but whose tabled value differs from 1 by as much as .0001 in some cases. In addition, the sum of all possible rank order probabilities for given m, n, and d was found to differ from 1 by as much as .0001.

b. Teichroew (1954, 1955b) has given 9- and 6-decimal-place tables of

$$(4.1) \qquad G(\delta; \alpha, \beta) = (\beta + 1) \int_{-\infty}^{\infty} F^{\alpha}(x + \delta)[1 - F(x)]^{\beta} f(x)\ dx,$$

where $F(x)$ is the standard normal distribution function, for $\beta = 0(1)4$, $\alpha = \beta'(1)9$ with $\beta' = \beta + 1$, and $\delta = -3.2(.1).00(.01)6.40$. In the notation of this paper, (4.1) may be written as

$$(4.2) \quad G(\delta; \alpha, \beta) = \begin{cases} P_{\alpha, \beta + 1}(\mathbf{z} \mid \delta) & \text{for } \delta > 0, \mathbf{z} = (\overbrace{0 \cdots 0}^{\alpha}\overbrace{1 \cdots 1}^{\beta+1}), \\ P_{\alpha, \beta + 1}(\mathbf{z} \mid -\delta) & \text{for } \delta < 0, \mathbf{z} = (\overbrace{1 \cdots 1}^{\beta+1}\overbrace{0 \cdots 0}^{\alpha}). \end{cases}$$

The 9-place tables are introduced by a detailed description of the method of calculation and are considered accurate to within 1 unit in the ninth decimal place.

c. Gupta (1963) and Milton (1963) give 5- and 8-decimal-place tables, respectively, of (in the notation of Milton)

$$(4.3) \qquad F(H; K, \rho) = \int_{-\infty}^{\infty} F^{K-1}\left(\frac{x\sqrt{\rho} + H}{\sqrt{1 - \rho}}\right) f(x)\ dx.$$

In the notation of this paper

$$F(d/\sqrt{2}; K, \tfrac{1}{2}) = P_{K-1,1}(\mathbf{z} \mid d),$$

where

$$\mathbf{z} = (\overbrace{0 \cdots 0 1}^{K-1}).$$

Gupta's table provides values for $H = -3.5(.10)3.5$, $K = 2(1)13$. Milton's table is for $H = .00(.05)5.15$, $K = 3(1)10(5)25$, and includes aids for interpolation. Both tables are considered to be accurate to the number of decimal places given within 1 unit in the last place.

5. METHOD OF COMPUTATION

a. Attempts at Use of Composite Quadrature Formulas

For the present purpose, a composite quadrature formula is defined to be a formula for approximation of a k-fold, or k-dimensional, integral based upon repeated application of a one-dimensional quadrature formula. For example, suppose that it is desired to approximate the integral

$$\iint_R g(x_1, x_2)\, dx_1\, dx_2, \quad R = \{a_1 \leq x_1 \leq b_1, a_2 \leq x_2 \leq b_2\}$$

by means of a particular one-dimensional formula

$$(5.1) \qquad \int_a^b h(y)\, dy \doteq \sum_{i=1}^n w_i h(y_i),$$

where the weights w_i and the abscissas y_i depend on a, b, and n. A typical composite formula based on (5.1) is

$$(5.2) \qquad \iint_R g(x_1, x_2)\, dx_1\, dx_2 \doteq \sum_{j=1}^n w_j \left[\sum_{i=1}^n w_i g(x_{1i}, x_{2j}) \right],$$

where

$$x_{1i} = \frac{(b_1 - a_1)(y_i - a)}{(b - a)} + a_1 \quad \text{and} \quad x_{2j} = \frac{(b_2 - a_2)(y_j - a)}{(b - a)} + a_2$$

for $i, j = 1, \ldots, n$. This method may be readily modified for integration over nonrectangular regions or regions with variable limits.

It is clear that the dimensionality of the integral $P(\mathbf{z} \mid d)$ in (1.1) can be reduced at least by a factor of 2 through the device of what might be called conditional integration; that is, regard a subset S of the variables of integration as fixed, integrate over the remaining variables, and then integrate with respect to the variables in S. One example of this is given in Section 4.b. A more general example is the following:

$$(5.3) \quad P(\overbrace{0 \cdots 0}^{\alpha}\overbrace{1 \cdots 1}^{\beta}\overbrace{0 \cdots 0}^{\gamma}\overbrace{1 \cdots 1}^{\delta})$$

$$= \frac{(\alpha + \gamma)! \, (\beta + \delta)!}{\alpha! \, (\beta - 2)! \, \gamma! \, (\delta - 1)!} \iiint_{-\infty < x \leq y \leq z < \infty} F^\alpha(x + d)[F(y) - F(x)]^{\beta - 2}$$

$$\cdot [F(z + d) - F(y + d)]^\gamma \cdot [1 - F(z)]^{\delta - 1} f(x) f(y) f(z)\, dx\, dy\, dz.$$

In this example a $(\alpha + \beta + \gamma + \delta)$-fold integral is written as a triple integral in terms of f and F. When composite integration formulas are used for k-fold integration, $k > 1$, the reduction of dimensionality is usually desirable. However, it was determined that this approach was not practical for evaluation of $P(\mathbf{z})$ because composite integration formulas used for three-fold integrals of the type in (5.3) took excessive computer time (at least one-half minute for 4- to 5-decimal-place accuracy) and computing time increased exponentially as the dimensionality of the integral increased. Evaluation of a seven-fold integral with minimal accuracy by the use of this method was estimated to require at least one-half hour. These considerations of computing time are based on the CDC 1604 computer which, for 36-bit mantissas, performs floating point operations in the following times: addition, 18.8 microseconds; multiplication, 36.0 microseconds; division, 56.0 microseconds.

b. A Modified Composite Midpoint Quadrature for Evaluation of $P_{m,n}(\mathbf{z})$

The method presented for evaluating rank order probability integrals of the type similar to (1.1) involves integration over a region R' (a truncation of the region R) by an efficient modified composite midpoint algorithm, followed by extrapolation to the limit. A typical composite quadrature formula for evaluation of a p-dimensional integral with N evaluation points per axis (dimension) requires performing on the order of N^p arithmetic operations. However, it should be noted that the algorithm used here for evaluation of (1.1) requires approximately Np arithmetic operations, and thus computing time increases linearly with N, making application of the algorithm practical.

First, it is useful to recall that in elementary calculus a one-dimensional definite integral is often defined as

$$(5.4) \qquad \int_a^b g(x)\, dx = \lim_{\substack{n \to \infty \\ \max\{\Delta x_i\} \to 0 \\ i}} \left[\sum_{i=1}^n g(x_i')\, \Delta x_i \right]$$

if the limit exists [i.e., when $g(x)$ is bounded and continuous almost everywhere], where $[a, b]$ is partitioned in n subintervals, x_i' is an arbitrary point in the ith subinterval, and Δx_i is the length of the ith subinterval. The midpoint approximation to $\int_a^b g(x)\, dx$, a natural application of (5.4), is

$$(5.5) \qquad \int_a^b g(x)\, dx \doteq h \sum_{i=1}^n g(x_i),$$

where the interval $[a, b]$ is divided into n subintervals each of length h, and x_i is the midpoint of the ith subinterval. For multidimensional integrals,

(5.4) and (5.5) may be extended to give an analogous approximation

$$(5.6) \quad \int \cdots \int_R g(x_1, \ldots, x_p) \, dx_1 \cdots dx_p$$

$$\doteq \left(\prod_{j=1}^{p} h_j \right) \left[\sum_{i_1=1}^{n_1} \cdots \sum_{i_p=1}^{n_p} g(x_{1 i_1}, \ldots, x_{p i_p}) \right],$$

where R is some rectangular region in p dimensions, and the interval of integration of the jth dimension is divided into n_j equal subintervals of length h_j with midpoint x_{ji_j} of the ith interval. Extension to general regions of integration is possible.

This relatively simple approximation often converges rather slowly (as the partition is increased) to the true value of the integral. Nevertheless, in cases where values of the integrand for large partitions are easy to obtain and convenient to use, the approximation has merit because of its simplicity.

The modified method of midpoint quadrature described here utilizes both the structure of the region of integration and the factoring of the integrand (the latter results from the statistical independence of the observations). The general p-dimensional form of the integral to be evaluated is

$$(5.7) \qquad I_p = \int \cdots \int_{R_p} \prod_{i=1}^{p} f_i(x_i) \, dx_i,$$

where R_p is the region $a \le x_1 \le \cdots \le x_p \le b$, and the functions $f_i(\cdot)$ are continuous in the interval $[a, b]$. [In this particular application, the $f_i(\cdot)$ are normal density functions.]

As an example let us first evaluate

$$I_2 = \int\int_{R_2} f_1(x_1) f_2(x_2) \, dx_1 \, dx_2,$$

where R_2 is the shaded area of Figure 1, by the modified midpoint method as follows. We divide the interval $[a, b]$ into, say, $N = 3$ equal subintervals of length h on each axis, and denote the values of f_1 and f_2 at the midpoint of the ith interval by f_{1i} and f_{2i}, respectively. The integrand at each midpoint is weighted by the area of the subregion of integration (rectangle or triangle) that includes the midpoint. Weighting the values that occur on the diagonal boundary of the region R_2 by $\frac{1}{2}$, we obtain as a natural midpoint approximation to I_2 (see Figure 1)

$$(5.8) \qquad I_2 \doteq h^2 \left(\frac{f_{11} f_{21} + f_{12} f_{22} + f_{13} f_{23}}{2} + f_{11} f_{22} + f_{11} f_{23} + f_{12} f_{23} \right),$$

where $h = (b - a)/3$.

Expression (5.8) may be written as

$$(5.9) \qquad I_2 \doteq h^2 \left(\sum_{i=1}^{N} \frac{f_{1i}f_{2i}}{2!} + \sum_{1 \le i < j \le N} f_{1i}f_{2j} \right)$$

for $N = 3 = (b - a)/h$. Similarly,

$$(5.10) \qquad I_3 = \iiint_{R_3} \prod_{i=1}^{3} f_i(x_i) \, dx_i$$

$$\doteq h^3 \left(\sum_{i=1}^{N} \frac{f_{1i}f_{2i}f_{3i}}{3!} + \sum_{1 \le i < j \le N} \frac{f_{1i}f_{2j}f_{3j} + f_{1i}f_{2i}f_{3j}}{2!} \right.$$

$$\left. + \sum_{1 \le i < j < k \le N} f_{1i}f_{2j}f_{3k} \right).$$

The terms whose members contain two equal second subscripts have denominator 2!, and those with three equal second subscripts have 3! as denominator. These divisors are weighting factors applied to compensate for including portions of squares and cubes on the edges of the region of integration.

Generalization to higher dimensions is immediate. Terms whose k members all have the same second subscripts have a multiplicative weighting factor $1/k!$ that is the proportion of the unit k-dimensional hypercube in the region $0 \le x_1 \le x_2 \le \cdots \le x_k \le 1$. Grouped sets of second subscripts, indicating subregions on lower-dimensional edges of the region of integration, are weighted in a geometrically analogous manner: $f_{11}f_{21}f_{31}f_{42}f_{52}f_{63}$

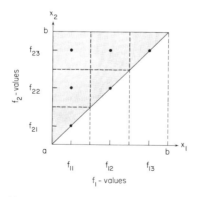

Figure 1 Region of integration and points at which the integrand is evaluated, for R_2.

has weight $1/(3!\,2!\,1!)$. We now write

(5.11)
$$I_p \doteq h^p \sum_{\alpha} \prod_{j=1}^{p} \frac{f_{ji(\alpha)}}{\pi_{\alpha}}, \quad \alpha \in S,$$

where S is the set of all possible inequalities of the form

$$1 \leq i_1 \mathcal{R} i_2 \mathcal{R} \cdots \mathcal{R} i_p \leq N;$$

the relation $\mathcal{R}$ is either " $<$ " or " $=$ ", and π_{α} is the product of factorials such that each string of $k - 1$ adjacent " $=$ " relations contributes $k!$ to the product. As $h \to 0$, the right-hand side of (5.11) approaches the true value of I_p.

c. p-Dimensional Midpoint Algorithm

A computing algorithm that economizes the number of arithmetic operations required to evaluate (5.11) was developed by I. R. Savage and the author. The algorithm is given below in terms of the vectors $\mathbf{f}_j$, $\mathbf{v}_{jk}$, and $\mathbf{s}_j$, $j = 1, \ldots, p$, consisting of N, N, and $N + 1$ elements, respectively. The elements of the vector $\mathbf{f}_j$ are values of $f_j(\cdot)$ evaluated at the midpoints of the N equal subintervals of length $h = (b - a)/N$ of the interval $[a, b]$; that is, $f_{ji} = f_j(x_i)$, where $x_i = a + (b - a)(2i - 1)/2N$, $i = 1, \ldots, N$. The approximations to $I_1, I_2, \ldots, I_p$ are successively developed. The vectors $\mathbf{s}_j$ and $\mathbf{v}_{jk}$ ($1 \leq k \leq j$) defined below are used for the storage of intermediary calculations; and their elements, evaluated during computation of the approximation to I_j, are saved and modified for computation of I_{j+1}. The $(N + 1)$st element of $\mathbf{s}_j = (s_{j1}, \ldots, s_{j,N+1})$ is seen to provide the approximation to I_j given in (2.5), that is, $I_j \doteq h^j s_{j,N+1}$.

The following definitions will be useful in describing the algorithm.

Definition 1. The product $\mathbf{a} * \mathbf{b}$ of the vectors $\mathbf{a} = (a_1, a_2, \ldots, a_N)$ and $\mathbf{b} = (b_1, b_2, \ldots, b_N)$ is the vector $\mathbf{c} = (c_1, c_2, \ldots, c_N)$, where $c_i = a_i b_i$.

Definition 2. The operator $\mathscr{A}$ applied to the vector $\mathbf{a} = (a_1, a_2, \ldots, a_N)$ gives a vector $\mathscr{A}\mathbf{a} = \mathbf{e}$, $\mathbf{e} = (e_1, e_2, \ldots, e_{N+1})$, where

$$e_r = \sum_{i=1}^{r-1} a_i,$$

$r = 1, \ldots, N + 1$ and $e_1 = 0$.

The algorithm works as follows.

Algorithm. The modified midpoint approximation to I_p is $I_p \doteq h^p s_{p,N+1}$, where $s_{p,N+1}$ is the $(N + 1)$st element of $\mathbf{s}_p$:

    $a.$ $p = 1: \mathbf{v}_{11} = \mathbf{f}_1,$

              $\mathbf{s}_1 = \mathscr{A}\mathbf{v}_{11};$

b. $p > 1$: for $j = 2, \ldots, p$ calculate the vector

$$\mathbf{v}_{ji} = \mathbf{v}_{j-1,i} * \mathbf{f}_j/(j + 1 - i), \quad i < j,$$

$$\mathbf{s}_j = \mathscr{A}\left(\sum_{i=1}^{j} \mathbf{v}_{ji}\right);$$

$$\mathbf{v}_{jj} = \mathbf{s}_{j-1} * \mathbf{f}_j,$$

as $h \to 0$, $h^p s_{p,N+1} \to I_p$.

The vector $\mathbf{s}_j$ is seen to be an indefinite integral representation of the approximation to I_j. This algorithm differs from an approach involving indefinite integration suggested by J. L. Hodges and used by Klotz (1964). The present algorithm corrects for effects of the edges of the region of integration and involves no high-degree quadrature formulas.

For the evaluation of (1.1), the range of integration was truncated to $(-8, 8)$. A single vector $\mathbf{f}$ was computed whose elements were $f(x)$ evaluated at increments of $h = .01$ from -8 to 8. When values of $f(x - d)$ were required for the vectors $\mathbf{f}_j$, they were found from $\mathbf{f}$ by using the translation property of $f(x)$. It was observed from preliminary computer results and confirmed by analysis that, for single-precision evaluation of (1.1) with $p = m + n$, the vectors $\mathbf{v}_{p1}, \ldots, \mathbf{v}_{p6}$ were sufficient for the approximation of I_p, $p > 6$, in the sense that errors introduced by this device were less than 10^{-10}. Consequently, only eight arrays (for the vectors $\mathbf{f}$, $\mathbf{v}_1$ through $\mathbf{v}_6$, $\mathbf{s}$) were necessary to implement the algorithm on the computer.

d. Improved Accuracy by Extrapolation to the Limit

Through comparison of values of $P(\mathbf{z})$ computed by the p-dimensional midpoint algorithm with values as given by Teichroew (1954) and by other checks it was seen that the error in the algorithm as implemented was still appreciable. Hence the technique of "extrapolation to the limit" or "deferred approach to the limit" was used. This technique is essentially one for improving the accuracy of an approximation by utilizing any knowledge (complete or incomplete) of the asymptotic behavior of the error [see Henrici (1964), pp. 239–241, and also Bauer, Rutishauser, and Stiefel (1963); Richardson and Gaunt (1927); Wilf (1967)].

To attain the desired accuracy, then, the results of the p-dimensional midpoint algorithm (Section 5.c) were improved by using this technique. For fixed p equal to $m + n$, the midpoint algorithm was used with subinterval lengths $h = h_i$, where $h_i = (2.56)/2^i$, $i = 0, 1, \ldots, 8$, to produce a series of approximations $A_{i,0}$ to $P_{m,n}(\mathbf{z} \mid d)$. The usual triangular extrapolation array

$$A_{0,0}$$

$$A_{1,0} \quad A_{1,1}$$

$$A_{2,0} \quad A_{2,1} \quad A_{2,2}$$

$$A_{3,0} \quad A_{3,1} \quad A_{3,2} \quad A_{3,3}$$

was created with

$$A_{i,k+1} = \frac{2^{2k+2}A_{i,k} - A_{i-1,k}}{2^{2k+2} - 1}, \quad k = 0, 1, 2, \ldots, 8 \text{ and } i = k, k+1, \ldots, 8.$$

Examination of this array showed that the error in approximating (1.1) decreased in a manner quite well described as an h^2 process as i increased, that is,

$$\frac{A_{i,k} - A_{i-1,k}}{A_{i+1,k} - A_{i,k}} \doteq 4^{k+1}$$

as might be expected from a midpoint method. Calculations were continued until column convergence in the array showed agreement to 9 decimal places or $|A_{i,k} - A_{i-1,k}| < 10^{-9}$. This convergence was always achieved at or before the $i = 8$ step in the extrapolation.

This method of calculation was used for $P_{7,7}(\mathbf{z} \mid d)$ and $P_{12,1}(\mathbf{z} \mid d)$; values of $P_{m,n}(\mathbf{z} \mid d)$ for smaller values of m and n were computed by means of Savage's back-recursive rule (1960), resulting in a considerable saving

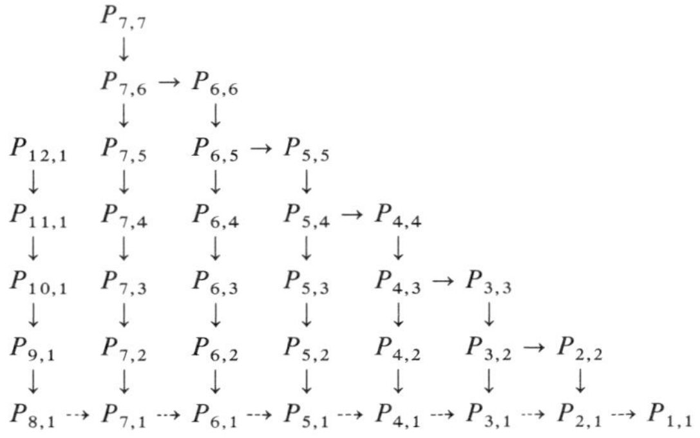

$$
\begin{array}{l}
P_{7,7} \\
\quad\downarrow \\
P_{7,6} \rightarrow P_{6,6} \\
\quad\downarrow \qquad\quad \downarrow \\
P_{12,1} \quad P_{7,5} \quad P_{6,5} \rightarrow P_{5,5} \\
\quad\downarrow \qquad\quad\downarrow \qquad\quad\downarrow \qquad\quad\downarrow \\
P_{11,1} \quad P_{7,4} \quad P_{6,4} \quad P_{5,4} \rightarrow P_{4,4} \\
\quad\downarrow \qquad\quad\downarrow \qquad\quad\downarrow \qquad\quad\downarrow \qquad\quad\downarrow \\
P_{10,1} \quad P_{7,3} \quad P_{6,3} \quad P_{5,3} \quad P_{4,3} \rightarrow P_{3,3} \\
\quad\downarrow \qquad\quad\downarrow \qquad\quad\downarrow \qquad\quad\downarrow \qquad\quad\downarrow \qquad\quad\downarrow \\
P_{9,1} \quad P_{7,2} \quad P_{6,2} \quad P_{5,2} \quad P_{4,2} \quad P_{3,2} \rightarrow P_{2,2} \\
\quad\downarrow \qquad\quad\downarrow \qquad\quad\downarrow \qquad\quad\downarrow \qquad\quad\downarrow \qquad\quad\downarrow \qquad\quad\downarrow \\
P_{8,1} \dashrightarrow P_{7,1} \dashrightarrow P_{6,1} \dashrightarrow P_{5,1} \dashrightarrow P_{4,1} \dashrightarrow P_{3,1} \dashrightarrow P_{2,1} \dashrightarrow P_{1,1}
\end{array}
$$

Figure 2

(about 80%) in computer time:

To compute $P_{m,n}(\mathbf{z})$ add all $[(m + n + 1)$ in number$]$ of the $P_{m+1,n}(\mathbf{z}^j)$ and divide by $m + 1$, where

$$\mathbf{z}^j = (z_1, \ldots, 0, z_j, \ldots, z_{m+n}), \quad j = 1, \ldots, m + n + 1.$$

The roles of m and n can be interchanged in the obvious manner.

Example.

$$P_{3,1}(0001) = \frac{P_{3,2}(10001) + P_{3,2}(01001) + P_{3,2}(00101) + 2P_{3,2}(00011)}{2}.$$

This procedure is shown schematically in Figure 2. The solid arrows indicate the paths of application of the back-recursive rule which result in tabled values. The dashed arrows indicate paths used for checking the accuracy of the computations (see Section 6.d).

6. ACCURACY OF COMPUTATIONS

Checks on the accuracy of the computations were of eight types.

a. During the final stages of program checkout, for the case of $m = n = 3$, both $P(\mathbf{z})$ and $P(\mathbf{z}^{tc})$ were computed directly (using the method of Section 5.d *without* the back-recursive rule) for $d = .25$ and 2.0 for the six values of $\mathbf{z}$ and $\mathbf{z}^{tc}$ such that $\mathbf{z} \neq \mathbf{z}^{tc}$. It was observed that $P_{3,3}(\mathbf{z})$ agreed with $P_{3,3}(\mathbf{z}^{tc})$ to at least 10 decimal places. This provided two independent calculations of the same quantity and is thus a valid check on the accuracy of the basic method of computation.

b. All values of $P(\mathbf{z})$ that correspond as in (4.2) to values in Teichroew's table were checked, and in all cases $[432$ of them: all d; $n = 1$, $m = 1(1)9$; $n = 2(1)5$, $m = n(1)7]$ Table A and Teichroew's table agree to within 5 units or less in the ninth decimal place. The frequencies of the magnitude of the absolute differences observed are given in units of the ninth decimal place as follows:

(Absolute Difference) $\times 10^9$	Frequency
0	317
1	98
2	3
3	9
4	3
5	2
	432

c. Spot checking that compared Table A with Klotz's table (1961) showed agreement to at least as many places as are considered accurate (see Section 4.*a*).

d. The values of $P_{7,1}(\mathbf{z})$ were computed in two ways from two different starting points, using Savage's back-recursive scheme. These two calculations, indicated by the dashed arrows in Figure 2, give results that in all cases (64 of them) agree to within 2 units or less in the ninth decimal place. In a similar manner, $P_{2,1}(\mathbf{z})$ was computed in two ways, and agreement to within 4 units or less in the ninth decimal place was observed in all 24 cases.

$P_{7,1}$		$P_{2,1}$	
(Abs. Difference) $\times 10^9$	Frequency	(Abs. Difference) $\times 10^9$	Frequency
0	36	0	9
1	24	1	10
2	4	2	1
	64	3	2
		4	2
			24

Using this scheme and denoting by ε_i the difference between the tabled value of $P_{m,n}(\mathbf{z}^i)$ and the true value, we find the maximum error possible under one step of the scheme to be

$$\text{From} \quad \text{To} \quad \text{Error} \qquad \text{Maximum Error}$$

$$(6.1) \qquad P_{m,n} \quad P_{m,n-1} \quad \sum_{i=1}^{m+n} \frac{\varepsilon_i}{n} \le \max_i |\varepsilon_i| \frac{m+n}{n} \equiv \frac{\varepsilon(m+n)}{n},$$

where the roles of *m* and *n* can be interchanged in the obvious manner. In accordance with Figure 2, application of (6.1) leads to the following selected numerical examples:

$$\text{From} \quad \text{To} \qquad \text{Maximum Error}$$

$$P_{7,7} \quad P_{7,6} \qquad 2\varepsilon$$

$$(6.2) \qquad P_{7,7} \quad P_{7,1} \qquad \frac{\displaystyle\prod_{i=0}^{5}(14-i)\varepsilon}{\displaystyle\prod_{i=0}^{5}(7-i)} = 429\varepsilon$$

$$P_{7,7} \quad P_{1,1} \qquad \frac{\displaystyle\prod_{i=0}^{11}(14-i)\varepsilon}{\left[\displaystyle\prod_{i=0}^{5}(7-i)\right]^2} = 1716\varepsilon.$$

However, these maximum errors will actually be obtained only if *all* values of $P(\mathbf{z})$ at the starting stage err by ε and in the same direction (i.e., all are too large or too small). Anything other than this situation will lead to an error less than the maximum one, and in fact if the errors at the starting stage are of both signs the result is very small error propagation because of cancellation.

e. Values of $P(\mathbf{z} \mid d)$ for $m = n = 1$ were checked by noting that $P_{1,1}(01) = F(d/\sqrt{2})$. Similarly, $P_{1,1}(10) = F(-d/\sqrt{2})$. Agreement to within 6 units or less in the ninth decimal place for all 16 cases was observed.

f. Savage (1956) and Savage, Sobel, and Woodworth (1966) present inequalities and a nonlinear relationship that may be used to verify the accuracy of Table A. The following statements of several of these relationships do not give the sufficient conditions exactly as they appear in the references; in each case the conditions of Section 1 are sufficient (and in some cases more than sufficient).

Theorem 6.1 [Savage (1956), p. 597]. *If* $\mathbf{z}$ *and* $\mathbf{z}'$ *differ only in their ith and jth components* $(i < j)$ *with* $(z_i, z_j) = (0, 1)$ *while* $(z_i', z_j') = (1, 0)$, *then* $P(\mathbf{z}) > P(\mathbf{z}')$.

Theorems 6.2 through 6.7 are from Savage, Sobel, and Woodworth (1966).

Theorem 6.2 (Corollary 2, p. 100). *If* $\mathbf{z}$ *and* $\mathbf{z}'$ *have the same number of zeros and ones and are such that* $\sum_{j=1}^{i} (z_j' - z_j) \geq 0$, $i = 1, \ldots, m + n$, *then* $P(\mathbf{z}) > P(\mathbf{z}')$.

Theorem 6.3 (Theorem 5, p. 101). *If* $d \neq 0$, *then* $P(1, 0, 0^r, 0, 1) > P(0, 1, 0^r, 1, 0)$ *or, equivalently,* $P(1, 0, 0^r, 0, 1) > P(1, 0, 1^r, 0, 1)$, *where* x^r *denotes a vector of* r x*'s.*

Theorem 6.4 (Theorem 6, p. 102). *If* $d \neq 0$, *then* $P(0^r, 0, 1, 1, 0, 0^r) > P(0^r, 1, 0, 0, 1, 0^r)$ *or, equivalently,* $P(1^r, 1, 0, 0, 1, 1^r) > P(0^r, 1, 0, 0, 1, 0^r)$.

Theorem 6.5 (Theorem 7, p. 103). *If* $d \neq 0$, *then* $P(0110\mathbf{z}) > P(1001\mathbf{z})$ *for any* $\mathbf{z}$ *or, equivalently,* $P(\mathbf{z}1001) > P(\mathbf{z}0110)$.

Theorem 6.6 (Theorem 8, p. 110, and Corollary 1, p. 111).

$$\text{(i)} \quad P(0110) = 2P(011) - 2P^2(01);$$
$$\text{(ii)} \quad P(0101) = 2P^2(01) - 2P(0011);$$
$$\text{(iii)} \quad P(1001) = 2P(100) - 2P^2(10);$$
$$\text{(iv)} \quad P(1010) = 2P^2(10) - 2P(1100).$$

Theorem 6.7 (p. 111). *For* $n = 2$ *and arbitrary fixed* $m = M$ *let*

$$\mathbf{z} = (\overbrace{0 \cdots 0}^{r_1} 1 0 \overbrace{\cdots 0}^{r_2} 1 0 \overbrace{\cdots 0}^{r_3}),$$

where $r_1 + r_2 + r_3 = M$. *Then*

$$P(\mathbf{z}) = \frac{M! \, 2!}{r_1! \, r_2! \, r_3!} \sum_{\substack{0 \le i_1 \le r_2 \\ 0 \le i_2 \le r_3}} \binom{r_2}{i_1}\binom{r_3}{i_2}(-1)^{r_2 + r_3 - (i_1 + i_2)}$$

$$\times \int_{-\infty}^{\infty} \int_{-\infty}^{y} F^{r_1 + r_2 - i_1}(x) F^{r_3 + i_1 - i_2}(y) f(x - d) f(y - d) \, dx \, dy.$$

Note that

$$A_i + A_{M-i} = \frac{P(\overbrace{0 \cdots 0}^{i} 1)}{i!} \cdot \frac{P(\overbrace{0 \cdots 0}^{M-i} 1)}{(M - i)!}$$

where

$$A_i = \int_{-\infty}^{\infty} \int_{-\infty}^{y} F^i(x) F^{M-i}(y) f(x - d) f(y - d) \, dx \, dy.$$

Therefore, if all rank order probabilities for $m < M$ and $n = 1$ and 2 have been computed, the only new integrals required to evaluate $P(\mathbf{z})$ are of the form A_i, $i = 0, \ldots, M$.

In general, the preceding inequalities are not capable of detecting other than gross errors; in all cases examined in a spot check, the inequality was verified. Some examples (for $d = .6$) are as follows.

1. Theorems 6.4 and 6.5:

$$P_{4,2}(100001) = .039412483 > .036066049$$
$$= P_{4,2}(010010) < P_{4,2}(001100) = .037706839.$$

2. Theorem 6.4:

$$P_{6,2}(00011000) = .028125870 > .027745063$$
$$= P_{6,2}(00100100).$$

3. Theorem 6.5:

$$P_{7,2}(011000000) = .008928414 > .007644742$$
$$= P_{7,2}(100100000).$$

4. Theorem 6.3:

$$P_{7,2}(100000001) = .022295925 > .021415866$$
$$= P_{7,2}(010000010).$$

Spot checking of Table A with respect to each of the above inequalities showed agreement in all cases checked.

Theorems 6.6 and 6.7 are more relevant to checking accuracy because they provide statements of equality rather than inequality. All tabled values to which Theorem 6.6 is applicable were checked (32 of them) and agreement to within less than 2 units in the eighth decimal place was observed. For example, for $d = .6$,

$$P(0101) = 2P^2(01) - 2P(0011) = 2(.556231457)^2 - 2(.214499821)$$
$$= .189787222 \quad \text{(tabled value is .189787226)}.$$

Two examples (both for $d = .6$) are given in application of Theorem 6.7. The first example is

$$P_{5,2}(0001010) = \frac{(3 + 1 + 1)! \, 2!}{3! \, 1! \, 1!} \sum_{\substack{0 \le i_1 \le 1 \\ 0 \le i_2 \le 1}} \binom{1}{i_1}\binom{1}{i_2}(-1)^{1 + 1 - (i_1 + i_2)}$$

(6.3)
$$\times \int_{-\infty}^{\infty} \int_{-\infty}^{y} F^{3 + 1 - i_1}(x) F^{1 + i_1 - i_2}(y) f(x - d) f(y - d) \, dx \, dy$$

$$= 40 \int_{-\infty}^{\infty} \int_{-\infty}^{y} \left[F^3(x)F(y) - F^3(x)F^2(y) - F^4(x) \right.$$

$$\left. + F^4(x)F(y) \right] f(x - d) f(y - d) \, dx \, dy.$$

Applying Theorem 6.7 again with $r_3 = 0$, one obtains

$$P(000101) = 8 \int_{-\infty}^{\infty} \int_{-\infty}^{y} \left[F^3(x)F(y) - F^4(x) \right] f(x - d) f(y - d) \, dx \, dy$$

(6.4.1)
$$= 8 \int_{-\infty}^{\infty} \int_{-\infty}^{y} F^3(x)F(y) f(x - d) f(y - d) \, dx \, dy - 4P(000011);$$

$$P(0000101) = 10 \int_{-\infty}^{\infty} \int_{-\infty}^{y} \left[F^4(x)F(y) - F^5(x) \right] f(x - d) f(y - d) \, dx \, dy$$

(6.4.2)
$$= 10 \int_{-\infty}^{\infty} \int_{-\infty}^{y} F^4(x)F(y) f(x - d) f(y - d) \, dx \, dy - 5P(0000011);$$

$$P(0001001) = 20 \int_{-\infty}^{\infty} \int_{-\infty}^{y} \left[F^3(x)F^2(y) - 2F^4(x)F(y) \right.$$

$$\left. + F^5(x) \right] f(x - d) f(y - d) \, dx \, dy$$

(6.4.3)
$$= 20 \int_{-\infty}^{\infty} \int_{-\infty}^{y} \left[F^3(x)F^2(y) - 2F^4(x)F(y) \right] f(x - d) f(y - d) \, dx \, dy$$

$$+ 10P(0000011).$$

Solving these equations for the integrals and substituting in (6.3), one obtains, after some simplification by combining like terms,

$$P(0001010) = 5P(000101) - 2P(0001001) - 4P(0000101)$$

(6.5)
$$= 5(.133383584) - 2(.086564516) - 4(.108517639)$$

$$= .059718332 \quad \text{(tabled value is .059718329)}.$$

The second example, after considerable simplification, is

$$P(00100100) = [15P(001001) + 54P(0001001) + 180P(0000101)$$

$$+ 360P(0000011) + 180P(001) \cdot P(00001)]$$

(6.6)
$$- [36P(00001001) + 90P(00000101) + 180P(00000011)$$

$$+ 360P(001) \cdot P(0001) + 9P(00010001)]$$

$$= 110.946032187 - 110.918287047$$

$$= .027745140 \quad \text{(tabled value is .027745063)}.$$

Various methods of treating digits beyond the ninth decimal place in (6.6) lead to slightly different results in the last three digits of the answer. However, since multiplication of an n-digit integer by a 9-digit decimal number can in general be expected to give a product with only 9 accurate leading digits (assuming the original n and 9 digits are accurate as given), one should really look only at the leading 9 digits when the difference is taken:

$$
\begin{array}{r}
110.946032187 \\
- \quad 110.918287047 \\
\hline
000.027745
\end{array}
$$

We see that *these* 9 digits agree as well as can be expected with the tabled value. Applying this method of error analysis to the preceding paragraph, one can say that in application of Theorem 6.7 agreement was also observed within 1 unit in the ninth leading decimal place.

g. All values of $P_{m,1}(\mathbf{z} \mid d)$ were computed for all d and for $m = 1(1)12$ by using a quadrature formula with $P_{m,1}(\mathbf{z} \mid d)$ expressed as a single integral:

$$P_{m,1}(\mathbf{z} \mid d) = P(\overbrace{0 \cdots 0}^{m-i} 1 \overbrace{0 \cdots 0}^{i}), \quad i = 0, \ldots, m$$

(6.7)

$$= \frac{m!}{(m-i)! \, i!} \int_{-\infty}^{\infty} F^{m-i}(x) [1 - F(x)]^i f(x - d) \, dx.$$

Quadrature was by one-dimensional Romberg integration [e.g., see Bauer (1961) and Wilf (1967)] and was performed in such a way that the error expected was of the order of 10^{-9}. The frequencies of the magnitude of the absolute differences between the tabled values of $P(\mathbf{z})$ and the values calculated by this one-dimensional quadrature are given below in units of 10^{-9}.

Frequency of absolute differences between $P_{m,1}(z)$ _as calculated by one-dimensional and_ $(m + 1)$-_dimensional quadrature_

m	$\left(\dfrac{\text{Absolute}}{\text{Difference}}\right) \times 10^9 =$ 0	1	2	3	4	5	Total Frequency
1	4	8			1	3	16
2	5	13	2	1	3		24
3	14	10	4	4			32
4	19	11	9	1			40
5	22	21	4	1			48
6	33	18	5				56
7	34	26	4				64
8	68	4					72
9	78	2					80
10	85	3					88
11	92	4					96
12	98	6					104
Total frequency:	552	126	28	7	4	3	720

Thus, in these 720 cases all differences were less than or equal to 5 units in the ninth decimal place.

h. Finally, $\sum_z P_{m,n}(z \mid d) = 1.0 \pm \delta$ for all tabled values of m, n, and d, where $|\delta| < 10^{-8}$.

In summary, the preceding checks on the accuracy of computations show errors of less than 10^{-8}, and it is reasonable to assign this degree of accuracy to all values in Table A.

Interpolation accuracy, with respect to the argument d, was briefly investigated and found to be difficult to specify because of the irregular spacing chosen for d and the physical size of the table. It is conjectured, on the basis of a preliminary investigation, that within the range $0 \leq d \leq 1$ linear and quadratic interpolation will be accurate to 3 and 4 decimal places, respectively.

7. AVAILABILITY OF TABLE A FOR MECHANICAL OR ELECTRONIC PROCESSING

The physical size of Table A discourages the use of a punched card form of the table for analysis. The present magnetic tape version is in CDC 1604 binary form and thus not universally usable. It would be possible, however, to make available a magnetic tape version in a form usable on common large-scale digital computers.

The set of computer programs used to compute Table A was written by the author in FORTRAN and FORTRAN Symbolic coding for the CDC 1604 computer. In part, the accuracy obtainable is dependent on the 48-bit word length of the 1604. Reproduction of the computations on another computer using the same programs would entail great difficulty. The computer time required on the 1604 for computation of Table A, excluding program checkout, was approximately 20 hours. The author will supply additional information on request.

CHAPTER 2

Small-Sample Power of Two-Sample Nonparametric Tests Against the Normal Shift Alternative

1. SUMMARY

Tables of the exact power of four two-sample nonparametric tests for location against the normal shift alternative are described and presented for small sample sizes:

Tables B-1 and B-2 : Wilcoxon rank-sum test;
Tables B-3 and B-4 : Terry-Hoeffding-Fisher-Yates normal scores (c_1) test;
Tables B-5 and B-6 : Mood-Brown median test;
Tables B-7 and B-8 : Kolmogorov-Smirnov $D_{m,n}$ test.

Odd-numbered tables are for one-sided tests, and even-numbered tables are for two-sided tests. Selected power and efficiency comparisons are made among these four tests and with the two-sample Student's t test. The most powerful two-sample rank test is also considered.

The power tables are derived from Table A, and values are presented to 8 decimal places for all sample sizes $2 \leq n \leq m \leq 7$ which yield non-trivial results. The sample of size m has location parameter μ_1, and the sample of size n has location parameter μ_2 $(\mu_2 > \mu_1)$, with common variance σ^2. Values are tabulated for the shift alternative

$$d = \frac{\mu_2 - \mu_1}{\sigma} = .2(.2)1.0,1.5,2.0,3.0.$$

2. INTRODUCTION

Asymptotic properties of nonparametric tests for location have been investigated by many authors. In practice, however, one often works with small samples for which the asymptotic results are not applicable. It is of interest, therefore, to investigate small-sample power and efficiency with the aid of Table A, as follows. (Assumptions and notation are the same as in Chapter 1.)

If $T(\mathbf{z})$ is a statistic defined for each value of $\mathbf{z}$ for a specified combination of sample sizes m and n, the following procedure can be used to construct critical regions based on this statistic. We arrange all of the $\mathbf{z}$'s in a list so that as we proceed down the list the corresponding values of $T(\mathbf{z})$ will be decreasing [or, strictly speaking, nonincreasing since several $\mathbf{z}$'s may give the same value of $T(\mathbf{z})$]. For a one-sided test we place in the critical region ω all $\mathbf{z}$ such that $T(\mathbf{z}) \geq T_0$. We will have a test with significance level

$$(2.1) \qquad a = \frac{K}{\binom{m + n}{n}},$$

where K is the number of $\mathbf{z}$'s such that $T(\mathbf{z}) \geq T_0$, for under the null hypothesis each rank order $\mathbf{z}$ is equally likely. When a nominal significance level of α is desired, we will use the smallest value of T_0 satisfying

$$(2.2) \qquad \frac{K}{\binom{m + n}{n}} \leq \alpha,$$

and the resulting T_0 is the critical value of the test. By using this procedure we ensure that the tests will always be conservative in the sense that the Type I error will never exceed the nominal level. However, the Type I error will be as close to the desired nominal level as the discrete character of the distributions permits.

Similarly, for a two-sided test (i.e., large and small values are significant), we place in the critical region all $\mathbf{z}$ such that $T(\mathbf{z}) \geq T'_0$ or $T(\mathbf{z}) \leq T''_0$ and have a test with significance level

$$(2.3) \qquad a = \frac{K'}{\binom{m + n}{n}}.$$

Here K' is the number of $\mathbf{z}$'s such that $T(\mathbf{z}) \geq T'_0$ plus the number of $\mathbf{z}$'s such that $T(\mathbf{z}) \leq T''_0$.

The following procedure is suitable for symmetric distributions and is the method that we shall employ in this chapter, even in cases where the distributions are not symmetric. When a nominal level of α is desired, we will use the smallest value of T'_0 and the largest value of T''_0 such that both

$$(2.4) \qquad \frac{K'}{\binom{m+n}{n}} \leq \alpha$$

and

$$(2.5) \qquad \frac{K'}{2} \text{ is the number of } \mathbf{z}\text{'s such that } T(\mathbf{z}) \geq T'_0.$$

This implies that K' is an even integer. It should be noted that Table A already has the $\mathbf{z}$'s arranged in decreasing order for the test statistic c_1.

The power of a test is then computed from Table A as

$$(2.6) \qquad \Pr(\omega_\alpha; d) = \sum_{\mathbf{z} \in \omega_\alpha} P_{m,n}(\mathbf{z} \mid d),$$

where ω_α is the critical region according to (2.2) or (2.4) and (2.5). Nominal levels of α used are .25,.10,.05,.025,.01,.005. Power values are given for $d = .2(.2)1.0,1.5,2.0,3.0$. In Tables B-1 through B-8 values are given to 8 decimal places and are believed to be accurate to within 1 unit in the eighth place.

3. POWER OF THE WILCOXON TWO-SAMPLE RANK-SUM TEST

The Wilcoxon two-sample rank-sum test [Wilcoxon (1945); Mann and Whitney (1947)] may be described as follows. For any rank order $\mathbf{z}$ the Wilcoxon statistic w is given by

$$(3.1) \qquad w = w(\mathbf{z}) = \sum_{i=1}^{m+n} i z_i,$$

the sum of the ranks of the Y sample when all the $m + n$ observations are ranked together. Note that $n(n + 1)/2 \leq w \leq n(n + 1)/2 + mn$ and thus that w may take on at most $mn + 1$ distinct values (it is seen in Table A and can be easily shown that for the present case of sampling from two non-degenerate normal distributions all $mn + 1$ values have nonzero probability). The null hypothesis of no difference between location parameters is rejected for large values of w if the translation alternative being considered is $d > 0$ (one-sided test). Similarly, for a two-sided test the null hypothesis is rejected for large or small values of w.

Tables B-1 and B-2 give the power of the Wilcoxon test against the normal shift alternative for one-sided and two-sided tests, respectively. The critical region ω_α for the one-sided Wilcoxon test at the nominal level of significance α was chosen in the following manner. From Table A an integer (critical value) T_0 was found such that $\Pr\{w(z) \geq T_0\} \leq \alpha$ and $\Pr\{w(z) \geq T_0 - 1\} > \alpha$ for the nominal levels of α. The critical region ω_α consists of all z such that $w(z) \geq T_0$. For example, for $m = 7$ and $n = 3$, a one-sided test of size $\alpha = .09167$ with $T_0 = 23$ determines the critical region $\omega_\alpha = \{z^1, \ldots, z^{10}, z^{14}\}$. Similarly, for the two-sided test, an integer T_0' was found such that $\Pr\{w(z) \geq T_0'\} \leq \alpha/2$ and $\Pr\{w(z) \geq T_0' - 1\} > \alpha/2$ for the nominal levels of α. The critical region ω_α consists of all z such that $w(z) \geq T_0'$ or $\leq n(n + 1)/2 + mn - T_0' = T_0''$. Power was calculated using (2.6).

Below each value of power for the Wilcoxon test, on the line labeled t, is given the power of the two-sample Student's t test for the same α's and shift alternatives. Let $\bar{x}_m$ and $\bar{y}_n$ denote two sample means which are to be tested by the statistic

$$t = \frac{\bar{x}_m - \bar{y}_n}{\left(\dfrac{(m - 1)s_x^2 + (n - 1)s_y^2}{m + n - 2} \cdot \dfrac{n + m}{nm} \right)^{1/2}},$$

using a test of size α for t, where m and n denote the sample sizes with sample variances s_x^2 and s_y^2, respectively. The power of a one-sided test based on t to detect a difference $d > 0$ between population means when the populations are normal with unit variance is given [Greenwood and Hartley (1962)] by

(3.3) $1 - \Pr\{t < t(\alpha, \gamma) \mid \gamma, \delta\}$

$$= 1 - 2^{(2 - \gamma)/2} \Gamma^{-1}\left(\frac{\gamma}{2}\right) z_0^{-\gamma} \int_0^\infty u^{\gamma - 1} e^{-u^2/2z_0^2} F(u - \delta) \, du,$$

where $\delta = d/(1/m + 1/n)^{1/2}$, $F(x)$ is the standard normal cumulative distribution function, $z_0 = t(\alpha, \gamma)/\sqrt{\gamma}$, $t(\alpha, \gamma)$ is the upper $100\,\alpha\%$ point of the t distribution with γ degrees of freedom, and $\gamma = m + n - 2$. Similarly, the power of the two-sided test is given by

$1 - \Pr\{-t(\alpha, \gamma) < t < t(\alpha, \gamma) \mid \gamma, \delta\}$

(3.4) $= 1 - \Pr\{t < t(\alpha, \gamma) \mid \gamma, \delta\} - \Pr\{t < -t(\alpha, \gamma) \mid \gamma, \delta\}$

$$= 2 - 2^{(2 - \gamma)/2} \Gamma^{-1}\left(\frac{\gamma}{2}\right) z_0^{-\gamma} \int_0^\infty u^{\gamma - 1} e^{-u^2/2z_0^2} [F(u - \delta) + F(u + \delta)] \, du.$$

The power of the t test was computed using a variation of Romberg quadrature to evaluate (3.3) and (3.4). Tables B-9 and B-10 give power values to 8 decimal places for one-and two-sided tests, respectively, for test sizes $\alpha = .25, .10, .05, .025, .01, .005$ and for normal shift alternatives $d = .2(.2)1.0, 1.5, 2.0, 3.0$.

Various numerical methods have been used previously to investigate the power of the Wilcoxon test. Tsao (1957) gives $P_{m,n}(\mathbf{z} \mid d)$ for $m = n = 2$ and 3, $d = .25(.25)1.50$, evaluated by means of interpolating polynomials, and uses this method to evaluate efficiency (asymptotic as $d \to 0$) for small sample sizes. Tsao's values of $P_{m,n}(\mathbf{z} \mid d)$ appear to be correct to within 1 or 2 units in the second leading nonzero digit. Dixon (1954), using numerical integration, gives values of power of the two-sided test to 3 decimal places for $m = n = 3, 4$, and 5; $d = .0(.25)2.0(.5)5.0$ for selected levels of significance. Most of Dixon's values are correct as given, but errors as large as .004 are to be found.

4. POWER OF THE TERRY-HOEFFDING-FISHER-YATES TWO-SAMPLE NORMAL SCORES (c_1) TEST

The two-sample normal scores (c_1) test [Terry (1952); Hoeffding (1951)] may be described as follows. For any rank order $\mathbf{z}$ the c_1 statistic is given by

$$(4.1) \qquad c_1 = c_1(\mathbf{z}) = \sum_{i=1}^{m+n} z_i \mathscr{E}(V_i),$$

where $V_1, \ldots, V_{m+n}$ are the order statistics of a sample of size $m + n$ drawn from a standard normal population. The null hypothesis of no difference between location parameters is rejected for large values of c_1 if the translation alternative being considered is $d > 0$ (one-sided test). Similarly, for a two-sided test the null hypothesis is rejected for large or small values of c_1.

Tables B-3 and B-4 give the power of the normal scores test against the normal shift alternative for one-sided and two-sided tests, respectively. The critical region ω_α for the one-sided test at the nominal significance level α was chosen as follows. From Table A the smallest number (critical value) T_0 such that $\Pr\{c_1(\mathbf{z}) \geq T_0\} \leq \alpha$ was found. The critical region ω_α consists of all $\mathbf{z}$ such that $c_1(\mathbf{z}) \geq T_0$. For example, for $m = 7$ and $n = 3$, a one-sided test of size $\alpha = .09167$ with $T_0 = 1.884 = 120(.0157)$ determines the critical region $\omega_\alpha = \{\mathbf{z}^1, \ldots, \mathbf{z}^{11}\}$. Similarly, for the two-sided test, the smallest number T_0' such that $\Pr\{c_1(\mathbf{z}) \geq T_0'\} \leq \alpha/2$ and $\Pr\{c_1(\mathbf{z}) \leq -T_0'\} \leq \alpha/2$ was found. The critical region consists of all $\mathbf{z}$ such that $c_1(\mathbf{z}) \geq T_0'$ or $\leq -T_0' = T_0''$. In both Tables B-3 and B-4 the power of the two-sample Student's t test is given for the same α's and shift alternatives as in Section 3.

The power of the c_1 test has been investigated by several authors. Terry (1952) gives some Monte Carlo results. Teichroew (1955a) presents Monte

Carlo values of $P_{m,n}(\mathbf{z} \mid d)$ for sample sizes $(m, n) = (3, 2)$, $(3, 3)$, $(4, 2)$, and $(4, 3)$; $d = .0(.25)1.5(.5)2.5$. His 4- and 5-decimal-place values are correct to at least within 1 unit in the second decimal place. Klotz (1964) gives $P_{m,n}(\mathbf{z} \mid d)$ for $m = n = 3$, 4, and 5; $d = .25,.5(.5)2(1)6$; and power values for $\alpha = .05714$ $(m = n = 4)$, $\alpha = .04762$ $(m = n = 5)$, for $d = .25,.5(.5)2.0,3.0,4.0$. Comparison with Table A shows differences as large as 1 unit in the sixth decimal place but generally high accuracy. The power values for $\alpha = .04762$ seem to be correct to within 1 unit in the fourth decimal place.

5. POWER OF THE MOOD-BROWN TWO-SAMPLE MEDIAN TEST

The median test [Mood and Graybill (1963); Brown and Mood (1951); Westenberg (1948), (1950)] may be described as follows. For any rank order $\mathbf{z}$ the median statistic M' is given by

(5.1) $$M' = M'(\mathbf{z}) = \sum_{i=[(m+n)/2+1]}^{m+n} z_i,$$

where $[x]$ is the smallest integer $\geq x$.
The statistic M'' is given by

(5.2) $$M'' = M''(\mathbf{z}) = \max \{M'(\mathbf{z}), M'(\mathbf{z}^t)\},$$

where $\mathbf{z}^t$ is the vector whose ith component is $z_i^t = z_{m+n+1-i}$. In the one-sided case, the null hypothesis of no difference between medians is rejected for large values of M' if the translation alternative being considered is $d > 0$. In the two-sided case, the null hypothesis is rejected for large values of M''. Note that $0 \leq M' \leq n$ and $[n/2] + 1 \leq M'' \leq n$.

Tables B-5 and B-6 give the power of the Mood-Brown test against the normal shift alternative for one-sided and two-sided tests, respectively. For Table B-5 the rejection region ω_α contains all $\mathbf{z}$ such that $M'(\mathbf{z}) \geq M'_\alpha$, where $\Pr\{M(\mathbf{z}) \geq M'_\alpha\} = \alpha$. For example, for $m = 7$ and $n = 3$, a one-sided test of size $\alpha = .08333$ with $M'_\alpha = 3$ determines the critical region

$$\omega_\alpha = \{\mathbf{z}^1, \ldots, \mathbf{z}^4, \mathbf{z}^6, \mathbf{z}^9, \mathbf{z}^{10}, \mathbf{z}^{14}, \mathbf{z}^{20}, \mathbf{z}^{26}\}.$$

For Table B-6 the rejection region contains all $\mathbf{z}$ such that $M''(\mathbf{z}) \geq M''_\alpha$, where $\Pr\{M''(\mathbf{z}) \geq M''_\alpha\} = \alpha$. In both tables the power of the two-sample Student's t test is given for the same α's and shift alternatives as before.

6. POWER OF THE KOLMOGOROV-SMIRNOV
TWO-SAMPLE TEST

The Kolmogorov-Smirnov two-sample test [Smirnov (1948); Kolmogorov (1941); Massey (1951)], often referred to as the $D_{m,n}$ test, may be described as follows. For any rank order $\mathbf{z}$ the statistic D^+ is given by

$$(6.1) \qquad D^+ = D^+(\mathbf{z}) = \max_{j=1,\ldots,m+n} \left\{ \sum_{i=1}^{j} [n(1 - z_i) - mz_i] \right\},$$

and the statistic D is given by

$$(6.2) \qquad D = D(\mathbf{z}) = \max_{j=1,\ldots,m+n} \left| \sum_{i=1}^{j} [n(1 - z_i) - mz_i] \right|.$$

In the one-sided case, the null hypothesis of no difference between location parameters is rejected for large positive values of D^+ if the translation alternative being considered is $d > 0$. In the two-sided case, the null hypothesis of no difference between distributions is rejected for large values of D.

Tables B-7 and B-8 give the power of the Kolmogorov-Smirnov test against the normal shift alternative for one-sided and two-sided tests, respectively. For Table B-7 the rejection region ω_α contains all $\mathbf{z}$ such that $D^+(\mathbf{z}) \geq D_\alpha^+$, where $\Pr\{D^+(\mathbf{z}) \geq D_\alpha^+\} = \alpha$. For example, for $m = 7$ and $n = 3$, a one-sided test of size $\alpha = .03333$ with $D_\alpha^+ = 18$ determines the critical region $\omega_\alpha = \{\mathbf{z}^1, \mathbf{z}^2, \mathbf{z}^4, \mathbf{z}^{10}\}$. Similarly, for Table B-8 the rejection region contains all $\mathbf{z}$ such that $D(\mathbf{z}) \geq D_\alpha$, where $\Pr\{D(\mathbf{z}) \geq D_\alpha\} = \alpha$. In both tables the power of the two-sample Student's t test is given for the same α's and shift alternatives as before.

7. MOST POWERFUL TWO-SAMPLE TEST
BASED ON RANKS

If, for a given d, the $\mathbf{z}$ values are arranged in order of decreasing values of $P_{m,n}(\mathbf{z} \mid d)$, it is easy to specify the critical region of size $\alpha = K \bigg/ \binom{m+n}{n}$ for the most powerful two-sample rank test (MPRT) for the normal shift alternative. The critical region consists of the $\mathbf{z}$ values corresponding to the K largest values of $P_{m,n}(\mathbf{z} \mid d)$, where $1 \leq K \leq \binom{m+n}{n}$. It is seen by inspection of Table A that the critical region for this test depends upon the value of d, the alternative hypothesis.

One reason for considering the MPRT is that it enables one to achieve the largest number of useful test sizes without resorting to randomization.

A second reason is the interest in theoretical statistics concerning the relative power of the MPRT, the Wilcoxon test, and the c_1 test. Various questions have been raised, such as the following:

(i) Is the c_1 test as powerful as or more powerful than the Wilcoxon test for all m, n, and d? [See, for example, Dwass (1956).]

(ii) Is the MPRT identical in power to either the c_1 or the Wilcoxon test?

The answer to both these questions is no. Inspection of Table A revealed 29 cases for which the (nonrandomized) test size α is less than .10 and the MPRT is more powerful than both the Wilcoxon and the c_1 tests. In 15 cases the Wilcoxon test is more powerful than the c_1 test; in 6 of these cases the MPRT is more powerful than the Wilcoxon test. In 14 cases the c_1 test is more powerful than the Wilcoxon test; in 3 of these cases the MPRT is more powerful than the c_1 test. When randomization is permitted, similar results can occur (see, for example, Table 4: $m = n = 6$, 7 and $\alpha = .01$). It is believed that these are all such cases in Table A except for a few that may have been overlooked for sample sizes $(m, n) = (6, 6)$, $(7, 6)$, and $(7, 7)$ when α is close to .10. It is interesting to note that in one case ($m = 7$, $n = 6$, $\alpha = 19/1716$) the Wilcoxon test is more powerful than the locally most powerful rank order test, c_1, for the relatively small value $d = .2$. These observations are summarized in Table 1.

8. SELECTED POWER COMPARISONS

Power comparisons of several kinds which are of interest may be made by using Table A. Comparisons are specified by fixing all but one of the following factors and varying that single factor for the comparison:

(i) type of test (t, Wilcoxon, c_1, median, Kolmogorov-Smirnov);
(ii) sample sizes;
(iii) size of test (Type I error);
(iv) one-sided or two-sided test;
(v) alternative hypothesis (d).

Perhaps of most interest are selected comparisons of the kind where factors ii through v are fixed and factor i is varied. Table 2 compares the power of the five tests for sample sizes $m = n = 5, 6, 7$ and $m = 7, n = 5, 6$; size of test $\alpha = .01$, .05 (achieved by linear randomization in Table A); one-sided tests; alternative hypotheses $d = .2(.2)1.0, 1.5, 2.0, 3.0$. Two immediate observations from these comparisons are that the c_1 and Wilcoxon tests have essentially the same power functions for these sample sizes, and that the t test shows an often negligible advantage in power over the c_1 and Wilcoxon tests.

Table 3, a rearrangement of Table 2, presents comparisons in which factor ii, sample size, is varied. In Table 4 comparisons are made wherein factor iv is varied. The power of one- and two-sided tests is compared for $m = n = 7$ and $m = 7, n = 5$ for $\alpha = .01$.

9. SELECTED EFFICIENCY COMPARISONS

The concept most frequently used in power-efficiency comparisons in large-sample theory is that of asymptotic relative efficiency (ARE) [Pitman (1949); Noether (1955)]. For small-sample theory Hodges and Lehmann (1956) have proposed a definition of efficiency that may be used in rough comparison with ARE. The Hodges-Lehmann (H-L) efficiency of test a relative to test b for alternative hypothesis d and test size α is defined to be

$$(8.1) \qquad e_{a,b} = e_{a,b}(d, \alpha) = \frac{N_b^*}{N_a},$$

where N_a is the sample size for test a, and N_b^* is the randomized sample size for test b needed to match the power of test a for the given α and d (randomization is done by linear interpolation in the sample size for test b with respect to power). The definition could be given equally well in terms of randomization on the sample size of test a.

Tables 5 through 8 present the efficiency of the one-sided Wilcoxon, c_1, median, and Kolmogorov-Smirnov tests, respectively, relative to the t test. Sample sizes with $m = n$ are used for these tables, and with this restriction sample size in the definition of $e_{a,b}$ for the two-sample case may be either $m (= n)$ or $m + n = 2m$. Comparisons are made for sample sizes $m = n = 5$, 6, 7; test sizes $\alpha = .01, .05$; and alternative hypotheses $d = .2(.2)1.0, 1.5, 2.0, 3.0$. The power of the t test required to bracket the power of each of the other tests is given, with the Hodges-Lehmann efficiency immediately below.

Example ($m = n = 5, \alpha = .05, d = .2$).

$$e_{\text{Wilcoxon},t} = .960 = \frac{4(1 - p) + 5p}{5},$$

where $p = (.086613 - .081918)/(.087800 - .081918) = .798$.

Table 9 presents the efficiency of the one-sided Wilcoxon test when there is no restriction to equal sample sizes. In this case sample size in the definition of $e_{a,b}$ is $m + n$. Two examples of unequal sample size are given: (i) Wilcoxon test sample size $m = n = 7$ and randomization between t-test sample sizes $m = n = 7$ and $m = 7, n = 6$; (ii) Wilcoxon test sample size $m = 7, n = 6$ and randomization between t-test sample sizes $m = 7, n = 6$ and $m = n = 6$.

Table 1. Power relationships among MPRT, Wilcoxon, and c_1 tests

Sample Size (m, n)	Shift Alternative, d	α	ω_α for Wilcoxon Test	Power Relationship	Power Values
7, 7	$.4 \leq d \leq 3.0$	$30/3432 = .0087$	$w \geq 71$	MPRT $= w > c_1$	MPRT $= w$: .990085* c_1: .988419* *for $d = 3.0$
7, 7	$d = 3.0$	$65/3432 = .0189$	$w \geq 69$	MPRT $> w > c_1$	MPRT: .997366 w: .997347 c_1: .996786
7, 6	$.2 \leq d \leq 3.0$	$19/1716 = .0111$	$w \geq 58$	MPRT $= w > c_1$	Not computed
7, 6	$.2 \leq d \leq 1.5$	$44/1716 = .0256$	$w \geq 56$	MPRT $= c_1 > w$	Not computed
7, 6	$d = 2.0, 3.0$	$44/1716 = .0256$	$w \geq 56$	MPRT $> w > c_1$	Not computed
6, 6	$d = 2.0, 3.0$	$30/924 = .0325$	$w \geq 51$	MPRT $= w > c_1$	MPRT $= w$: .885109† c_1: .884105† †for $d = 2.0$
7, 5	$d = 3.0$	$19/792 = .0240$	$w \geq 45$	MPRT $> w > c_1$	MPRT: .991364 w: .991255 c_1: .991089
7, 5	$.2 \leq d \leq 1.0$	$29/792 = .0366$	$w \geq 44$	MPRT $> c_1 > w$	MPRT: .068973‡ c_1: .068965‡ w: .068875‡ ‡for $d = .2$

30

7, 5	$1.5 \leq d \leq 3.0$	$29/792 = .0366$	$w \geq 44$	MPRT $> w > c_1$	Not computed
7, 5	$.2 \leq d \leq 1.5$	$59/792 = .0745$	$w \geq 42$	MPRT $= c_1 > w$	Not computed
7, 5	$d = 2.0, 3.0$	$59/792 = .0745$	$w \geq 42$	MPRT $> c_1 > w$	Not computed
6, 5	$.2 \leq d \leq 2.0$	$19/462 = .0411$	$w \geq 40$	MPRT $= c_1 > w$	Not computed
6, 5	$d = 3.0$	$19/462 = .0411$	$w \geq 40$	MPRT $= w > c_1$	Not computed
6, 5	$.2 \leq d \leq 1.5$	$29/462 = .0628$	$w \geq 39$	MPRT $= c_1 > w$	Not computed
6, 5	$d = 2.0, 3.0$	$29/462 = .0628$	$w \geq 39$	MPRT $= w > c_1$	Not computed
6, 5	$.2 \leq d \leq 1.5$	$41/462 = .0887$	$w \geq 38$	MPRT $= c_1 > w$	Not computed
6, 5	$d = 2.0$	$41/462 = .0887$	$w \geq 38$	MPRT $> c_1 > w$	Not computed
6, 5	$d = 3.0$	$41/462 = .0887$	$w \geq 38$	MPRT $> w > c_1$	Not computed
7, 4	$.2 \leq d \leq 3.0$	$18/330 = .0545$	$w \geq 33$	MPRT $= c_1 > w$	Not computed
7, 4	$.2 \leq d \leq 3.0$	$27/330 = .0818$	$w \geq 32$	MPRT $= c_1 > w$	Not computed
5, 5	$1.5 \leq d \leq 2.0$	$19/252 = .0754$	$w \geq 35$	MPRT $> c_1 > w$	Not computed
6, 4	$d = 1.5$	$12/210 = .0571$	$w \geq 30$	MPRT $> w > c_1$	Not computed
6, 4	$d = 2.0, 3.0$	$12/210 = .0571$	$w \geq 30$	MPRT $= w > c_1$	Not computed
6, 4	$.2 \leq d \leq 3.0$	$18/210 = .0857$	$w \geq 29$	MPRT $= c_1 > w$	Not computed
7, 3	$.2 \leq d \leq 3.0$	$7/120 = .0583$	$w \geq 24$	MPRT $= c_1 > w$	Not computed
7, 3	$.2 \leq d \leq 3.0$	$11/120 = .0917$	$w \geq 23$	MPRT $= c_1 > w$	Not computed
5, 4	$.8 \leq d \leq 3.0$	$12/126 = .0952$	$w \geq 26$	MPRT $= w > c_1$	Not computed
6, 3	$.2 \leq d \leq 2.0$	$7/84 = .0833$	$w \geq 21$	MPRT $= c_1 > w$	Not computed
6, 3	$d = 3.0$	$7/84 = .0833$	$w \geq 21$	MPRT $= w > c_1$	Not computed

Note: Power values are given to show the magnitudes involved. The values were calculated by hand, and those reported as "not computed" were not done to sufficient accuracy to report here.

Table 2. Comparison of power of two-sample tests (one-sided)

$d =$	.2	.4	.6	.8	1.0	1.5	2.0	3.0
$m = 5, n = 5, \alpha = .01$								
t	.019781	.036604	.063498	.103497	.158904	.364215	.616745	.941108
c_1 (normal scores)	.019451	.035430	.060585	.097515	.148161	.334549	.568186	.905400
Wilcoxon	.019451	.035430	.060585	.097515	.148161	.334549	.568186	.905400
Kolmogorov-Smirnov	.018186	.031279	.051015	.079107	.116956	.257104	.447080	.803946
Median	.017158	.028278	.044837	.068465	.100747	.225883	.408371	.781944
$m = 5, n = 5, \alpha = .05$								
t	.087800	.143539	.219105	.313330	.421448	.698489	.891583	.996038
c_1 (normal scores)	.086712	.140313	.212492	.302224	.405316	.673585	.870504	.992794
Wilcoxon	.086613	.140046	.211997	.301471	.404324	.672395	.869734	.992740
Kolmogorov-Smirnov	.082306	.128125	.188976	.264698	.353035	.597811	.806519	.981022
Median	.077591	.114247	.160195	.214768	.276485	.448204	.618812	.872697
$m = 6, n = 6, \alpha = .01$								
t	.021529	.042625	.077818	.131391	.205877	.469032	.745116	.982808
c_1 (normal scores)	.021201	.041394	.074652	.124771	.193997	.438808	.704358	.967803
Wilcoxon	.021180	.041325	.074497	.124499	.193600	.438418	.704767	.968874
Kolmogorov-Smirnov	.020094	.037505	.065212	.105975	.161541	.359811	.592286	.898963
Median	.018189	.031076	.050094	.076553	.111455	.237837	.411451	.770401
$m = 6, n = 6, \alpha = .05$								
t	.093207	.159116	.249725	.362029	.487576	.779866	.942029	.999225
c_1 (normal scores)	.091893	.155152	.241616	.348695	.468971	.755979	.927108	.998349
Wilcoxon	.091605	.153951	.239608	.345843	.465607	.752500	.925419	.998300
Kolmogorov-Smirnov	.086093	.138243	.207750	.293380	.391119	.645811	.840851	.985292
Median	.083807	.132579	.198098	.280010	.375315	.633004	.838271	.987563
$m = 7, n = 7, \alpha = .01$								
t	.023203	.048671	.092592	.160375	.254070	.564531	.836812	.995400
c_1 (normal scores)	.022986	.047321	.089051	.152967	.241011	.535269	.805793	.990138
Wilcoxon	.022801	.047134	.088646	.152294	.240128	.535000	.807526	.991453
Kolmogorov-Smirnov	.020808	.039954	.071048	.117447	.181230	.408721	.666931	.956515
Median	.020575	.038988	.068296	.111067	.168485	.365057	.582377	.871569

$m = 7, n = 7, \alpha = .05$

t	.098292	.174075	.279150	.407919	.547450	.840864	.969582	.999854
c_1 (normal scores)	.096909	.169828	.270484	.393980	.528795	.821065	.960526	.999652
Wilcoxon	.096541	.168787	.268527	.391084	.525236	.818054	.959479	.999646
Kolmogorov-Smirnov	.089200	.147312	.226064	.323762	.434871	.713343	.900265	.996589
Median	.083271	.130294	.192402	.269277	.358617	.605501	.814960	.983917

$m = 7, n = 6, \alpha = .01$

t	.022323	.045462	.084711	.144908	.228461	.515365	.792557	.990688
c_1 (normal scores)	.022006	.044243	.081553	.138317	.216759	.487511	.759330	.982500
Wilcoxon	.021969	.044123	.081268	.137780	.215906	.486111	.758507	.982911
Kolmogorov-Smirnov	.020694	.039655	.070550	.116883	.180934	.411295	.673236	.959926
Median	.019186	.034232	.057035	.089155	.131420	.278137	.460714	.794704

$m = 7, n = 6, \alpha = .05$

t	.095613	.166161	.263590	.383781	.516275	.810595	.957015	.999643
c_1 (normal scores)	.094251	.161994	.255031	.369774	.496999	.787583	.944323	.999131
Wilcoxon	.094004	.161336	.253863	.368148	.497769	.786447	.944320	.999198
Kolmogorov-Smirnov	.089492	.148319	.228354	.327874	.441059	.722669	.907000	.997124
Median	.082232	.128176	.189652	.266795	.357466	.609933	.821479	.985339

$m = 7, n = 5, \alpha = .01$

t	.021311	.041865	.075998	.127853	.199980	.456697	.731756	.980055
c_1 (normal scores)	.020975	.040620	.072833	.121296	.188296	.427249	.692024	.965204
Wilcoxon	.020957	.040553	.072671	.120976	.187750	.426010	.690489	.964678
Kolmogorov-Smirnov	.019103	.037119	.064422	.104706	.159909	.359892	.600334	.919765
Median	.019096	.034423	.058583	.094204	.143340	.326279	.558809	.900355

$m = 7, n = 5, \alpha = .05$

t	.092455	.156936	.245447	.355296	.478609	.769775	.936660	.999027
c_1 (normal scores)	.091187	.153114	.237615	.342359	.460421	.745654	.920757	.997878
Wilcoxon	.090865	.152211	.235892	.339703	.456934	.741783	.918695	.997845
Kolmogorov-Smirnov	.086007	.138981	.210221	.298725	.400431	.666165	.864097	.991826
Median	.079562	.119622	.170708	.232317	.302936	.502291	.696383	.935436

Table 3. Effect of change in sample size on power of two-sample tests (one-sided)

m	n	d = .2	.4	.6	.8	1.0	1.5	2.0	3.0
		t test							
		$\alpha = .01$							
5	5	.019781	.036604	.063498	.103497	.158904	.364215	.616745	.941108
7	5	.021311	.041865	.075998	.127853	.199980	.456697	.731756	.980055
6	6	.021529	.042625	.077818	.131391	.205877	.469032	.745116	.982808
7	6	.022323	.045462	.084711	.144908	.228461	.515365	.792557	.990688
7	7	.023203	.048671	.092592	.160375	.254070	.564531	.836812	.995400
		t test							
		$\alpha = .05$							
5	5	.087800	.143539	.219105	.313330	.421448	.698489	.891583	.996038
7	5	.092455	.156936	.245447	.355296	.478609	.769775	.936660	.999027
6	6	.093207	.159116	.249725	.362029	.487576	.779866	.942029	.999225
7	6	.095613	.166161	.263590	.383781	.516275	.810595	.957015	.999643
7	7	.098292	.174075	.279150	.407919	.547450	.840864	.969582	.999854
		c_1 (normal scores) test							
		$\alpha = .01$							
5	5	.019451	.035430	.060585	.097515	.148161	.334549	.568186	.905400
7	5	.020975	.040620	.072833	.121296	.188296	.427249	.692024	.965204
6	6	.021201	.041394	.074652	.124771	.193997	.438808	.704358	.967803
7	6	.022004	.044243	.081553	.138317	.216759	.487511	.759330	.982500
7	7	.022986	.047321	.089051	.152967	.241011	.535269	.805793	.990138
		c_1 (normal scores) test							
		$\alpha = .05$							
5	5	.086712	.140313	.212492	.302224	.405316	.673585	.870504	.992794
7	5	.091187	.153114	.237615	.342359	.460421	.745654	.920757	.997878
6	6	.091893	.155152	.241616	.348695	.468971	.755979	.927108	.998349
7	6	.094251	.161994	.255031	.369774	.496999	.787583	.944323	.999131
7	7	.096909	.169828	.270484	.393980	.528795	.821065	.960526	.999652
		Wilcoxon test							
		$\alpha = .01$							
5	5	.019451	.035430	.060585	.097515	.148161	.334549	.568186	.905400
7	5	.020957	.040554	.072671	.120976	.187750	.426010	.690489	.964678
6	6	.021180	.041325	.074497	.124499	.193600	.438418	.704767	.968874
7	6	.021970	.044123	.081268	.137780	.215906	.486111	.758507	.982910
7	7	.022801	.047134	.088646	.152294	.240128	.535000	.807526	.991453

34

Wilcoxon test

α = .05

m	n								
5	5	.086613	.140046	.211997	.301471	.404324	.672395	.869734	.992740
7	5	.090865	.152211	.235892	.339703	.456935	.741782	.918695	.997845
6	6	.091605	.153951	.239608	.345843	.465607	.752500	.925419	.998300
7	6	.094004	.161336	.253863	.368148	.497769	.786447	.944320	.999198
7	7	.096541	.168787	.268527	.391084	.525236	.818054	.959479	.999646

α = .01

m	n								
5	5	.018186	.031279	.051015	.079107	.116956	.257104	.447080	.803946
7	5	.019103	.037119	.064422	.104706	.159909	.359892	.600334	.919765
6	6	.020094	.037505	.065212	.105975	.161541	.359811	.592286	.898963
7	6	.020694	.039655	.070550	.116883	.180934	.411295	.673236	.959926
7	7	.020808	.039954	.071048	.117447	.181230	.408721	.666931	.956515

Kolmogorov-Smirnov test

α = .05

m	n								
5	5	.082306	.128125	.188976	.264698	.353035	.597811	.806519	.981022
7	5	.086007	.138981	.210221	.298725	.400431	.666165	.864097	.991826
6	6	.086093	.138243	.207750	.293380	.391119	.645811	.840851	.985292
7	6	.089492	.148319	.228354	.327874	.441059	.722669	.907000	.997124
7	7	.089200	.147312	.226064	.323762	.434871	.713343	.900265	.996589

Kolmogorov-Smirnov test

α = .01

m	n								
5	5	.017158	.028278	.044837	.068465	.100747	.225883	.408371	.781944
7	5	.019096	.034423	.058583	.094204	.143340	.326279	.558809	.900355
6	6	.018189	.031076	.050094	.076553	.111455	.237837	.411451	.770401
7	6	.019186	.034232	.057035	.089155	.131420	.278137	.460714	.794704
7	7	.020575	.038988	.068296	.111067	.168485	.365057	.582377	.871569

Median test

α = .05

m	n								
5	5	.077591	.114247	.160195	.214768	.276485	.448204	.618812	.872697
7	5	.079562	.119622	.170708	.232317	.302936	.502291	.696383	.935436
6	6	.083807	.132579	.198098	.280010	.375315	.633004	.838271	.987563
7	6	.082232	.128176	.189652	.266795	.357466	.609933	.821479	.985339
7	7	.083271	.130294	.192402	.269277	.358617	.605501	.814960	.983917

Median test

Table 4. Comparison of the power of one- and two-sided two-sample tests

Test	d = .2	.4	.6	.8	1.0	1.5	2.0	3.0
				$m = 7, n = 5, \alpha = .01$				
Student's t								
1-sided	.021311	.041865	.075998	.127853	.199980	.456697	.731756	.980055
2-sided	.013217	.023758	.044101	.077718	.128056	.334452	.607505	.950786
Normal scores (c_1)								
1-sided	.020975	.040620	.072833	.121296	.188296	.427249	.692024	.965204
2-sided	.013066	.023047	.042109	.073267	.119507	.308041	.562542	.921596
Wilcoxon								
1-sided	.020957	.040554	.072671	.120976	.187750	.426010	.690489	.964678
2-sided	.013066	.023046	.042106	.073261	.119496	.308000	.562461	.921512
Kolmogorov-Smirnov								
1-sided	.019103	.037119	.064422	.104706	.159909	.359892	.600334	.919765
2-sided	.012579	.020827	.036167	.060583	.096102	.240080	.446268	.822299
Median								
1-sided	.019096	.034423	.058583	.094204	.143340	.326279	.558809	.900355
2-sided	.012464	.020273	.034555	.056785	.088028	.208306	.363311	.592831
				$m = 7, n = 7, \alpha = .01$				
Student's t								
1-sided	.023203	.048671	.092592	.160375	.254070	.564531	.836812	.995400
2-sided	.014097	.027821	.055089	.101080	.170187	.439317	.739367	.986475
Normal scores (c_1)								
1-sided	.022986	.047321	.089051	.152967	.241011	.535269	.805793	.990138
2-sided	.013937	.027037	.052816	.095899	.160228	.411049	.700255	.974308
Wilcoxon								
1-sided	.022801	.047134	.088646	.152294	.240128	.535000	.807526	.991453
2-sided	.013927	.026995	.052709	.095699	.159929	.410892	.701223	.975789
Kolmogorov-Smirnov								
1-sided	.020808	.039954	.071048	.117447	.181230	.408721	.666931	.956515
2-sided	.013303	.024125	.044981	.079329	.130474	.337194	.605722	.944788
Median								
1-sided	.020575	.038988	.068296	.111067	.168485	.365057	.582377	.871569
2-sided	.012605	.020848	.035867	.059131	.091937	.218066	.393780	.755157

Table 5. *Hodges-Lehmann efficiency of the Wilcoxon test*

Test	$m = n$	α	$d = .2$	.4	.6	.8	1.0	1.5	2.0	3.0
Wilcoxon	5	.01	.019451	.035430	.060585	.097515	.148161	.334549	.568186	.905400
t	4	.01	.017923	.030556	.049637	.076968	.114140	.254781	.450658	.821336
t	5	.01	.019781	.036604	.063498	.103497	.158904	.364215	.616745	.941108
$e_{w,t} =$			.9645	.9612	.9580	.9549	.9520	.9458	.9415	.9404
Wilcoxon	6	.01	.021180	.041325	.074497	.124499	.193600	.438418	.704767	.968874
t	5	.01	.019781	.036604	.063498	.103497	.158904	.364215	.616745	.941108
t	6	.01	.021529	.042625	.077818	.131391	.205877	.469032	.745116	.982808
$e_{w,t} =$			.9667	.9640	.9613	.9588	.9564	.9513	.9476	.9443
Wilcoxon	7	.01	.022801	.047134	.088646	.152294	.240128	.535000	.807526	.991453
t	6	.01	.021203	.042625	.077818	.131391	.205877	.469032	.745116	.982808
t	7	.01	.023203	.048671	.092592	.160375	.254070	.564531	.836812	.995400
$e_{w,t} =$			.9657	.9637	.9618	.9602	.9587	.9558	.9544	.9552
Wilcoxon	5	.05	.086613	.140046	.211997	.301471	.404324	.672395	.869734	.992740
t	4	.05	.081918	.127045	.186890	.261375	.348431	.591364	.801534	.980600
t	5	.05	.087800	.143539	.219105	.313330	.421448	.698489	.891583	.996038
$e_{w,t} =$			.9597	.9576	.9559	.9543	.9531	.9513	.9515	.9573
Wilcoxon	6	.05	.091605	.153951	.239608	.345843	.465607	.752500	.925419	.998300
t	5	.05	.087800	.143539	.219105	.313330	.421448	.698489	.891583	.996038
t	6	.05	.093207	.159116	.249725	.362029	.487576	.779866	.942029	.999225
$e_{w,t} =$			.9506	.9447	.9449	.9446	.9440	.9440	.9451	.9516
Wilcoxon	7	.05	.096541	.168787	.268527	.391084	.525236	.818054	.959479	.999646
t	6	.05	.093207	.159116	.249725	.362029	.487576	.779866	.942029	.999225
t	7	.05	.098292	.174075	.279150	.407919	.547450	.840864	.969582	.999854
$e_{w,t} =$			.9508	.9495	.9484	.9476	.9470	.9466	.9476	.9528

Table 6. Hodges-Lehmann efficiency of the c_1 test

Test	$m = n$	α	$d = .2$	.4	.6	.8	1.0	1.5	2.0	3.0
c_1	5	.01	.019451	.035430	.060585	.097515	.148161	.334549	.568186	.905400
t	4	.01	.017923	.030556	.049637	.076968	.114140	.254781	.450658	.821336
t	5	.01	.019781	.036604	.063498	.103497	.158904	.364215	.616745	.941108
$e_{c_1,t} =$			.9645	.9612	.9580	.9549	.9520	.9458	.9415	.9404
c_1	6	.01	.021201	.041394	.074652	.124771	.193997	.438808	.704358	.967803
t	5	.01	.019781	.036604	.063498	.103497	.158904	.364215	.616745	.941108
t	6	.01	.021529	.042625	.077818	.131391	.205877	.469032	.745116	.982808
$e_{c_1,t} =$			.9687	.9659	.9632	.9604	.9578	.9519	.9471	.9400
c_1	7	.01	.022986	.047321	.089051	.152967	.241011	.535269	.805793	.990138
t	6	.01	.021529	.042625	.077818	.131391	.205877	.469032	.745116	.982808
t	7	.01	.023203	.048671	.092592	.160375	.254070	.564531	.836812	.995400
$e_{c_1,t} =$			.9815	.9681	.9658	.9635	.9613	.9562	.9517	.9403
c_1	5	.05	.086712	.140313	.212492	.302224	.405316	.673585	.870504	.992794
t	4	.05	.081918	.127045	.186890	.261375	.348431	.591364	.801534	.980600
t	5	.05	.087800	.143539	.219105	.313330	.421448	.698489	.891583	.996038
$e_{c_1,t} =$			.9630	.9609	.9589	.9572	.9558	.9535	.9532	.9580
c_1	6	.05	.091893	.155152	.241616	.348695	.468971	.755979	.927108	.998349
t	5	.05	.087800	.143539	.219105	.313330	.421448	.698489	.891583	.996038
t	6	.05	.093207	.159116	.249725	.362029	.487576	.779866	.942029	.999225
$e_{c_1,t} =$			.9595	.9576	.9559	.9544	.9531	.9511	.9507	.9542
c_1	7	.05	.096909	.169828	.270484	.393980	.528795	.821065	.960526	.999652
t	6	.05	.093207	.159116	.249725	.362029	.487576	.779866	.942029	.999225
t	7	.05	.098292	.174075	.279150	.407919	.547450	.840864	.969582	.999854
$e_{c_1,t} =$			.9611	.9594	.9579	.9566	.9555	.9536	.9530	.9542

Table 7. Hodges-Lehmann efficiency of the median test

Test	$m = n$	α	$d =$ .2	.4	.6	.8	1.0	1.5	2.0	3.0
Median	5	.01	.017158	.028278	.044837	.068465	.100747	.225883	.408371	.781944
t	3	.01	.015886	.024387	.036230	.052164	.072895	.149099	.261274	.550564
t	4	.01	.017923	.030556	.049637	.076968	.114140	.254781	.450658	.821336
$e_{M,t} =$			.7249	.7261	.7284	.7314	.7351	.7453	.7553	.7709
Median	6	.01	.018189	.031076	.050094	.076553	.111455	.237837	.411451	.770401
t	3	.01	.015886	.024387	.036230	.052164	.072895	.149099	.261274	.550564
t	4	.01	.017923	.030556	.049637	.076968	.114140	.254781	.450658	.821336
t	5	.01	.019781	.036604	.063498	.103497	.158904	.364215	.616745	.941108
$e_{M,t} =$			.6905	.6810	.6722	.6639	.6558	.6399	.6322	.6353
Median	7	.01	.020575	.038988	.068296	.111067	.168485	.365057	.582377	.871569
t	4	.01	.017923	.030556	.049637	.076968	.114140	.254781	.450658	.821336
t	5	.01	.019781	.036604	.063498	.103497	.158904	.364215	.616745	.941108
t	6	.01	.021529	.042625	.077818	.131391	.205877	.469032		
$e_{M,t} =$			.7792	.7708	.7622	.7531	.7434	.7154	.6847	.6313
Median	5	.05	.077591	.114247	.160195	.214768	.276485	.448204	.618812	.872697
t	2	.05						.270725	.383889	.617222
t	3	.05	.075244	.108982	.152139	.205016	.267114	.451838	.645202	.909800
t	4	.05	.081918	.127045	.186890	.261375	.348431	.591364	.801534	.980600
$e_{M,t} =$			.6703	.6583	.6464	.6346	.6230	.5960	.5798	.5746
Median	6	.05	.083807	.132579	.198098	.280010	.375315	.633004	.838271	.987563
t	4	.05	.081918	.127045	.186890	.261375	.348431	.591364	.801534	.980600
t	5	.05	.087800	.143539	.219105	.313330	.421448	.698489	.891583	.996038
$e_{M,t} =$			.7202	.7226	.7247	.7264	.7280	.7315	.7347	.7418
Median	7	.05	.083271	.130294	.192402	.269277	.358617	.605501	.814960	.983917
t	4	.05	.081918	.127045	.186890	.261375	.348431	.591364	.801534	.980600
t	5	.05	.087800	.143539	.219105	.313330	.421448	.698489	.891583	.996038
$e_{M,t} =$			.6043	.5996	.5959	.5932	.5914	.5903	.5927	.6021

39

Table 8. Hodges-Lehmann efficiency of the Kolmogorov-Smirnov test

Test	$m = n$	α	$d = .2$	.4	.6	.8	1.0	1.5	2.0	3.0
Kolmogorov-Smirnov	5	.01	.018186	.031279	.051015	.079107	.116956	.257104	.447080	.803946
t	3	.01							.261274	.550564
t	4	.01	.017923	.030556	.049637	.076968	.114140	.254781	.450658	.821336
t	5	.01	.019781	.036604	.063498	.103497	.158904	.364215		
$e_{\mathrm{KS},t} =$			.8283	.8239	.8199	.8161	.8126	.8042	.7962	.7872
Kolmogorov-Smirnov	6	.01	.020094	.037505	.065212	.105975	.161541	.359811	.592286	.898963
t	4	.01							.450658	.821336
t	5	.01	.019781	.036604	.063498	.103497	.158904	.254781	.616745	
t	6	.01	.021529	.042625	.077818	.131391	.205877	.364215		.941108
$e_{\mathrm{KS},t} =$			.8632	.8583	.8533	.8481	.8427	.8266	.8088	.7747
Kolmogorov-Smirnov	7	.01	.020808	.039954	.071048	.117447	.181230	.408721	.666931	.956515
t	5	.01	.019781	.036604	.063498	.103497	.158904	.364215	.616745	.941108
t	6	.01	.021529	.042625	.077818	.131391	.205877	.469032	.745116	.982808
$e_{\mathrm{KS},t} =$			.7982	.7938	.7896	.7857	.7822	.7749	.7701	.7671
Kolmogorov-Smirnov	5	.05	.082306	.128125	.188976	.264698	.353035	.597811	.806519	.981022
t	4	.05	.081918	.127045	.186890	.261375	.348431	.591364	.801534	.980600
t	5	.05	.087800	.143539	.219105	.313330	.421448	.698489	.891583	.996038
$e_{\mathrm{KS},t} =$			.8132	.8131	.8130	.8128	.8126	.8120	.8111	.8055
Kolmogorov-Smirnov	6	.05	.086093	.138243	.207750	.293380	.391119	.645811	.840851	.985292
t	4	.05	.081918	.127045	.186890	.261375	.348431	.591364	.801534	.980600
t	5	.05	.087800	.143539	.219105	.313330	.421448	.698489	.891583	.996038
$e_{\mathrm{KS},t} =$			.7850	.7798	.7746	.7693	.7641	.7514	.7394	.7173
Kolmogorov-Smirnov	7	.05	.089200	.147312	.226064	.323762	.434871	.713343	.900265	.996589
t	5	.05	.087800	.143539	.219105	.313330	.421448	.698489	.891583	.996038
t	6	.05	.093207	.159116	.249725	.362029	.487576	.779866	.942029	.999225
$e_{\mathrm{KS},t} =$			.7513	.7489	.7468	.7449	.7433	.7404	.7389	.7390

Table 9. Hodges-Lehmann efficiency of the Wilcoxon test: unequal sample sizes

Test	m	n	α	$d = .2$	.4	.6	.8	1.0	1.5	2.0	3.0
Wilcoxon	7	6	.01	.021969	.044123	.081268	.137780	.215906	.486111	.758507	.982911
t	6	6	.01	.021529	.042625	.077818	.131391	.205877	.469032	.745116	.982808
t	7	7	.01	.022323	.045462	.084711	.144908	.228461	.515365	.792557	.990688
$e_{\mathrm{W},t} =$				.9657	.9637	.9616	.9594	.9572	.9514	.9448	.9241
Wilcoxon	7	7	.01	.022801	.047134	.088646	.152294	.240128	.535000	.807526	.991453
t	7	6	.01	.022323	.045462	.084711	.144908	.228461	.515365	.792557	.990688
t	7	7	.01	.023203	.048671	.092592	.160375	.254070	.564531	.836812	.995400
$e_{\mathrm{W},t} =$				.9674	.9658	.9642	.9627	.9611	.9571	.9527	.9402
Wilcoxon	7	6	.05	.094004	.161336	.253863	.368148	.497769	.786447	.944320	.999198
t	6	6	.05	.093207	.159116	.249725	.362029	.487576	.779866	.942029	.999225
t	7	7	.05	.095613	.166161	.263590	.383781	.516275	.810595	.957015	.999643
$e_{\mathrm{W},t} =$				.9486	.9473	.9460	.9447	.9504	.9396	.9348	.9181
Wilcoxon	7	7	.05	.096541	.168787	.268527	.391084	.525236	.818054	.959479	.999646
t	7	6	.05	.095613	.166161	.263590	.383781	.516275	.810595	.957015	.999643
t	7	7	.05	.098292	.174075	.279150	.407919	.547450	.840864	.969582	.999854
$e_{\mathrm{W},t} =$				.9533	.9523	.9512	.9502	.9491	.9462	.9426	.9296

The H-L efficiencies for the Wilcoxon and c_1 tests are seen to be essentially independent of m, α, and d for the values used in Tables 5 and 6. The efficiencies of the two tests differ occasionally by as much as 1 in the second decimal place but usually by a lesser amount. The efficiency of the c_1 test is generally (but not always) greater than that of the Wilcoxon test. The Wilcoxon test H-L efficiency is seen to be very close to and occasionally greater than the asymptotic relative efficiency $3/\pi = .9580$ as given by Pitman (1949) and van der Vaart (1950). However, the H-L efficiency of the c_1 test is not as close to (differs by more than .03 from) the ARE of 1 given by Chernoff and Savage (1958). Witting (1960) gives another definition of efficiency for small parameter values and reports an efficiency for the Wilcoxon test ($m = n = 5$) of .9563.

The H-L efficiencies for the median and Kolmogorov-Smirnov tests are also seen to be essentially independent of m, α, and d for the values given in Tables 7 and 8, except that the efficiency of the Kolmogorov-Smirnov test is seen to decrease as d increases for α and m as in Table 8. The efficiency of the Kolmogorov-Smirnov test is generally (but not always) greater than that of the median test. The median test H-L efficiency is seen to take values straddling the ARE of $2/\pi = .6366$ given by Mood (1954).

CHAPTER 3

Sequential Two-Sample Rank Tests of the Normal Shift Hypothesis

1. SUMMARY

Two sequential two-sample rank tests, as developed by Wilcoxon, Rhodes, and Bradley (1963), are presented for testing the normal shift hypothesis. For both tests observations are taken in groups of $(m + n)$, m from one population and n from the other, and the observations are ranked within groups. Tables that facilitate the use of these tests are described, and values of the operating characteristic functions and average sample number functions are given.

2. INTRODUCTION

The general testing situation considered here is as follows. The random variables $X_1, \ldots, X_m$ and $Y_1, \ldots, Y_n$ are mutually independent. The X's (or Y's) have a common continuous cumulative distribution function $F(x)$ [or $G(x)$]. The null hypothesis is $H_0: G(x) = F(x)$. A particular alternative that has received rather extensive consideration is the Lehmann alternative:

$$H_L: G(x) \equiv F^k(x), \quad k \geq 0.$$

Savage (1956) has presented several nonsequential nonparametric tests of the null hypothesis against the Lehmann alternative in detail with a table, analogous to Table A, giving the distribution of rank orders under H_L and with a table of the power function of these tests. The sequential two-sample rank procedures of Wilcoxon, Rhodes, and Bradley (1963) are derived for this same alternative. In this book we treat both of these problems for the case of the normal shift alternative hypothesis; power functions of nonsequential tests were given in Chapter 2, and sequential tests are described in this chapter.

3. THE SEQUENTIAL TESTS

Consider the situation in which the random variables X and Y are normally and independently distributed with means μ_X and μ_Y, respectively, and variance 1. Denoting the cumulative distribution functions of X and Y by $F(x)$ and $G(y)$, respectively, we wish to use a sequential rank test of the null hypothesis H_0 against the alternative hypothesis H_1:

(3.1)
$$H_0: G(x) \equiv F(x),$$
$$H_1: G(x) \equiv F(x - d_1),$$

where $d_1 = \mu_Y - \mu_X \geq 0$, with Type I and Type II errors α and β, respectively. Applying the methods of Wald's (1947) sequential analysis, we use the following procedure. Two constants A and B $(0 < B < 1 < A)$ are to be chosen so that $A = (1 - \beta)/\alpha$ and $B = \beta/(1 - \alpha)$. At each stage or trial of the sequential procedure a group of $(m + n)$ independent observations is taken, consisting of m X-observations and n Y-observations. The $(m + n)$ observations are ranked within each group, and a probability (likelihood) ratio is computed for the group. The probability ratio for the tth group is denoted by

(3.2)
$$p_t = \frac{P_1}{P_0},$$

where P_0 (or P_1) is the probability density of the observed group ranking under the hypothesis H_0 (or H_1). If the vector of ranked observations is denoted by $\mathbf{u} = (u_1, u_2, \ldots, u_{m+n})$, a new vector $\mathbf{z} = (z_1, z_2, \ldots, z_{m+n})$ can be formed by letting $z_i = 0$ (or 1) if u_i is an X- (or Y-) observation $(i = 1, \ldots, m + n)$. The probability densities of the observed group ranking are then further denoted by

(3.3)
$$P_1 = P_{m,n}(\mathbf{z} \mid d_1) \quad \text{and} \quad P_0 = P_{m,n}(\mathbf{z} \mid 0),$$

where these quantities are to be found in Table A.

The probability ratio, denoted by r_t when t groups have been observed, is given by

(3.4)
$$r_t = \prod_{i=1}^{t} p_i = r_{t-1} p_t \quad (t = 1, 2, \ldots),$$

where $r_0 = 1$. The possible situations

(3.5)
(i) $B < r_t < A$,

(ii) $r_t \geq A$,

(iii) $r_t \leq B$,

determine the three respective actions to be taken, which are (i) continue the experiment by going to the $(t + 1)$st stage, (ii) terminate the process with the rejection of H_0, and (iii) terminate the process with the acceptance of H_0.

It is often computationally useful in the application of sequential tests of this type to use logarithms and to replace (3.5) by

(3.6)
$$\text{(i)} \quad \log B < \log r_t < \log A,$$
$$\text{(ii)} \quad \log r_t \geq \log A,$$
$$\text{(iii)} \quad \log r_t \leq \log B,$$

where in this presentation "log" refers to common logarithm (logarithm to the base 10). This procedure has actual Type I and II errors α' and β', which Wald [(1947), 3:27] has shown satisfy the inequality $\alpha' + \beta' \leq \alpha + \beta$, thus concluding that the differences $\alpha' - \alpha$ and $\beta' - \beta$ are small enough to be ignored for all practical purposes.

The two sequential rank tests considered by Wilcoxon, Rhodes, and Bradley provide as r_t the following ratios. The first test, the *SCR test* (sequential, configural rank test), is based upon the probability-ratio statistic $r_t^{(1)}$, whose value depends on the probabilities of the particular configuration of ordered observations obtained for each group. For this test

(3.7)
$$p_t = \frac{P_{m,n}(z^t \mid d_1)}{P_{m,n}(z^t \mid 0)} = P_{m,n}(z^t \mid d_1) \times \binom{m + n}{n}$$

and

(3.8)
$$r_t^{(1)} = \prod_{j=1}^{t} p_j = \binom{m + n}{n}^t \prod_{j=1}^{t} P_{m,n}(z^j \mid d_1);$$

here z^t denotes the **z** value observed at the tth stage of the experiment.

The *SRS test* (sequential, rank-sum test) is based on the probability-ratio statistic $r_t^{(2)}$, whose value depends on the probabilities of the within-group rank sum of the Y-observations for each group. For this test

(3.9)
$$p_t = \frac{\Pr\{w^t \mid d_1\}}{\Pr\{w^t \mid 0\}}$$

and

(3.10)
$$r_t^{(2)} = \prod_{j=1}^{t} p_j = \prod_{j=1}^{t} \frac{\Pr\{w^j \mid d_1\}}{\Pr\{w^j \mid 0\}},$$

where

$$w^j = w_Y^j = \sum_{i=1}^{m+n} i z_i^j,$$

the sum of the ranks of the Y-sample at the jth stage, $j = 1, 2, \ldots, t$. Clearly

$r_t^{(2)}$ could be written in terms of

$$w_X^j = \sum_{i=1}^{m+n} i(1 - z_i^j),$$

the sum of the ranks of the X-sample at the jth stage. Note that

$$w_X = \frac{(m + n)(m + n + 1)}{2} - w_Y.$$

4. COMPUTATION OF r_t

a. SCR test: $r_t^{(1)}$

Computation of $r_t^{(1)}$ may be done directly from Table A. Because of the large size of Table A no attempt has been made to provide values of $\log r_t^{(1)}$. In the notation of (3.4), (3.8) may be written as

$$(4.1) \qquad r_t^{(1)} = r_{t-1}^{(1)} \times P_{m,n}(z^t \mid d_1) \times \binom{m + n}{n} \equiv r_{t-1}^{(1)} r(z^t \mid d_1),$$

where $r_0^{(1)} \equiv 1$; $P_{m,n}(z \mid d_1)$ and $\binom{m + n}{n} = C(m, n)$ are found in Table A.

b. SRS test: $r_t^{(2)}$

Modifying slightly the notation of (3.4), we may write (3.10) as

$$(4.2) \qquad r_t^{(2)} = r_{t-1}^{(2)} \times \frac{\Pr\{w^t \mid d_1\}}{\Pr\{w^t \mid 0\}} \equiv r_{t-1}^{(2)} R(w^t \mid d_1)$$

or, taking logarithms, as

$$(4.3) \qquad \log r_t^{(2)} \equiv \log r_{t-1}^{(2)} + \log R(w^t \mid d_1),$$

where $r_0^{(2)} \equiv 1$ and $\log r_0^{(2)} \equiv 0$. Table C-1, derived from Table A, gives the distribution of w under the hypotheses $d = .0(.2)1.0, 1.5, 2.0, 3.0$. Both w_Y and w_X are given to facilitate use of the table. Table C-2 gives values of $\log R(w \mid d)$ for $d = .2(.2)1.0, 1.5, 2.0, 3.0$. Both tables are for sample sizes $1 \le n \le m \le 7$.

5. PROPERTIES OF THE SEQUENTIAL TESTS

Proof that the SCR and SRS tests terminate with probability 1 follows from Wald [(1947), p. 157]. In addition, the Wald sequential analysis provides two concepts that describe the properties of sequential tests and measure their performance: the OC function and the ASN function.

a. The Operating Characteristic Function (OC Function)

The OC function, $L(d)$, is the probability of accepting H_0 when d is the true value of the shift in location between the two populations. Wald's formulas (3:29a) and (3:43) provide approximations to $L(d)$ at two points: $L(0) \doteq 1 - \alpha$ and $L(d_1) \doteq \beta$. However, these points provide no way to compare performances of two tests. A third point at which an approximation to $L(d)$ may be evaluated with relative ease is at $d = d'$, where d' is such that $\mathscr{E}_{d'}(\log p_t) = 0$. [See Wald (1947), Appendix A.3.1.] From Wald's formula (A:98),

$$L(d') \doteq \frac{\log \left[(1 - \beta)/\alpha \right]}{\log \left[(1 - \beta)/\alpha \right] - \log \left[\beta/(1 - \alpha) \right]}.$$

Values of d' were obtained by inverse interpolation with a fifth-degree polynomial among values of $\mathscr{E}_{d'}(\log p_t)$.

As observed by Wilcoxon, Rhodes, and Bradley for the test with the Lehmann alternative, for both tests the value of d' is not very dependent on m and n when $1 \leq n \leq m \leq 7$. Furthermore, the values of d' are essentially the same for both tests, given m, n, and d_1. Table 1 gives the values of d' for $m = n = 1(1)7$ for the rank-sum test for $d_1 = .2(.2)1.0, 1.5, 2.0, 3.0$. For $m = n = 1(1)5$, only one case was found in which the configural rank test gave a result differing by more than 1 in the third decimal place.

b. The Average Sample Number Function (ASN Function)

The ASN function, generally of more interest in sequential tests than the OC function, gives the average (or expected) sample sizes required for termination of the test for various true values of d. Although the values of the ASN function depend upon d_1, α, β, and d, it is convenient to follow the notation of Wilcoxon, Rhodes, and Bradley and write the function as

Table 1. *Values of d' such that $\mathscr{E}_{d'}(\log p_t) = 0$, for alternative hypotheses $d_1 = .2(.2)1.0, 1.5, 2.0, 3.0$*

$m = n$	d_1							
	.2	.4	.6	.8	1.0	1.5	2.0	3.0
1	.100	.200	.299	.398	.495	.734	.961	1.374
2	.100	.200	.299	.399	.497	.740	.976	1.415
3	.100	.200	.300	.399	.498	.743	.982	1.434
4	.100	.200	.300	.399	.498	.744	.986	1.445
5	.100	.200	.300	.399	.499	.745	.988	1.452
6	.100	.200	.300	.399	.499	.746	.989	1.456
7	.100	.200	.300	.399	.499	.746	.990	1.459

ASN(d). Wald gives approximate formulas for ASN(d) for $d \neq d'$ (3:57) and $d = d'$ (A:99), which, in the notation of this chapter, are

$$(5.1) \quad \text{ASN}(d) \doteq \frac{L(d) \log \left[\beta/(1 - \alpha)\right] + \left[1 - L(d)\right] \log \left[(1 - \beta)/\alpha\right]}{\mathscr{E}_d(\log p_t)}$$

and

$$(5.2) \quad \text{ASN}(d') \doteq \frac{-\log \left[\beta/(1 - \alpha)\right] \log \left[(1 - \beta)/\alpha\right]}{\mathscr{E}_{d'}(\log^2 p_t)},$$

where p_t is $r(z \mid d_1)$ or $R(w \mid d_1)$, and $\mathscr{E}_d$ and $\mathscr{E}_{d'}$ denote expectation with respect to shift parameters d and d', respectively. Using Tables C-1 and C-2, one can compute these expectations in the usual way for discrete probability distributions by summing the products of the possible values of $\log p_t$ (and $\log^2 p_t$) and the corresponding probabilities of p_t based upon the appropriate value of d.

As noted by Wilcoxon, Rhodes, and Bradley, the values of ASN(d) from (5.1) and (5.2) are in terms of the number of groups of size $m + n$ of observations required (or in terms of the number of stages, t). In a slight modification of their procedure, ASN(d) is converted to the average total number of observations needed, rather than to the average number of Y-observations, thus providing for the case $m \neq n$. In Table 2 values of $(m + n)$ASN(d) are

Table 2. *Values for the ASN function converted to total sample sizes for the SRS (sequential, rank-sum) test*

$m = n$	$\alpha = \beta = .01$		
	$H_0 : d_1 = 0$	d'	$H_1 : d_1 = .2$
1	1396.9	3318.8	1396.9
2	1185.3	2746.6	1185.3
3	1096.7	2569.4	1107.0*
4	1057.1	2489.1*	1064.3
5	1034.8	2428.4*	1040.3*
6	1020.4	2401.5	1024.8
7	1010.3	2372.6	1014.1
	$H_0 : d_1 = 0$	d'	$H_1 : d_1 = .4$
1	355.6	829.7	358.8
2	294.1	692.6	299.7
3	274.8	647.6	279.4
4	265.2	624.7	269.3
5	259.4	610.8	263.6
6	255.4	601.1	259.6
7	252.6	594.4	256.6

* ASN for SRS test is greater than ASN for SCR test.

Table 2 **(Continued)**

$m = n$		$\alpha = \beta = .01$	
	$H_0 : d_1 = 0$	d'	$H_1 : d_1 = .6$
1	157.7	372.2	163.7
2	130.8	309.9	135.8
3	122.2	288.9	126.7
4	117.9	279.0	122.1
5	115.3	272.8	119.4
6	113.6	268.6	117.6
7	112.4	265.6	116.3
	$H_0 : d_1 = 0$	d'	$H_1 : d_1 = .8$
1	88.9	211.8	94.9
2	73.7	175.6	78.7
3	68.8	163.8	73.3
4	66.4	158.0	70.6
5	64.9	154.5	69.0
6	63.9	152.2	68.0
7	63.2	150.5	67.2
	$H_0 : d_1 = 0$	d'	$H_1 : d_1 = 1.0$
1	57.0	137.1	63.3
2	47.2	113.5	52.3
3	44.0	105.8	48.6
4	42.5	102.1	46.8
5	41.6	99.8	45.7
6	40.9	98.3	45.0
7	40.5	97.2	44.5
	$H_0 : d_1 = 0$	d'	$H_1 : d_1 = 1.5$
1	25.6	63.3	32.1
2	21.1	52.1	26.3
3	19.6	48.5	24.4
4	18.9	46.8	23.4
5	18.5	45.7	22.9
6	18.2	45.0	22.5
7	18.0	44.5	22.2
	$H_0 : d_1 = 0$	d'	$H_1 : d_1 = 2.0$
1	14.5	37.4	21.6
2	11.9	30.6	17.5
3	11.1	28.4	16.1
4	10.7	27.4	15.4
5	10.5	26.7	15.0
6	12.0†	26.3	14.7
7	14.0†	26.0	14.5

† For cases of large m, n, and d_1 marked † the tabled value is the minimum possible sample size since this value is greater than the ASN calculated from (5.1).

Table 2 **(Continued)**

$m = n$		$\alpha = \beta = .01$	
	$H_0 : d_1 = 0$	d'	$H_1 : d_1 = 3.0$
1	6.7	18.5	14.8
2	5.4	15.0	11.7
3	6.0†	13.9	10.7
4	8.0†	13.4	10.1
5	10.0†	13.1	10.0†
6	12.0†	12.9	12.0†
7	14.0†	14.0†	14.0†

$m = n$		$\alpha = \beta = .05$	
	$H_0 : d_1 = 0$	d'	$H_1 : d_1 = .2$
1	822.1	1362.7	822.1
2	697.5	1127.7	697.5
3	645.4	1055.0	651.4
4	622.1	1022.0	626.3
5	608.9	997.1	612.2
6	600.5	986.1	603.1
7	594.5	974.2	596.8
	$H_0 : d_1 = 0$	d'	$H_1 : d_1 = .4$
1	209.3	340.7	211.2
2	173.1	284.4	176.4
3	161.7	265.9	164.4
4	156.1	256.5	158.5
5	152.6	250.8	155.1
6	150.3	246.8	152.8
7	148.6	244.1	151.0
	$H_0 : d_1 = 0$	d'	$H_1 : d_1 = .6$
1	92.8	152.8	96.3
2	77.0	127.3	79.9
3	71.9	118.6	74.6
4	69.4	114.6	71.9
5	67.9	112.0	70.3
6	66.8	110.3	69.2
7	66.1	109.1	68.4
	$H_0 : d_1 = 0$	d'	$H_1 : d_1 = .8$
1	52.3	87.0	55.9
2	43.4	72.1	46.3
3	40.5	67.2	43.1
4	39.1	64.9	41.6
5	38.2	63.4	40.6
6	37.6	62.5	40.0
7	37.2	61.8	39.5

† For cases of large m, n, and d_1 marked † the tabled value is the minimum possible sample size since this value is greater than the ASN calculated from (5.1).

50

Table 2   *(Continued)*

$m = n$		$\alpha = \beta = .05$	
	$H_0: d_1 = 0$	d'	$H_1: d_1 = 1.0$
1	33.6	56.3	37.2
2	27.8	46.6	30.8
3	25.9	43.5	28.6
4	25.0	41.9	27.6
5	24.5	41.0	26.9
6	24.1	40.3	26.5
7	23.8	39.9	26.2
	$H_0: d_1 = 0$	d'	$H_1: d_1 = 1.5$
1	15.0	26.0	18.9
2	12.4	21.4	15.5
3	11.6	19.9	14.4
4	11.1	19.2	13.8
5	10.9	18.8	13.5
6	12.0†	18.5	13.2
7	14.0†	18.3	14.0†
	$H_0: d_1 = 0$	d'	$H_1: d_1 = 2.0$
1	8.6	15.3	12.7
2	7.0	12.6	10.3
3	6.5	11.7	9.5
4	8.0†	11.2	9.1
5	10.0†	11.0	10.0†
6	12.0†	12.0†	12.0†
7	14.0†	14.0†	14.0†
	$H_0: d_1 = 0$	d'	$H_1: d_1 = 3.0$
1	3.9	7.6	8.7
2	4.0†	6.1	6.9
3	6.0†	6.0†	6.3
4	8.0†	8.0†	8.0†
5	10.0†	10.0†	10.0†
6	12.0†	12.0†	12.0†
7	14.0†	14.0†	14.0†

† For cases of large m, n, and d_1 marked † the tabled value is the minimum possible sample size since this value is greater than the ASN calculated from (5.1).

given for the SRS test [based on $r_t^{(2)}$] when $\alpha = \beta = .05$ and $\alpha = \beta = .01$ for $d_1 = .2(.2)1.0, 1.5, 2.0, 3.0$ and $m = n = 1(1)7$. The function is computed for $d = 0$, d', and d_1. Table 3 provides values of $\mathscr{E}_d(\log p_t)$, $d = 0$, d_1, and $\mathscr{E}_{d'}(\log^2 p_t)$ for the SRS test to facilitate calculation of ASN(d) for other values of α and β.

In Table 2, four cases were found in which the ASN for the SRS test differs from (is greater than) that for the SCR test by one to three multiples

		$d_1 = .2$			$d_1 = .4$		
m	n	A	B	C	A	B	C
1	1	$-.0028$	.0024	.0028	$-.0110$	.0096	.0109
2	1	$-.0041$	.0036	.0041	$-.0166$	.0143	.0163
3	1	$-.0050$	.0043	.0050	$-.0199$	.0172	.0196
2	2	$-.0066$	.0058	.0066	$-.0266$	.0230	.0261
4	1	$-.0055$	.0048	.0055	$-.0222$	.0192	.0219
3	2	$-.0083$	.0072	.0083	$-.0332$	.0287	.0327
5	1	$-.0060$	.0052	.0059	$-.0238$	.0206	.0235
4	2	$-.0095$	.0082	.0094	$-.0379$	.0328	.0373
3	3	$-.0107$	.0093	.0106	$-.0427$	.0369	.0420
6	1	$-.0063$	.0054	.0062	$-.0251$	.0217	.0248
5	2	$-.0104$	.0090	.0103	$-.0415$	.0359	.0409
4	3	$-.0124$	.0108	.0124	$-.0498$	.0431	.0490
7	1	$-.0065$	.0057	.0065	$-.0260$	.0226	.0257
6	2	$-.0111$	.0096	.0110	$-.0443$	.0383	.0436
5	3	$-.0138$	.0120	.0138	$-.0553$	.0478	.0544
4	4	$-.0148$	.0128	.0147	$-.0590$	.0510	.0581
7	2	$-.0116$	.0101	.0116	$-.0465$	.0402	.0458
6	3	$-.0149$	.0130	.0149	$-.0597$	.0517	.0588
5	4	$-.0166$	.0144	.0165	$-.0664$	.0574	.0653
7	3	$-.0158$	.0137	.0158	$-.0633$	.0548	.0624
6	4	$-.0181$	.0157	.0180	$-.0724$	.0626	.0713
5	5	$-.0189$	.0164	.0188	$-.0754$	.0652	.0742
7	4	$-.0194$	.0168	.0193	$-.0774$	.0670	.0762
6	5	$-.0207$	.0180	.0207	$-.0829$	.0718	.0817
7	5	$-.0223$	.0194	.0222	$-.0893$	.0773	.0879
6	6	$-.0230$	.0199	.0229	$-.0919$	.0795	.0904
7	6	$-.0249$	.0216	.0248	$-.0995$	.0861	.0980
7	7	$-.0271$	.0235	.0270	$-.1084$	.0938	.1067

Table 3 **(Continued)**

m	n	$d_1 = .6$			$d_1 = .8$		
		A	B	C	A	B	C
1	1	−.0248	.0214	.0239	−.0440	.0376	.0412
2	1	−.0373	.0321	.0359	−.0662	.0566	.0621
3	1	−.0448	.0386	.0433	−.0796	.0681	.0749
2	2	−.0598	.0514	.0576	−.1061	.0907	.0994
4	1	−.0499	.0430	.0483	−.0887	.0759	.0838
3	2	−.0746	.0642	.0720	−.1326	.1133	.1244
5	1	−.0536	.0462	.0520	−.0952	.0816	.0902
4	2	−.0853	.0735	.0824	−.1516	.1296	.1425
3	3	−.0960	.0827	.0926	−.1706	.1459	.1601
6	1	−.0564	.0486	.0548	−.1002	.0860	.0952
5	2	−.0933	.0804	.0902	−.1657	.1418	.1561
4	3	−.1119	.0964	.1080	−.1989	.1701	.1868
7	1	−.0586	.0506	.0570	−.1041	.0894	.0991
6	2	−.0995	.0857	.0963	−.1768	.1514	.1669
5	3	−.1244	.1071	.1201	−.2210	.1890	.2078
4	4	−.1327	.1142	.1281	−.2357	.2016	.2215
7	2	−.1045	.0901	.1012	−.1857	.1590	.1755
6	3	−.1343	.1157	.1298	−.2386	.2042	.2246
5	4	−.1492	.1285	.1441	−.2652	.2268	.2492
7	3	−.1424	.1227	.1377	−.2531	.2166	.2385
6	4	−.1628	.1402	.1572	−.2893	.2474	.2721
5	5	−.1696	.1460	.1638	−.3013	.2578	.2833
7	4	−.1741	.1499	.1682	−.3093	.2647	.2912
6	5	−.1865	.1606	.1802	−.3315	.2835	.3117
7	5	−.2009	.1730	.1941	−.3570	.3054	.3358
6	6	−.2066	.1779	.1996	−.3672	.3141	.3453
7	6	−.2238	.1928	.2162	−.3978	.3403	.3741
7	7	−.2437	.2099	.2355	−.4331	.3705	.4074

Table 3 (Continued)

m	n	$d_1 = 1.0$			$d_1 = 1.5$		
		A	B	C	A	B	C
1	1	−.0686	.0581	.0618	−.1530	.1258	.1217
2	1	−.1032	.0875	.0934	−.2309	.1903	.1855
3	1	−.1242	.1054	.1131	−.2784	.2301	.2261
2	2	−.1656	.1403	.1497	−.3712	.3055	.2973
4	1	−.1384	.1176	.1267	−.3105	.2574	.2549
3	2	−.2069	.1754	.1875	−.4639	.3823	.3737
5	1	−.1487	.1265	.1367	−.3338	.2773	.2765
4	2	−.2366	.2007	.2150	−.5307	.4378	.4301
3	3	−.2664	.2258	.2413	−.5976	.4923	.4812
6	1	−.1565	.1333	.1444	−.3514	.2926	.2936
5	2	−.2588	.2196	.2358	−.5805	.4796	.4736
4	3	−.3105	.2633	.2818	−.6967	.5744	.5630
7	1	−.1626	.1387	.1506	−.3653	.3048	.3074
6	2	−.2761	.2345	.2523	−.6195	.5124	.5085
5	3	−.3450	.2927	.3136	−.7743	.6388	.6279
4	4	−.3681	.3121	.3341	−.8262	.6811	.6676
7	2	−.2899	.2464	.2657	−.6505	.5388	.5372
6	3	−.3726	.3162	.3393	−.8362	.6904	.6809
5	4	−.4140	.3512	.3761	−.9294	.7666	.7525
7	3	−.3952	.3355	.3605	−.8868	.7328	.7252
6	4	−.4517	.3832	.4107	−1.0138	.8366	.8229
5	5	−.4706	.3991	.4275	−1.0563	.8713	.8555
7	4	−.4830	.4099	.4398	−1.0842	.8952	.8825
6	5	−.5176	.4391	.4705	−1.1620	.9587	.9422
7	5	−.5574	.4729	.5070	−1.2514	1.0328	1.0164
6	6	−.5733	.4864	.5212	−1.2875	1.0622	1.0440
7	6	−.6211	.5269	.5648	−1.3941	1.1508	1.1320
7	7	−.6764	.5738	.6151	−1.5183	1.2533	1.2329

Table 3   (*Continued*)

		$d_1 = 2.0$			$d_1 = 3.0$		
m	*n*	*A*	*B*	*C*	*A*	*B*	*C*
1	1	−.2689	.2132	.1814	−.5881	.4299	.2637
2	1	−.4074	.3242	.2791	−.8992	.6618	.4128
3	1	−.4922	.3935	.3432	−1.0918	.8107	.5152
2	2	−.6564	.5208	.4483	−1.4560	1.0651	.6676
4	1	−.5497	.4414	.3895	−1.2231	.9160	.5925
3	2	−.8208	.6524	.5660	−1.8219	1.3366	.8508
5	1	−.5914	.4768	.4253	−1.3185	.9950	.6540
4	2	−.9394	.7481	.6544	−2.0874	1.5357	.9928
3	3	−1.0582	.8406	.7293	−2.3525	1.7229	1.1002
6	1	−.6230	.5041	.4539	−1.3911	1.0569	.7049
5	2	−1.0277	.8202	.7240	−2.2832	1.6852	1.1081
4	3	−1.2340	.9815	.8554	−2.7411	2.0134	1.2978
7	1	−.6479	.5259	.4776	−1.4483	1.1069	.7481
6	2	−1.0969	.8771	.7807	−2.4411	1.8039	1.2048
5	3	−1.3714	1.0919	.9568	−3.0174	2.2361	1.4607
4	4	−1.4637	1.1642	1.0150	−3.2040	2.3849	1.5435
7	2	−1.1519	.9229	.8281	−2.5574	1.8982	1.2877
6	3	−1.4810	1.1806	1.0408	−3.2408	2.4168	1.5988
5	4	−1.6464	1.3105	1.1457	−3.5516	2.6816	1.7488
7	3	−1.5706	1.2535	1.1118	−3.4000	2.5601	1.7183
6	4	−1.7960	1.4306	1.2553	−3.8351	2.9261	1.9245
5	5	−1.8707	1.4899	1.3029	−3.9604	3.0479	1.9919
7	4	−1.9215	1.5312	1.3492	−4.0172	3.1232	2.0778
6	5	−2.0569	1.6396	1.4364	−4.2377	3.3486	2.2019
7	5	−2.2127	1.7665	1.5518	−4.4397	3.6017	2.3862
6	6	−2.2730	1.8162	1.5920	−4.5153	3.7016	2.4430
7	6	−2.4600	1.9683	1.7275	−4.6940	3.9939	2.6561
7	7	−2.6690	2.1433	1.8818	−4.8644	4.3344	2.8958

Table 4.   *Sample size required for one-sided nonsequential t-test, and optimum ASN for SRS (sequential, rank-sum) test, for normal shift alternatives d_1*

| | $\alpha = \beta = .05$ | | | $\alpha = \beta = .01$ | | |
| | SRS Test | | | SRS Test | | |
d_1	H_0	H_1	t Test	H_0	H_1	t Test
.6	67	69	122			
.8	38	40	70	64	68	140
1.0	24	27	46	41	45	90
1.5	11	14	22	18	23	42
2.0	6	9	14	11	15	26
3.0	4	6	8	5	10	14

of the total group size, $m = n = 1(1)5$. These cases are marked with an asterisk. For both $\alpha = \beta = .05$ and $\alpha = \beta = .01$ it is seen that the approximately optimum group size shows a slight tendency to decrease as d_1, the alternative hypothesis, increases.

It is interesting to compare average sample numbers from Table 2 with the sample sizes required for a nonsequential t test with the same α and β, as given in Table 4 [see Davies (1956), Table E-1]. These sequential tests have optimum (best choice of group sizes) average sample numbers that are generally about 50–60 % of the sample size of the nonsequential t test.

Bradley, Martin, and Wilcoxon (1964) give Monte Carlo estimates of the ASN function for the rank-sum test based on the Lehmann alternative hypothesis when the underlying distributions are normal and differ in both location and dispersion. It is difficult to make any comparison between their results and the present work because of the difference in alternative hypotheses.

6. OTHER SEQUENTIAL RANK TESTS

Two additional sequential rank tests that might be considered are based upon the ratios

$$(6.1) \qquad r_t^{(3)} = \frac{P_{tm,tn}(\mathbf{z} \mid d)}{P_{tm,tn}(\mathbf{z} \mid 0)}$$

and

$$(6.2) \qquad r_t^{(4)} = \frac{\Pr\{w^t \mid d\}}{\Pr\{w^t \mid 0\}},$$

where $\mathbf{z}$ and w^t at the tth stage are based upon the ranks of all $t(m + n)$ observations (i.e., all observations are re-ranked with respect to each other at each stage). Thus

$$\mathbf{z} = [z_1, z_2, \ldots, z_{t(m+n)}] \quad \text{and} \quad w^t = \sum_{i=1}^{t(m+n)} i z_i.$$

Properties of these tests are of interest because it seems reasonable to expect that a test based on the ranks of all available observations relative to each other should be more economical than one in which only the ranks of observations relative to each stage are used. However, the nonindependence of the successive ratios indicates that in this case the usual Wald sequential procedure requires some justification.

Bradley, Merchant, and Wilcoxon (1966) discuss these modified sequential tests in detail for the case of Lehmann alternatives and present Monte Carlo results on a modified, sequential, configural rank test (MSCR test) corresponding to a ratio of the type $r_t^{(3)}$. That the MSCR test is a sequential probability ratio test in the Wald sense is shown by the theory presented by Hall, Wijsman, and Ghosh (1965). Berk and Savage (1968) show that the MSCR test $r_t^{(3)}$ terminates with probability 1. Furthermore, Savage and Savage (1965) have shown that the Wald bounds, A and B of Section 3, are appropriate for the MSCR test. For the modified, sequential, rank-sum test (MSRS test) $r_t^{(4)}$ no such results are available. Table A and Table C-2 provide the means for evaluation of $r_t^{(3)}$ and $r_t^{(4)}$ for several stages of an experiment (up to seven stages if $m = n = 1$). A survey of sequential rank tests and related results is given by Bradley (1967).

CHAPTER 4

Rank Order Probabilities
and Selection Procedures

1. INTRODUCTION

Single-sample multiple-decision procedures for ranking or selecting normal populations according to the magnitude of the population means have been presented by Bechhofer (1954). Since then many authors have contributed to "ranking and selection" and "multiple-decision" problems. Recent publications that contain exposition, theory, and rather complete references in this area include Gupta (1965), Barr and Rizvi (1966), and Bechhofer, Kiefer, and Sobel (1968).

Let X_i be normally and independently distributed random variables with unknown means μ_i and common known variance σ^2, $i = 1, \ldots, k$. The ranked μ_i are denoted as

$$\mu_{[1]} \le \mu_{[2]} \le \cdots \le \mu_{[k]};$$

it is assumed that it is not known which population is associated with $\mu_{[i]}$, $i = 1, \ldots, k$. On the basis of N independent observations from each of the k populations we wish to make an inference about the ordering of the populations according to the magnitude of the means; we may say the "best" population has the largest mean, the "second best" population has the second largest mean, etc. This inference is based on the sample means. The sample mean from the ith population is denoted by $\bar{X}_i$, and the ranked $\bar{X}_i$ are denoted by

$$\bar{X}_{[1]} < \bar{X}_{[2]} < \cdots < \bar{X}_{[k]}.$$

2. GOALS AND PROCEDURES

The goal of an experimenter might be Goal I of Mahamunulu (1967), which is to select a subset of size s that contains at least c of the t best

58

populations. The experimenter must specify two positive constants $d*$ and $P*$ (<1). He then desires a procedure for which the probability of selecting correctly the subset s of populations is at least equal to $P*$ whenever the distance between $\mu_{[k-t]}$ and $\mu_{[k-t+1]}$ is at least $d*$.

Two cases of Goal I are of special interest, corresponding to $c = t$ when $s \geq t$ and $c = s$ when $s \leq t$ [see Mahamunulu (1967)]:

Goal 1. To select a subset of size s that contains the t best populations, $s \geq t$.

Goal 2. To select a subset of size s that includes any s of the t best populations, $s \leq t$. This goal was first pointed out by Sobel [see the footnote in Bechhofer (1954), p. 22].

Clearly these goals coincide when $s = t (= c)$, in which case the common goal is to select the t best populations and no others.

The procedure, then, is to take N observations from each population and compute the k sample means, $\bar{X}_1, \bar{X}_2, \ldots, \bar{X}_k$. The means are ranked, and the populations associated with the s largest sample means are selected.

For Goal I, Mahamunulu has shown that, to guarantee a probability of correct selection at least equal to $P*$ whenever $\mu_{[k-t+1]} - \mu_{[k-t]} \geq d*$ and (to make the problem nontrivial)

$$\binom{k}{s}^{-1} \sum_{i=1}^{\min(s,t)} \binom{t}{i}\binom{k-i}{s-i} < P* < 1,$$

N should be at least equal to the smallest integer equal to or greater than the solution of

$$(2.1) \quad P* = \frac{t!}{(t-c)!\,(c-1)!} \sum_{\alpha=0}^{s-c} \binom{k-t}{\alpha} \int_{-\infty}^{\infty} F^{k-t-\alpha}(x+d)[1-F(x+d)]^{\alpha}$$
$$\times F^{t-c}(x)[1-F(x)]^{c-1}\,dF(x),$$

where $d = d*\sqrt{N}/\sigma$ and $F(x)$ is the standard normal cumulative distribution function.

For Goal 2 ($s \leq t$), the lower bound for $P*$ is $\binom{t}{s} / \binom{k}{s}$ and (2.1) reduces to

$$(2.2) \quad P* = \frac{t!}{(t-s)!\,(s-1)!} \int_{-\infty}^{\infty} F^{k-t}(x+d)F^{t-s}(x)[1-F(x)]^{s-1}\,dF(x)$$
$$\equiv Q(s,t).$$

Bechhofer (1954) gives (2.2) and a table of d for the case $t = s$, for various values of $P*$, k, and t. His table is based on an unpublished table by Teichroew (1954) of the integral in (2.2) with $t = s$, which was computed for Bechhofer's use.

For Goal 1 ($s \geq t$), applying the dual problem result of Desu and Sobel (1968) to (2.2), we get

(2.3) $\quad P^* = Q(k - s, k - t)$

$$= \frac{(k - t)!}{(s - t)! \, (k - s - 1)!} \int_{-\infty}^{\infty} F^t(x + d) F^{s-t}(x) [1 - F(x)]^{k-s-1} \, dF(x).$$

Desu and Sobel show that, if instead it is desired to fix the sample size N and determine the subset size s, the same expressions for P^* apply, namely, (2.2) and (2.3). Their Table 1 gives λ roots (d roots) for the equation $Q(s, t) = P^*$, which they show may then be used to determine s.

3. A TABLE FOR USE IN SELECTION PROCEDURES

Solutions for the procedures involving Goals I, 1, and 2 may be obtained from Table D, derived from Table A. Let the probability that at least n' of the ones in the random vector $\mathbf{z}$ (see Chapter 1, Section 1) are among the r rightmost elements $z_{m+n-r+1}, z_{m+n-r+2}, \cdots, z_{m+n-1}, z_{m+n}$ be denoted by $P(n'; r, m, n, d)$. Alternatively, this is the probability that, in a random sample of $m + n$ observations (m X-observations from a standard normal distribution and n Y-observations from a normal distribution with mean d and variance 1, all $m + n$ observations mutually independent), at least n' of the Y-observations are among the r largest observations. This probability is expressed in terms of $P_{m,n}(\mathbf{z} \mid d)$ as

(3.1) $$P(n'; r, m, n, d) = \sum_{z \in \mathscr{Z}} P_{m,n}(\mathbf{z} \mid d)$$

for $n' \leq n, r$; the summation is over the set $\mathscr{Z}$ of $(m + n)$-element $\mathbf{z}$ vectors with at least n' ones among the r rightmost elements.

Values of $P(n'; r, m, n, d)$ are given in Table D to 6 decimal places for $1 \leq n \leq m \leq 7$ and $n = 1$, $m = 8(1)12$; $r = 1(1)[m/2]$; $n' = 1(1)n$; and $d = .0(.2)1.0, 1.5, 2.0, 3.0$.

For Goal I, solution to the selection procedure may be obtained by solving (2.1), rewritten as

(3.2) $$P^* = P(c; s, k - t, t, d),$$

for N, where $d = d^* \sqrt{N}/\sigma$, using interpolation in Table D as necessary.

Similarly, for Goals 1 and 2 we have

(3.3) $$P^* = P(k - s; k - s, t, k - t, d) \quad \text{(Goal 1)},$$

and

(3.4) $$P^* = P(s; s, k - t, t, d) \quad \text{(Goal 2)},$$

and solution can be made for N or s, as desired.

CHAPTER 5

Exact Probabilities of Rank Orders for Two Widely Separated Normal Distributions

In this chapter we use the method of Table A to give numerical evidence of the adequacy of the Hodges-Lehmann (1962) asymptotic formula for the probability of rank orders from two widely separated normal distributions. A table is given of the probabilities of the most probable and second most probable rank orders from two normal distributions (with unit variances) with means differing by $d = 4$, 5, and 6 units.

The definitions and notation of Chapter 1, Section 1, should be recalled at this point, leading as before to (1.1). If $\mathbf{z} = (z_1, \ldots, z_{m+n})$ is a fixed vector of zeros and ones, the probability of rank order $\mathbf{z}$, $P_{m,n}(\mathbf{z} \mid d)$, is given by

$$(1.1) \qquad P_{m,n}(\mathbf{z} \mid d) = m! \, n! \int \cdots \int_R \prod_{i=1}^{m+n} f(t_i - z_i d) \, dt_i,$$

where the region of integration R is $-\infty < t_1 \leq \cdots \leq t_{m+n} < \infty$.

It is clear that $P_{m,n}(\mathbf{z}^0 \mid d) \to 1$ as $d \to \infty$ for $\mathbf{z}^0 = (0 \cdots 0 1 \cdots 1)$, and consequently that the probability of all other rank orders tends to zero. Hodges and Lehmann (1962) have presented a method describing this tendency for $\mathbf{z} \neq \mathbf{z}^0$, as follows. The rank order

$$\mathbf{z} = (\underbrace{0 \cdots 0}_{r_0} \underbrace{1 \cdots 1}_{s_1} \underbrace{0 \cdots 0}_{r_1} \underbrace{1 \cdots 1}_{s_2} \underbrace{0 \cdots 0}_{r_2} \cdots \underbrace{1 \cdots 1}_{s_c} \underbrace{0 \cdots 0}_{r_c} \underbrace{1 \cdots 1}_{s_0})$$

is characterized by the number of variables in the successive groups, a set of integers $(r_0, s_1, r_1, \ldots, s_c, r_c, s_0)$ with $\Sigma\, r_i = m$, $\Sigma\, s_i = n$. Here $r_0 = 0$ if $z_1 = 1$ and $s_0 = 0$ if $z_{m+n} = 0$. The "graph" of a rank order $\mathbf{z}$ may be obtained by representing each 0 by a horizontal and each 1 by a vertical

Table 1. $P_{m,n}(z^i \mid d)$, the probability of the rank order z^i $(i = 0, 1)$

m	n	$P_{m,n}(z^0 \mid d)$, $z^0 = (0\cdots01\cdots1)$			$P_{m,n}(z^1 \mid d)$, $z^1 = (0\cdots0101\cdots1)$			$P_{m,n}(z^1\mid d)^{1/d^2}$, Hodges-Lehmann Value: .7788		
		$d=4$	5	6	$d=4$	5	6	4	5	6
1	1	.99766113	.99979652	.99998895	.00233887	.00020348	.00001105	.6848	.7118	.7283
2	1	.99549652	.99959919	.99997803	.00432922	.00039466	.00002186	.7117	.7309	.7422
3	1	.99347006	.99940723	.99996720	.00607938	.00057588	.00003246	.7269	.7420	.7504
2	2	.99133986	.99921066	.99995628	.00797576	.00076486	.00004325	.7394	.7505	.7564
4	1	.99155761	.99922005	.99995649	.00764982	.00074871	.00004288	.7374	.7498	.7563
3	2	.98745787	.99883286	.99993476	.01115405	.00111521	.00006423	.7550	.7619	.7648
5	1	.98974182	.99903719	.99994586	.00907895	.00091431	.00005312	.7454	.7559	.7608
4	2	.98380234	.99846462	.99991344	.01398348	.00144887	.00008483	.7658	.7699	.7707
3	3	.98185533	.99827457	.99990264	.01554262	.00162491	.00009537	.7709	.7735	.7732
6	1	.98800962	.99885826	.99993533	.01039315	.00107356	.00006321	.7517	.7607	.7645
5	2	.98033865	.99810499	.99989230	.01653969	.00176816	.00010508	.7739	.7761	.7753
4	3	.97659023	.99773060	.99987082	.01942203	.00210972	.00012595	.7817	.7816	.7792
7	1	.98635083	.99868295	.99992488	.01161155	.00122717	.00007314	.7569	.7648	.7676
6	2	.97704073	.99775323	.99987135	.01887454	.00207485	.00012502	.7803	.7811	.7791
5	3	.97161075	.99719955	.99983929	.02290414	.00257309	.00015601	.7898	.7878	.7839
4	4	.96982514	.99701565	.99982861	.02419873	.00273755	.00016632	.7925	.7898	.7853
8	1	.98475726	.99851099	.99991451	.01274854	.00137571	.00008294	.7614	.7683	.7703
7	2	.97388823	.99740869	.99985057	.02102549	.00237033	.00014466	.7855	.7852	.7822
6	3	.96687784	.99668026	.99980802	.02606545	.00301767	.00018559	.7962	.7928	.7877
5	4	.96343809	.99631791	.99978677	.02846077	.00333694	.00020599	.8006	.7960	.7900

9	1	.98322223	.99834214	.99990422	.01381529	.00151965	.00009260	.7652	.7714	.7726
8	2	.97086486	.99707083	.99982995	.02302071	.00265573	.00016402	.7900	.7888	.7850
7	3	.96236108	.99617180	.99977702	.02896113	.00344556	.00021473	.8014	.7971	.7909
6	4	.95737714	.99563584	.99974530	.03230857	.00391144	.00024503	.8069	.8011	.7938
5	5	.95573461	.99545774	.99973474	.03339125	.00406546	.00025512	.8086	.8024	.7947
10	1	.98174016	.99817620	.99989401	.01482068	.00165938	.00010214	.7686	.7741	.7747
9	2	.96795727	.99673919	.99980949	.02488200	.00293202	.00018312	.7939	.7919	.7874
7	4	.95160183	.99496817	.99970417	.03581441	.00446382	.00028348	.8121	.8054	.7970
6	5	.94843554	.99461715	.99968315	.03781927	.00476302	.00030344	.8149	.8074	.7985
11	1	.98030635	.99801300	.99988387	.01577192	.00179526	.00011156	.7716	.7765	.7766
10	2	.96515438	.99641337	.99978918	.02662667	.00319997	.00020197	.7972	.7947	.7895
7	5	.94149041	.99379454	.99963199	.04183370	.00543312	.00035103	.8201	.8117	.8017
6	6	.93997547	.99362192	.99962154	.04274392	.00557768	.00036090	.8212	.8126	.8024
12	1	.97891677	.99785237	.99987379	.01667495	.00192757	.00012088	.7742	.7788	.7784
11	2	.96244679	.99609300	.99976901	.02826871	.00346027	.00022059	.8002	.7972	.7915
7	6	.93193653	.99264824	.99956045	.04718750	.00635958	.00041747	.8263	.8168	.8056
12	2	.95982651	.99577777	.99974898	.02981965	.00371350	.00023898	.8029	.7995	.7932
7	7	.92286990	.99152705	.99948950	.05199646	.00724805	.00048287	.8313	.8211	.8089

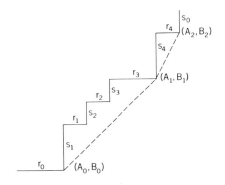

Figure 1

unit segment. Figure 1 illustrates the graph for

$$\mathbf{z} = (00110101001101)$$

with

$$(r_0, s_1, \ldots, r_c, s_0) = (2, 2, 1, 1, 1, 1, 2, 2, 1, 1)$$

The segments r_0 and s_0 are disregarded, and the lower convex hull of the graph is formed with $k + 1$ corner points $(A_0, B_0), (A_1, B_1), \ldots, (A_k, B_k)$. In Figure 1, $k = 2$ and the convex hull represented by the dotted line has three corner points: (2, 0), (6, 4), (7, 6). Hodges and Lehmann prove that, for $\mathbf{z} \neq \mathbf{z}^0$,

$$(1.2) \qquad \lim_{d \to \infty} \left[P_{m,n}(\mathbf{z} \mid d)^{d-2} \right] = \exp \left(-\frac{1}{2} \sum_{i=1}^{k} \frac{a_i b_i}{a_i + b_i} \right),$$

where

$$a_i = A_i - A_{i-1} \quad \text{and} \quad b_i = B_i - B_{i-1}, \quad i = 1, \ldots, k.$$

For the rank order shown in Figure 1, the right-hand side of (1.2) is $e^{-(4/3)} = .2636$. By using Table A of $P_{m,n}(\mathbf{z} \mid d)$, the values of $P_{m,n}(\mathbf{z} \mid d)^{d-2}$ for $d = 2$ and 3 are seen to be .0550 and .1341, respectively.

Table 1 gives values to 8 decimal places of $P_{m,n}(\mathbf{z} \mid d)$ for $\mathbf{z}^0 = (0 \cdots 01 \cdots 1)$ and $\mathbf{z}^1 = (0 \cdots 0101 \cdots 1)$; $1 \leq n \leq m \leq 7$ and $n = 1$ and 2, $m = 8(1)12$; $d = 4, 5, 6$. These values were computed as described in Chapter 1 and may be considered an abbreviated extension of Table A, since for fixed m and n the probability of any other rank order $\mathbf{z}^i$ ($i \neq 0, 1$) is less than both $1 - P_{m,n}(\mathbf{z}^0 \mid d) - P_{m,n}(\mathbf{z}^1 \mid d)$ and $P_{m,n}(\mathbf{z}^1 \mid d)$ [Savage, Sobel, and Woodworth (1966), Theorem 3]. The tabled values give an indication of the rate at which $P_{m,n}(\mathbf{z}^0 \mid d) \to 1$ and $P_{m,n}(\mathbf{z}^1 \mid d) \to 0$. The last three columns of the table contain values of $P_{m,n}(\mathbf{z}^1 \mid d)^{d-2}$. For sample sizes (m, n) as covered by the table, it is seen that, for example, $P_{m,n}(\mathbf{z}^1 \mid 6)^{1/36}$ ranges from .7283 to .8089, compared with the Hodges-Lehmann value $e^{-1/4} = .7728$ (which, it is interesting to note, is independent of m and n). The maximum departures above and below the Hodges-Lehmann value occur for the largest and smallest symmetric sample sizes tabled, namely $(m, n) = (7, 7)$ and $(1, 1)$, for all values of d.

References

Barr, D. R. and M. H. Rizvi (1966). An introduction to ranking and selection procedures. *J. Amer. Statist. Assoc.* **61,** 640–646.

Bauer, F. L. (1961). Algorithm 60, Romberg integration. *Comm. ACM* **4,** 255.

Bauer, F. L., H. Rutishauser, and E. Stiefel (1963). New aspects in numerical quadrature. *Proceedings of Symposia in Applied Mathematics: Experimental Arithmetic, High Speed Computing and Mathematics*, Vol. 15. American Mathematical Society, Providence, R.I. Pp. 199–218.

Bechhofer, R. E. (1954). A single-sample multiple decision procedure for ranking means of normal populations with known variances. *Ann. Math. Statist.* **25,** 16–39.

Bechhofer, R. E., J. Kiefer, and M. Sobel (1968). *Sequential Identification and Ranking Procedures.* University of Chicago Press, Chicago and London.

Berk, R. H. and I. R. Savage (1968). The information in a rank-order and the stopping time of some associated SPRT's. *Ann. Math. Statist.* **39,** 1661–1674.

Bradley, R. A. (1967). Topics in rank-order statistics. *Proceedings of the Fifth Berkeley Symposium on Mathematical Statistics*, Vol. 1 (L. le Cam and J. Neyman, editors). University of California Press, Berkeley and Los Angeles. Pp. 593–607.

Bradley, R. A., D. C. Martin, and F. Wilcoxon (1965). Sequential rank tests: I. Monte Carlo studies of the two-sample procedure. *Technometrics* **7,** 463–484.

Bradley, R. A., S. D. Merchant, and F. Wilcoxon (1966). Sequential rank tests: II. Modified two-sample procedures. *Technometrics* **8,** 615–623.

Brown, G. W. and A. M. Mood (1951). On median tests for linear hypotheses. *Proceedings of the Second Berkeley Symposium on Mathematical Statistics and Probability* (J. Neyman, editor). University of California Press, Berkeley. Pp. 159–166.

Chernoff, H. and I. R. Savage (1958). Asymptotic normality and efficiency of certain nonparametric test statistics. *Ann. Math. Statist.* **29,** 972–994.

Davies, O. L. (1956). *The Design and Analysis of Industrial Experiments* (2nd ed.). Oliver and Boyd, Edinburgh.

Desu, M. M. and M. Sobel (1968). A fixed subset-size approach to the selection problem. *Biometrika* **55,** 401–410.

Dixon, W. J. (1954). Power under normality of several nonparametric tests. *Ann. Math. Statist.* **25,** 610–614.

Dwass, M. (1956). The large-sample power of rank order tests in the two sample problem. *Ann. Math. Statist.* **27,** 352–374.

Greenwood, J. A. and H. O. Hartley (1962). *Guide to Tables in Mathematical Statistics.* Princeton University Press, Princeton, N.J.

Gupta, S. S. (1963). Probability integrals of multivariate normal and multivariate *t*. *Ann. Math. Statist.* **34**, 792–828.

Gupta, S. S. (1965). On some multiple decision (selection and ranking) rules. *Technometrics* **6**, 225–245.

Hall, W. J., R. A. Wijsman, and J. K. Ghosh (1965). The relationship between sufficiency and invariance with applications in sequential analysis. *Ann. Math. Statist.* **32**, 575–614.

Henrici, P. (1964). *Elements of Numerical Analysis*. John Wiley & Sons, New York.

Hodges, J. L., Jr. and E. L. Lehmann (1956). The efficiency of some nonparametric competitors of the *t*-test. *Ann. Math. Statist.* **27**, 324–335.

Hodges, J. L., Jr. and E. L. Lehmann (1962). Probabilities of rankings for two widely separated normal distributions. *Studies in Mathematical Analysis and Related Topics* (edited by Gilbarg, Solomon, and others). Stanford University Press, Stanford, Calif. Pp. 146–151.

Hoeffding, W. (1951). Optimum nonparametric tests. *Proceedings of the Second Berkeley Symposium on Mathematical Statistics and Probability* (J. Neyman, editor). University of California Press, Berkeley. Pp. 83–92.

Klotz, J. (1961). Private communication.

Klotz, J. (1964). On the normal scores two-sample test. *J. Amer. Statist. Assoc.* **59**, 652–664.

Kolmogorov, A. N. (1941). Confidence limits for an unknown distribution function. *Ann. Math. Statist.* **12**, 461–463.

Mahamunulu, D. M. (1967). Some fixed-sample ranking and selection problems. *Ann. Math. Statist.* **38**, 1079–1091.

Mann, H. B. and D. R. Whitney (1947). On a test of whether one of two random variables is stochastically larger than the other. *Ann. Math. Statist.* **18**, 50–60.

Massey, F. J. (1951). The distribution of the maximum deviation between two cumulative step functions. *Ann. Math. Statist.* **22**, 125–128.

Milton, R. C. (1963). Tables of the equally correlated multivariate normal probability integral. Technical Report No. 27, Department of Statistics, University of Minnesota. (Also Master's Thesis.)

Mood, A. M. (1954). On the asymptotic efficiency of certain nonparametric two-sample tests. *Ann. Math. Statist.* **25**, 514–522.

Mood, A. M. and F. A. Graybill (1963). *Introduction to the Theory of Statistics*. McGraw-Hill Book Company, New York.

Noether, G. E. (1955). On a theorem of Pitman. *Ann. Math. Statist.* **26**, 64–68.

Pitman, E. J. G. (1949). *Lecture Notes on Non-parametric Statistics*. Columbia University, New York.

Richardson, L. F. and J. A. Gaunt (1927). The deferred approach to the limit. *Trans. Roy. Soc. London Ser. A* **226**, 299–361.

Savage, I. R. (1956). Contributions to the theory of rank order statistics—the two-sample case. *Ann. Math. Statist.* **27**, 590–615.

Savage, I. R. (1960). Contributions to the theory of rank order statistics: computation rules for probabilities of rank orders. *Ann. Math. Statist.* **31**, 519–520.

Savage, I. R. (1964). Contributions to the theory of rank order statistics: applications of lattice theory. *Rev. Internatl. Statist. Inst.* **32**, 52–64.

Savage, I. R. and L. J. Savage (1965). Finite stopping time and finite expected stopping time. *J. Roy. Statist. Soc. Ser. B* **27**, 284–289.

Savage, I. R. and M. Sobel (1963). Contributions to the theory of rank order statistics— the two-sample case: fine structure of the ordering of probabilities of rank orders. Technical Report No. 29, Department of Statistics, University of Minnesota. (DDC—AD 414 707.)

Savage, I. R., M. Sobel, and G. Woodworth (1966). Fine structure of the ordering of probabilities of rank orders in the two-sample case. *Ann. Math. Statist.* **37**, 98–112.

Smirnov, N. V. (1948). Table for estimating the goodness of fit of empirical distributions. *Ann. Math. Statist.* **19**, 279–281.

Teichroew, D. (1954). Private communication.

Teichroew, D. (1955a). Empirical power functions for nonparametric two-sample tests for small samples. *Ann. Math. Statist.* **26**, 340–344.

Teichroew, D. (1955b). Probabilities associated with order statistics in samples from two normal populations with equal variance. *Engineering Agency Statistical Tables.* Chemical Corps Engineering Agency, Army Chemical Center, Maryland.

Teichroew, D. (1956). Tables of expected values of order statistics and products of order statistics for samples of size twenty and less from the normal distribution. *Ann. Math. Statist.* **27**, 410–426.

Terry, M. E. (1952). Some rank order tests which are most powerful against specific parameteric alternatives. *Ann. Math. Statist.* **23**, 346–366.

Tsao, C. K. (1957). Approximations to the power of rank tests. *Ann. Math. Statist.* **28**, 159–172.

van der Vaart, H. R. (1950). Some remarks on the power function of Wilcoxon's test for the problem of two samples, I and II. *Proc. Kon. Ned. Akad. Wetensch. Ser. A* **53**, 494–520.

Wald, A. (1947). *Sequential Analysis.* John Wiley & Sons, New York.

Westenberg, J. (1948). Significance test for median and interquartile range in samples from continuous populations of any form. *Proc. Kon. Ned. Akad. Wetensch. Ser. A* **51**, 252–261.

Westenberg, J. (1950). A tabulation of the median test for unequal samples. *Proc. Kon. Ned. Akad. Wetensch. Ser. A* **53**, 77–82.

Wilcoxon, F. (1945). Individual comparisons by ranking methods. *Biometrics* **1**, 80–83.

Wilcoxon, F., L. J. Rhodes, and R. A. Bradley (1963). Two sequential two-sample grouped rank tests with applications to screening experiments. *Biometrics* **19**, 58–84.

Wilf, H. S. (1967). Advances in numerical quadrature. *Mathematical Methods for Digital Computers*, Vol. II (A. Ralston and H. S. Wilf, editors). John Wiley & Sons, New York. Pp. 133–144.

Witting, H. (1960). A generalized Pitman efficiency for nonparametric tests. *Ann. Math. Statist.* **31**, 405–414.

Symbols

This short list of symbols is included to assist the reader in recalling the meaning of frequently used notation. Page numbers refer to the locations of definitions or first usage in the text.

Symbol	Page	Comment
$\mathbf{z}$	1, 73	rank order
$\mathbf{z}^c$, $\mathbf{z}^t$, $\mathbf{z}^{tc}$	2	conjugate, transpose, transpose-conjugate of $\mathbf{z}$
z^t	45	tth rank order in a sequential experiment
$P_{m,n}(\mathbf{z} \mid d)$, $P_{m,n}(\mathbf{z})$ $P(\mathbf{z} \mid d)$, $P(\mathbf{z})$	1, 2, 73	probability of rank order $\mathbf{z}$
$f(x - \theta)$	1, 73	normal density, $(2\pi)^{-\frac{1}{2}} \exp\left[-(x - \theta)^2/2\right]$
$F(x)$	5	normal distribution function, $\displaystyle\int_{-\infty}^{x} f(y)\, dy$
$C(m, n)$	4, 73	number of rank orders
c_1, c_2	2, 25, 74	ordering of the $\mathbf{z}$'s; also normal scores
w, w_X, w_Y	3, 23, 45, 258	Wilcoxon statistics
$e_{a,b}$	29	Hodges-Lehmann efficiency
●	4, 74	bullet marking "crossovers"
×	3, 74	indicates $\mathbf{z} = \mathbf{z}^{tc}$
$\doteq$	6	"approximately equal to"

Tables

Table A. Rank order probabilities

Let:

(1) $X_1, \ldots, X_m$ and $Y_1, \ldots, Y_n$ be normally distributed random variables with means μ_X and μ_Y, respectively, and common variance σ^2, all $m + n$ variables being mutually independent;

(2) $d = (\mu_Y - \mu_X)/\sigma$;

(3) $\mathbf{z} = (z_1, \ldots, z_{m+n})$ be a random vector of zeros and ones, where the ith component z_i is a 0 (or 1) if the ith member of an ordered (from smallest to largest) random sample from the X's and Y's is an X (or Y); and

(4) $f(x - \theta) = (2\pi)^{-\frac{1}{2}} \exp\left[-(x - \theta)^2/2\right]$, the normal density with mean θ and variance 1.

Then the quantity tabled is the probability of the rank order $\mathbf{z}$,

$$P_{m,n}(\mathbf{z} \mid d) = m! \, n! \int \cdots \int_R \prod_{i=1}^{m+n} f(t_i - z_i d) \, dt_i,$$

where the region of integration R is $-\infty < t_1 \le t_2 \le \cdots \le t_{m+n} < \infty$. Values are given to 9 decimal places for:

all $\mathbf{z}$ for $1 \le n \le m \le 7$ and $n = 1, m = 8(1)12$;
$d = .2(.2)1.0, 1.5, 2.0, 3.0$.

$C(m, n) = \dbinom{m + n}{n}$ is the total number of rank orders for a given m and n;

$1/C(m, n) = P_{m,n}(\mathbf{z} \mid 0)$.

The sections of Table A are arranged according to increasing values of $m + n$, from $2 \le m + n \le 14$. The values of $P_{m,n}(\mathbf{z} \mid d)$ for a small value of $m + n$ normally appear on one page. Values for large $m + n$ are listed on successive pages, and the heading of each page describes what portion of the table appears on that page.

73

The z's are ordered according to decreasing c_1 values; in case of ties in c_1 values, ordering is done according to decreasing c_2 values.

The column headed w contains the value of the Wilcoxon two-sample statistic for each z, $w = \sum_{i=1}^{m+n} iz_i$, the sum of the ranks of the Y's.

The symbol x in the rightmost column indicates cases where $z = z^{tc}$. The symbol ● in the body of the table indicates "crossovers."

Further explanation is found in Chapter 1, Sections 1–3. For $d > 3.0$, see Chapter 5.

Combined Sample Size $m + n$	m	n	Starting Page	Combined Sample Size $m + n$	m	n	Starting Page
2	1	1	75	9	8	1	84
					7	2	85
3	2	1	75		6	3	86
					5	4	88
4	3	1	75				
	2	2	75	10	9	1	90
					7	3	91
5	4	1	76		6	4	94
	3	2	76		5	5	99
6	5	1	76				
	4	2	77	11	10	1	102
	3	3	77		7	4	103
					6	5	110
7	6	1	78				
	5	2	78	12	11	1	119
	4	3	79		7	5	120
					6	6	136
8	7	1	80				
	6	2	80	13	12	1	146
	5	3	81		7	6	147
	4	4	83	14	7	7	182

Table A

m = 1, n = 1

Rank Orders 1 thru 2 of 2   **C(1, 1) = 2   1/C(1, 1) = .50000000**

z	w	c_1	c_2	d = .2	.4	.6	.8	1.0	1.5	2.0	3.0	z^{tc}
01	2	.2820948	.0000000	.556231457	.611351294	.664313379	.714196177	.760249938	.855577813	.921350391	.983052572	01x
10	1	-.2820948	.0000000	.443768541	.388648705	.336686620	.285803822	.239750060	.144422178	.078649599	.016947427	10x

m = 2, n = 1

Rank Orders 1 thru 3 of 3   **C(2, 1) = 3   1/C(2, 1) = .33333333**

z	w	c_1	c_2	d = .2	.4	.6	.8	1.0	1.5	2.0	3.0	z^{tc}
001	3	.2820948	.0918882	.391392382	.451874735	.513386681	.574469333	.633702045	.765811956	.865767172	.968795477	011
010	2	.0000000	-.1837763	.329678150	.318953118	.301853395	.279453687	.253095785	.179531716	.111166438	.028514190	101
100	1	-.2820948	.0918882	.278929466	.229172146	.184759922	.146076978	.113202167	.054656320	.023066380	.002690331	110

m = 3, n = 1

Rank Orders 1 thru 4 of 4   **C(3, 1) = 4   1/C(3, 1) = .25000000**

z	w	c_1	c_2	d = .2	.4	.6	.8	1.0	1.5	2.0	3.0	z^{tc}
0001	4	.2573438	.1378322	.304081133	.362695602	.424578451	.488231162	.552031438	.701862700	.822792953	.956374370	0111
0010	3	.0742529	-.1378322	.261933747	.267537398	.266424689	.258714514	.245011821	.191847768	.128922657	.037263319	1011
0100	2	-.0742529	-.1378322	.232583478	.210892279	.186355404	.160466017	.134631857	.077449805	.037827001	.005507966	1101
1000	1	-.2573438	.1378322	.201401640	.158874719	.122641454	.092588306	.068324881	.028839718	.010457379	.000854343	1110

m = 2, n = 2

Rank Orders 1 thru 5 of 5   **C(2, 2) = 6   1/C(2, 2) = .16666667**

z	w	c_1	c_2	d = .2	.4	.6	.8	1.0	1.5	2.0	3.0	z^{tc}
0011	7	.2210645	.1837763	.214499821	.269183507	.329757270	.394802532	.462545468	.630690497	.774915732	.942865689	0011x
0101	6	.1220607	-.0492427	.189787226	.209133795	.223109992	.230547296	.230869001	.202645805	.147941638	.047053344	0101x
1001	5	.0000000	-.1345336	.163997895	.156246659	.144148830	.128786307	.111144152	.067597113	.033761243	.004806232	0110
1010	4	-.1220607	-.0492427	.141573285	.116275121	.092309193	.070787463	.052434265	.021223401	.006868752	.000362572	1010x
1100	3	-.2210645	.1837763	.126143876	.092910255	.066535937	.046290093	.031262958	.010246063	.002751382	.000105929	1100x

75

Table A

m = 4, n = 1 — Rank Orders 1 thru 5 of 5 — C(4, 1) = 5 — 1/C(4, 1) = .20000000

z	w	c_1	c_2	d = .2	.4	.6	.8	1.0	1.5	2.0	3.0	z^{tc}
00001	5	.2325929	.1600041	.249631153	.305001348	.365088237	.428525999	.493698862	.652864757	.787838981	.945311326	01111
00010	4	.0990038	-.0866875	.217799918	.230077014	.237960860	.238820651	.233330303	.195991763	.139815891	.044252179	10111
00100	3	.0000000	-.1426333	.191767617	.188909274	.175908089	.159198052	.140028188	.089707884	.046121476	.008148369	11011
01000	2	-.0990038	-.0866875	.178666227	.155250190	.131201813	.107822654	.086157017	.043461151	.018355017	.001911709	11101
10000	1	-.2325929	.1600041	.156735083	.120062172	.089841001	.065632642	.046785627	.017974430	.005688625	.000376415	11110

m = 3, n = 2 — Rank Orders 1 thru 10 of 10 — C(3, 2) = 10 — 1/C(3, 2) = .10000000

z	w	c_1	c_2	d = .2	.4	.6	.8	1.0	1.5	2.0	3.0	z^{tc}
00011	9	.1657983	.1956624	.137155222	.182462731	.235739670	.296192132	.362427707	.540137485	.707166415	.920513927	00111
00101	8	.1162965	.0383150	.123375414	.147460313	.169912350	.187078390	.194475795	.200692974	.163484607	.060500394	01011
01001	7	.0667946	-.0583252	.111970957	.120367902	.124282776	.123314850	.117638019	.088366498	.032597262	.009676571	01101
01010	6	.0495019	-.0739733	.108258384	.112702016	.108770450	.108752809	.100877488	.070966060	.039763343	.006554889	10011
10001	6	.0000000	-.0956501	.098105450	.092637528	.084202436	.073684820	.062033647	.034390957	.015171208	.001543921	01110
10010	6	.0000000	-.1077077	.097868934	.091744958	.082397723	.070855407	.058379376	.029818472	.011661175	.000806101	10101
01100	5	-.0495019	-.0739733	.088080678	.075798494	.062172934	.048998949	.037099051	.015520372	.005047934	.000251409	11001
10100	5	-.0667946	-.0583252	.085706377	.070465492	.055556407	.041989612	.030413495	.011251970	.001299695	.000101364	11010
11000	5	-.1162965	.0383150	.077719084	.057074711	.041702440	.028763879	.019048215	.005673896	.001299695	.000030444	11100
11000	3	-.1657983	.1956624	.070636184	.048285854	.031910813	.020369150	.012547203	.003181307	.000635505	.000011978	11110

m = 5, n = 1 — Rank Orders 1 thru 6 of 6 — C(5, 1) = 6 — 1/C(5, 1) = .16666667

z	w	c_1	c_2	d = .2	.4	.6	.8	1.0	1.5	2.0	3.0	z^{tc}
000001	6	.2112011	.1702898	.212273864	.264316735	.322023644	.384213399	.449364877	.613354751	.758451568	.935304723	011111
000010	5	.1069592	-.0514288	.186786444	.203423065	.215322964	.221563001	.221669925	.196550030	.146937064	.050033015	101111
000100	4	.0335911	-.1188611	.170926906	.170099406	.164256221	.153925625	.139985909	.096879360	.055665600	.010564418	110111
001000	3	-.0335911	-.1188611	.144489889	.144752384	.128923927	.111404461	.093394404	.052633780	.024536861	.003016197	111011
010000	2	-.1069592	-.0514288	.144489889	.121686546	.095540303	.079076087	.060999070	.028009549	.010675341	.000881538	111101
100000	1	-.2112011	.1702898	.127837105	.095724863	.069932940	.049817425	.034585813	.012372520	.003733557	.000200108	111110

Table A

m = 4, n = 2   Rank Orders 1 thru 15 of 15   C(4, 2) = 15   1/C(4, 2) = .06666667

z	w	c_1	c_2	d = .2	.4	.6	.8	1.0	1.5	2.0	3.0	z^{tc}
000011	11	.1272641	.1837763	.095962013	.133481026	.179660622	.234328135	.296620749	.474302045	.653767460	.900793947	001111
000101	10	.0979169	.0732164	.087564870	.110314038	.133383584	.154900185	.172903111	.192396945	.170196102	.070837537	010111
001001	9	.0710440	.0001752	.080661296	.093316386	.103280825	.109420724	.111035869	.095804615	.064123743	.014111770	011011
000110	9	.0562201	-.0256720	.077207968	.085612781	.090932610	.092555801	.090324723	.070944815	.044399719	.008042385	100111
010001	8	.0416968	-.0505719	.073837763	.078138274	.079044207	.076475366	.070804764	.048371880	.025473348	.003405024	011101
001010	8	.0293472	-.0667399	.071084454	.072266364	.070056997	.064771157	.057122108	.033971830	.015102635	.001265427	101011
100001	7	.0000000	-.0703642	.065274353	.061271947	.055146613	.047599453	.039412483	.020551984	.008349849	.000680427	101101
001100	7	.0000000	-.0760686	.065162590	.060851679	.054294386	.046286138	.037706839	.018528983	.006875789	.000413488	011110
010010	7	.0000000	-.0819356	.065047948	.060424013	.053436834	.044991853	.036066049	.016741121	.005721522	.000267673	110011
010100	6	-.0293472	-.0667399	.059605466	.050799539	.041265615	.031947404	.023570596	.008923994	.002498194	.000079102	110101
100010	6	-.0416968	-.0505719	.057486628	.047324052	.037179085	.027866104	.019919894	.006984011	.001812085	.000048952	101110
011000	5	-.0562201	-.0256720	.055095106	.043563693	.032946384	.023825372	.016470644	.005366510	.001313988	.000033186	111001
100100	5	-.0710440	.0001752	.052658837	.039732899	.028625769	.019684142	.012914125	.003657362	.000766802	.000013465	110110
101000	4	-.0979169	.0732164	.048651065	.034011168	.022764203	.014579943	.008931338	.002152289	.000388994	.000005248	111010
110000	3	-.1272641	.1837763	.044699642	.028892138	.017983166	.010767822	.006196707	.001301607	.000209760	.000002370	111100

m = 3, n = 3   Rank Orders 1 thru 14 of 14   C(3, 3) = 20   1/C(3, 3) = .05000000

z	w	c_1	c_2	d = .2	.4	.6	.8	1.0	1.5	2.0	3.0	z^{tc}
000111	15	.1055254	.1734905	.074788958	.107703047	.149556342	.200574086	.260248910	.437209506	.623570405	.889891000	000111x
001011	14	.0853707	.0879087	.068751128	.090329844	.113496759	.136494134	.157260107	.189915315	.173689518	.076734934	001011x
010011	13	.0633603	.0182220	.062849402	.075222187	.085782693	.093277409	.096785426	.086948186	.059414267	.013010070	001101
001101	12	.0413499	-.0330715	.057430300	.062530360	.064561997	.063236057	.058779820	.039055519	.018894194	.001760967	010101x
100011	12	.0320878	-.0396151	.055498263	.058727022	.059270533	.057082594	.052490857	.034920437	.017684244	.002123779	001110
011001	11	.0211952	-.0570515	.053007804	.053231238	.050909320	.046151998	.039742286	.021901191	.008842711	.000587179	011001x
010101	11	-.0100773	-.0642450	.050694882	.048746516	.044456148	.038456029	.031556497	.015296401	.005352057	.000250208	010110
100101	10	.0100773	-.0642450	.046768161	.041486496	.034902305	.027849679	.021078813	.008345220	.002383638	.000074249	011010
011010	10	-.0211952	-.0570515	.044736082	.037956281	.030530909	.023277209	.016817972	.005892284	.001470973	.000033053	010101
100110	9	-.0320878	-.0396151	.042928391	.035112765	.027353895	.020291783	.014331959	.004845188	.001204720	.000029873	011100
101010	9	-.0413499	-.0330715	.041250417	.032238489	.023860265	.016718749	.011087690	.003116656	.000617943	.000008423	101010x
110010	8	-.0633603	.0182220	.037850834	.027249454	.018645494	.012119821	.007480177	.001779142	.000303295	.000003179	101100
110100	7	-.0853707	.0879087	.034718502	.023000120	.014526881	.008741783	.005008894	.001000764	.000145458	.000001151	110100x
111000	6	-.1055254	.1734905	.032136942	.019831741	.011735389	.006651355	.003608859	.000639609	.000084348	.000000577	111000

Table A

m = 6, n = 1 Rank Orders 1 thru 7 of 7 C(6, 1) = 7 1/C(6, 1) = .14285714

z	w	c_1	c_2	d = .2	.4	.6	.8	1.0	1.5	2.0	3.0	z^{tc}
0000001	7	.1931683	.1743292	.184978316	.233936117	.289188079	.349745146	.414216166	.580993534	.733156967	.926149168	0111111
0000010	6	.1081963	−.0242359	.163773291	.182283712	.197013388	.206809522	.210892268	.195367306	.151767605	.054933328	1011111
0000100	5	.0503867	−.0936966	.150926105	.154559915	.153345421	.147665198	.137779105	.101231825	.061392182	.012765724	1101111
0001000	4	.0000000	−.1127933	.140619006	.134112924	.123931882	.110964319	.096266345	.058782953	.029474957	.004107870	1110111
0010000	3	−.0503867	−.0936966	.131064428	.116543882	.100436979	.083883452	.067891847	.034463455	.014699073	.001443394	1111011
0100000	2	−.1081963	−.0242359	.120962096	.099406302	.079273571	.061337924	.046042145	.019666077	.006930780	.000480488	1111101
1000000	1	−.1931683	.1743292	.107676756	.079157146	.056720678	.039594438	.026912122	.009094841	.002578427	.000120026	1111110

m = 5, n = 2 Rank Orders 1 thru 21 of 21 C(5, 2) = 21 1/C(5, 2) = .04761905

z	w	c_1	c_2	d = .2	.4	.6	.8	1.0	1.5	2.0	3.0	z^{tc}
0000011	13	.1004549	.1662505	.071248681	.102725977	.142920839	.192163420	.250091374	.423868894	.610113936	.883110407	0011111
0000101	12	.0811850	.0848812	.065486141	.086155989	.108517639	.130955277	.151530524	.182620460	.172589702	.079084863	0101111
0001001	11	.0643895	.0298917	.060931090	.073448987	.086564516	.096228102	.102199249	.097761035	.071493952	.018065459	0110111
0000110	11	.0528610	.0027477	.058080516	.067619255	.075181277	.079868298	.081116351	.069545295	.045677917	.009332840	1001111
0010001	10	.0475939	−.0112523	.056741386	.064384202	.069607212	.071743024	.070539323	.052270339	.032791111	.005304715	0111011
0001010	10	.0360654	−.0332259	.054017605	.058232360	.059711329	.058223613	.053994963	.035980813	.017633801	.001717316	1010111
0100001	9	.0283240	−.0417962	.052333443	.054774610	.054625445	.051935364	.047102547	.030151015	.014605635	.001573126	0111101
0010010	9	.0192699	−.0417962	.050289050	.050382360	.047888940	.044190242	.036963626	.019928955	.007763526	.000448326	1011011
0001100	9	.0167956	−.0530624	.049850085	.049669310	.047109807	.042541100	.036581623	.020281040	.008326564	.000581625	1100111
1000001	8	.0000000	−.0536457	.046558306	.043517728	.038890797	.033238190	.027175363	.013568865	.005194864	.000360469	0111110
0010100	8	.0000000	−.0619689	.046395679	.042911158	.037676059	.031402232	.024849199	.011029402	.003542360	.000139267	1101011
0100010	8	−.0167956	−.0631701	.046372270	.042824528	.037504831	.031149325	.024534983	.010714664	.003361917	.000123185	1011101
0011000	7	−.0192699	−.0530624	.043289005	.037446436	.030820650	.022134719	.017980042	.006924132	.001952298	.000061099	1110011
0100100	7	−.0283240	−.0550419	.042772417	.036440081	.029445989	.022565404	.016399022	.006851851	.001497313	.000036080	1101101
1000010	7	−.0360654	−.0417962	.041246792	.033996482	.026653635	.019870949	.014083051	.004764578	.001169290	.000026662	1011110
0101000	6	−.0475939	−.0332259	.039897046	.031764540	.024028054	.017266583	.011785513	.003619154	.000803875	.000014939	1110101
1000100	6	−.0528610	−.0112523	.038036851	.028904903	.020888026	.014348700	.009366140	.002573540	.000510645	.000007464	1101110
0110000	5	−.0643895	.0027477	.037250118	.027790263	.019765641	.013397709	.008651905	.002341283	.000465367	.000007198	1111001
1001000	5	−.0811850	.0298917	.035470405	.025169636	.017006289	.010936402	.006691062	.001571446	.000268254	.000002951	1110110
1010000	4	−.0811850	.0848812	.033104367	.021988008	.013945095	.008440081	.004872263	.000999849	.000151148	.000001349	1111010
1100000	3	−.1004549	.1662505	.030628745	.018936484	.011241019	.006400264	.003491874	.000633381	.000086457	.000000660	1111100

Table A

m = 4, n = 3   Rank Orders 1 thru 35 of 35   C(4, 3) = 35   1/C(4, 3) = .02857143

z	w	c_1	c_2	d = .2	.4	.6	.8	1.0	1.5	2.0	3.0	z^{tc}
0000111	18	.0703503	.1410483	.045722947	.070060395	.102964294	.145399611	.197668388	.365664375	.559449096	.863100110	0001111
0001011	17	.0602729	.0902897	.042464615	.059995944	.080657613	.103287476	.126124465	.170058117	.174823092	.088777933	0010111
0010011	16	.0501956	.0486945	.039498371	.051707459	.064152187	.075497636	.084356929	.089298859	.069937705	.018741930	0011011
0001101	16	.0487110	.0434584	.039087363	.050629964	.062144366	.072342978	.079944991	.082268526	.062568777	.015762375	0100111
0100011	15	.0386337	.0115991	.036398051	.043836579	.049951869	.053900498	.055125876	.046426486	.028627881	.004625806	0011101
0010101	15	.0386337	.0096088	.036345775	.043577182	.049267529	.052551249	.052912850	.041983769	.023538842	.002723517	0101011
0001110	15	.0317166	-.0039792	.034712067	.039944699	.043566212	.045067447	.044252633	.033853884	.019093367	.002623253	1000111
1000011	14	.0285563	-.0142217	.033867182	.037830012	.039840830	.039581679	.037118390	.024658121	.011589038	.000952478	0011110
0100101	14	.0270717	-.0184535	.033484898	.036906624	.038270161	.037349794	.034321676	.021411113	.009261893	.000597181	0101101
0011001	14	.0216393	-.0223993	.032356161	.034721912	.035327314	.034099963	.031249812	.020129551	.009566411●	.000935842●	0110011
0010110	14	.0216393	-.0262321	.032268151	.034339443	.034442292	.032567587	.029040326	.016905423	.006874544	.000398559	1001011
1000101	13	.0169944	-.0345383	.031192232	.032000641	.030861856	.027989903	.023882152	.012332636	.004394244	.000189943	0101110
0101001	13	.0115619	-.0386530	.030055379	.029761649	.027748487	.024366096	.020157101	.009687155	.003233819	.000124682	0110101
0100110	13	.0100773	-.0390024	.029759981	.029206493	.027009102	.023538604	.019335896	.009142668	.003003835	.000111297	1001101
0011010	13	.0100773	-.0408449	.029721814	.029057469	.026700632	.023063076	.018728543	.008504456	.002636121	.000081819	1010011
1001001	12	.0069171	-.0414598	.029098087	.027904866	.025205926	.021452275	.017208757	.007677356	.002289981	.000081564	0110110
1000110	12	.0000000	-.0434905	.027766660	.025486448	.022096358	.018097175	.014004049	.005763758	.001674256	.000051059	1001110
0110001	12	.0000000	-.0455202	.027714774	.025296393	.021727057	.017562392	.013361966	.005183480	.001384655	.000033031	0111001
0101010	12	.0000000	-.0434905	.027675392	.025152549	.021438895	.017162334	.012885924	.004767038	.001186819	.000022585	1010101
0011100	12	-.0069171	-.0414598	.026410374	.022977368	.018811904	.014491420	.010502407	.003594388	.000839616	.000014644	1100011
1010001	11	-.0100773	-.0408449	.025843884	.022024670	.017685013	.013380585	.009541197	.003164208	.000728183	.000013206	0111010
1001010	11	-.0100773	-.0390024	.025810357	.021909689	.017475986	.013097907	.009223942	.002923819	.000628582	.000009123	1010110
0110010	11	-.0115619	-.0386530	.025562102	.021519846	.017044924	.012700227	.008901163	.002799404	.000600039	.000008746	1011001
0101100	11	-.0169944	-.0345383	.024585136	.019867177	.015073008	.010733723	.007172695	.001979528	.000365518	.000003697	1100101
1100001	10	-.0216393	-.0262321	.023896048	.018922230	.014182019	.010058290	.006749383	.001953247	.000400026●	.000006030	0111100
1010010	10	-.0216393	-.0223993	.023832891	.018725379	.013855688	.009654188	.006333620	.001695947	.000331466	.000003400	1011010
1001100	10	-.0270717	-.0184535	.022919214	.017278548	.012238207	.008141464	.005085576	.001188662	.000166448	.000001370	1100110
0110100	10	-.0285563	-.0142217	.022703126	.016984189	.011957192	.007918887	.004931713	.001150994	.000181704	.000001381●	1101001
1100010	9	-.0317166	-.0039792	.022261751	.016410980	.011440750	.007539456	.004694985	.001119294	.000186288●	.000001743	1011100
1010100	9	-.0386337	.0096088	.021187402	.014832602	.009796565	.006100840	.003580350	.000726200	.000100611	.000000602	1101010
0111000	9	-.0386337	.0115991	.021158788	.014754979	.009684867	.005981145	.003474147	.000681603	.000090510	.000000487	1110001
1100100	8	-.0487110	.0434584	.019758731	.012918205	.007979300	.004653379	.002560657	.000444525	.000053074	.000000240	1101100
1011000	8	-.0501956	.0486945	.019555780	.012655886	.007739044	.004468696	.002435071	.000412744	.000048157	.000000208	1110010
1101000	7	-.0602729	.0902897	.018256924	.011069449	.006362960	.003464803	.001785889	.000266672	.000027847	.000000100	1110100
1110000	6	-.0703503	.1410483	.017067591	.009732082	.005289637	.002736945	.001346476	.000181986	.000017546	.000000056	1111000

79

Table A

Rank Orders 1 thru 8 of 8 C(7, 1) = 8 1/C(7, 1) = .12500000

z	w	c_1	c_2	d = .2	.4	.6	.8	1.0	1.5	2.0	3.0	z^{tc}
00000001	8	.1779500	.1749419	.164121823	.210302303	.263200270	.322009814	.385480977	.553373951	.710998281	.917697168	01111111
00000010	7	.1065281	-.0042890	.145995447	.165436697	.181914663	.194147318	.201146318	.193299280	.155110803	.059164000	10111111
00000100	6	.0591028	-.0719588	.135220177	.145204529	.143802871	.141391373	.134683982	.103887730	.065854208	.014774650	11011111
00001000	5	.0190643	-.0986941	.126793950	.124501633	.118344529	.108899841	.097011275	.063061376	.033491411	.005162274	11101111
00010000	4	-.0190643	-.0986941	.119289309	.110195984	.098536264	.085287717	.071454829	.039808792	.018089764	.002026499	11110111
00100000	3	-.0591028	-.0719588	.111916614	.097043845	.081490012	.066264203	.052175688	.024923562	.009724844	.000804852	11111011
01000000	2	-.1065281	-.0042890	.103816907	.083626070	.065322496	.049472844	.036323940	.014635902	.004844295	.000292285	11111101
10000000	1	-.1779500	.1749419	.092845769	.067210564	.047388893	.032526888	.021722988	.007003998	.001886385	.000078271	11111110

Rank Orders 1 thru 28 of 28 C(6, 2) = 28 1/C(6, 2) = .03571429

z	w	c_1	c_2	d = .2	.4	.6	.8	1.0	1.5	2.0	3.0	z^{tc}
00000011	15	.0812795	.1487248	.055181624	.081976608	.117258665	.161741431	.215490212	.383789310	.573482510	.867059137	00111111
00000101	14	.0677294	.0865006	.050981523	.069484892	.090489805	.112689666	.134309271	.172904836	.172675855	.085838921	01011111
00000110	13	.0562898	.0440595	.047744147	.060746502	.073605004	.084988202	.093578197	.097180761	.076313485	.021580947	10011111
00001001	13	.0473231	.0186540	.045420819	.055011325	.063483240	.069842273	.073297695	.067572671	.047112701	.010468694	01101111
00001010	12	.0453959	.0116169	.044882409	.053594911	.060845879	.065715282	.067564426	.065866741	.038102269	.007143006	10101111
00001100	11	.0358836	-.0113352	.042520938	.048018466	.051446806	.052306930	.050480394	.036837061	.019531969	.002152676	11001111
00010001	12	.0339563	-.0143550	.042084130	.047090241	.050063478	.050600204	.048653268	.035629755	.019439585	.025546545•	01110111
00010010	11	.0249897	-.0317389	.039962729	.042323181	.042429446	.040269962	.036189432	.021840929	.009402964	.000639951	10110111
00010100	10	.0223335	-.0327244	.039429032	.041320221	.041112327	.038845748	.034864619	.021285677	.009486013	.000750447•	11010111
00011000	9	.0204063	-.0344512	.039020111	.040504670	.039967971	.037510673	.033504666	.020440496	.009310677	.000855503•	11100111
00100001	11	.0135501	-.0449443	.037464571	.037159518	.034851725	.030911488	.025929590	.013100776	.004681516	.000213094	01111011
00100010	10	.0114396	-.0452686	.037048210	.036384674	.033831927	.029787017	.024834670	.012427387	.004434502	.000206644	10111011
00100100	9	.0000000	-.0421620	.034881064	.032497802	.028886693	.024503404	.019842078	.009585187	.003507044	.000215139•	11011011
00101000	8	.0000000	-.0474027	.034778725	.032116769	.028125870	.023358863	.018398864	.008038714	.002526944	.000093773	11101011
00110000	7	.0000000	-.0498109	.034731844	.031943829	.027785861	.022858410	.017785415	.007444096	.002201258	.000068153	11110011
01000001	10	.0000000	-.0501010	.034726199	.031923031	.027745063	.022798551	.017712344	.007374307	.002163858	.000065409	01111101
01000010	9	-.0114396	-.0452686	.033592348	.028155621	.023023483	.017820413	.013055499	.004715212	.001207645	.000282283	10111101
01000100	8	-.0135501	-.0449443	.033188550	.027426407	.022091047	.016819666	.012104529	.004155023	.001001898	.000020153	11011101
01001000	7	-.0204063	-.0344512	.031043340	.025614889	.020056654	.014898043	.010495043	.003461572	.000816246	.000016474	11101101
01010000	6	-.0223335	-.0327244	.030701377	.025043172	.019380702	.014227829	.009907041	.003178359	.000731328	.000014248	11110101
01100000	5	-.0249897	-.0317389	.030205484	.024173334	.018304039	.013112107	.008885286	.002632475	.000550157	.000008401	11111001
10000001	9	-.0339563	-.0143550	.028766103	.021980310	.015926341	.010938534	.007118839	.001917757	.000366626	.000004725	01111110
10000010	8	-.0358836	-.0113352	.028446653	.021482332	.015378319	.010433614	.006707878	.001754887	.000326911	.000004067	10111110
10000100	7	-.0453959	.0116169	.026989381	.019360591	.013177250	.008506132	.005205650	.001205063	.000198490	.000001910	11011110
10001000	6	-.0473231	.0186540	.026730260	.019034680	.012890825	.008293991	.005077895	.001187218	.000200527•	.000002125	11101110
10010000	5	-.0562898	.0440595	.025412200	.017190677	.011050487	.006734697	.003911033	.000795253	.000115974	.000000894	11110110
10100000	4	-.0677294	.0865006	.023871003	.015212072	.009238256	.005342584	.002940727	.000530740	.000069605	.000000448	11111010
11000000	3	-.0812795	.1487248	.022195197	.013228575	.007552839	.004126600	.002155437	.000347054	.000041435	.000000231	11111100

Table A

m = 5, n = 3   Rank Orders   1 thru 50 of 56   C(5, 3) = 56   1/C(5, 3) = .01785714

d = .2

z	w	c_1	c_2	.2	.4	.6	.8	1.0	1.5	2.0	3.0	z^{tc}
00000111	21	.0490830	.1128405	.030195521	.048679746	.074958516	.110459366	.156093996	.313005782	.508294616	.839259621	00011111
00001011	20	.0433632	.0804447	.028230602	.042257909	.059959246	.080736107	.103292397	.154183295	.172086060	.098057134	00101111
00010011	19	.0379163	.0540216	.026515077	.037122709	.049076559	.061312190	.072461357	.086931890	.075621735	.023733000	00110111
00001101	18	.0365882	.0483052	.026114001	.036005955	.046850234	.057584558	.066927011	.076854769	.063808641	.007128316	01001111
00100011	18	.0321965	.0306136	.024838911	.032518171	.040103324	.046634821	.051186974	.050551523	.035712927	.007128316	01010111
00010101	18	.0311412	.0122889	.024516850	.031596296	.038245265	.043506364	.046541215	.044286939	.026573374	.003647975	01100111
01000011	17	.0263850	.0122889	.023292527	.028639383	.033219413	.036380561	.037652051	.032254898	.019877702	.003006042	10000111
00011001	17	.0254214	.0084650	.023016376	.027902914	.031844556	.034245974	.034739235	.028094643	.016198327	.002139697	10001111
00100101	17	.0254214	.0073435	.022985573	.027755028	.031450091	.034245091	.034739235	.013571253	.001323451	10101011	10010111
01000101	17	.0254214	.0065903	.022966678	.027655494	.031193457	.032972891	.032680257	.024200511	.012134768	.000986502	10100111
00010110	17	.0209381	−.0041850	.021862660	.025106373	.027052700	.027362207	.025988809	.017406685	.007954379	.000547155	11000111
00101001	16	.0197017	−.0083947	.021528622	.024271161	.025599345	.025271920	.023363583	.014459302	.006013974	.000329961	01011011
01001001	16	.0186464	−.0105466	.021278283	.023715240	.024732174	.024145119	.022076415	.013290585	.005372749	.000276059	01101011
00110001	16	.0152183	−.0133149	.020563016	.022336993	.022903285	.022183070	.020312169	.012827986	.005838911	.000494210	01110011
01010001	16	.0152183	−.0164274	.020490813	.022021648	.022173665	.020926188	.018517849	.010325066	.003893440	.000194210	10011011
10000011	15	.0142547	−.0170233	.020476711	.021959163	.022027828	.020674254	.018159394	.009843545	.003549771	.000139202	10101011
00011010	15	.0142547	−.0183440	.020263772	.021522665	.021406028	.019945892	.017421198	.009395049	.003428285	.000148272	10110011
00101010	15	.0129266	−.0213451	.019942433	.020798099	.020263128	.018450033	.015706487	.007849176	.002604698	.000087257	11000011
01001010	15	.0094985	−.0257823	.019187723	.019243711	.018016248	.015747529	.012852921	.005738401	.001676601	.000040733	01101101
00110010	15	.0084433	−.0242197	.019018737	.019018972	.017863522	.015763299	.013073153	.006259038	.002054105	.000074106	01110101
01010010	15	.0084433	−.0247118	.019007404	.018973209	.017763061	.015600126	.012854868	.006007219	.001898813	.000060450	01011110
10010001	15	.0074797	−.0265453	.018968647	.018819655	.017442101	.015103830	.012222073	.005366226	.001547495	.000037553	01101110
10001001	14	.0040516	−.0273646	.018766481	.018425952	.016908728	.014507699	.011643436	.005032432	.001449425	.000037079	01110110
00110100	14	.0027235	−.0297121	.018056256	.017048817	.015033775	.012382639	.009528062	.003678835	.000932603	.000018282	10011101
01010100	14	.0027235	−.0291270	.017811065	.016627931	.014490817	.011887503	.009105901	.003512359	.000903833	.000029184	10101101
01100001	14	.0027235	−.0293882	.017805815	.016608030	.014490728	.011467639	.008605649	.003440785	.000866721	.000016096	10110101
10000110	13	−.0015799	−.0311179	.017777417	.016480212	.014241115	.011467766	.008708781	.003084436	.000712283	.000021029	11001101
00111000	13	−.0017599	−.0291753	.017624273	.016291207	.014107501	.011448112	.007094113	.003322339	.000855031	.000008075	01110101
10010001	14	−.0027235	−.0291270	.016941947	.015048470	.012512528	.009738363	.007094113	.002405960	.000540360	.000008016	11010011
01011000	13	−.0027235	−.0291270	.016757011	.014718329	.012101402	.009315137	.006714348	.002225628	.000492739	.000007426	10111001
00110110	13	−.0027235	−.0293882	.016752110	.014700980	.012068822	.009269556	.006661517	.002182254	.000473484	.000006580	11011001
01010110	13	−.0027235	−.0311179	.016719604	.014586831	.011858153	.008982515	.006340312	.001948930	.000385601	.000004041	11100011
10001100	13	−.0040516	−.0297121	.016489145	.014214921	.011439743	.008593808	.006026030	.001835036	.000363147	.000003910	01111010
00111001	12	−.0074797	−.0273646	.015869878	.013168526	.010199629	.007372390	.004977789	.001369161	.000243509	.000002060	10011110
01011001	12	−.0084433	−.0242197	.015740082	.013022104	.010110559	.007366260	.005035842	.001472880	.000289587	.000002739	10101110
01101000	12	−.0084433	−.0247118	.015731835	.012996297	.010068192	.007315068	.004985271	.001446887	.000283319	.000003433	10110110
10010100	12	−.0084433	−.0265453	.015699317	.012888219	.009878442	.007069442	.004722828	.001274893	.000223784	.000001898	11011010
01100100	12	−.0094985	−.0257823	.015509735	.012570619	.009506723	.006708025	.004415927	.001145828	.000192563	.000001479	11100101
10011000	12	−.0129266	−.0213451	.014927210	.011645084	.008475757	.005754005	.003642592	.000854263	.000128860	.000000773	11101001
01110000	12	−.0142547	−.0183440	.014722779	.011352389	.008183739	.005513712	.003470861	.000809344	.000122817	.000000772	11001110
11000001	11	−.0152183	−.0133149	.014620785	.011281716	.008201217	.005615298	.003620682	.000929266	.000164033	.000001695	11111000
01011000	11	−.0152183	−.0164274	.014570286	.011127569	.007952319	.005316739	.003324779	.000766517	.000115965	.000000746	11110100
10100010	11	−.0152183	−.0170233	.014560859	.011099664	.007908853	.005266700	.003277455	.000743909	.000110375	.000000678	11101100
10010010	11	−.0186464	−.0105466	.014011576	.010275552	.007040767	.004506031	.002692772	.000549803	.000072739	.000000342	11011010
01101000	10	−.0197017	−.0083947	.013845105	.009974964	.006786763	.004287833	.002528637	.000498597	.000063530	.000000275	11011010
10011000	10	−.0209381	−.0041850	.013676260	.009813329	.006605456	.004163462	.002458076	.000493861	.000065683	.000000335	11100110
01100100	10	−.0254214	.0084650	.013020207	.008915068	.005728318	.003451782	.001949453	.000350678	.000041614	.000000164	11011010
10011000	10	−.0254214	.0073435	.013004348	.008872964	.005669329	.003390553	.001897105	.000331102	.000037796	.000000135	11100110
10100100	10	−.0254214	.0065903	.012993658	.008844542	.005629534	.003349352	.001862033	.000318223	.000035352	.000000118	11100110
01110000	9	−.0263850	.0122889	.012895511	.008766198	.005606214	.003371139	.001905097	.000347583	.000042654	.000000192	11110001

81

Table A

m = 5, n = 3 Rank Orders 51 thru 56 of 56 C(5, 3) = 56 1/C(5, 3) = .01785714

z	w	c_1	c_2	d = .2	.4	.6	.8	1.0	1.5	2.0	3.0	z^{tc}
10101000	9	−.0311412	.0258118	.012202227	.007818420	.004695007	.002641109	.001391156	.000209516	.000020741	.000000057	11101010
11000100	9	−.0321965	.0306136	.012072644	.007669484	.004574971	.002560799	.001344168	.000201797	.000020038	.000000056	11011100
10110000	8	−.0365882	.0483052	.011503112	.006974777	.003977629	.002131985	.001073285	.000146096	.000013268	.000000032	11110010
11001000	8	−.0379163	.0540216	.011335483	.006775591	.003810640	.002015008	.001001121	.000132066	.000011649	.000000026	11101100
11010000	7	−.0433632	.0804447	.010683791	.006039829	.003223264	.001622357	.000769493	.000091463	.000007378	.000000014	11110100
11100000	6	−.0490830	.1128405	.010051108	.005375922	.002728215	.001311849	.000596901	.000064957	.000004886	.000000009	11111000

82

Table A

m = 4, n = 4   **Rank Orders 1 thru 43 of 43**   **C(4, 4) = 70**   **1/C(4, 4) = .01428571**

z	w	c_1	c_2	d = .2	.4	.6	.8	1.0	1.5	2.0	3.0	z^{tc}
00001111	26	.0414452	.1015518	.024869263	.041162770	.064905061	.097698545	.140698265	.293082497	.488831240	.830373724	00001111x
00010111	25	.0370876	.0753955	.023322918	.035956418	.052410481	.072321434	.094602994	.147898909	.170905065	.101637833	00010111x
00100111	24	.0325118	.0516807	.021831430	.031373123	.042440077	.054101551	.065066296	.080647336	.071145572	.022461555	00011011
00101011	23	.0279360	.0307525	.020430775	.027347316	.034283556	.040281585	.044391427	.042980672	.028384674	.004159140	00011101
01000111	23	.0270918	.0282909	.021014049	.026837516	.033466145	.039238205	.043306049	.042882665	.029948458	.005650696	00011110
00101101	22	.0235784	.0146317	.019210770	.024152012	.028407980	.031285102	.032284660	.026429253	.014703054	.001526744	00100111x
01001011	22	.0225159	.0107119	.018913673	.023375229	.026982805	.029108987	.029366076	.022483989	.011493340	.000938368	00101011
00101110	22	.0189293	.0026191	.018046337	.021423443	.023920229	.025143076	.024905152	.018898601	.010167328	.001155458	00101101
01001101	21	.0181584	−.0022652	.017780047	.020623690	.022306923	.022511549	.021209581	.015589913	.005756607	.000314724	00101110
01001110	21	.0170959	−.0053194	.017506588	.019967546	.021205591	.020978127	.019340652	.011632885	.004554038	.000199552	00110011x
01010011	21	.0143534	−.0098533	.016882397	.018645732	.019252494	.018591927	.016798782	.009776407	.003798014	.000177615	00110101
01010101	20	.0135825	−.0120514	.016685630	.018182306	.018495321	.017572803	.015605070	.008667945	.003208230	.000137658	00110110
01010110	20	.0127383	−.0151527	.016453355	.017599968	.017488991	.016151014	.013866025	.006898780	.002194086	.000059617	00111001
01011001	20	.0099959	−.0180148	.015866832	.016435437	.015882076	.014322556	.012058499	.005824336	.001851916	.000036380	00111010
01011010	19	.0089334	−.0197928	.015624182	.015918289	.015110479	.013367081	.011022497	.005014560	.001481208	.000036380	00111100
01011100	19	.0081625	−.0215897	.015437890	.015504890	.014476811	.012570117	.010153527	.004350678	.001197072	.000022727	01000111x
01100011	19	.0054200	−.0228202	.014885076	.014470237	.013128672	.011121234	.008799560	.003648156	.000999741	.000022729	01001011
01100101	19	.0045758	−.0248156	.014680754	.014016890	.012434558	.010249736	.007851206	.002927147	.000692037	.000009988	01001101
01100110	18	.0035867	−.0252401	.014482207	.013648937	.011962557	.009753325	.007400143	.002715924	.000641848	.000009898	01001110
01101001	19	.0007709	−.0250380	.013942293	.012683491	.010754813	.008500273	.006262598	.002147500	.000476476	.000006489	01010011x
01101010	18	.0000000	−.0233831	.013825665	.013532956	.010642663	.008467456	.006313476	.002286752	.000557095	.000010473	01010101
01101100	18	.0000000	−.0261409	.013772363	.012340525	.010277661	.007956434	.005725923	.001829554	.000371677	.000004000	01010110
01110001	17	−.0007709	−.0262667	.013769935	.012331798	.010261239	.007933702	.005700710	.001810567	.000364584	.000003813	01011001
01110010	17	−.0035867	−.0250380	.013645012	.012150688	.010089099	.007812979	.005644175	.001850912	.000396107	.000005194	01011010
01110100	17	−.0045758	−.0248105	.013097323	.011157582	.008830169	.006138616	.004436082	.001232666	.000215325	.000001620	01011100
01111000	17	−.0054200	−.0228202	.012915285	.010674620	.008465853	.006026482	.004074824	.001125512	.000191834	.000001458	01100011x
10000111	17	−.0089334	−.0215897	.012788040	.009809524	.008307333	.005010330	.003202977	.000755002	.000200138	.000001697	01100101
10001011	16	−.0099959	−.0197928	.012282642	.009648960	.007275620	.004901014	.003138695	.000754195	.000116219	.000000592	01100110
10001101	16	−.0099959	−.0180148	.011992334	.009384297	.006844173	.004651622	.002945865	.000690128	.000103981	.000000736	01101001
10001110	16	−.0127383	−.0151527	.011516266	.008617128	.005983289	.003854213	.002302753	.000456470	.000056278	.000000204	01101010
10010011	16	−.0135825	−.0120514	.011405236	.008486802	.005883453	.003798356	.002282841	.000466412	.000060536	.000000258	01101100
10010101	15	−.0143534	−.0098533	.011294127	.008340163	.005751411	.003703204	.002226014	.000461414	.000062197	.000000313	01110001
10010110	15	−.0173674	−.0053194	.010845663	.007657776	.005026766	.003066686	.001738230	.000304171	.000033408	.000000098	01110010
10011001	15	−.0181584	−.0022652	.010691798	.007450239	.004831304	.002914401	.001634755	.000279579	.000030125	.000000086	01110100
10011010	14	−.0189293	.0026191	.010691760	.007407976	.004846611	.002972314	.001707960	.000320810	.000039867	.000000177	01101100
10011100	14	−.0235784	.0107119	.010060615	.006615333	.004051765	.002311956	.001228483	.000184633	.000017628	.000000040	10010100x
10100100	14	−.0235784	.0146317	.009924997	.006438031	.003896812	.002199658	.001157338	.000170304	.000015887	.000000036	10011000x
11001000	13	−.0270918	.0282909	.009464092	.005872204	.003409650	.001851269	.000939219	.000127418	.000011160	.000000022	11001100x
11010100	13	−.0279360	.0307525	.009342982	.005711739	.003262095	.001739519	.000865639	.000111391	.000009210	.000000016	11010010x
11100100	12	−.0325118	.0516807	.008782004	.005067070	.002741660	.001389976	.000659783	.000076423	.000005780	.000000009	11011000x
11101000	11	−.0370876	.0753955	.008253186	.004492141	.002301082	.001108194	.000501285	.000052135	.000003598	.000000005	11101000x
11110000	10	−.0414452	.1015518	.007787907	.004022235	.001964886	.000906507	.000394415	.000037790	.000002445	.000000003	11110000x

83

Table A

m = 8, n = 1 Rank Orders 1 thru 9 of 9 C(8, 1) = 9 1/C(8, 1) = .11111111

z	w	c_1	c_2	d = .2	.4	.6	.8	1.0	1.5	2.0	3.0	z^{tc}
000000001	9	.1650015	.1736242	.147642751	.191342970	.242045014	.299111534	.361433707	.529531734	.691318205	.909838545	011111111
000000010	8	.1035886	.0105417	.131832580	.151674664	.169242053	.183186240	.192378162	.190780934	.157440618	.062868986	101111111
000000100	7	.0635523	−.0540519	.122567759	.130885465	.135311466	.135437431	.131261707	.105463855	.069401052	.016614548	110111111
000001000	6	.0305029	−.0837863	.115451621	.116050141	.112851392	.106168799	.096633871	.066106240	.036809119	.006169971	111011111
000010000	5	.0000000	−.0926554	.109273375	.103940590	.095624819	.085086685	.073230213	.043489954	.020971425	.002612084	111101111
000100000	4	−.0305029	−.0837863	.103444196	.093161103	.081158168	.068389399	.055743556	.028902106	.012166484	.001152731	111110111
001000000	3	−.0635523	−.0540519	.097500054	.082811242	.068074266	.054157572	.041695806	.018780364	.006883219	.000496770	111111011
010000000	2	−.1035886	.0105417	.090790736	.071912297	.055204491	.041066801	.029599987	.011360929	.003569705	.000192106	111111101
100000000	1	−.1650015	.1736242	.081496927	.058221527	.040488332	.027393539	.018022990	.005583884	.001440172	.000054258	111111110

84

Table A

m = 7, n = 2   Rank Orders 1 thru 36 of 36   C(7, 2) = 36   1/C(7, 2) = .02777778

z	w	c_1	c_2	d = .2	.4	.6	.8	1.0	1.5	2.0	3.0	z^{tc}
000000011	17	.0671475	.1328536	.04110738	.067227684	.098476437	.138856650	.188785820	.351053998	.542130992	.852350784	001111111
000000101	16	.0571384	.0837892	.040913368	.057444304	.076944366	.098402983	.120258122	.163759763	.171465336	.091478716	001011111
000001001	16	.0488761	.0501807	.038491963	.050868815	.063466080	.075613329	.085777470	.095370061	.079481359	.024716008	011011111
000000110	15	.0417852	.0273093	.036582837	.045798163	.054531226	.061790480	.066672622	.065387417	.047995289	.011479762	100111111
000010001	15	.0412504	.0242834	.036404983	.045262302	.053416837	.059876927	.063793575	.060144938	.042019834	.008883664	011011011
000100001	14	.0336247	.0031155	.034444775	.040483107	.045123103	.047727792	.047938876	.038852374	.023209770	.003538425	011101011
000001010	14	.0335229	.0024081	.033440525	.040354795	.044830280	.047183104	.047061227	.037035151	.020976247	.002568137	101101011
000010010	13	.0258972	−.0153369	.032534259	.036004532	.037654118	.037219986	.034779301	.022980208	.010728096	.000835599	011101101
001000001	13	.0253623	−.0145685	.032452142	.035927099	.037712052	.037556744	.035509286	.024717377	.012666857	.001418125	101101101
000100010	13	.0235138	−.0178429	.032021092	.034967751	.036182521	.035484963	.032993454	.021824522	.010420461	.000917073	110111111
000100010	12	.0182714	−.0282775	.030777828	.032182739	.031761395	.029588094	.026021174	.014682633	.005797467	.000312682	101101111
000010100	12	.0158881	−.0303564	.030272713	.031173110	.030334605	.027896509	.024249009	.013361630	.005193178	.000278425	110101111
010000001	11	.0153532	−.0286463	.030208348	.031155564	.030489304	.028327845	.025004213	.014696258	.006382083	.000518319	011011111
000100100	11	.0100091	−.0369893	.028998373	.028546872	.026514773	.023234329	.019209816	.009264278	.003113501	.000119912	101011111
001000010	11	.0082624	−.0379961	.028634291	.027847960	.025552731	.022122753	.018072649	.008458884	.002757980	.000099607	110111011
001000100	10	.0076257	−.0364998	.028514683	.027743227	.025513227	.022199637	.018278683	.008838500	.003035535	.000139512	011111110
100000001	10	.0000000	−.0339773	.027106590	.025190046	.022295925	.018800769	.015108771	.007109933	.002509339	.000129781	011111111
000101000	10	.0000000	−.0398208	.026992679	.024768119	.021460669	.017559191	.013567120	.005542473	.001584281	.000044530	101101111
010000010	10	.0000000	−.0401438	.026986403	.024745103	.021415866	.017494157	.013488816	.005471170	.001548526	.000042356	101111101
010100000	9	.0000000	−.0409101	.026971511	.024690463	.021309399	.017339406	.013302212	.005300212	.001462104	.000036983	110111101
000110000	9	−.0076257	−.0364998	.025572915	.022269089	.018341878	.014288575	.010527548	.003843631	.000990546	.000023189	111100111
001001000	9	−.0082624	−.0379961	.025421995	.021948648	.017876350	.013734461	.009953981	.003447621	.000828918	.000016032	111101011
100000010	9	−.0100091	−.0369893	.025101575	.021393770	.017196074	.013034715	.009317097	.003112132	.000719584	.000012736	110101111
010001000	8	−.0153532	−.0303564	.024212966	.019998724	.015646074	.011591023	.008128937	.002630523	.000601852	.000011074	101111011
001100000	8	−.0158881	−.0282775	.024081103	.019721924	.015257418	.011149099	.007694792	.002370955	.000510710	.000008082	111101101
100000100	8	−.0182714	−.0287275	.023656844	.019009613	.014141411	.010306663	.006952990	.002012069	.000403580	.000004417	111111110
010010000	7	−.0235138	−.0178429	.022821959	.017754603	.013076372	.009115995	.006014370	.001669333	.000327390	.000003258	111110111
001010000	7	−.0253623	−.0145685	.022519592	.017283387	.012552133	.008623003	.005601431	.001488347	.000277041	.000002645	110111101
011000000	6	−.0258972	−.0153369	.022405967	.017071800	.012285336	.008348959	.005357517	.001373969	.000245629	.000001402	111110101
100001000	7	−.0335229	.0024081	.021230486	.015357775	.010512546	.006807837	.004170106	.000958873	.000155140	.000001349	111101101
010100000	6	−.0336247	.0031155	.021221062	.015350488	.010509215	.006800671	.004169208	.000956811	.000153864	.000001349	111011110
100010000	6	−.0412504	.0242834	.020096145	.013778142	.008948120	.005502395	.003202433	.000649033	.000092598	.000000647	111101110
011000000	5	−.0417852	.0273093	.020039569	.013729180	.008928914	.005509325	.003224403	.000668831	.000098962	.000000774	111110110
100100000	5	−.0488761	.0501807	.019038200	.012385702	.007644742	.004474539	.002482483	.000449141	.000057679	.000000334	111110100
101000000	4	−.0571384	.0837892	.017965053	.011060155	.006477627	.003606862	.001908336	.000309671	.000036123	.000000178	111111010
110000000	3	−.0671475	.1328536	.016765965	.009687242	.005351976	.002874288	.001422188	.000207268	.000022137	.000000096	111111100

85

z	w	c_1	c_2	d = .2	.4	.6	.8	1.0	1.5	2.0	3.0	z^{tc}
000000111	24	.0355867	.0906101	.021085737	.035483615	.056841584	.086852998	.126858289	.272643677	.466313506	.817772776	000111111
000001011	23	.0320457	.0687203	.019806737	.031103311	.046154984	.064802654	.086197708	.140119141	.167630091	.10597067	001011111
000010011	22	.0287775	.0509668	.018714750	.027664397	.038491602	.050463446	.062407681	.082870263	.078536553	.028072978	001101111
000100011	22	.0277561	.0459345	.018390676	.026697839	.036457099	.048677708	.056820475	.071533551	.063974224	.020075621	001110111
000001101	21	.0255093	.0353986	.017694982	.024675275	.032312609	.039776311	.046078009	.051420533	.040356717	.009504633	010011111
000010101	21	.0244879	.0304230	.017373105	.023724267	.030335182	.039634938	.040827771	.041406411	.028564368	.004519177	010101111
000100101	20	.0219683	.0209003	.016662487	.021852100	.026878186	.031039233	.033690745	.031788486	.020955225	.003429625	010110111
000011001	20	.0212197	.0171266	.016423963	.021147108	.025425376	.028860550	.032991621	.025339890	.014250201	.001390971	011001111
000101001	20	.0211761	.0187279	.016461256 •	.021375631	.026089508	.029957847	.032395909	.030519941	.020282342	.003448381	011010111
000110001	20	.0209469	.0165723	.016360549	.021016174	.025251219	.028398553	.029917554	.025743389	.015005431	.001688717	011100111
000001110	20	.0179080	.0067103	.015547155	.018977835	.021662170	.023132543	.023122257	.017357128	.008728429	.000692336	100011101
010000011	19	.0176787	.0065942	.015503390	.018916160	.021644956	.023251412 •	.023474624	.018477479	.010162114	.001161116	001110101
001010011	19	.0176787	.0051268	.015464179	.018718087	.021129654	.022248682	.021867125	.015554894	.007297315	.000483788	011011101
001100011	19	.0146398	.0051131	.015463782	.018718087	.021123585	.022235968	.021845445	.015509423	.007247826	.000471561	011101101
010001001	19	.0144106	−.0030601	.014695556	.016905733	.018129740	.018129874	.016911772	.010508798	.004263844 •	.000201219	101011101
000110010	19	.0143669	−.0035985	.014640132	.016789919	.017945836	.017907726	.016687035	.010412198	.004301870	.000224986 •	101101101
010010001	18	.0141377	−.0034087	.014636423	.016789919	.017978290	.017977762	.016796814	.010567670	.004405492	.000233286	101110101
001011001	18	.0133891	−.0048559	.014558192	.016560929	.017532351	.017287338	.015867338	.009433319	.003644715	.000156234	110011101
001101001	18	.0110988	−.0061607	.014386576	.016196165	.016993324	.016625030	.015173455	.008943319	.003458799	.000152515	110101101
100000011	18	.0110988	−.0082361	.013905284	.015267734	.015768906	.015332242	.014046914	.008760998	.003866310 •	.000293665	110110101
000100110	18	.0110988	−.0112284	.013834789	.014956712	.015045789	.014087093	.012279395	.006384172	.002138687	.000065934	100111001
010100001	18	.0110988	−.0113282	.013832397	.014946030	.015020802	.014044067	.012218714	.006306426	.002087740	.000061623	101011011
001010010	17	.0108696	−.0117303	.013780220	.014837889	.014868314	.013871834	.012056304	.006249445	.002108131 •	.000069356	011101011
010001010	17	.0100773	−.0116039	.013632651	.014595195	.014614239	.013691913	.012008191	.005397957	.001714709	.000048728	101101011
001100010	17	.0098481	−.0136449	.013542511	.014322900	.014089001	.012895105	.010986458	.004169602	.001202869	.000027088	011110011
010010010	17	.0078306	−.0165596	.013092664	.013388834	.012733848	.011266299	.009275094	.004298359	.001301139	.000035362	101110011
001001100	17	.0076014	−.0163896	.013052967	.013331129	.012687551	.011257502	.009317223	.003791273	.001025098	.000018950	110011011
010001100	17	.0075578	−.0174657	.013020449	.013215520	.012449157	.010885318	.008835672	.003791273	.001297740	.000037164	110101011
001100100	17	.0068092	−.0163096	.012902247	.013066750	.012367740	.010942740	.009052989	.003164082	.000837164	.000024761	110110011
100000101	17	.0068092	−.0172516	.012881917	.012984701	.012193618	.010669628	.008700783	.003843787	.001098563		110111001
000101010	17	.0068092	−.0178207	.012869754	.012936317	.012092856	.010515290	.008507349	.003360174	.001010140 •	.000020838 •	101011101
010100010	16	.0065800	−.0182475	.012817008	.012825424	.011932814	.010327202	.008317460	.003546868	.000976860	.000020873	101110110
001010100	16	.0042896	−.0208462	.012322149	.011830748	.010537466	.008707642	.006676536	.002478948	.000578077	.000007867 •	111001011
010011000	16	.0035410	−.0202410	.012190729	.011623256	.010320167	.008534708	.006575467	.002517043	.000622596	.000010520	111010011
100001001	16	.0033118	−.0206647	.012138299	.011550596	.010165596	.008356237	.006398407	.002414877	.000592282	.000010237	110111010
001101000	16	.0032682	−.0199486	.012143854 •	.011550038	.010243182	.008471693	.006535428	.002523792	.000635377	.000011523	111100011
100010001	16	.0032682	−.0209043	.012124390	.011476061	.010095613	.008254652	.006273957	.002295545	.000532967	.000007403	110111100
010010100	16	.0032682	−.0215732	.012110805	.011424809	.009994575	.008108450	.006100863	.002215472	.000476231	.000003870	111001101
001100100	16	.0010214	−.0219795	.011670148	.010623178	.008980145	.007050125	.005140838	.001690823	.000351418	.000004989	111010101
100001100	16	.0002292	−.0207574	.011540757	.010443264	.008793541	.006910770	.005063927	.001714366	.000375751		111100101
000111000	15	.0000000	−.0204000	.011502337	.011375031	.008736797	.008669269	.005043300	.001727877 •	.000387439 •	.000005568	111000111
100010010	15	.0000000	−.0210252	.011491628	.011336592	.008664589	.008769552	.004930678	.001645679	.000357127	.000004790	101011110
010101000	15	.0000000	−.0220428	.011471976	.011265821	.008530916	.008583527	.004718485	.001485342	.000295017	.000002946	111010110
011000100	15	.0000000	−.0226820	.011459659	.011221762	.008448624	.006470796	.004592491	.001396729	.000264051	.000002273	111100110
011000001	14	.0000000	−.0207574	.011452827	.010277596	.008605207	.006724340	.004905781	.001656025	.000245402	.000005388	111001110
001001010	14	−.0010214	−.0219795	.011276243	.009917596	.008098114	.006139591	.004321695	.001298422	.000213074 •	.000002191	110111011
100100001	14	−.0032682	−.0209043	.010880988	.009292953	.007369115	.005461760	.003776330	.001109906	.000177343	.000002272	111010111
010100100	14	−.0032682	−.0215732	.010863474	.009213057	.007261801	.005319461	.003621841	.001006392 •	.000157438	.000001417	111011011
001100010	14	−.0032682	−.0207574	.010851235	.009171459	.007187726	.005223172	.003519375	.000942458	.000174627	.000001069	111100111
100010010	15	−.0033118	−.0206647	.010858902	.009209264	.007259783	.005318408	.003620018	.001002006	.000174627	.000001318	101110110

86

Table A

m = 6, n = 3 Rank Orders 51 thru 84 of 84 C(6, 3) = 84 1/C(6, 3) = .01190476

z	w	c_1	c_2	d = .2	.4	.6	.8	1.0	1.5	2.0	3.0	z^{tc}
001011000	14	−.0035410	−.0202410	.010822795	.009158710	.007213791	.005288040	.003607484	.001012306	.000181229•	.000001514	111010011
010010100	14	−.0042896	−.0208462	.010667936	.008866140	.006833701	.004884507	.003237490	.000832097	.000133354	.000000830	110101101
100010010	13	−.0065800	−.0182475	.010273577	.008234759	.006128894	.004234485	.002715202	.000643993	.000095396	.000000506	101101110
101000001	13	−.0068092	−.0163096	.010263505	.008267602•	.006222850	.004376942	.002877389	.000751459	.000129580	.000001149	011011010
011101000	13	−.0068092	−.0172516	.010247183	.008214725	.006132770	.004263588	.002760179	.000681422	.000107805	.000000714	111010011
011000010	13	−.0068092	−.0178207	.010237271	.008182598	.006078273	.004195629	.002690884	.000642311	.000096695	.000000549	110111001
010101000	14	−.0075578	−.0174657	.010100151	.007950794	.005806779	.003934349	.002472846	.000557426	.000078517	.000000380	110110101
100001100	13	−.0076014	−.0163896	.010109620•	.007991007	.005877411	.004021319	.002558838	.000600893	.000089318	.000000498	110011110
010011000	14	−.0078306	−.0165596	.010063141	.007906540	.005773063	.003916755	.002468803	.000563922	.000081130	.000000418	111001110
100100010	13	−.0098481	−.0136449	.009726224	.007382962	.005205373	.003407927	.002071277	.000430236	.000055902	.000000229	101011010
001110000	12	−.0100773	−.0116039	.009715875	.007408525•	.005276600	.003509531	.002179372	.000489980	.000071543	.000000417	111100011
100010100	13	−.0108696	−.0117303	.009562850	.007139886	.004953139	.003191646	.001909697	.000381743	.000047761	.000000181	101101110
110000100	12	−.0110988	−.0082361	.009575201•	.007232001	.005127376	.003411498	.002129803	.000496324	.000077807	.000000596	011111110
010010100	12	−.0110988	−.0112284	.009527645	.007090553	.004905971	.003155355	.001886188	.000378002	.000047825	.000000192	110111010
101001000	12	−.0110988	−.0113282	.009526094	.007086068	.004899181	.003147796	.001879306	.000375079	.000047202	.000000187	110111001
100100100	12	−.0133891	−.0061607	.009173207	.006579461	.004391076	.002725944	.001573593	.000289338	.000033582	.000000112	101111010
011011000	12	−.0141377	−.0048559	.009052100	.006397791	.004201558	.002562998	.001451814	.000253080	.000027608	.000000079	101111001
011001000	12	−.0143669	−.0034087	.009030937	.006387211	.004210256	.002585761	.001479198	.000267892	.000030968	.000000107	110111010
110010000	12	−.0144106	−.0035985	.009019701	.006364402	.004180513	.002555252	.001452808	.000257442	.000028822	.000000090	111001110
011100000	11	−.0146398	−.0030601	.008985040	.006316472.	.004135074	.002520306	.001429879	.000253079	.000028524	.000000093•	111011001
110000010	11	−.0176787	.0065942	.008557175	.005753091	.003614987	.002121518	.001162104	.000190164	.000019994	.000000057	101111100
100101000	11	−.0176787	.0051268	.008536649	.005699468	.003541447	.002047186	.001100537	.000169366	.000016443	.000000038	110111010
101000100	11	−.0176787	.0051131	.008536472	.005699044	.003540914	.002046693	.001100164	.000169266	.000016430	.000000038	110111010
011010000	10	−.0179080	.0067103	.008516755	.005690379	.003549467•	.002066297	.001122258	.000179772	.000018559	.000000052	110111010
100110000	10	−.0209469	.0165723	.008091798	.005146442	.003039981	.001678405	.000863716	.000120250	.000012684	.000000021	111100110
011001000	10	−.0211761	.0187279	.008078725	.005140510•	.003064873	.001711143	.000894076	.000131575	.000012662	.000000032	111111001
101001000	9	−.0212197	.0171266	.008049393	.005074583	.002981523	.001631896	.000831744	.000131651	.000009697	.000000018	110111010
010011000	10	−.0219683	.0209003	.007962445	.004981447	.002912812	.001590763	.000810860	.000110799	.000009718	.000000019	111011100
101011000	9	−.0244879	.0304230	.007628796	.004669088	.002556230	.001335174	.000650767	.000079495	.000006241•	.000000010	110111110
110011000	9	−.0255093	.0353986	.007507323	.004433840	.002450565	.001266563	.000611738	.000073438	.000005700	.000000009	111011100
101100000	8	−.0277561	.0459345	.007235068	.004124616	.002203865	.001102833	.000516463	.000057738	.000004206	.000000006	111110010
010010000	8	−.0287775	.0509668	.007114124	.003990242	.002098887	.001034539	.000477472	.000051588	.000003643	.000000005	111101100
110010000	7	−.0320457	.0687203	.006745790	.003599904	.001807222	.000852719	.000377798	.000037268	.000002437	.000000003	111101110
111000000	6	−.0355867	.0906101	.006374511	.003231767	.001548889	.000700733	.000298845	.000027194	.000001668	.000000002	111111000

87

Table A

m = 5, n = 4   Rank Orders 1 thru 50 of 126   C(5, 4) = 126   1/C(5, 4) = .00793651

z	w	c_1	c_2	d = .2	.4	.6	.8	1.0	1.5	2.0	3.0	z^{tc}
000001111	30	.0259032	.0730097	.014812966	.026139287	.043703744	.069391508	.104882317	.241526004	.433710371	.801453863	000001111
000010111	29	.0237245	.0582323	.013979384	.023129786	.035999673	.052785275	.073026932	.128844319	.163925936	.111206075	000101111
000011011	28	.0215457	.0450372	.013206474	.020558490	.029975200	.040986967	.053672466	.075362247	.075222021	.028169512	000110111
000011101	28	.0213638	.0440077	.013144534	.020361206	.029533877	.040164179	.051277730	.072300868	.070929893	.025519033	001001111
000100111	27	.0191850	.0324152	.012427750	.018158124	.024783708	.031638184	.036709882	.044779617	.036352892	.008623343	001010111
000101011	27	.0191850	.0320465	.012415619	.018083678	.024543887	.031067923	.037824454	.044781058	.031142467	.005508894	001100111
000101101	27	.0185041	.0291397	.012215029	.017538193	.023518158	.029490642	.034625411	.039148078	.030316763	.005526991 ●	001101011
000110011	26	.0170062	.0217932	.011742938	.016151659	.020669194	.024631456	.027361046	.026340786	.016734569	.002098992	010001111
001000111	26	.0168243	.0207784	.011681914	.015962665	.020263228	.023913499	.026256869	.024320753	.014619227	.001531964	001101101
010000111	26	.0163253	.0193842	.011556796	.015683705	.019859639	.023493579	.025999375	.025260440	.016735569	.002615720	011000111
000010101	26	.0163253	.0186930	.011535907	.015566194	.019513256	.022741348	.024659189	.022140325	.012903990	.001272812	010010111
001001101	25	.0146456	.0175591	.011047151	.014247238	.017035805	.018899749	.019468788	.015238861	.007646155	.000544441	001101011
000110101	25	.0141465	.0099344	.010908707	.013891441	.016400049	.017962956	.018267313	.013842657	.006722157	.000444826	010011011
001001011	25	.0141174	.0113307	.010942538 ●	.014088233 ●	.016954876	.019095088	.020149068	.017489197	.010431756	.001347309	100001111
010001011	25	.0139646	.0094040	.010861923	.013780352	.016217305	.017715996	.017977943	.013578611	.006586765	.000435167	010101011
001010011	25	.0139646	.0090815	.010852897	.013733550	.016090765	.017465293	.017572607	.012858778	.005942841	.000335193	011001011
001100011	25	.0124668	.0043220	.010457190	.012767131	.014453708	.015367428 ●	.014816296	.010143470	.004467167	.000247499	100010111
010010011	25	.0119386	.0042545	.010359813	.012631731	.014400864	.015184792	.015368096	.011672489	.006100561	.000608385	100011011
010100011	25	.0119386	.0032347	.010332582	.012496094	.014046172	.014682867	.014282454	.009771392	.004331084	.000244135	011000111
000111001	24	.0117858	.0023032	.010280873	.012344024	.013747455	.014212161	.013649769	.008993776	.003818434	.000197554	011011101
010100101	24	.0117858	.0018991	.012270221	.012292008	.013615031	.013965111	.013273806	.008419412	.003375402	.000145359	011010111
010011001	24	.0117858	.0015568	.012261110	.012247165	.013500192	.013750037	.012945939	.007921652	.003000547	.000106088	010011101
011000011	24	.0111049	−.0002714	.012090065	.011850094	.012865309	.012915606	.011996988	.007128865	.002641891	.000092053	100011101
001011010	24	.0097599	−.0031947	.009768812	.011142352	.011782552	.011557246	.010521275	.006027921	.002199527	.000079937	011101011
010101010	24	.0096070	−.0040432	.009711721	.010968274	.011530591	.011183533	.010049448	.005536294	.001928029	.000063055	011110011
010101100	24	.0096070	−.0043657	.009711721	.011093170	.011438055	.011019099	.009811618	.005219975	.001719289	.000046184	011011110
100011010	24	.0095780	−.0031622	.009735734 ●	.011093170	.011745791	.011652828	.010588639	.006218064	.002358186	.000096219	100011110
100100110	24	.0095780	−.0038134	.009719664	.011019558	.011569668	.011253555	.010145063	.005631558	.001966054	.000062052	100111011
011010010	23	.0094252	−.0047206	.009669274	.010876974	.011302246	.010853494	.009636885	.005099055	.001673913	.000044926	011011101
011100010	23	.0089261	−.0062816	.009538261	.010561817	.010778280	.010140486	.008798759	.004341558	.001303010	.000027186	011010101
100010011	23	.0073992	−.0083164	.009204021	.008889124	.009847525	.009092166	.007787280	.003819914	.001187715	.000030785	001101110
001111100	23	.0073992	−.0083381	.009203772	.008889123	.009849878	.009099830	.007802561	.003850477	.001212227	.000033171	101000111
011010100	23	.0073992	−.0090089	.009187780	.008815576	.009687860	.008827626	.007403027	.003419309	.000964660	.000018117	100101011
011101000	22	.0072464	−.0087180	.009166258	.008804626	.009718685	.008933182	.007619673	.003710825	.001153933	.000031033	001111001
010101001	23	.0072464	−.0094091	.009150137	.008973047	.009562315	.008675996	.007275103	.003330599	.000944399	.000018916	011110011
010010101	23	.0067473	−.0107094	.009026013	.009451352	.009115562	.008099077	.006632471	.002823697	.000728235	.000011088	010101110
100010101	23	.0067182	−.0097167	.009042981	.009536111	.009309380	.008415743	.007047567	.003245135	.000935768	.000019871	010011110
100100101	23	.0067182	−.0100454	.009035572	.009505184	.009242135	.008308738	.006908005	.003104855	.000865575	.000016699	100011101
010011001	23	.0065654	−.0109024	.008987291	.008975851	.009014002	.007989167	.006530158	.002775217	.000717739	.000011168	011001101
100100011	22	.0052204	−.0119295	.008709076	.008847797	.008325130	.007258369	.005866867	.002484392	.000665288	.000012776	011101110
001100110	22	.0052204	−.0125807	.008694594	.008788181	.008197335	.007057921	.005610869	.002234657	.000544616	.000007622	100110011
100101100	22	.0050570	−.0124097	.008669504	.008755847	.008178374	.007068974	.005656783	.002312216	.000596163	.000010105	011110010
001011010	22	.0050385	−.0127984	.008655192	.008709352	.008088045	.006933418	.005487891	.002161648	.000520677	.000007081	101000101
001111100	22	.0045394	−.0139645	.008578183	.008618356	.008051519	.006997542	.005660399	.002443213	.000683389	.000011405	110001001
100010010	22	.0045394	−.0133762	.008547354	.008547354	.007788205	.006590479	.005147396	.001956467	.000452610	.000005562	010101110
001010010	22	.0045394	−.0137263	.008539856	.008463928	.007727314	.006499242	.005036441	.001862219	.000414039	.000005278	100111001
011001001	21	.0043866	−.0137277	.008510944	.008463708	.007674516	.006456874	.005013046	.001878931	.000429978	.000005278	011011100
010101001	21	.0043866	−.0140963	.008503096	.008384691	.007611881	.006364064	.004901938	.001747583	.000394049	.000004367	011011010
101001001	21	.0043866	−.0141165	.008210581	.007872692	.006999009	.005772154	.004418306	.001641283	.000388527	.000005980	001111010
101000011	21	.0028597										101001011

Table A

z	w	c_1	c_2	d = .2	.4	.6	.8	1.0	1.5	2.0	3.0	z^{tc}
100100101	21	.023607	−.0154748	.008086528	.007594482	.006574489	.005246759	.003860431	.001257680	.000247578	.000002153	010101110
010100110	22	.023607	−.0158034	.008079864	.007569556	.006526120	.005178351	.003781990	.001200440	.000227599	.000001764	100110101
100001110	22	.023316	−.0146825	.008097108	.007645572	.006682014	.005406090	.004049699	.001406921	.000304308	.000003556	011000110
010011100	20	.022079	−.0155823	.008055321	.007536759	.006501990	.005173595	.003798053	.001235903	.000245208	.000002258	111000101
001011100	21	.021788	−.0147683	.008065969	.007586668	.006604358	.005321440	.003969146	.001361893	.000289687	.000003185	110000101
100011001	21	.021788	−.0154335	.008052525	.007536458	.006508674	.005183226	.003809941	.001243265	.000246762	.000002236	011000111
010101001	20	.021788	−.0158593	.008043897	.007504188	.006444216	.005094500	.003708007	.001168284	.000222246	.000001697	101010101
011001001	20	.020810	−.0157606	.008018863	.007462035	.006401868	.005063888	.003694317	.001181816	.000230095	.000002031	011101001
001110001	20	−.0006810	−.0157606	.007757607	.007001026	.005834558	.004491145	.003193849	.000965094	.000179049	.000001467	110011010
110001010	21	−.0001528	−.0160128	.007651730	.006803808	.005578749	.004217517	.002939456	.000834869	.000142519	.000000904	100110011
110000011	20	.0000000	−.0142493	.007656606	.006875055	.005746546	.004472239	.003241648	.001065411	.000227117	.000002963	001111100
001101100	20	.0000000	−.0155917	.007630734	.006782438	.005573327	.004234449	.002975056	.000875250	.000158734	.000001253	110010011
101001100	20	.0000000	−.0160052	.007622782	.006754176	.005521103	.004163964	.002897758	.000824269	.000142385	.000000968	010011010
011000110	20	.0000000	−.0162582	.007617922	.006736933	.005489343	.004121296	.002851246	.000794259	.000133078	.000000823	100111001
100101010	20	.0000000	−.0162982	.007617158	.006734269	.005484576	.004115150	.002844905	.000790944	.000132368	.000000824	011101010
010101010	19	−.0001528	−.0167025	.007609394	.006706788	.005434144	.004047744	.002771019	.000744920	.000118551	.000000628	101010101
001011010	19	−.0006810	−.0160128	.007593545	.006703502	.005461501	.004107723	.002853055	.000811639	.000141645	.000001030	011101001
011010001	19	.0000000	−.0158077	.007495556	.006525751	.005259781	.003904169	.002675599	.000732309	.000121462	.000000749	110011100
001110010	19	−.0020260	−.0157606	.007238121	.006079935	.004702893	.003349282	.002195817	.000531667	.000076590	.000000334	100110110
110010001	19	−.0021788	−.0147683	.007227184	.006091047	.004751513	.003431157	.002293971	.000599359	.000097306	.000000631	110100110
100110001	19	−.0021788	−.0154335	.007215015	.006090909	.004679216	.003338371	.002197219	.000543130	.000081423	.000000416	111000011
010110001	19	−.0021788	−.0158593	.007207259	.006023832	.004633708	.003283060	.002137452	.000509553	.000074382	.000000311	011101101
001011001	20	−.0022079	−.0155823	.007206517	.006029536	.004647613	.003299895	.002157952	.000520402	.000074945	.000000331	101101010
100110010	19	−.0023316	−.0146825	.007199359	.006044093	.004696471	.003378064	.002249543	.000581986	.000093600	.000000598	110101010
010011010	19	−.0023607	−.0154748	.007179417	.005986008	.004600550	.003259248	.002128687	.000481986	.000075161	.000000360	110011001
001011100	19	−.0023607	−.0158034	.007173437	.005966030	.004565609	.003214913	.002083081	.000489316	.000068424	.000000284	011011100
101011000	19	−.0028597	−.0141165	.007107761	.005896021	.004529054	.003222080	.002120738	.000532032	.000082385	.000000468	011011001
100110100	19	−.0028597	−.0151363	.007089687	.005836771	.004428030	.003095249	.001993392	.000463440	.000064519	.000000269	110011010
010110100	19	−.0043866	−.0137277	.006822246	.005404746	.003945369	.002653315	.001643666	.000346298	.000043570	.000000147	100111010
001110100	19	−.0043866	−.0140963	.006815865	.005384402	.003911767	.002612939	.001604411	.000327530	.000039352	.000000115	110100110
100100110	18	−.0045394	−.0119645	.006823297	.005449854	.004043187	.002785869	.001782670	.000422540	.000063194	.000000361	111000011
010100110	18	−.0045394	−.0133762	.006799342	.005374541	.003919814	.002637951	.001638282	.000350774	.000045653	.000000177	011101101
001010110	18	−.0045394	−.0137263	.006793305	.005355320	.003888034	.002599623	.001600771	.000332286	.000041273	.000000137	101101001
100011100	18	−.0050385	−.0127984	.006713791	.005238729	.003770056	.002502003	.001531121	.000314472	.000038881	.000000130	110101010
010011001	18	−.0050676	−.0124097	.006714607	.005248815	.003770400	.002552428	.001553941	.000324845	.000041054	.000000144	110011001
001011001	18	−.0052204	−.0119295	.006693619	.005223309	.003770400	.002517176	.001554087	.000330498	.000043119	.000000171	011011100
100110100	18	−.0052204	−.0125807	.006682781	.005189939	.003716938	.002445489	.001494534	.000300498	.000036948	.000000120	100111100
011010100	18	−.0065654	−.0109024	.006445746	.004833401	.003331453	.002113064	.001233070	.000222421	.000023725	.000000055	101101010
101100100	17	−.0067182	−.0097167	.006445146	.004842278	.003365338	.002163408	.001286376	.000249360	.000029740	.000000100	101101010
110100010	17	−.0067182	−.0100454	.006439748	.004825869	.003339373	.002133372	.001258123	.000236641	.000026958	.000000078	011100001
101000110	18	−.0067473	−.0107094	.006423423	.004783952	.003277758	.002065523	.001196869	.000211628	.000022053	.000000049	101101010
110000101	18	−.0072464	−.0087180	.006360693	.004716281	.003233839	.002049966	.001200425	.000225566	.000025034	.000000070	100111100
011001100	18	−.0072464	−.0094091	.006349731	.004684143	.003184865	.001995201	.001151269	.000202819	.000021247	.000000048	011101100
100110001	17	−.0073992	−.0083164	.006338417	.004687306	.003209018	.002033604	.001187320	.000220413	.000025969	.000000083	011101010
010110001	17	−.0073992	−.0083381	.006337890	.004685201	.003204913	.002027980	.001186705	.000220415	.000024997	.000000073	110101001
001011001	17	−.0073992	−.0090089	.006327484	.004655345	.003160380	.001979480	.001143814	.000201973	.000021973	.000000056	101101010
100101100	16	−.0089261	−.0062816	.006028728	.004291765	.002787693	.001666367	.000916443	.000142544	.000013107	.000000023	110110010
010101100	16	−.0094252	−.0047206	.006012203	.004291810	.002708185	.001609200	.000881235	.000136574	.000012632	.000000046	011101010
001110001	16	−.0095780	−.0031622	.006007171	.004220329	.002751304	.001663992	.000933495	.000158275	.000016761	.000000032	110110100
011100100	16	−.0095780	−.0038134	.005997376	.004192994	.002711507	.001621538	.000896592	.000144512	.000014236		110110001

Table A

m = 5, n = 4 — Rank Orders 101 thru 126 of 126   C(5, 4) = 126   1/C(5, 4) = .00793651

z	w	c_1	c_2	d = .2	.4	.6	.8	1.0	1.5	2.0	3.0	z^{tc}
110001010	17	−.0096070	−.0040432	.005988189	.004172743	.002684031	.001592898	.000871832	.000135239	.000012546	.000000023	101011100
101001100	17	−.0096070	−.0043657	.005983472	.004160002	.002666164	.001574634	.000856697	.000130356	.000011801	.000000020	110110010
011011000	16	−.0097599	−.0031947	.005972489	.004161539•	.002683975	.001601768	.000884313	.000142265	.000014016	.000000031	110100001
011100010	16	−.0111049	−.0022714	.005764496	.003864570	.002390402	.001363649	.000771194	.000100596	.000008431	.000000012	011100110
100111000	16	−.0117858	−.0023032	.005674955	.003756256	.002300128	.001302309	.000681426	.000095274	.000008059	.000000013•	111000110
110010010	16	−.0117858	−.0018991	.005669280	.003741508	.002280193	.001282631	.000665655	.000090560	.000007387	.000000010	101101100
110011000	16	−.0117858	−.0015568	.005664525	.003729307	.002263933	.001266830	.000653204	.000087013	.000006908	.000000009	101010100
101010010	15	−.0119386	.0042545	.005674320•	.003783263	.002350666	.001360287	.000732771	.000113802	.000011222	.000000028	011111000
111000001	15	−.0119386	.0032347	.005660080	.003746219	.002300193	.001309715	.000691336	.000100290	.000009015	.000000017	111011010
110010100	16	−.0124668	.0043220	.005577525	.003627231	.002181585	.001212804	.000622881	.000082999	.000006676	.000000009	110011100
110100010	15	−.0139646	.0094040	.005372230	.003366842	.001952486	.001047171	.000519149	.000063491	.000004704	.000000006	101100100
101100100	15	−.0139646	.0090815	.005367962	.003356385	.001939158	.001034763	.000509766	.000061063	.000004403	.000000005	110110010
011101000	14	−.0141174	.0113307	.005370054•	.003386803	.001989399	.001087593	.000553026	.000073999	.000006229	.000000011	111000010
101011000	15	−.0141465	.0093344	.005346124	.003332276	.001920833	.001023418	.000503748	.000060364	.000004367	.000000004	101010100
100111000	15	−.0146456	.0117591	.005279501	.003249986	.001850338	.000973805	.000473509	.000055057	.000003867	.000000003	110110100
110011000	14	−.0163253	.0193842	.005073994	.003016865	.001666611	.000854672	.000406530	.000045420	.000003123	.000000003	111010010
101101000	14	−.0163253	.0186930	.005065488	.002997414	.001643382	.000834339	.000392028	.000042135	.000002761	.000000002	101110010
110101000	14	−.0168243	.0207784	.005002028	.002923020	.001582626	.000793506	.000368222	.000038371	.000002439	.000000002•	111001100
101010100	14	−.0170062	.0217932	.004982218	.002902845	.001568623	.000785616	.000364435	.000038049	.000002431	.000000002	110110010
111001000	13	−.0185041	.0291397	.004805262	.002708404	.001419717	.000691498	.000312690	.000030907	.000001890	.000000002	111100010
110100100	13	−.0191850	.0324152	.004724006	.002618139	.001349774	.000646734	.000287753	.000027341	.000001609	.000000001	101110000
110110000	13	−.0191850	.0324465	.004719832	.002609344	.001340078	.000638887	.000282571	.000026366	.000001520	.000000001	110110000
111010000	12	−.0213638	.0440077	.004767784	.002356625	.001155643	.000538643	.000224808	.000019255	.000001033	.000000001	111010000
110110100	12	−.0215457	.0450372	.004456821	.002334177	.001141896	.000519957	.000220358	.000018716	.000000997	.000000001	111011000
111010100	11	−.0237245	.0582323	.004226721	.002108844	.000984491	.000429529	.000174943	.000013613	.000000674	.000000001	111101000
111100000	10	−.0259032	.0730097	.004012142	.001910099	.000854856	.000359055	.000141312	.000010235	.000000479	.000000000	111110000

m = 9, n = 1 — Rank Orders 1 thru 10 of 10   C(9, 1) = 10   1/C(9, 1) = .10000000

z	w	c_1	c_2	d = .2	.4	.6	.8	1.0	1.5	2.0	3.0	z^{tc}
0000000001	10	.1538753	.1712104	.134279133	.175764636	.224440350	.279821724	.340935768	.508637579	.673645479	.902488438	0111111111
0000000010	9	.1001357	.0217240	.120772563	.140205008	.158441970	.173608296	.184481449	.188047399	.159054534	.066150971	1011111111
0000000100	8	.0656059	−.0394583	.112156358	.121715958	.127821359	.129904896	.127775933	.106324609	.072264646	.018306556	1101111111
0000001000	7	.0375765	−.0700862	.106005108	.108652493	.107684561	.103200870	.095641276	.068300810	.039585651	.007128347	1110111111
0000010000	6	.0122668	−.0833899	.100758485	.098134078	.092388790	.084078790	.073964294	.046287825	.023442042	.003189913	1111011111
0000100000	5	−.0122668	−.0833899	.095933590	.088958985	.079735884	.069081141	.057850089	.031994093	.014306524	.001511839	1111101111
0001000000	4	−.0375765	−.0700862	.091210567	.080435665	.068435665	.056530005	.045048608	.022023763	.008712044	.000721204	1111110111
0010000000	3	−.0656059	−.0394583	.086266970	.071999169	.058132630	.045040018	.034302348	.014707427	.005116120	.000329618	1111111011
0100000000	2	−.1001357	.0217240	.080572836	.062901542	.047571895	.034849147	.024724398	.009104189	.002736888	.000133715	1111111101
1000000000	1	−.1538753	.1712104	.072544390	.051232467	.035202566	.023521411	.015275835	.004572307	.001136074	.000039401	1111111110

Table A

m = 7, n = 3   Rank Orders 1 thru 50 of 120   C(7, 3) = 120   1/C(7, 3) = .00833333

z	w	c_1	c_2	d = .2	.4	.6	.8	1.0	1.5	2.0	3.0	z^{tc}
000000111	27	.0266347	.0734805	.01536168	.026825750	.04448857	.070120652	.105411390	.240749993	.431112422	.798213590	0001111111
000001011	26	.0242990	.0580130	.014480766	.023687727	.036522677	.053147530	.073094378	.127868738	.162449832	.111315250	0010111111
000001101	25	.0221898	.0455317	.013741931	.021254498	.030861783	.042138495	.054136415	.078303788	.079785291	.021803650	0011011111
000010011	25	.0214215	.0413481	.013485526	.020452498	.029100581	.038887205	.048860832	.066578324	.063512394	.031803650	0100111111
000010101	24	.0201454	.0346806	.013069212	.019180765	.024546089	.036369718	.041148224	.050730801	.043397927	.017714818	0011101111
000100011	24	.0193123	.0302550	.012795154	.018337187	.020940613	.033996536	.035945730	.039989935	.029855404	.005333603	0101011111
000011001	23	.0180362	.0247168	.012415164	.017284024	.020623340	.030712161	.031377796	.033341391	.024388323	.004722775	0110011111
000100101	23	.0172679	.0207662	.012167213	.016538602	.020490613	.027446789	.027200218	.025592309	.015812606	.001796110	0011110111
000101001	23	.0169765	.0198008	.012089520	.016348754	.020623340	.024712858	.026721014	.025347148	.016042907	.002047709	0111001111
000001110	24	.0169432	.0205462	.012105690	.016464834	.020985828	.025091687	.028173081	.028808721	.020445361	.003795403	1000111111
001000011	22	.0157004	.0151069	.011734022	.015428320	.018961799	.021807728	.023497912	.016673696	.013470238	.001907327	0011111011
000101010	22	.0151587	.0122201	.011557150	.014896766	.017855456	.019912648	.020673829	.016673618	.008724661	.000684132	0101110111
000110001	22	.0149321	.0114078	.011494650	.014736736	.017570019	.019492167	.020134097	.016044224	.008309285	.000644815	0110110111
001010001	23	.0148340	.0116458	.011483783	.014750052	.017667717	.019744244	.020599760	.017024045	.009285657	.000832443	0111010111
000011010	21	.0128229	.0053208	.010952648	.013445180	.015433533	.016581765	.016695605	.012896219	.006849572	.000700953	0011111101
000110010	21	.0128229	.0043641	.010926842	.013316034	.015090852	.015914139	.015626918	.010981074	.005025370	.000308157	0111100111
001001001	21	.0128229	.0041905	.010922175	.013292694	.015029774	.015796584	.015441693	.010667751	.004752466	.000265251	0110111011
000101100	21	.0128229	.0040018	.010917154	.013267908	.014965885	.015675737	.015255023	.010370773	.004513712	.000235273	0111101011
001100001	22	.0127896	.0043011	.010918745	.013295932	.015052531	.015849181	.015526799	.010788193	.004823034	.000226488	0011111110
000110100	22	.0124983	.0035899	.010847382	.013135188	.014803438	.015535617	.015190572	.010580680	.004803987	.000288422	1010011111
001010010	20	.0107138	-.0020238	.010376602	.011982761	.012839147	.012770904	.011799384	.007039272	.002687032	.000108014	0111010011
001000101	21	.0106804	-.0020197	.010370222	.011970999	.012822493	.012748203	.011768153	.006981346	.002627362	.000097931	1001101111
001100010	20	.0104871	-.0027541	.010316428	.011834899	.012586772	.012416042	.011365548	.006594394	.002429603	.000084378	0111011011
010100001	20	.0104538	-.0026621	.010312227	.011833259	.012586772	.012436711	.011397120	.006622136	.002429775	.000094416	1010101111
010001001	20	.0099455	-.0036382	.010194099	.011580996	.012120488	.011937470	.010946883	.006369345	.002389266	.000122693	1011001111
001010100	19	.0096208	-.0037087	.010131743	.011477588	.011119642	.011994020	.010971486	.006591270	.002607293	.000056951	0111110011
001001100	19	.0086693	-.0066228	.009883460	.010879124	.010877712	.011089151	.009324117	.004980564	.001712466	.000039388	1100011111
010010001	19	.0083780	-.0076395	.009804609	.010684060	.010773786	.011559302	.008766549	.004446739	.001428790	.000189498	0111101101
001001010	20	.0083446	-.0052506	.009854948	.010924345	.011353986	.011937470	.010172961	.006294382	.002714527	.000039388	0111011101
001100100	20	.0083446	-.0075131	.009801066	.010684612	.010799258	.010122913	.008802961	.004480537	.001438087	.000038190	1011010111
010010010	20	.0083446	-.0075736	.009799634	.010678337	.010784946	.010099044	.008770574	.004443997	.001417951	.000037163	1001111011
010001010	20	.0083446	-.0078428	.009793132	.010649236	.010717044	.009982996	.008608942	.004247594	.001299581	.000029419	1011011011
010100010	19	.0076097	-.0090025	.009694995	.010307541	.010299445	.009414803	.008039602	.003916784	.001209507	.000030694	0111101110
001101000	18	.0075763	-.0085984	.009630722	.010334730	.010298505	.009535754	.008203141	.004090576	.001297440	.000030438	1010110111
001011000	19	.0063355	-.0111459	.009343445	.009610570	.009178292	.008331357	.006902557	.003122436	.000898176	.000020122	0111011110
010010100	19	.0060089	-.0116756	.009306541	.009495735	.009006086	.008107574	.006625292	.002847996	.000755543	.000012671	1110001111
010100100	18	.0055005	-.0121229	.009353467	.009301696	.008759786	.007644980	.006398759	.002685315	.000691727	.000010722	1011011110
010001100	18	.0055005	-.0124703	.009149435	.009406215	.008990527	.008010606	.006186100	.002262423	.000704063	.000013350	0111111010
010011000	19	.0054672	-.0111494	.009171728	.009325505	.008818880	.007741855	.006655592	.003092206	.000937154	.000024834	1101101111
001101000	19	.0054672	-.0120612	.009151842	.009325505	.008818880	.007741855	.006310899	.002743539	.000756280	.000015068	1100111110
010001100	19	.0054672	-.0123411	.009145697	.009300495	.008765776	.007659215	.006206028	.002642299	.000707628	.000013019	1001111101
001101100	19	.0054672	-.0123614	.009145058	.009297085	.008756764	.007642348	.006180905	.002608591	.000686552	.000011683	1011110011
010011010	17	.0042244	-.0134844	.008883592	.008789085	.008073480	.006888720	.005462506	.002228183	.000582933	.000011077	0111110011
010101000	18	.0041910	-.0141306	.008863112	.008720895	.008073480	.006888416	.005216521	.001787761	.000474708	.000006454	1011100111
010110000	17	.0038997	-.0148156	.008797774	.008565904	.007705345	.006400535	.004910004	.001832190	.000396695	.000004453	1011101011
001110000	18	.0034561	-.0146143	.008712456	.008436616	.007560662	.006190243	.004865642	.001732629	.000437487	.000006646	1011101101
010101100	18	.0033580	-.0148061	.008689241	.008384782	.007488348	.006192091?	.004737024	.001908876	.000392418	.000004837	0111101111
011000000	18	.0031314	-.0140081	.008661642	.008372730	.007526164	.006292811	.004895290	.001892201	.000477279	.000008071	1101101111
100010000	18	.0031314	-.0140370	.008661642	.008369089	.007517782	.006279137	.004877553	.001641931	.000470730	.000007902	1110011110
001110100	18	.0031314	-.0150612	.008640428	.008287451	.007353587	.006036681	.004585123	.001641931	.000362596	.000004184	1100111011

91

Table A

z	w	c_1	c_2	d = .2	.4	.6	.8	1.0	1.5	2.0	3.0	z^{tc}
0100001010	18	.0031314	−.0153101	.008635275	.008267708	.007314230	.005979335	.004517179	.001587436	.000341316	.000003640	1010111101
0010100010	17	.0018553	−.0161779	.008372736	.007768838	.006657529	.005269503	.003852652	.001245075	.000245537	.000002186	1011101011
0101010001	16	.0013469	−.0155645	.008287644	.007643436	.006539645	.005192910	.003828805	.001296749	.000279958	.000003552	0111101101
1000010001	16	.0013136	−.0158988	.008274255	.007605068	.006471209	.005098309	.003719479	.001209193	.000244032	.000002411	1101100111
0001011000	17	.0010222	−.0152794	.008230334	.007549454	.006432094	.005090925	.003744093	.001262099	.000271132	.000003365	0111101110
0010100010	17	.0010222	−.0156128	.008223724	.007525011	.006384676	.005023146	.003650295	.001198131	.000244887	.000002523	1101100111
0100010010	17	.0010222	−.0163657	.008208901	.007470959	.006282011	.004880465	.003503295	.001082331	.000204185	.000001662	1101101101
0100001010	17	.0010222	−.0165852	.008204581	.007455251	.006252323	.004839503	.003457398	.001050507	.000193541	.000001469	1011101101
1000000110	18	.0009889	−.0150713	.008227832	.007551674	.006442695	.005109254	.003766476	.001277885	.000275758	.000003380	1001111101
0100000100	17	.0002539	−.0162911	.008062623	.007213683	.005968495	.004566766	.003231490	.000967147	.000177812	.000001402	1100111101
0011000010	16	−.0002539	−.0162911	.007964981	.007040114	.005754577	.004350114	.003041306	.000883469	.000157732	.000001175	1011110011
0110000100	15	−.0009889	−.0150713	.007846785	.006870330	.005594676	.004238463	.002988339	.000913435	.000180148	.000001969	0111111001
1000010000	16	−.0010222	−.0152794	.007836978	.006843978	.005552202	.004184641	.002930868	.000874924	.000166664	.000001637	0111110011
0001101000	16	−.0010222	−.0156128	.007829996	.006821952	.005511629	.004129557	.002869562	.000831377	.000150858	.000001236	1110100111
0010100010	16	−.0010222	−.0163657	.007815832	.006772600	.005422055	.004010596	.002740955	.000748952	.000124954	.000000796	1101101011
0100010010	16	−.0013136	−.0165852	.007811678	.006758089	.005396296	.003976643	.002704656	.000726572	.000118299	.000000700	1101101101
0010010100	16	−.0013469	−.0155645	.007768456	.006702621	.005352076	.003955003	.002734600	.000763586	.000126380	.000000851	0111101110
1000001100	17	−.0013469	−.0155645	.007767919	.006710781	.005371608	.003982975	.002735286	.000763586	.000131651	.000000920	1101101011
0100000110	17	−.0054672	−.0155645	.007659220	.006500429	.005094142	.003685991	.002462504	.000633459	.000098909	.000000539	1101101101
0001100100	15	−.0038997	−.0140081	.007453233	.006196833	.004789154	.003440235	.002296921	.000607531	.000101803	.000000733	1111000111
1000000010	15	−.0031314	−.0140370	.007452897	.006196404	.004789743	.003442885	.002301871	.000614304	.000105237	.000000852	0111111010
0011000100	15	−.0031314	−.0150612	.007434520	.006135134	.004682992	.003306336	.002159213	.000529068	.000079666	.000000421	1101110011
0101010000	15	−.0031314	−.0153101	.007430073	.006120437	.004657719	.003274576	.002126770	.000511140	.000074859	.000000364	1011110101
1000010000	16	−.0033580	−.0148061	.007379548	.006073095	.004614413	.003243945	.002109941	.000512314	.000076529	.000000399	1110110011
0001010010	16	−.0034561	−.0146143	.007379809	.006048921	.004587641	.003218590	.002088353	.000502455	.000073886	.000000362	1101101110
0100101000	15	−.0038997	−.0148156	.007291959	.005830937	.004391959	.003023034	.001902635	.000435172	.000059724	.000000272	1101101011
1000011000	15	−.0041910	−.0141306	.007247710	.005830416	.004337167	.002983574	.001897718	.000434349	.000060822	.000000272	1110111010
0100000010	16	−.0042244	−.0134844	.007253314	.005853937	.004380616	.003038838	.001953660	.000468213	.000068213	.000000351	1100111110
1000000100	17	−.0054672	−.0155645	.007054588	.005564084	.004088835	.002799856	.001788818	.000463683	.000066529	.000000457	0111110111
0010110000	14	−.0054672	−.0120612	.007038983	.005514401	.004005987	.002698223	.001684791	.000372373	.000051162	.000000232	1111000101
0110000010	14	−.0054672	−.0123411	.007034253	.005499584	.003981791	.002669292	.001656619	.000358456	.000047793	.000000198	1011111001
1000010010	14	−.0054672	−.0123614	.007034053	.005499398	.003982182	.002670565	.001658620	.000360337	.000048450	.000000208	1011110110
0001011000	15	−.0055005	−.0124703	.006934730	.005480719	.003955152	.002639620	.001628801	.000345233	.000044641	.000000168	1011101110
0101010000	14	−.0060089	−.0121229	.006934730	.005332097	.003787809	.002485803	.001506966	.000304677	.000037480	.000000127	1110110101
1000101000	14	−.0062355	−.0116756	.006899896	.005272696	.003735464	.002442192	.001475685	.000296431	.000036310	.000000123	1101101110
0010010010	15	−.0063335	−.0111459	.006888763	.005272696	.003735464	.002448732	.001484907	.000301882	.000037524	.000000131	1101101011
0110101000	13	−.0075763	−.0085984	.006694107	.004991332	.003453933	.002217842	.001321347	.000260708	.000032158	.000000119	1011101101
0100101000	14	−.0076097	−.0090025	.006681163	.004960105	.003408819	.002168063	.001275809	.000240861	.000027853	.000000085	1011111010
1100000010	13	−.0083446	−.0052506	.006601188	.004896112	.003398926	.002207886	.001341799	.000288662	.000041039	.000000246	1111111100
0101010000	13	−.0083446	−.0075131	.006566547	.004792534	.003240153	.002028592	.001175896	.000214507	.000024093	.000000070	1110110101
0110100100	13	−.0083446	−.0075736	.006564799	.004789800	.003236040	.002024070	.001171856	.000212913	.000023788	.000000069	1101111001
0101010100	13	−.0083446	−.0078428	.006560669	.004778064	.003218667	.002005258	.001155282	.000206568	.000022607	.000000062	1101101011
1000011000	14	−.0083780	−.0076395	.006557289	.004775111	.003216958	.002004337	.001154599	.000205963	.000022382	.000000060	1110101110
0011110000	14	−.0086228	−.0066228	.006518128	.004727560	.003160764	.001979849	.001142389	.000206269	.000022943	.000000066	1110110110
0011100000	12	−.0096208	−.0037087	.006383657	.004552688	.003021764	.001865990	.001071740	.000194837	.000022414	.000000075	1111100011
0101010000	12	−.0099455	−.0036382	.006323571	.004452462	.002907808	.001760778	.000988268	.000167456	.000017456	.000000045	1011110110
0101010000	12	−.0104538	−.0026621	.006243074	.004335056	.002789436	.001662945	.000918332	.000148732	.000014880	.000000035	1101101101
1000010000	13	−.0104871	−.0027541	.006233251	.004319929	.002773305	.001643868	.000902322	.000143840	.000013840	.000000029	1110111010
0110001000	12	−.0106804	−.0020197	.006210183	.004290768	.002747948	.001630911	.000896834	.000143851	.000014269	.000000033	1101111100
1000010010	13	−.0107138	−.0020238	.006203698	.004279254	.002734019	.001617310	.000885517	.000139762	.000013519	.000000029	1110111010

92

Table A

m = 7, n = 3   Rank Orders 101 thru 120 of 120   C(7, 3) = 120   1/C(7, 3) = .00833333

z	w	c_1	c_2	d = .2	.4	.6	.8	1.0	1.5	2.0	3.0	z^{tc}
0101100000	11	−.0124983	.035899	.005952806	.003952311	.002438015	.001396751	.000742935	.000110550	.000010277	.000000021	1111100101
0110000010	11	−.0127896	.043011	.005908989	.003891534	.002379661	.001350756	.000711533	.000103189	.000009338	.000000018	1111011001
1100000010	12	−.0128229	.053208	.005916717•	.003916811	.002415638	.001386964	.000740875	.000112227	.000010700	.000000024	1011111100
1000011000	12	−.0128229	.043641	.005903497	.003882793	.002369823	.001341622	.000704216	.000100726	.000008911	.000000016	1110011110
1010000100	12	−.0128229	.041905	.005901098	.003876635	.002361566	.001333506	.000697713	.000098748	.000008617	.000000015	1101111010
1001001000	12	−.0128229	.040018	.005898466	.003869826	.002352387	.001324446	.000690432	.000096531	.000008288	.000000014	1110110110
0110100000	10	−.0148340	.0116458	.005633582	.003546372	.002077930	.001132787	.000574344	.000076375	.000006409	.000000011	1111101001
1001001000	11	−.0149321	.0114078	.005612387	.003509723	.002037030	.001096822	.000547665	.000069196	.000005415	.000000008	1111101010
1010001000	11	−.0151587	.0122201	.005581873	.003471604	.002003904	.001073095	.000532896	.000066420	.000005128	.000000007	1110111010
1100000100	11	−.0157004	.0151069	.005521048	.003409363	.001960742	.001049388	.000522297	.000066164	.000005256•	.000000008	1101111100
0111000000	9	−.0169432	.0205462	.005366228	.003229918	.001815772	.000952748	.000466296	.000057405	.000004520	.000000007	1111110001
1001100000	10	−.0169765	.0198038	.005350793	.003198372	.001778643	.000919711	.000441961	.000051174	.000003710	.000000005	1111100110
1010010000	10	−.0172679	.0207662	.005310634	.003147388	.001733877	.000887460	.000421843	.000047416	.000003328	.000000004	1111011010
1100001000	10	−.0180362	.0247168	.005221993	.003052294	.001662789	.000843609	.000398323	.000044372	.000003114	.000000004	1110111110
1010001000	9	−.0193123	.0302550	.005062492	.002866868	.001512472	.000742941	.000339603	.000034902	.000002264	.000000002	1111101010
1100010000	9	−.0201454	.0346806	.004967808	.002766333	.001437738	.000696875	.000314787	.000031575	.000002011	.000000002	1111101100
1011000000	8	−.0214215	.0413481	.004821432	.002609398	.001319881	.000623453	.000274797	.000026061	.000001580	.000000001	1111011100
1100100000	8	−.0221898	.0455317	.004735165	.002518755	.001253092	.000582586	.000252913	.000023157	.000001361	.000000001	1111101100
1101000000	7	−.0242990	.0580130	.004509021	.002291314	.001092328	.000480022	.000204136	.000017214	.000000944	.000000001	1111110100
1110000000	6	−.0266347	.0734805	.004274985	.002070248	.000944891	.000405844	.000163812	.000012811	.000000662	.000000000	1111111000

93

Table A

m = 6, n = 4   Rank Orders 1 thru 50 of 210   C(6, 4) = 210   1/C(6, 4) = .00476190

z	w	c_1	c_2	d = .2	.4	.6	.8	1.0	1.5	2.0	3.0	z^{tc}
0000001111	34	.0170092	.0532682	.009430425	.017575582	.030896708	.051356297	.080922327	.203266396	.389369199	.775556980	0000111111
0000010111	33	.0158040	.0443064	.008939299	.015695392	.025814888	.039839031	.057786396	.112954145	.155983689	.118503837	0001011111
0000100111	32	.0146357	.0363965	.008495137	.014118527	.021888495	.031699785	.042948251	.069251914	.076267487	.033029512	0001101111
0001000111	32	.0144693	.0353286	.008434135	.013910315	.021436833	.030714775	.041236833	.064846527	.069403509	.028074502	0010011111
0000110111	31	.0134305	.0289853	.008065291	.012692153	.018581387	.025341558	.032241858	.043882059	.040089096	.011397546	0001110111
0001001011	31	.0133010	.0280432	.008013922	.012504738	.018108421	.024360766	.030475432	.039094488	.032630824	.006755935	0010101111
0001010111	31	.0128250	.0255739	.007860356	.012047972	.017169102	.022778673	.028175838	.035645673	.030159895	.006755935	0011001111
0001100111	30	.0120958	.0216294	.007619164	.011307804	.015586575	.019981092	.023857107	.027516263	.021118062	.004155730	0100011111
0010010111	30	.0120958	.0213953	.007611325	.011259129	.015428837	.019605907	.023127616	.025433226	.018036386	.002652620	0010110111
0010100111	30	.0120958	.0212834	.007607577	.011353908	.015353885	.019428735	.022786122	.024491574	.016722524	.002137493	0011010111
0000111011	30	.0116567	.0190669	.007467760	.010824427	.014513843	.018017583	.020727014	.021214655	.013783069	.001593149	0100101111
0001011011	29	.0108905	.0152174	.007224473	.010111171	.013064734	.015597560	.017220368	.015758638	.009034787	.000784257	0011100111
0001101011	29	.0107610	.0146554	.007186103	.010006388	.012866734	.015290412	.016807367	.015226464	.008649237	.000735365	0100110111
0010011011	29	.0104515	.0137777	.007108787	.009838310	.012639289	.015094178	.016781516	.016167548	.010348142	.001416498	0101010111
0010101011	29	.0104515	.0132123	.007091366	.009783998	.012344382	.014452723	.015643343	.013606066	.007400988	.000575671	0011110111
0001110011	30	.0104515	.0131172	.007088402	.009721963	.012293559	.014342089	.015447715	.013179002	.006942361	.000479032	1000011111
0011001011	28	.0102660	.0132402	.007061455	.009728542	.012466782	.014878789	.016560875	.016111305	.010499939	.001516888	0100111011
0010110011	28	.0097223	.0101016	.006876908	.009157658	.011254201	.012775175	.013407493	.011805040	.005459400	.000371851	0101011011
0011010011	28	.0095558	.0092408	.006823496	.009036735	.011936735	.012250820	.012660420	.009715752	.004595559	.000257482	0110010111
0011100011	28	.0092463	.0078445	.006730262	.008742207	.011441774	.011474372	.011607357	.008350163	.003630740	.000158595	0011111011
0000111110	28	.0091168	.0077254	.006704768	.008706375	.010435260	.011554886	.011831288	.008971415	.004269909	.000259017	0110001111
0100001011	28	.0091168	.0077067	.006704175	.008702923	.010424871	.011532152	.011791010	.008883918	.004177200	.000240062	0101001111
0100010111	28	.0091168	.0073922	.006695141	.008655133	.010293816	.011270285	.011366590	.008143594	.003548228	.000159738	0110011011
0010111010	29	.0090977	.0079584	.006707852	.008355537	.010524346	.011738572	.012130663	.009481819	.004688305	.000309156	1000101111
0101000111	27	.0083875	.0047495	.006494359	.008147410	.009409062	.010100467	.009819845	.006600403	.002728276	.000116713	0111001011
0100011011	27	.0080780	.0035072	.006405678	.007913688	.008983796	.009376890	.009004128	.005673087	.002154763	.000071635	0101101011
0011100111	27	.0079115	.0031023	.006365268	.007825073	.008852836	.009223692	.008856899	.005629707	.002189965	.000081135	0110101011
0011011001	27	.0079115	.0030090	.006362675	.007811802	.008817669	.009155863	.008750909	.005461907	.002061737	.000068322	0110110011
0011101001	27	.0079115	.0027709	.006356185	.007779317	.008733639	.008998010	.008511236	.005112704	.001820500	.000049829	0101110011
0010011011	28	.0078925	.0042723	.006393567	.007979946	.009268867	.010031161	.010128720	.007737195	.003970753	.000360064	0011111110
0011110010	28	.0078925	.0033568	.006368593	.007853644	.008935570	.009385898	.009108211	.006005168	.002457226	.000104839	1001001011
0100101011	28	.0078925	.0032803	.006366443	.007842503	.008905613	.009327162	.009014741	.005849076	.002330005	.000089862	1000110111
0011100101	27	.0074725	.0015621	.006245679	.007525410	.008334418	.008489314	.007957838	.004728349	.001698157	.000051033	0111011011
0011011000	26	.0071823	.0008085	.006175815	.007372706	.008107894	.008220369	.007690519	.004608661	.001708658	.000060101	0110110111
0100110101	26	.0067433	-.0006106	.006057500	.007079748	.007607603	.007521422	.006847219	.003795089	.001281582	.000035515	0111101001
0101000111	26	.0067433	-.0007033	.006055032	.007067744	.007577391	.007466189	.006765302	.003681974	.001206514	.000029903	0110111001
0010111001	26	.0067433	-.0009419	.006048833	.007038125	.007504329	.007335355	.006576270	.003439257	.001059422	.000021315	0101011101
0010101110	27	.0066877	-.0007394	.006043665	.007046756	.007550624	.007437511	.006737221	.003656206	.001186148	.000027530	0110010101
0101001001	26	.0065768	-.0013366	.006008651	.006948880	.007369070	.007168967	.006401099	.003324828	.001022494	.000020887	0110101101
0001001011	27	.0065577	-.0003840	.006029141	.007056473	.007638476	.007651906	.007098409	.004224231	.001582495	.000058894	0011111110
0011101010	27	.0065577	-.0007614	.006019729	.007013122	.007534692	.007470408	.006840344	.003896556	.001377227	.000044139	1010001011
0100101110	26	.0065577	-.0010300	.006012730	.006979546	.007451239	.007319365	.006618767	.003596414	.001180354	.000029393	1001011011
0100110101	26	.0062673	-.0022491	.005928898	.006761140	.007064385	.006765525	.005941197	.002946300	.000858155	.000015230	0111011101
0011110001	25	.0055380	-.0036627	.005759770	.006403273	.006546631	.006159102	.005335615	.002609918	.000775115	.000016210	0111101011
0101010001	25	.0055380	-.0036813	.005759435	.006402332	.006545585	.006160006	.005340143	.002624925	.000789169	.000017607	0111110011
0011100110	25	.0055380	-.0039959	.005751588	.006366377	.006546585	.006160006	.005136379	.002385766	.000655257	.000010631	1001110011
0010110110	25	.0055011	-.0038516	.005736606	.006337606	.006648683	.006065125	.005208839	.002469477	.000694123	.000012027	1010101011
0100110110	25	.0054085	-.0044251	.005717211	.006283446	.006324044	.005831233	.004928237	.002217721	.000585687	.000019328	0111011110
0101001010	25	.0053525	-.0037168	.005723779	.006342089	.006480522	.006109777	.005317616	.002659873	.000819845	.000019328	0110101110
0100101101	26	.0053525	-.0042061	.005711986	.006359717	.006359933	.005907483	.005042473	.002351352	.000652728	.000010972	1010010111

94

Table A

m = 6, n = 4   Rank Orders 51 thru 100 of 210   C(6, 4) = 210   1/C(6, 4) = .00476190

z	w	c_1	c_2	d = .2	.4	.6	.8	1.0	1.5	2.0	3.0	z^{tc}
0010010110	26	.0053525	−.0043982	.005707264	.006268429	.006310399	.005823901	.004928672	.022226194	.00588461	.00008413	1001011011
0100011001	25	.0050990	−.0051835	.005641518	.006113888	.006062993	.005503704	.004574889	.019965665	.00491563	.00006249	0101101011
0100011010	25	.0049325	−.0054536	.005604143	.006037468	.005956321	.005383318	.004459153	.019071114	.00477971	.00006264 ●	0110010111
0001110101	26	.0049134	−.0047794	.005625891 ●	.006142193	.006200371	.005790852	.005007207	.022497557	.00784042	.00020639	1100011101
1000001101	26	.0049134	−.0051109	.005616384	.006099868	.006102568	.005625992	.004781564	.022238531	.00638378	.00012312	0100111110
0110000011	24	.0042033	−.0061671	.005608661	.006066738	.006028954	.005506935	.004625609	.022080171	.00561223	.00009273	1000111101
0011010001	24	.0042033	−.0067324	.005437862	.005749863	.005690328	.005500784	.004134404	.021819093	.00494824	.00004527 ●	0011111001
0011010010	24	.0042033	−.0068275	.005435753	.005690328	.005458548	.004802095	.003876030	.021561909	.00371818	.00004527 ●	0111001011
0011100001	24	.0042033	−.0062045	.005435753	.005681425	.005439545	.004772711	.003883476	.021529726	.00358659	.00004195	0111011011
1000100011	25	.0041842	−.0066939	.005446331 ●	.005735088	.005562590	.004972036	.004097918	.021780171	.00473614	.00008213	0011110110
0001101010	25	.0041842	−.0068860	.005435093	.005687587	.005458615	.004806422	.003884290	.021571393	.00375697	.00004600	1010100111
0100110011	25	.0041842	−.0071371	.005430585	.005668201	.005415640	.004737414	.003794976	.021485279	.00337199	.00003460	1001100111
0100110100	24	.0040178	−.0074442	.005393567	.005592935	.005310683	.004618427	.003679296	.021422522	.00319787	.00003242	1010010011
0101101001	24	.0038938	−.0077637	.005363402	.005526661	.005211846	.004499658	.003557866	.021350048	.00298541	.00002942	0101011011
0101101010	25	.0037643	−.0070220	.005311600	.005456529	.005105572	.004369213	.003420971	.021260989	.00269131	.00002459	0101101011
0001011100	25	.0037082	−.0073190	.005337772 ●	.005506469	.005228931	.004571799	.003681450	.021497682	.00367031	.00003544	1100010111
1000000011	24	.0037082	−.0076691	.005330977	.005478081	.005167638	.004475701	.003559717	.021385257	.00318315	.00003544	0101011110
0100001010	24	.0037082	−.0076691	.005323190	.005446355	.005100824	.004373613	.003433886	.021277354	.00275235	.00002525	1001011101
0011010100	23	.0029981	−.0085531	.005169515	.005142786	.004690380	.003923516	.003011714	.021071266	.00225547	.00002170	0111101101
1001010010	24	.0029790	−.0080223	.005177434 ●	.005182768	.004778913	.004061215	.003182845	.021220288	.00287157	.00003940	0011101110
0001110010	24	.0029790	−.0083997	.005169150	.005149277	.004708737	.003954128	.003050401	.021103362	.00237373	.00002404	1011000111
0011000101	24	.0029790	−.0086683	.005163212	.005125200	.004658375	.003877761	.002956994	.021024892	.00206639	.00001691	1000110011
0110000101	23	.0028495	−.0090004	.005131413	.005054599	.004551808	.003747907	.002822074	.020940058	.00179843	.00001266	1010101011
0101000011	23	.0025590	−.0090268	.005075707	.004958785	.004403165	.003592792	.002684672	.020930911	.00185454	.00001544	0101111011
0101010010	23	.0025590	−.0092609	.005070811	.004939633	.004384714	.003566118	.002653403	.020883527	.00169302	.00001269	0111011011
1001101000	24	.0025590	−.0093728	.005068444	.004930404	.004384714	.003712852	.002827810	.020994654	.00161636	.00001144	1011010111
1000110001	24	.0025400	−.0087294	.005078420 ●	.004976991	.004483214	.003633326	.002732038	.020999083	.00200909	.00002041	0101010011
0110100010	24	.0025400	−.0090282	.005071926	.004951099	.004429928	.003563162	.002643701	.020917150	.00179783	.00001404	0101101101
0010100110	24	.0025400	−.0093783	.005064515	.004922343	.004372403	.003549993	.002634255	.020844912	.00155091	.00000992	1001101101
0010110001	24	.0023735	−.0090481	.005039817	.004894918	.004363187	.003569710 ●	.002680942	.020902079	.00178499	.00001446	0111101110
1000011001	24	.0023735	−.0091088	.005038559	.004890165	.004353978	.003556870	.002666617	.020892945	.00176037	.00001436	0110011110
0100011110	24	.0023735	−.0095049	.005030208	.004857863	.004289521	.003463646	.002557897	.020812076	.00148161	.00000951	1000111101
0011100010	25	.0023544	−.0087209	.005043163	.004916062	.004411652	.003645523	.002774763	.020981452	.00209834	.00002216	1011000111
1000110100	22	.0018298	−.0094718	.004927399	.004682333	.004087502	.003279197	.002418651	.020784800	.00152344	.00001283	0111110011
1010010001	23	.0016443	−.0091193	.004898848	.004646948	.004066348	.003299638	.002499868	.020835446	.00176466	.00002005	1011001011
0101010001	23	.0016443	−.0100548	.004898822	.004574248	.003922638	.003077778	.002209844	.020655503	.00112137	.00000639	1010010111
0011010100	23	.0016443	−.0101312	.004878926	.004564477	.003911613	.003062561	.002192963	.020644582	.00108934	.00000602	1010111010
0111010000	22	.0013538	−.0102883	.004819769	.004461266	.003777551	.002927048	.002076236	.020599393	.00100594	.00000577	0111101101
0011100001	22	.0013347	−.0096420	.004829166 ●	.004503196	.003862239	.003047593	.002213210	.020693585	.00130641	.00001045	1010101101
0110110000	23	.0013347	−.0100359	.004821043	.004472449	.003802154	.002962414	.002115735	.020624035	.00107395	.00000646	0111101110
0101000110	23	.0013347	−.0103674	.004814365	.004447793	.003755229	.002897743	.002043937	.020577033	.00093241	.00000463	1001110101
0010101100	22	.0012243	−.0102276	.004796203	.004422315	.003733742	.002887773	.002045791	.020590930	.00099652	.00000583	0101111100
0011011000	23	.0012052	−.0101330	.004794287	.004421663	.003735700	.002891343	.002050194	.020592115	.00099176	.00000549	1110010011
0101100001	23	.0012052	−.0101937	.004793082	.004417307	.003727636	.002880610	.002038775	.020585662	.00097626	.00000542	1010101101
0101011000	23	.0011492	−.0105897	.004785122	.004388014	.003672118	.002804485	.001954771	.020531730	.00081804	.00000352	1001011110
1000011100	23	.0007292	−.0100075	.004785990	.004411473	.003728814	.002890303	.002053828	.020598311	.00101438	.00000580	1010101101
0011010110	23	.0004391	−.0102974	.004699961	.004250419	.003522087	.002674338	.001860808	.020513214	.00082228	.00000419	1110010011
0110100100	22	.0001855	−.0106038	.004638543	.004132317	.003367439	.002510387	.001712378	.020446703	.00067205	.00000297	1010101101
0101100010	21	.0001855	−.0104285	.004593493	.004059942	.003288569	.002441910	.001662687	.020436380	.00067216 ●	.00000333	0111011101
0110110001	21	.0000191	−.0104916	.004560446	.004000002	.003214057	.002366558	.001597338	.020409638	.00061565 ●	.00000290	0111011001

95

Table A

m = 6, n = 4 Rank Orders 101 thru 150 of 210 C(6, 4) = 210 1/C(6, 4) = .00476190

z	w	c_1	c_2	d = .2	.4	.6	.8	1.0	1.5	2.0	3.0	z^{tc}
1100000011	22	.0000000	−.0091472	.004582450•	.004083867	.003371087	.002578137	.001827420	.000556198	.000106702	.00001044	0011111100
0001111100	22	.0000000	−.0096178	.004573392	.004051568	.003311083	.002496534	.001737031	.000494376	.000085747	.00000603	1110000011
1010000101	22	.0000000	−.0102498	.004561265	.004008735	.003232746	.002392320	.001624830	.000424908	.000065313	.00000320	0101111010
0010110100	22	.0000000	−.0103940	.004558501	.003998997	.003215012	.002368870	.001599785	.000409860	.000061092	.00000271	0111001100
1000110001	22	.0000000	−.0104193	.004557359	.003997351	.003212150	.002365336	.001599346	.000408460	.000060938	.00000275•	1100110011
0011001100	22	.0000000	−.0104537	.004555796	.003994996	.003207792	.002359446	.001589886	.000396329	.000059639	.00000257	0110110110
0110000110	22	.0000000	−.0105353	.004555522	.003989585	.003197867	.002346443	.001574484	.000395892	.000057539	.00000236	1110110101
1000101001	22	.0000000	−.0105498	.004555083	.003988585	.003196304	.002344596	.001574162	.000375539	.000057613	.00000242	1001110110
0110010010	22	.0000000	−.0108340	.004550083	.003963513	.003161841	.002299525	.001527022	.000368737	.000050498	.00000174	0111011010
0101011000	22	.0000000	−.0109272	.004548301	.003963294	.003150704	.002285135	.001512105	.000360659	.000048541	.00000159	1010110101
1000100110	23	−.0000191	−.0104916	.004552825	.003984949	.003192527	.002340762	.001570525	.000392620	.000056317	.00000219	1001101110
1000001010	23	−.0001855	−.0104285	.004522193	.003933507	.003133425	.002285746	.001526794	.000378658	.000054171	.00000214	1010011101
0011110001	21	−.0004391	−.0106038	.004470430	.003837471	.003011848	.002161150	.001417685	.000333655	.000044812	.00000150	1100001110
0111100001	20	−.0007292	−.0102974	.004420741	.003760555	.002931411	.002094139	.001371151	.000325167	.000044894	.00000172	0111110001
0011101000	21	−.0011492	−.0100075	.004345815	.003638485	.002795263	.001971011	.001236941	.000296773	.000040822	.00000165	0111110010
1001101001	21	−.0012052	−.0101330	.004332710	.003611398	.002757737	.001929427	.001236941	.000277978	.000036315	.00000124	1100110101
0101010001	21	−.0012052	−.0105897	.004331596	.003607671	.002751349	.001921545	.001229153	.000274272	.000035539	.00000120	1010101010
0101000110	22	−.0012243	−.0102276	.004327185•	.003583052	.002709975	.001870113	.001177697	.000248419	.000029604	.00000076	1001110101
1000011000	21	−.0013347	−.0096420	.004316692	.003598252	.002737607	.001905411	.001213112	.000265562	.000033279	.00000098	1001110011
0101000110	21	−.0013347	−.0096420	.004316692	.003596702	.002754089	.001937870	.001258288	.000290725	.000039751	.00000153	1110101011
1010010001	21	−.0013347	−.0100359	.004309682	.003573813	.002715532	.001890802	.001206574	.000268269	.000034740	.00000118	1011011010
0110010100	21	−.0013347	−.0103674	.004303652	.003553712	.002680994	.001847839	.001163533	.000246507	.000029693	.00000079	1010011001
1000101010	22	−.0013538	−.0102883	.004301300	.003551180	.002679165	.001846671	.001162685	.000245850	.000029429	.00000076	0101011010
1100000101	21	−.0016443	−.0091393	.004266365	.003521907	.002679281	.001875796	.001209818	.000281727	.000039203	.00000164	0101110011
0101101000	21	−.0016443	−.0100548	.004250169	.003469316	.002589784	.001767835	.001103472	.000229613	.000027309	.00000072	1101101010
0101100010	21	−.0016443	−.0101312	.004248793	.003467871	.002582127	.001758473	.001094283	.000225252	.000026377	.00000067	1010101001
1000011100	22	−.0018298	−.0094718	.004224995	.003440704	.002571314	.001763103	.001109074	.000238593	.000029924	.00000094	1100011010
0110010001	19	−.0023544	−.0087209	.004138086	.003309289	.002435497	.001649625	.001028444	.000219902	.000028165	.00000102•	0111010110
0101011001	20	−.0023735	−.0090481	.004128799	.003285009	.002398413	.001606994	.000988108	.000201384	.000024075	.00000071	1101100101
0110100001	20	−.0023735	−.0091088	.004127730	.003281575	.002392752	.001600224	.000981693	.000198574	.000023528	.00000068	0111010101
0101100100	20	−.0023735	−.0095049	.004120841	.003259596	.002356587	.001557110	.000940268	.000179652	.000019547	.00000043	1011100101
0011011010	20	−.0025400	−.0087294	.004102477	.003246867	.002360433	.001576158	.000966666	.000196317	.000023462	.00000069	1110011010
1001010010	20	−.0025400	−.0090282	.004097432	.003231248	.002335489	.001547300	.000939764	.000184936	.000021250	.00000057	0111011001
1001100001	20	−.0025400	−.0093783	.004091370	.003212006	.002304005	.001509997	.000904161	.000168972	.000017962	.00000037	1001111010
1000010100	21	−.0025590	−.0090268	.004093682	.003224133	.002323955	.001536794	.000929832	.000180058	.000020091	.00000048	1011100110
0101001010	20	−.0025590	−.0092609	.004089627	.003212262	.002304924	.001511942	.000906210	.000169654	.000018014	.00000037	1010101100
0110100100	20	−.0028495	−.0093728	.004038967	.003205189	.002295088	.001500440	.000895409	.000165075	.000017144	.00000033	1101010101
1001010001	20	−.0029790	−.0090004	.004030739	.003130865	.002217826	.001435591	.000849083	.000153894	.000015863	.00000031	0101010110
1100010010	20	−.0029790	−.0080223	.004024407	.003188930	.002248310	.001480886	.000896848	.000177225	.000020726	.00000060	1101010101
0110110001	20	−.0029790	−.0083397	.004019952	.003113970	.002217034	.001444531	.000862680	.000162261	.000017657	.00000040	1010101101
0110001100	20	−.0029790	−.0086683	.004019952	.003105825	.002195807	.001420449	.000840685	.000153457	.000016051	.00000033	1100111001
1000101100	21	−.0029981	−.0085531	.004018134	.003104761	.002196099	.001421571	.000841902	.000153762	.000016034	.00000032•	1100101010
0011101100	19	−.0037082	−.0070202	.003904105	.002952094	.002028928	.001310653	.000770419	.000141597	.000015452	.00000039	1110110010
1010010100	19	−.0037082	−.0073190	.003904105	.002937774	.002025928	.001286227	.000748557	.000133159	.000013943	.00000032	0111101001
1001010100	19	−.0037082	−.0076691	.003898320	.002920232	.002001526	.001255097	.000720013	.000121558	.000011761	.00000021	1001111010
1010011000	20	−.0037643	−.0077637	.003886070	.002896982	.001972318	.001226033	.000695707	.000113088	.000010355	.00000015	1010110110
1001100010	20	−.0038938	−.0074442	.003866783	.002870928	.001948414	.001208418	.000684745	.000111326	.000010248	.00000016	1101011010
1100001100	19	−.0040178	−.0071371	.003848422	.002846546	.001926796	.001193363	.000676238	.000110731	.000010388•	.00000017	0111011010
1010010010	19	−.0041842	−.0062045	.003831886	.002837082	.001932236	.001210342	.000697227	.000121716	.000012603	.00000029	1100101101
0110010010	19	−.0041842	−.0066939	.003824045	.002813973	.001897005	.001171324	.000662315	.000108235	.000010174	.00000017	0111101010

Table A

m = 6, n = 4   Rank Orders 151 thru 200 of 210   C(6, 4) = 210   1/C(6, 4) = .00476190

z	w	c_1	c_2	d = .2	.4	.6	.8	1.0	1.5	2.0	3.0	z^{tc}
0110101100	19	−.0041842	−.0068860	.003821045	.002805392	.001884364	.001157853	.000650766	.000104304	.000009568	.000000015	1101011001
1100000110	20	−.0042033	−.0061671	.003828738	.002831860	.001925868	.001203759	.000691279	.000119012	.000011999	.000000024	1001111100
1000110100	20	−.0042033	−.0067324	.003819783	.002805811	.001886722	.001161092	.000653729	.000105271	.000009683	.000000015	1101011010
1001001100	20	−.0042033	−.0068275	.003818241	.002801236	.001879734	.001153374	.000646916	.000102732	.000009259	.000000014	1110011010
1011000001	18	−.0049134	−.0043855	.003722876	.002687190	.001790010	.001100232	.000623738	.000105868	.000010813	.000000025	1111000011
1001100010	18	−.0049134	−.0047794	.003716858	.002670244	.001765383	.001073986	.000601192	.000097956	.000009508	.000000019	1011110010
1010100001	18	−.0049134	−.0051109	.003711621	.002655019	.001742477	.001048940	.000579056	.000089651	.000008050	.000000013	1011110001
1001100010	18	−.0049325	−.0054436	.003702703	.002633709	.001713325	.001018982	.000553920	.000081306	.000006770	.000000009	1011011010
1010100001	19	−.0050990	−.0051835	.003675682	.002593831	.001672913	.000985874	.000530738	.000075859	.000006132	.000000007	0111101010
1100100001	18	−.0053525	−.0037168	.003650679	.002578324	.001677167	.001004666	.000554157	.000087271	.000008216	.000000016	1111011100
0101010000	18	−.0053525	−.0042061	.003643199	.002557262	.001646456	.000972097	.000526219	.000077511	.000006614	.000000009	1110100101
0110100100	18	−.0053525	−.0043982	.003640331	.002549414	.001635381	.000960781	.000516906	.000074637	.000006210	.000000006	1101110110
1001010100	19	−.0054085	−.0044251	.003629317	.002530258	.001613089	.000940096	.000500705	.000069788	.000005511	.000000006	1101010110
1000110010	18	−.0055190	−.0038516	.003616596	.002520662	.001610713	.000943656	.000506775	.000073115	.000006105	.000000008	1111001001
1000011000	19	−.0055380	−.0036627	.003616242	.002522034	.001614127	.000947718	.000510281	.000074151	.000006229	.000000008	1110001110
1010000100	19	−.0055380	−.0036813	.003611613	.002520859	.001612175	.000945421	.000508137	.000073288	.000006078	.000000006	1011011010
1011000001	19	−.0062673	−.0039959	.003501806	.002508361	.001594769	.000927865	.000493879	.000069037	.000005502	.000000006	1100111010
1001000010	17	−.0065577	−.0022491	.003475062	.002357046	.001451696	.000817880	.000421397	.000054221	.000003970	.000000004	1011111010
0101000001	17	−.0065577	−.0003840	.003473840	.002342226	.001457590	.000837275	.000443855	.000063880	.000005562	.000000006	0111110010
0101010000	17	−.0065577	−.0007614	.003469475	.002326957	.001435943	.000814914	.000425140	.000057693	.000004594	.000000006	1111000101
0111010000	17	−.0065577	−.0010300	.003465681	.002317099	.001422699	.000801995	.000414964	.000054850	.000004226	.000000005	1101110001
1001010100	17	−.0065768	−.0013366	.003453667	.002299365	.001399907	.000780027	.000397650	.000049909	.000003572	.000000003	1101100110
0110101100	17	−.0066872	−.0007394	.003445421	.002289600	.001396497	.000781722	.000401525	.000052059	.000003931	.000000004	1101010010
1000110100	18	−.0067433	−.0006106	.003437214	.002276760	.001383176	.000770314	.000393084	.000049738	.000003613	.000000004	1011011010
1001001100	18	−.0067433	−.0007039	.003432565	.002273458	.001378829	.000766167	.000389896	.000046630	.000003514	.000000003	1011011010
1010000100	18	−.0067433	−.0009419	.003377378	.002264796	.001367299	.000755062	.000381288	.000045330	.000003242	.000000003	1101011010
1001000100	17	−.0071823	.0008855	.003333529	.002201943	.001318849	.000725271	.000366011	.000039689	.000003250	.000000003	1100111001
1000100010	17	−.0074725	.0015621	.003293723	.002141580	.001262025	.000681889	.000337669	.000047028	.000002684	.000000003	1011111010
1010000001	16	−.0078925	.0042723	.003293723	.002113911	.001257915	.000693584	.000354156	.000039689	.000003838	.000000006	0111111000
0110110000	16	−.0078925	.0033568	.003281290	.002082651	.001217014	.000654484	.000323789	.000038607	.000002715	.000000003	1111011001
0111001000	16	−.0078925	.0032803	.003280288	.002080229	.001213986	.000651732	.000321766	.000038129	.000002662	.000000003	1110110001
1100100010	17	−.0079115	.0031023	.003274250	.002068040	.001199186	.000638079	.000311398	.000035393	.000002323	.000000002	1011010100
1010100100	17	−.0079115	.0030090	.003273017	.002055041	.001185317	.000634649	.000308876	.000034803	.000002259	.000000002	1101010010
1001001000	17	−.0079115	.0027709	.003269810	.002057116	.001167930	.000625319	.000301940	.000033143	.000002079	.000000002	1011001010
1001100100	16	−.0080780	.0035072	.003249383	.002034292	.001167030	.000614642	.000296381	.000032560	.000002055	.000000002	1110011010
1010101000	15	−.0083875	.0047495	.003209260	.001987429	.001129861	.000589410	.000281997	.000030487	.000001901	.000000001	1011001010
0111010000	15	−.0090977	.0079584	.003123231	.001891451	.001057094	.000544925	.000258977	.000028180	.000001830	.000000002	1111010001
1001100000	16	−.0091168	.0077254	.003116650	.001878392	.001041777	.000531319	.000249065	.000025805	.000001560	.000000001	1011110010
1100100000	16	−.0091168	.0077067	.003116435	.001877938	.001041287	.000530959	.000248830	.000025773	.000001558	.000000001	0111111000
1011010000	16	−.0091168	.0073922	.003112403	.001868430	.001029691	.000520693	.000241494	.000024168	.000001397	.000000001	1101110010
1010101000	16	−.0092463	.0078445	.003094934	.001846879	.001011454	.000508145	.000234092	.000023020	.000001307	.000000001	1110101010
1001010000	16	−.0095558	.0092408	.003057119	.001804406	.000978577	.000487369	.000222793	.000022569	.000001210	.000000001	1101101100
1100100000	16	−.0097223	.0091016	.003038142	.001784666	.000964558	.000479338	.000218894	.000021231	.000001200	.000000000	1110001100
0111100000	14	−.0102660	.0132402	.002980116	.001729076	.000929228	.000464213	.000212555	.000021556	.000001322	.000000001	1111100001
1010000010	15	−.0104515	.0137777	.002953573	.001694356	.000898037	.000439349	.000198197	.000018904	.000001069	.000000001	1011001010
0110101000	15	−.0104515	.0132123	.002946790	.001679327	.000880755	.000424855	.000188382	.000016999	.000000896	.000000001	1011001010
1001010000	15	−.0104515	.0131172	.002945665	.001676785	.000877990	.000422588	.000186880	.000016726	.000000873	.000000000	1110110100
1100100000	15	−.0107610	.0146554	.002909507	.001637970	.000849087	.000405027	.000177682	.000015643	.000000807	.000000000	1101101010
1001100000	16	−.0108905	.0152174	.002893403	.001616963	.000834542	.000395649	.000172475	.000014937	.000000757	.000000000	1101110100
1011010000	14	−.0116567	.0190669	.002804358	.001524127	.000763891	.000352831	.000150087	.000012305	.000000595	.000000000	1111010010

97

Table A

z	w	c_1	c_2	d = .2	.4	.6	.8	1.0	1.5	2.0	3.0	z^{tc}
								C(6, 4) = 210		1/C(6, 4) = .00476190		
				Rank Orders 201 thru 210 of 210								
1110000100	14	−.0120958	.0216294	.002757256	.001477436	.000731846	.000334799	.000141321	.000011442	.000000551	.000000000	1101111000
1100110000	14	−.0120958	.0213953	.002754668	.001472143	.000726219	.000330434	.000138575	.000010998	.000000517	.000000000	1111001100
1101001000	14	−.0120958	.0212834	.002753432	.001469622	.000723548	.000328373	.000137286	.000010793	.000000502	.000000000	1111011010
1011100000	13	−.0128250	.0255739	.002675530	.001392646	.000670832	.000298742	.000122880	.000009367	.000000427	.000000000	1111100010
1010010000	13	−.0133010	.0280432	.002621117	.001335315	.000629085	.000273853	.000110071	.000007915	.000000340	.000000000	1111010100
1110001000	13	−.0134305	.0289853	.002609141	.001325198	.000623260	.000271163	.000109032	.000007873	.000000341•	.000000000	1110111000
1011000000	13	−.0144693	.0353286	.002500435	.001219634	.000551993	.000231560	.000089945	.000006003	.000000243	.000000000	1111100100
1110010000	12	−.0146357	.0363965	.002483531	.001203685	.000541518	.000225890	.000087281	.000005757	.000000231	.000000000	1111011000
1110100000	11	−.0158040	.0443064	.002368912	.001098960	.000474768	.000190757	.000071193	.000004353	.000000164	.000000000	1111101000
1111000000	10	−.0170092	.0532682	.002257977	.001003380	.000417053	.000161853	.000058550	.000003355	.000000120	.000000000	1111110000

m = 6, n = 4

z	w	c_1	c_2	d = .2	.4	.6	.8	1.0	1.5	2.0	3.0	z^{tc}
0000011111	40	.0146611	.0476993	.008044765	.015314531	.027441933	.04640468	.074264344	.192358015	.376637460	.768259534	0000011111x
0000101111	39	.0136876	.0402036	.007637902	.013720539	.023042018	.036239334	.053473114	.108316337	.153633357	.120643991	0000101111x
0000110111	38	.012832	.0331392	.007246323	.012298087	.019425621	.028600552	.033310598	.065143827	.072914025	.031724835	0000110111
0001001111	37	.0116789	.0265595	.006873899	.011016458	.016351821	.022503115	.028743658	.038553720	.033320058	.007224730	0001001111
0001010111	37	.0115709	.0260620	.006841572	.010931069	.016207264	.022331694	.028639063	.039384845	.035912261	.009872794	0001010111x
0001100111	36	.0107053	.0208720	.006538294	.009947331	.013988821	.018204003	.021946585	.025275681	.018574310	.002858521	0001100111
0010001111	36	.0105666	.0200252	.006489226	.009787095	.013626140	.017529008	.020856002	.023040076	.015991788	.002056420	0010001111
0010010111	36	.0102007	.0183884	.006380041	.009491226	.013082638	.016733357	.019891411	.022494325	.016742444	.003022603	0010010111
0010100111	36	.0095930	.0148581	.006171553	.008832068	.011640484	.014141915	.015851948	.014925675	.008696586	.000756026	0010100111
0100010111	35	.0094543	.0140975	.006125517	.008691488	.011343997	.013629657	.015086586	.013658794	.007548037	.000556832	0010110011x
0100100111	35	.0091963	.0130268	.006050855	.008494237	.010987542	.013108140	.011436421	.013038335	.007310289	.000591666	0001011011
0011000111	34	.0085887	.0102067	.005865516	.007974337	.009980722	.011511441	.012247409	.010118985	.005179377	.000351705	0001101011
0011001111	34	.0084807	.0094507	.005824841	.007838664	.009676748	.010965142	.011412401	.008733813	.003990788	.000186612	0010011011x
0011010011	34	.0082228	.0085083	.005753922	.007661339	.009434742	.010550797	.010931023	.008365592	.003900126	.000206317	0010101011
1000001111	35	.0080841	.0078516	.005711244	.007540754	.009139882	.010176810	.010417094	.007681903	.003408963	.000155204	0010110101
0101000111	33	.0080682	.0087590	.005735166	.007682865	.009551518	.011035262	.011865677	.010500611	.006162470	.000707422	0011000111
0101010011	33	.0074764	.0053422	.005535407	.007074238	.008288202	.008907092	.008788793	.005872653	.002338134	.000082973	0011001101
1000010011	34	.0072184	.0045180	.005437672	.006912504	.008025139	.008653118	.008405744	.005607759	.002273773	.000091045	0011010101
0110000111	33	.0071105	.0041202	.005430219	.006912319	.007786638	.008167639	.007848758	.004861102	.001763946	.000048699	0011011001x
1000100011	34	.0070638	.0038535	.005438683	.006872812	.008013552	.008627670	.008584044	.006036460	.002646068	.000131926	0011100101
0100011011	33	.0067138	.0025377	.005324600	.006540418	.007358967	.007589238	.007178656	.004300500	.001526703	.000042454	0100001111x
0011010011	32	.0064720	.0017186	.005259717	.006381209	.007090887	.007221915	.006746442	.003919171	.001351222	.000035759	0011010011x
0110010011	32	.0061062	.0008928	.005172458	.006193120	.006817087	.006904830	.006028383	.003804372	.001376975	.000045178	0011110001
0101001011	32	.0061062	.0004061	.005159898	.006131931	.006663254	.006624294	.006441505	.003247964	.001021210	.000021022	0100100011x
0101010011	33	.0061062	.0004061	.005159517	.006130032	.006658387	.006615273	.006246726	.003229542	.001009571	.000020336	0011110001x
1000000111	33	.0060903	.0009595	.005171114	.006195575	.006828827	.006928792	.006476226	.003839578	.001388613	.000044105	0100101001
0101100011	32	.0059516	.0004607	.005132995	.006099924	.006660713	.006688684	.006179985	.003537553	.001221818	.000033822	0100110001
1001000011	32	.0057403	.0008117	.005050461	.005892315	.006261251	.006075888	.005386449	.002691595	.000771858	.000012371	0101000101
0110001011	31	.0054985	-.0011202	.005010305	.005796307	.006148034	.005983168	.005346445	.002793268	.000872699	.000019456	0101100011x
1001010011	31	.0051018	-.0025425	.004902072	.005577266	.005691123	.005335805	.004604029	.002140045	.000575377	.000008439	0101010101x
1000100011	32	.0050859	-.0019867	.004912980	.005586001	.005836279	.005607222	.004957224	.002547224	.000794517	.000018636	0011110110
0110100011	31	.0049939	-.0026126	.004880510	.005490070	.005651266	.005325862	.004597749	.002177512	.000604194	.000009787	0101110001
1010000011	31	.0049780	-.0025164	.004879779	.005494821	.005661341	.005355645	.004638737	.002217712	.000620681	.000010079	0101010110
1001100011	31	.0047359	-.0035840	.004809148	.005311830	.005349023	.004912626	.004116446	.001775988	.000435922	.000004985	0100110110
1010010011	32	.0045813	-.0033954	.004785139	.005288547	.005359319	.004982205	.004251114	.001972662	.000543562	.000009134	0101011001
1100000011	30	.0041283	-.0047325	.004669103	.005018187	.004928098	.004426614	.003637076	.001514074	.000366644	.000004448	0100110101
0111000011	31	.0039895	-.0050183	.004636516	.004948219	.004825880	.004303245	.003510308	.001434674	.000340876	.000003973	0101100101
1010100011	31	.0039736	-.0044713	.004646455	.004999335	.004946050	.004502755	.003775032	.001704471	.000469932	.000008802	0011110110
1100010011	30	.0039736	-.0044207	.004636399	.004955053	.004845402	.004336811	.003554124	.001473640	.000355853	.000004258	0101101001x
1011000011	31	.0039736	-.0049197	.004635685	.004891890	.004838185	.004324889	.003538261	.001457292	.000348070	.000004000	0101011010
0101010101	30	.0037316	-.0058715	.004568400	.004786062	.004564181	.003963192	.003134437	.001161208	.000241791	.000001910	0101010101
1000110101	31	.0036237	-.0058501	.004548728	.004755777	.004556692	.003950014	.003140253	.001191985	.000258753	.000002356	0100110110
0110101011	29	.0036078	-.0057239	.004548501	.004761867	.004554001	.003979189	.003177395	.001222588	.000269455	.000002514	0101011110
1010110011	30	.0030160	-.0066860	.004415614	.004490271	.004157047	.003551345	.002765162	.001008933	.000215058	.000002055	0011110110
1101000011	30	.0029693	-.0068086	.004403935	.004463158	.004131526	.003494368	.002701199	.000959772	.000196137	.000001614	0101010110
1011010011	29	.0028614	-.0067977	.004383810	.004343101	.004096579	.003494368	.002687080	.000968576	.000203130	.000001820	0100110110
1100100011	29	.0027580	-.0074128	.004350761	.004343101	.003948411	.003270087	.002468000	.000815159	.000151829	.000000970	0101100101x
1110000011	29	.0026193	-.0075947	.004320574	.004283269	.003867382	.003181197	.002384675	.000774493	.000141877	.000000878	0101011010
1001110011	30	.0026034	-.0063609	.004345685	.004386735	.004086770	.003154827	.002792480	.001116133	.000276899	.000004382	0011110100
1100010101	30	.0026034	-.0074365	.004320904	.004291019	.003886964	.003212104	.002422002	.000801944	.000150550	.000000988	0101001110

99

m = 5, n = 5 Rank Orders 51 thru 100 of 142   **C(5, 5) = 252**   **1/C(5, 5) = .00396825**

z	w	c_1	c_2	d = .2	.4	.6	.8	1.0	1.5	2.0	3.0	z^{tc}
1001001101	30	.0026034	−.0074521	.004320557	.004289610	.003884015	.003207649	.002416595	.000797576	.000148955	.000000959	0100110110
1000011101	31	.0024488	−.0070524	.004300005	.004275039	.003900571	.003267212	.002513400	.000901930	.000192538	.000001949	100001110x
0111000011	28	.0020116	−.0077222	.004202626	.004076884	.003624082	.002953431	.002207670	.000733569	.000144019	.000001204	0011110001
1010010011	29	.0019957	−.0079776	.004194113	.004050086	.003574279	.002883678	.002127606	.000674694	.000123606	.000000831	0011011010
0110110011	29	.0018570	−.0081437	.004164223	.003991251	.003494899	.002796646	.002045793	.000633863	.000113138	.000000715	0101101001
0101010101	28	.0016458	−.0086157	.004114222	.003884909	.003441931	.002619784	.001872075	.000540194	.000088113	.000000436	0101010101
1001011001	28	.0015991	−.0086799	.004103852	.003863427	.003310989	.002583275	.001835031	.000518050	.000081581	.000000355	0110101001x
0110010101	28	.0015070	−.0087114	.004085848	.003832031	.003274571	.002550253	.001810719	.000514406	.000082663	.000000887	0011011100
1001100101	28	.0014911	−.0079348	.004098677	.003886048	.003382383	.002722961	.001983506	.000631127	.000118605	.000000887	0101011001x
1000110011	29	.0014911	−.0085237	.004086591	.003840614	.003294389	.002579611	.001844260	.000536169	.000088792	.000000461	0110011110
1010011001	29	.0014911	−.0085776	.004085478	.003836410	.003286222	.002568151	.001831337	.000527463	.000086132	.000000427	0100111010
1000011110	30	.0014752	−.0083598	.004086814	.003847149	.003310109	.002603099	.001871089	.000553328	.000093500	.000000505	100001011x
0101010011	28	.0008835	−.0087749	.003965318	.003620027	.003020115	.002303316	.001606427	.000442592	.000070474	.000000359	0011110010
0111000101	27	.0006414	−.0089749	.003915406	.003525938	.002898474	.002175458	.001491154	.000390694	.000058454	.000000176	0011011100
1010110001	27	.0005335	−.0092120	.003907850	.003504105	.002861104	.002127317	.001440480	.000361214	.000050618	.000000193	0101101001
0110101001	27	.0004868	−.0086057	.003890211	.003473880	.002826466	.002095999	.001417096	.000356340	.000050729	.000000193	0110101001x
1100001011	28	.0004868	−.0092242	.003893015	.003499779	.002883286	.002177086	.001506874	.000411106	.000065538	.000000343	0011011100
1001011010	27	.0004868	−.0092625	.003880908	.003456443	.002803498	.002070947	.001393374	.000344049	.000047513	.000000161	0110101010
1010011100	28	.0004868	−.0089749	.003880160	.003453779	.002798632	.002064539	.001386608	.000340226	.000046551	.000000153	0101011010
1000110101	29	.0004709	−.0090283	.003881615	.003464194	.002820498	.002094857	.001419293	.000358736	.000051096	.000000187	1000110110
1100000111	28	.0001209	−.0087826	.003819480	.003364290	.002711639	.001999872	.001349599	.000342476	.000049988	.000000205	0100111100
1011000011	27	.0001209	−.0087826	.003773399	.003284556	.002617725	.001910679	.001277619	.000319186	.000046554	.000000205	0011110010
0111001001	26	−.0004709	−.0090283	.003701858	.003151695	.002449392	.001738050	.001126345	.000256590	.000033414	.000000109	0110110010
1010100011	27	−.0004868	−.0086057	.003766639	.003172868	.002489215	.001790126	.001180355	.000285624	.000040498	.000000170	0011011100
1001100110	27	−.0004868	−.0092625	.003695085	.003133392	.002419844	.001702035	.001090423	.000238269	.000029134	.000000077	0101011010x
1001010101	27	−.0005335	−.0092625	.003694369	.003130952	.002415576	.001696655	.001084986	.000235514	.000028512	.000000073	1001010110
1100101001	28	−.0006255	−.0092002	.003886254	.003117686	.002400227	.001681766	.001072264	.000230130	.000027314	.000000065	1001010110x
1100011100	27	−.0006414	−.0089749	.003669056	.003088902	.002367879	.001654152	.001050691	.000224912	.000026893	.000000068	0110101010
1010101010	28	−.0006414	−.0089749	.003670118	.003090799	.002384687	.001675116	.001073497	.000236256	.000029336	.000000082	1000110110
1010010110	27	−.0008835	−.0087924	.003627194	.003027630	.002298389	.001605352	.001019537	.000219786	.000026774	.000000072	1010011100
0111010001	25	−.0014752	−.0083598	.003522230	.002858185	.002120724	.001439072	.000893276	.000183977	.000021866	.000000061	0111110001x
1001000111	26	−.0014911	−.0079348	.003526697	.002877171	.002154924	.001481941	.000935947	.000204800	.000026619	.000000095	0011110010
1001110001	26	−.0014911	−.0085237	.003516185	.002842805	.002097045	.001411396	.000866718	.000171568	.000019186	.000000044	0111000010
1011000101	26	−.0014911	−.0085776	.003515233	.002839727	.002091931	.001405261	.000860808	.000168893	.000018641	.000000031	1001110010x
1010101001	26	−.0015991	−.0087114	.003509751	.002826295	.002071749	.001382157	.000839091	.000159336	.000016735	.000000032	0110110010
1010101010	26	−.0016458	−.0086157	.003492925	.002798744	.002041421	.001355558	.000819501	.000154566	.000016260	.000000027	1010101010●
1010011010	26	−.0018570	−.0074128	.003485036	.002786190	.002027302	.001342291	.000808573	.000150435	.000015455	.000000034	0101101010
1100010101	26	−.0019957	−.0067977	.003453117	.002743410	.001989637	.001317078	.000795733	.000151335	.000016256	.000000032	0110101100
1100001101	26	−.0019957	−.0079776	.003429594	.002706989	.001951029	.001288323	.000777304	.000144817	.000015386	.000000032	0011011100
1100001110	27	−.0020116	−.0077222	.003430917	.002715308	.001966586	.001303161	.000789951	.000152952	.000016948	.000000039	1000110110
0111100001	24	−.0024488	−.0070524	.003359262	.002608204	.001857199	.001212791	.000726341	.000138189	.000015367	.000000039	0111110001x
1110000101	25	−.0026034	−.0063609	.003341416	.002593068	.001854204	.001221534	.000741372	.000148666	.000017910	.000000058	0011110010
1011001001	25	−.0026034	−.0074365	.003323483	.002538205	.001767586	.001122404	.000649886	.000110752	.000010619	.000000017	0110110010
1010010101	25	−.0026193	−.0074521	.003323228	.002537442	.001766412	.001121100	.000648724	.000110319	.000010547	.000000017	0101101010
1010011010	26	−.0026193	−.0075947	.003317739	.002524470	.001747800	.001069833	.000608379	.000103366	.000009339	.000000013	1010011100x
1001011100	26	−.0027580	−.0074128	.003294517	.002489017	.001710973	.001069833	.000608597	.000097839	.000008665	.000000011	0101101010
1100101010	25	−.0028614	−.0067977	.003285170	.002486778	.001720920	.001051110	.000618773	.000103611	.000010528	.000000019	0110101100
1100011010	25	−.0029693	−.0068086	.003264579	.002451594	.001680368	.001051110	.000600025	.000098963	.000009213	.000000014	0110101100
1011011000	25	−.0029693	−.0068567	.003263509	.002441732	.001674447	.001043733	.000593596	.000096887	.000008842	.000000013	1011011100

Table A

The eight probability columns correspond to $d = .2,\ .4,\ .6,\ .8,\ 1.0,\ 1.5,\ 2.0,\ 3.0$.

z	w	c_1	c_2	.2	.4	.6	.8	1.0	1.5	2.0	3.0	z^{tc}
1011000110	25	−.0036237	−.0058501	.003156559	.002288794	.001512375	.000910464	.000499247	.000073780	.000006058	.000000007	1001110010
1010101010	25	−.0037316	−.0058715	.003135958	.002254074	.001473301	.000875474	.000472859	.000066680	.000005157	.000000005	1010101000
1100110001	24	−.0039736	−.0044713	.003112337	.002240537	.001479758	.000896236	.000497612	.000077987	.000007060	.000000008	0101111000x
1101001001	24	−.0039736	−.0048901	.003105939	.002222628	.001453927	.000869277	.000474972	.000070663	.000005986	.000000008	0110110100
1101001010	24	−.0039736	−.0049197	.003105492	.002221388	.001452157	.000867451	.000473456	.000070187	.000005918	.000000008	0110101100
1100011010	25	−.0039895	−.0050183	.003100855	.002211368	.001438716	.000853637	.000461760	.000066189	.000005293	.000000008	1010110100x
1011001010	24	−.0041283	−.0047325	.003079273	.002180704	.001408884	.000830092	.000445866	.000062774	.000004927	.000000006	1010100100
1011010001	24	−.0045813	−.0033954	.003015005	.002098061	.001336805	.000779715	.000416236	.000058775	.000004762	.000000006	1001001100
1011001001	24	−.0047359	−.0035840	.002983311	.002042832	.001273726	.000722952	.000373439	.000047344	.000003314	.000000003	1011010010x
1101010001	23	−.0049780	−.0025164	.002954246	.002013037	.001255278	.000716228	.000373888	.000049870	.000003814	.000000004	0111010100
1101000110	24	−.0049939	−.0026126	.002949753	.002003738	.001243354	.000704512	.000364396	.000046949	.000003400	.000000003	1001011100
1110000101	23	−.0050859	−.0019867	.002941871	.001973908	.001247817	.000713350	.000373696	.000050569	.000003947	.000000003	0110110100x
1100011100	24	−.0051018	−.0025425	.002930730	.001910054	.001211943	.000678081	.000345601	.000042529	.000002903	.000000002	1010101100
1101010010	24	−.0054985	−.0011202	.002837898	.001850752	.001159517	.000643506	.000326341	.000040176	.000002791	.000000002	1100110010x
1101001100	23	−.0057403	−.0008117	.002816898	.001835345	.001100138	.000596030	.000294200	.000033446	.000002112	.000000002	1011010010x
1110010010	22	−.0059516	−.0004607	.002798329	.001812498	.001097123	.000601677	.000302570	.000037103	.000002640	.000000003	0111010100
1101100010	23	−.0060903	.0009595	.002794345	.001804785	.001077842	.000588267	.000294578	.000035808	.000002531	.000000002	1010110100
1101010100	23	−.0061062	.0008928	.002787825	.001788314	.001068427	.000579387	.000287639	.000033831	.000002268	.000000002	1001110100
1011100010	23	−.0061062	.0004201	.002787637	.001787853	.001047007	.000559254	.000277232	.000030084	.000001854	.000000001	1100110000
1101100100	23	−.0061062	.0004061	.002777103	.001787637	.001046427	.000558726	.000272047	.000029998	.000001846	.000000001	1011011000
1100101100	23	−.0064720	.0017186	.002738748	.001728121	.000996403	.000524709	.000252244	.000027052	.000001627	.000000001	1011101000x
1110100010	22	−.0067138	.0025377	.002705802	.001687112	.000961441	.000500524	.000237934	.000024840	.000001457	.000000001	1100111000x
1110010100	22	−.0070636	.0044570	.002677919	.001651804	.000941297	.000493482	.000237909	.000026546	.000001743	.000000002	1011010100
1101101000	22	−.0071105	.0038535	.002651402	.001619026	.000903131	.000460037	.000213901	.000021106	.000001169	.000000001	1010110100
1110100100	22	−.0072184	.0045180	.002640551	.001609817	.000898575	.000458939	.000214350	.000021534	.000001224	.000000001	1100110100
1110011000	22	−.0074764	.0053422	.002604794	.001564692	.000859680	.000431798	.000198177	.000019005	.000001028	.000000001	1101010000
1011100100	21	−.0080682	.0087590	.002542078	.001506136	.000824552	.000416758	.000194329	.000020192	.000001252	.000000001	1100111000
1101110000	21	−.0080841	.0078516	.002527820	.001475399	.000788572	.000385750	.000178219	.000016219	.000000801	.000000001	1011011000
1101101000	21	−.0082228	.0085083	.002511265	.001457278	.000774935	.000377393	.000172615	.000015095	.000000770	.000000001	1011101000
1101011000	21	−.0084807	.0094507	.002477103	.001416071	.000740970	.000354715	.000155282	.000013276	.000000643	.000000001	1010110100x
1110011100	21	−.0085887	.0102067	.002467175	.001408495	.000737792	.000354343	.000155931	.000013607	.000000679	.000000000	1100111000
1110101000	20	−.0091963	.0130268	.002393933	.001327426	.000676033	.000315990	.000135466	.000011121	.000000525	.000000000	1011011000
1110100100	20	−.0094543	.0140975	.002361379	.001289900	.000646396	.000296984	.000125030	.000009775	.000000438	.000000000	1011100100x
1110011000	20	−.0095930	.0148581	.002346004	.001274243	.000635415	.000290702	.000121940	.000009468	.000000422	.000000000	1011101000x
1110011010	19	−.0102007	.0183884	.002280799	.001209925	.000591724	.000266507	.000110431	.000008425	.000000375	.000000000	1011100100x
1110010110	19	−.0105666	.0200252	.002236140	.001160201	.000553809	.000243025	.000097971	.000006944	.000000286	.000000000	1011110000
1101110010	19	−.0107053	.0208720	.002221668	.001146247	.000544569	.000238007	.000095621	.000006735	.000000276	.000000000	1011101000x
1110110000	18	−.0115709	.0260620	.002130278	.001057162	.000484418	.000204741	.000079737	.000005244	.000000203	.000000000	1011101000x
1110101000	18	−.0116789	.0265595	.002117374	.001043233	.000474189	.000198666	.000076654	.000004918	.000000186	.000000000	1011110000x
1110110000	17	−.0126832	.0331392	.002016950	.000950264	.000414491	.000167186	.000062290	.000003704	.000000131	.000000000	1011110000x
1111010000	16	−.0136876	.0402036	.001921085	.000865256	.000362037	.000140527	.000050534	.000002782	.000000093	.000000000	1111010000x
1111100000	15	−.0146611	.0476993	.001833904	.000792351	.000319411	.000119911	.000041848	.000002165	.000000069	.000000000	1111100000x

Table A

m = 10, n = 1 Rank Orders 1 thru 11 of 11 C(10, 1) = 11 1/C(10, 1) = .09090909

z	w	c_1	c_2	d = .2	.4	.6	.8	1.0	1.5	2.0	3.0	z^{tc}
00000000001	11	.1442215	.1681843	.123214096	.162715971	.209528709	.263305376	.323201250	.490114473	.657630645	.895579862	01111111111
00000000010	10	.0965379	.0302604	.110650372	.130486645	.149116414	.165163478	.177345181	.185231059	.160148334	.069085756	10111111111
00000000100	9	.0662581	−.0275517	.103436142	.113835136	.121185984	.124805831	.124353931	.106697229	.074605167	.019868950	11011111111
00000001000	8	.0419980	−.0580565	.098024815	.102159498	.102908571	.100200769	.094309296	.069889420	.041935039	.008037987	11101111111
00000010000	7	.0204446	−.0736166	.093469345	.092852111	.089121404	.082650829	.074061923	.048445539	.025577808	.003754392	11110111111
00000100000	6	.0000000	−.0784398	.089353756	.084845623	.077831894	.068975989	.059054281	.034441003	.016190714	.001874555	11111011111
00001000000	5	−.0204446	−.0736166	.085427853	.077560288	.068033229	.057655243	.047204914	.024622652	.010351945	.000957603	11111101111
00010000000	4	−.0419980	−.0580565	.081484893	.070587927	.059095291	.047811296	.037380918	.017392432	.006530380	.000483089	11111110111
00100000000	3	−.0662581	−.0275517	.077276877	.063528488	.050505053	.038825787	.028860090	.011862122	.003946257	.000230863	11111111011
01000000000	2	−.0965379	.0302604	.072352734	.055773161	.041634316	.030093322	.021058200	.007479738	.002164041	.000097269	11111111101
10000000000	1	−.1442215	.1681843	.065309117	.045655151	.031039134	.020512079	.013170015	.003824333	.000919669	.000029674	11111111110

102

C(7, 4) = 330   1/C(7, 4) = .00303030

z	w	c_1	c_2	d = .2	.4	.6	.8	1.0	1.5	2.0	3.0	z^{tc}
0000001111	38	.116339	.0395960	.006318093	.012348633	.022679237	.039239193	.064130023	.173920246	.352870092	.752125343	0000111111
0000010111	37	.109154	.0338507	.006009019	.011104140	.019154255	.030896508	.046688602	.099749953	.147955298	.124184861	0001011111
0000011011	36	.102339	.0288246	.005734273	.010073049	.016451917	.025017511	.035477089	.063286056	.075800262	.037184677	0001101111
0000100111	36	.101067	.0279277	.005684692	.009894086	.015999433	.024073283	.033755092	.058349220	.067383976	.030237701	0010011111
0000101011	35	.095524	.0241829	.005475052	.009159312	.014207023	.020461454	.027404780	.041853653	.042163738	.013935220	0010101111
0001000111	35	.094252	.0232488	.005424110	.008970709	.013722905	.019437845	.025522646	.036470129	.033321428	.007900044	0011001111
0000110011	35	.090974	.0213087	.005310474	.008608441	.012955538	.018002376	.023293186	.032489061	.029705209	.007950732•	0100011111
0001001011	34	.088340	.0196873	.005216790	.008300555	.012233354	.016683417	.021137992	.027888693	.024031447	.005676729	0011010111
0001010011	34	.087438	.0189580	.005178425	.008153300	.011838186	.015865777	.019646573	.023878198	.018119806	.005733819	0011100111
0000111110	34	.087068	.0187995	.005167769	.008126667	.011796948	.015825368	.019641221	.024182099•	.018839940	.003180335	0011101011
0001100011	34	.084159	.0170679	.005066467	.007801192	.011072638	.014500744	.017539209	.020068717	.014308013	.001896610	0100101111
0010000111	33	.080253	.0151030	.004942891	.007443611	.010363341	.013359117	.015970196	.018148688	.013406556	.002307985•	0001111101
0001011001	33	.080253	.0148354	.004933779	.007386415	.010176996	.012915966	.015113431	.015793561	.010162475	.001056520•	0010111101
0010010011	33	.077344	.0148174	.004931185	.007382811	.010165723	.012890378	.015066528	.015684639	.010043661	.001035519	0011011101
0000111101	33	.076975	.0132190	.004836564	.007087662	.009543354	.011815756	.013462595	.013039036	.007657682	.000625519	0110011111
0001100101	33	.075079	.0130753	.004826281	.007062458	.009503517	.011770269	.013429849	.013132221	.007872223	.000774717	0101011011
0001101001	34	.073438	.0126179	.004782145	.006981980	.009443074	.011847557	.013809245	.014834813	.008744267	.001668161	1000011011
0010011001	32	.073066	.0113012	.004714392	.006737471	.008853059	.010706129	.011927699	.010994841	.006449267	.000744724	0011110011
0010100101	32	.072166	.0110233	.004699728	.006685993	.008731712	.010476744	.011559015	.010301855	.005553165	.000384005•	0010110111
0100000111	32	.070160	.0106733	.004674433	.006621976	.008622482	.010328864	.011393506	.010217373	.005599347	.000413383	0011010111
0000111011	32	.070160	.0100833	.004624058	.006513556	.008481733	.010225429	.011431808	.010988365	.006870617	.000852307	0011111101
0010001011	32	.070160	.0095649	.004607748	.006419120	.008198463	.009606378	.010334095	.008577992	.004248512	.000234763	0001111011
0001010101	32	.068888	.0095350	.004606795	.006413569	.008181777	.009570026	.010270198	.008444677	.004117687	.000214821	0101101111
0010010101	33	.068265	.0091587	.004574760	.006343877	.008088028	.009489714	.010257276	.008764996	.004601474	.000302701	1000101011
0001001101	31	.066254	.0090747	.004561856	.006323959	.008079307	.009520559•	.010357129	.009068335	.004924068	.000371893	0011011011
0011000011	31	.064982	.0078158	.004490839	.006099003	.007596712	.008685536	.009123569	.007167820	.003392912	.000145736	0101110111
0010100011	31	.063345	.0072342	.004452358	.005992198	.007392587	.008366880	.008694635	.006627397	.003025912	.000099933	0101011101
0001011010	31	.062975	.0064571	.004401996	.005885059	.007117547	.007932276	.008103541	.005869840	.002510217	.000042394	0110101011
0010001101	31	.062073	.0062095	.004388475	.005806481	.007025534	.007767393	.007861154	.005712651	.002247470	.000050955	0111010011
0100100011	31	.062073	.0061837	.004372650	.005793082	.007025589•	.007896771•	.008139946	.006155098	.002839989	.000148096	0010111011
0010110001	31	.062073	.0059693	.004366323	.005758841	.006956930	.007702858	.007822412	.005596613	.002372826	.000093902	0110101111
0010011010	31	.061450	.0059329	.004365278	.005753359	.006942133	.007637943	.007777034	.005525938	.002332014	.000090038	0101011111
0001011100	32	.061080	.0059272	.004354452	.005743360	.006957272	.007745040	.007928806	.005854007	.002602397	.000118724	1000011101
0010101001	30	.059439	.0058113	.004344866	.005721147	.006923185	.007704878	.007892339	.005863825	.002647178	.000129879	0011100111
0010010110	30	.058167	.0049965	.004293565	.005575733	.006642090	.007264720	.007302360	.005148117	.002194356	.000095908	0010101111
0001100110	30	.056161	.0043776	.004254092	.005546143	.006426264	.006925707	.006845665	.004583476	.001827648	.000066365	0011010011
0101000011	30	.055258	.0034403	.004171996	.005294236	.006102916	.006445715	.006183370	.003801186	.001348014	.000033946	0111101011
0100101001	30	.054889	.0032386	.004171751	.005250652	.006045715	.006372590	.006153620	.003862846	.001427927	.000042394	0101101011
0001110001	30	.054889	.0032131	.004164627	.005240649	.006040426	.006388987	.006169554	.003973228	.001520522	.000050955	0111010111
0011001001	30	.054889	.0030167	.004159098	.005212165	.005967063	.006243903	.005975117	.003634970	.001282564	.000032667	0111011101
0010010101	30	.054889	.0029504	.004157237	.005313808	.005941955	.006196167	.005902175	.003529623	.001212952	.000028232	0101010111
1000000111	31	.054265	.0038060	.004170218	.005313808	.006276858	.006882397	.007015001	.005403580	.002741403	.000231313	0001111110
0001011011	31	.054265	.0030159	.004148140	.005200068	.005972747	.006289077	.006074396	.003833476	.001431941	.000044742	1000011110
0010110010	31	.054265	.0029725	.004146889	.005193474	.005954830	.006253802	.006018396	.003743222	.001364004	.000037828	0101010111
0100011001	31	.053993	.0027047	.004117384	.005133376	.005887271	.005988140	.005598140	.003588414	.001504828	.000037828	1010010111
0101000101	29	.051980	.0021074	.004083229	.005032054	.005675224	.005861467	.005547823	.003317094	.001170045	.000031776	0110010111
0011100001	29	.051352	.0019055	.004066760	.004993082	.005612887	.005781892	.005462796	.003268932	.001166800	.000034278	0011110011
0101001001	29	.049346	.0010627	.004008435	.004838334	.005331686	.005366722	.004919082	.002713819	.000861740	.000017536	0101101011
0111000001	29	.048074	.0008506	.003980192	.004783534	.005263918	.005307532	.004907124	.002775160•	.000930492	.000023268	0111010011
0100100111	29	.048074	.0007983	.003978848	.004777224	.005248720	.005281139	.004870356	.002733512	.000909495	.000022661	0110111101

103

Table A

z	w	c_1	c_2	d = .2	.4	.6	.8	1.0	1.5	2.0	3.0	z^{tc}
0010101001	29	.048074	.005947	.003973356	.004750160	.005180127	.005155475	.004685125	.002487535	.000758283	.000013955	0101101011
0001011001	29	.048074	.005320	.003971673	.004741916	.005159410	.005117926	.004630500	.002418184	.000718350	.000012099	1011101101
0001100110	30	.047451	.005920	.003962074	.004735086	.005174671	.005173465	.004734039	.002583082	.000820428	.000017034	1001011011
0010010110	30	.047081	.003898	.003939701	.004700443	.005107505	.004975605	.004597090	.002431182	.000737216	.000013121	1001011110
0010011001	30	.046802	.001799	.003943228	.004668933	.005045954	.004975605	.004478612	.002318243	.000687191	.000011831	0010111011
1000000101	30	.046179	.007452	.003929611	.004724761	.005215918	.005308809	.004985422	.003012294	.001122438	.000039153	1000111011
0010011010	30	.046179	.002227	.003929418	.004659576	.005054643	.005064643	.004564250	.002461006	.000777951	.000016576	1010111011
0010101010	30	.046179	.002168	.003929418	.004658456	.005051363	.005012107	.004552836	.002441223	.000762716	.000015330	1010111101
0101011010	30	.044795	-.000265	.003888361	.004549056	.004854276	.004201674	.004201674	.002107687	.000604810	.000009668	0011101111
0110000001	28	.044168	-.002839	.003880360	.004550380	.004890561	.004821459	.004364246	.002357171	.000764725	.000018985	0111110011
0000111100	30	.042900	-.003083	.003856602	.004514526	.004864132	.004827364	.004414523	.002491842	.000863947	.000026052	1100001111
0101000001	28	.041259	.012355	.003803244	.004363540	.004580011	.004401390	.003876057	.001937565	.000574221	.000011410	0011110011
0001100101	28	.041259	-.014319	.003798173	.004339612	.004521967	.004299633	.003732047	.001766985	.000480356	.000007063	0101100011
0010100101	28	.041259	-.014983	.003796465	.004331594	.004506772	.004266093	.003685802	.001713705	.000452826	.000006031	0101010111
0010110010	28	.040889	-.016647	.003790598	.004326117	.004506772	.004287771	.003685802	.001777415	.000489007	.000007465	0110101011
0010011001	29	.040266	-.016820	.003773644	.004283719	.004433774	.004185378	.003604363	.001663553	.000435758	.000005607	1000111011
0101010010	28	.039987	-.018912	.003763423	.004253497	.004376646	.004101536	.003502226	.001575643	.000400736	.000004866	0110101101
0011011001	28	.039364	-.018164	.003753871	.004245304	.004384767	.004137959	.003569635	.001671309	.000451830	.000006638	0111101110
1000011001	29	.038994	-.014826	.003755360	.004272932	.004465932	.004289875	.003789571	.001935540	.000595808	.000013189	0011011111
0110000001	29	.038994	-.019930	.003742648	.004248920	.004329453	.004057266	.003469946	.001576703	.000407666	.000005154	1010110111
0010010110	29	.038994	-.020306	.003741698	.004210539	.004319088	.004039600	.003445806	.001550658	.000395069	.000004761	1001011011
0101000001	28	.037981	-.023831	.003714492	.004144815	.004216674	.003898603	.003288809	.001435989	.000354397	.000004029	0111011101
0100100101	28	.036709	-.026470	.003684682	.004081459	.004120132	.003739281	.003183601	.001379755	.000340650	.000003999	0110011110
1000101001	28	.036085	-.022811	.003682014	.004103112	.004195031	.003936808	.003392830	.001617998	.000459733	.000007201	1100001011
0010011100	29	.036085	-.023543	.003680236	.004095157	.004176575	.003905608	.003350074	.001569281	.000433140	.000005800	1000110111
0101010010	28	.034075	-.025528	.003675499	.004074478	.004130127	.003800190	.003251509	.001473078	.000390150	.000006139	0111011011
0101000010	27	.034075	-.029683	.003628394	.003974382	.003985508	.003661993	.003085756	.001386346	.000371286	.000003714	0111110111
0011010001	27	.034075	-.031827	.003623128	.003946032	.003939868	.003570703	.002962972	.001256311	.000307207	.000003268	1010101111
0011100010	28	.034075	-.032191	.003622202	.003946442	.003920680	.003553269	.002939045	.001230117	.000294232	.000002970	0111110011
0011001010	28	.033451	-.033580	.003607256	.003913456	.003870410	.003490569	.002871563	.001181806	.000276118	.000002970	1100110011
0010101001	27	.033172	-.035706	.003597092	.003884076	.003816467	.003413911	.002781326	.001110582	.000250090	.000002368	0101010011
0101010001	27	.032803	-.035977	.003590068	.003872687	.003805268	.003407251	.002781409	.001121088	.000256488	.000002557	0111001110
1000010001	27	.032179	-.031959	.003587520	.003893704	.003876091	.003540984	.002970469	.001324917	.000353165	.000005745	1100110110
0001011010	28	.032179	-.034752	.003580924	.003865147	.003812282	.003437588	.002853236	.001189986	.000289762	.000003514	1010101011
0010011010	28	.032179	-.037176	.003575078	.003839417	.003754159	.003342889	.002711407	.001069692	.000236900	.000002099	1001011011
0010100101	28	.032179	-.037514	.003574260	.003835824	.003746079	.003329822	.002694496	.001053910	.000230388	.000001958	1100110001
0101010010	27	.031166	-.039751	.003550255	.003784546	.003671991	.003244036	.002610595	.001011098	.000220754	.000001951	1010101101
0101011010	28	.030907	-.039910	.003544992	.003774638	.003658930	.003229587	.002596310	.001001322	.000216675	.000001832	0111101110
0100110010	28	.029894	-.042763	.003519522	.003717017	.003567067	.003121318	.002483642	.000932280	.000196381	.000001589	1010101101
0011001100	28	.029270	-.039493	.003515526	.003730413	.003622052	.003219102	.002619835	.001068178	.000254003	.000002970	1100100111
1000001001	28	.028901	-.040523	.003506035	.003708156	.003585694	.003170617	.002564311	.001022967	.000235366	.000002460	0101010111
0100011010	28	.028901	-.040956	.003505076	.003704258	.003577503	.003158110	.002548870	.001009556	.000229869	.000002319	1100011101
0100010010	26	.027638	-.043375	.003449949	.003680748	.003562657	.003078786	.002449544	.000923043	.000195900	.000001611	1001011110
0011110010	26	.027260	-.044419	.003466501	.003622137	.003458698	.003099667	.002411836	.000935340	.000211036	.000002229	0111001011
0110010000	26	.025988	-.044001	.003443645	.003537708	.003343081	.003009803	.002426206	.000984201	.000240605	.000003416	0011110001
0010100001	26	.025988	-.049185	.003431742	.003537230	.003321101	.002840316	.002213971	.000794096	.000161392	.000001264	0111010111
0011010010	26	.025988	-.049484	.003431070	.003534986	.003315366	.002831688	.002203583	.000784201	.000158563	.000001219	1010101101
0110100010	27	.025364	-.045322	.003433072	.003554933	.003378085	.002945935	.002350235	.000933517	.000220656	.000002828	0111110110
1000010101	26	.025274	-.048536	.003417150	.003506351	.003274472	.002783802	.002155028	.000755635	.000148526	.000001035	1010110111

Table 11

m = 7, n = 4 Rank Orders 101 thru 150 of 330 C(7, 4) = 330 1/C(7, 4) = .00303030

z	w	c_1	c_2	d = .2	.4	.6	.8	1.0	1.5	2.0	3.0	z^{tc}
0010110010	27	.024995	−.0050478	.003410088	.003494412	.003261395	.002773014	.002148459	.000757179	.000150202	.000001078	1011001011
0010010010	27	.024093	−.0053646	.003385970	.003437063	.003169893	.002656467	.002023116	.000674454	.000124062	.000000713	1010101101
0101010001	26	.023981	−.0052393	.003386805	.003446050	.003193244	.002695784	.002074224	.000719949	.000141397	.000001024	0101110101
0101001001	26	.023079	−.0055174	.003363615	.003392728	.003110662	.002593385	.001966769	.000652695	.000121017	.000000751	0110110101
0101000101	27	.022086	−.0054942	.003357181	.003394510	.003102786	.002590285	.001969279	.000660973	.000124912	.000000827	0110111001
0000110100	27	.022086	−.0051515	.003353022	.003394510	.003145187	.002668767	.002074383	.000755660	.000161188	.000001542	1011001101
0001000101	27	.022086	−.0053273	.003348964	.003377432	.003108503	.002611324	.002002577	.000693253	.000136565	.000001001	1100101110
0001010100	27	.022086	−.0053819	.003347793	.003377793	.003099166	.002597605	.001986326	.000680697	.000132087	.000000920	1010011101
0001010010	27	.022086	−.0056165	.003342618	.003352031	.003056342	.002534113	.001910928	.000623721	.000112962	.000000642	1001011101
1000010010	27	.020814	−.0054539	.003342179	.003229478	.003041784	.002541304	.001940196	.000667604	.000131873	.000001007	0110011110
0010011100	27	.020814	−.0055635	.003319778	.003316571	.003021920	.002511637	.001904533	.000639234	.000121553	.000000816	1100111011
0100001100	27	.020814	−.0057679	.003315328	.003298912	.002986003	.002459031	.001842803	.000593902	.000106734	.000000610	1010011101
1000001110	28	.020191	−.0052834	.003314147	.003215535	.003051383	.002570306	.001985979	.000716648	.000152661	.000001504	1000111110
0011001001	25	.019173	−.0058567	.003282467	.003239865	.002914883	.002391408	.001789833	.000582306	.000107972	.000000717	0111100011
0011001100	26	.018803	−.0059678	.003273096	.003218783	.002883092	.002353140	.001750913	.000559826	.000101664	.000000643	0111011001
1001000011	26	.018180	−.0055358	.003270649	.003235903	.002936331	.002445307	.001870208	.000662853	.000140840	.000001481	1001111010
0011010010	26	.018180	−.0060583	.003259238	.003190318	.002842613	.002305972	.001703482	.000531368	.000092882	.000000510	1110010011
0010101010	26	.018180	−.0060642	.003259135	.003190017	.002842212	.002305688	.001703498	.000531996	.000093277	.000000521	0110100111
0101010010	26	.017278	−.0063424	.003236001	.003137400	.002761842	.002207725	.001602722	.000472593	.000076556	.000000337	1010100111
1000110010	26	.016908	−.0063219	.003229478	.003127307	.002751989	.002200956	.001600000	.000475252	.000077996	.000000359	1100101011
0110000101	25	.015894	−.0061411	.003214169	.003110041	.002745939	.002112943	.001628346	.000509248	.000091013	.000000560	0101111001
0101010010	25	.015894	−.0064087	.003208532	.003088400	.002703060	.002126510	.001555839	.000462814	.000076930	.000000387	0111010101
1010010001	25	.015894	−.0064267	.003208156	.003086977	.002700605	.002148782	.001555500	.000460135	.000076192	.000000381	0111010110
0011100100	26	.015271	−.0062219	.003200525	.003082469	.002707438	.002169000	.001585183	.000485088	.000084125	.000000472	0111011101
1001100010	26	.015271	−.0062652	.003199652	.003079247	.002701312	.002160563	.001575818	.000478936	.000082254	.000000447	1001111011
0011010001	26	.015271	−.0065071	.003194539	.003059568	.002662516	.002105527	.001513313	.000436779	.000069658	.000000304	1110010111
0001000011	26	.014902	−.0062554	.003199841	.003068057	.002689114	.002150253	.001568875	.000478559	.000092796	.000000462	1110100111
1000110010	26	.013999	−.0059007	.003183171	.003067549	.002712641	.002220890	.001641181	.000543523	.000106742	.000000895	0101010111
0010101010	26	.013999	−.0063816	.003172984	.003028093	.002634023	.002088662	.001510025	.000448564	.000075225	.000000391	0111010101
1000101010	26	.013999	−.0064988	.003170526	.003018675	.002615481	.002062311	.001479954	.000427782	.000068766	.000000308	1100101011
0100010010	26	.013999	−.0066997	.003166344	.003002837	.002584799	.002019602	.001432434	.000397571	.000060367	.000000229	1010101101
0011010100	27	.013006	−.0063254	.003155211	.002999644	.002603674	.002063351	.001492932	.000446705	.000075866	.000000406	1001011110
0011010001	24	.011988	−.0065674	.003131045	.002949006	.002532857	.001984618	.001419302	.000413052	.000068768	.000000373	0111100011
0110010001	25	.011365	−.0066259	.003117890	.002923381	.002498313	.001946445	.001382862	.000393700	.000063403	.000000305	1011100011
1001010001	26	.010721	−.0066763	.003104542	.002897984	.002465046	.001910903	.001350211	.000378215	.000059637	.000000270	1100111101
1010000001	25	.010093	−.0061576	.003103364	.002918366	.002525519	.002002419	.001462913	.000465631	.000089713	.000000798	0011111101
0011010010	25	.010093	−.0069476	.003087055	.002856815	.002401812	.001834680	.001273493	.000336537	.000049040	.000000177	1010010101
0010010010	24	.010093	−.0069911	.003086911	.002856553	.002395605	.001826415	.001262336	.000331076	.000047619	.000000165	0101011011
1001010001	25	.009080	−.0068932	.003068932	.002828372	.002372730	.001812465	.001261126	.000339496	.000051513	.000000220	0111011101
0101010010	24	.008710	−.0069894	.003059999	.002809198	.002345362	.001781387	.001231394	.000324915	.000048086	.000000193	1101011011
1001001010	25	.008087	−.0067568	.003052694	.002805033	.002351241	.001798152	.001254901	.000342686	.000053155	.000000241	0101111010
0011010100	25	.008087	−.0068301	.003051021	.002799908	.002334816	.001785624	.001245114	.000334779	.000051028	.000000157	1101100101
0101000110	25	.008087	−.0070285	.003047247	.002785123	.002313912	.001747723	.001200292	.000309802	.000044313	.000000210	1001110101
0110010001	24	.007808	−.0068882	.003044800	.002799908	.002322444	.001765025	.001192358	.000326337	.000049249	.000000179	0110110101
1001010010	25	.007185	−.0069253	.003032081	.002754620	.002292265	.001732223	.001192358	.000312009	.000045814	.000000143	0111011011
0010100100	25	.007185	−.0070350	.003029885	.002771195	.002277195	.001711812	.001173113	.000298478	.000042120	.000000105	1010110101
0101010010	25	.007185	−.0072393	.003025808	.002739841	.002249812	.001675387	.001131475	.000276637	.000036758	.000000105	1010101101
0010111000	25	.006815	−.0066318	.003030890	.002771751	.002317645	.001772028	.001238970	.000342761	.000054408	.000000263	1110001111
1000011001	25	.006815	−.0066651	.003026418	.002755523	.002287447	.001731528	.001195420	.000316972	.000047583	.000000263	0111001011
0010011100	25	.006815	−.0069996	.003023561	.002745097	.002267881	.001705025	.001166595	.000299313	.000042716	.000000151	1101011011

105

Table A

m = 7, n = 4   Rank Orders 151 thru 200 of 330   C(7, 4) = 330   1/C(7, 4) = .00303030

z	w	c_1	c_2	d = .2	.4	.6	.8	1.0	1.5	2.0	3.0	z^{tc}
0100110010	25	.0006815	−.0071884	.003019802	.002731491	.002242691	.001671515	.001130965	.000279089	.000037695	.000000114	1010011101
1000011100	26	.0006192	−.0069029	.003013499	.002730768	.002254591	.001695802	.001161920	.000300184	.000043297	.000000158	1001101110
0011100010	23	.005174	−.0067753	.002996794	.002707563	.002235467	.001687190	.001164462	.000312517	.000048479	.000000235	0111100011
1000010100	26	.0004920	−.0069169	.002988878	.002687803	.002203305	.001646400	.001121445	.000286458	.000041094	.000000152	0100011110
0101001100	25	.0003906	−.0071699	.002964717	.002636758	.002131739	.001566629	.001046561	.000251490	.000033307	.000000098	1100111101
0011010010	24	.0003278	−.0072019	.002952168	.002614121	.002104090	.001539527	.001024071	.000243745	.000032062	.000000095	1011001011
0011100100	24	.0002909	−.0072677	.002943862	.002597527	.002081869	.001515801	.001002714	.000234780	.000030284	.000000085	1010110011
0101100010	23	.0001895	−.0071472	.002926962	.002572924	.002058964	.001500483	.000956181	.000238654•	.000032268	.000000110	0111011001
0011011000	23	.0001272	−.0069536	.002918732	.002564889	.002056741	.001505255•	.001005662	.000246658	.000034438	.000000127	0110100111
0110001001	23	.0000623	−.0070960	.002903650	.002534014	.002014596	.001459507	.000963842	.000228380	.000030624	.000000104	0110011001
1100000011	24	.0000000	−.0061307	.002910150•	.002577679	.002106214	.001588045	.001105320	.000316018	.000055638	.000000431	0011111100
0010100101	24	.0000000	−.0068486	.002896386	.002529164	.002017795	.001470972	.000980001	.000239778	.000033607	.000000127	0101111110
0011011010	24	.0000000	−.0068555	.002896252	.002528685	.002016894	.001469734	.000978618	.000238840	.000033315	.000000123	1100100111
1000100110	24	.0000000	−.0070874	.002891825	.002513271	.001983377	.001434344	.000942141	.000219443	.000028713	.000000085	0101100110
0111000010	24	.0000000	−.0070915	.002891749	.002512966	.001988755	.001433408	.000940996	.000218508	.000028387	.000000089	1001111101
0110101000	24	.0000000	−.0070918	.002891740	.002512980•	.001988870	.001433714	.000941522	.000219169	.000028666	.000000074	0111011110
0010101010	24	.0000000	−.0071716	.002890617	.002507645	.001979269	.001421233	.000928480	.000211920	.000026850	.000000068	1101011101
0101010100	24	.0000000	−.0072270	.002889158	.002503975	.001977249	.001412905	.000919974	.000207542	.000025861	.000000053	1011101011
0110010100	23	.0000000	−.0073879	.002886091	.002493350	.001953936	.001388988	.000895691	.000195306	.000023179	.000000050	1010101001
0101010010	24	.0000000	−.0074194	.002885499	.002491272	.001950269	.001384349	.000891011	.000192998	.000022689	.000000050	1011010001
1001000110	25	−.0000623	−.0070960	.002879671	.002491374	.001962065	.001406436	.000917544	.000208866	.000026392	.000000073	1001110110
0010011110	24	−.0001272	−.0069536	.002870024	.002479389	.001953559	.001403798	.000919947•	.000213697	.000027915	.000000085	1100010101
0100101010	25	−.0001895	−.0071472	.002854403	.002405056	.001905663	.001350126	.000869588	.000190641	.000022983	.000000055	1010110101
0101001010	24	−.0002909	−.0072677	.002832903	.002405056	.001854139	.001297966	.000825037	.000174213	.000020121	.000000044	1100101101
0011101000	24	−.0003278	−.0072019	.002827104	.002397148	.001847299	.001293798	.000823641	.000175264•	.000020540	.000000047	1011101001
0111010000	22	−.0003906	−.0071699	.002815770	.002378728	.001817699	.001276507	.000815770	.000172093	.000020243	.000000048•	0111100110
0101101000	23	−.0004920	−.0069169	.002801176	.002361328	.001815613	.001273648	.000815385•	.000179651	.000022552	.000000068	1100111110
1100100001	22	−.0006192	−.0067753	.002798786	.002360648	.001818083	.001278427•	.000820748	.000182017	.000022881	.000000067	0111010011
1101000010	23	−.0006815	−.0069029	.002777185	.002320166	.001767409	.001227963	.000778412	.000167174	.000020438	.000000059	0111101001
0011101010	23	−.0006815	−.0066318	.002770171	.002315529	.001769904	.001237234	.000791075	.000175266	.000022345	.000000071	1110101001
1000100001	23	−.0006815	−.0068568	.002766041	.002301705	.001746159	.001207834	.000761883	.000161083	.000019257	.000000051	0111001110
0010101001	23	−.0006815	−.0069996	.002764432	.002292994	.001731216	.001189333	.000743492	.000152088	.000017279	.000000039	1101100101
0100110010	23	−.0006815	−.0071884	.002759981	.002281529	.001711733	.001165547	.000720285	.000141474	.000015160	.000000028	0111100110
0011010100	23	−.0007185	−.0069253	.002755756	.002285415•	.001724838	.001185662	.000742470	.000153463	.000017825	.000000044	1010011001
0101010010	23	−.0007185	−.0070350	.002752033	.002278749	.001713430	.001171577	.000728014	.000146702	.000016356	.000000035	1010110101
1010001001	24	−.0007808	−.0072393	.002752033	.002266423	.001692574	.001146240	.000703930	.000135643	.000014192	.000000025	1001110011
0101001001	24	−.0007808	−.0068882	.002746442	.002266486	.001702082	.001164625	.000725530	.000146952	.000016556	.000000036	0110111010
1010100001	23	−.0008087	−.0067568	.002743586	.002265544	.001706312	.001172216	.000734634	.000153020	.000018044	.000000047	0110101011
0010101010	23	−.0008087	−.0068301	.002742217	.002260865	.001698100	.001161811	.000724053	.000147551	.000016779	.000000038	1110101011
0100001010	23	−.0008087	−.0070285	.002738645	.002249154	.001678439	.001138072	.000701119	.000137273	.000014756	.000000028	1010101001
1000010010	24	−.0008710	−.0069894	.002727399	.002230453	.001656968	.001117937	.000684883	.000131764	.000013823	.000000025	1010101110
1000100001	24	−.0009080	−.0068932	.002722106	.002224065	.001652435	.001116240	.000685458•	.000133358	.000014265	.000000027	1010111010
1100010100	23	−.0010093	−.0061576	.002715219	.002233431	.001684769•	.001165563	.000739446	.000162234	.000020727	.000000068	0101111110
0101010001	23	−.0010093	−.0069476	.002701919	.002188131	.001609403	.001070044	.000652144	.000122606	.000012620	.000000022	1100101010
0011100100	22	−.0010721	−.0069911	.002701143	.002185617	.001605238	.001070044	.000667445	.000120609	.000012253	.000000021	1010101101
0111011000	23	−.0011365	−.0066763	.002694810	.002183235	.001611348	.001083405	.000663607	.000129635	.000014162	.000000030	1110011001
1000011010	24	−.0011988	−.0066259	.002683393	.002164650	.001590515	.001064351	.000634581	.000124886	.000013366	.000000026	1100111101
1000101100	23	−.0012000	−.0065674	.002672490	.002147063	.001570913	.001046480	.000634480	.000120299	.000012630	.000000024	1110101010
0011011100	21	−.0013006	−.0063254	.002657564	.002128535	.001557243	.001040900	.000635856	.000125353	.000014199	.000000036	

Table A

m = 7, n = 4 Rank Orders 201 thru 250 of 330 **C(7, 4) = 330** **1/C(7, 4) = .00303030**

z	w	c_1	c_2	d = .2	.4	.6	.8	1.0	1.5	2.0	3.0	z^{tc}
1001010001	22	−.0013999	−.0063816	.002637633	.002092493	.001513099	.000997418	.000599473	.000112248	.000011876	.000000025	0111010110
0110100100	22	−.0013999	−.0064989	.002635588	.002085077	.001502427	.000984800	.000587489	.000107010	.000010844	.000000019	1011101001
0101010100	22	−.0014902	−.0066997	.002632082	.002074841	.001484338	.000963688	.000567796	.000098957	.000009405	.000000014	1101010010
1010011000	22	−.0015271	−.0062554	.002622614	.002069951	.001488932	.000967570	.000567666	.000107666	.000011135	.000000021	1101101001
0110011000	22	−.0015271	−.0062219	.002616211	.002059991	.001478530	.000967968	.000578011	.000106641	.000011136	.000000023	0111011011
0011001000	22	−.0015271	−.0062652	.002615424	.002056937	.001474058	.000962449	.000572541	.000106641	.000010562	.000000019	1110011001
1100000100	23	−.0015894	−.0065071	.002611257	.002043904	.001453068	.000938207	.000550144	.000095015	.000008983	.000000013	0110111001
1010010010	23	−.0015894	−.0061411	.002605674	.002043090	.001460560	.000951833	.000565420	.000104412	.000010412	.000000019	1001011010
1001001010	23	−.0015894	−.0064087	.002601047	.002028517	.001437056	.000924566	.000540114	.000092250	.000008589	.000000012	1010110110
0011010010	23	−.0015894	−.0064267	.002600733	.002027526	.001435456	.000922714	.000538404	.000091576	.000008475	.000000012	0110101110
0101100100	22	−.0016908	−.0063219	.002583392	.002000819	.001407686	.000899614	.000522197	.000087984	.000008121	.000000011	1101100101
0101010010	22	−.0017278	−.0063424	.002576044	.001987841	.001392331	.000885139	.000510703	.000084473	.000007618	.000000010	1110111010
1100001001	22	−.0018180	−.0055358	.002572490	.002000198	.001424012	.000928064	.000553606	.000103256	.000011079	.000000025	0101111100
0101011000	22	−.0018180	−.0060583	.002563741	.001973413	.001381858	.000880163	.000509890	.000085974	.000008004	.000000011	1100111100
1000011000	23	−.0018803	−.0060642	.002563620	.001972981	.001381080	.000879173	.000508895	.000085504	.000007911	.000000010	1100101101
1000010100	23	−.0019173	−.0059678	.002553374	.001957484	.001364909	.000865386	.000498776	.000082762	.000007534	.000000010	0111001110
1001000110	23	−.0018894	−.0058567	.002548260	.001951406	.001364058	.000863291	.000498512	.000083528	.000007739	.000000011	1010101110
1010000110	20	−.0020191	−.0052834	.002538648	.001943368	.001369844	.000882289	.000520628	.000095054	.000010094	.000000023	0111110001
1001100001	21	−.0020814	−.0054539	.002524003	.001919599	.001333052	.000845322	.000489534	.000084094	.000008236	.000000015	0111010010
0011100010	21	−.0020814	−.0055635	.002522168	.001913979	.001324188	.000835211	.000480254	.000080345	.000007547	.000000012	1011100110
0101100010	21	−.0020814	−.0057679	.002518748	.001903553	.001307903	.000816905	.000463782	.000074149	.000006520	.000000008	1011110001
1010100001	21	−.0022086	−.0051515	.002504824	.001893116	.001307885	.000825757	.000476364	.000081045	.000007822	.000000013	0111010101
1000100100	21	−.0022086	−.0053273	.002502048	.001885131	.001296127	.000813313	.000465843	.000077708	.000007374	.000000013	0111111100
0101001010	21	−.0022086	−.0053819	.002501115	.001882210	.001291416	.000807819	.000460689	.000075515	.000006952	.000000010	0111110100
1000101000	21	−.0022709	−.0056165	.002497238	.001870545	.001273431	.000775866	.000442971	.000069066	.000005918	.000000007	1001110010
1010001001	22	−.0022709	−.0054942	.002487398	.001856211	.001258973	.000775903	.000434412	.000063858	.000005250	.000000006	1011101110
0101000110	21	−.0023079	−.0055174	.002480051	.001843366	.001244027	.000762105	.000437713	.000063806	.000005220	.000000006	1010101110
1100010001	22	−.0023981	−.0052393	.002467561	.001828120	.001231898	.000754860	.000420510	.000061314	.000005364	.000000006	1010111010
1001010010	20	−.0024093	−.0053646	.002463513	.001819082	.001219959	.000743028	.000420510	.000061314	.000005005	.000000006	0101110010
0011000110	21	−.0024995	−.0050478	.002451608	.001805441	.001210073	.000738027	.000409550	.000062098	.000005206	.000000006	1110101101
1100001000	21	−.0025364	−.0045322	.002452903	.001817925	.001233269	.000765673	.000433995	.000071765	.000006808	.000000012	0110111100
0101011000	21	−.0025364	−.0050426	.002444725	.001793974	.001197215	.000726494	.000400809	.000059696	.000004892	.000000005	1110110101
1100000110	21	−.0025364	−.0050802	.002444135	.001792289	.001194752	.000723911	.000398644	.000055021	.000004801	.000000005	1101111010
1010000011	22	−.0025988	−.0044001	.002443175	.001803808	.001218942	.000753621	.000426131	.000069192	.000006398	.000000010	1001111110
1000101001	21	−.0025988	−.0049185	.002434926	.001779805	.001180926	.000714987	.000392742	.000057750	.000004654	.000000005	1100110010
1000011010	21	−.0025988	−.0049484	.002434440	.001778373	.001180788	.000712646	.000390717	.000057067	.000005017	.000000005	1110101110
1010100100	22	−.0027260	−.0044419	.002418581	.001761746	.001170788	.000709620	.000390717	.000059340	.000005078	.000000008	1101010110
1001001100	20	−.0028901	−.0040523	.002394010	.001728684	.001140924	.000688293	.000379589	.000057965	.000004816	.000000008	0111101101
0101101000	21	−.0028901	−.0040956	.002393921	.001728684	.001140924	.000684384	.000376010	.000056527	.000004816	.000000006	0111110101
1001010100	21	−.0028901	−.0043375	.002389465	.001715454	.001121134	.000666929	.000361074	.000051541	.000004076	.000000004	1101111001
0011011000	20	−.0029270	−.0039493	.002388600	.001721247	.001133678	.000682343	.000375247	.000056663	.000004859	.000000007	1111011100
1001010100	21	−.0029894	−.0042763	.002371749	.001686974	.001090159	.000639892	.000341083	.000046345	.000003433	.000000003	1101101010
0101010010	20	−.0030907	−.0039910	.002357265	.001668410	.001074255	.000629157	.000335122	.000045785	.000003450	.000000003	1101110110
1000011001	20	−.0031166	−.0039751	.002352611	.001660803	.001065773	.000621545	.000329332	.000044174	.000003239	.000000003	1111100101
1100011010	20	−.0031959	−.0034752	.002345487	.001663458	.001080461	.000642678	.000350047	.000052138	.000004487	.000000007	0111011000
0101011010	20	−.0032179	−.0037176	.002341349	.001651111	.001062356	.000623466	.000333643	.000046604	.000003639	.000000004	1110101010
0101001100	20	−.0032179	−.0037685	.002337685	.001639609	.001048076	.000608934	.000321801	.000042801	.000003190	.000000003	1101101001
0101000110	20	−.0032179	−.0037514	.002337174	.001639609	.001046114	.000606958	.000320211	.000042678	.000003135	.000000003	1110101011
1001001010	20	−.0032803	−.0035777	.002328115	.001627463	.001034831	.000598347	.000314517	.000041456	.000002997	.000000003	1101100111
1001010100	21	−.0033172	−.0035706	.002321322	.001616420	.001022839	.000587988	.000306986	.000039616	.000002787	.000000002	1101011010

Table A

m = 7, n = 4   Rank Orders 251 thru 300 of 330   C(7, 4) = 330   1/C(7, 4) = .00303030

z	w	c_1	c_2	d = .2	.4	.6	.8	1.0	1.5	2.0	3.0	z^{tc}
01100011000	20	−.0033451	−.0033580	.002319406	.001617137	.001026841	.000593692	.000312495	.000041612	.000003072	.000000003	1110011001
11000011000	21	−.0034075	−.0029683	.002313547	.001613653	.001027598	.000597145	.000316492	.000043144	.000003279	.000000003	1010111110
10001011000	21	−.0034075	−.0031827	.002310358	.001604991	.001015520	.000585009	.000306720	.000040353	.000002928	.000000003	1100101110
10110001000	21	−.0034075	−.0032191	.002309837	.001603633	.001013712	.000583280	.000305401	.000040030	.000002895	.000000003	1100111010
10110000001	19	−.0036085	−.0022811	.002286537	.001581443	.001002380	.000582188	.000309831	.000043680	.000003574	.000000005	0111110010
00111010000	19	−.0036085	−.0023543	.002285425	.001578329	.000997866	.000577428	.000305772	.000042287	.000003351	.000000004	1011110010
01110010000	19	−.0036085	−.0025528	.002282386	.001569823	.000985626	.000564716	.000295172	.000038951	.000002881	.000000003	1011111100
10011001000	20	−.0036709	−.0026470	.002269375	.001547050	.000959696	.000541574	.000277938	.000034555	.000002360	.000000003	1011110010
10011010000	20	−.0037981	−.0023831	.002249689	.001519409	.000933291	.000521231	.000264609	.000031958	.000002116	.000000002	1011111010
11001000001	19	−.0038994	−.0014826	.002244184	.001524857	.000950364	.000543206	.000284713	.000038758	.000003074	.000000004	0111110100
01011001000	19	−.0038994	−.0019930	.002236675	.001504629	.000922302	.000515049	.000261986	.000032100	.000002188	.000000002	1110100101
01101001000	19	−.0038994	−.0020306	.002236135	.001503211	.000920396	.000513206	.000260560	.000031734	.000002146	.000000002	1101110010
01011010000	19	−.0039364	−.0018164	.002232378	.001500290	.000919543	.000513845	.000261722	.000032255	.000002217	.000000002	1111001010
01101010000	20	−.0039987	−.0018912	.002219724	.001478686	.000895529	.000492926	.000246525	.000028624	.000001816	.000000001	1101010010
01101000010	19	−.0040266	−.0016820	.002217669	.001478773	.000898290	.000497010	.000250426	.000029943	.000001990	.000000002	1110110010
10001011010	20	−.0040889	−.0014647	.002209221	.001468202	.000889112	.000490448	.000246348	.000029180	.000001913	.000000001	1010101110
10001100010	20	−.0041259	−.0012355	.002205632	.001465604	.000888546	.000491230	.000247522	.000029643	.000001970	.000000001	1010111110
10001010010	20	−.0041259	−.0014319	.002202855	.001458437	.000879057	.000482182	.000240611	.000027919	.000001782	.000000001	1100110010
10101000010	20	−.0041259	−.0014983	.002201915	.001456013	.000875855	.000479138	.000238298	.000027352	.000001721	.000000001	1100110110
00111110000	18	−.0042900	−.0003083	.002188695	.001452304	.000884878	.000494842	.000253883	.000032766	.000002453	.000000003	1011000011
11000111000	20	−.0044168	−.0002839	.002165558	.001415232	.000845254	.000461102	.000229628	.000026888	.000001764	.000000001	1001111000
10100011000	19	−.0044795	−.0004065	.002152359	.001392825	.000820888	.000440491	.000215142	.000023720	.000001444	.000000001	1011110010
01110001000	18	−.0046179	−.0007452	.002143153	.001394302	.000834128	.000458723	.000231853	.000029084	.000002151	.000000002	0111111010
01110010000	18	−.0046179	−.0002227	.002135793	.001375250	.000808738	.000434117	.000212730	.000023920	.000001511	.000000001	1101111010
01011100000	18	−.0046179	−.0002168	.002135728	.001375147	.000808625	.000434114	.000212711	.000023930	.000001513	.000000001	1011110010
10011000010	19	−.0046802	−.0001799	.002132747	.001355629	.000778839	.000416715	.000200540	.000021266	.000001241	.000000001	1011110110
01101100000	18	−.0047081	−.0003898	.002121625	.001355337	.000789813	.000419768	.000203433	.000022190	.000001355	.000000001	1110101001
01000111000	18	−.0047451	−.0005920	.002117629	.001351632	.000787711	.000419028	.000203407	.000022338	.000001378	.000000001	1011001001
10001110010	19	−.0048074	−.0008506	.002109719	.001342353	.000780147	.000413935	.000200414	.000021839	.000001331	.000000001	1100110010
11000110010	18	−.0048074	−.0007983	.002108978	.001340438	.000777609	.000411514	.000198565	.000021378	.000001281	.000000001	1011110100
10010101000	19	−.0048074	−.0005947	.002106220	.001333619	.000768957	.000403604	.000192772	.000020074	.000001152	.000000001	1110101010
10100101000	19	−.0048074	−.0005320	.002105369	.001331512	.000766285	.000401168	.000190994	.000019680	.000001114	.000000001	1101101010
10101001000	18	−.0049346	−.0010627	.002089419	.001313412	.000752369	.000392593	.000186542	.000019227	.000001096	.000000001	1100111010
01110000010	18	−.0051352	−.0019055	.002064244	.001284633	.000729904	.000378413	.000178914	.000018816	.000001043	.000000001	1101011110
01011000010	18	−.0051980	−.0021074	.002055589	.001273838	.000720765	.000372188	.000175316	.000017810	.000001010	.000000001	1011110010
01011110000	17	−.0052993	−.0027047	.002045174	.001264961	.000716590	.000371598	.000176307	.000018474	.000001101	.000000001	0111111110
11000110000	17	−.0054265	−.0036060	.002036707	.001264973	.000725424	.000383849	.000187299	.000021792	.000001517	.000000002	1011111010
01110110000	17	−.0054265	−.0030159	.002026289	.001239678	.000693639	.000354823	.000165881	.000016686	.000000950	.000000001	1110110010
10110000010	17	−.0054265	−.0029725	.002025735	.001238388	.000692091	.000353481	.000164945	.000016497	.000000933	.000000001	1011110010
11001000010	18	−.0054889	−.0032131	.002017539	.001228316	.000683431	.000347325	.000161142	.000015810	.000000869	.000000001	1011111100
10011010010	18	−.0054889	−.0030167	.002014988	.001222259	.000676041	.000340821	.000156552	.000014864	.000000783	.000000000	1110110010
10101010000	18	−.0054889	−.0029504	.002011425	.001220209	.000673544	.000338631	.000155013	.000014552	.000000755	.000000000	1110010110
10010110000	18	−.0055258	−.0032386	.002011202	.001218924	.000674272	.000340250	.000156555	.000014969	.000000797	.000000000	1101011110
10100110000	18	−.0055258	−.0034403	.001997594	.001200608	.000657650	.000328189	.000149153	.000013764	.000000703	.000000000	1011110010
10110010000	18	−.0056161	−.0043776	.001973583	.001174441	.000638189	.000316492	.000143166	.000013135	.000000672	.000000000	1010111110
11000011010	18	−.0058167	−.0049965	.001958677	.001158585	.000626732	.000309845	.000139912	.000012850	.000000662	.000000000	1100111110
10001011110	16	−.0059439	−.0058113	.001939692	.001138824	.000613837	.000302264	.000135509	.000012599	.000000676	.000000000	1111110100
01101100010	16	−.0061080	−.0059272	.001934566	.001133433	.000607475	.000298516	.000134319	.000012375	.000000652	.000000000	1011110010
01101100100	16	−.0061450	−.0059272	.001926675	.001122835	.000599469	.000292985	.000131002	.000011820	.000000604	.000000000	1011110100

Table A

m = 7, n = 4   Rank Orders 301 thru 330 of 330   C(7, 4) = 330   1/C(7, 4) = .00303030

z	w	c_1	c_2	d = .2	.4	.6	.8	1.0	1.5	2.0	3.0	z^{tc}
10110000100	17	−.0062073	.0059329	.001923449	.001115698	.000591083	.000285868	.000126149	.000010897	.000000526	.000000000	11011110010
10101001000	17	−.0062975	.0062095	.001910865	.001099903	.000577666	.000276709	.000120845	.000010141	.000000474	.000000000	11101010010
10011001000	17	−.0063345	.0064571	.001907349	.001097088	.000576357	.000276417	.000120964	.000010235	.000000484	.000000000	11100101010
10001010000	17	−.0064982	.0072342	.001887866	.001075871	.000560514	.000266807	.000115973	.000009677	.000000453	.000000000	11011101010
01110100000	17	−.0066254	.0078158	.001872453	.001058763	.000547476	.000258729	.000111683	.000009167	.000000453	.000000000	11011011100
10011100000	15	−.0068265	.0090747	.001852235	.001041032	.000537652	.000254999	.000110999	.000009501	.000000471	.000000000	11110101100
11100000010	16	−.0068888	.0091587	.001842186	.001026971	.000524599	.000245374	.000105019	.000008499	.000000390	.000000000	11101110100
10101010000	16	−.0070160	.0100833	.001830776	.001018450	.000521082	.000244950	.000105689•	.000008832	.000000424	.000000000	10111111000
10110010000	16	−.0070160	.0095649	.001824718	.001005447	.000506676	.000233387	.000098214	.000007574	.000000329	.000000000	11011110010
10101100000	16	−.0070160	.0095350	.001824373	.001004720	.000505888	.000232769	.000097826	.000007514	.000000325	.000000000	11101010010
11010100100	16	−.0072166	.0106733	.001802633	.000983240	.000491390	.000224859	.000094151	.000007214	.000000313	.000000000	11011110100
11001001000	16	−.0073069	.0110233	.001790981	.000969595	.000480512	.000217855	.000090309	.000006729	.000000283	.000000000	11101101100
11000110000	14	−.0073438	.0113012	.001787752	.000967253	.000479574	.000217743	.000090472•	.000006802	.000000290	.000000000	11110011100
01111000000	15	−.0075079	.0126179	.001774386	.000959018	.000477942	.000219422•	.000092717	.000007447	.000000351	.000000000	11110011010
10101100000	15	−.0076975	.0130753	.001746735	.000923644	.000444719	.000198808	.000080817	.000005763	.000000234	.000000000	11111001010
11100100000	15	−.0077344	.0132190	.001741996	.000918184	.000443463	.000196139	.000079396	.000005598	.000000224	.000000000	11111010010
11000010000	15	−.0080253	.0151030	.001712904	.000891650	.000426951	.000187856	.000075877	.000005379	.000000219•	.000000000•	11011111000
11001010000	15	−.0080253	.0148354	.001710009	.000885890	.000421026	.000183435	.000073316	.000005003	.000000195	.000000000	11011101010
11010010000	15	−.0080253	.0148174	.001709810	.000885483	.000420466	.000183112	.000073019	.000004974	.000000193	.000000000	11110101010
10110100000	14	−.0084159	.0170679	.001667718	.000843820	.000392211	.000167340	.000065490	.000004281	.000000161	.000000000	11101010010
11001100000	14	−.0087068	.0187995	.001636926	.000813808	.000372028	.000156251	.000060244	.000003806	.000000139	.000000000	11111001100
11010100000	14	−.0087438	.0189580	.001632386	.000808813	.000368314	.000154029	.000059114	.000003688	.000000133	.000000000	11110101100
10111000000	14	−.0088340	.0196873	.001624798	.000803199	.000365653	.000153137	.000058943	.000003723•	.000000136	.000000000	11101111000
10101010000	13	−.0090974	.0213087	.001597452	.000777026	.000348351	.000143783	.000054585	.000003341	.000000119	.000000000	11110110010
10110010000	13	−.0094252	.0232488	.001562664	.000743100	.000325558	.000131287	.000048692	.000002813	.000000095	.000000000	11110111000
11011000000	13	−.0095524	.0241829	.001550944	.000733144	.000319718	.000128487	.000047536	.000002739	.000000092	.000000000	11111010010
11100100000	12	−.0101067	.0279277	.001496683	.000684042	.000288947	.000112676	.000040519	.000002189	.000000070	.000000000	11101111000
11010010000	12	−.0102339	.0288246	.001484587	.000673393	.000282445	.000109417	.000039106	.000002085	.000000066	.000000000	11111011000
11011000000	11	−.0109154	.0338507	.001421765	.000619663	.000250511	.000093806	.000032493	.000001618	.000000048	.000000000	11111011000
11110000000	10	−.0116339	.0395960	.001359365	.000569130	.000221945	.000080469	.000027077	.000001270	.000000036	.000000000	11111110000

109

Table A

m = 6, n = 5 Rank Orders 1 thru 50 of 462 C(6, 5) = 462 1/C(6, 5) = .00216450

z	w	c_1	c_2	d = .2	.4	.6	.8	1.0	1.5	2.0	3.0	z^{tc}
00000011111	45	.0087967	.0319033	.004710814	.009572649	.018209476	.032515197	.054658438	.156928170	.331555773	.738711759	00000011111
00000101111	44	.0083099	.0277775	.004491574	.008652043	.015500250	.025876756	.040337846	.091939499	.143147298	.127563706	00000101111
00000110111	43	.0078231	.0239466	.004285798	.007845843	.013300914	.020916421	.030564588	.057923453	.072351003	.036902953	00000110111
00000111011	43	.0077967	.0237461	.004274631	.007804631	.013192177	.020679900	.030116446	.056529241	.069816309	.034621996	00000111011
00001001111	42	.0073099	.0202080	.004081563	.007095418	.011388744	.016905173	.023245888	.037244792	.038816309	.012945098	00001001111
00001010111	42	.0073099	.0201357	.004078656	.007073894	.011306037	.016673185	.022716770	.036123898	.034131049	.009781113	00001010111
00001011011	42	.0072191	.0195782	.004046690	.006972057	.011086678	.016297417	.022184419	.034671741	.034942804	.010931776	00001011011
00001100111	41	.0068232	.0168404	.003895045	.006437269	.009782474	.013686802	.017653691	.023585576	.019570975	.003581478	00001100111
00001101011	41	.0067968	.0166318	.003883948	.006394785	.009671127	.013449487	.017219012	.022413020	.017872414	.002842626	00001101011
00001110011	41	.0067323	.0163536	.003865602	.006351833	.009617691	.013441297	.017368498	.023683300	.020660124	.004763646	00001110011
00010001111	41	.0067323	.0162184	.003860514	.006316636	.009491669	.013112813	.016674161	.021321571	.016685841	.002549823	00100101011
00010010111	41	.0064982	.0149091	.003782247	.006079948	.009003767	.012304237	.015543482	.020013096	.016469635	.003374820	01000011011
00010011011	40	.0063100	.0135570	.003708717	.005811651	.008358669	.011017124	.013333394	.014904185	.010037604	.001087037	00110110111
00010100111	40	.0062455	.0131712	.003686289	.005745231	.008202423	.010738529	.012905925	.014162774	.009356612	.000960299	00101101011
00010101011	40	.0062191	.0130437	.003678162	.005722715	.008160480	.010676657	.012830520	.014113384	.009373414	.000990298	00101110011
00010110011	40	.0062191	.0129808	.003675941	.005708350	.008112628	.010561221	.012605890	.013509835	.008607299	.000787949	00011110111
00100001111	40	.0060114	.0120590	.003614831	.005549493	.007844005	.010225178	.012314846	.014043449	.010214978	.001633568	10000001011
00100010111	40	.0060114	.0118649	.003608074	.005506298	.007701261	.009882169	.011647197	.012204183	.007722626	.000741906	00110011011
00100011011	39	.0058232	.0107772	.003543935	.005306581	.007274388	.009139410	.010536298	.010407374	.006183568	.000523746	00110101011
00100100111	39	.0057323	.0102860	.003514131	.005217058	.007089922	.008829820	.010089524	.009743299	.005566745	.000457003	00110110011
00100101011	39	.0057324	.0102196	.003511899	.005203336	.007046517	.008730484	.009906365	.009314079	.005183063	.000361969	00101101111
00100110011	39	.0057324	.0101541	.003480951	.005189585	.007002759	.008629891	.009720370	.008878958	.004710996	.000276117	00110101101
00101000111	39	.0056415	.0096945	.003480951	.005107012	.006840909	.008373444	.009373887	.008462179	.004466425	.000266597	00110110101
00101001011	39	.0055246	.0091308	.003444862	.005005856	.006647792	.008110608	.008983836	.008034495	.004250183	.000267913	00111001011
00101010011	39	.0054982	.0090621	.003439307	.004998319	.006650626	.008110600	.009075794	.008315868	.004572927	.000325902	00111010011
00101100011	39	.0054982	.0087652	.003435330	.004974574	.006577711	.007948697	.008788311	.007689887	.003939612	.000222449	01000111011
01000000111	40	.0053628	.0087652	.003408088	.004942613	.006611867	.008117319	.009345247	.009425496	.006114071	.000794349	10000011011
01000001011	38	.0052456	.0076878	.003355498	.004744783	.006125560	.007227385	.007801631	.006448836	.003140388	.000166945	00111100011
01000010011	38	.0052456	.0076250	.003353457	.004732693	.006088801	.007146716	.007659374	.006154253	.002858695	.000127412	00111010101
01000100011	38	.0052192	.0075036	.003345485	.004710800	.006047766	.007084462	.007578692	.006065621	.002809472	.000124927	00111011001
00101101011	38	.0051547	.0071184	.003323205	.004641078	.005899425	.006829795	.007206134	.005515935	.002402075	.000088298	00110101011
01000101011	38	.0050114	.0065753	.003283563	.004543076	.005737976	.006621875	.006989556	.005453009	.002496803	.000115082	00111011101
01001000011	38	.0050114	.0065717	.003283485	.004542820	.005737790	.006622676	.006992949	.005469113	.002519550	.000119974	01010010011
00110001011	38	.0050114	.0064484	.003279586	.004459437	.005571936	.006354063	.006619058	.004964589	.002113233	.000073187	00110100101
00110010011	38	.0049206	.0061131	.003254707	.004449436	.005543420	.006295473	.006481070	.004797128	.002161032	.000085953	01000100101
00110100011	38	.0049206	.0060543	.003252029	.004520037	.005784558	.006840664	.007487781	.007487781	.002028783	.000073939	01010010101
01000110011	39	.0048760	.0064799	.003258691	.004520037	.005650062	.006651193	.006982157	.005703971	.002822126	.000177130	10000111110
01001001011	39	.0048760	.0062221	.003250704	.004774504	.005318138	.005989327	.006173575	.004557132	.002282157	.000166664	00111101011
00110101011	37	.0047324	.0053301	.003200364	.004317060	.005253481	.005856762	.005955249	.004176464	.001990261	.000086702	00111101101
00110110011	37	.0047324	.0051949	.003196256	.004294309	.005253481	.005856762	.005955249	.004176464	.001833568	.000057751	00111110011
00111000011	37	.0046680	.0048373	.003174920	.004230431	.005123589	.005643855	.005658181	.003788797	.001431247	.000038672	01010101011
01010001011	37	.0046416	.0047342	.003167505	.004211620	.005091281	.005599365	.005600442	.003748405	.001417990	.000038805	01100110011
01010010011	37	.0045246	.0043565	.003136995	.004140879	.004982939	.005471553	.005487993	.003748273	.001492206	.000051292	00111101011
01010100011	37	.0045246	.0042348	.003133345	.004120950	.004927193	.005359238	.005306535	.003449286	.001267999	.000032814	01011010011
01100001011	37	.0044982	.0041240	.003125684	.004100718	.004890684	.005380894	.005239726	.005239726	.001232711	.000031163	01010100101
01001010011	37	.0044338	.0040876	.003113902	.004088608	.004904608	.005380894	.005405288	.005405288	.001541016	.000059963	01010101101
01001100011	37	.0044338	.0038527	.003105527	.004051128	.004800987	.005174357	.005074864	.003213827	.001145869	.000027824	01100011011
01010101011	37	.0044338	.0037912	.003105150	.004041556	.004774962	.005123530	.004995494	.003095606	.001146664	.000022826	01010011011
01011000011	38	.0043893	.0039861	.003103290	.004041556	.004732800?	.005342599	.005367248	.003724440	.001528812	.000057751	10000111110
01010110011	38	.0043620	...	.003102025	.004090814?	.004900817	.005418394	.005505610	.003988058	.001746827	.000080428	00101100111110

z	w	c_1	c_2	d = .2	.4	.6	.8	1.0	1.5	2.0	3.0	z^{tc}
0010011110	38	.0043629	.0038498	.003094841	.004041185	.004823367	.005262663	.005256289	.003576409	.001426441	.000048880	1000011011
0111001101	36	.0042456	.0031987	.003056351	.003928445	.004600093	.004911559	.004786080	.003027459	.001106985	.000030916	1000011101
0100011101	37	.0041996	.0029474	.003041320	.003884048	.004512844	.004722885	.004599044	.002805206	.000975956	.000023558	0110001111
0110010011	36	.0041548	.0027459	.003027913	.003847897	.004445036	.004670797	.004662734	.002662734	.000897637	.000016431	0100111011
0010100111	36	.0041548	.0026735	.003025866	.003837379	.004417432	.004618784	.004389315	.002556280	.000833213	.000016431	0011110011
0101001011	36	.0040115	.0023249	.002991524	.003765356	.004320269	.004523224	.004326037	.002625261	.000930525	.000016431	0011110101
0001101011	36	.0040115	.0021309	.002986008	.003736840	.004244806	.004379514	.004106764	.002313209	.000729265	.000013463	0011110110
0010011101	36	.0039470	.0018896	.002967994	.003691030	.004165263	.004267664	.003972338	.002190651	.000672626	.000011477	0101010011
0100101011	36	.0039470	.0018257	.002966346	.003682662	.004143583	.004227386	.003912613	.002113036	.000628434	.000009663	0101010101
0010011011	36	.0039206	.0019063	.002964035	.003689586	.004177522	.004305262	.004041364	.002304578	.000746001	.000015138	0110100011
0100011011	36	.0039206	.0018032	.002961182	.003675247	.004140692	.004237299	.003941075	.002174622	.000671088	.000011799	0101100111
0010011001	36	.0039206	.0017332	.002959218	.003665241	.004114671	.004188752	.003868753	.002079321	.000615853	.000009414	0011111101
0000111101	37	.0038761	.0020277	.002959568	.003697767	.004224672	.004416927	.004229444	.002599436	.000942591	.000027591	0110100101
1000011011	37	.0038761	.0019429	.002957432	.003687328	.004198467	.004369420	.004160200	.002510307	.000890051	.000024780	1001001111
0100110011	37	.0038761	.0018343	.002954180	.003670366	.004153065	.004281768	.004024350	.002307690	.000751888	.000015298	0011100111
0010011110	37	.0037853	.0016637	.002933638	.003630098	.004103508	.004240382	.004008496	.002371517	.000819609	.000020874	0010011110
0100010111	36	.0037129	.0015411	.002930316	.003613669	.004061780	.004163883	.003895823	.002223175	.000730072	.000016048	1000011101
0101001101	36	.0036680	.0009999	.002902924	.003527016	.003883865	.003877979	.003512688	.001797120	.000505949	.000006936	1000011011
0100110101	36	.0036680	.0009280	.002893166	.003509185	.003864814	.003642729	.003518980	.001842160	.000541713	.000008827	0111001011
0010110011	35	.0036416	.0008169	.002885528	.003489240	.003829524	.003817494	.003458323	.001787612	.000517672	.000008119	1001010111
0011100011	35	.0035247	.0004478	.002855002	.003416986	.003713159	.003666063	.003290963	.001665371	.000473355	.000007209	0101100111
0100101101	35	.0034338	.0004207	.002838202	.003394401	.003706160	.003697983	.003375644	.001835062	.000588196	.000013658	0001111101
0101001110	35	.0034338	.0001638	.002831340	.003366121	.003622956	.003549559	.003160395	.001568359	.000437218	.000005548	0101011011
0110001101	35	.0034338	.0001193	.002830151	.003355257	.003608663	.003523095	.003124111	.001525281	.000414449	.000004787	0100110111
0010011011	35	.0034338	.0000690	.002828462	.003343743	.003592842	.003494578	.003083798	.001478790	.000390962	.000004762	0011100111
0101010101	35	.0034338	.0000555	.002828119	.003347151	.003588837	.003488349	.003075554	.001471041	.000388025	.000003804	0110101011
1000011101	36	.0033893	−.0000119	.002826652	.003338348	.003567018	.003449623	.003020771	.001408152	.000356686	.000011944	1000011011
0011001101	36	.0033893	.0003072	.002827135	.003368834	.003664887	.003642729	.003310393	.001774563	.000556564	.000011944	0001111110
0100110110	36	.0033893	.0001218	.002822199	.003322170	.003544873	.003535503	.003156282	.001581939	.000446960	.000006715	0011111101
0001110110	36	.0033629	.0000308	.002815088	.003317423	.003572423	.003575930	.003109952	.001542800	.000430113	.000006198	1001010111
0010111001	35	.0033430	−.0001828	.002806003	.003296932	.003512622	.003395249	.002978942	.001411551	.000370011	.000004496	0110011011
0000111110	35	.0032985	−.0000424	.002803988	.003315376	.003580030	.003536432	.003196499	.001697643	.000532097	.000011797	1010001111
0010001110	36	.0032985	−.0000957	.002800202	.003296484	.003532642	.003447548	.003065321	.001568359	.000437548	.000006397	0101011010
0100110110	36	.0032261	−.0002237	.002796876	.003280761	.003480761	.003381233	.002973055	.001525338	.000429125	.000006397	1001001111
0100101110	35	.0031997	−.0006527	.002772860	.003212364	.003366751	.003193429	.002742473	.001216960	.000377201	.000004689	1001010101
0010101010	35	.0031548	−.0007234	.002766307	.003197713	.003347151	.003166680	.002715196	.001201562	.000293101	.000002798	0101011011
1000110011	35	.0031548	−.0007082	.002758720	.003189612	.003347616	.003191589	.002766713	.001281703	.000333648	.000004286	0110101011
0100001101	36	.0030643	−.0007521	.002741143	.003161266	.003321848	.003182361	.002781383	.001336096	.000368347	.000005670	0100111110
0100101110	36	.0030643	−.0008158	.002739553	.003154041	.003305014	.003154103	.002732049	.001296429	.000349610	.000005073	1000111101
0010101110	34	.0029471	−.0012713	.002706828	.003072029	.003166202	.002965559	.002526134	.001122916	.000281158	.000003367	0110101011
0101000111	34	.0029471	−.0013744	.002704200	.003059883	.003137544	.002917048	.002460551	.001055004	.000250091	.000002508	0111001011
0101001011	34	.0029471	−.0014444	.002701493	.003051493	.003117616	.002883144	.002414573	.001017473	.000228725	.000001972	1011001101
0100100111	34	.0029207	−.0014750	.002696879	.003041642	.003106860	.002875687	.002413367	.001018729	.000236678	.000002245	0101011010
1001010011	35	.0028761	−.0015339	.002695370	.003034633	.003090280	.002847611	.002375486	.000980137	.000219631	.000001836	0101011011
0011011001	34	.0028761	−.0012035	.002689050	.003061383	.003173015	.003003867	.002599838	.001232289	.000341904	.000005785	0001110110
0110000111	34	.0028562	−.0014613	.002684931	.003023919	.003101082	.002881010	.002431607	.001049352	.000251721	.000002606	0010100111
0100101011	34	.0028562	−.0014833	.002680069	.003032919	.003093842	.002877143	.002433439	.001061576	.000259823	.000002891	0010111010
1001001101	36	.0030643	−.0015656	.002674654	.003001827	.003042628	.002792014	.002320519	.000950591	.000212066	.000001770	0110111001
0010101110	34	.0028117	−.0016773	.002671613	.003002050	.003061574	.002837932	.002392004	.001032228	.000248519	.000002601	0101101010
0010100111	35	.0028117	−.0016882	.002671613	.002988343	.003003011	.002785752	.002323089	.000964994	.000219733	.000001937	1001010111

Table A

$C(6, 5) = 462$

z	w	c_1	c_2	d = .2	.4	.6	.8	1.0	1.5	2.0	3.0	z^{tc}
00001011010	35	.0027853	−.0016084	.002668833	.002990843	.003047698	.002825330	.002384047	.001037179	.000254058	.000002851	10100101011
00101011001	35	.0027853	−.0016146	.002668634	.002989767	.003044837	.002820012	.002376304	.001028082	.000249607	.000002727	01100110110
00100110110	35	.0027853	−.0017609	.002664992	.002973288	.003006715	.002756669	.002292181	.000944653	.000213139	.000001833	10010111011
00100110101	34	.0027129	−.0021555	.002643900	.002919322	.002914980	.002633189	.002113050	.000843527	.000179075	.000001330	01001101011
01001011010	34	.0027129	−.0019383	.002642096	.002911187	.002896332	.002602648	.002113050	.000806862	.000164672	.000001075	00111000111
00111000011	34	.0026681	−.0022609	.002639312	.002923692	.002853781	.002564465	.002249049	.000952107	.000228423	.000002558	01100011101
01001011001	34	.0026220	−.0017353	.002622930	.002920537	.002974335	.002777666	.002380759	.000813137	.000172840	.000001321	11000001111
00001111100	35	.0025775	−.0021985	.002627570	.002869337	.002856320	.002581416	.002118685	.000850197	.000188400	.000001606	10010111010
01000101101	35	.0025775	−.0022651	.002616168		.002862423	.002557007	.002087656	.000822736	.000177773	.000001409	01000101110
00100101110	35	.0025775		.002614575		.002840986		.002087656				
00011110010	33	.0024339	−.0024495	.002583779	.002768168	.002487662	.002036516		.000825478	.000190328	.000001989	00110101110
01010100101	33	.0024339	−.0025845	.002578150	.002779518	.002400971	.002026054	.001826574	.000726574	.000151047	.000001157	10011100111
00110110010	33	.0024339	−.0027459	.002576641	.002772808	.002376318	.001894200	.000697446	.000139491	.000000932		10011000111
00011001110	34	.0023894	−.0026462	.002570779	.002770363	.002407400	.001943219	.000750883	.000161655	.000001363		01101001011
01100010101	33	.0023694	−.0028617	.002561984	.002742708	.002329554	.001499945	.000676139	.000134609	.000000898		10100011011
01101010010	33	.0023430	−.0028112	.002555819	.002740505	.002345792	.001876512	.000706218	.000147340	.000001152		10101101011
01100100111	33	.0023430	−.0029329	.002555417	.002727973	.002302244	.001821721	.000659176	.000129732	.000000843		01101101010
10010100011	33	.0022985	−.0025292	.002556819	.002757148	.002716295	.002302244	.002017526	.000851394	.000129732	.000002961	00011110010
10010100011	34	.0022985	−.0028100	.002550026	.002727210	.002648661	.002336862	.001873796	.000712329	.000150708	.000001212	01010010111
00110101010	34	.0022985	−.0028498	.002549121	.002723481	.002640815	.002325026	.001859553	.000701407	.000147094	.000001167	10100100111
10010101001	34	.0022985	−.0028641	.002548728	.002721567	.002636144	.002316956	.001848544	.000690181	.000142247	.000001059	00110101110
00110101001	34	.0022985	−.0029143	.002547558	.002716620	.002625448	.002300347	.001827926	.000672903	.000135862	.000000948	10001110011
01010101010	34	.0022985	−.0030049	.002545377	.002707105	.002604998	.002266690	.001785264	.000635826	.000121986	.000000717	10011001101
01010010101	34	.0022261	−.0032724	.002525750	.002660155	.002529512	.002172283	.001685349	.000574347	.000104606	.000000544	01011000111
00100111010	34	.0022077	−.0030983	.002526376	.002671835	.002561021	.002225570	.001754035	.000632346	.000124878	.000000826	10101101011
01100101001	34	.0021997	−.0033513	.002524716	.002505686	.002142896	.001654770	.000555978	.000099453	.000000493		00110101011
01100100101	33	.0021353	−.0033072	.002508093	.002630428	.002497827	.002148418	.001674461	.000585194	.000111629	.000000685	00110110011
01001001101	33	.0021353	−.0034423	.002504938	.002617114	.002469231	.002104500	.001620792	.000542812	.000099295	.000000493	10101101011
10010010011	33	.0020907	−.0033528	.002498659	.002612116	.002473325	.002121550	.001648934	.000571439	.000107609	.000000627	01011001101
10010001110	34	.0020907	−.0034165	.002497205	.002606106	.002460659	.002102435	.001625951	.000553937	.000101880		01011000111
00010111110	34	.0020643	−.0030513	.002500749	.002633925	.002529819	.002216567	.001848544	.000681195	.000149482	.000001396	11000010111
10000110101	34	.0020643	−.0033912	.002492866	.002600684	.002458131	.002105466	.001634668	.000566070	.000106897	.000000637	01010010101
01001000011	34	.0020643	−.0034675	.002491117	.002593418	.002442733	.002082083	.001606362	.000544073	.000099510	.000000537	00111100011
01011000011	32	.0019471	−.0036016	.002466302	.002546731	.002384497	.002005352	.001561943	.000536609	.000102354	.000000675	11010111001
10000011110	35	.0019471	−.0036604	.002464915	.002546731	.002384497	.002005352	.001536837	.000515573	.000094654	.000000545	01101110011
01001010101	32	.0019290	−.0033311	.002469012	.002566297	.002431581	.002101269	.001657043	.000615963	.000131453	.000001214	10000111011
01101110001	32	.0018563	−.0037700	.002445487	.002503516	.002323498	.001956093	.001494726	.000501624	.000093388	.000000586	00110110001
01011010101	32	.0018563	−.0037736	.002445346	.002502688	.002321222	.001951837	.001488590	.000494872	.000093391	.000000520	01110110010
00110101010	34	.0018563	−.0038969	.002442588	.002491568	.002298188	.001917847	.001488652	.000466166	.000081496	.000000418	10101010011
00011010110	34	.0018117	−.0037867	.002436739	.002488153	.002305463	.001939403	.001481984	.000498817	.000093441	.000000597	01101000111
10001100011	33	.0018117	−.0037929	.002436544	.002487142	.002302916	.001934948	.001475906	.000492818	.000090994	.000000549	00110110010
00110100110	33	.0018117	−.0039392	.002433211	.002487364	.002273862	.001891034	.001422969	.000452012	.000077334	.000000362	00110110011
00110100011	33	.0017853	−.0038737	.002429767	.002471563	.002278799	.001905087	.001444660	.000473554	.000085316	.000000480	10101101010
00110100110	33	.0017853	−.0039963	.002426984	.002460115	.002254782	.001868900	.001401412	.000440799	.000074579	.000000341	01101110110
10101000011	33	.0017209	−.0038339	.002418584	.002455532	.002266600	.001910410	.001451365	.000489842	.000092565	.000000605	10101010011
00100110101	32	.0017209	−.0040917	.002412717	.002431757	.002216380	.001827139	.001362686	.000422983	.000070612	.000000313	00110101011
01010001101	32	.0017130	−.0042477	.002407852	.002415712	.002186381	.001785669	.001316459	.000393413	.000062489	.000000243	01101001101
01001010101	32	.0017130	−.0043112	.002394372	.002399400	.002154394	.001753463	.001289219	.000384036	.000061141	.000000244	01101011101
10001101001	32			.002389449	.002382315	.002145708	.001746340	.001284696	.000384047	.000061506	.000000249	

m = 6, n = 5 Rank Orders 151 thru 200 of 462 C(6, 5) = 462 1/C(6, 5) = .00216450

z	w	c_1	c_2	d = .2	.4	.6	.8	1.0	1.5	2.0	3.0	z^{tc}
11000000111	33	.0015776	−.0035977	.002396862	.002432317	.002263210	.001932139	.001514565	.000570590	.000128915	.00000 1547	0001111100
00110111001	33	.0015776	−.0039772	.002388386	.002397707	.002190518	.001821772	.001379744	.000458691	.000085903	.00000 0563	1100011011
10100001101	33	.0015776	−.0042625	.002382068	.002373205	.002138344	.001744858	.001289192	.000392480	.000061829	.00000 0286	0100110011
01010010110	33	.0015776	−.0043165	.002380896	.002367696	.002129120	.001731651	.001274136	.000382481	.000064668	.00000 0235	1000111010
10010010110	33	.0015776	−.0043985	.002379046	.002360154	.002126299	.001726950	.001268016	.000377128	.000059934	.00000 0199	0101010101
00100111100	33	.0014867	−.0041627	.002367129	.002352449	.002113485	.001708497	.001246886	.000363016	.000055941	.00000 0400	1100001011
00100011110	33	.0014867	−.0043295	.002363518	.002335838	.002124753	.001703558	.001302153	.000414013	.000073075	.00000 0298	1010010110
01000011010	33	.0014867	−.0044264	.002361408	.002329934	.002079538	.001679355	.001255237	.000382014	.000063588	.00000 0239	1010001110
01000010011	34	.0014422	−.0042793	.002356125	.002328126	.002088296	.001700623	.001257530	.000388182	.000065908	.00000 0324	1000101110
01010000111	31	.0013695	−.0044391	.002339149	.002292921	.002039776	.001647746	.001209445	.000368793	.000062825	.00000 0338 ●	0011110001
00110100101	31	.0013695	−.0045608	.002336480	.002288311	.002018258	.001616453	.001173130	.000343230	.000054935	.00000 0241	0111001101
00110110001	31	.0013431	−.0045740	.002331204	.002272785	.002006601	.001604994	.001163582	.000339880	.000054374	.00000 0239	0110110001
10110100010	32	.0012986	−.0044440	.002325602	.002269602	.002012905	.001623036	.001190435	.000363860	.000062426	.00000 0343	1011000101
00110101010	32	.0012986	−.0045821	.002322594	.002257781	.001988950	.001588405	.001150478	.000336129	.000053987	.00000 0243	1011010110
01010101001	32	.0012341	−.0047101	.002319811	.002246823	.001966950	.001556770	.001114224	.000311577	.000046795	.00000 0168	1001110011
00110011001	32	.0012262	−.0047779	.002306195	.002220805	.001933216	.001521578	.001083056	.000298733	.000044234	.00000 0154	1010010110
00110010011	31	.0012077	−.0048493 ●	.002303203	.002212659	.001919618	.001548259	.001065260	.000289064	.000041941	.00000 0139	0101110011
01010001011	32	.0012077	−.0048168	.002300385	.002209262	.001917693	.001504828	.001067759	.000292031	.000051987	.00000 0146	1010110100
01001110001	31	.0011353	−.0048884	.002285216	.002180647	.001811235	.001467651	.001035758	.000280105	.000040810	.00000 0139	1110001101
01010100101	31	.0011353	−.0049547	.002283817	.002175313	.001870899	.001453340	.001020005	.000270569	.000038355	.00000 0121	0101010101
00111011000	31	.0011353	−.0050202	.002282445	.002170121	.001860923	.001439669	.001005135	.000261895	.000036227	.00000 0107	0101010111
10011000011	32	.0010908	−.0046010	.002282860 ●	.002189669	.001910711	.001517374	.001097137	.000324691	.000054156	.00000 0284	1101001101
01011000101	32	.0010908	−.0049410	.002275621	.002161705	.001855581	.001455647	.000991634	.000256707	.000037766	.00000 0117	1001001110
10011001010	32	.0010644	−.0050172	.002274023	.002155647	.001843904	.001423359	.000988909	.000257394	.000035114	.00000 0098	1010010110
01011001001	32	.0010644	−.0050638	.002269379	.002143498	.001827225	.001418158	.000974655	.000249075	.000033522	.00000 0103	1001011010
00110101001	31	.0010445	−.0050278	.002268051	.002145064	.001827194	.001409228	.000982316	.000256577	.000035903	.00000 0090	0110011011
11000001011	32	.0010000	−.0045239	.002267143	.002165724	.001887013	.001499915	.001087868	.000327106	.000055930	.00000 0311	0011111100
00101011100	32	.0010000	−.0048405	.002260503	.002140379	.001837482	.001430295	.001009567	.000275658	.000040977	.00000 0149	1100101011
01001100101	32	.0010000	−.0049079	.002259153	.002135479	.001828435	.001418356	.000997041	.000268954	.000039470	.00000 0141	1001110011
10001011010	33	.0010000	−.0049413	.002258461	.002132893	.001823504	.001411623	.000989721	.000264627	.000038373	.00000 0132	1001100011
01001011010	33	.0010000	−.0050128	.002256929	.002126952	.001811792	.001395126	.000971245	.000252940	.000035228	.00000 0106	1001101110
10001001110	33	.0009554	−.0049325	.002254973	.002119589	.001797690	.001375824	.000950240	.000240593	.000032155	.00000 0085	1010010110
01110010001	30	.0009290	−.0049290	.002250126	.002118713	.001806744	.001395500	.000976286	.000259360	.000037224	.00000 0123	1001011010
01101100001	30	.0008563	−.0048234	.002244740	.002108713	.001794355	.001383095	.000965778	.000255577	.000036598	.00000 0121	0011111110
01101110000	31	.0008563	−.0050175	.002233723	.002092636	.001789377	.001390194	.000983481	.000275887	.000043731	.00000 0137	0111011010
10011010001	31	.0008118	−.0050010	.002229731	.002081217	.001760864	.001351041	.000940489	.000249302	.000040731	.00000 0139 ●	0111111110
00110100011	31	.0008118	−.0051236	.002221584	.002067792	.001745943	.001337810	.000930649	.000246931	.000036436		0111110011
01000111100	31	.0007658	−.0050164	.002219030	.002058137	.001727322	.001312061	.000902245	.000229367	.000031530	.00000 0097	1011000011
01010011100	32	.0007210	−.0050891	.002212443	.002051006	.001724520	.001315303 ●	.000910125 ●	.000237066	.000033777	.00000 0115	1100011011
00101011010	31	.0007210	−.0051689	.002202533	.002031601	.001699214	.001289185	.000887629	.000229101	.000032639	.00000 0117 ●	1001011010
10010100110	31	.0006486	−.0052826	.002200855	.002016812	.001686759	.001271810	.000868285	.000216892	.000029292	.00000 0070	1010100101
01110100010	30	.0006486	−.0053082	.002198559	.001997059	.001671170	.001251112	.000846438	.000204992	.000026543	.00000 0068	1011101010
01101010010	30	.0006222	−.0053711	.002184330	.001987059	.001641172	.001222519	.000823470	.000197854	.000025542	.00000 0060 ●	0110011011
00101110010	30	.0005776	−.0053626	.002183065	.001979078	.001632682	.001211327	.000811750	.000191615	.000024139	.00000 0060 ●	0111011010
10010101001	31	.0005776	−.0049322	.002178220	.001996352	.001663843	.001203533	.000805991	.000190220	.000024005	.00000 0161	1101010011
00110100101	31	.0005776	−.0053954	.002178442	.001978430	.001667398	.001267483	.000788869	.000235567	.000035880	.00000 0050	1101010110
				.002169045	.001961775	.001601474	.001180078	.000784988	.000181214	.000022142		

113

Table A

m = 6, n = 5   Rank Orders 201 thru 250 of 462   C(6, 5) = 462   1/C(6, 5) = .00216450

z	w	c_1	c_2	d = .2	.4	.6	.8	1.0	1.5	2.0	3.0	z^{tc}
0101010010110	31	.0005776	−.0054620	.002167716	.001956985	.001592929	.001168607	.000773082	.000175032	.000020796	.000000043	1001101010101
0110010100111	30	.0005777	−.0054255	.002164739	.001953696	.001591558	.001170641	.000777666	.000179551	.000022115	.000000072	0110110010011
0010101110001	31	.0005132	−.0052137	.002160434	.001953928	.001601328	.001189296	.000800544	.000193729	.000025458	.000000052	1100100011101
1000110101010	31	.0005132	−.0053806	.002157121	.001942022	.001579547	.001160801	.000770889	.000168973	.000019929	.000000041	0110100011101
1001101101010	31	.0005132	−.0054774	.002155189	.001935060	.001566786	.001144092	.000753503	.000189745	.000021948	.000000120	1010100111100
1100010010011	31	.0004868	−.0049866	.002159931	.001961642	.001621519	.001220115	.000835869	.000215795	.000031233	.000000069	0011001111001
1010010100101	31	.0004868	−.0053791	.002155070	.001944058	.001588987	.001176874	.000789930	.000189745	.000024744	.000000050	0101110010011
1010001010101	31	.0004868	−.0053791	.002152124	.001933453	.001569545	.001151370	.000763299	.000175551	.000021513	.000000050	0110011011010
1000011101010	31	.0004868	−.0054214	.002151287	.001930458	.001564106	.001144324	.000756057	.000171894	.000020741	.000000046	0010110111001
0110010100110	31	.0004868	−.0054222	.002151266	.001930368	.001563899	.001143989	.000755629	.000171538	.000020626	.000000045	1011001110011
0101010010110	31	.0004868	−.0055057	.002149614	.001924461	.001553177	.001130106	.000741365	.000164342	.000019109	.000000038	1010010010101
1001010101110	31	.0004423	−.0053116	.002144978	.001923578	.001561031	.001146401	.000761922	.000177418	.000022211	.000000055	1000111001011
1000101011100	32	.0004423	−.0053840	.002143500	.001918115	.001550758	.001132596	.000747172	.000169145	.000020250	.000000043	1001100010110
0011110000001	29	.0003695	−.0051509	.002144413	.001911812	.001555863	.001150779	.000769754	.000189754	.000025937	.000000088	0111100000101
1000011010110	31	.0003514	−.0053109	.002127729	.001894019	.001526756	.001114542	.000736898	.000169992	.000021207	.000000053	1010011001010
1100001101011	31	.0002790	−.0051234	.002117582	.001883053	.001521929	.001179966	.000746415	.000179131	.000023744	.000000072	1100100111101
1010010001101	31	.0002790	−.0053812	.002112551	.001865228	.001489684	.001070131	.000703131	.000156376	.000018575	.000000040	0011001011101
0110111000011	31	.0002342	−.0052599	.002106522	.001859874	.001490248	.001084083	.000716292	.000167453	.000021737	.000000066	0011101101010
1010010010110	30	.0002342	−.0054454	.002102869	.001846828	.001466488	.001053073	.000684019	.000150235	.000017747	.000000040	1101001010110
0011101101010	30	.0002078	−.0054388	.002097970	.001838601	.001457108	.001044463	.000677307	.000148218	.000017452	.000000039	1011001011001
0111000001010	29	.0001354	−.0053579	.002085744	.001820247	.001438852	.001030347	.000668517	.000147462	•.000017662	.000000042	0101110001001
0101010010110	29	.0001354	−.0054930	.002083186	.001811489	.001423585	.001011311	.000649632	.000138612	.000015891	.000000033	0111010010101
0101101000110	29	.0000908	−.0054987	.002074543	.001797609	.001404723	.000992508	.000633577	.000132414	.000014721	.000000028	0111011001010
0110110000101	30	.0000908	−.0055625	.002073325	.001791882	.001397324	.000983220	.000626554	.000132044	.000013852	.000000024	1001110010101
1010001010110	29	.0000709	−.0055284	.002070266	.001788154	.001395095	.000977324	.000622681	.000130713	.000014614	•.000000029	0111011001010
0110101001001	31	.0000709	−.0055016	.002065750	.001781243	.001388012	.000977885	.000626554	.000130021	.000014583	.000000029	0110101001101
0001111000011	30	.0000000	−.0048875	.002068932	.001806871	.001441945	.001051701	.000701246	.000152337	.000024407	.000000098	0011011011001
1100101000110	30	.0000000	−.0051288	.002064326	.001790828	.001413288	.001014810	.000663167	.000151947	.000019499	.000000059	1101101001100
0010111010010	30	.0000000	−.0052786	.002061467	.001780895	.001395613	.000992182	.000639980	.000139862	.000016709	.000000040	1101001010011
0011101100100	30	.0000000	−.0053837	.002058744	.001773996	.001383488	.000976929	.000624698	.000132503	.000015191	.000000032	1101010100011
1000110001010	30	.0000000	−.0054272	.002058638	.001771141	.001378478	.000970637	.000618409	.000129497	.000014577	.000000029	0110101011010
1010011010110	30	.0000000	−.0055238	.002056800	.001764816	.001367401	.000956772	.000604608	.000122998	.000013278	.000000023	1010101011010
0101011010110	30	.0000000	−.0055337	.002056611	.001764159	.001366236	.000955285	.000603093	.000122230	.000013111	.000000023	0110101010110
1100001010110	30	.0000000	−.0055687	.002055948	.001761895	.001362322	.000950475	.000598418	.000120208	.000012755	.000000022	1001010010110
0101011001010	30	.0000000	−.0055698	.002055925	.001761809	.001362150	.000950222	.000598119	.000119994	.000012693	.000000017	1010101100110
0110100101010	30	−.0000445	−.0056559	.002054291	.001756208	.001352409	.000938148	.000586251	.000114649	.000011692	.000000021	0110010101011
1000110010110	31	.0000000	−.0055016	.002048695	.001751302	.001351902	.000942275	.000592947	.000119041	.000012603	.000000021	1001100110011
1010010010110	30	−.0000709	−.0055284	.002043163	.001740900	.001338732	.000928957	.000581596	.000114927	.000011591	.000000018	0110110010101
1001011010110	30	−.0000908	−.0054987	.002040011	.001736818	.001335807	.000928182	.000582727	.000116867	.000012478	.000000018	0110111001010
0110100111010	30	.0000000	−.0055625	.002038805	.001732695	.001328637	.000919273	.000573932	.000112824	.000011694	.000000019	1010110111001
1010010111010	31	−.0001354	−.0053579	.002034172	.001731136	.001334095	.000931039	.000588444	.000121225	.000013491	.000000027	1000111101010
1000110101110	31	−.0001354	−.0054930	.002031586	.001722185	.001318327	.000911185	.000565565	.000111734	.000011572	.000000018	1010101011010
0101011101010	30	−.0002078	−.0054388	.002018864	.001722161	.001297192	.000893500	.000556234	.000109266	.000011412	.000000019	1100100111001
1100001001110	30	−.0002342	−.0052599	.002017129	.001704462	.001305672	.000906584	.000570506	.000116500	.000012859	•.000000025	1100110010101
0101011010110	29	−.0002790	−.0051234	.002013717	.001693085	.001286342	.000883076	.000547750	.000106471	.000010978	.000000018	0011110010010
1011001000110	31	−.0002790	−.0053812	.002011268	.001699241	.001305621	.000912613	.000580525	.000108180	.000010960	.000000040	1011001110010
1000110010110	31	−.0003514	−.0053109	.002006437	.001683037	.001277654	.000877997	.000544826	.000105065	.000011231	.000000022	1011100111010
0101110000110	31	−.0003514	−.0053109	.002004161	.001663518	.001256973	.000860065	.000533525	.000105065	.000011231	.000000026	0111101000101
0110111001010	28	−.0003695	−.0051509	.001993974	.001666925	.001256549	.000872283	.000545818	.000110792	.000012321	.000000031	1001111000101
0101001110010	31	−.0003695	−.0051509	.001993395	.001663518	.001265549	.000872283	.000545818	.000110792	.000012321	.000000031	1011001111001

114

z	w	c_1	c_2	.2	.4	.6	.8	1.0	1.5	2.0	3.0	z^{tc}
				$d = .2$								
0110101001	28	−.0004423	−.0053840	.001975413	.001629726	.001215717	.000820218	.000500653	.000094251	.000009569	.000000016	0111010101
1100110001	29	−.0004868	−.0049866	.001974144	.001638838	.001238494	.000852209	.000534093	.000110624	.000012906	.000000033	0011101100
0011100100	29	−.0004868	−.0052309	.001966978	.001624000	.001213273	.000821405	.000540016	.000096794	.000010105	.000000018	1100111010
1010010100	29	−.0004868	−.0053791	.001966679	.001615089	.001198297	.000803395	.000486775	.000089396	.000008748	.000000013	0101110110
1001100010	29	−.0004867	−.0054214	.001966208	.001612557	.001194067	.000798358	.000481872	.000087448	.000008415	.000000012	1010011010
0101100110	29	−.0004868	−.0054222	.001966194	.001612509	.001193980	.000798235	.000482014	.000087346	.000008386	.000000012	1001100110
0011011000	29	−.0004868	−.0055057	.001964673	.001607502	.001185611	.000788259	.000472436	.000083475	.000007723	.000000010	1010110101
0101011001	29	−.0005132	−.0052137	.001964969	.001616384	.001204899	.000813983	.000498842	.000095242	.000008894	.000000018	0101101100
0110110010	29	−.0005132	−.0053806	.001961928	.001606353	.001188054	.000793756	.000479098	.000087017	.000007438	.000000012	0111001010
0101011100	29	−.0005132	−.0054774	.001960171	.001600594	.001178455	.000782329	.000468292	.000082560	.000007630	.000000009	1011001101
1001010010	30	−.0005577	−.0054255	.001952603	.001588852	.001165966	.000771523	.000460266	.000080339	.000007321	.000000009	0011010110
0010101110	29	−.0005776	−.0049322	.001957718	.001611470	.001206969	.000822463	.000509855	.000101968	.000011296	.000000024	1100111100
1010101010	30	−.0005776	−.0053954	.001949436	.001584616	.001162565	.000769863	.000460205	.000081309	.000007692	.000000010	0110101010
0110101010	30	−.0006222	−.0054620	.001948230	.001580671	.001155607	.000762083	.000452877	.000078326	.000007092	.000000008	1010101010
1000111010	30	−.0006486	−.0053626	.001941516	.001571700	.001148083	.000756643	.000449854	.000078080	.000007108	.000000009	1001010110
1001101010	30	−.0006486	−.0053082	.001937477	.001566267	.001143321	.000753484	.000448253	.000078132	.000007168	.000000009	1010011100
0101101010	30	−.0006486	−.0053711	.001936349	.001562613	.001137310	.000746434	.000441695	.000075560	.000006749	.000000008	0101110010
1100101010	29	−.0007210	−.0050891	.001927633	.001555324	.001136755	.000752489	.000451104	.000081322	.000007888	.000000012	0111001100
0011010011	29	−.0007210	−.0051689	.001926254	.001551009	.001129865	.000744606	.000443916	.000078580	.000007438	.000000010	0011100011
0101010101	29	−.0007210	−.0052826	.001924221	.001544434	.001119054	.000731914	.000432081	.000073876	.000006654	.000000008	1101110011
0011110010	28	−.0007658	−.0050164	.001920484	.001545376	.001127845	.000746612	.000448358	.000081911	.000008179	.000000014	0111100011
1100011001	29	−.0008118	−.0050010	.001911942	.001530792	.001110723	.000730277	.000435035	.000077373	.000007430	.000000011	0110001110
0110001110	29	−.0008118	−.0051236	.001909808	.001524057	.001099885	.000717784	.000423564	.000072929	.000006894	.000000009	1100011100
1100011100	30	−.0008563	−.0048234	.001906594	.001526187	.001109962	.000733308	.000440041	.000080565	.000008089	.000000014	1000111100
1000101110	30	−.0008563	−.0050175	.001903167	.001515193	.001091956	.000712158	.000420227	.000072455	.000006660	.000000009	0111001110
0111001010	27	−.0009290	−.0049512	.001890664	.001496142	.001072726	.000697003	.000410489	.000071204	.000006728	.000000010	0011100011
1101010010	28	−.0009554	−.0049325	.001885987	.001488652	.001064591	.000689900	.000405227	.000069823	.000006554	.000000010	1101110010
1010000111	28	−.0010000	−.0045239	.001884622	.001496547	.001083923	.000716177	.000431777	.000081803	.000008821	.000000020	1010001110
0011011010	28	−.0010000	−.0048405	.001879176	.001479176	.001055814	.000675588	.000401293	.000069272	.000006534	.000000008	0011011100
0100110101	28	−.0010000	−.0049079	.001877910	.001475203	.001049203	.000666155	.000393837	.000066155	.000005978	.000000008	1101110110
0111000010	28	−.0010000	−.0049413	.001877323	.001473338	.001046185	.000672092	.000390615	.000064899	.000005769	.000000007	1001110001
0101101001	29	−.0010000	−.0050128	.001876118	.001469669	.001040505	.000665819	.000385115	.000063033	.000005506	.000000006	0111001100
0101101010	29	−.0010000	−.0051068	.001874485	.001464544	.001032293	.000656434	.000376591	.000059848	.000005003	.000000005	1010110010
1001101001	29	−.0010445	−.0050278	.001867369	.001454166	.001021941	.000648025	.000370719	.000058438	.000004829	.000000005	1001110010
1010101001	28	−.0010644	−.0050000	.001864146	.001449689	.001018021	.000645581	.000369739	.000058820	.000004967	.000000005	0110101010
0011010001	28	−.0010644	−.0050638	.001863041	.001446226	.001012503	.000639303	.000364065	.000056732	.000004643	.000000005	1010101010
0110110001	28	−.0010908	−.0046010	.001865951	.001462226	.001041439	.000674082	.000396489	.000069223	.000006656	.000000010	1100110010
0100110110	28	−.0010908	−.0049410	.001860160	.001444439	.001013257	.000642271	.000367872	.000058665	.000004981	.000000005	0101110010
0110011010	28	−.0010908	−.0050160	.001858844	.001440230	.001006733	.000634872	.000361203	.000056222	.000004604	.000000005	0111001100
1000110101	29	−.0011353	−.0048884	.001852581	.001432503	.001000537	.000631134	.000359501	.000056297	.000004651	.000000005	1100100110
0110101010	29	−.0011353	−.0049547	.001851444	.001428980	.000994996	.000624913	.000353959	.000054341	.000004363	.000000004	1010101010
1001010101	29	−.0011353	−.0050202	.001850317	.001425482	.000989485	.000618729	.000348460	.000052421	.000004087	.000000003	1010101010
1100101010	29	−.0012077	−.0046313	.001843269	.001422832	.000995365	.000630990	.000362442	.000058863	.000005163	.000000006	0101110010
0101101001	28	−.0012077	−.0048168	.001840135	.001413239	.000989404	.000604844	.000343718	.000051191	.000004035	.000000004	1100110010
1011001010	29	−.0012341	−.0048493	.001836076	.001405521	.000970764	.000609144	.000339905	.000052875	.000004299	.000000004	1010011010
0101011010	28	−.0012986	−.0047779	.001835806	.001406833	.000973989	.000616726	.000344117	.000057137	.000005006	.000000006	1110001011
1001001110	28	−.0012986	−.0044440	.001829208	.001403148	.000976675	.000604996	.000353211	.000053621	.000004496	.000000005	0101010110
1100001101	28	−.0012986	−.0045821	.001826910	.001396213	.000966014	.000594083	.000342941	.000050346	.000004020	.000000005	0101011100
0110010110	28	−.0012986	−.0047101	.001824775	.001389768	.000956099	.000590444	.000333383	.000050365	.000004054	.000000004	1100100110
1000101100	29	−.0013431	−.0045740	.001818604	.001382236	.000950093	.000586297	.000331683	.000047000	.000004054	.000000004	1001001110

115

Table A

z	w	c_1	c_2	d = .2	.4	.6	.8	1.0	1.5	2.0	3.0	z^{tc}
11000010110	29	−.0013695	−.0044391	.001815844	.001380470	.000950669	.000592828	.000334643	.000051727	.000004268	.000000004	1010111100
10010011100	29	−.0013695	−.0045608	.001813832	.001374446	.000941481	.000582805	.000325948	.000048830	.000003861	.000000004	1100011100
01111010000	26	−.0014422	−.0042793	.001804914	.001366181	.000938728	.000585596	.000331702	.000052587	.000004587	.000000006	0111010001
00111100100	27	−.0014867	−.0041627	.001798419	.001357690	.000931290	.000580445	.000328748	.000052256	.000004585	.000000006	0111100011
10011100001	27	−.0014867	−.0043295	.001795622	.001349183	.000918087	.000565757	.000315719	.000047603	.000003868	.000000004	0011110001
11100000011	27	−.0014867	−.0044264	.001794010	.001344318	.000910607	.000557522	.000308501	.000045118	.000003503	.000000003	1100100011
11011011000	27	−.0015776	−.0035977	.001790501	.001356269	.000940522	.000597014	.000346877	.000060647	.000006125	.000000012	0111011000
00110011001	27	−.0015776	−.0039772	.001784291	.001337700	.000911994	.000565381	.000318717	.000050192	.000004375	.000000006	1010110010
01110010010	27	−.0015776	−.0042625	.001779613	.001323776	.000890836	.000542328	.000298683	.000043385	.000003375	.000000003	0111101100
01011001100	27	−.0015776	−.0043165	.001778711	.001321052	.000886646	.000537719	.000294650	.000042009	.000003176	.000000003	1010110010
10101010010	27	−.0015776	−.0043251	.001778610	.001320873	.000886563	.000537834	.000294927	.000042247	.000003230	.000000003	0111010010
01101010010	27	−.0015776	−.0043985	.001777396	.001317245	.000881043	.000531833	.000289740	.000040533	.000002990	.000000003	1011010101
01100110100	27	−.0016221	−.0043089	.001770435	.001307292	.000871270	.000524017	.000284369	.000039309	.000002850	.000000002	1001110110
10010110010	28	−.0016221	−.0043718	.001769398	.001304199	.000866591	.000518972	.000280058	.000037942	.000002671	.000000002	1010100110
10100110010	28	−.0016485	−.0043112	.001765419	.001298814	.000861687	.000516414	.000277891	.000037615	.000002651	.000000002	1010011010
10101100100	28	−.0017130	−.0042477	.001754348	.001281728	.000843912	.000500640	.000267617	.000035323	.000002418	.000000002	1010010110
11010010010	27	−.0017209	−.0038339	.001759565	.001298943	.000870599	.000529678	.000292491	.000043348	.000003502	.000000004	0111100001
01101001100	27	−.0017209	−.0040917	.001755419	.001286865	.000851528	.000510528	.000276246	.000038187	.000002918	.000000002	1100011010
11001001010	27	−.0017853	−.0038737	.001746806	.001276840	.000845210	.000506615	.000274945	.000038624	.000002918	.000000003	0111011010
01011101100	27	−.0017853	−.0039963	.001744848	.001271176	.000836860	.000497805	.000267544	.000036335	.000002616	.000000002	1101011000
01010111000	27	−.0018117	−.0037867	.001743198	.001272377	.000841439	.000504012	.000273376	.000038318	.000002877	.000000002	0111001010
01100110010	27	−.0018117	−.0037929	.001743136	.001272314	.000841525	.000504300	.000273791	.000038592	.000002935	.000000003	1011010010
01100110010	27	−.0018117	−.0039392	.001740810	.001265606	.000831665	.000493922	.000265093	.000035913	.000002583	.000000002	1101101001
10001110100	28	−.0018563	−.0037700	.001735076	.001259081	.000826777	.000491090	.000263771	.000035838	.000002581	.000000002	1101000111
10010100100	28	−.0018563	−.0037736	.001735059	.001259148	.000827040	.000491538	.000264286	.000036098	.000002627	.000000002	0111110100
10001110100	28	−.0018563	−.0038969	.001733065	.001253314	.000818360	.000482313	.000256499	.000033688	.000002315	.000000002	1100110101
11100000101	27	−.0019471	−.0033311	.001728479	.001256972	.000832327	.000501653	.000275256	.000040700	.000003351	.000000004	1101000101
11001001010	28	−.0019471	−.0036016	.001720713	.001238524	.000806808	.000475493	.000253443	.000033803	.000002392	.000000002	0110110101
10100110010	28	−.0019471	−.0036604	.001719808	.001236006	.000803240	.000471879	.000250532	.000033001	.000002299	.000000002	1101111010
00110111000	26	−.0020643	−.0030513	.001707486	.001227380	.000803816	.000479563	.000260629	.000037738	.000003055	.000000004	1101111000
10101100001	26	−.0020643	−.0033912	.001702158	.001212203	.000781719	.000456467	.000241352	.000031788	.000002255	.000000002	0111001010
01101100010	26	−.0020643	−.0034675	.001700951	.001208743	.000776664	.000451178	.000236945	.000030450	.000002082	.000000002	1011010010
10110010010	26	−.0020907	−.0033528	.001697826	.001205798	.000775317	.000451334	.000237870	.000031062	.000002184	.000000002	0111011010
01110010010	26	−.0020907	−.0034165	.001696814	.001202885	.000771045	.000446851	.000234123	.000029918	.000002035	.000000001	1001110010
10010100010	27	−.0021353	−.0033072	.001690149	.001193742	.000764216	.000440206	.000229720	.000028995	.000001937	.000000001	1010011010
01110000100	27	−.0021353	−.0034423	.001688054	.001187872	.000754055	.000431705	.000222859	.000027100	.000001719	.000000001	1010101010
10110010100	26	−.0021997	−.0033513	.001677445	.001172173	.000738390	.000419212	.000214458	.000025418	.000001565	.000000001	1101000101
01011100100	27	−.0022077	−.0030983	.001679890	.001180781	.000751515	.000433023	.000225852	.000028677	.000001950	.000000001	1100111000
01011100010	26	−.0022261	−.0032724	.001673738	.001167518	.000734474	.000416605	.000213014	.000025257	.000001559	.000000001	1110011010
11100000101	27	−.0022985	−.0025292	.001671594	.001167923	.000755044	.000441159	.000236660	.000032042	.000002419	.000000002	1110111000
11011001001	26	−.0022985	−.0028100	.001667368	.001165348	.000738823	.000424828	.000221525	.000028345	.000001968	.000000002	0110111000
01101101001	26	−.0022985	−.0028498	.001666732	.001163506	.000736101	.000421947	.000219094	.000027583	.000001865	.000000001	1110100101
10001100001	26	−.0022985	−.0028641	.001666549	.001163103	.000735685	.000421689	.000219024	.000027667	.000001889	.000000001	0111010001
01110011000	27	−.0022985	−.0029143	.001665765	.001160881	.000732470	.000418354	.000216264	.000026840	.000001783	.000000001	1001110010
01010101000	27	−.0022985	−.0030049	.001664403	.001157176	.000727337	.000413268	.000212258	.000025790	.000001667	.000000001	1010010110
11000010110	27	−.0023430	−.0028112	.001658993	.001151422	.000722373	.000411223	.000211467	.000025814	.000001675	.000000001	1001010110
10011001100	27	−.0023430	−.0029329	.001657144	.001146334	.000716250	.000404094	.000205796	.000024296	.000001504	.000000001	1100100101
10010111000	26	−.0023894	−.0028617	.001653316	.001141330	.000711831	.000400972	.000203936	.000024022	.000001486	.000000001	1001011010
01100111100	27	−.0023894	−.0026462	.001652899	.001144396	.000718146	.000408353	.000210349	.000025965	.000001722	.000000001	1110011001
01110111000	27	−.0024339	−.0024495	.001647548	.001138858	.000714571	.000406797	.000210038	.000026187	.000001762	.000000001	1110111000

116

Table A

m = 6, n = 5   Rank Orders 351 thru 400 of 462   1/C(6, 5) = .00216450

C(6, 5) = 462

z	w	c_1	c_2	d = .2	.4	.6	.8	1.0	1.5	2.0	3.0	z^{tc}
1100101010	27	−.0024339	−.0026845	.001643990	.001129087	.000700894	.000393095	.000199112	.000023230	.000001424	.000000001	1010101100
1010100100	27	−.0024339	−.0027459	.001643086	.001126677	.000697629	.000389937	.000196687	.000022639	.000001364	.000000001	1100110001
0011111100	25	−.0025775	−.0017353	.001631564	.001124754	.000708793	.000408153	.000214702	.000028966	.000002213	.000000002	1110000011
1011010000	25	−.0025775	−.0021985	.001624701	.001106223	.000683144	.000382600	.000194322	.000023294	.000001517	.000000001	0111110010
0111010001	26	−.0026220	−.0022651	.001623690	.001090434	.000679218	.000378639	.000191135	.000023402	.000001196	.000000001	1011010001
1001011100	26	−.0026220	−.0022609	.001615445	.001088946	.000664696	.000366210	.000182230	.000020352	.000001196	.000000001	1100011110
1100001100	27	−.0026861	−.0013283	.001611669	.001070067	.000667315	.000370721	.000186626	.000021752	.000001359	.000000001	1010111000
1011000010	26	−.0027129	−.0020832	.001601265	.001070067	.000645745	.000351786	.000172958	.000018680	.000001057	.000000000	1101110100
0101101010	26	−.0027129	−.0021555	.001600217	.001067326	.000642103	.000348337	.000170366	.000018085	.000001000	.000000000	1101011100
0101111000	26	−.0027853	−.0016084	.001594858	.001067204	.000648310	.000357431	.000178794	.000020641	.000001291	.000000001	1101010101
1100100001	25	−.0027853	−.0016146	.001594794	.001067112	.000648300	.000357540 •	.000178978	.000020760	.000001314	.000000001	1110010010
0101100100	25	−.0027853	−.0017609	.001592661	.001061460	.000640651	.000350116	.000173230	.000019296	.000001153	.000000001	1100100101
0111010100	25	−.0028117	−.0015656	.001590638	.001061228	.000642672 •	.000353225	.000176178	.000020236	.000001267	.000000001	0111011100
0111001010	25	−.0028117	−.0016882	.001588844	.001056459	.000636200	.000346928	.000170179	.000018945	.000001129 •	.000000001	1010110001
1001101000	26	−.0028562	−.0014833	.001583546	.001050877	.000632320	.000344837	.000170383	.000018986	.000001129 •	.000000001	1010010110
1011001010	26	−.0028562	−.0016773	.001580739	.001043518	.000622494	.000335453	.000170524	.000017240	.000000958	.000000001	1100011110
1100000011	25	−.0028761	−.0012035	.001583913 •	.001055640	.000640448	.000353540	.000177506	.000020900	.000001354	.000000001	0111111010
0101110010	26	−.0028761	−.0014613	.001580222	.001046025	.000627637	.000341281	.000168136	.000018582	.000001105	.000000001	1010110100
1001010110	26	−.0029207	−.0014750	.001571767	.001032181	.000612742	.000328687	.000159261	.000016641	.000000931	.000000000	1101101010
1010110000	26	−.0029207	−.0015339	.001570936	.001030061	.000609990	.000326136	.000157383	.000016230	.000000877	.000000000	0111101010
1010011100	26	−.0029471	−.0012713	.001569820	.001032015 •	.000614659	.000331494	.000161812	.000017364	.000000991	.000000000	1011010010
1000110010	26	−.0029471	−.0013744	.001568352	.001028231	.000609689	.000328826	.000158324	.000016566	.000000914	.000000000	1101001100
1010010110	26	−.0029471	−.0014444	.001567368	.001025726	.000604443	.000323823	.000156177	.000016083	.000000870	.000000000	1110011000
1011000001	24	−.0030643	−.0007521	.001555728	.001017369	.000605117	.000327255 •	.000160885	.000017945	.000001100	.000000000	0111011100
0111100010	24	−.0030643	−.0008158	.001554798	.001014897	.000601754	.000323969	.000158319	.000017272	.000001024	.000000000	1001111010
1100100010	26	−.0031548	−.0007082	.001539648	.000992312	.000579168	.000305965	.000146234	.000014859	.000000800	.000000000	1100011110
1011001001	25	−.0031997	−.0007234	.001531261	.000979082	.000565615	.000291988	.000139050	.000013529	.000000691	.000000000	1101100010
0101110001	25	−.0032261	−.0006527	.001527427	.000974065	.000561744	.000291988	.000137169	.000013245	.000000672	.000000000	1010111000
0101011100	25	−.0032985	−.0004424	.001523846	.000977725 •	.000571352	.000303912	.000147069	.000015793	.000000931	.000000001	0101110100
1010100001	24	−.0032985	−.0000957	.001522014	.000973311	.000565921	.000299124	.000143706	.000015131	.000000876	.000000000	0111010100
0110100100	24	−.0032985	−.0002237	.001520221	.000968745	.000559974	.000293560	.000139548	.000014157	.000000777	.000000000	1011010001
1001101000	25	−.0033430	−.0001828	.001512597	.000957189	.000548285	.000284174	.000133222	.000012895	.000000662	.000000000	1101011000
0101101100	24	−.0033629	−.0000308	.001511955	.000959140 •	.000552396	.000288750	.000136958	.000013852	.000000760	.000000000	1110001001
1100010001	24	−.0033893	−.0003072	.001510925	.000961086 •	.000556810	.000293697	.000140990	.000014884	.000000869	.000000000	0111011100
1100100010	24	−.0033893	.0001218	.001504386	.000954751	.000548714	.000286254	.000135518	.000013647	.000000746	.000000000	1001111010
1001011000	25	−.0034338	.0004207	.001504286	.000951946	.000548108	.000286876	.000136530	.000013970	.000000777	.000000000	1100111110
1010010010	25	−.0034338	.0001638	.001500787	.000943279	.000537158	.000277012	.000129312	.000012419	.000000635	.000000000	1010111010
0110100010	25	−.0034338	.0001193	.001500182	.000941787	.000535281	.000275321	.000128099	.000012167	.000000613	.000000000	1010110100
1100110000	25	−.0034338	.0000690	.001499499	.000940110	.000533182	.000273444	.000126764	.000011897	.000000590	.000000000	1101110100
0101111100	24	−.0034338	.0000555	.001499309	.000939625	.000532557	.000272868	.000126343	.000011807	.000000582	.000000000	1101010100
1010101010	25	−.0034338	−.0000119	.001498400	.000937409	.000529807	.000270430	.000124627	.000011469	.000000554	.000000000	1101010010
1010011100	25	−.0035247	.0004478	.001488126	.000927275	.000523403	.000267491	.000123716	.000011596 •	.000000577	.000000000	1100011010
1100011000	25	−.0036416	.0008169	.001471893	.000906864	.000503179	.000255981	.000116754	.000010603	.000000511	.000000000	1100110010
0111010100	24	−.0036680	.0009280	.001468604	.000903179	.000503200	.000255860	.000115914	.000010527	.000000508 •	.000000000	1010011100
1011100000	24	−.0037129	.0009999	.001461475	.000893226	.000493971	.000247088	.000109974	.000009874	.000000462	.000000000	1010011010
0111110000	23	−.0037853	.0016637	.001457260	.000890815 •	.000500913	.000255544	.000117753	.000011618	.000000632	.000000000	0111110010
1110001000	24	−.0037853	.0020277	.001455610	.000890764	.000495815	.000250920	.000115421	.000010885	.000000562	.000000000	1101111000
1101110000	23	−.0038761	.0019479	.001445628	.000878212	.000490022	.000245296	.000110885 •	.000010571	.000000598	.000000000	0111011000
0111011100	23	−.0038761	.0019479	.001444538	.000876050	.000486565	.000245296	.000112508	.000010571	.000000546	.000000000	1110010100
0111101000	23	−.0038761	.0018343	.001443091	.000875146	.000482579	.000241894	.000110186	.000010138	.000000511	.000000000	1110101001

117

Table A
m = 6, n = 5

z	w	c_1	c_2	d = .2	.4	.6	.8	1.0	1.5	2.0	3.0	z^{tc}
1001110000	24	−.039206	.0019063	.001435946	.000864916	.000472748	.000234359	.000105322	.000009256	.00000436	.00000000	1110100010
1100110000	24	−.039206	.0018032	.001434593	.000861694	.000468835	.000230956	.000102966	.000008807	.00000400	.00000000	1011001100
1010110000	24	−.039206	.0017332	.001433689	.000859580	.000466314	.000228806	.000101507	.000008544	.00000380	.00000000	1011001010
1010010000	24	−.039470	.0018845	.001430913	.000857079	.000464926	.000228304	.000101446	.000008601	.00000387	.00000000	1011010100
1011010000	24	−.039470	.0018257	.001430151	.000855292	.000463777	.000226479	.000100206	.000008377	.00000370	.00000000	1010101000
1100000010	24	−.040115	.0023249	.001425010	.000852930	.000465947	.000228926	.000102513	.000008947	.00000420	.00000000	1010110010
1010001000	24	−.040115	.0021309	.001422529	.000844768	.000441785	.000223852	.000098886	.000008239	.00000366	.00000000	1100110010
1011000000	24	−.041548	.0027459	.001404768	.000827245	.000439230	.000211685	.000093784	.000007731	.00000341	.00000000	1010101000
1010100100	24	−.041548	.0026735	.001403840	.000825089	.000437236	.000211083	.000092321	.000007471	.00000321	.00000000	1010101010
1011000100	23	−.041996	.0029474	.001399312	.000821210	—	—	.000092349	.000007576	.00000332	.00000000	1011110010
1100111000	24	−.042456	.0031987	.001394281	.000816329	.000434117	.000209555	.000091765	.000007576	.00000336	.00000000	1100001100
1110100001	22	−.043629	.0040352	.001383988	.000809858	.000433422	.000212027	.000094761	.000008499	.00000430	.00000000	0111000001
0111011000	22	−.043629	.0038498	.001381651	.000804470	.000427037	.000206564	.000091011	.000007775	.00000368	.00000000	1110010001
0111011000	23	−.043893	.0039861	.001378644	.000801353	.000424830	.000205302	.000090396	.000007718	.00000365	.00000000	1110000110
1001110000	23	−.044338	.0040876	.001371916	.000792194	.000416402	.000199085	.000086516	.000007062	.00000313	.00000000	1011101100
1010100000	23	−.044338	.0038527	.001369024	.000785720	.000408990	.000192993	.000082520	.000006391	.00000265	.00000000	1011010100
1010100100	23	−.044338	.0037912	.001368263	.000784009	.000407027	.000191380	.000081465	.000006217	.00000253	.00000000	1010101010
1010010100	23	−.044982	.0041240	.001360956	.000776534	.000401863	.000188528	.000080139	.000006115	.00000250	.00000000	1010110100
1011001000	23	−.044982	.0043565	.001359123	.000776020	.000402605	.000189661	.000081075	.000006308	.00000264	.00000000	1010110100
1010011000	23	−.045246	.0042348	.001357646	.000772765	.000398938	.000186698	.000079166	.000006003	.00000243	.00000000	1110011010
1001101000	23	−.046416	.0047342	.001343111	.000756367	.000386355	.000178918	.000075080	.000005548	.00000219	.00000000	1011001100
1010100100	23	−.046680	.0048373	.001339714	.000752418	.000383244	.000176949	.000074024	.000005426	.00000213	.00000000	1010101100
1100001000	22	−.047324	.0053301	.001334306	.000749046	.000382611	.000177692	.000074976	.000005675	.00000233	.00000000	1100101100
1110001000	22	−.047324	.0051949	.001332685	.000745510	.000378666	.000174531	.000072955	.000005356	.00000211	.00000000	1110001100
1100000001	21	−.047324	.0064799	.001322892	.000743074	.000383196	.000181250	.000078567	.000006605	.00000318	.00000000	0111110000
0111010000	21	−.048760	.0062221	.001318834	.000736408	.000375698	.000175141	.000074559	.000005904	.00000262	.00000000	1011101000
1011010000	21	−.048760	.0061131	.001310667	.000722113	.000361745	.000164667	.000068069	.000004887	.00000190	.00000000	1011110100
1011000010	22	−.049206	.0060543	.001309965	.000720592	.000360060	.000163327	.000067219	.000004756	.00000181	.00000000	1011011000
1010110000	22	−.049206	.0065753	.001300237	.000711203	.000353953	.000160166	.000065854	.000004675	.00000180	.00000000	1011010010
1010010010	22	−.050114	.0065717	.001300182	.000711061	.000353770	.000160001	.000065736	.000004652	.00000178	.00000000	1110001010
1011010000	22	−.050114	.0064484	.001298763	.000708086	.000350581	.000157547	.000064229	.000004437	.00000165	.00000000	1110101010
1101010000	22	−.051547	.0071184	.001281641	.000689506	.000336852	.000149365	.000060083	.000004014	.00000144	.00000000	1011101010
1100110000	22	−.052192	.0075036	.001274903	.000683153	.000332810	.000147313	.000059211	.000003962	.00000143	.00000000	1010111010
1110000100	22	−.052456	.0076878	.001272430	.000681119	.000331730	.000146892	.000059097	.000003972	.00000145	.00000000	1100101010
1010110000	22	−.052456	.0076250	.001271725	.000679678	.000330223	.000145760	.000058418	.000003880	.00000139	.00000000	1111011000
0111101000	20	−.053628	.0087652	.001264496	.000678664	.000334256	.000150918	.000062407	.000004647	.00000197	.00000000	1010111000
1011100100	21	−.054982	.0090838	.001244689	.000653394	.000312837	.000136497	.000055231	.000003564	.00000128	.00000000	1011101000
1011010000	21	−.054982	.0089621	.001243329	.000650623	.000309941	.000134320	.000052922	.000003386	.00000118	.00000000	1011101010
1010110000	21	−.055246	.0091308	.001240675	.000648205	.000308447	.000133580	.000052612	.000003367	.00000117	.00000000	1011010010
1011001000	21	−.056415	.0096945	.001226915	.000633505	.000297757	.000127310	.000049485	.000003060	.00000103	.00000000	1110100010
1100111000	21	−.057324	.0102860	.001217898	.000625412	.000292844	.000124926	.000048515	.000003009	.00000102	.00000000	1111001100
1010100100	21	−.057324	.0102196	.001217178	.000623984	.000291393	.000123866	.000047895	.000002930	.00000098	.00000000	1011011100
1101001000	21	−.057324	.0101541	.001216474	.000622604	.000290007	.000122865	.000047317	.000002859	.00000094	.00000000	1011001100
1110010000	20	−.058232	.0107772	.001207788	.000615144	.000285713	.000120916	.000046590	.000002835	.00000095	.00000000	1100111100
1110000010	20	−.060114	.0120590	.001189661	.000599355	.000276428	.000116571	.000044898	.000002761	.00000084	.00000000	1011110100
1011011000	20	−.060114	.0118649	.001187617	.000595414	.000272517	.000113769	.000043287	.000002563	.00000094	.00000000	1011101000
1010010100	20	−.062191	.0130437	.001165089	.000573098	.000257383	.000105450	.000039380	.000002227	.00000069	.00000000	1011101010
1101001000	20	−.062191	.0129808	.001164439	.000571869	.000256192	.000104617	.000038914	.000002173	.00000067	.00000000	1011101010
1010110000	20	−.062455	.0131712	.001161995	.000569821	.000255030	.000104090	.000038714	.000002164	.00000067	.00000000	1011010100
1110010100	20	−.063100	.0135570	.001155223	.000563321	.000250760	.000101816	.000037679	.000002082	.00000063	.00000000	1110101000

118

Table A

m = 6, n = 5 Rank Orders 451 thru 462 of 462 C(6, 5) = 462 1/C(6, 5) = .00216450

z	w	c_1	c_2	d = .2	.4	.6	.8	1.0	1.5	2.0	3.0	z^{tc}
10111100000	19	−.0064982	.0149091	.001137728	.000548551	.000242291	.000097922	.000036170	.000002010	.000000062	.000000000	11111000010
11110000100	19	−.0067323	.0163536	.001113512	.000525579	.000227319	.000089985	.000032565	.000001721	.000000051	.000000000	11011110000
11101010000	19	−.0067323	.0162184	.001112193	.000523221	.000225146	.000088537	.000031790	.000001641	.000000047	.000000000	11110101000
11101101000	19	−.0067968	.0166318	.001105698	.000517228	.000221354	.000086588	.000030933	.000001578	.000000045●	.000000000	11101001000
11100110000	19	−.0068232	.0168404	.001103416	.000515444	.000220411	.000086192	.000030794	.000001573	.000000045	.000000000●	11110011000
11011010000	18	−.0072191	.0195782	.001065393	.000481909	.000200062	.000076134	.000026529	.000001284	.000000035	.000000000	11110100100
11110001000	18	−.0073099	.0202080	.001056682	.000474232	.000195404	.000073832	.000025552	.000001217	.000000033	.000000000	11101110000
11101001000	18	−.0073099	.0201357	.001056019	.000473116	.000194434	.000073221	.000025241	.000001189	.000000032	.000000000	11101101000
11110010000	17	−.0077967	.0237461	.001011506	.000435641	.000172676	.000062908	.000021037	.000000928	.000000023	.000000000	11110110000
11110011000	17	−.0078231	.0239466	.001009134	.000433675	.000171552	.000062382	.000020826	.000000916	.000000023	.000000000	11100110000
11111010000	16	−.0083099	.0277775	.000966516	.000399206	.000152265	.000053546	.000017335	.000000713	.000000017	.000000000	11110100000
11111100000	15	−.0087967	.0319033	.000926298	.000368356	.000135820	.000046339	.000014601	.000000568	.000000013	.000000000	11111000000

m = 11, n = 1 Rank Orders 1 thru 12 of 12 C(11, 1) = 12 1/C(11, 1) = .08333333

z	w	c_1	c_2	d = .2	.4	.6	.8	1.0	1.5	2.0	3.0	z^{tc}
000000000001	12	.1357690	.1648349	.113894932	.151612550	.196712980	.248972988	.307667755	.473531903	.643007391	.889058873	011111111111
000000000010	11	.0929777	.0368436	.102510805	.122137632	.140973022	.157656269	.170868453	.182408273	.160855793	.071730874	101111111111
000000000100	10	.0660699	−.0177857	.096023023	.106988387	.115275170	.120117783	.121056233	.106729456	.076536870	.021317290	110111111111
000000001000	9	.0447369	−.0476657	.091196787	.096430336	.098523099	.097268033	.092795714	.071034804	.043941670	.008900946	111011111111
000000010000	8	.0260207	−.0643239	.087174666	.088077949	.085952372	.081016049	.073759135	.050126298	.027438018	.004302570	111101111111
000000100000	7	.0085491	−.0719031	.083588026	.080965515	.075733769	.068409356	.059673442	.036403369	.017857953	.002236064	111110111111
000001000000	6	−.0085491	−.0719031	.080227194	.074584793	.066958037	.058046625	.048592740	.026738470	.011825023	.001200620	111111011111
000010000000	5	−.0260207	−.0643239	.076938631	.068605607	.059082191	.049139222	.039470050	.019593831	.007820897	.000647219	111111101111
000100000000	4	−.0447369	−.0476657	.073572413	.062755600	.051714929	.041170922	.031663737	.014117679	.005068823	.000340638	111111110111
001000000000	3	−.0660699	−.0177857	.069925379	.056727397	.044490088	.033730099	.024718865	.009792256	.003133596	.000168620	111111111011
010000000000	2	−.0929777	.0368436	.065602931	.050004997	.036899730	.026356634	.018220247	.006269261	.001753726	.000073272	111111111101
100000000000	1	−.1357690	.1648349	.059345214	.041109242	.027684613	.018116021	.011513629	.003254400	.000760240	.000023013	111111111110

119

Table A
m = 7, n = 5   Rank Orders 1 thru 50 of 792   C(7, 5) = 792   1/C(7, 5) = .0012263

z	w	c_1	c_2	d = .2	.4	.6	.8	1.0	1.5	2.0	3.0	z^{tc}
000000011111	50	.0055390	.0218965	.002923978	.006291547	.012613182	.023628852	.041490197	.130480906	.295069858	.712125630	000001111111
000000101111	49	.0052743	.0194748	.002796759	.005724202	.010847978	.019075867	.031194711	.078821224	.133179607	.132637624	000010111111
000001001111	48	.0050152	.0172507	.002679739	.005233656	.009302633	.015684528	.024154936	.051051466	.051362955	.041241160	000011011111
000000110111	48	.0049907	.0170462	.002668568	.005189618	.009302633	.015398595	.023582200	.049348459	.066448806	.037012202	000100111111
000010001111	47	.0047505	.0151180	.002565854	.004786021	.008207846	.012964613	.018895920	.034532517	.039942824	.015711564	000101011111
000001010111	47	.0047316	.0149335	.002531440	.004742893	.008074898	.012633233	.018188921	.031766187	.034139770	.010184056	000010111111
000000111011	47	.0046675	.0144767	.002456069	.004655143	.007864821	.012229001	.017532150	.030639509	.033721504	.011839016 ●	000011101111
000010010111	46	.0046469	.0129835	.002506869	.004356869	.007106933	.010655345	.014710943	.023274553	.023038889	.064459464	000100101111
000001100111	46	.0044669	.0129338	.002448597	.004341506	.007047327	.010487594	.014329404	.021736960	.020023804	.004259997	000110101111
000000111101	46	.0044669	.0129155	.002447842	.004335805	.007025162	.010425202	.014187755	.021174855	.018960903	.003604920	000101101111
000010100111	46	.0044669										000111001111
000001011011	46	.0044084	.0124921	.002424792	.004252875	.006820589	.010014606	.013479669	.019539322	.016954826	.003009366	001001011111
000000111110	46	.0042598	.0115292	.002219422	.004074332	.006420183	.009292206	.012376230	.017849128	.016048144	.003682714 ●	010000111111
000010011011	45	.0042022	.0110221	.002217207	.003967420	.006125665	.008639476	.011144014	.014370116	.010923021	.001423293	000111010111
000001101011	45	.0041833	.0109048	.002180687	.003921861	.006057187	.008587364	.011196777	.014179147	.010779105	.001410530	000101110111
000010101011	45	.0041437	.0107347	.002174486	.003891204	.005946058	.008295955	.010580915	.015277701	.012964077	.002641976	000110110111
000001110011	45	.0041437	.0106220	.002173981	.003399901	.005830736	.008253120	.010490333	.013237421	.011176405	.001176405	010001010111
000010110011	45	.0041437	.0106060	.002147437	.003886745	.005929787	.008253120	.010490333	.012942220	.009288320	.001019674	001001101111
000001111001	45	.0040007	.0097073	.002132624	.003720990	.005563109	.007265963	.008948176	.011294716	.007932465	.000844433	001011010111
000010111001	44	.0039431	.0093134	.002126264	.003646344	.005388329	.007265963	.008952404	.010257981	.006904537	.000705141	001011100111
000011011001	45	.0039186	.0091262	.002140397	.003609587	.005296538	.007081206	.008635667	.009561898	.006128058	.000535179	100000111111
000010101010	44	.0038790	.0088421	.002123437	.003554889	.005164700	.006824372	.008208683	.008689403	.005224386	.000370527	001010101111
000100101011	44	.0038601	.0087880	.002219422	.003551117	.005178214	.006888399	.008371731	.009258395	.005996778	.000556627	001011011011
000011100011	44	.0038601	.0087907	.002219167	.003549528	.005178214	.006876967	.008351183	.009217162	.005965974	.000561162 ●	001100011111
000100110011	44	.0038601	.0087373	.002217207	.003538441	.005128073	.006766844	.008133853	.008625836	.005229781	.000388226	000111011011
010000101011	44	.0037360	.0081756	.002180687	.004980607	.004980607	.006610136	.008065801	.009316300	.006689281 ●	.000979900	001001111011
000011101001	44	.0037360	.0080023	.002174486	.003403111	.004844891	.006281364	.007424778	.007558090	.004476128	.000328880	010001001111
000100111001	44	.0037360	.0079885	.002173981	.003399901	.004833509	.006253538	.007370273	.007443083	.004300776	.000291138	001010011111
000011110001	43	.0036596	.0075288	.002147437	.003316274	.004655004	.005945823	.006918903	.006779264	.003821027	.000255370	001101010111
000100101101	43	.0036319	.0072614	.002132624	.003266104	.004539368	.005730620	.006577455	.006160937	.003256326	.000176122	001011100111
000000111110	45	.0036114	.0075205	.002140397	.003244419	.004738447	.006208742	.007492390	.008472527	.006003346	.000871346	100000111111
001000101011	43	.0035954	.0071512	.002125409	.003247545	.004508681	.005693586	.006547115	.006211388	.003373776	.000204882	001101101111
000101010011	43	.0035954	.0071255	.002124505	.003241767	.004489801	.005649265	.006463872	.006013330	.003161662	.000170146	001100101111
000100110101	43	.0035954	.0070882	.002123200	.003233494	.004463025	.005587229	.006348984	.005751868	.002898063	.000134223	001011010111
001000110011	43	.0035369	.0067931	.002104851	.003181943	.004365968	.005442491	.006169866	.005611364	.002888281	.000150232 ●	001001011111
010000011011	43	.0034713	.0063898	.002080038	.003108152	.004205509	.005160725	.005748104	.004952780	.002373508	.000099561	010010111111
000101100011	43	.0034524	.0064190	.002080038	.003117359 ●	.004255267	.005296900	.006020413	.005619882	.003064662	.000200411	000100111011
000100111010	43	.0034524	.0063459	.002077596	.003102528	.004209218	.005194081	.005836562	.005231001	.002690939	.000149177	001011101111
000011110011	43	.0034524	.0062999	.002076016	.003092687	.004177918	.005122601	.005706052	.004942401	.002404532	.000108982	010010011111
000110011001	42	.0033948	.0060392	.002058929	.003047704	.004099716	.005017628	.005594389	.004927449	.002500967 ●	.000137219	001111001111
000000111110	44	.0033524	.0059596	.002049891	.003035019	.004102338	.005067946	.005729014	.005323012	.002923452	.000200024	010000111111
001000101101	42	.0033363	.0056818	.002038430	.002982701	.003959608	.004774056	.005233703	.004379137	.002080794	.000095008	001110011011
000101010101	42	.0033363	.0056562	.002037563	.002977382	.003942971	.004736719	.005166766	.004238248	.001948420	.000078649	001101001111
000100111101	42	.0033363	.0056188	.002036307	.002969733	.003919239	.004683973	.005073274	.004048645	.001779396	.000060992	001100110111
001000110101	42	.0033118	.0054955	.002028608	.002947948	.003877559	.004620071	.004990402	.003962410	.001738140	.000060161	001101010111
001001010011	42	.0032722	.0052758	.002015480	.002908574	.003797292	.004488146	.004805939	.003721009	.001581901	.000050070	001011010111
010000011101	42	.0032122	.0049719	.001996573	.002854715	.003693618	.004327747	.004595699	.003492423	.001463687	.000046017	001011011101
000101100101	42	.0031877	.0049395	.001991810	.002850149	.003703125 ●	.004373309	.004699788	.003753359	.001709841	.000072074	001100101111
000100111110	42	.0031877	.0048481	.001988849	.002832751	.003650991	.004261273	.004507534	.003391564	.001406498	.000043103	001011010101
000110011010	41	.0031877	.0048159	.001987798	.002828530	.003632292	.004221077	.004438756	.003265045	.001305798	.000035227	011000110111
001000111001	42	.0031292	.0046722	.001974269	.002802751	.003616582	.004246860	.004544067	.003614834	.001659967	.000075038	001000011111

120

Table A

m = 7, n = 5 Rank Orders 51 thru 100 of 792 C(7, 5) = 792 1/C(7, 5) = .00126263

z	w	c_1	c_2	d = .2	.4	.6	.8	1.0	1.5	2.0	3.0	z^{tc}
01000001101	42	.031292	.046200	.001972573	.002792751	.003586459	.004181660	.004431157	.003395291	.001467208	.000081654	00100111101
00100011110	42	.031292	.045617	.001970726	.002782158	.003555506	.004116859	.004322938	.003205902	.001321029	.000041702	01000111101
10000010111	43	.030876	.047603	.001970545	.002819230	.003703252	.004173854	.004979950	.004611452	.002685057	.000268877	00001111110
00001101110	43	.030876	.045199	.001962874	.002774487	.003568941	.004181991	.004468500	.003548313	.001624736	.000071427	10000111011
00101000011	43	.030716	.045087	.001962501	.002772246	.003562037	.004166684	.004441308	.003491683	.001571220	.000066450	10000110111
00110100011	41	.030716	.043218	.001954243	.002740043	.003484022	.004021916	.004220215	.003162093	.001345224	.000049535	00111000111
00111000111	41	.030716	.043144	.001953963	.002738389	.003478695	.004009682	.004197880	.003113765	.001299439	.000043992	00110110011
00110110011	41	.030716	.042610	.001952256	.002728507	.003449550	.003948010	.003928092	.002928092	.001153053	.000043869	00101110011
00101010011	41	.030528	.041374	.001945423	.002706524	.003402249	.003867157	.003977602	.002771312	.001053853	.000026427	00110110011
01000101101	41	.030131	.039345	.001932847	.002670456	.003332059	.003757215	.003831301	.002603618	.000959762	.000021995	00101011110
01000100011	41	.029886	.038251	.001925587	.002651147	.003297476	.003707806	.003711959	.002554130	.000942297	.000022067	00110011011
00101101011	41	.029286	.036329	.001910159	.002616924	.003249882	.003662320	.003749253	.002633127	.001044188	.000032652	00010011111
00110101101	41	.029286	.035416	.001907314	.002610897	.003203900	.003567860	.003594561	.002375980	.000856193	.000019233	00011110011
00111001011	41	.029286	.035094	.001906302	.002600897	.003187333	.003533761	.003538778	.002284898	.000792480	.000015492	01010110111
00110011011	41	.029041	.033981	.001898977	.002595150	.003151450	.003481420	.003474198	.002224875	.000766060	.000014811	01010010111
01001010011	41	.028645	.033062	.001899818	.002575431	.003135005	.003480125	.003502670	.002332384	.000859276	.000017271	01100010111
01000110011	41	.028645	.032701	.001888704	.002558370	.003117423	.003444478	.003445116	.002240606	.000795689	.000012938	01110010111
00100110110	41	.028645	.032095	.001886860	.002552166	.003089105	.003388001	.003355591	.002105515	.000708430	.000020995	01100111011
00101100111	42	.028229	.031751	.001879034	.002542027	.003095903	.003431167	.003452467	.002304224	.000852754	.000053222	00011110011
10000110011	42	.028041	.032915	.001879489	.002533302	.003162149	.003581249	.003710614	.002769484	.001219229		01011001110
00000111011	42	.028041	.031453	.001875130	.002527877	.003096330	.003449337	.003498517	.002423067	.000959807	.000031065	10010001111
00110011111	42	.028041	.031078	.001873962	.002521288	.003097340	.003410042	.003433523	.002311005	.000873390	.000032839	10000111011
00110111011	40	.027880	.030231	.001868910	.002507498	.003052855	.003376010	.003397358	.002299663	.000888890	.000027583	00011110011
00101110011	40	.027880	.029105	.001865477	.002485565	.002999648	.003269569	.003225269	.002026031	.000696161	.000014420	00111100101
00110010111	40	.027880	.028844	.001864998	.002485989	.002992604	.003255816	.003203925	.001995436	.000767307	.000013543	00101101101
01010101011	40	.027484	.027062	.001852885	.002452543	.002992964	.003161429	.003083144	.001870086	.000613735	.000011102	01101110101
01001010101	40	.027295	.025951	.001846483	.002423324	.002891022	.003090018	.003101335	.001775168	.000564021	.000009331	01010011101
01000010111	40	.027215	.026493	.001846799	.002444447	.002919566	.003160938	.003015645	.001935094	.000668241	.000014689	01100111011
00011011011	41	.026639	.024380	.001831079	.002403102	.002857035	.003160453	.003101335	.001890253	.000667879	.000016603	00011110011
00011110011	40	.026639	.023649	.001828884	.002391183	.002824074	.003015204	.002912694	.001734398	.000564116	.000010438	01110001111
00011001011	40	.026639	.023189	.001827509	.002383772	.002803783	.002975544	.002851093	.001646049	.000509787	.000007953	01010011011
00010101011	40	.026450	.022029	.001820969	.002363753	.002763014	.002909811	.002762340	.001546088	.000458248	.000006238	01010101011
00011001101	40	.026054	.021277	.001812209	.002348257	.002749173	.002910021	.002787574	.001626212	.000518524	.000009230	00111001011
01001001011	40	.026054	.020916	.001811136	.002342513	.002733542	.002879143	.002740632	.001559610	.000477885	.000007383	01001101011
01010001011	40	.026054	.020310	.001809364	.002333177	.002708589	.002832213	.002668815	.001464354	.000424555	.000005483	01100101011
00010011011	40	.025809	.019994	.001804376	.002325813	.002704721	.002842317	.002700326	.001533235	.000471322	.000007516	01001110011
00010011001	40	.025809	.019982	.001804327	.002325476	.002704721	.002840653	.002695769	.001524727	.000464961	.000007139	01101100101
00100011011	40	.025809	.019337	.001802456	.002315692	.002678784	.002791644	.002622308	.001429181	.000412467	.000005332	01010101011
01000110110	40	.025639	.020183	.001802008	.002326022	.002711042	.002873911	.002754433	.001616813	.000521164	.000009504	10001100111
10000100111	41	.025393	.020798	.001799649	.002332199	.002750130	.002953449	.002891388	.001841591	.000674275	.000018745	00110111110
00001011010	41	.025393	.019153	.001794910	.002307408	.002683868	.002826248	.002696154	.001562318	.000495907	.000008658	10010010111
00010011110	41	.025393	.018891	.001794132	.002303243	.002672253	.002804188	.002662000	.001513571	.000465997	.000007306	01010110111
00110110011	39	.025233	.017737	.001788220	.002285085	.002636633	.002749260	.002592843	.001452386	.000443959	.000007269	00110111011
10000001111	41	.024808	.018215	.001782297	.002281516	.002654890	.002810008	.002708344	.001668161	.000580535	.000014001	10010011110
00000111010	41	.024808	.017888	.001781459	.002281516	.002654789	.002810008	.002708344	.001653429	.000579115	.000014895	10010011110
00011000011	41	.024808	.016985	.001778834	.002277697	.002566039	.002773473	.002598236	.001495277	.000477818	.000009053	10000111101
00110001011	39	.024648	.015496	.001771902	.002243904	.002547461	.002651717	.002478126	.001354776	.000402467	.000006058	01011110011
00101010111	39	.024648	.014999	.001770496	.002236729	.002547461	.002617411	.002427839	.001292700	.000369880	.000005003	00110011011
01010100011	39	.024648	.014816	.001769764	.002234051	.002540446	.002604640	.002408786	.001269016	.000357419	.000004603	00101101101
01000101101	40	.024568	.014644	.001768099	.002230146	.002535020	.002598286	.002402690	.001264456	.000354599	.000004413	01010111101

Table A
m = 7, n = 5

C(7, 5) = 792   1/C(7, 5) = .00126263

z	w	c_1	c_2	d = .2	.4	.6	.8	1.0	1.5	2.0	3.0	z^{tc}
01010000111	39	.023803	.0013142	.001750912	.002198003	.002500513	.002580851	.002419703	.001364608	.000435591	.000008984	00111110101
00011011001	39	.023803	.0011410	.001745970	.002172535	.002433803	.002455976	.002233707	.001122703	.000298789	.000003288	01011001011
00011001011	39	.023803	.0011272	.001745585	.002170610	.002428925	.002447284	.002221165	.001108437	.000293125	.000003125	01101100111
00011101001	39	.023407	.0010806	.001737525	.002158100	.002412210	.002456676	.002253554	.001181041	.000338647	.000004921	01101010101
01010010011	39	.023407	.0010284	.001736045	.002150528	.002401538	.002419195	.002199790	.001113404	.000302124	.000003623	01001110101
01001010001	39	.023407	.0009701	.001734408	.002142258	.002380367	.002380598	.002144014	.001047447	.000269345	.000002713	01001101110
01000101001	39	.023218	.0009322	.001730124	.002133983	.002370596	.002373197	.002142360	.001059791	.000277958	.000003082	01101010111
01000101101	39	.023218	.0009310	.001730078	.002133682	.002369649	.002371140	.002138896	.001054168	.000275454	.000002934	01101011001
00110101001	39	.023218	.0008665	.001728279	.002124661	.002346752	.002329800	.002079816	.000986649	.000243371	.000002151	00111011110
10000100111	40	.022803	.0010343	.001725736	.002140661	.002411748	.002470019	.002301786	.001285732	.000408389	.000008319	00111011101
00011010110	40	.022803	.0008698	.001721185	.002117843	.002353390	.002363019	.002145216	.001088564	.000298775	.000003762	10010100111
00011010011	40	.022803	.0008436	.001720437	.002113999	.002343361	.002344347	.002117607	.001053703	.000280069	.000003137	10001101011
00011100001	38	.022642	.0008196	.001717097	.002108854	.002340827	.002350315	.002136500	.001096469	.000309168	.000004405	00111100011
00100001101	39	.022577	.0006964	.001712538	.002090557	.002299089	.002306113	.002032338	.000976785	.000249145	.000002543	01110010011
00111000011	40	.022558	.0007531	.001713705	.002097324	.002315409	.002306743	.002074765	.001020896	.000268065	.000002916	10000101011
10000011001	40	.022161	.0007393	.001706405	.002087942	.002314069	.002324181	.002116575	.001095714	.000295946	.000004420	10001101110
10000010111	40	.022161	.0006907	.001705138	.002081974	.002299751	.002299603	.002082976	.001060423	.000251587	.000004019	10100101011
01100010111	40	.022161	.0006116	.001702947	.002071008	.002271873	.002248992	.002009951	.000973367	.000233377	.000002655	01110001101
01100011110	38	.022001	.0004927	.001697004	.002052987	.002236976	.002196871	.001945955	.000919732	.000233377	.000002419	00111010101
01000010111	39	.021977	.0004487	.001695356	.002046291	.002221156	.002169154	.001906798	.000874666	.000210255	.000001816	00110110011
00110100011	38	.021812	.0004488	.001692519	.002043527	.002223764	.002182477	.001933187	.000916136	.000232907	.000002476	01010100011
01000010101	39	.021732	.0003687	.001688923	.002031163	.002137254	.002139083	.001875040	.000855456	.000205185	.000001791	01110100011
01001100011	38	.021215	.0001716	.001673561	.001994391	.002137989	.002062841	.001792434	.000801106	.000188791	.000001615	01000011111
00000011101	40	.020731	.0004099	.001672441	.002016437	.002214944	.002133868	.002029008	.001097746	.000345658	.000007478	10000111110
10000011100	40	.020731	.0002665	.001666899	.001989637	.002167712	.002103203	.001906936	.000948582	.000263415	.000003867	00111111001
10000000111	40	.020731	.0002018	.001665005	.001991092	.002161463	.002102393	.001862122	.000900670	.000240899	.000003171	01110000111
01100000111	38	.020571	.0002350	.001662041	.001978069	.002128980	.002074280	.001914124	.000876106	.000283969	.000005018	00110110101
00011110001	38	.020571	.0001362	.001659951	.001966546	.002100744	.002024834	.001831090	.000800461	.000231329	.000002927	01000110111
01010010011	38	.020571	.0000467	.001658820	.001961189	.002080782	.002002504	.001762206	.000769172	.000195413	.000001915	00110110101
01000100011	38	.020571	.0000044	.001658820	.001961189	.002080782	.002002504	.001731716	.000769172	.000181912	.000001631	01000110111
00110100011	38	.020571	.0000013	.001658664	.001960329	.002085515	.001998259	.001725435	.000761377	.000177899	.000001515	01101010011
01001001011	39	.020571	−.0000065	.001658521	.001959623	.002083756	.001995141	.001721055	.000756562	.000175682	.000001463	01100111001
00011001011	38	.020571	−.0000128	.001658358	.001958869	.002081996	.001992229	.001717258	.000753183	.000174485	.000001450	01101010011
00011100011	38	.020571	−.0000161	.001658279	.001958531	.002081284	.001991197	.001716199	.000752820	.000174716	.000001474	01010101011
00100000111	38	.020571	−.0000635	.001656996	.001952294	.002065935	.001964332	.001678916	.000713496	.000157434	.000001111	01010101101
00100100101	38	.020156	−.0000796	.001656568	.001950247	.002060986	.001955842	.001667411	.000702101	.000152813	.000001034	10001111011
00110010011	38	.020156	−.0001046	.001654118	.001964936	.002118240	.002074156	.001846645	.000917702	.000258572	.000004136	10011101101
00011000011	39	.020156	−.0000416	.001650236	.001946266	.002072452	.001995675	.001733780	.000790011	.000194832	.000001591	10011100111
01001001001	39	.020126	−.0000791	.001649222	.001941318	.002060201	.001972040	.001703444	.000756492	.000179096	.000001328	10011001101
00100101101	38	.019986	−.0002427	.001641938	.001917500	.002011778	.001897011	.001608399	.000670742	.000145490	.000000998	01110011011
00010101001	39	.019967	−.0001811	.001643173	.001924290	.002028424	.001925219	.001645800	.000705361	.000158555	.000001190	10010101011
10000101001	39	.019571	−.0001782	.001636191	.001915795	.002027593	.001904494	.001680144	.000758836	.000185599	.000001845	10101101111
00011011010	39	.019571	−.0002268	.001634978	.001910331	.002015102	.001920116	.001653751	.000734981	.000176519	.000001096	10001011001
01001010011	39	.019571	−.0003059	.001632871	.001900220	.001900489	.001877414	.001594986	.000673313	.000149236	.000001438	10001101011
00101100101	37	.019410	−.0003333	.001629389	.001894087	.001984647	.001875763	.001600327	.000691191	.000160481	.000000872	01110011001
01010001001	39	.019330	−.0004471	.001624968	.001878116	.001950248	.001820287	.001527833	.000747606	.000131815	.000001903	01010011101
00011001010	39	.019325	−.0002436	.001630128	.001892000	.002007084	.001916845	.001658808	.000690303	.000183858	.000001328	00110011110
00011011010	39	.019325	−.0003296	.001627910	.001880007	.001983007	.001875121	.001600437	.000653843	.000158803	.000001036	10011001111
00011101001	39	.019325	−.0003824	.001623658	.001892429	.001967248	.001848207	.001564017	.000611651	.000143551	.000001285	10100111011
00100101101	39	.019325	−.0003824	.001623658	.001892429	.001967248	.001848207	.001552920	.000611651	.000150510	.000001285	10011011011

122

z	w	c_1	c_2	d = .2	.4	.6	.8	1.0	1.5	2.0	3.0	z^{tc}
010001010101	38	.0019141	−.0005357	.001619298	.001863024	.001923494	.001782755	.001483804	.000587206	.000119677	.000000699	010110101111
000110010101	37	.0018566	−.0006197	.001606883	.001840037	.001896847	.001761350	.001474100	.000602038	.000130367	.000000959	010111010111
010000111001	38	.0018500	−.0006626	.001604600	.001839631	.001880978	.001742884	.001451677	.000583335	.000123464	.000000848	011100110011
000001011100	39	.0018084	−.0005210	.001600696	.001824161	.001917573	.001813863	.001557906	.000701203	.000175287	.000001978	110000101111
100001010110	39	.0018084	−.0005506	.001597367	.001816591	.001863634	.001800899	.001474481	.000616816	.000138428	.000001119	100010111110
010000101110	39	.0018084	−.0007178	.001595689	.001799370	.001831642	.001677934	.001437648	.000583220	.000125448	.000000899	001110100111
000101100101	37	.0017924	−.0008378	.001589821	.001799145	.001831010	.001676702	.001384235	.000543636	.000112834	.000000733	011101010011
010001100101	37	.0017924	−.0008389	.001588781	.001791458	.001813032	.001646832	.001382365	.000541274	.000111651	.000000521	010101010101
001011010011	37	.0017924	−.0009035	.001588120	.001791458	.001813032	.001646832	.001343129	.000505144	.000097910	.000000521	011101101011
000011100011	37	.0017735	−.0008218	.001586802	.001786373	.001832807	.001686237	.001399757	.000562665	.000120661	.000000885	011100010111
010100010011	37	.0017735	−.0008741	.001585420	.001790554	.001819616	.001664963	.001372591	.000539206	.000112221	.000000766	001101110101
001100101011	37	.0017735	−.0009324	.001583996	.001783561	.001803193	.001637540	.001336355	.000505204	.000098963	.000000547	010101111001
011000101011	38	.0017339	−.0008700	.001578433	.001781081	.001815091	.001671627	.001392176	.000570171	.000126624	.000001053	001011011101
001001010011	37	.0017339	−.0010432	.001574067	.001761762	.001769502	.001596941	.001295188	.000482341	.000093174	.000000504	011010010011
001010100011	37	.0017339	−.0010570	.001573716	.001759664	.001765824	.001590934	.001287440	.000475562	.000093747	.000000472	011100101111
100100000111	38	.0017320	−.0007373	.001581469	.001796206	.001851566	.001733921	.001476689	.000657651	.000166531	.000002180	100110101010
000111000110	38	.0017320	−.0009777	.001575331	.001767879	.001785043	.001622197	.001327306	.000509103	.000102251	.000000611	100101010011
000011001011	38	.0017320	−.0009889	.001575048	.001766598	.001782113	.001617436	.001321192	.000503793	.000102025	.000000587	010101010101
100001101011	37	.0016923	−.0013025	.001588189	.001758046	.001779652	.001627423	.001346039	.000538507	.000116082	.000000883	001011101011
010010110010	38	.0016923	−.0010087	.001567432	.001754885	.001772982	.001617435	.001333242	.000530365	.000113935	.000000883	011100100111
001010011110	38	.0016923	−.0010990	.001565126	.001744281	.001748273	.001576425	.001279286	.000479880	.000094074	.000000535	100011101011
100001010001	38	.0016735	−.0010497	.001562920	.001744869	.001757510	.001597749	.001311552	.000514016	.000107762	.000000760	001101110110
001010101010	38	.0016735	−.0011357	.001560790	.001735355	.001735927	.001562821	.001266680	.000474261	.000092886	.000000525	101010010011
000111010010	38	.0016574	−.0011401	.001559456	.001729301	.001735037	.001540174	.001237494	.000448694	.000083697	.000000404	100101011011
001101000011	36	.0016494	−.0012488	.001557861	.001731574	.001735037	.001568381	.001280008	.000495362	.000103297	.000000755	001110010011
001010100101	37	.0016494	−.0012984	.001553666	.001717091	.001705182	.001522268	.001222268	.000445570	.000084762	.000000455	010011110010
001011100011	37	.0016494	−.0013167	.001552420	.001711480	.001692418	.001501653	.001195966	.000423142	.000076920	.000000356	010011101011
010011001001	38	.0016093	−.0013025	.001551961	.001709415	.001687730	.001494107	.001186385	.000415127	.000074201	.000000325	010101011011
010000110010	38	.0016093	−.0013025	.001545019	.001700527	.001684095	.001450570	.001204942	.000440814	.000084585	.000000466	010101111011
010010110001	37	.0015909	−.0014390	.001538335	.001681079	.001648286	.001450570	.001146225	.000398211	.000071233	.000000321	011100111011
000110111100	38	.0015494	−.0012757	.001534724	.001687486	.001679148	.001512803	.001234670	.000484059	.000103814	.000000825	110001001111
100001001110	38	.0015494	−.0014053	.001531524	.001673222	.001646843	.001460578	.001167596	.000424446	.000081271	.000000449	010011101110
010001001110	38	.0015494	−.0014724	.001529914	.001666267	.001631622	.001436860	.001138317	.000401198	.000073597	.000000360	001111001101
010011000011	36	.0015333	−.0015009	.001526368	.001659632	.001624006	.001431069	.001136388	.000406958	.000077391	.000000446	001011100011
000111100011	36	.0015333	−.0015021	.001526331	.001659431	.001623468	.001430065	.001134925	.000405279	.000077621	.000000430	011101010011
010110000101	38	.0015248	−.0015666	.001524732	.001652309	.001607422	.001404368	.001102368	.000377814	.000067010	.000000303	110000100111
100001001100	38	.0015248	−.0013612	.001528123	.001671441	.001652728	.001477865	.001195508	.000455172	.000093587	.000000646	010100111011
010000110110	38	.0015248	−.0014585	.001525753	.001661039	.001629554	.001441047	.001149090	.000416003	.000079656	.000000449	101001011011
001101101010	36	.0015088	−.0015317	.001524000	.001653465	.001612965	.001415165	.001117083	.000390409	.000071108	.000000345	100101111110
000111001000	36	.0015088	−.0015447	.001520804	.001648244	.001608215	.001413454	.001119681	.000396634	.000075283	.000000423	011100100111
010101000011	36	.0015088	−.0015809	.001519949	.001644609	.001600402	.001401527	.001105283	.000387905	.000071987	.000000390	001101110101
001100010011	36	.0015088	−.0016415	.001518446	.001637913	.001585311	.001377360	.001074674	.000362123	.000062987	.000000272	011101010101
010101010011	36	.0014692	−.0017513	.001508546	.001615763	.001552000	.001337433	.001034427	.000340140	.000057516	.000000232	010101010011
001011001010	37	.0014673	−.0016794	.001509901	.001621357	.001571139	.001361051	.001063264	.000361812	.000064128	.000000295	011101011011
001100001011	36	.0014503	−.0016487	.001507564	.001621357	.001571470	.001373411	.001082990	.000382977	.000072248	.000000417	001110111001
001100111001	36	.0014503	−.0017218	.001505793	.001613626	.001554370	.001346476	.001049365	.000354413	.000058505	.000000291	011101011001
000110111001	36	.0014503	−.0017678	.001504698	.001608948	.001544269	.001330984	.001030587	.000341300	.000094683	.000000248	011100111011
100000001111	39	.0014248	−.0015184	.001505966	.001627583	.001594745	.001417519	.001143840	.000440655	.000106761	.000000823	100011111110
101000000111	37	.0014088	−.0014994	.001503536	.001625704	.001597487	.001427985	.001162427	.000465176	.000115235	.000001179	000111110011
000111110010	37	.0014088	−.0016712	.001499390	.001607568	.001557015	.001363199	.001079612	.000390522	.000076843	.000000511	101100100111

Table A

m = 7, n = 5   Rank Orders 201 thru 250 of 792   C(7, 5) = 792   1/C(7, 5) = .00126263

z	w	c_1	c_2	d = .2	.4	.6	.8	1.0	1.5	2.0	3.0	z^{tc}
100100001011	37	.0014088	−.0016760	.001499234	.001606717	.001554763	.001359082	.001073762	.000384225	.000074098	.000000453	001011101110
100000110011	37	.0014088	−.0017098	.001498420	.001603185	.001546990	.001346891	.001058597	.000371879	.000069869	.000000398	001100111110
100010100011	37	.0014088	−.0017328	.001497860	.001600733	.001541547	.001338279	.001047891	.000362875	.000066708	.000000354	001101110011
001100001110	37	.0014088	−.0017878	.001496566	.001595244	.001529727	.001320129	.001025673	.000345728	.000061101	.000000286	100011100011
000101011010	37	.0014088	−.0018028	.001496205	.001593675	.001526271	.001314704	.001018923	.000340243	.000059238	.000000263	101001100111
000100101010	37	.0014088	−.0018118	.001495990	.001592754	.001524277	.001311642	.001015208	.000337459	.000058388	.000000255	101001010111
001010000110	37	.0014088	−.0018534	.001494979	.001588335	.001514519	.001296341	.000996255	.000322452	.000053526	.000000203	100101100011
010100010110	37	.0014088	−.0018669	.001494654	.001586926	.001511439	.001291570	.000990424	.000318021	.000052166	.000000191	100101110011
001110000011	35	.0013927	−.0017252	.001495162●	.001597661	.001541489	.001343867	.001059394	.000378392	.000073478	.000000478	011111000011
010010100101	36	.0013847	−.0019733	.001487747	.001569858	.001483862	.001256786	.000954122	.000297530	.000045968	.000000161	011010110101
010100100101	36	.0013658	−.0019946	.001483776	.001562479	.001474829	.001248209	.000947536	.000296296	.000047338●	.000000166	010101110101
001010011001	37	.0013502	−.0019806	.001481201	.001559210	.001473060	.001249229●	.000951169●	.000300717	.000048730	.000000175	101011010101
011000001101	36	.0013262	−.0019594	.001477303	.001554881	.001472742	.001255859●	.000964576	.000316625	.000054697	.000000248	011101001101
010000011011	36	.0013262	−.0020157	.001474659	.001543737	.001448994	.001219912	.000914982	.000285304	.000045307	.000000147	011101010011
010010111001	36	.0013262	−.0020882	.001474276	.001542092	.001445439	.001214465	.000909331	.000280441	.000045852	.000000160	011101101110
000111011100	37	.0012846	−.0018924	.001471186	.001549623●	.001476239	.001272600	.000993321	.000347773	.000066523	.000000422	110001001111
100000111110	37	.0012846	−.0020358	.001467784	.001535061	.001446607	.001223590	.000933046	.000299864	.000050387	.000000210	110100111110
010101001010	37	.0012846	−.0021004	.001466300	.001528928	.001431777	.001204503	.000910580	.000284111	.000045831	.000000170	110001101111
001101011001	37	.0012658	−.0020045	.001465013	.001532614●	.001445852	.001230279	.000944441	.000311382	.000054314	.000000251	110101011001
100001010101	36	.0012658	−.0021019	.001462736	.001523029	.001425424	.001193305	.000907268	.000283986	.000045968	.000000170	110101011110
010001010110	37	.0012658	−.0021750	.001461053	.001516069	.001410864	.001177664	.000881844	.000266340	.000050963	.000000130	100101011101
001011010001	35	.0012497	−.0020966	.001459994	.001519793●	.001424998	.001204195	.000917454	.000297454	.000051380	.000000246	101011010101
001101100001	35	.0012497	−.0021328	.001459169	.001517989	.001417989	.001193864	.000905667	.000289188	.000049029	.000000225	011110101010
001101000101	35	.0012497	−.0021933	.001457725	.001510221	.001404559	.001173156	.000880404	.000269767	.000039101	.000000156	011011001011
000110000110	35	.0012101	−.0022096	.001448229	.001489833	.001375110	.001139230	.000847512	.000253419	.000039101	.000000133	011011010011
000110000111	36	.0012082	−.0021281	.001449630●	.001496456	.001389435	.001160743	.000872826	.000270823	.000049669	.000000173	011011001001
001000011100	37	.0012016	−.0021609	.001450251	.001502034	.001403074	.001183284	.000899371	.000290039	.000047769	.000000225	110100011011
010010001010	37	.0012016	−.0022566	.001449020	.001499020	.001396882	.001173340	.000888727	.000283080	.000041721	.000000211	110000111110
010000011010	37	.0012016	−.0022458	.001447572	.001491050	.001380259	.001148653	.000859673	.000262566	.000046181	.000000154	011000111010
011000011010	35	.0011856	−.0022238	.001445149	.001488861	.001381449●	.001155079	.000870894	.000274709	.000046181	.000000210	001110110011
001010010001	35	.0011856	−.0023151	.001443019	.001479935	.001362507	.001126485	.000836721	.000249708	.000038554	.000000132	011100101011
001100101001	35	.0011856	−.0023473	.001442291	.001476978	.001356449	.001117684	.000826628	.000243146	.000036817	.000000120	101011100111
100000101110	38	.0011601	−.0021702	.001441534	.001485611●	.001355880	.001161348	.000881341	.000283905	.000048809	.000000227	100010101011
000101011010	36	.0011440	−.0022866	.001435933	.001470278	.001339179	.001126616	.000843848	.000261148	.000042822	.000000181	001110101110
000110101010	36	.0011440	−.0023726	.001433975	.001463244	.001328304	.001085601	.000814840	.000240818	.000036864	.000000124	101010101011
010010011010	36	.0011256	−.0024255	.001432743	.001457094	.001314737	.001070522	.000795584	.000227319	.000033019	.000000093	001101010101
010011000101	35	.0011256	−.0024736	.001428264	.001447509	.001314737	.001075460	.000781688	.000221629	.000032068	.000000092	011100101101
100110010010	36	.0011252	−.0023089	.001431914●	.001462595	.001345999	.001114278	.000835095	.000257888	.000042262	.000000178	011101011011
000101110010	37	.0011252	−.0023415	.001431172	.001459543	.001339609	.001106919	.000823724	.000249519	.000039642	.000000149	011010010111
001100010110	36	.0011252	−.0024318	.001429098	.001450986	.001321763	.001080480	.000792766	.000248195	.000033619	.000000100	100101011001
010101000101	35	.0011011	−.0025363	.001422289	.001434533	.001295886	.001126451	.000760707	.000211504	.000029889	.000000081	010110101011
101000001001	36	.0010855	−.0022754	.001425286●	.001453916	.001340795	.001118330	.000844080	.000270842	.000047186	.000000241	001101111010
001001011010	35	.0010855	−.0025158	.001419822	.001431543	.001294368	.001049720	.000763727	.000214757	.000030799	.000000086	101010110011
000101011010	36	.0010855	−.0025271	.001413140	.001430482	.001292169	.001046490	.000759988	.000212284	.000030142	.000000081	101010101011
010010101001	35	.0010615	−.0026131	.001410776	.001419697	.001269697	.001019900	.000733949	.000199776	.000027580	.000000071	010101011101
011000010001	36	.0010426	−.0025633	.001410629	.001415640	.001273190●	.001028891	.000746765	.000209895	.000030373	.000000089	010101111001
011000011001	35	.0010426	−.0010629	.001409433	.001414067	.001272295	.001027825	.000745812	.000199714	.000030444●	.000000091	011000111101
001010111010	35	.0010426	−.0026240	.001412425●	.001414255	.001262531	.001013775	.000729858	.000199604	.000027839	.000000074	011011100001
000110011001	36	.0010111	−.0021425	.001409255	.001409255	.001263980	.001137895	.000727839	.000317378	.000046617	.000000638	001111111100
110000001001	36	.0010111	−.0024509	.001414425	.001444437	.001321763	.001070717	.000882291	.000254769	.000044509	.000000241	110101000111

124

Table A

m = 7, n = 5   Rank Orders 251 thru 300 of 792   C(7, 5) = 792   1/C(7, 5) = .00126263

z	w	c_1	c_2	d = .2	.4	.6	.8	1.0	1.5	2.0	3.0	z^{tc}
00110111101100	36	.0010011	−.0025065	.001404263	.001406981	.001270220	.001033520	.000758092	.000222365	.000034368	.000000124	11010010011
00011011001101	36	.0010011	−.0025133	.001404110	.001406362	.001268948	.001031665	.000755952	.000220944	.000033981	.000000106	11000100011
10010000011001	36	.0010011	−.0025556	.001403173	.001402637	.001261465	.001020948	.000743826	.000213244	.000031964	.000000106•	01001110011
10000011000110	36	.0010011	−.0025968	.001402235	.001398791	.001253484	.001009211	.000730163	.000203961	.000029374	.000000085	01101110010
01010000010110	36	.0010011	−.0026143	.001401885	.001397546	.001251277	.001006478	.000727429	.000203061	.000029336	.000000088•	10001110101
01000011001100	36	.0010011	−.0026177	.001401761	.001396859	.001249502	.001003384	.000723429	.000199489	.000028166	.000000077	10110101110
01000010010110	36	.0010011	−.0026725	.001400563	.001392154	.001241079	.000990248	.000708827	.000190704	.000026020	.000000064	10110011101
00111110000011	34	−.0009850	−.0026882	.001401451	.001390699	.001237183	.000985904	.000703852	.000187511	.000025200	.000000159	01110010011
01110000011010	34	−.0009850	−.0025006	.001400205	.001403102	.001267688	.001034123	.000762106	.000228789	.000036957	.000000159	01111001101
—	34	−.0009850	−.0025528	.001400313	.001398633	.001258793	.001021510	.000747929	.000219842	.000034587	.000000140	00111100101
00111001011100	34	−.0009850	−.0026111	.001398974	.001393102	.001247216	.001004264	.000727572	.000205357	.000030268	.000000097	01011100011
10000101110010	36	−.0009426	−.0026434	.001390247	.001376810	.001226437	.000982744	.000708444	.000196977	.000028378	.000000083	11000010110
10000011000011	36	−.0009426	−.0026763	.001389535	.001374035	.001220983	.000975149	.000700104	.000192145	.000027256	.000000078	01100101110
01100100000111	36	−.0009265	−.0027611	.001387675	.001366707	.001206387	.000954501	.000677025	.000178027	.000023734	.000000056	10100101110
00101010000001	34	−.0009265	−.0027262	.001387262	.001372431	.001223264	.000982926	.000712506	.000204147	.000031291	.000000120	00111101000
00110111010001	34	−.0009265	−.0027389	.001385215	.001364178	.001206414	.000958445	.000684338	.000185368	.000026054	.000000075	01110101011
00111011000111	34	−.0009020	−.0027711	.001384514	.001361436	.001201005	.000950874	.000675973	.000180402	.000024851	.000000068	01101010011
10000110000011	34	−.0009020	−.0027873	.001379567	.001352527	.001190270	.000940625	.000667801	.000178029	.000024570	.000000068•	01100101011
—	37	−.0009010	−.0026638	.001382001	.001362223	.001209242	.000966800	.000696282	.000194255	.000028844	.000000088	01110011001
—	35	−.0008850	−.0026888	.001378515	.001355687	.001201381	.000959762	.000691547	.000194769•	.000029189	.000000104	00111011110
00111101010010	35	−.0008850	−.0027748	.001376633	.001343267	.001186545	.000938627	.000651677	.000179521	.000025101	.000000072	10101100011
01011000110010	35	−.0008850	−.0028276	.001375449	.001343499	.001176850	.000924641	.000651744	.000169315	.000022435	.000000053	10100011101
10000100000110	37	−.0008765	−.0026963	.001372696	.001351908	.001195730	.000952680	.000683849	.000189428	.000027503	.000000086	11000111011
01101010110010	35	−.0008605	−.0027479	.001372622	.001343036	.001183158	.000938856	.000671376	.000184844	.000026931	.000000089•	01110011010
01110110010010	35	−.0008605	−.0027966	.001371556	.001338822	.001174703	.000926763	.000657653	.000175953	.000024509	.000000070	01110001101
01010010000100	34	−.0008757	−.0028757	.001369813	.001331921	.001160906	.000907188	.000635721	.000162472	.000021129	.000000048	10010100011
01010010101010	35	−.0008421	−.0029254	.001365315	.001322312	.001147371	.000892232	.000622013	.000156867	.000020153	.000000045	01010101101
01001011010010	34	−.0008208	−.0029561	.001360632	.001312923	.001134547	.000878110	.000608812	.000150735	.000018837	.000000038	01010101101
10010100010011	34	−.0008024	−.0029875	.001361217	.001304932	.001124249	.000871841	.000600225	.000148215	.000018760	.000000040•	01110111100
01100011001001	35	−.0008020	−.0027664	.001361217•	.001323226	.001160211	.000917918	.000655579	.000181219	.000026760	.000000093	00110111110
00100111010010	35	−.0008020	−.0029126	.001358040	.001310784	.001135532	.000883065	.000616588	.000157004	.000020494	.000000048	10100010111
01100001110010	36	−.0008020	−.0029501	.001357233	.001357678	.001129493	.000874743	.000607543	.000151887	.000019327	.000000042	10100011110
01000011000110	36	−.0007939	−.0028107	.001358726•	.001316802	.001149549	.000904184	.000640959	.000172453	.000024430	.000000074	11000111101
01101011001001	34	−.0007779	−.0029732	.001352235	.001298446	.001118133	.000863723	.000598676	.000149364	.000019074	.000000043	01110011010
00011010010011	34	−.0007779	−.0029989	.001351692	.001296386	.001114191	.000858380	.000592967	.000146275	.000018401	.000000040	01110011010
01010100010010	35	−.0007779	−.0030362	.001350896	.001293343	.001108334	.000850397	.000584395	.000141583	.000017369	.000000035	01110100101
00011010011010	35	−.0007363	−.0029090	.001345780	.001290696	.001087415	.000865453	.000605418	.000156909	.000021276	.000000058	11000101101
10010101001001	35	−.0007363	−.0030064	.001343676	.001282492	.001097941	.000843150	.000580876	.000142464	.000017800	.000000037	01010101010
01010011011001	34	−.0007363	−.0030795	.001342128	.001276612	.001086668	.000827831	.000564457	.000133498	.000015829	.000000028	01101011101
10000011001001	35	−.0007194	−.0030301	.001340031	.001275642	.001089504	.000835100	.000574635	.000141101	.000017800	.000000040	01100111100
00010111010100	35	−.0007175	−.0028703	.001343058•	.001287946	.001113724	.000868648	.000611250	.000162206	.000022804	.000000069	10100010111
10100010010110	35	−.0007175	−.0030137	.001339978	.001276046	.001090463	.000836325	.000575725	.000141253	.000017716	.000000038	01010111011
00011111000100	35	−.0007175	−.0030783	.001338606	.001270886	.001080597	.000822944	.000561404	.000133438	.000015992	.000000030	10010111010
11000001011010	35	−.0006778	−.0027090	.001339011	.001288656	.001125857	.000893311	.000640303	.000187614	.000030465	.000000147	11100000111
10100101001110	35	−.0006778	−.0027126	.001338887	.001287996	.001124205	.000893490	.000640489	.000184286	.000029252	.000000129	00101111100
00100110000011	34	−.0006778	−.0029491	.001333905	.001269019	.001087415	.000839490	.000584058	.000149931	.000018870	.000000056	10001111101
00100110011010	35	−.0006778	−.0029961	.001332920	.001265296	.001076062	.000829787	.000573591	.000144006	.000018262	.000000048	10001111011
00010011000110	35	−.0006778	−.0030135	.001332615	.001263620	.001075699	.000824628	.000567595	.000140002	.000017788	.000000040	11001011101
10001011001001	35	−.0006778	−.0030226	.001332320	.001262857	.001075294	.000822530	.000565281	.000138638	.000017463	.000000028	11000101101
10000010011001	35	−.0006778	−.0030416	.001331933	.001261458	.001072748	.000819267	.000562007	.000137188	.000017230	.000000038	01010011110

125

Table A
m = 7, n = 5 **Rank Orders 301 thru 350 of 792** **C(7, 5) = 792** **1/C(7, 5) = .00126263**

z	w	c_1	c_2	d = .2	.4	.6	.8	1.0	1.5	2.0	3.0	z^{tc}
100011011001	35	.0006778	−.0030602	.001331531	.001259894	.001069680	.000814996	.000557311	.000134438	.000016576	.000000034	110011011110
010001101010	35	.0006778	−.0031290	.001330093	.001254500	.001059457	.000801252	.000542730	.000126649	.000014892	.000000027	101010011101
001010011010	35	.0006778	−.0031425	.001253222	.001253222	.001057259	.000798221	.000539439	.000124808	.000014482	.000000025	101011001101
011101100001	35	.0006618	−.0029380	.001331187	.001265400	.001085132	.000839981	.000587326	.000154628	.000021949	.000000074	001110010011
001011100000	33	.0006618	−.0030111	.001329592	.001259135	.001072658	.000822294	.000567449	.000142093	.000018627	.000000048	011101100011
001110100001	33	.0006618	−.0030571	.001328637	.001255571	.001065933	.000813276	.000557888	.000136943	.000017484	.000000042	011110010011
011101010001	33	.0006429	−.0030998	.001324198	.001246338	.001053157	.000799299	.000545110	.000131585	.000016479	.000000038	100011101110
100010001110	33	.0006363	−.0030786	.001324398	.001249414	.001060112	.000809072	.000555474	.000136708	.000017451	.000000041	100111101110
100001001101	36	.0006014	−.0030786	.001319725	.001239153	.001045091	.000791663	.000538597	.000128500	.000015670	.000000031	100111011010
010010101010	34	.0006014	−.0030388	.001317631	.001237967	.001047587	.000798792	.000549114	.000137253	.000018216	.000000051	011111011110
000111010010	34	.0006014	−.0030874	.001316608	.001234087	.001040116	.000788530	.000537925	.000130679	.000016586	.000000040	101101000111
010001000010	34	.0006014	−.0031665	.001314930	.001227698	.001027821	.000771731	.000519795	.000120515	.000014255	.000000027	100111110011
010111000010	33	.0005773	−.0031781	.001310168	.001219319	.001018007	.000762681	.000512885	.000118820	.000012699	.000000028	010111110011
001011011010	34	.0005617	−.0032303	.001306125	.001210210	.001004515	.000747020	.000497802	.000111789	.000016077	.000000038	101001011011
000000111010	36	.0005533	−.0030805	.001307593	.001218887	.001022803	.000772720	.000525702	.000127109	.000016958	.000000046	101110011010
101000100001	34	.0005372	−.0030736	.001304776	.001201850	.001010439	.000777166	.000529476	.000129476	.000017484	.000000021	101011101110
010010101010	34	.0005372	−.0032381	.001301339	.001199886	.000994720	.000737895	.000490673	.000109761	.000012443	.000000032	101101010011
010010110110	34	.0005372	−.0032643	.001300802	.001199886	.000991100	.000733175	.000485826	.000107400	.000011985	.000000020	101010101101
100010110110	34	.0005349	−.0031632	.001302411	.001206635	.001004062	.000750435	.000503797	.000116388	.000013777	.000000026	110001011110
010100010110	33	.0005188	−.0032342	.001297975	.001196540	.000989391	.000734022	.000488746	.000110409	.000012790	.000000024	011101101101
010101010001	33	.0005188	−.0032599	.001297453	.001194637	.000985892	.000729463	.000484063	.000108111	.000012336	.000000022	011101010101
010101000001	33	.0005188	−.0032972	.001296686	.001191817	.000980662	.000722597	.000476959	.000104566	.000011623	.000000019	011101011010
010100110100	34	.0004943	−.0032947	.001292083	.001184133	.000972120	.000715112	.000471522	.000103371	.000011532	.000000020	110011100111
001110010100	34	.0004773	−.0031483	.001291877	.001189738	.000986424	.000736497	.000495343	.000116615	.000014415	.000000033	010111001010
010001001101	34	.0004773	−.0032457	.001289853	.001182180	.000972150	.000717352	.000475047	.000105714	.000012015	.000000021	110011001101
100110110110	34	.0004773	−.0033188	.001288365	.001176749	.000962141	.000704270	.000461558	.000099017	.000010674	.000000016	011011110111
010001000101	33	.0004547	−.0033105	.001284301	.001170444	.000955934	.000699860	.000459436	.000099758	.000011041	.000000018	011100111001
010011010110	34	.0004547	−.0031580	.001284368	.001181145	.000976117	.000726600	.000487313	.000114013	.000014019	.000000032	110011011010
100100010100	34	.0004528	−.0032876	.001283007	.001171251	.000957592	.000695423	.000451423	.000100440	.000011106	.000000020	011001101110
010100100110	34	.0004528	−.0033548	.001283007	.001166311	.000948541	.000690219	.000449411	.000094598	.000009966	.000000014	100110101101
010101010110	34	.0004131	−.0032865	.001276891	.001158650	.000943354	.000689186	.000451817	.000097948	.000010817	.000000018	110010101011
100010010010	34	.0004131	−.0033194	.001276232	.001156284	.000939076	.000683715	.000446312	.000095409	.000010353	.000000016	011010101101
010100010100	34	.0004131	−.0034042	.001274520	.001150086	.000927771	.000669082	.000431398	.000088249	.000008976	.000000011	101010101101
110100011000	34	.0003943	−.0030234	.001278640	.001171997	.000972382	.000730366	.000496742	.000122999	.000016519	.000000050	111010101101
001010111100	34	.0003943	−.0030572	.001277960	.001169511	.000967731	.000724114	.000490025	.000119086	.000015522	.000000042	110100001110
001001111100	34	.0003943	−.0032353	.001274372	.001156470	.000943675	.000692474	.000457017	.000101793	.000011732	.000000022	011110011001
100001011100	33	.0003943	−.0032523	.001274034	.001155268	.000941523	.000689756	.000454322	.000100613	.000011545	.000000020	011100111001
100100011000	34	.0003943	−.0032652	.001273764	.001154243	.000939389	.000687072	.000451423	.000098978	.000011158	.000000020	010110111010
001110010110	35	.0003943	−.0032655	.001273760	.001154234	.000939550	.000687073	.000451432	.000098987	.000011159	.000000017	110011101110
100110110011	34	.0003943	−.0033142	.001272779	.001150692	.000933097	.000678740	.000442947	.000094911	.000010371	.000000017	100111110011
011000101100	34	.0003943	−.0033181	.001276698	.001150380	.000932486	.000677887	.000441987	.000094319	.000010223	.000000016	100101111001
010100110110	34	.0003943	−.0033456	.001272153	.001148455	.000929049	.000673535	.000437655	.000092351	.000009865	.000000015	101110011101
010100101100	34	.0003943	−.0033905	.001271244	.001145154	.000923012	.000665722	.000429693	.000088544	.000009012	.000000012	001111110001
011110000011	32	.0003782	−.0030839	.001274445	.001162850	.000959521	.000716335	.000484108	.000118080	.000015703	.000000048	011101110100
001101110000	32	.0003782	−.0032571	.001270945	.001150087	.000935893	.000685125	.000451374	.000100578	.000011727	.000000023	100111110110
011100010001	34	.0003782	−.0032710	.001270676	.001149146	.000934227	.000683030	.000449291	.000099620	.000011545	.000000015	100101011110
100100110110	32	.0003527	−.0032586	.001266031	.001141388	.000925868	.000674515	.000442252	.000099692	.000011017	.000000015	110011011010
100100110011	35	.0003527	−.0033083	.001265005	.001137583	.000918132	.000665012	.000435012	.000091692	.000008894	.000000014	100111010110
001101100110	35	.0003527	−.0033265	.001264629	.001136195	.000915546	.000661595	.000428694	.000089881	.000009523	.000000014	110011101110
100110110011	33	.0003367	−.0031935	.001264356	.001141368	.000929188	.000682479	.000452478	.000103864	.000012753	.000000031	001111011010

Table A

m = 7, n = 5   Rank Orders 351 thru 400 of 792   C(7, 5) = 792   1/C(7, 5) = .00126263

z	w	c_1	c_2	d = .2	.4	.6	.8	1.0	1.5	2.0	3.0	z^{tc}
00111100010	33	.0003367	−.0032261	.001263686	.001138878	.000924484	.000676133	.000445674	.000100013	.000011830	.000000025	10110000111
00111001010	35	.0003367	−.0033165	.001261857	.001132215	.000912199	.000660025	.000428966	.000091472	.000010027	.000000017	10011100011
10000101010	35	.0002942	−.0033357	.001253385	.001116862	.000893253	.000641176	.000413035	.000085680	.000009041	.000000013	10110101011
10100100001	33	.0002782	−.0032364	.001252413	.001119424	.000901956	.000655412	.000429741	.000095803	.000011403	.000000026	00111101110
00101101010	33	.0002782	−.0034009	.001249109	.001105665	.000880010	.000626988	.000406469	.000081081	.000008333	.000000012	10101001011
11000001010	33	.0002782	−.0034272	.001248593	.001105665	.000876884	.000622936	.000396463	.000079305	.000008019	.000000012	01101111010
01000101100	34	.0002701	−.0031256	.001252993	.001107166	.000881432	.000629049	.000442637	.000101811	.000012519	.000000030	01001111100
01000100110	34	.0002701	−.0033660	.001248253	.001107166	.000881057	.000624018	.000402968	.000082394	.000008571	.000000012	10011011101
01001100001	34	.0002701	−.0033773	.001248023	.001106323	.000879497	.000629049	.000400853	.000081341	.000008361	.000000012	11000110101
01101000101	32	.0002541	−.0033573	.001245436	.001102962	.000877186	.000626650	.000402238	.000083313	.000008908	.000000014	01011110101
01001110001	32	.0002541	−.0033647	.001245284	.001102406	.000876158	.000625306	.000400856	.000082639	.000008777	.000000014	01111000101
01101001001	32	.0002541	−.0034181	.001244239	.001098736	.000869652	.000617127	.000392741	.000078970	.000008103	.000000013	01101100101
00110001001	32	.0002537	−.0034224	.001244047	.001098137	.000868531	.000615580	.000391040	.000077984	.000007879	.000000011	10110110101
01101010001	32	.0002352	−.0033458	.001240136	.001090840	.000859547	.000606912	.000384124	.000076040	.000007665	.000000011	01110101011
01101001001	32	.0001956	−.0034434	.001232676	.001078264	.000845298	.000594047	.000374360	.000073420	.000007345	.000000010	10101010101
00111010100	33	.0001937	−.0032861	.001235391	.001088452	.000863771	.000617589	.000397975	.000084400	.000009436	.000000018	11010100011
10101001010	33	.0001937	−.0034156	.001232821	.001079290	.000847261	.000596456	.000376589	.000074185	.000007434	.000000010	11010101010
01010100110	33	.0001937	−.0034828	.001231514	.001074731	.000839497	.000586445	.000366749	.000069849	.000006856	.000000008	10011010110
01001101001	32	.0001711	−.0034283	.001228328	.001071411	.000838109	.000588114	.000370309	.000072678	.000007305	.000000010	01110010110
00101101001	33	.0001541	−.0033978	.001225655	.001067721	.000834747	.000585710	.000368870	.000072410	.000007254	.000000010	01110101101
10001100101	33	.0001541	−.0034307	.001225021	.001065531	.000830938	.000581022	.000364329	.000070504	.000006937	.000000009	01101100101
01001100110	33	.0001541	−.0035155	.001223376	.001059804	.000820887	.000568522	.000353076	.000065163	.000006003	.000000008	10101001010
11000100011	35	.0001456	−.0032307	.001227284	.001076268	.000851546	.000607909	.000391623	.000083350	.000009391	.000000018	11000011110
00110110100	33	.0001295	−.0031633	.001227072	.001075707	.000853889	.000613101	.000398290	.000087655	.000010425	.000000024	00110111100
00110110010	33	.0001295	−.0032132	.001224585	.001072288	.000847739	.000605176	.000390149	.000083469	.000009497	.000000019	10011000111
10100101010	33	.0001295	−.0033935	.001221094	.001060113	.000826215	.000578086	.000363143	.000070917	.000007079	.000000010	11010101010
01001100101	33	.0001295	−.0034073	.001220829	.001059199	.000824630	.000576141	.000361621	.000070141	.000006952	.000000009	11010101010
00010011010	33	.0001295	−.0034146	.001220686	.001058699	.000823746	.000575029	.000360159	.000069631	.000006856	.000000009	01100111010
01000110001	33	.0001295	−.0034153	.001220674	.001058662	.000823682	.000574996	.000337934	.000069688	.000006880	.000000009	01110101110
00101101001	33	.0001295	−.0034586	.001219838	.001055769	.000818644	.000568740	.000354057	.000067050	.000006421	.000000008	01110011010
01000100110	33	.0001295	−.0034628	.001219756	.001055473	.000818099	.000568021	.000353301	.000066649	.000006334	.000000007	10011011001
01010010101	33	.0001295	−.0035061	.001218920	.001052583	.000813060	.000561802	.000347258	.000064081	.000005900	.000000006	10110101010
01010101010	33	.0001295	−.0035375	.001218315	.001050496	.000809441	.000557369	.000342990	.000062325	.000005616	.000000006	10101010101
00110101010	31	.0001135	−.0033221	.001219488	.001062115	.000829731	.000584905	.000371510	.000075939	.000008182	.000000015	01111011101
10000101010	33	.0001135	−.0034500	.001212097	.001047218	.000803695	.000549944	.000336992	.000060352	.000005942	.000000013	10011001110
00110101010	33	.0000880	−.0032597	.001212549	.001049940	.000819652	.000578658	.000365992	.000074561	.000007954	.000000013	11100011011
01001010100	33	.0000710	−.0034031	.001208589	.001040671	.000803653	.000557228	.000346946	.000066354	.000006507	.000000009	11100111010
10100010010	33	.0000710	−.0034677	.001208589	.001036887	.000796188	.000547996	.000337934	.000062442	.000005957	.000000007	01110101110
01100001010	32	.0000691	−.0034348	.001208825	.001037701	.000798622	.000550961	.000340700	.000063444	.000005957	.000000007	10011110110
01010010010	34	.0000295	−.0033232	.001203476	.001032468	.000797292	.000554249	.000346898	.000068014	.000006959	.000000011	10011111010
10001101010	34	.0000295	−.0034359	.001201298	.001024904	.000784009	.000537683	.000330576	.000060734	.000005636	.000000006	10100111010
10101000110	34	.0000295	−.0034519	.001200985	.001023804	.000782064	.000535245	.000328169	.000059667	.000005448	.000000006	10100111010
10101010010	32	.0000134	−.0032616	.001201665	.001031712	.000799403	.000559242	.000353413	.000072204	.000007932	.000000016	01111111010
00110101010	32	.0000134	−.0034078	.001198665	.001031712	.000791839	.000537390	.000331278	.000061833	.000005905	.000000012	00110111010
01010101010	33	.0000134	−.0034453	.001198126	.001014780	.000771837	.000532281	.000326722	.000060006	.000005610	.000000007	10101010101
11000101010	33	.0000054	−.0034739	.001196030	.001014780	.000771133	.000524792	.000319834	.000057218	.000005234	.000000005	11000111000
10010101100	32	−.0000054	−.0032616	.001194021	.001011588	.000767903	.000522381	.000318508	.000057345	.000005234	.000000006	01010111100
10101010100	32	.0000134	−.0034348	.001196442	.001022160	.000787162	.000547996	.000341708	.000067032	.000006861	.000000007	10101111010
01010011010	33	−.0000134	−.0034078	.001193728	.001013085	.000711729	.000527828	.000324069	.000059746	.000005627	.000000007	11010111010
10100100110	33	−.0000134	−.0034453	.001193009	.001010614	.000767511	.000522559	.000318963	.000057579	.000005257	.000000006	11000111010

z	w	c_1	c_2	d = .2	.4	.6	.8	1.0	1.5	2.0	3.0	z^{tc}
01110000101	31	−.0000295	−.0033232	.001192291	.001013380	.000775307	.000533992	.000331146	.000063447	.000006342	.00000009	010111110001
01010100001	31	−.0000295	−.0034359	.001190196	.001106370	.000761455	.000517756	.000317632	.000057943	.000005425	.00000007	011110010101
01010110001	31	−.0000295	−.0034519	.001189901	.001005401	.000761843	.000517865	.000315865	.000057264	.000005319	.00000005	011011100101
01101001001	31	−.0000691	−.0034348	.001182710	.000993705	.000749046	.000506688	.000307663	.000055230	.000005089	.00000011	110111010001
00011110100	32	−.0000710	−.0034597	.001185629	.001094144	.000767198	.000509100	.000329139	.000064368	.000006693	.00000007	010111100100
10010000101	32	−.0000710	−.0034031	.001182906	.000994836	.000751092	.000509100	.000309830	.000055927	.000005168	.00000006	010111101110
01010000110	32	−.0000710	−.0034677	.001181698	.000990780	.000744214	.000500826	.000301973	.000052746	.000004647	.00000005	011110011001
01100100001	31	−.0000880	−.0034600	.001178859	.000988720	.000740891	.000498582	.000301164	.000053268	.000004825	.00000005	011101011001
01100101100	32	−.0001135	−.0033221	.001176320	.000985888	.000743180	.000503795	.000307080	.000055917	.000005234	.00000005	011011101110
01110001100	34	−.0001135	−.0035375	.001176291	.000981616	.000755160	.000506229	.000307080	.000055917	.000005234	.00000010	011011011001
10001000011	32	−.0001295	−.0031633	.001176291	.000991017	.000741969	.000506582	.000304764	.000064677	.000006979	.00000014	001110110011
00011011000	32	−.0001295	−.0032132	.001175362	.000987875	.000749742	.000513836	.000318167	.000061613	.000006364	.00000011	111010000111
00101001100	32	−.0001295	−.0033935	.001172004	.000976609	.000730581	.000490683	.000295908	.000052207	.000004713	.00000005	110101001011
10100100100	32	−.0001295	−.0034073	.001171748	.000975759	.000729160	.000489002	.000294344	.000051615	.000004624	.00000005	110011010010
00011010001	32	−.0001295	−.0034146	.001171611	.000975301	.000728384	.000488006	.000293451	.000051246	.000004562	.00000005	110010101001
10010101000	32	−.0001295	−.0034153	.001171600	.000975269	.000728347	.000488049	.000293466	.000051295	.000004579	.00000004	011010100110
01100100010	32	−.0001295	−.0034586	.001170795	.000972587	.000723835	.000482670	.000288410	.000049307	.000004264	.00000004	010101011010
01100101010	32	−.0001295	−.0034628	.001170718	.000972326	.000723382	.000482105	.000287848	.000049042	.000004213	.00000004	100110110010
01001010010	32	−.0001295	−.0035061	.001169914	.000969652	.000718899	.000476783	.000282872	.000047122	.000003918	.00000003	011110110101
01010101010	32	−.0001295	−.0035375	.001169331	.000967718	.000715671	.000472977	.000279345	.000045803	.000003724	.00000003	101011010101
00111000001	30	−.0001456	−.0032307	.001171997	.000981616	.000741969	.000506229	.000311846	.000059535	.000006054	.00000010	011101100011
00110110100	32	−.0001541	−.0033978	.001167280	.000968448	.000721102	.000477589	.000289081	.000050264	.000004464	.00000005	011010010011
01010110001	32	−.0001541	−.0034307	.001166670	.000966420	.000717705	.000477883	.000285325	.000048826	.000004244	.00000004	101000110110
01010010110	32	−.0001541	−.0035155	.001165102	.000960857	.000709018	.000467589	.000275714	.000045118	.000003671	.00000005	100100110010
10001000011	33	−.0001711	−.0034283	.001163453	.000960857	.000717113	.000471555	.000280153	.000047099	.000003979	.00000006	111000010011
00101011011	32	−.0001937	−.0032861	.001161787	.000962237	.000717263	.000471136	.000290413	.000051789	.000004774	.00000006	101101000011
10100101000	32	−.0001937	−.0034156	.001159439	.000954594	.000704691	.000466461	.000276890	.000046708	.000004002	.00000004	101010011010
01100010011	33	−.0001937	−.0034828	.001158200	.000950493	.000697846	.000458363	.000269339	.000043802	.000003554	.00000004	101010101010
01001010010	33	−.0001956	−.0034434	.001158533	.000952064	.000700607	.000461604	.000272269	.000044797	.000003684	.00000004	100110110110
01010110100	32	−.0002352	−.0034458	.001150984	.000939191	.000685867	.000448174	.000260339	.000042024	.000003352	.00000003	101010101101
01011100100	32	−.0002537	−.0034224	.001147956	.000934868	.000681899	.000445462	.000260618	.000042111	.000003415	.00000003	110011000011
10100000110	33	−.0002541	−.0033573	.001149037	.000938475	.000687821	.000452311	.000266848	.000044352	.000003737	.00000003	100101111010
10000111000	33	−.0002541	−.0033647	.001148909	.000938077	.000687203	.000451635	.000266269	.000044181	.000003719	.00000003	101100111010
10001100001	33	−.0002541	−.0034181	.001147930	.000934857	.000681859	.000445352	.000260448	.000041981	.000003387	.00000003	101001110110
01010100011	31	−.0002701	−.0031256	.001150282	.000947588	.000706046	.000479979	.000290443	.000054807	.000005620	.00000010	001111010010
00111011000	31	−.0002701	−.0033660	.001145685	.000933088	.000681789	.000447059	.000263099	.000043600	.000003644	.00000004	101101010101
00111010010	31	−.0002701	−.0033773	.001145688	.000932467	.000680803	.000445950	.000262115	.000043265	.000003644	.00000004	011110110100
11000010001	32	−.0002782	−.0032364	.001146656	.000937786	.000690544	.000457745	.000273117	.000047374	.000004256	.00000005	110101011100
01001010010	33	−.0002782	−.0034009	.001143110	.000928294	.000675076	.000436853	.000256736	.000041306	.000003342	.00000004	110101011010
01010101100	32	−.0002782	−.0034272	.001143231	.000926732	.000672509	.000436853	.000254005	.000040313	.000003199	.00000003	110010110110
01010100001	30	−.0002942	−.0033357	.001141901	.000927293	.000676278	.000443052	.000260806	.000043510	.000003748	.00000004	011110100011
11000011001	32	−.0003367	−.0031935	.001136342	.000921436	.000673069	.000442826	.000262392	.000044857	.000003990	.00000005	011001111000
01001011100	32	−.0003367	−.0032261	.001135769	.000919614	.000670126	.000439434	.000259287	.000043681	.000003803	.00000004	111000111101
01001001100	32	−.0003367	−.0033165	.001134164	.000914483	.000661831	.000429907	.000250640	.000040539	.000003339	.00000004	011001101001
01110011000	30	−.0003527	−.0032586	.001132187	.000912823	.000661775	.000431769	.000253012	.000041966	.000003610	.00000004	011011010001
01100110001	30	−.0003527	−.0033083	.001131323	.000910117	.000657494	.000424648	.000248741	.000040496	.000003403	.00000004	011110011001
00111010010	33	−.0003527	−.0033265	.001131003	.000909115	.000655880	.000424685	.000263115	.000039952	.000003327	.00000004	011110111100
11000000111	33	−.0003782	−.0030839	.001130426	.000914345	.000668028	.000440798	.000262675	.000046108	.000004289	.00000006	100111111100
10000101100	33	−.0003782	−.0032571	.001127333	.000904383	.000651755	.000421864	.000245221	.000038438	.000003234	.00000003	110010101101
01010011100	33	−.0003782	−.0032710	.001127078	.000903536	.000650335	.000420176	.000243635	.000038815	.000003136	.00000003	110011011110
1_000000___	31	−.0003943	−.0030234	.001128529	.000913067	.000668835	.000443682	.000266629	.000048586	.000004830	.00000008	011110100100

128

Table A

m = 7, n = 5 Rank Orders 451 thru 500 of 792 C(7, 5) = 792 1/C(7, 5) = .00126263

z	w	c_1	c_2	d = .2	.4	.6	.8	1.0	1.5	2.0	3.0	z^{tc}
00011110100	31	−.0003943	−.0030572	.001127918	.000911059	.000665461	.000439610	.000262702	.000046846	.000004494	.000000007	11101000111
00101110001	31	−.0003943	−.0032353	.001124727	.000900744	.000648538	.000419815	.000244337	.000039669	.000003320	.000000003	11011000101
10101000001	31	−.0003943	−.0032523	.001124422	.000899763	.000646952	.000418004	.000242710	.000039107	.000003244	.000000003	01011000011
00111000110	31	−.0003943	−.0032652	.001124203	.000899086	.000645894	.000416825	.000241670	.000038748	.000003193	.000000003	01011110010
10011000100	31	−.0003943	−.0032655	.001124194	.000899051	.000645822	.000416723	.000241557	.000038687	.000003180	.000000003	11001100011
01100000010	31	−.0003943	−.0033142	.001123326	.000896274	.000641343	.000411603	.000236944	.000037057	.000002949	.000000003	01011110110
01011000010	31	−.0003943	−.0033181	.001123259	.000896061	.000640995	.000411191	.000236552	.000036889	.000002918	.000000002	01111110010
01011010100	31	−.0003943	−.0033456	.001122758	.000894429	.000638311	.000408065	.000233684	.000035830	.000002762	.000000002	10011101001
01011010010	31	−.0003943	−.0033905	.001121964	.000891916	.000634304	.000403545	.000229670	.000034474	.000002580	.000000002	10101100101
01010101000	31	−.0004131	−.0032865	.001120262	.000891874	.000637165	.000408509	.000235071	.000036829	.000002947	.000000002	11011010011
01010101001	31	−.0004131	−.0033194	.001119675	.000889998	.000634143	.000405065	.000231981	.000035756	.000002798	.000000002	01101010110
01011010010	31	−.0004131	−.0034042	.001118169	.000885198	.000626434	.000396292	.000224114	.000033014	.000002416	.000000002	10101010101
01101011001	31	−.0004528	−.0031580	.001115003	.000886205	.000633872	.000407987	.000226293	.000034011	.000003163	.000000003	11100100101
01101001001	31	−.0004528	−.0032876	.001112744	.000879131	.000622682	.000395424	.000225158	.000034216	.000002640	.000000002	01011010101
00101000110	31	−.0004528	−.0033548	.001111554	.000877349	.000616619	.000388537	.000221764	.000032073	.000002342	.000000002	10101001101
10010100110	32	−.0004547	−.0033105	.001110533	.000877033	.000619440	.000391719	.000218993	.000032936	.000002446	.000000003	11100110110
01011001010	31	−.0004773	−.0031483	.001108849	.000878871	.000625759	.000400812	.000230945	.000036629	.000003000	.000000003	01101110110
01011011010	31	−.0004773	−.0032457	.001107356	.000873639	.000617543	.000391658	.000222891	.000033933	.000002635	.000000002	01111011110
01011010010	31	−.0004773	−.0033188	.001093159	.000869553	.000611011	.000384255	.000216273	.000031636	.000002314	.000000002	10101011010
01010101010	32	−.0004943	−.0032961	.001091104	.000865197	.000606441	.000380357	.000213428	.000030904	.000002226	.000000001	10111001110
10100010010	32	−.0005188	−.0032342	.001101167	.000860819	.000603074	.000378582	.000212921	.000031205	.000002296	.000000002	10010110110
10010110010	32	−.0005188	−.0032599	.001100715	.000859390	.000600800	.000376019	.000210647	.000030437	.000002193	.000000001	11100101110
10011010010	32	−.0005188	−.0032972	.001100063	.000857354	.000597597	.000372455	.000207529	.000029425	.000002064	.000000002	10101011010
00111010010	30	−.0005349	−.0031632	.001099421	.000863686	.000604103	.000381271	.000216200	.000034759	.000002558	.000000002	01110110110
10001101010	31	−.0005372	−.0030736	.001100482	.000854920	.000610254	.000388139	.000222202	.000030267	.000002817	.000000002	11011011010
00100101010	31	−.0005372	−.0032381	.001097190	.000854474	.000596531	.000377877	.000208791	.000029524	.000002206	.000000002	01011101101
01010011010	31	−.0005372	−.0032643	.001097651	.000854396	.000594244	.000370320	.000206540	.000033994	.000002109	.000000001	11100101110
01010010010	29	−.0005533	−.0030815	.001079239	.000830942	.000604384	.000373055	.000209524	.000029008	.000002070	.000000001	11010101010
01010011010	31	−.0005617	−.0030571	.001077236	.000827363	.000588437	.000365802	.000218571	.000029274	.000002757	.000000003	01110111100
10100011010	32	−.0005773	−.0031781	.001091104	.000838184	.000587020	.000365349	.000203764	.000029008	.000002115	.000000001	10111001011
11000001010	31	−.0006014	−.0030388	.001088950	.000845034	.000589898	.000370377	.000209134	.000031486	.000002442	.000000002	01101011100
01010111000	31	−.0006014	−.0030874	.001088115	.000842449	.000588841	.000365843	.000205120	.000030103	.000002247	.000000001	11100101101
01110100010	31	−.0006014	−.0031665	.001086776	.000838377	.000579593	.000359044	.000199287	.000028281	.000002019	.000000001	11001110001
01110001100	29	−.0006174	−.0030786	.001085317	.000838304	.000581964	.000363197	.000203826	.000030291	.000002335	.000000001	01111101110
10000101010	32	−.0006363	−.0030265	.001082631	.000834949	.000579236	.000361519	.000203033	.000030291	.000002352	.000000001	11100101110
11000001010	31	−.0006429	−.0030998	.001080053	.000828555	.000569865	.000351300	.000194058	.000027193	.000001911	.000000001	11001011010
01110010010	31	−.0006618	−.0029380	.001079033	.000830942	.000576600	.000353707	.000201810	.000027982	.000002291	.000000002	10010101110
10010001010	32	−.0006618	−.0030111	.001077033	.000827363	.000570674	.000349358	.000196975	.000028527	.000002114	.000000001	10010111100
01011111000	32	−.0006618	−.0030571	.001072806	.000816320	.000566775	.000349358	.000193147	.000027246	.000001942	.000000002	11000111010
00101110100	30	−.0006778	−.0027090	.001069932	.000838184	.000589880	.000376469	.000217889	.000036222	.000003279	.000000004	11110000111
11010000011	30	−.0006778	−.0027126	.001080144	.000838246	.000590184	.000377057	.000218648	.000036734	.000003403	.000000005	00111111010
01110000110	30	−.0006778	−.0029491	.001076079	.000825614	.000570209	.000354482	.000198359	.000029327	.000002250	.000000002	01011111010
01110110100	30	−.0006778	−.0029961	.001075272	.000823119	.000566303	.000350132	.000194527	.000028029	.000002070	.000000002	10011111000
00111010010	30	−.0006778	−.0030135	.001075004	.000822385	.000565299	.000349171	.000193816	.000027896	.000002065	•.000000002	11011011010
10011100000	30	−.0006778	−.0030226	.001074855	.000821942	.000564634	.000348464	.000193222	.000027719	.000002044	.000000002	11011101110
00100110010	30	−.0006778	−.0030416	.001074210	.000820884	.000562958	.000346586	.000192170	.000027170	.000001972	.000000001	01111001101
01011001010	30	−.0006778	−.0030602	.001073029	.000819962	.000561569	.000345102	.000190315	.000026793	.000001926	.000000001	01110111100
01010110010	30	−.0006778	−.0031290	.001072806	.000816320	.000555899	.000338842	.000184863	.000026793	.000001693	.000000001	11011100110
00101110100	30	−.0006778	−.0031425	.001072806	.000815658	.000554910	.000337798	.000183996	.000024772	.000001665	.000000001	10101010101
00101101010	30	−.0007175	−.0028703	.001069932	.000817093	.000562298	.000348660	.000194782	.000028789	.000002218	.000000002	11010000111

129

Table A

z	w	c_1	c_2	$d=.2$	.4	.6	.8	1.0	1.5	2.0	3.0	z^{tc}
101010001001	30	−.0007175	−.0030137	.001067530	.000809861	.000551284	.000336744	.000184596	.000025603	.000001812	.000000001	011011101100
011010000010	30	−.0007175	−.0030783	.001066430	.000806499	.000546097	.000331067	.000179696	.000024043	.000001612	.000000001	101011101001
001101011000	31	−.0007194	−.0030301	.001066860	.000808206	.000548830	.000334030	.000182181	.000024752	.000001691	.000000001	110111100110
101001100001	30	−.0007363	−.0029090	.001065726	.000809160●	.000552593	.000339419	.000187509	.000026754	.000001971	.000000001	111010101001
011010011000	30	−.0007363	−.0030064	.001064106	.000804321	.000545288	.000331595	.000180894	.000024750	.000001725	.000000001	101101011010
101001010110	30	−.0007363	−.0030795	.001062867	.000800551	.000539502	.000325300	.000175494	.000022855	.000001513	.000000001	101101011010
101000100010	31	−.0007779	−.0029732	.001056812	.000792752	.000532867	.000320905	.000173123	.000022287	.000001514●	.000000001	100111011010
100011000010	31	−.0007779	−.0029989	.001056378	.000791433	.000530850	.000318722	.000171263	.000021529	.000001446	.000000001	101101011010
001010100100	31	−.0007779	−.0030362	.001055750	.000789546	.000527993	.000315664	.000168688	.000020635	.000001322	.000000001	110110101011
001111000010	29	−.0007939	−.0028107	.001056524●	.000795962	.000539853	.000329615	.000181168	.000023596	.000001710	.000000001	101111000011
110010000101	30	−.0008020	−.0027664	.001055736	.000795556	.000540187●	.000330461	.000182122	.000026006	.000001935	.000000002	011011101100
010011011000	30	−.0008020	−.0029126	.001053314	.000788331	.000529274	.000318740	.000172165	.000022928	.000001546	.000000001	110111100101
001011001100	30	−.0008020	−.0029501	.001052688	.000786455	.000526443	.000315712	.000169614	.000022169	.000001456	.000000001	110111100101
010110001100	31	−.0008208	−.0029875	.001051959	.000786330	.000523217	.000312206	.000166613	.000021243	.000001343	.000000001	101010101101
010101010100	30	−.0008421	−.0029561	.001049050	.000780308	.000519577	.000309686	.000165219	.000021146	.000001353●	.000000001	101010101101
101000010010	31	−.0008605	−.0029254	.001045544	.000775043	.000514190	.000305239	.000162102	.000020442	.000001280	.000000001	011011011100
110001001000	30	−.0008605	−.0027479	.001045023	.000778027	.000520949	.000313664	.000169809	.000022995	.000001603	.000000001	110101101101
010110010000	31	−.0008605	−.0027966	.001044222	.000775651	.000517374	.000309834	.000166560	.000021989	.000001475	.000000001	110010110010
010111000001	30	−.0008605	−.0028757	.001042934	.000775817	.000511818	.000304024	.000161770	.000020635	.000001322	.000000001	110110101011
011011000001	28	−.0008765	−.0026963	.001042908	.000775817●	.000519809	.000313724	.000170585	.000023596	.000001710	.000000001	011111000011
110000110001	30	−.0008850	−.0026888	.001041387	.000773235	.000516754	.000310787	.000168193	.000022833	.000001603	.000000001	011100111100
010100111000	30	−.0008850	−.0027748	.001039971	.000769034	.000510444	.000304045	.000162494	.000021093	.000001386	.000000001	110001111001
011100000110	30	−.0008850	−.0028276	.001039125	.000766604	.000506921	.000300426	.000159563	.000020301	.000001301	.000000001	101010111001
011010000110	31	−.0009010	−.0026638	.001038836	.000769738	.000513689	.000307798	.000167241	.000022908	.000001645	.000000001	100111011100
100110001000	28	−.0009020	−.0027873	.001036559	.000763070	.000503572	.000297832	.000157830	.000019930	.000001262	.000000001	110110101010
100001001100	31	−.0009265	−.0026476	.001034218	.000762016	.000504828●	.000300595	.000160816	.000021035	.000001401	.000000001	100110111100
010010110100	31	−.0009265	−.0027389	.001032751	.000757782	.000498649	.000294194	.000155581	.000019578	.000001238	.000000001	110101101110
001010100100	31	−.0009265	−.0027171	.001032221	.000756214	.000496313	.000291730	.000153535	.000018992	.000001172	.000000001	110010110110
001110100100	29	−.0009426	−.0026434	.001031356	.000757691●	.000500649	.000297547	.000159071	.000019071	.000001426	.000000001	011011011010
011011000001	29	−.0009426	−.0026963	.001030813	.000756078	.000498234	.000294981	.000156921	.000020298	.000001350	.000000001	011111000011
010110100001	29	−.0009426	−.0027611	.001029422	.000751981	.000492134	.000288534	.000151540	.000018717	.000001162	.000000001	101110100101
100000111100	31	−.0009850	−.0025006	.001025645	.000750899	.000495319	.000294295	.000157460	.000020821	.000001426	.000000001	101000111110
110000011100	31	−.0009850	−.0025528	.001024782	.000748332	.000491453	.000290157	.000153957	.000019751	.000001293	.000000001	100011111010
110000011100	31	−.0009850	−.0026111	.001023867	.000745752	.000487774	.000286430	.000150977	.000018965	.000001209	.000000001	100111111000
110100000011	29	−.0010011	−.0021425	.001028448	.000763018	.000515492	.000317088	.000177582	.000027674	.000002412	.000000003	111111110000
001011110000	29	−.0010011	−.0023800	.001024582	.000751585	.000498213	.000298349	.000160360	.000022201	.000001611	.000000001	111010010011
001101101001	30	−.0010011	−.0025065	.001022580	.000745866	.000499919	.000289775	.000154333	.000020197	.000001374	.000000001	111010010011
101101000010	30	−.0010011	−.0025133	.001022472	.000745556	.000489470	.000289312	.000153099	.000020093	.000001362	.000000001	111010111000
100101010010	30	−.0010011	−.0025556	.001021782	.000743556	.000486478	.000286167	.000151343	.000019332	.000001272	.000000001	101011100110
101011000001	29	−.0010011	−.0025968	.001021133	.000741697	.000483845	.000283489	.000149194	.000018759	.000001210	.000000001	011011100110
011100001010	29	−.0010011	−.0026143	.001020821	.000740703	.000482255	.000281695	.000147601	.000018222	.000001139	.000000001	101011110001
101001000110	29	−.0010011	−.0026177	.001020803	.000740759●	.000482498	.000280498	.000148089	.000018463	.000001178	.000000001	011011110010
011001100010	31	−.0010011	−.0026725	.001019902	.000738098	.000478530	.000277918	.000144584	.000017434	.000001056	.000000000	011001110011
011010010010	30	−.0010011	−.0026882	.001019654	.000737398	.000471534	.000276915	.000143788	.000017231	.000001036	.000000000	101110110010
101011100001	30	−.0010426	−.0025633	.001013866	.000730237	.000471755	.000273273	.000141929	.000017101	.000001039●	.000000000	100111100110
100011110000	30	−.0010426	−.0025706	.001013735	.000729863	.000471101	.000272547	.000141296	.000016901	.000001015	.000000000	101101100110
100101010010	30	−.0010426	−.0026240	.001012881	.000727412	.000467556	.000268917	.000138370	.000016124	.000000933	.000000000	101011010110
100101010010	30	−.0010615	−.0026151	.001012887	.000726219	.000461938	.000263370	.000135159	.000015472	.000000876	.000000000	011011110010
001101010010	30	−.0010615	−.0026751	.001009503	.000722066	.000476448	.000264262	.000135472	.000015159	.000001356	.000000000	011011110110
101100000101	29	−.0010855	−.0022754	.001009422●	.000732209	....	.000280607	.000149603	....	....	....	011101110010

130

m = 7, n = 5   Rank Orders 551 thru 600 of 792   1/C(7, 5) = .00126263

C(7, 5) = 792

z	w	c_1	c_2	d = .2	.4	.6	.8	1.0	1.5	2.0	3.0	z^{tc}
01010010100	29	−.0010855	−.0025271	.001006436	.000718852	.000460038	.000263762	.000135477	.000015827	.000000929	.000000000	11010110010101
01010010010	30	−.0011011	−.0025363	.001003355	.000713539	.000460407	.000258399	.000131527	.000014896	.000000836	.000000000	10101101010110
11001010001	29	−.0011252	−.0023089	.001002472	.000716484	.000460889	.000266798	.000138972	.000017146	.000001094	.000000001	01101110110110
00011101000	29	−.0011252	−.0023415	.001001954	.000714997	.000458719	.000264539	.000137106	.000016602	.000001028	.000000001	11011000011101
01010000010	30	−.0011256	−.0024318	.001000540	.000711021	.000258849	.000255670	.000132582	.000015430	.000000906	.000000001	10111011000110
10100100010	30	−.0011256	−.0024736	.000999774	.000708891	.000449992	.000255670	.000132992	.000014720	.000000829	.000000001	10110011001010
11000110010	29	−.0011440	−.0022866	.000999298	.000711693	.000456026	.000262862	.000136294	.000016601	.000001044	.000000001	01111011000110
01010101000	29	−.0011440	−.0023726	.000997938	.000707819	.000450440	.000257136	.000131652	.000015325	.000000901	.000000001	11100101011010
01010101010	29	−.0011440	−.0024255	.000997125	.000705574	.000447312	.000254050	.000129252	.000014741	.000000845	.000000000	11100101010110
01110100000	27	−.0011601	−.0021702	.000998935	.000712170	.000458569	.000266492	.000139783	.000017808	.000001201	.000000001	01111010111100
11000010010	30	−.0011856	−.0022238	.000992492	.000701577	.000445803	.000254550	.000130565	.000015360	.000000919	.000000000	10011011010
10001010010	30	−.0011856	−.0023151	.000991082	.000697666	.000440320	.000249095	.000126281	.000014284	.000000811	.000000000	11010100110
10010110000	30	−.0011856	−.0023473	.000990572	.000696214	.000438240	.000246984	.000124595	.000013846	.000000766	.000000000	11011001110
00111100010	28	−.0012016	−.0021281	.000991049	.000701199	.000447162	.000257001	.000133118	.000016309	.000001044	.000000001	10111110000
11001000010	28	−.0012016	−.0021609	.000990526	.000699700	.000444991	.000254768	.000131303	.000015808	.000000988	.000000000	11111000110
01011000010	30	−.0012016	−.0022458	.000989188	.000695900	.000439525	.000249176	.000126777	.000014568	.000000849	.000000000	10111100011
01100011100	29	−.0012082	−.0022096	.000988524	.000695468	.000439581	.000249562	.000127237	.000014737	.000000869	.000000000	11100111001
10100110100	29	−.0012101	−.0022865	.000986958	.000691462	.000434062	.000244078	.000122909	.000013627	.000000754	.000000000	11010101110
10000111010	30	−.0012497	−.0020966	.000982506	.000687586	.000432575	.000244548	.000124187	.000014242	.000000830	.000000000	11100101010
11000010010	30	−.0012497	−.0021328	.000981935	.000685971	.000430266	.000242206	.000122313	.000013751	.000000778	.000000000	10101011100
10100100010	30	−.0012497	−.0021933	.000981023	.000683502	.000426891	.000238936	.000119816	.000013169	.000000724	.000000000	11010111010
00111010110	28	−.0012658	−.0020045	.000981006	.000687044	.000433691	.000246740	.000126520	.000015121	.000000945	.000000001	11101010010
01101010000	28	−.0012658	−.0021019	.000979507	.000682891	.000427869	.000240937	.000121946	.000013948	.000000822	.000000000	01110101010
01011011000	28	−.0012658	−.0021750	.000978364	.000679680	.000423306	.000236333	.000118276	.000012985	.000000719	.000000000	10111011010
00110111000	28	−.0012846	−.0018924	.000979185	.000685928	.000433980	.000247774	.000127707	.000015539	.000000994	.000000001	11100010011
01110001001	28	−.0012846	−.0020358	.000977005	.000679953	.000425589	.000239562	.000121274	.000013908	.000000824	.000000000	10111100010
01110000100	28	−.0012846	−.0021004	.000975992	.000677099	.000421520	.000235439	.000117972	.000013030	.000000729	.000000000	10111101100
10110000010	30	−.0013262	−.0019594	.000970415	.000670463	.000416292	.000232225	.000116360	.000012912	.000000730	.000000000	10011101100
10010000010	29	−.0013262	−.0020882	.000968690	.000668449	.000409709	.000225747	.000111333	.000011691	.000000611	.000000000	01110111100
10011001010	29	−.0013262	−.0020882	.000968449	.000665074	.000408834	.000224909	.000110702	.000011550	.000000599	.000000000	10101110110
01011010100	28	−.0013502	−.0019806	.000965650	.000662504	.000407688	.000225005	.000111363	.000011910	.000000644	.000000000	11010100101
10100010010	29	−.0013658	−.0019946	.000962524	.000657148	.000401718	.000219828	.000107660	.000011106	.000000570	.000000000	10101101010
01010010010	29	−.0013847	−.0019733	.000959555	.000652425	.000397031	.000216153	.000105257	.000010677	.000000537	.000000000	10110101010
11000001100	30	−.0013927	−.0017252	.000961594	.000660099	.000408241	.000227366	.000113990	.000012778	.000000738	.000000000	10000111100
11000100010	28	−.0014088	−.0014994	.000962212	.000649772	.000415992	.000235821	.000121016	.000014734	.000000957	.000000001	01011111100
01011110000	28	−.0014088	−.0016712	.000959440	.000657554	.000406277	.000226212	.000113462	.000012775	.000000744	.000000000	11110000010
10011110001	28	−.0014088	−.0017760	.000958891	.000657512	.000404330	.000226569	.000115683	.000012892	.000000763	.000000000	01110110010
10001100001	28	−.0014088	−.0017098	.000958548	.000656149	.000404468	.000224569	.000112293	.000012557	.000000730	.000000000	01111011100
01100010010	28	−.0014088	−.0017328	.000958548	.000655231	.000403224	.000223368	.000111376	.000012339	.000000709	.000000000	01101101100
01110010010	29	−.0014088	−.0017878	.000957705	.000652901	.000399959	.000220111	.000108804	.000011674	.000000638	.000000000	11001110100
01010101000	28	−.0014088	−.0018028	.000957481	.000652300	.000399318	.000219318	.000108197	.000011530	.000000624	.000000000	11010100101
01011001100	28	−.0014088	−.0018118	.000957346	.000651934	.000398642	.000218836	.000107829	.000011444	.000000617	.000000000	11000100110
01010001010	28	−.0014088	−.0018534	.000956732	.000650306	.000396462	.000216768	.000106281	.000011100	.000000586	.000000000	10101101001
01110100100	26	−.0014248	−.0018669	.000958531	.000649772	.000395744	.000216083	.000105767	.000010986	.000000576	.000000000	01111110110
01111000010	29	−.0014503	−.0015184	.000958797	.000659190	.000410204	.000230995	.000117695	.000014075	.000000901	.000000001	10011101110
10011100010	29	−.0014503	−.0016487	.000957295	.000646277	.000394597	.000216509	.000106850	.000011453	.000000628	.000000000	10110000110
10001100100	29	−.0014503	−.0017218	.000950991	.000643171	.000390473	.000212740	.000103988	.000010790	.000000566	.000000000	11100101110
01100101000	29	−.0014503	−.0017678	.000950296	.000641288	.000387897	.000210242	.000102079	.000010344	.000000524	.000000000	11001100110
01100101100	28	−.0014673	−.0016794	.000948501	.000639923	.000387654	.000208114	.000102901	.000010665	.000000561	.000000000	11001011001
10010101000	29	−.0014692	−.0017513	.000947059	.000636391	.000382989	.000206366	.000099533	.000009887	.000000489	.000000000	11010101110

131

Table A

m = 7, n = 5   Rank Orders 601 thru 650 of 792 $C(7, 5) = 792$ $1/C(7,5) = .00126263$

z	w	c_1	c_2	d = .2	.4	.6	.8	1.0	1.5	2.0	3.0	z^{tc}
100110110000	29	-.0015088	-.0015447	.000942797	.000632866	.000381748	.000206847	.000100642	.000010357	.000000541	.000000000	111001001110
110010010100	29	-.0015088	-.0015809	.000942248	.000631367	.000376693	.000204834	.000099093	.000009989	.000000506	.000000000	101011011010
101001001100	29	-.0015088	-.0016415	.000941371	.000629087	.000376693	.000202052	.000097054	.000009561	.000000470	.000000000	110011011010
001111000001	27	-.0015248	-.0013612	.000942603	.000635660	.000387155	.000212935	.000105747	.000011754	.000000690	.000000000	111110010100
101111000010	27	-.0015248	-.0014585	.000941160	.000631802	.000381926	.000207888	.000101888	.000010831	.000000599	.000000000	011111000110
110010000010	29	-.0015333	-.0015317	.000940060	.000628825	.000377843	.000203902	.000098809	.000010079	.000000524	.000000000	101100111010
110001011000	29	-.0015333	-.0015009	.000938907	.000627311	.000376387	.000202742	.000097990	.000009883	.000000503	.000000000	101001111100
100100111000	27	-.0015021	-.0015021	.000938894	.000627288	.000376403	.000202740	.000097998	.000009889	.000000504	.000000000	110100111010
100110011000	29	-.0015333	-.0015666	.000937333	.000624852	.000373183	.000199753	.000095801	.000009424	.000000464	.000000000	110100110110
001110110000	27	-.0015494	-.0012757	.000939317	.000631688	.000383908	.000210809	.000104571	.000011608	.000000682	.000000000	111010010011
101100100001	27	-.0015494	-.0014053	.000937423	.000626690	.000377216	.000204424	.000099743	.000010482	.000000573	.000000000	011110110010
011010010010	27	-.0015494	-.0014724	.000936413	.000623956	.000373469	.000200770	.000096923	.000009796	.000000504	.000000000	011011011001
100101100010	28	-.0015909	-.0014390	.000929228	.000623003	.000362388	.000191929	.000091044	.000008700	.000000416	.000000000	101110100110
011011000100	27	-.0016093	-.0013025	.000921227	.000612939	.000363908	.000194134	.000093038	.000009245	.000000469	.000000000	010111000101
010011100010	28	-.0016494	-.0012488	.000920516	.000603205	.000354333	.000186663	.000088159	.000008363	.000000398	.000000000	101011011010
101011000010	28	-.0016494	-.0012984	.000920255	.000601380	.000351969	.000184494	.000086691	.000008046	.000000373	.000000000	011011011010
101010010010	28	-.0016574	-.0013167	.000921303	.000600711	.000351105	.000183705	.000086240	.000007933	.000000364	.000000000	110011011100
110000110010	29	-.0016574	-.0011401	.000919720	.000604861	.000357105	.000189497	.000090337	.000008845	.000000439	.000000000	100101111100
011010010001	27	-.0016735	-.0010497	.000919467	.000603921	.000357414	.000190623	.000091614	.000009302	.000000493	.000000000	110100110110
010101010100	27	-.0016735	-.0011357	.000918467	.000600626	.000353021	.000186451	.000088475	.000008581	.000000425	.000000000	111010100101
011010100100	27	-.0016735	-.0011886	.000917715	.000598704	.000350537	.000184174	.000086828	.000008244	.000000397	.000000000	110101101001
110100010001	27	-.0016923	-.0009761	.000917310	.000601175	.000355330	.000189379	.000091001	.000009255	.000000493	.000000000	011110110010
010101110000	28	-.0016923	-.0010087	.000916805	.000599769	.000353348	.000187395	.000089432	.000008850	.000000450	.000000000	101010111001
011110010010	27	-.0016923	-.0010990	.000915530	.000596533	.000349192	.000183601	.000086695	.000008291	.000000404	.000000000	110110011001
101100010010	27	-.0016923	-.0009797	.000913400	.000598302	.000354643	.000190129	.000092151	.000009672	.000000537	.000000000	111010011001
011100101000	28	-.0017320	-.0007373	.000909996	.000589597	.000343337	.000179658	.000084460	.000008002	.000000387	.000000000	111010011001
010101101000	28	-.0017320	-.0009996	.000909836	.000589185	.000342804	.000179168	.000084104	.000007929	.000000381	.000000000	100111110001
011010011000	29	-.0017320	-.0009889	.000911163	.000592890	.000347704	.000183704	.000087405	.000008612	.000000438	.000000000	110110100110
011001011000	28	-.0017339	-.0008700	.000908689	.000586530	.000339433	.000176063	.000081829	.000007441	.000000339	.000000000	110100100110
100101010100	28	-.0017339	-.0010570	.000908494	.000586034	.000338799	.000175490	.000081420	.000007361	.000000333	.000000000	111010100110
100011101000	28	-.0017735	-.0008218	.000904579	.000583236	.000338291	.000176447	.000082731	.000007797	.000000375	.000000000	111010001110
110100010010	28	-.0017735	-.0008741	.000903824	.000581276	.000335716	.000174047	.000080965	.000007420	.000000343	.000000000	101011011010
101010010100	28	-.0017735	-.0009324	.000903012	.000579245	.000333152	.000171752	.000079344	.000007110	.000000319	.000000000	100111011010
100110101000	28	-.0017924	-.0008378	.000900885	.000577169	.000331907	.000171231	.000079221	.000007144	.000000324	.000000000	101011001010
100101011000	28	-.0017924	-.0009035	.000899974	.000574917	.000329052	.000168687	.000077435	.000006808	.000000299	.000000000	111011000110
001110101000	26	-.0018084	-.0005210	.000902480	.000584289	.000342528	.000181766	.000081766	.000009003	.000000496	.000000000	111010001010
101101000100	28	-.0017924	-.0006506	.000900656	.000579645	.000336519	.000176217	.000083222	.000008119	.000000416	.000000000	111110100011
100110010010	28	-.0018084	-.0007178	.000899685	.000577113	.000333169	.000173058	.000080861	.000007584	.000000366	.000000000	011111100100
011101000100	26	-.0018084	-.0007178	.000898904	.000576645	.000333169	.000173058	.000080861	.000007584	.000000366	.000000000	011111011010
110011000010	27	-.0018500	-.0006626	.000892842	.000567122	.000323446	.000165571	.000076045	.000006749	.000000302	.000000000	111010100110
101000111000	28	-.0018566	-.0006197	.000892242	.000566767	.000323482	.000165822	.000076316	.000006826	.000000309	.000000000	111000111010
101010010100	27	-.0019141	-.0005357	.000882934	.000553553	.000311061	.000156629	.000070653	.000005958	.000000251	.000000000	101010111100
110010100100	28	-.0019165	-.0004254	.000884404	.000549258	.000314928	.000160081	.000073063	.000006401	.000000284	.000000000	110001011010
100110110001	26	-.0019325	-.0002436	.000883655	.000558603	.000318874	.000164338	.000077207	.000007207	.000000358	.000000000	011111001100
011011100100	26	-.0019325	-.0003296	.000882450	.000553550	.000314945	.000160729	.000073822	.000006645	.000000308	.000000000	111110000100
101100010100	27	-.0019325	-.0003824	.000881726	.000553765	.000312717	.000158753	.000072436	.000006381	.000000288	.000000000	101110100011
011011010010	28	-.0019330	-.0004471	.000880723	.000551285	.000309549	.000155867	.000070358	.000005963	.000000253	.000000000	101110110001
110010110000	28	-.0019410	-.0003333	.000880819	.000552859	.000313255	.000158315	.000072159	.000006317	.000000281	.000000000	101100111100
110010110000	27	-.0019410	-.0003333	.000880819	.000552067	.000314801	.000161478	.000074759	.000006960	.000000341	.000000000	101010110100
001110100100	28	-.0019571	-.0002702	—	—	—	—	—	—	—	—	—

z	w	c_1	c_2	d = .2	.4	.6	.8	1.0	1.5	2.0	3.0	z^{tc}
011100100100	26	-.0019571	-.0003059	000878323	000549520	000309148	000156357	000071082	000006207	00000278	000000000	110101100 01
011010101000	26	-.0019967	-.0001811	000872836	000542693	000303421	000152520	000066916	000005928	00000261	000000000	110101001 01
100101010000	27	-.0019986	-.0002427	000871622	000539964	000300082	000149560	000066829	000005522	00000229	000000000	110101000 10
111000010001	26	-.0020156	-.0001046	000873290	000546940	000310012	000158953	000073696	000006931	00000346	000000000	011101000 11
011100110000	26	-.0020156	-.0000416	000871296	000540816	000303832	000153410	000069746	000005974	00000267	000000000	111100110 01
111000110000	26	-.0020156	-.0000791	000870795	000539496	000302368	000152142	000070850	000006461	00000308	000000000	111001110 01
100011100000	27	-.0020571	-.0002350	000867478	000536277	000300471	000154288	000068395	000005988	00000271	000000000	100111000 11
110000010010	27	-.0020571	-.0001362	000866152	000536277	000299495	000150781	000068930?	000005988	00000238	000000000	111000011 10
110100001000	27	-.0020571	-.0000467	000864941	000533320	000295837	000147558	000067993	000005978	00000238	000000000	011111000 01
110011000000	27	-.0020571	-.0000044	000864367	000531920	000294109	000146045	000065102	000005373	00000224	000000000	110100111 00
100101010100	27	-.0020571	-.0000013	000864297	000531767	000293939	000145912	000065021	000005362	00000224	000000000	110100100 10
101101001000	27	-.0020571	-.0000065	000864228	000531604	000293745	000145747	000064911	000005344	00000222	000000000	110011100 10
101100110010	27	-.0020571	-.0000128	000864140	000531384	000293467	000145500	000064737	000005311	00000220	000000000	111011010 10
110011000010	27	-.0020571	-.0000161	000864092	000531257	000293301	000145346	000064626	000005289	00000218	000000000	101101011 00
101001100010	27	-.0020571	-.0000635	000863462	000529752	000291485	000143793	000063577	000005044	00000206	000000000	110101100 10
101001001010	27	-.0020571	-.0000796	000863245	000529229	000290848	000143243	000063203	000005109	00000201	000000000	111010100 11
001111000000	25	-.0020731	.0004099	000866977	000541047	000306674	000157793	000073663	000007148	00000375	000000000 ●	101101010 01
011010100100	25	-.0020731	.0002665	000843817	000536322	000300784	000152537	000067993	000006930	00000310	000000000	011111100 01
010110010010	25	-.0020731	.0002018	000838550	000534053	000297876	000149873	000067993	000005978	00000273	000000000	101111001 10
110000111000	27	-.0021156	.0001716	000837854	000521328	000285053	000139882	000061582	000004914	00000197	000000000	111001111 00
101011001000	26	-.0021732	-.0003687	000848301	000531983	000277473	000134985	000058924	000004608	00000182	000000000	101111001 00
100110010000	27	-.0021812	-.0004488	000847908	000512290	000278325	000135908	000059622	000004743	00000191	000000000	110011101 00
101101001000	26	-.0021977	-.0004487	000844953	000507887	000274106	000132781	000057713	000004464	00000174	000000000	011101101 10
110100100010	27	-.0022001	.0004927	000845090	000508540	000274989	000133555	000057552	000004548	00000179	000000000	101010101 00
101010100010	25	-.0022161	.0007393	000845523	000512254	000280690	000139075	000062297	000005377	00000247	000000000	111010100 10
101010010010	26	-.0022161	.0006907	000844847	000510555	000275634	000137097	000060866	000005077	00000221	000000000	101011010 11
011011010000	26	-.0022161	.0006116	000843817	000508133	000270539	000134646	000059218	000004792	00000200	000000000	111000010 01
011011001000	25	-.0022558	.0007531	000838550	000501830	000267688	000131349	000057418	000004577	00000189	000000000	101011110 10
110001100100	26	-.0022577	.0006964	000837435	000499420	000267688	000128901	000057742	000004272	00000166	000000000	101110001 10
110000111000	27	-.0022642	.0008196	000837854	000501392	000270411	000131356	000057440	000004570	00000186	000000000	101011111 00
111000100001	25	-.0022803	.0010343	000837817	000503888	000274535	000135440	000060481	000005202	00000239	000000000	011101111 00
011101010000	25	-.0022803	.0008698	000835664	000498782	000268375	000130127	000056839	000004531	00000187	000000000	111100110 01
011101001000	25	-.0022803	.0008436	000835231	000498022	000267494	000129401	000056366	000004455	00000182	000000000	101101101 10
110010100010	26	-.0023218	.0009322	000829004	000489450	000259734	000123831	000053019	000003963	00000150	000000000	110101010 10
101010010100	26	-.0023218	.0009310	000828994	000489441	000259739	000123847	000053037	000003969	00000151	000000000	110100100 10
101010110000	26	-.0023218	.0008665	000828162	000487512	000257479	000124586	000051805	000003772	00000138	000000000	111000010 10
100101110010	26	-.0023407	.0008006	000827537	000488847	000260108	000124586	000053688	000004107	00000161	000000000	111011010 10
110100010010	26	-.0023407	.0010284	000826873	000487324	000258340	000123126	000052734	000003955	00000151	000000000	101011010 10
011001010010	26	-.0023407	.0009701	000826125	000485598	000256325	000121456	000051642	000003782	00000140	000000000	101011011 00
110001011000	26	-.0023803	.0013142	000824335	000485208	000258106	000123850	000053567	000004166	00000167	000000000	110100111 00
101001011000	26	-.0023803	.0011410	000821244	000480208	000252324	000119091	000050469	000003674	00000135	000000000	110101101 10
101010101000	25	-.0023803	.0011272	000821065	000479794	000251838	000118687	000050204	000003632	00000133	000000000	110101010 10
101010010010	25	-.0024568	.0014644	000811750	000469615	000244365	000114320	000048664	000003444	00000126	000000000	101101110 10
110010010010	26	-.0024648	.0015496	000811393	000469942	000245169	000115149	000047759	000003549	00000133	000000000	110011101 00
110001010100	26	-.0024648	.0014999	000810760	000468493	000243487	000113761	000047441	000003406	00000123	000000000	101101101 10
110010010100	26	-.0024648	.0014816	000810529	000467968	000242883	000113268	000047968	000003357	00000120	000000000	101101101 10
101110000001	24	-.0024808	.0018215	000812022	000473819	000250722	000118918	000052268	000004221	00000183	000000000	111111000 01
011110000100	24	-.0024808	.0017888	000812022	000472668	000249249	000116525	000051299	000004020	00000166	000000000	111110010 01
110111000010	24	-.0024808	.0016985	000810439	000470125	000246330	000116525	000049742	000003769	00000149	000000000	101111000 01
110001011000	26	-.0025233	.0017737	000803809	000461068	000238142	000110691	000046275	000003278	00000118	000000000	110100111 00

133

z	w	c_1	c_2	d = .2	.4	.6	.8	1.0	1.5	2.0	3.0	z^{tc}
111001001000001	24	-.0025393	.0020798	.000804850	.000465832	.000244665	.000116555	.000050341	.000004011	.000000172	.000000000	01111101101000
011011010010000	24	-.0025393	.0019153	.000802779	.000461101	.000239155	.000111962	.000047293	.000003490	.000000135	.000000000	11110101001001
011100101010000	24	-.0025393	.0018891	.000802458	.000460394	.000238365	.000111332	.000046895	.000003430	.000000131	.000000000	11110101010001
011100110101000	24	-.0025639	.0020183	.000799715	.000457638	.000236503	.000110331	.000046445	.000003399	.000000130	.000000000	11111101010001
100111001010000	25	-.0025809	.0019994	.000796433	.000452582	.000231568	.000106627	.000044162	.000003058	.000000108	.000000000	10111110011000
100101101010000	25	-.0025809	.0019982	.000796425	.000452576	.000231575	.000106644	.000044180	.000003063	.000000109	.000000000	11011001001010
101011001001000	25	-.0025809	.0019337	.000795623	.000450784	.000229548	.000105014	.000043144	.000002909	.000000099	.000000000	11011100101000
100101001001000	25	-.0026054	.0021277	.000793686	.000449832	.000229729	.000105657	.000043738	.000003033	.000000108	.000000000	11011100101000
100101101001000	25	-.0026054	.0020916	.000793246	.000448867	.000228655	.000104807	.000043206	.000002956	.000000103	.000000000	11011101101000
100101001001000	25	-.0026054	.0020310	.000792500	.000447213	.000226799	.000103327	.000042273	.000002820	.000000095	.000000000	11011101101000
101010101010000	25	-.0026450	.0022029	.000787625	.000441820	.000222778	.000100931	.000041072	.000002706	.000000090	.000000000	11010101010010
101000110010010	25	-.0026639	.0024380	.000787156	.000443280	.000225303	.000103274	.000042662	.000002956	.000000095	.000000000	10110111011000
101100111000000	25	-.0026639	.0023649	.000786277	.000441369	.000223195	.000101617	.000041631	.000002809	.000000097	.000000000	11100011011000
101110000010000	24	-.0026639	.0023189	.000785721	.000440159	.000221861	.000100572	.000040982	.000002718	.000000092	.000000000	10111110100010
100100011010000	24	-.0027215	.0026493	.000779605	.000434411	.000218301	.000098868	.000040332	.000002703	.000000093	.000000000	10111101101100
100100010010000	25	-.0027295	.0025951	.000775546	.000430989	.000214920	.000096371	.000038841	.000002505	.000000082	.000000000	11011010101100
100100010100100	25	-.0027484	.0027062	.000775581	.000429191	.000213847	.000095887	.000038672	.000002505	.000000083	.000000000	11011011011000
110000101100000	25	-.0027880	.0030231	.000772425	.000427483	.000213833	.000096597	.000039378	.000002653	.000000093	.000000000	10011110101100
110010101101000	25	-.0027880	.0029105	.000771087	.000424613	.000210705	.000094167	.000037881	.000002446	.000000081	.000000000	11011101101000
110010011000000	25	-.0027880	.0028944	.000770893	.000424189	.000210238	.000093800	.000037653	.000002413	.000000079	.000000000	11100011011000
110101000000001	23	-.0028041	.0032915	.000772896	.000430751	.000218410	.000100688	.000042179	.000003141	.000000128	.000000000	01111101101000
011011000001000	23	-.0028041	.0031453	.000771112	.000426794	.000213926	.000096282	.000039817	.000002758	.000000101	.000000000	11110101001001
011101100001000	23	-.0028041	.0031078	.000770675	.000425871	.000212935	.000096282	.000039352	.000002693	.000000097	.000000000	11110101000001
100111001010000	23	-.0028229	.0031751	.000768186	.000422945	.000210628	.000094835	.000038588	.000002611	.000000093	.000000000	10111110011000
100101000010000	24	-.0028645	.0033062	.000762448	.000415837	.000204744	.000090897	.000036386	.000002335	.000000077	.000000000	11011010101100
101010000010000	24	-.0028645	.0032701	.000761306	.000414938	.000203737	.000090155	.000035935	.000002274	.000000074	.000000000	11011101101000
101011001000000	24	-.0028645	.0032095	.000761306	.000413404	.000202076	.000088875	.000035154	.000002169	.000000068	.000000000	11100111001000
110010100100000	24	-.0029041	.0033981	.000756627	.000408429	.000198503	.000086821	.000034159	.000002081	.000000065	.000000000	10011110101100
101101001001000	24	-.0029286	.0036329	.000755103	.000408214	.000199328	.000087846	.000034917	.000002205	.000000072	.000000000	11011100101000
101010011000000	24	-.0029286	.0035416	.000754042	.000405989	.000196959	.000086047	.000033835	.000002063	.000000064	.000000000	11100011011000
101101010100000	24	-.0029286	.0035094	.000753672	.000405224	.000196157	.000085449	.000033482	.000002019	.000000062	.000000000	11010110011000
110011001000000	24	-.0029886	.0038251	.000746946	.000398420	.000191505	.000082902	.000032306	.000001927	.000000059	.000000000	10110110011000
110010010101000	24	-.0030131	.0039345	.000743972	.000395180	.000189127	.000081509	.000031618	.000001864	.000000056	.000000000	11010110101100
110000101101000	24	-.0030528	.0041374	.000743097	.000390533	.000185892	.000079704	.000030768	.000001793	.000000054	.000000000	10101111001000
101010101000000	24	-.0030716	.0043218	.000738297	.000390341	.000186440	.000080383	.000031261	.000001869	.000000058	.000000000	10101111011000
110000110100100	24	-.0030716	.0043144	.000738223	.000390205	.000186317	.000080305	.000031222	.000001867	.000000058	.000000000	11100011011000
110000011100000	24	-.0030716	.0042610	.000737619	.000388970	.000185033	.000079354	.000030664	.000001798	.000000054	.000000000	11011101101000
111011000000001	22	-.0030876	.0047603	.000740591	.000397136	.000194170	.000088870	.000036367	.000002480	.000000096	.000000000	01111101101000
011101000001000	22	-.0030876	.0045199	.000737840	.000391392	.000188324	.000082139	.000032454	.000002059	.000000070	.000000000	11011011001000
011011100001000	22	-.0030876	.0045087	.000737717	.000391146	.000188074	.000081956	.000032348	.000002046	.000000069	.000000000	11011101000001
100111100010000	23	-.0031292	.0046722	.000732368	.000384891	.000183102	.000077780	.000030630	.000001844	.000000058	.000000000	11110000110
100110000010000	23	-.0031292	.0046200	.000731785	.000383712	.000181888	.000077787	.000030108	.000001780	.000000055	.000000000	10111101000100
101110000010000	23	-.0031292	.0045617	.000731119	.000382336	.000180444	.000076806	.000029467	.000001699	.000000051	.000000000	11011011001000
100101010010000	23	-.0031877	.0049395	.000725333	.000377266	.000177516	.000075503	.000028102	.000001692	.000000052	.000000000	11011010010100
101011010100000	23	-.0031877	.0048481	.000724312	.000375203	.000175397	.000073949	.000028102	.000001583	.000000046	.000000000	11011010010100
101101010010000	23	-.0031877	.0048159	.000723956	.000374493	.000174677	.000073429	.000027805	.000001549	.000000044	.000000000	11011010010100
110011001001000	23	-.0032192	.0049719	.000721500	.000372829	.000173353	.000072799	.000027569	.000001537	.000000044	.000000000	11011100101000
110011000100100	23	-.0032722	.0052758	.000714640	.000365213	.000168419	.000068419	.000026259	.000001430	.000000040	.000000000	11011101101000
110011011000000	23	-.0033118	.0054955	.000710306	.000360928	.000165213	.000065501	.000025551	.000001376	.000000038	.000000000	11011010110100

Table A
m = 7, n = 5   Rank Orders 751 thru 792 of 792   C(7, 5) = 792   1/C(7, 5) = .00126263

z	w	c_1	c_2	d = .2	.4	.6	.8	1.0	1.5	2.0	3.0	z^{tc}
11001011000	23	−.0033363	.0056562	.000707894	.000358801	.000164290	.000067911	.000025325	.000001365	.000000038	.000000000	11100101100
11011010100	23	−.0033363	.0056188	.000707493	.000358020	.000163518	.000067367	.000025021	.000001332	.000000037	.000000000	11101011010
01111011000	21	−.0033524	.0059596	.000708526 •	.000361877 •	.000168076	.000066489	.000027177	.000001611	.000000051	.000000000	11110100001
11100011000	23	−.0033948	.0060392	.000702131	.000353706	.000161267	.000066491	.000024769	.000001339	.000000038 •	.000000000	11110101010
11110100000	22	−.0034524	.0064190	.000695698	.000348768	.000158381	.000065164	.000024269	.000001321	.000000034	.000000000	10111111000
10101101000	22	−.0034524	.0063459	.000695212	.000347207	.000156816	.000064041	.000023630	.000001247	.000000033	.000000000	11011100010
10110101000	22	−.0034524	.0062999	.000693000	.000346278	.000155912	.000063413	.000023283	.000001211	.000000031	.000000000	11101110010
10111100100	22	−.0034713	.0063898	.000693360	.000343959	.000154280	.000062498	.000022852	.000001175	.000000030	.000000000	11101101010
11010110000	22	−.0035369	.0067931	.000686245	.000337659	.000150292	.000060468	.000021976	.000001117	.000000030	.000000000	11101011010
11011001000	22	−.0035954	.0071512	.000680203	.000332002	.000146694	.000058628	.000021178	.000001062	.000000028	.000000000	11110011100
11001101000	22	−.0035954	.0071255	.000679938	.000331509	.000146226	.000058309	.000021006	.000001045	.000000027	.000000000	11101001100
01111010010	20	−.0035954	.0070882	.000679551	.000330784	.000145533	.000057837	.000020751	.000001019	.000000026	.000000000	11101011010
10101110000	22	−.0036114	.0075205	.000677260	.000335970	.000151144	.000061969	.000023130	.000001297	.000000040	.000000000	11111010010
11100101000	22	−.0036199	.0072614	.000673417	.000328865	.000144456	.000057359	.000020574	.000001012	.000000026	.000000000	11100110100
11110000010	21	−.0036596	.0075288	.000673346	.000325493	.000142452	.000056403	.000020188	.000000990	.000000025	.000000000	11110011010
10101111000	21	−.0037360	.0081756	.000667346	.000321487	.000140961	.000056180	.000020327 •	.000001041 •	.000000028	.000000000	10111111010
10110110000	21	−.0037360	.0080023	.000665583	.000318223	.000137872	.000054084	.000019194	.000000925	.000000023	.000000000	11011101010
11010100100	21	−.0037360	.0079885	.000665446	.000317975	.000137644	.000053934	.000019115	.000000917	.000000023	.000000000	11101101010
11011000100	21	−.0038601	.0087980	.000653120	.000306823	.000130823	.000050565	.000017699	.000000827	.000000020	.000000000	11101011010
11100110000	21	−.0038601	.0087907	.000653043	.000306714	.000130683	.000050464	.000017646	.000000822	.000000020	.000000000	11111001100
11010001000	21	−.0038601	.0087373	.000652514	.000305764	.000129812	.000049893	.000017348	.000000794	.000000019	.000000000	11101011010
11101010010	21	−.0038790	.0088421	.000650464	.000303761	.000128491	.000049197	.000017038	.000000772	.000000018	.000000000	11101101010
11100110000	21	−.0039186	.0091262	.000646778	.000300656	.000126717	.000048383	.000016722	.000000756	.000000018	.000000000	11110101010
11101001000	20	−.0039431	.0093134	.000644611	.000298936	.000125802	.000047997	.000016587	.000000751	.000000017 •	.000000000	11110011100
11011010000	20	−.0040007	.0097073	.000639075	.000294111	.000122944	.000046636	.000016037	.000000720	.000000017	.000000000	11100111100
11011001100	20	−.0041437	.0107347	.000625761	.000282809	.000116392	.000043564	.000014811	.000000651	.000000015	.000000000	11101110100
10111101000	20	−.0041437	.0106220	.000624697	.000280977	.000114779	.000042704	.000014297	.000000606	.000000014	.000000000	11101101100
11011010010	20	−.0041437	.0106060	.000624547	.000280723	.000114559	.000042408	.000014229	.000000601	.000000014	.000000000	11110101100
11010100010	20	−.0041833	.0109048	.000620997	.000277830	.000112958	.000041695	.000013960	.000000588	.000000013	.000000000	11101011100
11100111000	20	−.0042022	.0110221	.000619064	.000276047	.000111839	.000041130	.000013719	.000000572	.000000013	.000000000	11110011100
10111111000	19	−.0042598	.0115292	.000614573	.000272978	.000110492	.000040702	.000013629	.000000578 •	.000000013	.000000000	11101011010
11010010100	19	−.0044084	.0124921	.000599765	.000259603	.000102281	.000036643	.000011926	.000000471	.000000010	.000000000	11111010100
11100101000	19	−.0044669	.0129835	.000594951	.000256010	.000100462	.000035904	.000011672	.000000461	.000000010	.000000000	11110110010
11101010000	19	−.0044669	.0129338	.000594509	.000255293	.000099864	.000035418	.000011501	.000000448	.000000009	.000000000	11101111000
11011100000	19	−.0044669	.0129155	.000594347	.000255032	.000099649	.000035546	.000011440	.000000444	.000000009	.000000000	11110011100
11101100100	18	−.0046675	.0144767	.000576733	.000240892	.000088933	.000031956	.000010123	.000000377	.000000008	.000000000	11110101100
11110010000	18	−.0047316	.0149335	.000570732	.000235798	.000088592	.000030584	.000009579	.000000347	.000000007	.000000000	11101111010
11101011000	18	−.0047505	.0151180	.000569395	.000234960	.000079849	.000030481	.000009557	.000000348 •	.000000007	.000000000	11101011100
11110100000	17	−.0049907	.0170462	.000548779	.000218752	.000079849	.000026652	.000008123	.000000278	.000000005	.000000000	11110111000
11111000000	17	−.0050152	.0172507	.000546740	.000217194	.000079032	.000026303	.000007995	.000000272	.000000005	.000000000	11101111000
11110100000	16	−.0052743	.0194748	.000525673	.000201444	.000070926	.000022904	.000006773	.000000217	.000000004	.000000000	11111010000
11111000000	15	−.0055390	.0218965	.000505274	.000186922	.000063779	.000020027	.000005776	.000000176	.000000003	.000000000	11111110000

Table A

m = 6, n = 6   Rank Orders 1 thru 50 of 494   C(6, 6) = 924   1/C(6, 6) = .00108225

z	w	c_1	c_2	d = .2	.4	.6	.8	1.0	1.5	2.0	3.0	z^{tc}
000000111111	57	.0048587	.0197022	.002555816	.005597888	.011403860	.021673081	.038548395	.124428252	.286651625	.706096058	000000111111x
000001011111	56	.0046367	.0176228	.002446868	.005102747	.009836541	.017566847	.029131213	.075774556	.130843046	.133920609	000001011111x
000010011111	55	.0044098	.0156254	.002341909	.004655610	.008518243	.014372246	.022406132	.048779249	.067929044	.039978732	000010101111x
000011001111	54	.0041829	.0137190	.002241256	.004245953	.007368997	.011733830	.017168666	.031031066	.034253274	.010586410	000010110111x
000100011111	54	.0041667	.0136206	.002235747	.004230289	.007343891	.011718874	.017221974	.031875081	.036953031	.014228373	000001101111
000101001111	53	.0039608	.0119858	.002148653	.003893175	.006446947	.009771400	.013576224	.021338570	.022020676	.004452164	000011001111
000110001111	53	.0039398	.0118125	.002139499	.003856781	.006347783	.009551642	.013157895	.021002212	.020226632	.003495017	000010011111
001000101111	53	.0038897	.0114991	.002121837	.003801856	.006236412	.009382763	.012972718	.020244821	.019913769	.005279342	000100101111
001001001111	52	.0037178	.0101748	.002050920	.003534973	.005548231	.007939466	.013071864	.013702184	.010565979	.001382047	000011010111
001010001111	52	.0036968	.0100123	.002042237	.003502335	.005464454	.007765227	.010061823	.012942481	.009605412	.001105159	000100110111
001000111111	52	.0036628	.0098041	.002030361	.003465178	.005386670	.007636868	.009887567	.012795371	.009691424	.001231033	000010111011
010000011111	52	.0035402	.0090674	.001987954	.003333319	.005111120	.007180423	.009260959	.012210108	.009924032	.001820386	000011011011
000110101111	51	.0034937	.0086302	.001965507	.003242831	.004865502	.006655505	.008291458	.009708152	.006612216	.000675060	000110001111
000101101111	51	.0034747	.0084699	.001957511	.003208844	.004771511	.006442224	.007905187	.008738492	.005439743	.000404815	001000101111
001001101111	51	.0034407	.0082761	.001946147	.003174970	.004704322	.006337948	.007773557	.008660764	.005526249	.000465946	000101010111
001000011111	51	.0034197	.0081255	.001937955	.003145986	.004634472	.006201905	.007547454	.008199185	.004848515	.000378639	000110010111
010001001111	51	.0033133	.0075158	.001902131	.003037341	.004411808	.005836435	.007043483	.007603379	.004769783	.000408036	001000110111
001010101111	50	.0032478	.0070237	.001875856	.002942761	.004182149	.005389591	.006304719	.006152607	.003359881	.000190121	001001010111
001010011111	50	.0032138	.0068426	.001864103	.002911346	.004122134	.005299775	.006195046	.006087929	.003404396	.000218112	001010010111
001100011111	50	.0031976	.0066929	.001857408	.002881367	.004043389	.005136948	.005913537	.005494409	.002812836	.000131419	001000111011
010000111111	50	.0031426	.0063781	.001838910	.002825327	.003928694	.004949954	.005653273	.005175775	.002635326	.000127575	010000111101
010001011111	50	.0030912	.0061273	.001823100	.002782056	.003850036	.004837297	.005524596	.005115674	.002687216	.000149320	000101101011
001100101111	50	.0030209	.0059915	.001815648	.002756937	.003793795	.004735656	.005368069	.004853041	.002465653	.000123064	001001101011
001101001111	49	.0029845	.0056686	.001797463	.002697844	.003662717	.004502422	.005010442	.004306869	.002052274	.000466764	001100001111
100000101111	49	.0029707	.0057709	.001795674	.002724749	.003787848	.004831545	.005667261	.005932089	.003861971	.000378639	000111010111
001100011111	49	.0029707	.0054480	.001782832	.002660353	.003600251	.004424255	.004943606	.004344897	.002179265	.000114123	000110011011
001110001111	49	.0029707	.0053597	.001779817	.002641745	.003541755	.004292588	.004707218	.003849869	.001720844	.000060155	001010101011
010010001111	49	.0029296	.0053255	.001779749	.002641308	.003540332	.004289280	.004701172	.003836391	.001707820	.000058690	001001110111
010001101111	49	.0028644	.0050553	.001762340	.002586782	.003424706	.004092372	.004416940	.003441076	.001442879	.000041688	001010011011
010010101111	49	.0028644	.0048353	.001746830	.002550034	.003370414	.004037958	.004390473	.003571441	.001633818	.000017590	001100110111
010010011111	49	.0028482	.0046985	.001739881	.002524229	.003307377	.003916742	.004195720	.003231292	.001355531	.000040994	001010110111
010100011111	48	.0027988	.0044725	.001725150	.002485117	.003238337	.003821308	.004087918	.003175263	.001373519	.000048707	000111001111
001100111111	49	.0027932	.0044207	.001722605	.002474575	.003214668	.003775988	.004015847	.003045589	.001280281	.000101232	010001011101
100000011111	50	.0027576	.0044489	.001718035	.002482075	.003267051	.003922520	.004301137	.003673689	.001835359	.000025963	000110101111
010100101111	48	.0027438	.0041156	.001705265	.002420829	.003098618	.003578866	.003732985	.002671203	.001031838	.000029887	001101010111
001101101111	48	.0027217	.0040694	.001701318	.002412989	.003090259	.003576669	.003744543	.002724854	.001082735	.000018059	000111011011
010101001111	48	.0026937	.0038331	.001688549	.002370633	.002996711	.003412969	.003504418	.002390234	.000866881	.000034095	001110010111
010100111111	48	.0026213	.0035807	.001669727	.002323925	.002940033	.003363275	.003490978	.002527453	.001010709	.000018423	001101001111
010101011111	48	.0026213	.0035087	.001667101	.002313741	.002895408	.003269543	.003334290	.002255048	.000823048	.000017603	001100111011
001110011111	48	.0026213	.0035049	.001666974	.002312988	.002893137	.003264655	.003325938	.002229920	.000811379		010010011101
010010111111	48	.0025711	.0032374	.001646753	.002265731	.002800080	.003117763	.003129762	.002017236	.000691071	.000012801	010010011011
100000111111	49	.0025355	.0031255	.001646517	.002272706	.002848792	.003246021	.003365728	.002457641	.001022727	.000036007	000111101011
001110101111	47	.0025218	.0030293	.001639668	.002229241	.002737578	.003033331	.003035312	.001958695	.000683381	.000014173	001110001111
100000011111	49	.0025145	.0031713	.001639648	.002252437	.002807913	.003179377	.003272992	.002336288	.000942377	.000030058	000110110111
011000101111	47	.0025008	.0029256	.001630007	.002210955	.002702963	.002980784	.002967789	.001888853	.000648746	.000012963	001100101011
100100101111	47	.0024668	.0027020	.001617717	.002165244	.002620042	.002841921	.002773216	.001649035	.000513532	.000007496	001010110011
011000011111	47	.0024506	.0026665	.001614053	.002165244	.002614434	.002843325	.002787175	.001691685	.000545887	.000009045	010010101101
100100011111	48	.0024437	.0026867	.001613589	.002168576	.002639406	.002879066	.002849657	.001798497	.000491494	.000013074	010100110111
011001001111	47	.0023944	.0026061	.001597114	.002119385	.002530087	.002792066	.002636309	.001552814	.000486014	.000007608	001101101011
010011001111	47	.0023782	.0024061	.001593700	.002119385	.002493151	.002720988	.002632095	.001557280	.000457868	.000017603	001110011011

136

z	w	c_1	c_2	d = .2	.4	.6	.8	1.0	1.5	2.0	3.0	z^{tc}
011000001111	47	.0023442	.0023500	.001587287	.002070132	.002534416•	.002765755	.002741901	.001779609	.000655431	.000018228	000001111001
010001101011	47	.0023442	.0021587	.001581540	.002075893	.002448725	.002597377	.002478855	.001395230	.000411776	.000005428	001010001101
010010011011	47	.0023086	.0021568	.001581481	.002075561	.002447788	.002595496	.002475866	.001390788	.000409034	.000005315	001010101101
100010010111	48	.0022925	.0022195	.001577467	.002082264•	.002491445	.002704821	.002667264	.001704529	.000614340	.000015836	000101011001x
001101000111	46	.0022925	.0021020	.001571270	.002061603	.002445927	.002625659	.002551993	.001546137	.000511867	.000009677	000111011110
001011000111	46	.0022787	.0019347	.001564186	.002035315	.002386267	.002523101	.002408169	.001378641	.000426194	.000006958	001010101110
100001110011	48	.0022447	.0017254	.001552400	.001999279	.002313026	.002405359	.002249781	.001202269	.000336277	.000003967	001011010011x
001101110011	47	.0022375	.0018824	.001555722	.002022087	.002378243	.002533171	.002445578	.001466600	.000487110	.000009927	001000111110
001100101011	48	.0022237	.0016337	.001546241	.001983075	.002284311	.002364716	.002201296	.001161485	.000320379	.000003664	001000101110
011000101111	47	.0022217	.0016430	.001546166	.001984183•	.002288479	.002373579	.002215213	.001179513	.000329461	.000003891	010001011101x
010011000111	46	.0021723	.0014594	.001532635	.001951079	.002234312	.002303409	.002139448	.001133473	.000319216	.000004073•	000111011101
010101001011	46	.0021513	.0013661	.001526424	.001934575	.002204720	.002260950	.002087994	.001087983	.000300243	.000003629	001010101101
010110001011	46	.0021173	.0011691	.001515049	.001900992	.002138834	.002158807	.001955673	.000955014	.000240015	.000002145	001010101101
011000010111	46	.0021012	.0012475	.001514554	.001910259•	.002173071	.002231360	.002069846	.001106146	.000319633	.000004507	001011011110
100010001011	46	.0021012	.0011464	.001511690	.001895662	.002135384	.002162102	.001968992	.000984603	.000257963	.000002688	001010101101
100100000111	46	.0021012	.0011401	.001511506	.001894692	.002132810	.002162102	.001961756	.000975447	.000253165	.000002547	001010101101
001100011011	47	.0020656	.0012165	.001507616	.001900592•	.002170423	.002247591	.002113041	.001196385	.000381589	.000007735	001000011110
010001110011	47	.0020656	.0011418	.001505446	.001889201	.002140006	.002189505	.002024712	.001075012	.000308747	.000004287	001010101110
001110010011	45	.0020518	.0010524	.001500669	.001875187	.002137444	.002184554	.002017110	.001064416	.000302429	.000004120•	001110101101
100101000011	47	.0020154	.0009229	.001490707	.001850115	.002069834	.002088369	.001901226	.000962423	.000259592	.000002986	001010011110
001101001011	45	.0020017	.0007526	.001483689	.001824948	.002015123	.001998762	.001782007	.000842666	.000207945	.000001901	001010101101
011001010011	46	.0019948	.0007058	.001481193	.001817188	.001999373	.001973808	.001749262	.000809593	.000193300	.000001578	010100101101
011010001011	45	.0019806	.0006716	.001477839	.001810339	.001990565	.001965810	.001744762	.000815353•	.000199686	.000001903•	001101100011x
010100110011	45	.0019293	.0005147	.001464679	.001780354	.001945043	.001911350	.001690796	.000790171	.000155400	.000001768	001110101101
100000111011	47	.0018953	.0003321	.001453771	.001749466	.001886912	.001824903	.001583383	.000692972	.000155770	.000001110	010100101011
001100110011	45	.0018880	.0003601	.001457201	.001771083	.001914944	.001930118	.001733616	.000861689	.000234904	.000003149	001011010011x
010011010011	45	.0018743	.0002552	.001450896	.001749445	.001899521	.001855649	.001634847	.000760356	.000188442	.000001874	001110101101
010101010011	45	.0018743	.0002508	.001447914	.001734437	.001863894	.001794789	.001550262	.000670519	.000148856	.000001034	001110101101
100010100011	46	.0018387	.0002665	.001442091	.001729745	.001868539•	.001818775	.001596102	.000734916	.000179798	.000001706	001101011110
011001000111	45	.0018241	.0001605	.001436736	.001713475	.001836833	.001770943	.001536559	.000682269	.000159130	.000001320	001101011001
100100100011	46	.0018225	.0002389	.001438527	.001723061	.001860863	.001813087	.001594749	.000743037	.000185695	.000001894	001110101110
010100101011	46	.0017885	.0002477	.001432852	.001717973	.001867305	.001841961	.001650953	.000663047	.000153571	.000003979	000111011110
100010110011	46	.0017885	.0000737	.001428112	.001694655	.001809029	.001737929	.001503263	.000663047	.000142846	.000001245	001110101110
001101010011	44	.0017885	.0000696	.001427998	.001694086	.001807590	.001735341	.001499574	.000658846	.000153571	.000001196	001011100011
011100010011	45	.0017748	−.0000167	.001423354	.001680952	.001745285	.001701569	.001460362	.000630367	.000142846	.000001123	010101010101x
100100010011	44	.0017679	−.0001404	.001418835	.001663658	.001754649	.001638650	.001377455	.000551433	.000127739	.000000605	001101011101
010100011011	45	.0017586	−.0001141	.001417938	.001665430	.001754649	.001659285	.001409318	.000588493	.000118966	.000000897	010101011010x
100100011011	45	.0017517	−.0001566	.001415585	.001658498	.001741288	.001639157	.001384185	.000566060	.000118966	.000000746	010001101011
010110000111	44	.0017024	−.0002260	.001405071	.001639668	.001721146	.001626380	.001384619•	.000590241	.000133923	.000001120	001110101101
011000110011	46	.0016660	.0003152	.001396196	.001619922	.001690982	.001588785	.001344228	.000561452	.000123279	.000000901	010001011110
011010011011	44	.0016522	−.0003815	.001392105	.001609521	.001673837	.001566952	.001321545	.000550134	.000121846	.000000970•	001101011101
010011000111	44	.0016522	−.0004864	.001389237	.001596018	.001641721	.001515097	.001252401	.000484053	.000095703	.000000523	001110100101
010101001011	44	.0016312	−.0004908	.001383792	.001593395	.001622085	.001512724	.001249188	.000480903	.000094453	.000000504	001011010101
100100110011	45	.0016166	−.0005570	.001383761	.001591883•	.001648558	.001448418	.001223932	.000466139	.000080569	.000000473	001110011110
011000101011	44	.0015972	−.0004565	.001376206	.001568522	.001602994	.001537701	.001292820	.000534203	.000112234	.000000899	001011100011
100101001011	45	.0015956	−.0006159	.001378057	.001577965	.001615650	.001469609	.001209220	.000464232	.000091730	.000000508	001110101110
100100101011	44	.0015616	−.0005319	.001367923	.001551186	.001602994	.001507660	.001259294	.000510378	.000109282	.000000782	001010110101
010010101011	45	.0015616	−.0006886	.001367517	.001551186	.001578468	.001441755	.001182382	.000450464	.000088213	.000000473	001010110101

137

Table A

m = 6, n = 6   Rank Orders 101 thru 150 of 494   C(6, 6) = 924   1/C(6, 6) = .00108225

z	w	c_1	c_2	d = .2	.4	.6	.8	1.0	1.5	2.0	3.0	z^{tc}
010011001101	44	.0015458	−.0008378	.001361349	.001530590	.001538647	.001383441	.001112963	.000398163	.000071710	.000000305	01001100101x
101000010011	45	.0015454	−.0006082	.001367117	.001557150	.001600427	.001484395	.001243965	.000515185	.000115007	.000000952	00010111010
100001110011	45	.0015454	−.0006920	.001364965	.001547198	.001576813	.001445515	.001192589	.000466676	.000095785	.000000612	001100011110
001110010011	45	.0015454	−.0007056	.001364610	.001545535	.001572863	.001438675	.001183675	.000458066	.000092331	.000000552	001101100011
001010001101	43	.0015317	−.0007850	.001360189	.001533623	.001552200	.001411127	.001153115	.000438705	.000087228	.000000523	010001111001
011100100111	44	.0015248	−.0009044	.001355895	.001517933	.001518947	.001359058	.001087643	.000383429	.000067871	.000000277	010000111100
100000011111	44	.0014747	−.0009596	.001345502	.001499615	.001499071	.001344725	.001083051	.000395041	.000074495	.000000384	000011111100
100001100101	45	.0014391	−.0006961	.001345644	.001520434	.001562443	.001461673	.001246127	.000561453	.000145707	.000002085	010010001011
100001101101	45	.0014391	−.0010229	.001337438	.001483190	.001476461	.001319743	.001059629	.000383792	.000071783	.000000359	010100001101
010010010111	45	.0014391	−.0010251	.001337382	.001482929	.001475853	.001318734	.001058307	.000382580	.000071326	.000000352	010001101110
011010000111	43	.0014253	−.0010112	.001335328	.001481804	.001480000	.001331530	.001080023	.000409196	.000082937	.000000564	00001111001
011011100011	43	.0014253	−.0011123	.001332780	.001470271	.001453610	.001288627	.001024864	.000360924	.000065364	.000000305	001110001101
011010100011	43	.0014253	−.0011186	.001332619	.001469527	.001451888	.001285800	.001021201	.000357700	.000064210	.000000290	001110011101
010101001011	43	.0014091	−.0012032	.001327605	.001456203	.001428807	.001254651	.000986465	.000334609	.000057602	.000000233	001101010101
011100100011	43	.0013751	−.0012477	.001320340	.001442592	.001412170	.001239122	.000975067	.000333708	.000058522	.000000252	001101100101
100010010111	44	.0013735	−.0011595	.001322208	.001451708	.001433144	.001272852	.001017689	.000368985	.000070540	.000000402	010100100110
100010011011	43	.0013541	−.0013019	.001315194	.001431314	.001395571	.001219670	.000955921	.000323910	.000056258	.000000239	001101010110
100011010111	43	.0013395	−.0013019	.001312492	.001427091	.001391562	.001217229	.000955427	.000325432	.000056806	.000000240	001101100101
100011011011	46	.0013323	−.0011092	.001315906	.001446182	.001437438	.001293092	.001053635	.000412072	.000088823	.000000755	100000011110x
010100111010	44	.0013185	−.0012661	.001309566	.001425670	.001397067	.001232097	.000980652	.000352464	.000067182	.000000386	001101101110
100011001101	44	.0013185	−.0013558	.001307348	.001415804	.001374907	.001197642	.000936057	.000315367	.000054437	.000000225	001100101110
100100100111	44	.0013185	−.0013653	.001307112	.001414752	.001372539	.001193858	.000931288	.000311442	.000053124	.000000211	001110010110
001111001001	42	.0013048	−.0013649	.001304695	.001411843	.001372251	.001198791	.000941954	.000325947	.000059243	.000000303	001110100101x
010101010011	42	.0013027	−.0015064	.001300852	.001396046	.001338064	.001145742	.000876646	.000275308	.000043173	.000000136	010100110101
010100010111	43	.0012817	−.0015623	.001295673	.001384645	.001321237	.001126040	.000857422	.000265575	.000040996	.000000124	010100111010x
010100011011	44	.0012684	−.0014133	.001288689	.001372173	.001306902	.001113759	.000849576	.000266489	.000042167	.000000139	010100111001
010010010111	43	.0012121	−.0016417	.001280993	.001357273	.001287490	.001093556	.000831827	.000259415	.000040797	.000000132	010010101101
110000101011	43	.0011960	−.0013977	.001283845	.001377822	.001338257	.001176324	.000937655	.000346120	.000069886	.000000491	000101011100
010100111010	44	.0011960	−.0016386	.001278129	.001353379	.001285267	.001095074	.000837405	.000267594	.000043969	.000000168	010100011110
100100001101	44	.0011960	−.0016459	.001277953	.001352616	.001283596	.001092476	.000834215	.000265126	.000043187	.000000160	010100110110
011100001011	42	.0011822	−.0015394	.001278012	.001360101	.001305352	.001130639	.000844518	.000308508	.000058158	.000000352	000111110010
010101100011	42	.0011822	−.0017307	.001273460	.001340639	.001263275	.001065352	.000805899	.000248554	.000039078	.000000134	001110001101
011010001011	42	.0011822	−.0017326	.001273415	.001340441	.001262847	.001065320	.000805105	.000247970	.000038904	.000000132	010011001110
011011000011	42	.0011483	−.0017644	.001264412	.001327781	.001247889	.001051791	.000795437	.000247117	.000039491	.000000143	010110010101
011100010011	43	.0011466	−.0016729	.001268268	.001336455	.001267049	.001081405	.000831430	.000274221	.000047928	.000000233	001101100110
010010110011	44	.0011321	−.0018384	.001261700	.001322285	.001228466	.001026892	.000769920	.000231531	.000035510	.000000116	010110101010
011000100111	45	.0011102	−.0017055	.001260710	.001322286	.001249983	.001062749	.000818471	.000266709	.000045982	.000000207	100000111110x
100010001101	43	.0010965	−.0017839	.001256403	.001311212	.001230809	.001039770	.000791032	.000253751	.000044044	.000000195	001101100110
011001000111	43	.0010965	−.0018737	.001254269	.001302102	.001211167	.001009710	.000754687	.000226627	.000034712	.000000112	001101001110
100101001011	43	.0010965	−.0018832	.001254041	.001301125	.001209052	.001006462	.000750753	.000223705	.000033833	.000000104	001101011010
011111000001	41	.0010827	−.0018044	.001253391	.001305334	.001223147	.001031871	.000784248	.000251409	.000042726	.000000196	001110000011x
010110001011	42	.0010758	−.0019908	.001247764	.001285105	.001182309	.000972947	.000715681	.000204078	.000029168	.000000077	010110010101
100100110011	43	.0010755	−.0019264	.001249158	.001290971	.001194881	.000990728	.000736094	.000217206	.000032548	.000000099	001101100110
010010101011	42	.0010597	−.0020689	.001242972	.001277960	.001162955	.000947809	.000689521	.000189521	.000025776	.000000059	010101011010x
010011010011	43	.0010415	−.0019601	.001242092	.001277960	.001179009	.000975584	.000724239	.000214361	.000032446	.000000102	001010110110
011100001011	42	.0010257	−.0020868	.001236284	.001261620	.001150500	.000937839	.000668901	.000190544	.000032446	.000000067	010101011001
011001001101	42	.0010047	−.0021306	.001231389	.001251431	.001136276	.000922065	.000668901	.000188405	.000026623	.000000062	010010110110
010011001101	42	.0010047	−.0021154	.001222994	.001136276	.001134741	.000922065	.000677064	.000186465	.000026018	.000000062	010011001110

138

m = 6, n = 6   Rank Orders 151 thru 200 of 494   C(6, 6) = 924   1/C(6, 6) = .00108225

z	w	c_1	c_2	d = .2	.4	.6	.8	1.0	1.5	2.0	3.0	z^{tc}
100101011010	43	.0009691	−.0021591	.001224102	.001238117	.001119976	.000906186	.000655926	.000179830	.000024691	.000000060	01010110110
100100010011	43	.0009691	−.0021641	.001223985	.001237635	.001118975	.000904715	.000654225	.000178713	.000024399	.000000058	01001010101
010110100011	41	.0009553	−.0021691	.001223389	.001233705	.001118827	.000904113	.000656622	.000183539	.000026237	.000000075	00110100100
110000101001	43	.0009189	−.0020262	.001217767	.001235416	.001130204	.000932607	.000694330	.000212233	.000034415	.000000140	00100111001
100000111001	43	.0009189	−.0021597	.001223507	.001235416	.001130204	.000887780	.000649177	.000186376	.000027446	.000000086	01100011110
011100000110	41	.0009052	−.0021770	.001214794	.001223928	.001102823	.000893080	.000650569	.000182825	.000026486	.000000079	01011011001
011100100011	41	.0009052	−.0021992	.001214404	.001221928	.001102823	.000892344	.000650569	.000185726	.000027533	.000000088	00110111001
011001100101	41	.0009052	−.0022529	.001211709	.001218476	.001085606	.000871718	.000627432	.000171511	.000023980	.000000066	00110111001
011101000101	41	.0009052	−.0022567	.001210174	.001211244	.001085606	.000870705	.000626284	.000170791	.000023797	.000000065	01011011110
010010011110	44	.0008833	−.0021858	.001207524	.001210897	.001084901	.000870705	.000640401	.000180972	.000026395	.000000080	10001110110
101010000111	42	.0008696	−.0021863	.001205035	.001206669	.001087276	.000882119	.000644838	.000187693	.000029014	.000000111	01010110010
100011100011	42	.0008696	−.0022700	.001203119	.001198793	.001070919	.000857995	.000616714	.000168485	.000023593	.000000065	01001011010
100110010011	42	.0008696	−.0022836	.001202808	.001197521	.001068291	.000834141	.000612252	.000165518	.000022790	.000000059	00111000110
010101000011	42	.0008534	−.0023489	.001198341	.001186882	.001051724	.000834007	.000592016	.000155139	.000020507	.000000047	00110100101
100110100011	41	.0008328	−.0024549	.001192166	.001171543	.001027312	.000804055	.000561865	.000139950	.000017287	.000000033	01010101010
101010010101	42	.0008194	−.0023682	.001191573	.001174970	.001037871	.000821448	.000582719	.000153290	.000020497	.000000050	01010110010
101000110001	41	.0007988	−.0024601	.001185720	.001160938	.001016113	.000795284	.000562719	.000140599	.000019520	.000000038	01000110110
101100110011	42	.0007984	−.0023995	.001186960	.001165911	.001025831	.000808930	.000577710	.000140599	.000019778	.000000047	01010110010
011001010011	41	.0007826	−.0025275	.001181216	.001150205	.000999476	.000775237	.000536888	.000130792	.000015806	.000000029	01011010011
110001000011	42	.0007470	−.0025904	.001179726	.001159630	.001028245	.000822808	.000594446	.000169394	.000025875	.000000099	00110110001
100011010101	42	.0007470	−.0025373	.001174315	.001138332	.000985878	.000762956	.000527667	.000128511	.000015571	.000000029	01010110010
100011100101	42	.0007470	−.0025424	.001174202	.001137885	.000984985	.000761697	.000526269	.000127686	.000015378	.000000028	01001110010
010111000011	40	.0007333	−.0024690	.001173279	.001140313	.000993956	.000777352	.000545734	.000141016	.000018792	.000000048	00111100010
101000100101	41	.0007260	−.0025575	.001170244	.001130839	.000971206	.000764533	.000521320	.000127299	.000015614	.000000031	01000111001
100100100101	42	.0007260	−.0025753	.001169557	.001128710	.000972840	.000748922	.000515015	.000123375	.000014648	.000000025	01010101010
110110001001	42	.0006920	−.0024296	.001166318	.001130328	.000985163	.000748926	.000544532	.000143204	.000019520	.000000052	01010111100
100011001011	41	.0006920	−.0025672	.001163373	.001119023	.000963245	.000742119	.000511824	.000124714	.000015285	.000000030	01000110110
010110001001	42	.0006920	−.0025823	.001163048	.001117769	.000960809	.000738767	.000508191	.000122681	.000014831	.000000028	01001110001
011101000011	40	.0006783	−.0025802	.001160682	.001114220	.000958109	.000738093	.000510078	.000126040	.000015969	.000000036	01010101001
110001000011	40	.0006621	−.0027122	.001158025	.001110973	.000955967	.000738093	.000511566	.000128009	.000016515	.000000039	00110111001
100010101110	43	.0006564	−.0025751	.001156486	.001107100	.000949379	.000729260	.000501779	.000121756	.000014865	.000000028	10001010110
100010001011	43	.0006403	−.0025599	.001153803	.001103559	.000947041	.000729325	.000504131	.000125046	.000015901	.000000036	10001110110
101100000111	41	.0006265	−.0024728	.001153147	.001106948	.000957812	.000747473	.000526444	.000140577	.000020123	.000000068	00111110010
100110010001	41	.0006265	−.0026468	.001149396	.001092421	.000929370	.000707913	.000482941	.000114873	.000013836	.000000026	00110110010
100100101011	41	.0006059	−.0026509	.001149308	.001092084	.000928719	.000707023	.000481982	.000114344	.000013719	.000000026	00110110110
101010101001	41	.0005557	−.0027498	.001143352	.001077836	.000906897	.000681262	.000457035	.000102931	.000011518	.000000018	01010100101
110010101101	41	.0005764	−.0026576	.001142785	.001081009	.000916279	.000696110	.000474146	.000112859	.000013719	.000000027	01011101010
100010011011	42	.0005764	−.0027122	.001138589	.001071581	.000902391	.000680116	.000458893	.000105975	.000013690	.000000022	01001111100
110001001101	41	.0004700	−.0027122	.001140338	.001080848	.000921667	.000707192	.000488374	.000122320	.000015983	.000000040	01001101010
101001001101	40	.0004700	−.0027122	.001134713	.001065165	.000895404	.000674223	.000454876	.000105408	.000012391	.000000023	01001101110
010110010101	40	.0005557	−.0028005	.001132869	.001058275	.000882455	.000657019	.000436897	.000096282	.000010548	.000000016	01101001001
110010100011	41	.0005557	−.0028024	.001132829	.001058127	.000882181	.000656659	.000436526	.000096102	.000010514	.000000016	00111011010
100011100011	41	.0005201	−.0025512	.001131393	.001066809	.000907670	.000697185	.000497185	.000124831	.000017339	.000000056	00111110110
100110010011	41	.0005201	−.0027922	.001126230	.001047614	.000850950	.000647283	.000430103	.000094970	.000010465	.000000015	01011000110
010110111001	40	.0005056	−.0027995	.001126166	.001047032	.000869844	.000645794	.000428825	.000094139	.000010291	.000000015	00111101010
010101010101	41	.0005040	−.0028467	.001122479	.001039506	.000855848	.000633606	.000413264	.000090046	.000009656	.000000014	01101011010
110001010011	41	.0004700	−.0028582	.001121895	.001037378	.000855650	.000629579	.000413264	.000087605	.000009120	.000000012	01100110001
100110100011	42	.0004700	−.0026981	.001118803	.001038952	.000866707	.000649563	.000437421	.000102051	.000012259	.000000025	00101011010
100011001101	41	.0004700	−.0028358	.001115974	.001028532	.000847345	.000624072	.000410966	.000087740	.000009565	.000000012	01001100110
101001001101	41	.0004700	−.0028509	.001115661	.001027371	.000845177	.000621209	.000407990	.000087248	.000009267	.000000013	01001101110

139

Table A

m = 6, n = 6   Rank Orders 201 thru 250 of 494   C(6, 6) = 924   1/C(6, 6) = .00108225

z	w	c_1	c_2	d = .2	.4	.6	.8	1.0	1.5	2.0	3.0	z^{tc}
011011000011	39	.0004562	−.0027703	.001114788	.001029594•	.000852964	.000634139	.000423299	.000096541	.000011381	.000000023	001111001001
100100111001	41	.0004490	−.0027146	.001114497	.001031033•	.000856007	.000639894	.000399368	.000099368	.000011856	.000000024	001110111100
100100011010	41	.0004490	−.0028618	.001111478	.001019330	.000836310	.000612819	.000401343	.000085306	.000009019	.000000013	011000110110
101000110101	41	.0004490	−.0028718	.001111274	.001019183	.000834937	.000611035	.000399519	.000084432	.000008854	.000000012	010101011010
011110100011	39	.0004352	−.0027821	.001110579	.001022027•	.000843803	.000625286	.000416092	.000094214	.000009401	.000000014	001110110001
100011001110	42	.0004344	−.0028251	.001109452	.001017897	.000835769	.000614103	.000403801	.000087113	.000011035	.000000022	100011001110x
100110100011	42	.0004134	−.0028508	.001104958	.001009292	.000824709	.000602792	.000394094	.000083611	.000008839	.000000012	100010110010
010111000011	40	.0003996	−.0028579	.001102283	.001004871	.000820139	.000599442	.000392490	.000083611	.000009197	.000000015	001110100010
101000011010	39	.0003838	−.0029101	.001098237	.000995960	.000807027	.000585077	.000379173	.000078762	.000008190	.000000011	010111010010x
101100001011	42	.0003632	−.0028092	.001093377	.000996752•	.000813342	.000595802	.000391902	.000086045	.000009718	.000000018	100011111010
101100001011	40	.0003495	−.0028224	.001093521	.000991465	.000806786	.000589337	.000386648	.000084246	.000009435	.000000017	001011110010
101001000011	40	.0003495	−.0028971	.001092021	.000986068	.000796999	.000576766	.000373822	.000078232	.000008292	.000000013	001110001110
101010000011	40	.0003495	−.0029032	.001091898	.000985622	.000796181	.000575704	.000372736	.000077706	.000008190	.000000012	001101111010
110000101101	41	.0003426	−.0028284	.001092051•	.000988484	.000802555	.000584441	.000381735	.000081795	.000008893	.000000014	010010111100
011101001011	39	.0003288	−.0029484	.001084037	.000973265	.000777001	.000555865	.000354576	.000070460	.000007261	.000000009	011001010010
011010100011	39	.0003127	−.0029484	.001081040	.000970622	.000774300	.000556217•	.000356137	.000071834	.000007261	.000000010	010010010010
101011000011	39	.0002787	−.0030086	.001076435	.000955655	.000755489	.000536205	.000339072	.000065962	.000006389	.000000008	001111010010
110010000011	40	.0002678	−.0026678	.001082543•	.000978200	.000798824	.000589882	.000393983	.000093079	.000011946	.000000034	000111110010
100101100011	40	.0002771	−.0030146	.001075970	.000954445	.000755430	.000533522	.000336244	.000064463	.000006080	.000000007	001111100010
100110100011	40	.0002771	−.0030168	.001075927	.000954292	.000755158	.000533181	.000335909	.000064317	.000006056	.000000006	010101100110
110010001011	40	.0002431	−.0028460	.001072899	.000955540•	.000764601	.000549755	.000355243	.000074844	.000008137	.000000014	001011110010
000111001000	40	.0002431	−.0029795	.001070264	.000946214	.000747958	.000528703	.000334822	.000065271	.000006378	.000000008	011010001110
101010001101	40	.0002431	−.0029968	.001069919	.000944988	.000745752	.000525903	.000331449	.000063997	.000006147	.000000007	011101101100
110000101001	40	.0002269	−.0028877	.001066104	.000947354	.000753416	.000537416	.000344089	.000070385	.000007359	.000000011	001101101010
101010101010	40	.0002269	−.0030349	.001065907	.000937119	.000733843	.000513007	.000321420	.000059659	.000005568	.000000006	011001010110
010101010101	40	.0002269	−.0030449	.000995	.000733843	.000511932	.000319932	.000059461	.000005461	.000000014	010101010010
011101001011	38	.0002132	−.0028768	.001066678	.000944136	.000750753	.000536543	.000344797	.000071962	.000007831	.000000014	001111010010
100101001110	41	.0001913	−.0030054	.001059931	.000927645	.000725421	.000506831	.000316354	.000059449	.000005521	.000000006	010101010010
100111000011	40	.0001775	−.0029340	.001058783	.000928526	.000730091	.000514825	.000325660	.000064578	.000006549	.000000009	001111100010
100110100101	40	.0001719	−.0030188	.001056042	.000920798	.000717483	.000499652	.000311018	.000058250	.000005431	.000000006	011000111010
100100110110	41	.0001703	−.0030204	.001055664	.000919900	.000715998	.000497723	.000308979	.000057165	.000005210	.000000005	100100110110x
101101010011	41	.0001363	−.0029871	.001049897	.000901373	.000707848	.000491875	.000305793	.000057263	.000005351	.000000005	100100111010
101010100011	39	.0001226	−.0029971	.001047160	.000906697	.000702724	.000487646	.000303084	.000057134	.000005459•	.000000007	011011011010
011011000110	40	.0001205	−.0029574	.001047484	.000908292	.000705538	.000490914	.000305962	.000057962	.000005523	.000000006	010101011010
011011000101	39	.0001068	−.0030350	.001044440	.000899087	.000692600	.000476896	.000293714	.000053743	.000004928	.000000005	010101110001
011100100101	39	.0001064	−.0029618	.001044769	.000903780	.000700695	.000486978	.000303467	.000057836	.000005616	.000000007	001101010010
110010010011	40	.0000995	−.0029635	.001043388	.000901133	.000697128	.000483044	.000299775	.000056160	.000005288	.000000006	001101111010
011011000101	38	.0000858	−.0030319	.001039523	.000892259	.000685203	.000470409	.000284878	.000052517	.000004789	.000000005	010101010010
010110100101	40	.0000518	−.0030796	.001032201	.000878680	.000667790	.000451299	.000274779	.000048112	.000004196	.000000004	010101010001x
100010100101	39	.0000502	−.0030822	.001031806	.000877738	.000666663	.000451299	.000272806	.000047136	.000004009	.000000003	011010100110
011100110001	38	.0000356	−.0030329	.001030022	.000876460	.000666905•	.000453873	.000276342	.000049200	.000004398	.000000005	011001110001
110000011110	41	.0000138	−.0028423	.001029464	.000881527•	.000679504	.000471519	.000294579	.000057477	.000005851	.000000005	100011111100
110000101011	39	.0001226	−.0026542	.001030452•	.000889502	.000696221	.000449229	.000318298	.000069413	.000008338	.000000021	000111111010
110001000011	39	.0000000	−.0028609	.001026522	.000875986	.000672578	.000464676	.000288923	.000055612	.000005569	.000000008	001011011010
110010010011	39	.0001068	−.0029293	.001025074	.000871571	.000664101	.000454983	.000279979	.000051838	.000004921	.000000006	010101110110
110010100011	39	.0000000	−.0029372	.001025225	.000871059	.000664101	.000454325	.000278842	.000051402	.000004846	.000000006	011101101100
110011110000	39	.0000000	−.0029849	.001024171	.000867985	.000658824	.000444898	.000272767	.000048828	.000004412	.000000004	011100111010
101100001101	38	.0000000	−.0030073	.001023746	.000866354	.000656354	.000451299	.000269895	.000047647	.000004217	.000000004	011011110010

140

z	w	c_1	c_2	d = .2	.4	.6	.8	1.0	1.5	2.0	3.0	z^{tc}
101001100101	39	.0000000	-.0030884	.001022214	.000861364	.000647551	.000434332	.000259930	.000043759	.000003621	.000000003	010110011010
101001100011	37	-.0000138	-.0030925	.001022136	.000861101	.000647105	.000433799	.000259430	.000043567	.000003592	.000000003	010110110100
011100000111	37	-.0000356	-.0028423	.001024285	.000857875	.000646136	.000446136	.000288531	.000056202	.000005770	.000000009	001111100001
101001001001	40	-.0000502	-.0030329	.001016505	.000853368	.000640271	.000429299	.000257220	.000043691	.000003675	.000000003	100011100001
101010010110	39	-.0000518	-.0030822	.001012855	.000845775	.000636073	.000418964	.000248560	.000040964	.000003320	.000000003	011100101010x
101001010011	40	-.0000858	-.0030796	.001012561	.000845184	.000629268	.000417830	.000247386	.000040375	.000003210	.000000004	100010101010x
101011000011	40	-.0000995	-.0030319	.001007039	.000837392	.000622195	.000413048	.000244975	.000040521	.000003313	.000000003	001011011100
110001110010	38	-.0001064	-.0029635	.001005764	.000837578	.000625186	.000414367	.000251071	.000043594	.000003865	.000000004	010001101100
101001110011	38	-.0001068	-.0029618	.001004447	.000835169	.000622073	.000415064	.000248081	.000042346	.000003639	.000000003	101000110110
101000110110	40		-.0030350	.001003008	.000830479	.000614245	.000405785	.000239415	.000039023	.000003135	.000000002	100101110010
101100100011	38	-.0001205	-.0029574	.001001899	.000831204	.000618115	.000412111	.000246415	.000042402	.000003726	.000000004	010110110010
110001000101	39	-.0001226	-.0029971	.001000722	.000827732	.000612338	.000405104	.000239648	.000039533	.000003240	.000000003	011011110100
011101110001	37	-.0001363	-.0028871	.000998360	.000824296	.000609239	.000403168	.000238932	.000039939	.000003372	.000000003	001111010001
101101001001	37	-.0001703	-.0030204	.000991332	.000811487	.000593760	.000388430	.000227252	.000036575	.000002951	.000000002	010111010100
101100010011	38	-.0001719	-.0030188	.000991016	.000810819	.000592725	.000387144	.000225941	.000036938	.000002833	.000000002	011011001001x
011000111001	37	-.0001775	-.0029340	.000991489	.000814015	.000598858	.000394768	.000233173	.000038681	.000003235	.000000003	010111001100
011100010110	40	-.0001913	-.0030054	.000987632	.000805664	.000587613	.000383272	.000223622	.000035778	.000002873	.000000002	001011001100
110001010011	40	-.0002132	-.0028768	.000985764	.000801136	.000591050	.000388892	.000229499	.000038113	.000003200	.000000003	100010011010
101010100011	38	-.0002269	-.0028877	.000983020	.000801136	.000588875	.000384525	.000226544	.000037700	.000003217	.000000003	100011101010
110010110010	38	-.0002269	-.0030349	.000980345	.000792427	.000571591	.000367945	.000211391	.000032178	.000002413	.000000002	010101010110
101010000011	38	-.0002269	-.0030449	.000980162	.000798134	.000570621	.000366825	.000210375	.000031818	.000002363	.000000003	011110110010
110100010011	38	-.0002431	-.0028460	.000980710	.000798359	.000583672	.000368172	.000226552	.000038076	.000003299	.000000003	010011110100
100101111001	38	-.0002431	-.0029795	.000978359	.000790446	.000570870	.000368516	.000212635	.000032923	.000002532	.000000002	011100110100
101100100101	38	-.0002431	-.0029968	.000978282	.000790446	.000569244	.000366651	.000210954	.000032333	.000002450	.000000002	010111011100
111000011001	38	-.0002771	-.0026878	.000977970	.000796412	.000585951	.000389082	.000233155	.000041366	.000003869	.000000005	101000110110x
011100110011	38	-.0002771	-.0030146	.000977082	.000777687	.000555606	.000354188	.000201481	.000029894	.000002182	.000000002	010011011100
110010010011	38	-.0002787	-.0030168	.000971245	.000777558	.000555396	.000353948	.000201264	.000029818	.000002119	.000000001	100100111010
101010110010	38	-.0003127	-.0030086	.000971205	.000777295	.000555051	.000353501	.000200768	.000029540	.000002185	.000000001	010101011010
100110100110	39	-.0003288	-.0029484	.000971011	.000770051	.000548699	.000340647	.000198717	.000029624		.000000001	100011011010
110010100011	39		-.0029844	.000961983	.000762791	.000539457	.000340206	.000191288	.000027422	.000001913	.000000001	100011101010
101010000011	37	-.0003426	-.0028284	.000962218	.000767619	.000549574	.000353195	.000203813	.000032255	.000002626	.000000002	011110010010
110000110011	38	-.0003495	-.0028224	.000960982	.000765487	.000546964	.000350554	.000201520	.000031379	.000002479	.000000001	010011110100
100101110001	38	-.0003495	-.0028971	.000959648	.000761219	.000540085	.000342706	.000194472	.000028749	.000002139	.000000001	011011001100
110000010011	38	-.0003495	-.0029032	.000959541	.000760883	.000539551	.000342107	.000193942	.000028434	.000002116	.000000001	010111011100
101101001010	36	-.0003632	-.0028092	.000958628	.000761909	.000543356	.000347704	.000199671	.000031100	.000002472	.000000002	101000011010x
101100110001	38	-.0003838	-.0029101	.000952936	.000749616	.000526742	.000330596	.000185311	.000026575	.000001879	.000000001	011011011100
011100101001	36	-.0003996	-.0028579	.000950874	.000747482	.000525651	.000330647	.000186022	.000027096	.000001964	.000000001	010011011010
011001100011	36	-.0004134	-.0028508	.000948456	.000743839	.000522147	.000328135	.000184656	.000027115	.000002011	.000000001	100100110110
100110010110	36	-.0004344	-.0028251	.000944938	.000738567	.000516841	.000323883	.000181796	.000026551	.000001961	.000000001	011100100001x
110000110011	39	-.0004352	-.0027821	.000945458	.000740209	.000519209	.000326214	.000183546	.000026870	.000001969	.000000001	100011011100
110011001001	37	-.0004490	-.0027146	.000944900	.000739877	.000520950	.000329598	.000187416	.000028691	.000002269	.000000002	001110011100
100111001001	37	-.0004490	-.0028618	.000941523	.000731811	.000508196	.000315310	.000174799	.000024451	.000001696	.000000001	011011001100
110000011011	38	-.0004490	-.0028718	.000941347	.000731259	.000507325	.000314338	.000173946	.000024172	.000001660	.000000001	010111100110
011001010011	39	-.0004521	-.0027703	.000941694	.000734162	.000512676	.000320590	.000179478	.000027842	.000001867	.000000002	100101110010
100110110010	37	-.0004700	-.0026981	.000940415	.000734089	.000514821	.000324417	.000183728	.000023923	.000002179	.000000001	011100110100
101100110010	37	-.0004700	-.0028358	.000938912	.000726568	.000502948	.000311137	.000170816	.000021661	.000001652	.000000002	010111011100
101010110001	37	-.0004700	-.0028509	.000937751	.000725759	.000501688	.000309749	.000170816	.000021604	.000001604	.000000001	010110110010
100110100011	37	-.0005040	-.0028548	.000931235	.000714740	.000489334	.000298848	.000162821	.000021661	.000001415	.000000001	101010100110
100110010110	38	-.0005056	-.0028467	.000931095	.000714642	.000489252	.000298714	.000162627	.000021521	.000001388	.000000001	100110100110x
110000010011	37	-.0005201	-.0025512	.000933430	.000725699	.000508466	.000321026	.000182624	.000022588	.000002284	.000000002	001101111000

141

Table A

m = 6, n = 6   Rank Orders 301 thru 350 of 494   C(6, 6) = 924   1/C(6, 6) = .00108225

z	w	c_1	c_2	d = .2	.4	.6	.8	1.0	1.5	2.0	3.0	z^{tc}
101001110001	37	−.0005201	−.0027922	.000929318	.000713093	.000488942	.000299568	.000164008	.000022240	.000001495	.000000001	0111000110010
101100011001	37	−.0005201	−.0027995	.000929195	.000712717	.000488368	.000298946	.000163479	.000022080	.000001476	.000000001	0110011100010
101100001110	38	−.0005557	−.0027122	.000923966	.000705986	.000482710	.000295308	.000161631	.000022005	.000001493•	.000000001	1000111001010
101011001110	38	−.0005557	−.0028005	.000922453	.000701354	.000475578	.000287555	.000155011	.000019996	.000001255	.000000001	1001100110010
101001010110	38	−.0005695	−.0028024	.000922420	.000701252	.000475421	.000287385	.000154867	.000019953	.000001250	.000000001	0011111000010
101110000011	36	−.0005695	−.0025665	.000923884•	.000709406	.000489976	.000304437	.000170186	.000025079	.000001921	.000000002	0101011100010
101010010101	37	−.0005764	−.0027122	.000920087	.000699526	.000475554	.000289052	.000157074	.000020942	.000001386	.000000001	0101011011010
101001011010	37	−.0005925	−.0026576	.000917968	.000697240	.000474161	.000288669	.000157313•	.000021232	.000001433	.000000001	0101011101010x
101001011010	38	−.0006059	−.0027498	.000913890	.000688304	.000462273	.000276799	.000147742	.000018576	.000001135	.000000001	0101011110010
110011001101	37	−.0006265	−.0024728	.000914660•	.000659836	.000434344	.000293294	.000162469	.000023294	.000001717	.000000001	1001111111000
110001100011	37	−.0006265	−.0026468	.000911760	.000687158	.000463235	.000279299	.000150604	.000019723	.000001285	.000000001	0101001110010
110011001001	37	−.0006265	−.0026509	.000911691	.000686951	.000462921	.000278964	.000150321	.000019639	.000001275	.000000001	0110011010100
011110101001	35	−.0006403	−.0025599	.000910658	.000681393•	.000465466	.000282770	.000154132	.000021064	.000001469	.000000001	0110111000001
110101010001	35	−.0006564	−.0025751	.000907386	.000681691	.000468970	.000276993	.000149878	.000020057	.000001366	.000000001	0111010100010x
110010101110	38	−.0006621	−.0025570	.000906555	.000680446	.000457529	.000275537	.000148594	.000019564	.000001290	.000000001	1001011011010
110010101011	38	−.0006783	−.0025802	.000903113	.000674135	.000449970	.000268478	.000143150	.000018135	.000001131	.000000001	0101011011100
101011010010	36	−.0006920	−.0025109	.000903109	.000677757•	.000457480	.000277778	.000151751	.000021120	.000001529	.000000001	0111011010010
110011010001	36	−.0006920	−.0025672	.000900798	.000670795	.000446889	.000266350	.000142022	.000018123	.000001155	.000000001	0111011001010
101010010101	36	−.0006920	−.0025823	.000900546	.000670044	.000445759	.000266147	.000141013	.000017827	.000001121	.000000001	0101011101010
101001001001	36	−.0007260	−.0025753	.000894293	.000659879	.000434788	.000255815	.000134407	.000016399	.000000988	.000000000	0111011011010
100111000110	37	−.0007276	−.0025596	.000894217	.000659958•	.000434964	.000255956	.000134454	.000016347	.000000974	.000000000	1001110001100x
100110011010	38	−.0007333	−.0024690	.000894645•	.000662570	.000439490	.000261032	.000138795	.000017612	.000001116	.000000000	1010001110100
110001100011	36	−.0007470	−.0022904	.000895028•	.000667135	.000448122	.000271237	.000147934	.000020616	.000001503	.000000000	0111011100010
101101101001	36	−.0007470	−.0025373	.000890978	.000654972	.000430038	.000252162	.000132342	.000016064	.000000965	.000000000	0111010110010
101001001001	36	−.0007470	−.0025424	.000890896	.000654972	.000430038	.000242605	.000132055	.000015988	.000000957	.000000000	0111010010010
101001010110	36	−.0007826	−.0025275	.000884449	.000644572	.000418832	.000242605	.000125250	.000014484	.000000813	.000000000	0101011011100
110011000011	36	−.0007984	−.0023995	.000883591	.000645848•	.000452573	.000247465	.000129697	.000015866	.000000970	.000000000	0101011001010
101100011010	37	−.0007988	−.0024601	.000882517	.000642765	.000418031	.000242712	.000125780	.000014771	.000000849	.000000000	1001011110010
101100100110	36	−.0008194	−.0023682	.000880166	.000640853	.000417685	.000243662•	.000127216	.000015416	.000000933	.000000000	0101011101010
101010010101	37	−.0008328	−.0024549	.000876253	.000632623	.000407178	.000233595	.000119428	.000013469	.000000737	.000000000	1010010110010
110010101001	36	−.0008534	−.0023489	.000874121	.000631315	.000407662•	.000235340	.000121452	.000014228	.000000827	.000000001	0101010101100
110001101001	36	−.0008696	−.0021863	.000873676	.000633676	.000412764	.000241378	.000126683	.000015723	.000000988	.000000001	1010011100100
110001100110	36	−.0008696	−.0022700	.000872361	.000629975	.000407505	.000236084	.000122483	.000014648	.000000878	.000000001	0111011100010
101001001001	36	−.0008696	−.0022836	.000872147	.000629371	.000406646	.000235217	.000121794	.000014470	.000000860	.000000001	1001011110010
011110010001	34	−.0008951	−.0021858	.000871172	.000628817•	.000408951	.000238509	.000124958	.000015547	.000000994	.000000001	0111011001100
101010100110	37	−.0009052	−.0021709	.000867280	.000623546	.000402109	.000232536	.000120566	.000014463	.000000875	.000000001	1000111101010
101001100110	37	−.0009052	−.0022529	.000865954	.000619705	.000396502	.000226745	.000115860	.000013190	.000000739	.000000001	1001010110010
101001100110	37	−.0009052	−.0022567	.000865893	.000619531	.000396253	.000226493	.000115659	.000013139	.000000734	.000000001	1001011101010
101011000011	35	−.0009189	−.0020262	.000867054•	.000626158	.000407595	.000239164	.000126508	.000016372	.000001113	.000000001	0111011110100
100111001010	35	−.0009189	−.0021597	.000864849	.000619881	.000398362	.000225513	.000118534	.000014076	.000000843	.000000001	0111110001010
101111000001	35	−.0009189	−.0021770	.000864619	.000619090	.000397218	.000228339	.000117584	.000013818	.000000815	.000000000	0101110110010
110010101010	37	−.0009553	−.0021691	.000857927	.000608206	.000385453	.000218306	.000110459	.000012261	.000000669	.000000000	1001010111100
110111010001	35	−.0009691	−.0019121	.000859464•	.000615785	.000397999	.000232040	.000122032	.000015602	.000001051	.000000000	0111011100010
101101110001	35	−.0009691	−.0021591	.000855568	.000604741	.000382168	.000215925	.000109073	.000012064	.000000664	.000000000	0101110100010
101011001001	35	−.0009691	−.0021641	.000855489	.000604520	.000381858	.000215616	.000108832	.000012064	.000000664	.000000000	1001010110010
101010100110	35	−.0009901	−.0021154	.000852308	.000600297	.000378064	.000212910	.000107215	.000011830	.000000650	.000000000	0111011001010
101010010101	35	−.0009901	−.0021306	.000849358	.000595082	.000372124	.000207215	.000103376	.000010961	.000000568	.000000000	1001011001010
101101110010	36	−.0010257	−.0020868	.000846134	.000590663	.000368067	.000204649	.000101573	.000010687	.000000549	.000000000	1001011001010

142

m = 6, n = 6   Rank Orders 351 thru 400 of 494   C(6, 6) = 924   1/C(6, 6) = .00108225

z	w	c_1	c_2	d = .2	.4	.6	.8	1.0	1.5	2.0	3.0	z^{tc}
110011001001	35	-.0010755	-.0019264	.000839388	.000582707●	.000362020	.000201262	.000100118	.000010741	.000000575	.000000000	011011001100
101110001100	36	-.0010758	-.0019908	.000838305	.000579747	.000357872	.000197136	.000096886	.000009946	.000000498	.000000000	101001111010
110000011100	37	-.0010827	-.0018044	.000839888●	.000558548●	.000366580	.000206004	.000103881	.000011657	.000000661	.000000000	110000111100x
111000100101	35	-.0010965	-.0017839	.000837651	.000582303	.000363432	.000203583	.000102325	.000011389	.000000762	.000000000	101100110100
101010010001	35	-.0010965	-.0018737	.000836295	.000578631	.000359265	.000198625	.000098625	.000010532	.000000562	.000000000	011100100100
110101001001	35	-.0010965	-.0018832	.000836153	.000578249	.000357899	.000198233	.000098250	.000010447	.000000555	.000000000	011101100100
011110100001	33	-.0011102	-.0017055	.000836345●	.000584613●	.000364192	.000205197	.000104055	.000012025	.000000722	.000000000	011110100001x
100100100110	36	-.0011321	-.0018384	.000830202	.000569331●	.000348919	.000191041	.000093421	.000009512	.000000475	.000000000	011101100100
111000001100	35	-.0011466	-.0016729	.000830031	.000571901●	.000353868	.000196567	.000097995	.000010697	.000000592	.000000000	011101111000
101000010110	36	-.0011483	-.0017644	.000828333	.000567593	.000348070	.000190963	.000093701	.000009678	.000000495	.000000000	100101110100
111000001110	36	-.0011822	-.0015394	.000825437	.000566713	.000350065●	.000194468	.000097116	.000010718	.000000604	.000000000	100011111000
100001101010	36	-.0011822	-.0017307	.000822549	.000558882	.000339351	.000184080	.000089178	.000008866	.000000432	.000000000	100010111000
110011001010	36	-.0011822	-.0017326	.000825081●	.000558806●	.000339250	.000183983	.000089106	.000008851	.000000431	.000000000	101010111000
101011000011	34	-.0011960	-.0013977	.000821422	.000568725	.000354372	.000199576	.000101582	.000012062	.000000762	.000000000	001111100010x
101011100001	34	-.0011960	-.0016386	.000821313	.000558717	.000340505	.000185907	.000090919	.000009391	.000000487	.000000000	011101100010
110111000011	34	-.0011960	-.0016459	.000818400	.000558427	.000340115	.000185534	.000090639	.000009327	.000000481	.000000000	011110110010
011010100011	34	-.0012121	-.0016417	.000799927	.000545892	.000335492	.000181854	.000088215	.000008896	.000000449	.000000000	100111010010
101101000110	35	-.0012478	-.0015945	.000812500	.000545151	.000326812	.000175010	.000083698	.000008061	.000000381	.000000000	011101100100
110110000101	34	-.0012684	-.0014133	.000811400	.000546343●	.000330296	.000179246	.000087300	.000008994	.000000470	.000000000	100111010010
101010000110	35	-.0012817	-.0015623	.000806711	.000536480	.000318221	.000168323	.000079378	.000007326	.000000329	.000000000	100110111000
101100101010	35	-.0013027	-.0015064	.000803667	.000532544	.000314814	.000165984	.000078036	.000007152	.000000319	.000000000	101010101100
110010101001	36	-.0013048	-.0013649	.000805359●	.000537350	.000321300	.000172086	.000084523	.000008082	.000000393	.000000000	100101011100x
111000100011	34	-.0013185	-.0012661	.000804308	.000537323	.000322560●	.000173945	.000084240	.000008578	.000000444	.000000000	011100101100
101110100001	34	-.0013185	-.0013558	.000803004	.000533922	.000317619	.000169768	.000081167	.000007925	.000000389	.000000000	011110101100
011111000001	34	-.0013185	-.0013653	.000802866	.000533567	.000317266	.000178533	.000080853	.000007859	.000000383	.000000000	001111100001x
101101100001	34	-.0013323	-.0011092	.000804089●	.000539461	.000326693	.000172266	.000080025	.000009572	.000000547	.000000000	011110110010
101100101100	34	-.0013395	-.0013019	.000799210	.000529884	.000314523	.000167266	.000079691	.000007715	.000000375	.000000000	011101100100
111010010010	35	-.0013541	-.0013019	.000797210	.000525483	.000309836	.000163362	.000076998	.000007178	.000000330	.000000000	101010111000
110110010001	34	-.0013735	-.0011595	.000795729	.000525332	.000313110●	.000165525	.000078955	.000007704	.000000380	.000000000	101010111000
110100100110	35	-.0013751	-.0012477	.000794139	.000521468	.000306317	.000160910	.000075567	.000006983	.000000318	.000000000	110001101100
110010101010	35	-.0014091	-.0012032	.000788541	.000513326	.000298456	.000154932	.000071786	.000006368	.000000276	.000000000	101001101100
110011001100	36	-.0014253	-.0011123	.000786330	.000515759●	.000302888	.000159578	.000075387	.000007154	.000000340	.000000000	101100111000
110010110010	35	-.0014253	-.0011186	.000786890	.000512080	.000298147	.000155123	.000072278	.000006535	.000000292	.000000000	101100011100
111010000011	35	-.0014253	-.0006961	.000786802	.000511859	.000297868	.000155002	.000072103	.000006503	.000000289	.000000000	001111001000
011111000010	35	-.0014391	-.0010229	.000790391	.000523837	.000314899	.000176609	.000084760	.000009435	.000000566	.000000000	001111110010
101101000110	33	-.0014391	-.0010251	.000785689	.000511619	.000298770	.000156443	.000073444	.000006875	.000000325	.000000000	011100010100
101110000110	34	-.0014391	-.0010229	.000785658	.000511541	.000298671	.000156354	.000073380	.000006863	.000000324	.000000000	011111011000
011110010010	34	-.0014747	-.0009596	.000780038	.000503576	.000291076	.000150584	.000069700	.000006232	.000000276	.000000000	011110010010
110110010010	34	-.0015248	-.0009044	.000771666	.000491367	.000279346	.000141750	.000064187	.000005373	.000000220	.000000000	101101110000
110010100110	35	-.0015317	-.0007850	.000772079●	.000493597	.000282650	.000144908	.000066485	.000005818	.000000252	.000000000	110001101100
111010010010	33	-.0015454	-.0006082	.000772087	.000496144	.000287052	.000149470	.000067021	.000006610	.000000320	.000000000	011101101000
110011100001	33	-.0015454	-.0006920	.000770914	.000493189	.000283288	.000146065	.000067589	.000006127	.000000282	.000000000	011011101000
110011001001	33	-.0015454	-.0007056	.000770726	.000492723	.000282704	.000145544	.000063134	.000006057	.000000276	.000000000	011011100010
110100110010	34	-.0015458	-.0008378	.000768772	.000487806	.000276409	.000139829	.000063134	.000006251	.000000213	.000000000	101101001010
111001001001	33	-.0015616	-.0006886	.000768023	.000487792●	.000278943	.000142725	.000065470	.000005784	.000000258	.000000000	011101100100
110101000110	34	-.0015972	-.0005319	.000763909	.000484613	.000271828	.000141268	.000064887	.000005780	.000000262●	.000000000	100111010010
110110100010	34	-.0016166	-.0006159	.000762520	.000481161	.000271470	.000137441	.000062177	.000005251	.000000220	.000000000	011101110010
110011010010	34	-.0016312	-.0004565	.000761218	.000481308●	.000273470	.000139582	.000063983	.000005679	.000000257	.000000000	100111010010
110011001010	34	-.0016522	-.0005570	.000757149	.000473654	.000264854	.000132333	.000059062	.000004787	.000000190	.000000000	101101110000
110011001100	34	-.0016522	-.0003815	.000755744	.000473742●	.000266428	.000134355	.000060723	.000005148	.000000218	.000000000	100110111000

143

Table A

m = 6, n = 6   Rank Orders 401 thru 450 of 494   C(6, 6) = 924   1/C(6, 6) = .00108225

d = .2

z	w	c_1	c_2	.2	.4	.6	.8	1.0	1.5	2.0	3.0	z^{tc}
110110110010	34	-.0016522	-.0004864	000754308	000470217	000262065	000130534	000058089	000004677	000000185	.00000000	10110110100
110101001010	34	-.0016522	-.0004908	000754249	000470076	000261897	000130392	000057994	000004662	000000184	.00000000	10101010100
101110100001	32	-.0016660	-.0003152	000754202	000472378	000265838	000134389	000061016	000005291	000000234	.00000000	01111010000
110100011010	34	-.0017024	-.0002260	000748772	000465036	000259109	000129448	000057957	000004798	000000198	.00000000	10011011000
110100101100	33	-.0017517	-.0001566	000740792	000453792	000248678	000121862	000053383	000004142	000000158	.00000000	10101110010
101101011000	34	-.0017586	-.0001141	000740116	000453260	000248464	000121856	000053446	000004166	000000160	.00000000	110010101100x
101010001010	33	-.0017679	-.0001404	000738102	000449953	000245506	000119277	000051840	000003930	000000146	.00000000	101101010100x
111000101010	33	-.0017748	.0000167	000738517	000452047	000248077	000122003	000053746	000004266	000000169	.00000000	110001110100
110110010101	32	-.0017885	-.0002477	000739590	000456899	000254966	000128432	000058361	000005158	000000237	.00000000	010111101000
110110100001	32	-.0017885	.0000737	000737285	000451395	000248304	000122693	000054449	000004460	000000186	.00000000	101101010100
110101000001	32	-.0017885	.0000696	000737231	000451270	000248156	000122568	000054366	000004447	000000185	.00000000	011101101000
111010001001	33	-.0018225	.0002389	000733351	000447374	000245599	000121288	000053865	000004442	000000188	.00000000	100111011000
111010000101	33	-.0018241	.0001605	000731991	000444322	000241937	000118123	000052492	000004046	000000159	.00000000	100110110000
110101010001	32	-.0018387	.0002665	000730808	000434867	000242400	000118992	000049896	000004246	000000176	.00000000	011101011000
110110010010	33	-.0018743	.0003601	000725604	000437012	000236307	000114655	000047710	000003861	000000150	.00000000	100111001100
111000110010	33	-.0018743	.0002552	000724222	000433745	000232407	000111368	000047631	000003503	000000127	.00000000	101110011000
101110101010	31	-.0018743	.0002508	000724166	000436615	000232256	000111244	000051535	000003491	000000126	.00000000	011110011000
111010000110	33	-.0018880	.0005048	000725059	000437918	000238316	000116802	000046872	000004198	000000176	.00000000	100111101010
110110110001	32	-.0018953	.0003321	000721465	000430508	000229857	000109771	000046872	000003415	000000123	.00000000	100111101000
111000110001	33	-.0019293	.0005147	000717751	000426981	000227708	000108804	000046557	000003430	000000126	.00000000	101110001100
110011001100	32	-.0019806	.0006716	000710603	000418121	000220365	000103981	000043908	000003122	000000110	.00000000	110010101100x
101110100010	32	-.0019948	.0007058	000708528	000415447	000218087	000102455	000043060	000003022	000000105	.00000000	101101010100x
110101001100	33	-.0020017	.0007526	000707896	000414997	000217943	000102486	000043134	000003043	000000106	.00000000	110011101100
101101010001	31	-.0020154	.0009229	000707676	000416635	000220077	000105271	000045155	000003421	000000133	.00000000	011101011000
110110100010	31	-.0020518	.0010524	000702792	000410635	000215730	000101838	000045271	000003143	000000115	.00000000	110011110000
111000101001	31	-.0020656	.0011418	000702451	000411912	000218061	000104149	000044852	000003460	000000138	.00000000	010111101000
111001001100	31	-.0020656	.0011357	000701521	000409825	000215684	000102221	000043613	000003269	000000126	.00000000	011110011000
110110000110	33	-.0021012	.0012475	000701447	000409662	000215501	000102074	000043521	000003255	000000126	.00000000	011110011000
111010010010	32	-.0021012	.0011464	000696493	000403423	000210179	000098434	000041417	000002967	000000108	.00000000	100111101100
110011100001	32	-.0021012	.0011464	000695210	000400496	000206804	000095674	000039638	000002695	000000091	.00000000	101110011000
110110001010	32	-.0021012	.0011401	000695132	000400325	000206614	000095524	000039545	000002682	000000090	.00000000	101011100100
101011001010	32	-.0021173	.0011691	000692635	000396995	000203706	000093542	000038428	000002548	000000084	.00000000	101010101000
110101010010	32	-.0021513	.0013661	000689076	000393760	000201825	000092740	000038185	000002562	000000086	.00000000	101010110000
101101100001	32	-.0021723	.0014594	000680090	000390971	000201539	000091539	000035720	000002508	000000084	.00000000	101110100100x
110110011000	32	-.0022217	.0016430	000679615	000383487	000193941	000087927	000035373	000002318	000000075	.00000000	110011001000
101111001000	31	-.0022237	.0016337	000680286	000386147	000193165	000087349	000038000	000002272	000000073	.00000000	011111010000
111001100010	30	-.0022375	.0017254	000677043	000379882	000197820	000091417	000033768	000002708	000000100	.00000000	011100110000
101101100010	32	-.0022447	.0018824	000673614	000373614	000189432	000086124	000034768	000002217	000000070	.00000000	110011010100x
110110010001	32	-.0022787	.0019347	000673289	000378143	000191644	000085473	000034601	000002236	000000073	.00000000	110011011000
111010100001	30	-.0022925	.0021020	000673289	000378143	000191644	000087607	000036110	000002499	000000090	.00000000	110011101000
111100010010	30	-.0023086	.0022195	000671856	000377117	000191262	000087592	000036203	000002531	000000092	.00000000	011011110000
111100001001	31	-.0023442	.0023500	000667143	000371441	000186609	000084521	000034485	000002311	000000079	.00000000	100111011000
101011001100	31	-.0023442	.0021587	000664846	000366429	000181171	000080609	000031884	000001955	000000060	.00000000	101101010100
110110011010	31	-.0023442	.0021568	000664824	000366429	000181123	000080254	000031863	000001953	000000060	.00000000	101101100100
110110101000	31	-.0023782	.0023664	000663413	000363413	000179423	000079548	000031652	000001963	000000061	.00000000	101010100100x
111001101000	31	-.0023944	.0024061	000659042	000360440	000176947	000077931	000030777	000001866	000000057	.00000000	110101010100
101111000100	30	-.0024437	.0026867	000653761	000355407	000173753	000074952	000030120	000001833	000000056	.00000000	101100110100x
110111001010	30	-.0024506	.0026045	000653252	000353252	000171764	000074749	000029337	000001742	000000052	.00000000	110011011000
111000110100	31	-.0024742	.0026332	000650214	000350214	000169241	000073316	000029337	000001742	000000052	.00000000	110011100100x
111010101000	31	-.0024668	.0027020	000649943	000350214	000169241	000073316	000028460	000001649	000000048	.00000000	110101010100x

144

Table A

m = 6, n = 6   Rank Orders 451 thru 494 of 494   C(6, 6) = 924   1/C(6, 6) = .00108225

z	w	c_1	c_2	d = .2	.4	.6	.8	1.0	1.5	2.0	3.0	z^{tc}
11101100001	29	−.0025145	.0031713	.000647187•	.000350388	.000171632	.000076018	.000030428	.000001975	.000000067	.000000000	01111001000
11000110100	31	−.0025218	.0030293	.000644224	.000344941	.000166020	.000071781	.000027863	.000001626	.000000048	.000000000	10100011000
11100100001	30	−.0025355	.0032841	.000644848•	.000348026	.000170012	.000075124	.000030009	.000001940	.000000066	.000000000	01111010000
11011010000	30	−.0025711	.0032374	.000638106	.000338169	.000160997	.000068821	.000026402	.000001494	.000000043	.000000000	10100110000
11100100010	30	−.0026213	.0035907	.000633491	.000334540	.000159237	.000068258	.000026328	.000001525	.000000045	.000000000	10110111000
11001010010	30	−.0026213	.0035087	.000632566	.000332662	.000157306	.000066844	.000025510	.000001429	.000000040	.000000000	10101110000
11100010010	30	−.0026213	.0035049	.000632524	.000332580	.000157224	.000066786	.000025478	.000001426	.000000040	.000000000	10110101000
11010010100	30	−.0026937	.0038331	.000623812	.000323812	.000150411	.000062853	.000023571	.000001261	.000000034	.000000000	10110111000
11100011000	30	−.0027277	.0040694	.000620644	.000323183	.000150630	.000062381	.000023462	.000001273	.000000035	.000000000	11001101000
11010101000	30	−.0027438	.0041156	.000618395	.000320630	.000149110	.000061046	.000022774	.000001206	.000000032	.000000000	11011011000
11101010001	28	−.0027576	.0044489	.000619804•	.000322400	.000152182	.000065133	.000025258	.000001529	.000000049	.000000000	01111010100
11011100010	29	−.0027932	.0044207	.000613361	.000313330	.000144163	.000059699	.000022235	.000001179	.000000032	.000000000	01011010000
11000111000	30	−.0027988	.0044725	.000612970	.000313165	.000144200•	.000059795	.000022313	.000001191	.000000032	.000000000	11000011000x
11010110000	29	−.0028482	.0046985	.000607071	.000306867	.000139692	.000057230	.000021089	.000001106	.000000029	.000000000	10110101000
11100010010	29	−.0028644	.0048353	.000605809	.000306097	.000139478	.000057268•	.000021170	.000001089	.000000030	.000000000	10110101000
11011001000	29	−.0029206	.0050553	.000598689	.000298157	.000133601	.000053833	.000019495	.000000961	.000000026	.000000000	10110101000
11100001100	29	−.0029707	.0054480	.000594435	.000295107	.000132289	.000053497	.000019499	.000000987	.000000024	.000000000	10110100100x
11010101000	29	−.0029707	.0053597	.000593499	.000293318	.000130553	.000052297	.000018843	.000000920	.000000023	.000000000	11001101000
11101010100	29	−.0034197	.0053578	.000593480	.000293282	.000130520	.000052275	.000018831	.000000919	.000000023	.000000000	11010011000
11111100001	27	−.0029845	.0057709	.000595605•	.000298883	.000136582	.000056770	.000021440	.000001228	.000000038	.000000000	01111111000
11101011000	29	−.0030209	.0056686	.000588353	.000288554	.000127571	.000050803	.000018211	.000000880	.000000022	.000000000	11101011000x
11011010010	28	−.0030702	.0059915	.000583491	.000284259	.000125024	.000049604	.000017740	.000000857	.000000021	.000000000	10111101000
11100110000	28	−.0030912	.0061273	.000581396	.000282370	.000123869	.000049038	.000017505	.000000843	.000000021	.000000000	11011100010x
11101010100	28	−.0031426	.0063781	.000575435	.000276189	.000119569	.000046657	.000016398	.000000756	.000000018	.000000000	11010101000
11110010010	28	−.0031976	.0066929	.000569547	.000270516	.000115881	.000044740	.000015558	.000000699	.000000016	.000000000	10110101000
11011001000	28	−.0032138	.0068426	.000568388	.000269883	.000115747	.000044798•	.000015634	.000000712	.000000017	.000000000	11010101000
11100011000	28	−.0032478	.0070237	.000564617	.000266139	.000113250	.000043472	.000015042	.000000670	.000000015	.000000000	11001101000
11101010100	27	−.0033133	.0075158	.000558773	.000261499	.000110806	.000042468	.000014708	.000000662	.000000016	.000000000	11011101000
11101001000	27	−.0034197	.0081255	.000547388	.000250523	.000103658	.000038744	.000013071	.000000549	.000000012	.000000000	11011101000
11110010100	27	−.0034407	.0082761	.000545434	.000248877	.000102718	.000038312	.000012903	.000000540	.000000012	.000000000	11011011000
11101010100	27	−.0034747	.0084699	.000541800	.000245397	.000100477	.000037162	.000012406	.000000508	.000000011	.000000000	11101010100x
11100011000	27	−.0034909	.0086302	.000540721	.000244864	.000100398	.000037233	.000012478	.000000518	.000000011	.000000000	11100110100
11101000100	26	−.0035402	.0090674	.000536924	.000242369	.000099398	.000036984	.000012469	.000000530	.000000012	.000000000	10111101000
11101000100	26	−.0036628	.0098041	.000524175	.000230424	.000091835	.000033149	.000010827	.000000423	.000000009	.000000000	10110101000
11110010100	26	−.0036968	.0100123	.000520686	.000227205	.000089833	.000032154	.000010410	.000000398	.000000008	.000000000	10110101000
11100011000	26	−.0037178	.0101748	.000518838	.000225730	.000089033	.000031804	.000010280	.000000392	.000000008	.000000000	10111001000x
11110001000	25	−.0038897	.0114991	.000503652	.000213525	.000082347	.000028850	.000009170	.000000339	.000000007	.000000000	10111110000
11101010100	25	−.0039398	.0118125	.000498582	.000208940	.000079559	.000027495	.000008615	.000000306	.000000006	.000000000	11011100000
11101001000	25	−.0039608	.0119858	.000496824	.000207600	.000078862	.000027203	.000008511	.000000302	.000000006	.000000000	11101010100
11101010100	24	−.0041667	.0136206	.000479035	.000193579	.000071311	.000023913	.000007290	.000000245	.000000004	.000000000	11011011000
11110010100	24	−.0041829	.0137190	.000477393	.000192108	.000070432	.000023497	.000007124	.000000236	.000000004	.000000000	11101010100x
11101000100	23	−.0044098	.0156254	.000458650	.000177948	.000063104	.000020422	.000006023	.000000188	.000000003	.000000000	11101100000
11101010000	22	−.0046367	.0176228	.000440614	.000164794	.000056513	.000017738	.000005087	.000000149	.000000002	.000000000	11101010000x
11111000000	21	−.0048587	.0197022	.000423884	.000153173	.000050944	.000015560	.000004355	.000000122	.000000002	.000000000	11111000000x

145

Table A

m = 12, n = 1   Rank Orders 1 thru 13 of 13   C(12, 1) = 13   1/C(12, 1) = .07692308

| | | | | d = .2 | .4 | .6 | .8 | 1.0 | 1.5 | 2.0 | 3.0 | |
z	w	c_1	c_2									z^{tc}
0000000000001	13	.1283069	.1613382	.105933964	.142039170	.185563603	.236395242	.293920763	.458563281	.629568129	.882881290	0111111111111
0000000000010	12	.0895444	.0419607	.095531605	.114880562	.133792526	.150932955	.164963897	.179623462	.161271145	.074130995	1011111111111
0000000000100	11	.0653719	−.0097223	.089641002	.100982702	.109979240	.115806362	.117909281	.106520599	.078143464	.022664774	1101111111111
0000000001000	10	.0463731	−.0387354	.085288751	.091344542	.094503213	.094449923	.091193993	.071849160	.045669266	.009719911	1110111111111
0000000010000	9	.0298713	−.0558424	.081690671	.083765788	.082937067	.079291770	.073200659	.051443801	.029069161	.004833039	1111011111111
0000000100000	8	.0146557	−.0650295	.078514126	.077361815	.073586386	.067571684	.059900870	.037993033	.019340585	.002593306	1111101111111
0000001000000	7	.0000000	−.0679385	.075576238	.071675580	.065616753	.057985080	.049462535	.028481533	.013151890	.001446604	1111110111111
0000010000000	6	−.0146557	−.0650295	.072752700	.066423435	.058542274	.049807002	.040905382	.021424635	.008898419	.000818259	1111111011111
0000100000000	5	−.0298713	−.0558424	.069937508	.061393755	.052034365	.042579457	.033639212	.016000350	.006107333	.000459416	1111111101111
0001000000000	4	−.0463731	−.0387354	.067013214	.056388020	.045826855	.035970359	.027267555	.011712305	.004044061	.000250000	1111111110111
0010000000000	3	−.0653719	−.0097223	.063806490	.051156470	.039640050	.029685011	.021482371	.008237015	.002547097	.000127344	1111111111011
0100000000000	2	−.0895444	.0419607	.059965654	.045249730	.033046969	.023355417	.015970748	.005341554	.001450047	.000056779	1111111111101
1000000000000	1	−.1283069	.1613382	.054348076	.037338431	.024930699	.016169737	.010182733	.002809271	.000639403	.000018281	1111111111110

146

z	w	c_1	c_2	d = .2	.4	.6	.8	1.0	1.5	2.0	3.0	z^{tc}
0000000111111	63	.028343	.0125612	.001478765	.003460227	.007489070	.015040787	.028125839	.100768973	.250874338	.676902080	0000000111111111
0000001011111	62	.027232	.0114353	.001420066	.003174353	.006524568	.012361311	.021641859	.063224069	.119597313	.13391354	0000001011111111
0000001101111	61	.026122	.0103673	.001364448	.002919509	.005720006	.012083904	.017004207	.042079052	.063491465	.044377006	0000010011111111
0000010011111	61	.026080	.0103272	.001362406	.002910167	.005691214	.010211523	.016847318	.041423637	.063491465	.042448440	0000100011111111
0000010101111	60	.024969	.0093160	.001309703	.002683919	.005013501	.010013501	.013438812	.028678490	.037404756	.017152396	0000010101111111
0000011001111	60	.024969	.0093012	.001308990	.002675631	.004985046	.008480132	.013194210	.027284640	.033624666	.012186410	0000101011111111
0000011010111	60	.024830	.0091918	.001303220	.002654544	.004984387	.008387291	.013063841	.027139010	.020284831	.015303806	0000011101111111
0000100011111	59	.023859	.0083365	.001258546	.002467370	.004398146	.007139930	.010574365	.019116288	.019005078	.005257366	0000110011111111
0000100101111	59	.023817	.0082940	.001256357	.002457215	.004366274	.007059514	.010401895	.018456214	.021798759	.004398213	0000101101111111
0000100110111	59	.023719	.0082394	.001253310	.002453044	.004363873	.007092091	.010543978	.019517532	.021798759	.007110240	0001000011111111
0000010110111	59	.023719	.0082118	.001252045	.002439913	.004318986	.006955040	.010204804	.017900828	.018201329	.004093005	0001000101111111
0000011100111	59	.023390	.0079669	.001238959	.002393572	.004209983	.006755161	.009909195	.017707669	.019043062	.005721200	0001000110111111
0000100111011	58	.022706	.0073714	.001207857	.002265254	.003849258	.005934849	.008314809	.012835257	.011277761	.001795748	0001001001111111
0000101011011	58	.022609	.0072927	.001203704	.002249239	.003807302	.005846225	.008155387	.012441198	.010786414	.001664421	0001001010111111
0000101101011	58	.022567	.0072655	.001202198	.002244612	.003798294	.005833798	.008144900	.012487566	.010929702	.001739700	0001001101111111
0000110011011	58	.022567	.0072527	.001201639	.002240223	.003780724	.005783413	.008029609	.012042364	.010163193	.001418658	0001010011111111
0000110101011	58	.022280	.0070796	.001191939	.002212790	.003735819	.005747615	.008075999	.012908039	.012359588	.002923966	0001000111101111
0000011110011	58	.022280	.0070403	.001190240	.002193481	.003682586	.005594602	.007723775	.011500431	.009760502	.001455867	0001001110111111
0000101110011	57	.021596	.0065065	.001161872	.002093123	.003411927	.005039771	.006756108	.009295759	.007242084	.000902132	0001011001111111
0000110110011	58	.021559	.0065336	.001162665	.002104934	.003464861	.005196378	.007116095	.010658435	.009557327	.001982714	0001000111110111
0000011100111	57	.021456	.0064006	.001156204	.002072370	.003360383	.004936692	.006580782	.008920848	.006839745	.000822734	0001100011111111
0001001011011	57	.021456	.0063889	.001155718	.002068712	.003346366	.004898275	.006496887	.008634212	.006406016	.000683494	0001010101111111
0001001101011	57	.021456	.0063757	.001155163	.002064506	.003330185	.004853797	.006399599	.008302819	.005912561	.000538705	0001010110111111
0001010011011	57	.021316	.0062766	.001149787	.002045996	.003287353	.004774895	.006277661	.008114055	.005793912	.000546856	0001011001111111
0001010101011	57	.021170	.0061742	.001144204	.002026982	.003249965	.004722659	.006157806	.007940895	.005704753	.000567391	0001100101111111
0001011001011	57	.021127	.0061614	.001143275	.002026497	.003249965	.004722659	.006227775	.008219073	.006131797	.000690898	0001011010111111
0001000111011	57	.021127	.0061368	.001142268	.002019057	.003227037	.004647720	.006067676	.007703338	.005401099	.000491453	0001011101111011
0001100011011	57	.020449	.0057261	.001118926	.001948645	.003084205	.004445862	.005849928	.007920803	.006416137	.001088804	0001100110111101
0001010110011	57	.020449	.0056757	.001117883	.001933779	.003029005	.004298672	.005535833	.006884814	.004841283	.000485682	0001011110011111
0001011010011	56	.020346	.0055697	.001111636	.001910911	.002963910	.004153516	.005266267	.006213878	.004061010	.000331061	0001011101110111
0001100110011	56	.020346	.0055568	.001111117	.001907128	.002949933	.004116664	.005189054	.005979017	.003750979	.000260657	0001100011011111
0001011100011	56	.020304	.0055275	.001109514	.001901685	.002937466	.004093810	.005153655	.005921011	.003705768	.000256826	0001101001111111
0001101000011	56	.020206	.0054456	.001105249	.001885008	.002893692	.004002023	.004992605	.005559685	.003324523	.000198040	0001101010111111
0001010111001	56	.020017	.0053381	.001099013	.001867246	.002861637	.003960999	.004958628	.005655617	.003358442	.000263690	0001101101111111
0001011011001	56	.020017	.0053365	.001098954	.001866877	.002860497	.003958558	.004954860	.005651877	.003356874	.000269062	0001110011111111
0001100111001	56	.019877	.0052332	.001098011	.001860117	.002835937	.003894952	.004823890	.005274337	.003094446	.000177211	0001100111101111
0001101011001	56	.019877	.0052214	.001093390	.001846861	.002811616	.003862404	.004793787	.005322540	.003228590	.000213277	0001101110111111
0001110011001	56	.019338	.0048770	.001092934	.001843689	.002800428	.003843648	.004737795	.005173814	.003059757	.000184651	0001110100111111
0001101110001	56	.019338	.0048791	.001073612	.001781602	.002666325	.003603482	.004440855	.004726995	.002797105	.000183137	0001110101011111
0001110110001	56	.019296	.0048791	.001073194	.001784234	.002681796	.003649650	.004504089	.005022317	.003154568	.000250171	0001110111011101
0001110111000	56	.019296	.0048430	.001071820	.001774808	.002648976	.003568036	.004342657	.004594867	.002659658	.000160861	0010000111101111
0010001011011	55	.019193	.0047783	.001068191	.001763461	.002625510	.003530263	.004293334	.004561171	.002684928	.000179257	0001111110011111
0010001101011	55	.019193	.0047508	.001067138	.001756224	.002600316	.003467749	.004170213	.004241796	.002327213	.000091052	0010010101111111
0010010011011	55	.019096	.0046724	.001063027	.001740755	.002561292	.003389384	.004037861	.003976983	.002079662	.000091763	0010010110111111
0010010101011	55	.019054	.0046456	.001061519	.001735973	.002551143	.003372313	.004013885	.003948387	.002066283	.000092228	0010011101111101
0010011001011	55	.018907	.0045719	.001057023	.001734319	.002552862	.003353408	.004011012	.004048830	.002232414	.000124194	0010100111111111
0010000111011	55	.018907	.0045473	.001056086	.001717919	.002510719	.003299962	.003904605	.003781161	.001945513	.000083630	0010011110111011
0010100011011	55	.018864	.0045191	.001054526	.001712755	.002499178	.003279291	.003873266	.003732019	.001908246	.000083466	0010101101111011
0010010110011	55	.018767	.0044978	.001052572	.001711955	.002510664	.003324743	.003981574	.004067678	.002300964	.000141719	0011000011111111
0010011010011	55	.018767	.0044555	.001050977	.001701132	.002473477	.003233759	.003805017	.003628268	.001832307	.000074557	0010101110111101

147

Table A

m = 7, n = 6 Rank Orders 51 thru 100 of 1716 C(7, 6) = 1716 1/C(7, 6) = .00058275

z	w	c_1	c_2	d = .2	.4	.6	.8	1.0	1.5	2.0	3.0	z^{tc}
0000101011011	55	0018767	0044434	001050524	001698094	002463191	003209037	003758054	003519346	001727295	000063747	0010010110111
0000001111110	57	0018623	0045325	001052037	001726057	002578517	003514394	004379585	005211040	003741587	000510335	1000000111111
0100001011011	55	0018438	0042480	001039287	001663992	002394307	003099527	003614076	003385170	001696901	000071943	0010010011011
0100001001111	55	0018186	0041241	001031589	001643646	002360079	003057937	003580223	003441092	001815176	000093271	0010001001111
0000111101101	55	0018186	0041093	001031063	001640275	002349152	003032732	003534142	003342584	001725332	000084027	0100010010111
0000101011101	55	0018083	0040869	001030221	001634616	002329897	002986112	003444693	003127115	001504570	000055744	0001110001111
0000111000111	54	0018046	0040349	001027050	001626229	002316059	002970430	003435126	003169519	001580221	000068768	0011100001111
0100001100111	55	0018046	0040283	001026326	001625805	002319180	002982695	003462910	003242763	001651658	000076164	0100100111101
0100010011101	55	0018046	0040051	001025486	001620394	002301486	002941541	003387079	003077830	001499283	000060364	0100100110111
0110001100111	54	0017943	0039292	001021430	001606089	002267687	002877964	003286487	002906661	001362794	000048415	0100011100111
0001010100111	54	0017943	0039144	001020900	001602668	002256548	002852242	003239580	002808741	001278092	000041397	0011010010111
0010100001111	54	0017754	0038350	001015623	001590688	002211769	002846664	003261695	002953601	001458601	000065287	0001011101101
0000110101011	54	0017754	0037957	001014194	001581369	002211060	002774748	003128397	002660538	001186150	000036766	0010101000111
0100100100111	54	0017657	0037353	001010760	001570497	002187839	002735037	003070984	002581004	001133308	000033429	0010010110111
0100101001011	54	0017657	0037235	001010337	001567775	002179003	002722533	003034129	002505538	001069936	000028687	0011010100111
0010010111011	54	0017614	0037286	001009968	001570286	002192405	002751552	003107512	002674529	001221354	000034318	0011001010111
0010010010111	54	0017614	0037112	001009349	001566337	002179696	002722533	003055173	002568099	001131468	000034154	0001001101011
1000001101111	54	0017614	0036978	001008868	001563229	002169560	002699107	003012445	002479214	001055375	000028152	0010101010111
0100100101101	56	0017512	0038482	001012844	001560262	002302269	003023782	003631369	003943613	002589268	000297493	0100010010111
0000010111110	56	0017844	0037844	001010549	001585293	002252741	002903089	003400296	003348660	001875072	000121076	1000001011111
0010001011011	54	0017328	0035233	000998903	001532215	002104725	002591110	002861136	002290320	000946775	000023651	0010001011011
0100000001111	54	0017076	0034258	000992109	001517416	002087583	002586630	002894464	002451902	001128538	000042932	0100011101101
0010010101101	54	0017076	0033896	000999833	001509322	002061641	002527536	002783922	002233342	000939117	000025837	0010001101011
0100101101101	54	0017033	0033632	000989334	001504559	002054559	002509986	002758426	002197628	000915249	000024360	0100010010101
0100000010111	54	0016936	0033458	000987426	001497173	002059465	002541770	002830433	002386349	001097381	000042694	0010010110101
0100000101011	54	0016936	0033190	000986461	001491990	002039138	002494667	002743890	002200523	000928616	000025884	0010101011101
0010010000111	53	0016833	0032950	000985628	001483672	002022882	002458482	002680448	002080510	000835065	000019755	0011000101011
0010011001011	53	0016790	0032446	000982449	001478342	002008695	002440546	002663758	002089243	000860113	000022958	0010110101011
0011001100111	53	0016644	0032153	000980897	001463902	001996794	002419450	002632483	002044718	000831235	000021439	0010110010111
	53	0016644	0031328	000979029	001463902	001968044	002373868	002571737	001977931	000797456	000020346	0011001100111
0010000011011	54	0016607	0031327	000975508	001464614	001974309	002392142	002608241	002057908	000864713	000025489	0100001111101
0100000111111	54	0016607	0031211	000975118	001462252	001967083	002376457	002581337	002009562	000828433	000023126	0010000111101
0110001100111	53	0016504	0031033	000973085	001460634	001973477	002403457	002642540	002163612	000970587	000036528	0011011100011
0010100010111	53	0016504	0030511	000971284	001449480	001938562	002325790	002505697	001898890	000753323	000018524	0010100100111
0010100101011	53	0016504	0030465	000971123	001448471	001935375	002318643	002493033	001874217	000733326	000017054	0010101100101
0101001001011	53	0016504	0030359	000970758	001446217	001926641	002303200	002466223	001825177	000696669	000014893	0010110101011
0010010101011	53	0016504	0030330	000970210	001445635	001926741	002299667	002466348	001816201	000691411	000014774	0010101010111
0010010011101	53	0016402	0030200	000970210	001442840	001917903	002280283	002426695	001754455	000645493	000012176	0010101010111
0000101110110	55	0016402	0030959	000971324	001377179	001788550	002077739	002160641	001561766	000576517	000012011	0100011010011
0000110111010	53	0016364	0029597	000966199	001432940	001902442	002262960	002413793	001777118	000671383	000014234	0011001110011
1000000101111	55	0016359	0031160	000971397	001464927	002001222	002479850	002791467	002490098	001262341	000066704	0000101011110
0010010011011	55	0016359	0030669	000969727	001454668	001969155	002408126	002663616	002227234	001023518	000103499	1000101011101
0010100110011	53	0016217	0028564	000960625	001414214	001860955	002189083	002304919	001621185	000579520	000102700	0010101001101
0100100101011	53	0016175	0028335	000959250	001410252	001852978	002177507	002290255	001607672	000574385	000002448	0011001101011
0101001011011	53	0015923	0027589	000953169	001399325	001846164	002191809	002344944	001825177	000729768	000022018	0100100010111
0010100101101	53	0015923	0027081	000951444	001388830	001813959	002121692	002224318	001560221	000564258	000010881	0010101001011
0100100110101	53	0015825	0026666	000948686	001381816	001810024	002110188	002160641	001557653	000568805	000011301	0100101010111
0010010111101	53	0015825	0026434	000947873	001377179	001788550	002077739	002160641	001475265	000513219	000008707	0100110101101
0000101111110	53	0015783	0026478	000947415	001378711	001797503	002101525	002205424	001561766	000576517	000012011	0011001010111
	53			000947747	001378290	001796191	002098645	002200463	001553017	000570350	000011709	

TABLE A

m = 7, n = 6   Rank Orders 101 thru 150 of 1716   C(7, 6) = 1716   1/C(7, 6) = .00058275

z	w	c_1	c_2	d = .2	.4	.6	.8	1.0	1.5	2.0	3.0	z^{tc}
000100110101	53	.0015783	.0026194	.000946462	.001372980	.001780157	.002064376	.002142749	.001455325	.000502990	.000008371	0100100110101
000110100011	52	.0015680	.0025875	.000943933	.001368218	.001776402	.002067953	.002160835	.001514867	.000555378	.000011703	0001110010011
000101111001	53	.0015496	.0025566	.000945566	.001364263	.001782632	.002105070	.002235517	.001690870	.000702774	.000022920	0110000101011
010000101011	53	.0015496	.0024763	.000937550	.001348312	.001734552	.001997630	.002061184	.001385757	.000477687	.000008259	0010010011101
001100110001	53	.0015394	.0024644	.000937162	.001346050	.001727918	.001983877	.002038742	.001350861	.000455796	.000007259	0100010011011
001011011011	52	.0015394	.0024339	.000934669	.001341427	.001724337	.001987437	.002056018	.001405696	.000502624	.000010084	0011101000111
001011001011	52	.0015394	.0024164	.000934091	.001338003	.001714143	.001965809	.002019961	.001346606	.000462074	.000008054	0001110010011
001011010111	52	.0015351	.0024031	.000933645	.001335330	.001706091	.001948657	.001991205	.001298102	.000429781	.000006553	0010110010011
001101001011	52	.0015351	.0023883	.000932543	.001332997	.001703373	.001947493	.001993948	.001313280	.000443868	.000007373	0010101011011
	52		.0023765	.000932150	.001330656	.001696354	.001932618	.001969147	.001272607	.000416773	.000006178	0010100110011
001100010011	52	.0015254	.0023672	.000930409	.001329801	.001703009	.001956047	.002017009	.001368224	.000484630	.000009362	0001011100111
100000010111	52	.0015254	.0023279	.000929121	.001322262	.001680802	.001909771	.001941057	.001247902	.000407130	.000006016	0011010010111
000110101011	54	.0015249	.0024712	.000933683	.001349163	.001760131	.002075592	.002215093	.001699077	.000720582	.000024363	0000110111110
000011101011	54	.0015249	.0024416	.000932736	.001343713	.001744214	.002042428	.002160207	.001606742	.000653461	.000019798	1000100011111
000101010111	54	.0015249	.0024205	.000932020	.001339337	.001730664	.002012565	.002108021	.001507047	.000572332	.000013240	0010101001011
100010001011	54	.0015109	.0023537	.000927769	.001346614	.001729920	.002025160	.002143631	.001603338	.000657537	.000020074	0001100111110
001100100111	52	.0015065	.0022234	.000927912	.001327877	.001710172	.001983904	.002075221	.001488213	.000574498	.000014698	1000001110111
001011001101	52	.0015065	.0022086	.000922912	.001304080	.001645147	.001854236	.001868542	.001173121	.000372404	.000005351	0010110010011
001010101011	52	.0014925	.0021560	.000922229	.001301261	.001638678	.001837128	.001840746	.001130795	.000346812	.000004253	0010101001011
	52	.0014812	.0021131	.000918643	.001292506	.001624240	.001824425	.001833116	.001145632	.000363733	.000005127	0011000110111
000111001011	52	.0014812	.0021131	.000915581	.001285320	.001613585	.001813110	.001825222	.001154944	.000376328	.000005907	0100011001111
010100100111	53	.0014775	.0021199	.000915246	.001286874	.001621539	.001833120	.001861393	.001219478	.000420909	.000008152	0100000111101
001011000111	52	.0014673	.0021023	.000913138	.001284297	.001623434	.001846922	.001893877	.001293790	.000479501	.000011977	0001011110101
001010010111	52	.0014673	.0020479	.000911369	.001273997	.001593173	.001788327	.001789914	.001124562	.000364033	.000005607	0010100110111
001001010111	52	.0014673	.0020386	.000911017	.001272350	.001588605	.001784505	.001775259	.001103899	.000352360	.000005244	0010101001111
001110100011	52	.0014673	.0020356	.000910973	.001271721	.001586605	.001774038	.001768441	.001092644	.000344919	.000004916	0001101110111
001100100101	52	.0014673	.0020270	.000910701	.001271172	.001582173	.001761471	.001754162	.001077730	.000332520	.000004471	0001010101011
001010011011	52	.0014673	.0020099	.000910145	.001266964	.001577813	.001742262	.001742192	.001025386	.000304895	.000003525	0011010010111
001001100111	51	.0014533	.0020175	.000908845	.001268998	.001587937	.001782673	.001797492	.001156322	.000391058	.000007151	0010101010111
	52		.0019564	.000906320	.001257908	.001559135	.001717028	.001710893	.001030755	.000314721	.000004053	0010111001110
010001011011	52	.0014386	.0018773	.000901563	.001244125	.001532345	.001685608	.001657134	.000975907	.000289431	.000003428	0010110111101
001001011101	52	.0014386	.0018657	.000901202	.001242105	.001526677	.001674390	.001639707	.000952227	.000276693	.000003081	0100010010111
001000111011	52	.0014344	.0019143	.000902084	.001240760	.001555182	.001735282	.001739476	.001102757	.000366663	.000006328	0110000010111
010000111011	52	.0014344	.0018576	.000900293	.001240891	.001526429	.001705245	.001647825	.000969258	.000288106	.000003468	0100100110111
001011000101	51	.0014344	.0018445	.000899881	.001238439	.001519879	.001664394	.001627441	.000969476	.000272479	.000003011	0100010011101
001010010101	51	.0014344	.0018561	.000898682	.001240891	.001535697	.001705245	.001700868	.001065400	.000351098	.000006112	0011110001111
001101010011	51	.0014241	.0018138	.000897353	.001233422	.001514565	.001662930	.001634097	.000969111	.000294155	.000003943	0011101010101
001010110011	51	.0014241	.0018017	.000896962	.001231178	.001508083	.001649708	.001612880	.000937604	.000275320	.000003264	0010110101011
100100011011	53	.0014241	.0017566	.000894042	.001223383	.001494184	.001630667	.001589670	.000919170	.000268848	.000003176	0011101010011
		.0014144	.0018826	.000897883	.001245173	.001555830	.001753500	.001784797	.001205056	.000444192	.000010988	0000111011110
000110011110	53	.0014139	.0018335	.000896333	.001236380	.001530568	.001701830	.001701024	.001072495	.000354956	.000006049	1000011011011
001100100011	51	.0014139	.0017571	.000893403	.001232545	.001498175	.001641576	.001611063	.000955686	.000291495	.000003992	0011011110011
001100110010	51	.0014101	.0017325	.000892630	.001219214	.001485972	.001617276	.001573032	.000902435	.000261419	.000003014	0110010001011
001011110010	53	.0014097	.0018102	.000894941	.001232272	.001522384	.001688776	.001683426	.001052129	.000343803	.000005600	1001010001011
100000101110	53	.0014012	.0018000	.000893137	.001230906	.001527212	.001707067	.001724027	.001138024	.000409776	.000009744	1001010011101
001010011110	53	.0013999	.0017908	.000892799	.001228711	.001520136	.001691934	.001695533	.001083213	.000364463	.000006657	1000101011011
001010101110	53	.0013999	.0017533	.000891638	.001222260	.001502008	.001655754	.001638445	.000999968	.000316695	.000004639	0010101010111
001101011011	51	.0013954	.0016451	.000887618	.001204031	.001455343	.001568438	.001508008	.000833443	.000229194	.000002244	0010110101011
001010011011	51	.0013912	.0016196	.000886166	.001199863	.001446513	.001564455	.001489617	.000814538	.000229194	.000002072	0010101011011
001100110011	51	.0013815	.0016047	.000884184	.001197048	.001447089	.001563205	.001510120	.000854202	.000244935	.000002792	0010101110110

149

z	w	c_1	c_2	d = .2	.4	.6	.8	1.0	1.5	2.0	3.0	z^{tc}
0010010110011	51	.0013815	.0015772	.000883329	.001192321	.001433978	.001537555	.001470724	.000802127	.000217615	.000002074	0011001010111
1000000111011	53	.0013670	.0016382	.000882905	.001202107	.001472164	.001623198	.001613035	.001015394	.000343045	.000006675	0010010011110
0100000111110	53	.0013670	.0016137	.000882165	.001198102	.001461195	.001601843	.001580123	.000969812	.000316444	.000005601	1000010111001
0010010011101	52	.0013562	.0015314	.000879972	.001184331	.001423579	.001528642	.001467464	.000815764	.000229194	.000002476	0101001011101
0000110100101	51	.0013562	.0014893	.000876673	.001177178	.001412089	.001514561	.001453884	.000813369	.000232990•	.000002789	0101010001111
0000101001101	51	.0013562	.0014873	.000876608	.001176789	.001398232	.001512136	.001449906	.000807198	.000229184	.000002645	0001110010111
0001011001001	51	.0013520	.0014609	.000875787	.001172231	.001401574•	.001487165	.001411328	.000755272	.000201275	.000001856	0001011001011
0010110001011	51	.0013520	.0014609	.000875124	.001172213	.001405574	.001497057	.001429668	.000785337	.000218788	.000002380	0100101010111
0001100101101	51	.0013520	.0014376	.000874403	.001168233	.001390542	.001475490	.001396558	.000741559	.000195754	.000001763	0010011001101
0100000010111	51	.0013423	.0014437	.000873055	.001169053	.001400340	.001501869	.001443586	.000814993	.000237400	.000002984	0001011110101
0010010110101	51	.0013423	.0013929	.000871492	.001160479	.001376720	.001455933	.001373373	.000722818	.000188948	.000001660	0101001010111
0000011110011	51	.0013233	.0014339	.000869748	.001166801	.001408824	.001533123	.001505925	.000929703	.000314223	.000006711	0001011110011
0010011011011	51	.0013233	.0013585	.000867439	.001154124	.001373681	.001463934	.001398186	.000776836	.000222717	.000002723	0110001110011
0001010011011	51	.0013233	.0013142	.000866094	.001146827	.001353782	.001425602	.001340109	.000701994	.000183883	.000001667	0100010111011
0010010001101	51	.0013233	.0013039	.000865784	.001145168	.001349331	.001417192	.001327654	.000687063	.000178857	.000001524	0101001001101
0010001011011	51	.0013233	.0013007	.000865677	.001144543	.001347505	.001413447	.001321653	.000678430	.000177060	.000001393	0110001011011
0010001001101	51	.0013233	.0012880	.000865292	.001142463	.001341886	.001402760	.001305734	.000659152	.000162967	.000001214	0011001001101
0001001100101	50	.0013131	.0012968	.000863939	.001143669	.001353040	.001431906	.001356870	.000736583	.000205484	.000002364	0001110010101
0001110010111	51	.0013131	.0012858	.000863574	.001141643	.001347391	.001420777	.001339621	.000713019	.000192480	.000001958	0001011110100
1000100111001	51	.0013094	.0012835	.000862921	.001141088	.001348452•	.001425096	.001347985	.000725977	.000199232	.000002112	0011000110101
0100001110011	51	.0013094	.0012576	.000862150	.001136994	.001337544	.001404589	.001317696	.000689561	.000181771	.000001722	0011000111101
0010001110011	51	.0013094	.0012421	.000861681	.001134459	.001330671	.001391448	.001297976	.000664992	.000169641	.000001441	0011100110011
0010010011011	51	.0012991	.0012287	.000859641	.001131890	.001331429•	.001400420	.001313411	.000702949	.000192155	.000002109	0011010011011
0001010010011	50	.0012991	.0012270	.000859577	.001131466	.001330068	.001397401	.001313225	.000694253	.000186555	.000001893	0101010011101
0001101011001	50	.0012991	.0012033	.000858873	.001127737	.001320159	.001378825	.001285886	.000661746	.000171177	.000001559	0000111101011
1001010000111	52	.0012986	.0012653	.000862558	.001147823	.001374699	.001483525	.001444408	.000869358	.000284407	.000005511	0001011101110
1000011001011	52	.0012986	.0012482	.000860615	.001137210	.001345405	.001426073	.001355267	.000744091	.000210428	.000002457	1000010101101
1000101001101	52	.0012889	.0012120	.000858813	.001133694	.001342771	.001427446•	.001362068	.000763051	.000222424	.000002839	0000010111110
0001001011110	52	.0012889	.0012482	.000857433	.001127933	.001327279	.001397928	.001318607	.000706418	.000192931	.000002011	0010010101110
1000010011011	52	.0012889	.0012324	.000857356	.001130923	.001338431	.001422242	.001347985	.000762350	.000229977	.000002980	0011000111110
0001011101010	52	.0012846	.0012167	.000856903	.001128569	.001332259	.001410753	.001341140	.000741589	.000213345	.000002665	1001000101101
0010011010110	52	.0012846	.0011911	.000856124	.001124305	.001320531	.001387920	.001306120	.000694746	.000187917	.000001898	1000100110101
0010101001011	50	.0012802	.0010982	.000852667	.001109701	.001285280	.001325572	.001218101	.000597613	.000144643	.000001069	0011010011001
0010110100011	50	.0012704	.0010592	.000849382	.001102910	.001274087	.001311071	.001202254	.000588448	.000142566	.000001072•	0010110111001
0010101010011	50	.0012662	.0010491	.000848939	.001100955	.001271772	.001309346	.001202201	.000591082•	.000144272	.000001106	0101010111001
0010110010011	50	.0012662	.0010362	.000848986	.001099986	.001266626	.001299883	.001188572	.000575928	.000137707	.000000996	0011010111001
1001000010111	52	.0012560	.0011495	.000850259	.001116044	.001319460	.001406899	.001354845	.000797658	.000257612	.000005036	1010000101111
0010100100111	52	.0012560	.0010920	.000848499	.001106370	.001293615	.001353998	.001272403	.000680608	.000187759	.000002071	0111001010111
1001001010011	52	.0012560	.0010667	.000847762	.001102535	.001282564	.001335302	.001245021	.000647924	.000171915	.000001692	0010010101110
0100011101001	51	.0012555	.0010007	.000845765	.001092566	.001256943	.001288563	.001177826	.000572039	.000137680	.000001020	0101001110101
0010110110010	51	.0012512	.0009827	.000844544	.001089447	.001251638	.001281440	.001169866	.000566679	.000136166	.000001654	0101001110101
0000111110010	50	.0012410	.0009815	.000842847	.001089576	.001256593•	.001297667	.001200044	.000613265	.000161216	.000001654	0001011100101
0101010100010	50	.0012410	.0009547	.000842056	.001084416	.001245601	.001277158	.001169960	.000577538	.000144121	.000001253	0101011011010
0100101010010	50	.0012410	.0009307	.000842047	.001080601	.001235422	.001258014	.001141706	.000543798	.000128150	.000000910	0101010110010
0010110010010	50	.0012312	.0008895	.000838533	.001073400	.001223068	.001241200	.001122547	.000529804	.000123572	.000000853	0011010110010
0010101010010	50	.0012270	.0009048	.000838293	.001075436	.001259336	.001259336	.001151432	.000567214	.000141756	.000001246	0101010110010
0101010001010	50	.0012270	.0008687	.000837928	.001069841	.001231960	.001231960	.001111535	.000521037	.000120567	.000000818	0011010001101
0000111110001	50	.0012270	.0008621	.000834629	.001067521	.001220413•	.001248202	.001143239	.000570351	.000136360	.000001430	0110010001110
0100011110001	50	.0012123	.0008054	.000832964	.001058798	.001197476	.001205602	.001081247	.000498368	.000113307	.000000723	0110011011101

150

m = 7, n = 6   Rank Orders 201 thru 250 of 1716   C(7, 6) = 1716   1/C(7, 6) = .00058275

z	w	c_1	c_2	d=.2	.4	.6	.8	1.0	1.5	2.0	3.0	z^{tc}
0101001010101	50	.0012081	.0007817	.000831573	.001054778	.001189792	.001194111	.001066858	.000485669	.000108225	.000000657	0010101011101
0010010001011	50	.0012081	.0007702	.000831238	.001053048	.001185319	.001185983	.001055307	.000473310	.000103121	.000000582	0100101010101
0110000010011	50	.0011983	.0008360	.000831537	.001061927	.001214755	.001245811	.001146552	.000583978	.000154712	.000001648	0010101111001
0101000011001	50	.0011983	.0007734	.000829717	.001052477	.001190083	.001200283	.001080448	.000507266	.000118798	.000000850	0110001010101
0101001010011	50	.0011983	.0007707	.000829422	.001052186	.001189472	.001199427	.001079594	.000507357	.000118295	.000000918	0100011110101
0110010001011	50	.0011983	.0007628	.000829001	.001051020	.001186484	.001194037	.001071974	.000499232	.000115900	.000000821	0010010101101
0110100101101	50	.0011983	.0007467	.000828951	.001044862	.001180061	.001182216	.001054925	.000480137	.000107514	.000000673	0011001011011
0011011001011	50	.0011983	.0007316	.000828511	.001046278	.001174129	.001171386	.001039448	.000463305	.000100420	.000000563	0101010101011
0101101011011	49	.0011881	.0007572	.000827576	.001049241	.001188718	.001204591	.001092963	.000532411	.000133439	.000001250	0101101011011
0011011000111	49	.0011881	.0007326	.000826856	.001045489	.001178891	.001186409	.001066515	.000501641	.000119007	.000000924	0011011000111
0000110001110	51	.0011876	.0007800	.000828115	.001052147	.001196007	.001217251	.001150092	.000548591	.000139351	.000001313	0010110001111
1000010111110	52	.0011839	.0008089	.000828333	.001056073	.001208738	.001243199	.001150215	.000600699	.000166542	.000002126	0100001111101
0100000111011	52	.0011839	.0007959	.000827970	.001064255	.001204144	.001234952	.001138512	.000587571	.000160377	.000001967	1000011111101
0011011001101	49	.0011838	.0007177	.000825719	.001042780	.001174583	.001180988	.001060816	.000498280	.000118137	.000000918	0011100100111
1001000100111	51	.0011736	.0007945	.000826219	.001052902	.001207404	.001247464	.001162784	.000627875	.000184590	.000002921	0001011111010
1000010100011	51	.0011736	.0007429	.000824695	.001044826	.001185795	.001206392	.001101041	.000547738	.000140871	.000001374	0011101101110
0011010110110	49	.0011736	.0007319	.000824376	.001043160	.001181417	.001189259	.001089143	.000533649	.000134135	.000001219	1001001000111
0001010011010	51	.0011736	.0007178	.000823987	.001041247	.001176683	.001189969	.001077718	.000522002	.000129352	.000001139	1000010100111
0011000101011	51	.0011736	.0007104	.000823768	.001040093	.001173627	.001184251	.001069300	.000511900	.000124470	.000001024	0011000101011
1000101010111	51	.0011736	.0006918	.000823223	.001037227	.001166058	.001170138	.001048619	.000487511	.000113017	.000000783	1000101010111
0101011001011	49	.0011691	.0006345	.000820874	.001028721	.001147606	.001140309	.001019858	.000452447	.000102271	.000000628	0101011001011
1100000111011	50	.0011654	.0006394	.000820959	.001028959	.001150495	.001147536	.001021752	.000467946	.000107024	.000000746	1100000111011
1100001010111	51	.0011596	.0006501	.000819698	.001029497	.001155102	.001158458	.001039112	.000487946	.000115776	.000000888	1100001010111
0111001001011	49	.0011551	.0005853	.000817133	.001019861	.001133681	.001123105	.000992253	.000432183	.000098436	.000000631	0111001001011
0110100101011	49	.0011551	.0005736	.000816801	.001018184	.001129426	.001115504	.000981614	.000443246	.000093926	.000000562	0110100101011
0101010101101	51	.0011449	.0006028	.000816427	.001016296	.001124673	.001107089	.000969959	.000420474	.000093368	.000000502	0101010101101
1000001011110	51	.0011449	.0005783	.000815184	.001020565	.001141120	.001141035	.001020861	.000446926	.000112683	.000000756	1000001011110
1000010111010	49	.0011412	.0005181	.000812881	.001017146	.001132546	.001125845	.000999708	.000454958	.000103608	.000000706	1000010111010
0011000110011	51	.0011407	.0006219	.000815727	.001023057	.001113195	.001094178	.000958231	.000416940	.000089604	.000000527	0011000110011
1010110101011	51	.0011407				.001150491	.001161053	.001052413	.000517823	.000133088	.000001347	1010110101011
0101001111011	51	.0011407	.0005881	.000814741	.001017914	.001137005	.001136051	.001015929	.000474877	.000112641	.000000874	0101001111011
1000100111101	51	.0011407	.0005602	.000813956	.001013975	.001127059	.001118303	.000991020	.000448417	.000101398	.000000677	1000100111101
0100011110101	50	.0011402	.0005089	.000812455	.001006937	.001100002	.001088955	.000951307	.000410247	.000086980	.000000489	0100011110101
0101010101101	49	.0011299	.0004941	.000812027	.001004730	.001100308	.001078625	.000936644	.000394594	.000080519	.000000394	0101010101101
0001110101011	49	.0011299	.0005056	.000810650	.001005246	.001112110	.001098581	.000969888	.000437336	.000080394	.000000740	0001110101011
0101010100111	50	.0011262	.0004823	.000809980	.001001807	.001103258	.001082516	.000947002	.000412270	.000089096	.000000539	0101010100111
0011100101011	49	.0011160	.0004578	.000808663	.000997823	.001095440	.001070622	.000932128	.000399386	.000088435	.000000480	0011100101011
0101100111010	49	.0011160	.0004447	.000806571	.000999458	.001093745	.001073334	.000940847	.000415674	.000092787	.000000647	0101100111010
0111000100111			.0004299	.000806139	.000992406	.001087862	.001062504	.000925207	.000398008	.000084893	.000000496	0111000100111
0110100010101			.0004074	.000805512	.000989294	.001080140	.001049026	.000906799	.000380119	.000078141	.000000411	0110100010101
0000111011001	49	.0010970	.0004089	.000802359	.000986853	.001085250	.001068128	.000941853	.000428322	.000100608	.000000838	0101000011101
0101010011011	49	.0010970	.0003285	.000800096	.000975491	.001056642	.001017348	.000871106	.000355045	.000070408	.000000333	1001011001011
0101010101011	49	.0010970	.0003166	.000799763	.000973832	.001052518	.001010144	.000861275	.000345582	.000066915	.000000292	0100110101011
0010101011001	49	.0010873	.0003230	.000798287	.000973250	.001056639	.001022181	.000881664	.000370422	.000077177	.000000431	0101101010011
0100100101011	49	.0010873	.0002970	.000797569	.000969723	.001047961	.001007160	.000861311	.000351011	.000069977	.000000342	0100111001101
0100100010111	49	.0010873	.0003527	.000797134	.000967548	.001042526	.000997548	.000848207	.000338168	.000065126	.000000283	0101010101101
0001100100111	49	.0010830	.0003048	.000797061	.000976715	.001067729	.001043865	.000913457	.000405354	.000091939	.000000681	0101000110111
0110100101101	49	.0010830	.0002863	.000796543	.000970105	.001051266	.001014929	.000873503	.000364767	.000075482	.000000415	0101011101001
0011000101011	49	.0010830		.000796543	.000967536	.001044882	.001003765	.000858209	.000349727	.000069704	.000000338	0110101110011
0110100101101	49	.0010830	.0002771	.000796288	.000966276	.001041767	.000998350	.000850848	.000342683	.000067103	.000000308	0100101110011

Table A

$m = 7$, $n = 6$ Rank Orders 251 thru 300 of 1716 $C(7, 6) = 1716$ $1/C(7, 6) = .00058275$

$d = .2$

z	w	c_1	c_2	.2	.4	.6	.8	1.0	1.5	2.0	3.0	z^{tc}
0101010011011	49	.0010830	.0002758	.000796255	.000966130	.001041457	.000997902	.000850367	.000342553	.000067189•	.00000312	0011001101101
0011010011011	49	.0010830	.0002619	.000795866	.000964195	.001036645	.000989501	.000838905	.000331522	.000063115	.00000265	0101010011101
0000011111100	51	.0010729	.0004650	.000799743•	.000990431	.001108075	.001121233	.001027458	.000540339	.000158219	.000002646	1100000011111
1000010011101	51	.0010729	.0003306	.000795998	.000969932	.001060534	.001035539	.000905119	.000399937	.000089819	.000000628	0100001011110
0100010011110	51	.0010729	.0003172	.000795637	.000969637	.001056323	.001028331	.000895408	.000390592	.000086202	.000000574	1000001011101
0011110000111	48	.0010728	.0003285	.000795977•	.000971867	.001061675	.001038661	.000910962	.000409880	.000095631	.000000795	0001110010011
1000110000111	48	.0010728	.0002891	.000794890	.000966493	.001048329	.001015230	.000878853	.000376584	.000081723	.000000540	0011101000011
0001110100011	50	.0010626	.0002939	.000793232	.000965068	.001050071	.001022517	.000891723	.000393036	.000088659	.000000649	1000110010011
1001010011001	50	.0010626	.0002782	.000792813	.000963076	.001045296	.001014428	.000880941	.000382938	.000084853	.000000595	1001100111101
0101011001001	50	.0010626	.0002526	.000792087	.000959398	.001035940	.000997631	.000857252	.000357583	.000074047	.000000405	1001100101011
1001010010011	50	.0010583	.0002706	.000791857	.000961157•	.001042703	.001011597	.000878256	.000381287	.000084155	.000000574	0011011011110
0011010010111	50	.0010583	.0002343	.000790856	.000956221	.001030475	.000990200	.000844822	.000351609	.000072212	.000000387	1001011110011
0110000011011	49	.0010544	.0002330	.000790169	.000955633	.001031499	.000994353	.000856667	.000362779	.000077452	.000000480	0110001111001
0010010011011	49	.0010544	.0001822	.000788769	.000948769	.001014627	.000965135	.000816990	.000324450	.000062894	.000000286	0110001011011
1000010101011	50	.0010486	.0002607	.000789017•	.000958241	.001041013	.001013548	.000885077	.000394066	.000090696	.000000711	0011011110110
0100010101110	50	.0010486	.0001968	.000788151	.000949521	.001019361	.000975557	.000832654	.000340672	.000068903	.000000353	1001010101101
0010110010011	48	.0010441	.0001556	.000785940	.000941624	.001007830	.000958566	.000812440	.000326002	.000064701	.000000326	1010110100111
1011010010001	48	.0010441	.0001427	.000784877	.000941917	.001003683	.000951478	.000802963	.000172963	.000061582	.000000291	1011100110011
0101010100011	48	.0010399	.0001305	.000784877	.000939563	.001003683	.000947512	.000799199	.000317274	.000061334	.000000291•	1011000110011
0011010101011	48	.0010301	.0000883	.000782087	.000932324	.000987911	.000930990	.000780785	.000303035	.000057765	.000000261	1011001100111
1010000011011	50	.0010297	.0002648	.000786793•	.000955791	.001045915	.001032315	.000920258	.000446757	.000119002	.000001585	0001111110010
0000011011011	50	.0010297	.0001763	.000784351	.000943690	.001015617	.000978422	.000844413	.000363151	.000080129	.000000566	1010001101110
1000010101101	50	.0010297	.0001531	.000783694	.000944380	.001007246	.000963482	.000813101	.000341055	.000070822	.000000404	0010010101110
1000010101011	49	.0010297	.0001410	.000783355	.000938688	.001003008	.000955998	.000808486	.000330493	.000066586	.000000341•	0010101011110
0010100101101	50	.0010297	.0001312	.000783108	.000937584	.001000550	.000952176	.000808486	.000327400	.000065922	.000000346•	1000011101011
0101010101101	50	.0010292	.0001138	.000782623	.000935162	.000994495	.000944514	.000793771	.000312646	.000060145	.000000267	1000101011011
0101001001101	49	.0010249	.0000668	.000781290	.000929204	.000980725	.000918962	.000764798	.000284495	.000052399	.000000200	0101011011001
0101010101011	50	.0010249	.0000463	.000780002	.000925759	.000974634	.000910510	.000755079	.000281553	.000050320	.000000182	0101011011101
0011001110011	49	.0010157	.0001159	.000780279	.000933095	.000997125	.000957247	.000814487	.000339784	.000071850	.000000442	1010001110011
1000110011011	50	.0010157	.0001129	.000780189	.000932633	.000995952	.000950669	.000811623	.000337002	.000070833	.000000431	0110001110011
0010001110011	50	.0010157	.0000798	.000779295	.000928317	.000985487	.000932754	.000787522	.000314059	.000062140	.000000312	1001001011011
1001010101101	49	.0010152	.0000428	.000778239	.000923699	.000974942	.000915621	.000765626	.000295844	.000056208	.000000255	0101001011011
0100010101011	49	.0010152	.0000152	.000777481	.000919997	.000965906	.000900128	.000747056	.000276673	.000049390	.000000377	1010101011101
0101011010101	48	.0010049	.0000412	.000776435	.000921786•	.000975923	.000922240	.000778619	.000313568	.000063962	.000000377	0001110101101
0011110010011	48	.0010049	-.0000050	.000775444	.000916948	.000964083	.000901832	.000751034	.000287186	.000053987	.000000240	0101011011011
0001110010101	48	.0010007	.0000082	.000774352	.000914446	.000960255	.000897190	.000746326	.000284626	.000053365	.000000236	0010110101110
0101001101101	48	.0009860	-.0000673	.000770213	.000904069	.000934901	.000874234	.000720976	.000277677	.000048468	.000000192	0010111100101
0101101001101	48	.0009860	-.0000788	.000769902	.000902576	.000939323	.000868210	.000713050	.000261721	.000046114	.000000169	1011001011011
0110110010001	48	.0009823	-.0000555	.000769888	.000904910	.000946837	.000882499	.000733188	.000280278	.000053125	.000000247	0111000111101
0001110101011	48	.0009720	-.0000099	.000769321	.000908784•	.000961439	.000911797	.000775938	.000325057	.000070995	.000000525	0111001110110
0110001000111	48	.0009720	-.0000541	.000768142	.000903167	.000947975	.000888971	.000745464	.000296246	.000059897	.000000350	0011110111001
0001110101101	48	.0009720	-.0000822	.000767384	.000895519	.000939154	.000873913	.000725285	.000277120	.000052650	.000000247	0110101010111
0110110010001	48	.0009720	-.0001035	.000766817	.000896832	.000932784	.000862292	.000711400	.000265071	.000048573	.000000205	0011101010011
0100110010101	48	.0009720	-.0001057	.000766757	.000896551	.000932120	.000862185	.000709950	.000264722	.000048123	.000000200	0111100011011
0101001010101	48	.0009720	-.0001209	.000766348	.000894937	.000927451	.000854344	.000699849	.000254722	.000045035	.000000168	0010110101101
0010100110101	48	.0009720	-.0001224	.000766306	.000894379	.000926892	.000853341	.000698249	.000253317	.000044502	.000000162	0101011010101
0101001010101	48	.0009720	-.0001304	.000766060	.000893381	.000924379	.000849516	.000692901	.000249316	.000043258	.000000152	0011010100111
0101011010001	48	.0009720	-.0001318	.000765675	.000893842	.000924266	.000849096	.000692901	.000249152	.000043285	.000000154•	0011010110011
0101101010001	48	.0009720	-.0001461		.000891195	.000919851	.000841687	.000683195	.000240758	.000040519	.000000129	0110101010011

152

Table A

m = 7, n = 6   Rank Orders 301 thru 350 of 1716   C(7, 6) = 1716   1/C(7, 6) = .00058275

z	w	c_1	c_2	d = .2	.4	.6	.8	1.0	1.5	2.0	3.0	z^{tc}
0100010011110	50	.0009618	−.0001033	.000765010	.000894560	.000932147	.000865932	.000717487	.000272809	.000051396	.000000232	1000011011110
0011111000011	47	.0009618	−.0000789	.000765681	.000897958	.000940671	.000880945	.000738189	.000293533	.000059561	.000000352	0011110000011
0101000110011	48	.0009580	−.0001672	.000762681	.000886098	.000914260	.000838248	.000683305	.000246046	.000043076	.000000159	0011001100011
0011100110011	48	.0009580	−.0001788	.000762372	.000884595	.000910747	.000832373	.000675627	.000239416	.000048880	.000000139	0101000110011
0000010111100	50	.0009576	−.0000086	.000767229	.000907746	.000924596	.000867361	.000797121	.000349732	.000081276	.000000697	1100000110011
1000001101101	50	.0009576	.0001033	.000764261	.000891646	.000932014	.000867361	.000720616	.000276972	.000053064	.000000253	1001010111110
0000011100011	50	.0009576	−.0001181	.000763875	.000891856	.000927847	.000860514	.000717182	.000269426	.000050516	.000000226	0100100111101
0000111100111	49	.0009473	−.0001162	.000762150	.000890254	.000929415	.000867886	.000725497	.000287315	.000058219	.000000346	1001110011101
1000110001111	49	.0009473	−.0001254	.000761892	.000888960	.000926182	.000862203	.000717670	.000279480	.000055112	.000000298	0001110101110
0011100011110	49	.0009473	−.0001628	.000760895	.000884230	.000914907	.000843236	.000692609	.000256718	.000046900	.000000196	1000010100011
0010110011001	48	.0009434	−.0002149	.000758848	.000877147	.000900365	.000821084	.000665554	.000235901	.000040504	.000000141	0110010011011
0110000100101	48	.0009391	−.0001957	.000758603	.000878591	.000905736	.000831518	.000680178	.000249357	.000045015	.000000183	0010101111001
0010101000111	49	.0009391	−.0002319	.000757656	.000874181	.000895460	.000814690	.000658620	.000231562	.000039346	.000000133	0110000110011
0010101010110	49	.0009376	−.0001968	.000758288	.000878010	.000904975	.000830652	.000679256	.000248564	.000044681	.000000178	1001010010111
0010101010011	49	.0009333	−.0001635	.000757127	.000881324	.000914952	.000849131	.000704963	.000273325	.000053801	.000000292	0001101011010
0011011110010	49	.0009333	−.0002127	.000756236	.000873880	.000900408	.000824819	.000673051	.000244853	.000043733	.000000172	1001001010011
0010101110011	47	.0009289	−.0002175	.000755521	.000870586	.000899742	.000825895	.000672327	.000250247	.000046117	.000000205	0010111100001
1010011110010	49	.0009289	−.0002451	.000752798	.000863852	.000892140	.000813549	.000660624	.000237435	.000042036	.000000167	0010101100011
0010001010011	47	.0009191	−.0002838	.000752628	.000870586	.000881087	.000799256	.000645250	.000227989	.000039539	.000000149	0110101010011
0001011110011	49	.0009187	−.0002103	.000754628	.000872612	.000910672	.000833185	.000689055	.000265362	.000052194	.000000291	1010010100011
0010110011110	49	.0009187	−.0002441	.000753709	.000868164	.000890873	.000814708	.000646251	.000242006	.000039488	.000000175	0010110011110
0111001011001	49	.0009187	−.0002720	.000752982	.000864795	.000883048	.000801917	.000647854	.000228381	.000039087	.000000135	0111001011001
0111100110011	47	.0009149	−.0002935	.000751796	.000864772	.000887241	.000796225	.000642611	.000227156	.000039481	.000000150	0111100110011
1000010101011	49	.0009144	−.0002647	.000752420	.000861801	.000878070	.000806331	.000640731	.000223854	.000037861	.000000156	1000010101011
0010101101011	49	.0009144	−.0002892	.000751785	.000855931	.000864497	.000795383	.000615314	.000204808	.000032416	.000000126	0010101101011
0001011110010	48	.0009139	−.0003389	.000752193	.000869715	.000901421	.000838121	.000699622	.000279530	.000058112	.000000379	0001011110010
0010101110011	49	.0009047	−.0002084	.000750615	.000862308	.000883921	.000808885	.000661178	.000244664	.000045345	.000000206	0010101110011
1010011110010	49	.0009047	−.0002685	.000750616	.000860415	.000879486	.000801545	.000651627	.000236296	.000042431	.000000173	1010011110010
1010001010011	49	.0009047	−.0003843	.000750188	.000860364	.000879488	.000801758	.000652190	.000237446	.000043069	.000000185	1010001010011
1000011110011	49	.0009047	−.0002852	.000750188	.000860364	.000879488	.000801758	.000652190	.000237446	.000043069	.000000185	1000011110011
0011001011011	49	.0009047	−.0002883	.000750090	.000859828	.000878049	.000799089	.000648362	.000233391	.000041453	.000000164	0011001011011
1001001011011	49	.0009047	−.0003206	.000749251	.000855943	.000869038	.000784372	.000629532	.000217810	.000036446	.000000118	1001001011011
0101010110011	48	.0009042	−.0003665	.000747998	.000850585	.000857210	.000765892	.000606913	.000201220	.000031812	.000000089	0101010110011
0100101010011	48	.0008999	−.0003716	.000747116	.000849056	.000855624	.000764877	.000606782	.000202256	.000032265	.000000093	0100101010011
0100100110011	49	.0008999	−.0003845	.000746781	.000847512	.000852073	.000759155	.000599592	.000196692	.000030643	.000000082	0100100110011
1000000111110	51	.0008902	−.0003176	.000749314	.000865043	.000897313	.000837079	.000702968	.000288996	.000063317	.000000511	1000000111110
0011110111101	47	.0008897	−.0003685	.000745400	.000847363	.000870970	.000794150	.000647627	.000240299	.000045473	.000000235	0011110111101
0101110011101	47	.0008897	−.0003843	.000741793	.000841269	.000856609	.000766671	.000617206	.000214347	.000036668	.000000136	0101110011101
1010000011101	49	.0008713	−.0004054	.000741179	.000838825	.000845211	.000758680	.000607051	.000211184	.000036391	.000000139	1010000011101
0101100011011	48	.0008713										0101100011011
0100010111001	48	.0008713	−.0004447	.000740162	.000834145	.000834440	.000741248	.000584968	.000193409	.000030839	.000000091	0110010101101
0001110000111	47	.0008610	−.0004046	.000739395	.000836793	.000845399	.000763115	.000616046	.000222807	.000041086	.000000202	0011110111001
0101011001011	47	.0008610	−.0004525	.000738156	.000831080	.000832190	.000744581	.000588494	.000199750	.000033413	.000000119	0101011001011
0011011010101	47	.0008610	−.0004711	.000737675	.000828867	.000827096	.000733336	.000587049	.000191340	.000030781	.000000096	0010101001011
0101101010010	47	.0008610	−.0004802	.000737438	.000827784	.000824616	.000729351	.000573046	.000187435	.000029614	.000000082	0010101010010
0010101010101	47	.0008610	−.0004816	.000737407	.000827651	.000728961	.000728961	.000572625	.000187252	.000029608	.000000088	0010101001011
0011011010101	47	.0008610	−.0004955	.000737045	.000825980	.000722724	.000722724	.000564747	.000181030	.000027735	.000000074	0101101010011
0001101110101	47	.0008568	−.0004646	.000737092	.000828706	.000828647	.000737390	.000584345	.000197625	.000032928	.000000116	0111001100111
0101011000011	47	.0008568	−.0004905	.000736425	.000825656	.000821678	.000726199	.000570291	.000186586	.000029574	.000000089	0011011001011
0010101100101	47	.0008568	−.0005060	.000736023	.000823801	.000817405	.000719294	.000561577	.000179715	.000027506	.000000073	1010110101011

153

Table A
m = 7, n = 6   Rank Orders 351 thru 400 of 1716   C(7, 6) = 1716   1/C(7, 6) = .00058275

z	w	c_1	c_2	d = .2	.4	.6	.8	1.0	1.5	2.0	3.0	z^{tc}
000101111001	47	.0008470	−.0004484	.000735755	.000828216•	.000831680	.000745468	.000596812	.000209730	.000036961	.000000153	011100010111
010100010011	47	.0008470	−.0005288	.000733717	.000819027	.000810938	.000712536	.000555878	.000178263	.000027545	.000000077	001101011001
110000000111	47	.0008470	−.0005407	.000733409	.000817613	.000807700	.000707334	.000549356	.000173215	.000026059	.000000066	010101101101
110000001111	49	.0008466	−.0002687	.000740260	.000849145	.000880418	.000825984	.000702025	.000306212	.000074444	.000000860	000011111100
100011011110	49	.0008466	−.0003687	.000737680	.000837152	.000852156	.000778514	.000638750	.000244554	.000048683	.000000290	110000010011
100010001100	49	.0008466	−.0004688	.000735131	.000825549	.000825593	.000735530	.000584028	.000198914	.000033384	.000000118	100011101101
100010011001	49	.0008466	−.0004804	.000734844	.000824288	.000822814	.000731220	.000578792	.000195114	.000032314	.000000110	010010110111
100101010111	49	.0008466	−.0004821	.000734779	.000822899	.000821711	.000729208	.000575709	.000191833	.000031048	.000000096	100010101101
001101011010	49	.0008466	−.0004966	.000734420	.000822300	.000818154	.000723540	.000568889	.000186795	.000029616	.000000086	100101010011
001101100110	48	.0008363	−.0004648	.000734423	.000822828•	.000826410	.000740844	.000593956	.000210880	.000038013	.000000174	100111001110
000111001110	48	.0008363	−.0005010	.000732492	.000819571	.000816628	.000724976	.000573736	.000194078	.000032452	.000000115	100011000111
000100111100	49	.0008326	−.0004199	.000733861•	.000827799	.000836744	.000758254	.000616328	.000229000	.000043716	.000000227	011001111110
010000011100	49	.0008326	−.0005034	.000731759	.000818354	.000815405	.000724201	.000573607	.000194791	.000032787	.000000113	011001110111
010000110110	49	.0008326	−.0005211	.000731319	.000816388	.000811010	.000717274	.000565045	.000188252	.000030826	.000000103	101000100011
001100100111	47	.0008281	−.0005496	.000729809	.000812191	.000803570	.000707154	.000553797	.000181094	.000028936	.000000090	001101110010
100101101001	48	.0008281	−.0005643	.000729440	.000810548	.000799916	.000692231	.000546808	.000175956	.000027476	.000000085	101001001111
100010101001	48	.0008281	−.0005868	.000728867	.000807964	.000794091	.000692231	.000535440	.000167481	.000025070	.000000064	010101010011
100101011001	48	.0008223	−.0005135	.000729684•	.000814933	.000812267	.000723018	.000575157	.000199600	.000034991	.000000149	100111011010
010101011010	48	.0008223	−.0005431	.000728922	.000811451	.000804256	.000710009	.000558561	.000185787	.000030424	.000000101	100110011001
001101011010	48	.0008223	−.0005642	.000728391	.000809051	.000798853	.000701461	.000547983	.000177783	.000028090	.000000084	100101001101
001110000011	46	.0008178	−.0005598	.000727722	.000808680	.000800136•	.000705329	.000554299	.000184486	.000030460	.000000107	001110001011
010000110011	47	.0008141	−.0005886	.000726327	.000800495	.000792246	.000694001	.000541072	.000175091	.000027727	.000000085	100110111001
001100111001	47	.0008141	−.0006119	.000725743	.000801894	.000786452	.000684931	.000529988	.000166915	.000025405	.000000069	011001110011
011010100011	46	.0008038	−.0006235	.000723989	.000798855	.000786265	.000688130•	.000536537	.000174574	.000028062	.000000093	001101110001
001011101010	48	.0008038	−.0005502	.000723828	.000798282	.000782828	.000682882	.000530250	.000170176	.000026868	.000000085	101011011010
100010010011	48	.0008034	−.0005381	.000725381•	.000806362	.000801117	.000711891	.000566319	.000198216	.000033540	.000000167	101010101011
100100011010	48	.0008034	−.0005935	.000722897	.000799535	.000785377	.000686318	.000533732	.000171319	.000026771	.000000078	100101010101
001010111010	48	.0008034	−.0006330	.000723265	.000796735	.000779163	.000676629	.000521915	.000162712	.000024365	.000000062	100011101110
010101010010	47	.0008029	−.0006664	.000722321	.000793095	.000773095	.000671509	.000503852	.000153950	.000022178	.000000051	100110011010
001101011010	48	.0007936	−.0006246	.000721737	.000795402•	.000780319	.000681731	.000530644	.000177697	.000027306	.000000085	101010101010
100001010011	48	.0007936	−.0005555	.000721652	.000794994	.000773348	.000680128	.000528591	.000170061	.000026822	.000000082	001010011110
011010010110	48	.0007936	−.0006608	.000720823	.000791297	.000771099	.000667182	.000512689	.000158235	.000023428	.000000058	100101101011
101000011010	48	.0007894	−.0005769	.000722175•	.000799759	.000791931	.000701625	.000556569	.000193323	.000034288	.000000154	101011001010
100110011011	48	.0007894	−.0006378	.000720646	.000792915	.000776540	.000677198	.000526115	.000169383	.000026802	.000000083	010110011010
100100110011	48	.0007894	−.0006381	.000720625	.000792768	.000774096	.000676325	.000524830	.000169038	.000026304	.000000078	011100100111
001010010111	48	.0007894	−.0006458	.000720146	.000791917	.000774211	.000673392	.000521266	.000165478	.000025603	.000000074	011100101110
010100101101	48	.0007894	−.0006575	.000719960	.000790660	.000771452	.000669127	.000516099	.000161742	.000024552	.000000066	100011101010
001010011010	48	.0007894	−.0006755	.000719693	.000788635	.000766928	.000662033	.000507408	.000155375	.000022774	.000000055	100011011001
010100111001	47	.0007889	−.0006992	.000719039	.000786126	.000761836	.000654677	.000499076	.000150366	.000021645	.000000051	010110101011
010010111010	47	.0007889	−.0007109	.000718747	.000784829	.000758961	.000650215	.000493676	.000146559	.000021645	.000000045	101101101101
010101010010	47	.0007889	−.0007241	.000718414	.000783349	.000755682	.000645131	.000487539	.000142285	.000019521	.000000039	101010101101
100001011110	50	.0007792	−.0005889	.000720009•	.000795649	.000786679	.000696231	.000551827	.000191185	.000033698	.000000145	100010111101
001111000101	46	.0007786	−.0007500	.000719965	.000787073	.000768192	.000667941	.000517736	.000166459	.000026548	.000000087	011011000011
010100101011	47	.0007749	−.0007508	.000715266	.000777158	.000747733	.000637271	.000481278	.000140910	.000019549	.000000041	010101110101
100000011011	48	.0007607	−.0006674	.000714743	.000782531•	.000765138	.000668475	.000522133	.000173160	.000028934	.000000109	001001011011
001010101110	47	.0007313	−.0007313	.000713156	.000775506	.000749539	.000644093	.000492186	.000150729	.000022348	.000000057	011001011010
011000110101	47	.0007602	−.0007735	.000712047	.000771007	.000740093	.000630022	.000478517	.000140373	.000019720	.000000044	011011001001
011001101001	48	.0007560	−.0007642	.000711506	.000770899	.000741457	.000633286	.000480510	.000140140	.000019720	.000000050	100101011101
010010110011	47	.0007560	−.0007888	.000710904	.000768287	.000735801	.000624698	.000470851	.000143946	.000020771	.000000046	011011001101
010010011011	46	.0007457	−.0007889	.000709204	.000763670	.000727020	.000612836	.000454714	.000137125	.000019039	.000000041	011001110110

154

Table A

m = 7, n = 6 Rank Orders 401 thru 450 of 1716 C(7, 6) = 1716 1/C(7, 6) = .00058275

z	w	c_1	c_2	.2	.4	.6	.8	1.0	1.5	2.0	3.0	z^{tc}
0001110100101	46	.0007457	−.0007715	.000709497	.000767757	.000738869	.000632800	.000482644	.000148419	.000022519	.000000066	0110101000111
0101110000101	46	.0007457	−.0007742	.000709418	.000767347	.000737834	.000630990	.000480196	.000146184	.000021756	.000000059	0101011100101
0011100001101	46	.0007457	−.0007821	.000709220	.000766471	.000735896	.000627980	.000476538	.000143543	.000021022	.000000054	0110110001111
0100110100101	46	.0007457	−.0007982	.000708835	.000764838	.000732437	.000622840	.000470582	.000139763	.000020114	.000000050	0111010010011
0101101010001	46	.0007457	−.0008133	.000708457	.000763155	.000728699	.000617018	.000463500	.000134670	.000018716	.000000043	0110110100011
0110100100101	46	.0007360	−.0008329	.000706224	.000758673	.000722803	.000611021	.000458556	.000133418	.000018700	.000000043	0011101010011
0000110011011	46	.0007360	−.0008444	.000705937	.000757402	.000719998	.000606680	.000453316	.000129727	.000017712	.000000047	0101101010101
0110011001101	48	.0007360	−.0006874	.000709681	.000773908	.000756137	.000662321	.000520471	.000178146	.000031546	.000000147	1100010001111
1000011010101	48	.0007355	−.0007993	.000706925	.000761788	.000729372	.000620605	.000469401	.000139794	.000020088	.000000048	1000100011110
0110011010011	48	.0007355	−.0008141	.000706568	.000760260	.000726104	.000615694	.000463629	.000135969	.000019119	.000000043	1000110011101
0001101110001	46	.0007317	−.0007827	.000706678	.000762875	.000733515	.000628551	.000480306	.000149195	.000023017	.000000071	0111000100111
0101001100101	46	.0007317	−.0008394	.000705300	.000756915	.000720624	.000608948	.000456980	.000133150	.000018734	.000000044	0011100100101
0110010011001	46	.0007317	−.0008525	.000704974	.000755473	.000717444	.000604029	.000451041	.000128965	.000017611	.000000037	0110100110011
1001001010001	48	.0007313	−.0008166	.000705731	.000758838	.000724560	.000614455	.000462894	.000136030	.000019178	.000000043	0101001010011
0010010011110	48	.0007313	−.0008296	.000705420	.000757516	.000721755	.000610273	.000458023	.000132886	.000018407	.000000039	1000101011101
1100000010111	48	.0007215	−.0006644	.000707679	.000763162	.000758516	.000670167	.000533156	.000191119	.000036185	.000000203	0000111100111
0010101001101	48	.0007215	−.0007516	.000705543	.000763162	.000737678	.000637167	.000492429	.000158761	.000025671	.000000088	1010010101011
1001000011101	48	.0007215	−.0008129	.000704075	.000756893	.000722419	.000617116	.000468453	.000142365	.000021286	.000000060	1010010101011
0101010010101	48	.0007215	−.0008217	.000703868	.000756026	.000722439	.000614446	.000465378	.000140422	.000020813	.000000057	1000011110010
1000011001110	48	.0007215	−.0008363	.000703490	.000754290	.000718458	.000608048	.000457355	.000134197	.000018968	.000000043	1000110011110
0100001110110	48	.0007215	−.0008535	.000703077	.000752523	.000714681	.000602376	.000450696	.000129799	.000017859	.000000038	1010010111101
0110100101001	46	.0007171	−.0008468	.000702458	.000752329	.000716227	.000606415	.000456878	.000135608	.000019674	.000000051	1001011001011
0010101010011	46	.0007171	−.0008829	.000701578	.000746819	.000707992	.000593919	.000442063	.000125592	.000017080	.000000036	0110100110101
1010010011001	46	.0007128	−.0008922	.000700582	.000746449	.000705130	.000590799	.000439234	.000124461	.000016887	.000000035	0110101001011
1000011100101	47	.0007113	−.0008072	.000702307	.000755012	.000724324	.000620540	.000475155	.000149992	.000023960	.000000086	0011110010011
0011110001001	47	.0007070	−.0008564	.000701161	.000749714	.000712657	.000602496	.000453225	.000133908	.000019304	.000000049	1010010001111
0010110110001	47	.0007031	−.0008664	.000700147	.000747554	.000709583	.000599028	.000449953	.000132426	.000019007	.000000047	1001001000111
0101001101001	46	.0007031	−.0008700	.000699347	.000746184	.000708133	.000597981	.000449535	.000132883	.000019217	.000000049	0111000110011
0110001010101	46	.0007031	−.0008968	.000698707	.000743469	.000702381	.000589423	.000439586	.000126472	.000017634	.000000041	0110011001101
0110100110010	46	.0007031	−.0009208	.000698126	.000740977	.000697041	.000581392	.000430154	.000120253	.000016067	.000000033	0110100110110
0011011000011	45	.0006928	−.0008922	.000696963	.000741360	.000701776	.000591559	.000444353	.000131878	.000019384	.000000054	0011110100011
1000110010011	47	.0006924	−.0008937	.000696810	.000740787	.000700396	.000589124	.000441000	.000128729	.000018311	.000000044	0101110010011
0011011001101	47	.0006924	−.0009182	.000696220	.000738270	.000695019	.000581047	.000431521	.000124433	.000016707	.000000036	1000111001011
0010011110010	48	.0006886	−.0008307	.000697641	.000746237	.000711436	.000599855	.000463309	.000146841	.000023313	.000000079	1100001110011
1000001111001	48	.0006886	−.0008728	.000696646	.000742047	.000704576	.000599639	.000450836	.000136489	.000020580	.000000061	1010000111110
0100100110110	47	.0006886	−.0008957	.000696100	.000739736	.000699652	.000589246	.000442126	.000130581	.000019002	.000000051	1010000111101
0011100111001	47	.0006886	−.0009044	.000695899	.000738994	.000698282	.000587488	.000440403	.000129993	.000018994	.000000102	0011101100011
1010001010011	45	.0006784	−.0009737	.000695419	.000742185	.000708912	.000600649	.000454473	.000149711	.000024813	.000000078	1011000011010
1010000101110	47	.0006784	−.0008735	.000694770	.000739357	.000702720	.000596915	.000453485	.000140755	.000022182	.000000078	0110011001110
1000001101101	47	.0006784	−.0009157	.000693739	.000734871	.000692922	.000581826	.000435246	.000127525	.000018380	.000000048	0011001110110
0010110011010	47	.0006784	−.0009164	.000693730	.000734860	.000692951	.000581940	.000435449	.000127740	.000018443	.000000048	1010010010111
1001011010010	47	.0006784	−.0009351	.000693268	.000732830	.000688498	.000575081	.000427195	.000121940	.000016879	.000000038	0010111011010
0011011010001	47	.0006784	−.0009363	.000693250	.000732798	.000688527	.000575265	.000427574	.000122453	.000017075	.000000040	0110110101010
0010011011110	47	.0006784	−.0009432	.000693073	.000731197	.000684989	.000569839	.000424090	.000119929	.000016385	.000000036	0011100010011
0010101001110	47	.0006784	−.0009514	.000692879	.000731197	.000684187	.000568669	.000421067	.000117919	.000015863	.000000033	1001100101011
0010101010110	47	.0006784	−.0009553	.000692787	.000730794	.000680045	.000562421	.000419740	.000117132	.000015692	.000000032	1001010101011
0101010011010	47	.0006779	−.0009737	.000692340	.000728868	.000680804	.000562421	.000412386	.000122284	.000014484	.000000026	0011101100110
1001010010101	47	.0006779	−.0009928	.000691818	.000728860	.000676502	.000557616	.000407285	.000109711	.000014011	.000000025	0101011010101
0100101100101	46	.0006779	−.0010056	.000691506	.000726571	.000673645	.000553343	.000403211	.000106548	.000013260	.000000022	0101010010101
0100101100111	46	.0006736	−.0010128	.000690562	.000723822	.000671254	.000550926	.000400298	.000105931	.000013188	.000000022	0011100101101

155

Table A

m = 7, n = 6   Rank Orders 451 thru 500 of 1716   C(7, 6) = 1716   1/C(7, 6) = .00058275

z	w	c_1	c_2	d = .2	.4	.6	.8	1.0	1.5	2.0	3.0	z^{tc}
1000010011110	49	.0006682	−.0008996	.000692241	.000733592	.000693702	.000585507	.000441239	.000132843	.000019892	.000000057	1000011011110
1001000110011	47	.0006644	−.0009669	.000689962	.000725894	.000678815	.000564416	.000417421	.000117933	.000016184	.000000036	0011001110010
0011000110110	47	.0006644	−.0009914	.000689376	.000723409	.000673538	.000565545	.000408242	.000111961	.000014695	.000000056	1001001110110
1000100101110	49	.0006639	−.0009094	.000691232	.000731454	.000690677	.000582120	.000438074	.000131462	.000019633	.000000056	1000100111110
0101001010101	46	.0006639	−.0010436	.000688056	.000718150	.000662777	.000547223	.000399688	.000101518	.000012355	.000000028	1010010011101
0010011011010	47	.0006497	−.0009814	.000686156	.000719292	.000664656●	.000558485	.000400114●	.000109033	.000014250	.000000041	1010011101101
0101001001101	47	.0006455	−.0009814	.000686052	.000716742●	.000675514	.000543204	.000413932	.000118560	.000016685	.000000041	0010101111010
0100100110101	47	.0006455	−.0010306	.000686156	.000714394	.000661195	.000537898	.000396231	.000107219	.000013892	.000000026	1010001110011
0100101001101	46	.0006450	−.0010501	.000684454	.000712436	.000657382	.000543204	.000390427	.000103995	.000013212	.000000024	1010011011001
0100111001001	46	.0006450	−.0010517	.000684423	.000712343	.000657265	.000537840	.000390491●	.000104239	.000013317	.000000025	0101001110011
0101010101001	46	.0006450	−.0010754	.000683855	.000709932	.000652163	.000530530	.000381762	.000098779	.000012034	.000000019	0101101010101
0011111001001	45	.0006347	−.0010187	.000683341	.000712958●	.000662254	.000548029	.000404300	.000114929	.000016227	.000000042	0111011000011
0011011000011	45	.0006347	−.0010446	.000682719	.000710310	.000656609	.000539563	.000394362	.000108305	.000014507	.000000032	0011011000101
0011011000101	45	.0006347	−.0010601	.000681148	.000708705	.000653166	.000534377	.000388254	.000104238	.000013467	.000000026	0111100010101
0110110010101	46	.0006310	−.0010819	.000680877	.000705513	.000647742	.000527316	.000380768	.000100078	.000012539	.000000022	0101000110011
0101000010101	46	.0006310	−.0010936	.000681962●	.000714089	.000645483	.000545483	.000377185	.000098041	.000012104	.000000081	0110101101001
0110100010011	45	.0006207	−.0009682	.000680185	.000706618	.000669536	.000562561	.000423954	.000130703	.000020896	.000000040	0011111110001
1110101000111	45	.0006207	−.0010437	.000679140	.000702230	.000653743	.000538972	.000396238	.000111770	.000015678	.000000026	0111010100111
0011011001001	45	.0006207	−.0010879	.000678892	.000702230	.000644524	.000525350	.000380484	.000101664	.000013156	.000000023	0110101111100
0011100010101	45	.0006207	−.0010982	.000678892	.000703176	.000642260	.000521961	.000376522	.000099081	.000012512	.000000023	1100101100111
0101010100011	45	.0006207	−.0011015	.000678829	.000700961	.000641941	.000521664	.000376380	.000099304●	.000012640●	.000000024	0010101010101
0011010100011	47	.0006207	−.0011142	.000678524	.000699660	.000639177	.000517548	.000371596	.000096241	.000011894	.000000019	0101010001011
0000111101100	47	.0006203	−.0009573	.000682121●	.000714928	.000671358	.000565246	.000427020	.000132598	.000021359	.000000084	1101000011111
1000100100110	47	.0006203	−.0010917	.000678932	.000701435	.000640709	.000522359	.000376638	.000098602	.000012274	.000000021	1000101010101
0010100101110	47	.0006105	−.0011051	.000678623	.000700169	.000640126	.000518664	.000372512	.000096221	.000011757	.000000019	1100010010111
0010111011100	46	.0006105	−.0010243	.000678723●	.000705322	.000654138	.000541576	.000400403	.000114983	.000016427	.000000043	0010100101101
1000010101101	47	.0006105	−.0011078	.000676773	.000699216	.000637216	.000516878	.000372909	.000099266	.000012142	.000000020	0101010111010
0011010101010	47	.0006063	−.0011254	.000676364	.000695535	.000633815	.000511898	.000366491	.000093949	.000011399	.000000018	1001010010111
1100010101010	47	.0006063	−.0009617	.000679408	.000710291	.000666192	.000560647	.000423696	.000131871	.000021264	.000000042	0011011011110
0001100111100	47	.0006063	−.0010341	.000677712	.000703176	.000651094	.000538157	.000397196	.000113570	.000016160	.000000023	0011010011001
1001000101101	47	.0006063	−.0011126	.000675883	.000695583	.000635303	.000515045	.000370708	.000096960	.000012125	.000000021	0101011101110
1010010010011	47	.0006063	−.0011226	.000675645	.000694581	.000633204	.000511964	.000367180	.000094798	.000011627	.000000019	0101001011110
1001000010110	47	.0006063	−.0011228	.000675650	.000694638	.000633392	.000512333	.000367702	.000095253	.000011758	.000000020	1000110101001
0110100010110	47	.0006063	−.0011384	.000675281	.000693092	.000630163	.000507608	.000362312	.000091981	.000011013	.000000017	1001110011011
0101010000011	45	.0006018	−.0010953	.000675496	.000694277	.000638498	.000521178	.000378875	.000103236	.000013826	.000000031	0010110111011
0011011010101	45	.0006018	−.0011462	.000675432●	.000691355	.000628316	.000506375	.000362056	.000092996	.000011432	.000000020	0101010010111
0011110101010	46	.0005960	−.0010526	.000675371	.000692554	.000645895	.000533617	.000394389	.000114438	.000016908	.000000053	0001011100110
0011011010011	45	.0005960	−.0011165	.000672484	.000688847	.000632665	.000514004	.000371573	.000099406	.000012985	.000000027	1000111111001
0110100110011	45	.0005921	−.0011479	.000671941	.000688511	.000625698	.000504956	.000362081	.000094348	.000011905	.000000023	0011011011001
0101001011001	45	.0005921	−.0011712	.000671712	.000686268	.000621069	.000498253	.000354503	.000089809	.000010867	.000000018	0011011010011
0010101110001	45	.0005878	−.0011498	.000671649	.000687046●	.000623958	.000503278	.000360671	.000093743	.000011755	.000000022	0110101010101
0110100100011	45	.0005878	−.0011517	.000671618	.000686974	.000623932	.000503417●	.000361033	.000094276	.000011953	.000000019	0111000110111
1100100101011	47	.0005878	−.0011781	.000671003	.000686433	.000618687	.000495823	.000352445	.000089124	.000010772	.000000018	0110101010111
1000010111100	47	.0005776	−.0010340	.000672423●	.000685015	.000643644	.000534234	.000397571	.000118180	.000017961	.000000058	0010111111100
0000101110001	47	.0005776	−.0011094	.000670674	.000686758	.000628496	.000511886	.000371641	.000101092	.000013479	.000000029	1100010010111
0110100000111	47	.0005776	−.0011224	.000670404	.000686776	.000626736	.000509721	.000369631	.000100576	.000013541●	.000000031	0111000011101
0110101001010	46	.0005776	−.0011260	.000670325	.000686462	.000626118	.000508867	.000368709	.000100088	.000013444●	.000000031	1000111111001
0101001011010	47	.0005776	−.0011523	.000669697	.000683803	.000620478	.000500456	.000358892	.000093656	.000011813	.000000022	0110101011110
0110001110100	47	.0005776	−.0011747	.000669652	.000681711	.000616216	.000494936	.000352065	.000089652	.000010914	.000000018	1010010011001
0010001010011	44	.0005775	−.0011311	.000670216	.000686141	.000625667	.000508497	.000366860	.000100490	.000013651	.000000033	0011110100011

m = 7, n = 6   Rank Orders 501 thru 550 of 1716   C(7, 6) = 1716   1/C(7, 6) = .00058275

z	w	c_1	c_2	d = .2	.4	.6	.8	1.0	1.5	2.0	3.0	z^{tc}
1010010000111	46	.0005673	-.0011143	.000668720	.000684821	.000626268•	.000511805	.000374069	.000105424	.000015104	.000000044	000111011010
0001110011010	46	.0005673	-.0011752	.000667299	.000678915	.000613952	.000493700	.000353186	.000091972	.000011676	.000000023	101011100011
0001011001011	46	.0005673	-.0011755	.000667101	.000678789	.000613602	.000493064	.000352321	.000091234	.000011455	.000000021	101011001110
1000110010011	46	.0005673	-.0011832	.000667101	.000678054	.000612090	.000490884	.000349870	.000089791	.000011134	.000000018	011011001110
0010110001110	46	.0005673	-.0011949	.000666836	.000676980	.000609918	.000487794	.000346430	.000087793	.000010687	.000000015	100101100011
0001011010110	46	.0005673	-.0012129	.000666414	.000675225	.000606273	.000482489	.000340405	.000084160	.000009858	.000000015	100101011110
1000111001011	46	.0005631	-.0011841	.000666312	.000676869•	.000610119	.000490592	.000350336	.000090774	.000011455	.000000022	001110011011
1000101111010	46	.0005631	-.0011872	.000666230	.000676487	.000610250	.000489222	.000348666	.000089604	.000011151	.000000020	001110101011
0010101101010	46	.0005631	-.0012202	.000665464	.000673341	.000603789	.000479905	.000338171	.000083367	.000009738	.000000014	100110101011
0101110001101	45	.0005626	-.0012246	.000665289	.000672932	.000603258	.000479462	.000337973	.000083615	.000009860	.000000015	010101010111
0100110100101	45	.0005626	-.0012507	.000664656	.000700357	.000598041	.000472063	.000329798	.000079043	.000008902	.000000012	010110100101
0001011110010	46	.0005534	-.0011607	.000665047•	.000676254	.000612946	.000495416	.000357232	.000096046	.000012804	.000000029	101100010111
0011001010011	46	.0005534	-.0012182	.000663720	.000670809	.000601767	.000479288	.000339033	.000085116	.000010274	.000000017	100101011011
0010101010110	46	.0005534	-.0012435	.000663137	.000668427	.000596908	.000472331	.000331255	.000080595	.000009277	.000000013	100011110011
1000100011110	48	.0005529	-.0011634	.000664880•	.000675697	.000611782	.000493613	.000355023	.000094436	.000012367	.000000026	100001101110
0101010010101	45	.0005529	-.0011781	.000664527	.000674196	.000608578	.000488801	.000349362	.000090647	.000011386	.000000021	101010101001
0101001010101	45	.0005529	-.0012795	.000662239	.000665083	.000590442	.000461486	.000321795	.000075693	.000008325	.000000013	010101110101
1000101110100	45	.0005486	-.0012842	.000661350	.000663451	.000588474	.000461645	.000320389	.000075372	.000008305	.000000011•	100100111010
0010101100101	48	.0005389	-.0011876	.000661744•	.000669447•	.000603480	.000484903	.000347409	.000091651	.000011935	.000000025	100110111010
1010000100101	46	.0005344	-.0012251	.000660072	.000664768	.000595433	.000474471	.000336468	.000085764	.000010645	.000000020	001101100101
0010011101010	46	.0005344	-.0012547	.000659401	.000662064	.000589983	.000466750	.000327915	.000080866	.000009568	.000000015	101010010011
0010101010101	46	.0005344	-.0012758	.000658915	.000660074	.000585921	.000460938	.000321430	.000077134	.000008915	.000000013	101010010111
0110010010101	45	.0005339	-.0012790	.000658708	.000659784	.000585636	.000460847	.000321624•	.000077598	.000008922	.000000013	010011010011
0010101100101	45	.0005339	-.0013036	.000658208	.000657525	.000581097	.000454464	.000314628	.000073762	.000008134	.000000010	011001110011
0010101110011	45	.0005339	-.0013096	.000657288	.000655765	.000578868	.000452422	.000312805	.000073206	.000008065	.000000010	011010011011
1010000110011	46	.0005297	-.0012495	.000656933	.000658516•	.000587141	.000465792	.000328907	.000083033	.000010228	.000000019	101010111010
0010011010101	46	.0005205	-.0012858	.000656112	.000655209	.000580485	.000456379	.000318499	.000077104	.000008932	.000000014	010100111011
0100011110100	45	.0005205	-.0012799	.000656164•	.000655684	.000581667	.000458260	.000320757	.000078583	.000009286	.000000015	011100111101
0101011101001	45	.0005200	-.0013222	.000655209	.000651807	.000573872	.000444927...	.000308676	.000071873	.000007879	.000000010	010101110101
1010010110011	45	.0005200	-.0013343	.000654933	.000650745	.000571789	.000444416	.000305633	.000070326	.000007587	.000000009	011001011001
0001110010001	44	.0005097	-.0012452	.000655066	.000656020	.000585644	.000466352	.000331479	.000086326	.000011253	.000000026	011101000001
0101101000011	44	.0005097	-.0013019	.000653782	.000650845	.000575203	.000451530	.000315004	.000076732	.000009072	.000000015	010111000011
0011011000101	44	.0005097	-.0013150	.000653479	.000649600	.000572650	.000447875	.000310871	.000074287	.000008518	.000000013	010110100011
1000110001101	46	.0005092	-.0013094	.000653494•	.000649755	.000572883	.000447983	.000310754	.000073838	.000008342	.000000011	100111001110
0010110001011	46	.0005092	-.0013224	.000653205	.000648620	.000570653	.000444912	.000307455	.000072110	.000008002	.000000011	100110111011
0101100010101	47	.0005092	-.0012279	.000654648•	.000656092	.000586791	.000468508	.000334075	.000087745	.000011506	.000000026	010101001011
0010100100011	44	.0005054	-.0013124	.000652762	.000648680	.000572161	.000448170	.000311924	.000075477	.000008855	.000000015	011101100011
0011100100011	44	.0005054	-.0013329	.000652495	.000647582	.000569908	.000444927	.000308276	.000073317	.000008366	.000000013	011101010101
0001011110100	46	.0005054	-.0011789	.000653873•	.000657750	.000593557	.000480844	.000349967	.000099567	.000014816	.000000051	101011100101
1100000110111	46	.0004952	-.0011901	.000653594	.000650518	.000590835	.000476619	.000344828	.000095782	.000013701	.000000011	110111111001
0001011011100	46	.0004952	-.0012461	.000652333	.000651431	.000580500	.000461756	.000327988	.000085231	.000011029	.000000024	110100000111
0001110101100	46	.0004952	-.0012648	.000651906	.000649899	.000576974	.000456699	.000322304	.000081823	.000010234	.000000020	110010100101
1000110100101	46	.0004952	-.0013258	.000650540	.000644260	.000566136	.000441512	.000305657	.000072520	.000008231	.000000012	010110001110
0010110010101	46	.0004952	-.0013416	.000650181	.000642811	.000563219	.000437397	.000301126	.000069994	.000007699	.000000010	100110011011
1000101011110	46	.0004952	-.0013448	.000650116	.000642580	.000562815	.000436904	.000300664	.000069837	.000007684	.000000010•	100100110111
0001010101110	46	.0004952	-.0013523	.000649947	.000641843	.000561266	.000434639	.000298094	.000068323	.000007356	.000000009	100010110101
0010110101010	46	.0004952	-.0013683	.000649581	.000641898•	.000561446	.000434983	.000298565	.000068697	.000007453	.000000009	100101011101
0010101010110	46	.0004952	-.0013400	.000649415	.000641870•	.000562886	.000430424	.000293993	.000066204	.000006948	.000000008	100110101101
0010110011001	44	.0004908	-.0013012	.000649908	.000639901	.000559912	.000438145	.000302850	.000071766	.000008185	.000000012	010110010101
0110000111001	45	.0004871	-.0013554	.000648378	.000639383	.000559007	.000433455	.000298195	.000069527	.000007745	.000000011	011000111001

157

Table A

m = 7, n = 6   Rank Orders 551 thru 600 of 1716   C(7, 6) = 1716   1/C(7, 6) = .00058275

z	w	c_1	c_2	d = .2	.4	.6	.8	1.0	1.5	2.0	3.0	z^{tc}
000111000110	45	.0004850	−.0013069	.000649079	.000643101	.000567020	.000445088	.000311193	.000076938	.000009348	.000000017	100111000111
100100110110	46	.0004813	−.0013662	.000647043	.000636572	.000554975	.000428816	.000293705	.000067376	.000007311	.000000009	010100111101
011010100011	46	.0004768	−.0013792	.000646756	.000635450	.000552788	.000425827	.000290518	.000065742	.000006997	.000000008	100101110001
011100000101	44	.0004768	−.0012940	.000647841	.000641792	.000566788	.000446401	.000313693	.000079112	.000009902	.000000020	100101111001
011010010011	44	.0004768	−.0013485	.000646641	.000637088	.000557552	.000433645	.000299899	.000071629	.000008325	.000000014	001110010011
011100110001	44	.0004768	−.0013577	.000646426	.000636201	.000555728	.000431011	.000296930	.000069862	.000007923	.000000013	001110010011
011100011001	44	.0004768	−.0013608	.000646171	.000636030	.000555492	.000430831	.000296898	.000070074	.000008018	.000000013	001110011001
010101011100	46	.0004768	−.0013693	.000645794	.000635213	.000553813	.000428409	.000294171	.000068457	.000007652	.000000011	011001110001
010101011100	46	.0004666	−.0013865	.000645484	.000633736	.000550933	.000424476	.000289981	.000066308	.000007235	.000000011	011011001011
100001011010	46	.0004666	−.0013287	.000645484	.000635484	.000550933	.000434487	.000301657	.000072881	.000008545	.000000014	110001001011
011011011010	46	.0004666	−.0013708	.000644221	.000631894	.000550026	.000424951	.000291481	.000067606	.000007505	.000000010	011010011110
001111100001	46	.0004666	−.0013937	.000643716	.000629915	.000546150	.000419622	.000285763	.000064605	.000006908	.000000009	001011110011
011100110011	43	.0004665	−.0012999	.000645813	.000638334	.000563029	.000443407	.000311986	.000079503	.000010183	.000000023	001111000011
001011001011	44	.0004628	−.0013832	.000643265	.000629866	.000547278	.000422120	.000289142	.000067000	.000007474	.000000011	011100100011
010101010011	46	.0004623	−.0012768	.000645500	.000638738	.000564667	.000446022	.000314840	.000080715	.000010311	.000000021	110000110011
010101011001	46	.0004623	−.0013375	.000644160	.000633455	.000554214	.000431441	.000298887	.000071754	.000008348	.000000013	001101111010
010101011010	46	.0004623	−.0013664	.000643532	.000631027	.000549522	.000425070	.000292122	.000068277	.000007664	.000000011	110010111010
101000010111	46	.0004623	−.0013713	.000643423	.000630606	.000548708	.000423965	.000290952	.000067682	.000007550	.000000011	100101110010
011000011001	46	.0004623	−.0013830	.000643161	.000629560	.000546624	.000424476	.000291056	.000065974	.000007203	.000000009	100011101110
010010110010	46	.0004623	−.0014040	.000642699	.000627758	.000543113	.000416260	.000282674	.000063343	.000006693	.000000008	100011101101
101010000111	45	.0004521	−.0013052	.000643020	.000633675	.000558083	.000439583	.000309917	.000080127	.000010592	.000000027	001111001010
100110010011	45	.0004521	−.0013653	.000641659	.000628178	.000546941	.000423659	.000292057	.000069387	.000008034	.000000014	001011110010
001110101010	45	.0004521	−.0013808	.000641330	.000626938	.000544615	.000420602	.000288923	.000067933	.000007784	.000000012	101010001011
011110010011	45	.0004521	−.0013820	.000641289	.000626742	.000544083	.000419713	.000287794	.000067094	.000007560	.000000012	100011110001
100101100011	45	.0004521	−.0013851	.000641224	.000626476	.000543622	.000419115	.000287198	.000066855	.000007532	.000000011	001101010110
010101001011	44	.0004516	−.0014174	.000640504	.000623625	.000537981	.000411281	.000278699	.000062261	.000006586	.000000008	100101110101
010011000101	45	.0004423	−.0014307	.000640126	.000622402	.000535809	.000408509	.000275914	.000061011	.000006374	.000000008	010110011010
011010010110	45	.0004423	−.0014126	.000638803	.000621196	.000535016	.000410527	.000279168	.000063442	.000006923	.000000009	100110110010
100001110010	45	.0004423	−.0014371	.000638260	.000619058	.000531817	.000404741	.000272946	.000060162	.000006268	.000000007	100110110010
010010110010	46	.0004419	−.0013855	.000639289	.000623185	.000539822	.000416260	.000284301	.000065762	.000007307	.000000010	100011101110
001101110010	45	.0004381	−.0013803	.000638733	.000622771	.000540332	.000417334	.000287067	.000068070	.000007915	.000000013	011000100111
100110010110	45	.0004381	−.0014141	.000637983	.000619820	.000534525	.000409308	.000278397	.000063417	.000006957	.000000010	001110110011
011011010110	45	.0004376	−.0014420	.000637365	.000617387	.000529748	.000402726	.000271317	.000059679	.000006208	.000000007	100010110011
011100100110	47	.0004376	−.0013978	.000638224	.000620832	.000536380	.000411596	.000280534	.000064081	.000006993	.000000010	010101101110
011101000101	44	.0004376	−.0014780	.000636495	.000614269	.000526268	.000398214	.000266702	.000057524	.000005828	.000000006	010101010011
010101010011	45	.0004279	−.0013827	.000636765	.000619401	.000535945	.000413882	.000284488	.000067448	.000007858	.000000013	100001110110
100011001100	47	.0004279	−.0014103	.000636140	.000616882	.000531408	.000406670	.000276484	.000062838	.000006837	.000000009	001011110110
010011001011	47	.0004234	−.0014125	.000635288	.000615590	.000530360	.000406394	.000277063	.000063947	.000007181	.000000011	100011011011
010101011010	45	.0004234	−.0014616	.000634207	.000611366	.000522115	.000395098	.000264978	.000057643	.000005930	.000000007	100101001011
001011011010	45	.0004192	−.0014678	.000633282	.000610516	.000519817	.000392768	.000263012	.000056985	.000005836	.000000007	101010101011
010100101101	44	.0004187	−.0014564	.000633455	.000610516	.000521873	.000395791	.000266417	.000058924	.000006242	.000000006	100111101001
010110101001	44	.0004187	−.0014957	.000632597	.000607201	.000515489	.000387184	.000257374	.000054460	.000005415	.000000006	011001010010
010011110010	45	.0004094	−.0014383	.000632113	.000609170	.000521684	.000397104	.000268748	.000060665	.000006616	.000000009	011000101010
010100101010	45	.0004094	−.0014475	.000631915	.000608413	.000520244	.000395188	.000266765	.000059729	.000006453	.000000009	011010111010
001110010110	45	.0004094	−.0014850	.000631097	.000605245	.000514113	.000386865	.000257944	.000055245	.000005588	.000000007	101010101011
011101010010	44	.0004089	−.0015050	.000630578	.000603491	.000510954	.000382808	.000253856	.000053409	.000005278	.000000005	010101010011
010101010101	45	.0004089	−.0015168	.000630326	.000602534	.000509047	.000380425	.000251411	.000052282	.000005086	.000000005	011010101010
010101011001	44	.0004047	−.0014897	.000630120	.000603491	.000512066	.000386394	.000256630	.000055033	.000005595	.000000006	111000101101
011001011011	44	.0004047	−.0015071	.000629745	.000601061	.000509349	.000381419	.000252885	.000053249	.000005277	.000000005	010101101101

Table A

m = 7, n = 6   Rank Orders 601 thru 650 of 1716   C(7, 6) = 1716   1/C(7, 6) = .00058275

z	w	c_1	c_2	d = .2	.4	.6	.8	1.0	1.5	2.0	3.0	z^{tc}
0101001101010	44	.0004047	−.0015204	.000629459	.000600978	.000507304	.000378720	.000250114	.000051969	.000005058	.000000005	0110100110101
1000001111010	47	.0003950	−.0014183	.000628854●	.000606503	.000520454	.000398137	.000271614	.000063469	.000007338	.000000012	1010001111110
1100001111101	46	.0003945	−.0013748	.000630689●	.000600893	.000527075	.000407177	.000281254	.000068545	.000006568	.000000017	1000010111101
0100010111100	46	.0003945	−.0014387	.000629300	.000604500	.000516563	.000392715	.000265651	.000060042	.000006568	.000000009	1100010101101
0011110100001	43	.0003904	−.0013904	.000630372●	.000608819	.000525245	.000404994	.000279258	.000067921	.000008350	.000000017	0111100000011
0101110100101	43	.0003944	−.0014707	.000628619	.000602161	.000512419	.000387584	.000260722	.000058169	.000006329	.000000009	0011110100011
1100010100011	43	.0003944	−.0014826	.000628373	.000601109	.000510335	.000384684	.000257569	.000056447	.000005968	.000000008	0111010100011
0001110011100	45	.0003842	−.0013629	.000629049●	.000607948	.000526287	.000408304	.000277548	.000058169	.000009289	.000000023	1100011000011
0101010011001	45	.0003842	−.0014353	.000627473	.000601811	.000514269	.000391667	.000265947	.000061322	.000006973	.000000011	0101100110110
1000110001011	45	.0003842	−.0015137	.000625775	.000595280	.000501700	.000374673	.000247982	.000052182	.000005190	.000000006	0101101010101
1000110101010	45	.0003842	−.0015238	.000625553	.000594413	.000500014	.000372379	.000245552	.000050963	.000004962	.000000005	1001010001110
0101010001110	45	.0003842	−.0015240	.000625659●	.000594471	.000500189	.000372698	.000245970	.000051265	.000005034	.000000005	1001011001011
1001011001101	45	.0003842	−.0015396	.000625215	.000593132	.000497593	.000369174	.000242249	.000049414	.000004692	.000000004	1001011001101
0001101100110	45	.0003800	−.0014425	.000626524●	.000599922	.000511752	.000389011	.000263602	.000060424	.000006822	.000000011	1100100101101
1000101100101	45	.0003800	−.0015260	.000624715	.000592957	.000498352	.000370901	.000244478	.000050743	.000004949	.000000005	0110100101110
0100101011001	44	.0003760	−.0015436	.000624338	.000591530	.000495656	.000367336	.000240806	.000049021	.000004648	.000000004	0110101001101
0010101110100	45	.0003702	−.0015427	.000625291●	.000599225	.000513015	.000392505	.000241290●	.000049710	.000004825	.000000005	1101010010111
1001010010011	45	.0003702	−.0014152	.000623639	.000590551	.000494945	.000367235	.000268471	.000063804	.000007605	.000000014	1101000001111
0101010101011	45	.0003702	−.0015496	.000622391	.000588095	.000491225	.000363015	.000238030	.000048348	.000004585	.000000004	0101011001110
0101010010110	45	.0003702	−.0015630	.000622106	.000587021	.000489632	.000361026	.000235337	.000047119	.000004377	.000000004	1100101010101
0101011010001	43	.0003658	−.0015059	.000622520●	.000590516	.000497588	.000372595	.000248036	.000053767	.000005649	.000000007	0111010001011
0110110101001	43	.0003658	−.0015079	.000622490	.000590443	.000497535	.000372644●	.000248217	.000054028	.000005735	.000000008	0111011011001
0100101000011	43	.0003615	−.0015343	.000621918	.000588404●	.000493322	.000366967	.000242241	.000051024	.000005155	.000000006	0110101001011
0011100101001	43	.0003615	−.0015172	.000621498	.000588467	.000494766	.000369691	.000245606	.000053059	.000005582	.000000006	0011101010001
0111000101011	43	.0003615	−.0014947	.000620994	.000587144●	.000491046	.000364679	.000240331	.000050408	.000005071	.000000009	0110101101001
0011010011001	43	.0003518	−.0014947	.000620142	.000583085	.000494758	.000371198	.000248069	.000054779	.000005942	.000000006	0110101011001
0111000110100	43	.0003518	−.0015455	.000619064	.000590551	.000487113	.000361078	.000237594	.000049712	.000004997	.000000006	0111000111111
0001111110000	45	.0003513	−.0015350	.000623403●	.000599798	.000519249	.000404674	.000284090	.000074626	.000010389	.000000032	1100000001111
1000010011000	45	.0003513	−.0013380	.000620931	.000590204	.000500473	.000378636	.000255619	.000058240	.000006550	.000000006	1010110111100
0100101000110	45	.0003513	−.0014521	.000620031	.000586819	.000494088	.000370151	.000235850	.000053850	.000005702	.000000007	1001110001011
0010101101001	45	.0003513	−.0015147	.000619598	.000585141	.000490839	.000365732	.000242081	.000051438	.000005230	.000000006	1100101010111
1000001101001	45	.0003513	−.0015344	.000619199	.000583706	.000488269	.000362513	.000238949	.000050182	.000005048	.000000006	1101001001110
1010000100101	45	.0003513	−.0015421	.000619027	.000583028	.000486934	.000360650	.000236959	.000049126	.000004837	.000000005	1010011011010
0110100111010	45	.0003513	−.0015475	.000618913	.000582605	.000486150	.000359650	.000235928	.000048663	.000004759	.000000005	1010001101111
1000101001011	45	.0003410	−.0015586	.000618683	.000581759	.000484597	.000357656	.000233932	.000047790	.000004616	.000000004	1010011101101
0001110101010	45	.0003410	−.0015595	.000618653	.000581602	.000484223	.000357059	.000233217	.000047344	.000004520	.000000004	1010101011011
0011111000011	44	.0003271	−.0015809	.000618198	.000579895	.000481027	.000352870	.000228938	.000045380	.000004184	.000000008	1011001011010
1000111000011	44	.0003271	−.0015182	.000617564	.000582025●	.000487838	.000363858	.000241615	.000052411	.000005573	.000000007	1011010001110
1001011000011	44	.0003271	−.0015213	.000617564	.000581686	.000487113	.000362785	.000240389	.000051677	.000005410	.000000007	1011011011011
0010101010011	44	.0003410	−.0015544	.000616853	.000578969	.000481922	.000355817	.000233076	.000048024	.000004711	.000000005	0011101010110
1100010110011	45	.0003373	−.0014674	.000617987●	.000584686	.000493567	.000371797	.000249969	.000056435	.000006304	.000000010	1100111111100
0011000111001	45	.0003373	−.0015152	.000616979	.000580896	.000484417	.000362295	.000240068	.000051521	.000005354	.000000008	1010011101110
1010010100101	45	.0003373	−.0015593	.000616048	.000577403	.000479872	.000353683	.000231214	.000047338	.000004603	.000000005	0101001111010
0110000110011	45	.0003373	−.0015659	.000615911	.000576893	.000478929	.000352460	.000229981	.000046792	.000004513	.000000004	1011000111011
0110000110010	45	.0003373	−.0015671	.000615883	.000576786	.000478715	.000352162	.000229656	.000046611	.000004475	.000000004	1010100111100
1011010101011	45	.0003373	−.0015841	.000615527	.000575462	.000476263	.000348982	.000226443	.000045176	.000004236	.000000004	1010010111101
0011110000111	44	.0003271	−.0014425	.000616637●	.000583887	.000449036	.000376047	.000256134	.000061192	.000007547	.000000017	1011010110111
1000110110010	44	.0003271	−.0015310	.000614751	.000576707	.000481304	.000357502	.000236449	.000050807	.000005355	.000000007	1011001000011
1001010000011	44	.0003271	−.0015496	.000614247	.000574769	.000481686	.000352469	.000231137	.000048118	.000004832	.000000007	1011101010110
1010010100011	44	.0003271	−.0015664	.000613994	.000573832	.000475845	.000350225	.000228869	.000047096	.000004658	.000000005	0011101010110

159

Table A

m = 7, n = 6   Rank Orders 651 thru 700 of 1716   C(7, 6) = 1716   1/C(7, 6) = .00058275

z	w	c₁	c₂	d = .2	.4	.6	.8	1.0	1.5	2.0	3.0	z^tc
0011100010110	44	.0003271	−.0015762	0013779	000572991	000474205	000347978	000226466	000045845	000004411	000000004	1001011100011
1101010100110	44	.0003271	−.0015935	0013415	000571641	000471705	000344736	000223190	000045380	000004165	000000004	1000110101011
1000110100110	46	.0003266	−.0015631	0013944	000573695	000475459	000349469	000227806	000046233	000004437	000000004	1000110101110
0101011000101	43	.0003266	−.0016105	0012970	000570201	000469192	000341601	000220117	000043085	000003959	000000003	0101110010101
0101010100101	43	.0003223	−.0016178	0011935	000568352	000466800	000339176	000218083	000042433	000003873	000000003	0101101001101
1000101010110	46	.0003169	−.0015720	0011206	000570031•	000471026	000345254	000224487	000045573•	000004336	000000004	1000100011110
1001000101110	46	.0003126	−.0015690	0011206	000568986	000470129	000344778	000224435	000045573•	000004404	000000004	1001011101110
1001000101011	44	.0003126	−.0015819	0010930	000567935	000468130	000342114	000221666	000044242	000004164	000000004	1001001101110
1001010100011	44	.0003126	−.0015477	0010846	000569508•	000469508	000348761	000229242	000048442	000005003	000000006	0010011101010
1010100011011	44	.0003081	−.0015477	0010846	000569508•	000472408	000348761	000210014	000040436	000003672	000000006	0101011101010
1010101010010	44	.0003081	−.0016116	0009497	000564452	000472938	000336312	000216460	000039651	000003548	000000003	1010101100101
0011011001001	43	.0003076	−.0016212	0009212	000609212	000461596	000334740	000215012	000041907	000003857	000000004	0101101001001
1010100011001	44	.0002984	−.0015889	0008153	000608153	000463132•	000338146	000219310	000044361	000004321	000000005	0010110011010
0011010011010	44	.0002984	−.0016252	0007390	000607390	000457832	000331222	000212253	000041111	000003754	000000005	1011101011010
1010010011010	44	.0002942	−.0015877	0007386	000607386	000461938	000337266	000218833	000044406	000004354	000000005	1011000101011
0010101110010	44	.0002942	−.0016034	0007047	000607047	000459440	000333900	000215290	000042628	000004014	000000003	1010100101011
0111001101010	44	.0002942	−.0015832	0006519	000606519	000455912	000329393	000220804	000040698	000003704	000000003	1010011110011
0111000000101	43	.0002937	−.0015775	0007499•	000607499•	000463112	000338814	000220392	000045071	000004456	000000005	1001111110001
0110101100001	44	.0002937	−.0016297	0006416	000606416	000455784	000339367	000210886	000040830	000003739	000000005	0111001011010
0101010011000	43	.0002937	−.0015477	0006320	000606320	000455120	000328505	000210014	000040436	000003672	000000006	0101011101010
0111001011001	43	.0002937	−.0016449	0006101	000606101	000453659	000326642	000208166	000039651	000003548	000000003	1110000111101
0110101010001	43	.0002937	−.0016479	0006039	000557288	000453237	000326100	000207623	000039416	000003510	000000003	0101101011001
0101001011001	43	.0002937	−.0016608	0005771	000556310	000451459	000323840	000205392	000038481	000003366	000000003	0101010011010
1010000011110	46	.0002840	−.0015482	0006290•	000561940	000464111	000341674	000224296	000047573	000004986	000000007	1000011111010
1000010111010	45	.0002840	−.0015876	0005451	000558752	000458047	000333559	000215796	000043325	000004166	000000004	1100010101011
0100011001101	42	.0002835	−.0015898	0005711•	000560066	000457569	000337643	000215287	000043118	000004132	000000004	0011110000101
0111110000011	42	.0002834	−.0015717	0005463	000559113	000455035	000335172	000217734	000044533	000004697	000000006	1101111001001
0011001011001	43	.0002834	−.0015832	0003417	000552983	000452983	000322234	000205003	000038988	000004432	000000005	0111100011100
0101001101100	43	.0002797	−.0016480	0006287	000560287	000448602	000339310	000222028	000046411	000004724	000000006	0101011100101
0100100111101	45	.0002792	−.0015477	0006564	000556564	000455111	000330457	000213072	000042326	000004011	000000004	1100001101101
1100100000111	44	.0002689	−.0014838	0004826•	000562374	000468686	000350308	000234974	000054211	000006484	000000014	0011111011100
0011110101100	44	.0002689	−.0015710	0003003	000555559	000455868	000333268	000217151	000045115	000004618	000000006	1100100100111
0011010001110	44	.0002689	−.0016323	0001722	000550769	000447019	000321706	000205337	000039587	000003623	000000003	1000111100100
0011011010010	44	.0002689	−.0016411	0001643	000550153	000445857	000320241	000201860	000038986	000003530	000000003	1000111100101
0100110101010	44	.0002689	−.0016557	0001235	000549000	000443709	000317443	000198097	000037710	000003316	000000002	1001101001101
0100110100110	43	.0002650	−.0016728	0000881	000547708	000441360	000314453	000198174•	000036828	000003215	000000003•	0111010011001
0110010011001	44	.0002608	−.0016723	0000170	000546661	000440472	000314042	000197259	000036653	000003205	000000002	0111010011100
1001010101001	44	.0002592	−.0016732	0009358	000545296	000438962	000312742	000195695	000035897	000003066	000000002	1010101010101
0101010100110	44	.0002592	−.0016757	0008994	000544479	000437698	000311193	000193550	000035013	000002931	000000002	1010110010110
0101001110110	44	.0002592	−.0016887	0008728	000543517	000435964	000309006	...	...	...	...	...
0010110110100	44	.0002550	−.0015636	0000524•	000551738	000462062	000330412	000215504	000045107	000004681	000000006	1101001000111
1001010100100	44	.0002550	−.0016755	0008205	000543198	000436341	000310088	000194972	000035802	000003068	000000002	0101100110110
0101001100110	44	.0002550	−.0016903	0007902	000542099	000433461	000307591	000192521	000034789	000002913	000000003	1000111100110
0010111100001	42	.0002505	−.0015954	0009027•	000546102	000446095	000322347	000208732	000042199	000004173	000000005	0111101010100
0101101001011	42	.0002505	−.0016222	0008500	000546102	000442916	000319478	000205061	000040796	000003975	000000005	0111101010100
0011101101001	42	.0002407	−.0016463	0007999	000544244	000439473	000314997	000200501	000038685	000003598	000000003	0111010100011
1100100010011	44	.0002403	−.0016477	0006138	000541095	000435936	000311868	000198202	000038145	000003545	000000004	0010110011011
0010110011100	44	.0002403	−.0016323	0007580•	000546412	000435546	000313204	000199243	000043304	000004399	000000006	1100110011011
...	44	.0002403	−.0016323	0006335	000541859	000437204	000313204	000199346	000038352	000003527	000000003	1100110001101
...	44	.0002402	−.0016612	0005753	000539780	000433498	000308632	000194829	000036495	000003240	...	0110111001010

160

m = 7, n = 6 — Rank Orders 701 thru 750 of 1716 C(7, 6) = 1716 1/C(7, 6) = .00058275

z	w	c_1	c_2	d = .2	.4	.6	.8	1.0	1.5	2.0	3.0	z^{tc}
0110010001110	44	.0002403	−.0016662	.00595652	.00539423	.00432866	.00307849	.000194073	.000036194	.000003196	.000000003	1000111011001
1000110011001	44	.0002403	−.0016778	.00595408	.00538513	.00431178	.00305660	.000191863	.000035213	.000003034	.000000002	1010011001110
1001100110011	44	.0002365	−.0016989	.00594979	.00542379	.00428379	.00302126	.000188390	.000033767	.000002810	.000000003	0011101110001
0111000100001	42	.0002365	−.0016154	.00596003	.00542183	.00438877	.00316248	.000202962	.000040497	.000003971	.000000005	0111001100011
0010101011011	42	.0002360	−.0016515	.00595265	.00539501	.00434001	.00311264	.000197717	.000037852	.000003462	.000000004	1100100100011
1000110101001	44	.0002360	−.0016368	.00595446	.00540206	.00435158	.00304291	.000190872	.000034998	.000003016	.000000003	1010010010110
0100101101001	44	.0002360	−.0016789	.00594591	.00537125	.00429620	.00300458	.000187106	.000033431	.000002783	.000000003	1010110010111
0001011111000	44	.0002263	−.0017018	.00594124	.00535439	.00426582	.00287106	.000183431	.000033223	.000002773	.000000002	1110110001011
1000010100111	44	.0002263	−.0015137	.00596130	.00546345	.00448633	.00330121	.000217718	.000047712	.000005310	.000000009	1011011000111
1100001010011	43	.0002263	−.0015971	.00594422	.00540093	.00437137	.00315195	.000202495	.000040488	.000003328	.000000005	0011011010110
0010011110100	44	.0002263	−.0016034	.00594305	.00539707	.00436493	.00314436	.000201791	.000040217	.000003915	.000000004	1101010010011
1000010111000	44	.0002263	−.0016377	.00593609	.00537196	.00431967	.00308704	.000196115	.000037769	.000003511	.000000004	1001110001110
0011001011100	46	.0002263	−.0016465	.00593419	.00536461	.00430552	.00306797	.000194112	.000036779	.000003328	.000000003	1101000110011
0100101010101	44	.0002263	−.0016648	.00593058	.00535203	.00428360	.00304117	.000191557	.000035785	.000003182	.000000003	1011000111110
1010010101000	44	.0002263	−.0016894	.00592546	.00533317	.00424901	.00299675	.000187115	.000033852	.000002868	.000000002	0101010101010
1010010100110	44	.0002263	−.0016963	.00592406	.00532817	.00424009	.00298566	.000186042	.000033427	.000002806	.000000003	1001110010111
0100101011010	44	.0002263	−.0016976	.00592379	.00532714	.00423816	.00298310	.000185776	.000033298	.000002783	.000000002	1001101110110
0011110011010	44	.0002263	−.0017150	.00592028	.00531457	.00421577	.00295522	.000183077	.000032223	.000002625	.000000002	1000101101110
0001111100001	43	.0002160	−.0016228	.00592005	.00535257	.00430891	.00308982	.000197457	.000039088	.000003817	.000000005	1011010101110
1001011000011	43	.0002160	−.0016566	.00591306	.00532688	.00426170	.00302887	.000191304	.000036293	.000003328	.000000002	0011110001110
0010101000110	43	.0002160	−.0016845	.00590732	.00530592	.00422347	.00297992	.000186408	.000034134	.000002965	.000000002	0110000111010
1000011011110	45	.0002156	−.0016706	.00590906	.00531305	.00423587	.00294439	.000187694	.000034515	.000002999	.000000006	1001001111010
1000011110010	46	.0002118	−.0015671	.00592324	.00537854	.00436379	.00316413	.000205099	.000042596	.000004436	.000000002	1001000111110
0011100100110	43	.0002118	−.0016640	.00590360	.00530835	.00423772	.00300453	.000189258	.000035625	.000003236	.000000002	1001101100010
0101100100110	42	.0002113	−.0016885	.00589856	.00528993	.00420411	.00296149	.000184954	.000033728	.000002918	.000000002	1011011001010
0101010010010	44	.0002113	−.0017082	.00589371	.00527454	.00417805	.00292998	.000181962	.000032565	.000002747	.000000002	0110010100101
1000110011010	45	.0002016	−.0016767	.00588159	.00526534	.00418089	.00294463	.000183964	.000033612	.000002911	.000000002	1001110001101
1001000110110	45	.0002016	−.0016883	.00587919	.00525659	.00416496	.00292429	.000181940	.000032740	.000002771	.000000002	1001110001101
0011010101010	44	.0002016	−.0017036	.00587645	.00524651	.00414651	.00290065	.000179582	.000031723	.000002608	.000000002	1001101010110
0101011000011	43	.0001971	−.0016957	.00586954	.00523902	.00414468	.00290673	.000180750	.000032614	.000002791	.000000002	0110110011011
1010000011001	44	.0001934	−.0016890	.00586387	.00523203	.00413996	.00290558	.000180910	.000032822	.000002833	.000000002	0110001111010
1001000011100	44	.0001934	−.0017020	.00586125	.00522266	.00412324	.00288465	.000178866	.000031977	.000002702	.000000002	1001001110001
1011000010101	45	.0001876	−.0016485	.00585043	.00520414	.00410138	.00286356	.000177120	.000031368	.000002607	.000000002	1001100111010
1010110010011	43	.0001831	−.0016244	.00585760	.00524679	.00418888	.00298156	.000189261	.000036903	.000003548	.000000004	0011011110010
1010010010011	43	.0001831	−.0016761	.00584727	.00521005	.00412345	.00289962	.000181231	.000033508	.000002995	.000000003	0011011110010
0110110011001	43	.0001831	−.0016870	.00584509	.00520234	.00410986	.00288280	.000179607	.000032855	.000002895	.000000003	1001011010110
0011011100110	43	.0001831	−.0017011	.00584215	.00519145	.00409145	.00285665	.000176955	.000031646	.000002686	.000000002	1001101010110
0011010100110	43	.0001831	−.0017086	.00584068	.00518634	.00408083	.00284584	.000175932	.000031260	.000002632	.000000002	1001110001101
0111010010011	43	.0001831	−.0017271	.00583697	.00517319	.00405760	.00281712	.000173166	.000030164	.000002470	.000000002	1010101010110
0101011010010	42	.0001826	−.0017096	.00583960	.00518453	.00407911	.00284490	.000175931	.000031331	.000002652	.000000002	1110010101011
0110110000101	42	.0001826	−.0017271	.00583611	.00517208	.00405702	.00281746	.000173275	.000030264	.000002491	.000000002	0101111001001
0101011001100	42	.0001826	−.0017404	.00583345	.00516271	.00404058	.00279728	.000171348	.000029519	.000002384	.000000002	0110110010011
0101100100101	42	.0001784	−.0017331	.00582693	.00515466	.00403521	.00279609	.000171545	.000029757	.000002430	.000000002	0110010110100
0101101001010	44	.0001784	−.0017331	.00582459	.00514639	.00402069	.00277826	.000169843	.000029101	.000002336	.000000002	0110100101001
1000010011010	45	.0001729	−.0016964	.00582375	.00516175	.00405744	.00282857	.000174885	.000031108	.000002620	.000000004	1000101111010
0101100101100	42	.0001691	−.0017121	.00581368	.00514010	.00402816	.00279851	.000172392	.000030383	.000002539	.000000003	1001011011010
1010010001110	45	.0001687	−.0017035	.00581928	.00516104	.00406521	.00284392	.000176710	.000032036	.000002776	.000000002	1000101101110
1000100010110	45	.0001687	−.0017018	.00581432	.00514322	.00403331	.00280388	.000172792	.000030410	.000002523	.000000002	1010011010110
0111010001010	42	.0001686	−.0017031	.00581452	.00514475	.00403729	.00281034	.000173560	.000030861	.000002612	.000000002	0101011100011
0101010110010	42	.0001686	−.0017424	.00580677	.00511771	.00399026	.00275306	.000168126	.000028787	.000002316	.000000002	1110010101001

Table A

m = 7, n = 6 Rank Orders 751 thru 800 of 1716 C(7, 6) = 1716 1/C(7, 6) = .00058275

z	w	c_1	c_2	d = .2	.4	.6	.8	1.0	1.5	2.0	3.0	z^{tc}
1100001001101	44	.0001682	−.0016551	.000582301	.000517566	.000409161	.000287678	.000179885	.000333306	.000002968	.000000003	01001101011100
1100011101100	44	.0001682	−.0016847	.000581725	.000515574	.000405717	.000283495	.000175916	.000331773	.000002743	.000000002	11001000011101
0100101011100	44	.0001579	−.0017058	.000581292	.000513998	.000402853	.000279846	.000172290	.000330213	.000002492	.000000002	11001100001101
0001111001100	43	.0001579	−.0016405	.000580696	.000515641	.000408141	.000288065	.000181369	.000344721	.000003294	.000000004	11001100001110
1000111000101	43	.0001579	−.0017240	.000579013	.000509615	.000397348	.000274480	.000168001	.000329032	.000002364	.000000001	10011100000111
0100001100110	43	.0001542	−.0016631	.000578663	.000508384	.000395188	.000271825	.000165458	.000328036	.000002219	.000000002	10010011111100
1000000110101	44	.0001542	−.0016994	.000578796	.000512602	.000403279	.000282171	.000175588	.000332121	.000002829	.000000002	01010011111100
0101000011100	44	.0001542	−.0017459	.000579506	.000510127	.000398971	.000276901	.000170550	.000330137	.000002532	.000000002	11000011111001
0101001011001	42	.0001497	−.0017459	.000577057	.000505658	.000392195	.000269318	.000163785	.000327822	.000002229	.000000002	01010110110010
0110100010001	42	.0001497	−.0017606	.000576765	.000504637	.000390415	.000267149	.000161728	.000327043	.000002119	.000000001	11010101011001
1100100000111	43	.0001439	−.0015460	.000579946	.000517977	.000415384	.000299254	.000193801	.000341243	.000004594	.000000004	01011111110100
0001111011000	43	.0001439	−.0016460	.000577947	.000510827	.000402502	.000282826	.000177297	.000333584	.000003153	.000000002	10100101000011
0101110101010	43	.0001439	−.0017461	.000575943	.000503697	.000389821	.000266987	.000161836	.000327151	.000002129	.000000001	01011111100110
1001010101010	43	.0001439	−.0017578	.000575715	.000502899	.000388434	.000265300	.000160238	.000326545	.000002044	.000000001	10010101010110
1001010101001	43	.0001439	−.0017595	.000575681	.000502782	.000388234	.000265064	.000160025	.000326478	.000002037	.000000001	01010101010110
1001011000011	43	.0001439	−.0017739	.000575397	.000501794	.000386522	.000262989	.000158068	.000325748	.000001936	.000000001	00111111010010
1001011000011	41	.0001395	−.0016702	.000576637	.000507807	.000393950	.000278078	.000173087	.000332091	.000002938	.000000003	10011111010001
0011111010000	41	.0001395	−.0016935	.000576167	.000506124	.000395132	.000274259	.000169310	.000330453	.000002662	.000000002	10101010110101
0111001010001	42	.0001357	−.0017401	.000574540	.000501677	.000380651	.000265961	.000165551	.000327455	.000002210	.000000001	01110101011001
0111001000011	41	.0001255	−.0016744	.000573918	.000503103	.000392717	.000273099	.000169277	.000331074	.000002819	.000000001	11001101011010
0011011100001	41	.0001255	−.0016891	.000573614	.000501987	.000390673	.000270466	.000166627	.000329876	.000002610	.000000002	01111100010011
1001010100100	41	.0001255	−.0017116	.000573184	.000500531	.000388203	.000267525	.000163888	.000328862	.000002466	.000000003	01011010100011
1100100000111	43	.0001250	−.0016369	.000574542	.000505305	.000396604	.000277765	.000173642	.000332699	.000003049	.000000003	00101111100011
0101010100100	43	.0001250	−.0017123	.000573053	.000500117	.000387375	.000266323	.000162535	.000328110	.000002318	.000000002	10011111001011
1010100100100	43	.0001250	−.0017253	.000572738	.000499285	.000385975	.000264673	.000161018	.000327574	.000002247	.000000002	00111111000011
1010101000011	43	.0001250	−.0017289	.000572213	.000499051	.000385580	.000264205	.000160588	.000327422	.000002227	.000000001	10011111001101
1001101010001	41	.0001250	−.0017552	.000572213	.000497202	.000383231	.000260210	.000156762	.000325929	.000002010	.000000001	10101010110110
0011101010010	41	.0001153	−.0017776	.000571772	.000495666	.000379667	.000256975	.000153705	.000324776	.000001848	.000000001	10101011011010
0111001001001	42	.0001153	−.0016706	.000572035	.000499900	.000388944	.000269454	.000166244	.000329984	.000002638	.000000002	01110101011001
0011101000011	43	.0001153	−.0017184	.000573918	.000496638	.000383255	.000262476	.000159548	.000327304	.000002227	.000000001	11001100111010
1010010010101	43	.0001153	−.0017626	.000570234	.000493637	.000378058	.000256169	.000153584	.000025041	.000001905	.000000001	01010110110010
1001010011001	43	.0001153	−.0017691	.000570107	.000493198	.000377306	.000255271	.000152750	.000024744	.000001866	.000000001	01011011001011
0101010101010	43	.0001153	−.0017873	.000569751	.000491966	.000375187	.000252720	.000150361	.000023867	.000001852	.000000001	10101010101011
0011101111000	43	.0001110	−.0016063	.000572494	.000498878	.000395184	.000277718	.000174592	.000033660	.000001747	.000000001	10011110100011
1100000110100	43	.0001110	−.0016661	.000571325	.000498878	.000384260	.000268842	.000161468	.000028185	.000003257	.000000004	00111100101011
0011101010100	43	.0001110	−.0016972	.000570714	.000496739	.000381541	.000260898	.000158339	.000028185	.000002364	.000000002	10101010101011
1000101110100	43	.0001110	−.0017204	.000570260	.000495169	.000379914	.000259918	.000156460	.000026986	.000002191	.000000001	11011101011010
0011001110001	43	.0001110	−.0017342	.000569991	.000494231	.000376962	.000255354	.000153107	.000026264	.000002087	.000000001	00111100101100
0011110100101	43	.0001110	−.0017596	.000569495	.000492518	.000376962	.000255354	.000153107	.000025010	.000001911	.000000001	10110111011010
0101001011010	43	.0001110	−.0017598	.000569490	.000492494	.000376910	.000255276	.000153018	.000024958	.000001901	.000000001	01110010101101
1001010101010	43	.0001110	−.0017677	.000569337	.000491974	.000376027	.000254234	.000152065	.000024635	.000001862	.000000001	01010110110101
0101010101010	43	.0001008	−.0017692	.000569307	.000491865	.000375834	.000253990	.000151823	.000024531	.000001845	.000000001	01110011011101
1001010101010	43	.0001008	−.0017874	.000568945	.000490618	.000373694	.000251424	.000149432	.000023666	.000001730	.000000004	10010011011110
1100000111110	45	.0001008	−.0016172	.000570353	.000498966	.000390202	.000272808	.000170554	.000032382	.000003083	.000000003	10000111111100
0001111100100	45	.0001008	−.0016681	.000571325	.000495401	.000383864	.000264854	.000162717	.000028977	.000002502	.000000001	10111101011010
0011111000011	42	.0001008	−.0016546	.000569634	.000495565	.000386148	.000267767	.000166009	.000030623	.000002813	.000000003	11110010101100
1001110000011	42	.0001008	−.0017121	.000568511	.000492657	.000379338	.000259613	.000157983	.000027378	.000002302	.000000001	00111110100110
0011110100101	42	.0001008	−.0017374	.000568009	.000490891	.000376629	.000255354	.000154277	.000025872	.000002068	.000000001	11110100111010

162

z	w	c_1	c_2	d = .2	.4	.6	.8	1.0	1.5	2.0	3.0	z^{tc}
1001000001110	44	.0000906	−.0017502	.000565818	.000486678	.000370810	.000250263	.000149616	.000024331	.000001857	.00000001	1000111011010
1000110001100	44	.0000906	−.0017631	.000565561	.000485768	.000369205	.000248281	.000147712	.000023584	.000001749	.00000001	1001100110111
1000101100010	44	.0000863	−.0017621	.000564782	.000484504	.000367850	.000247145	.000146925	.000023431	.000001737	.00000001	1001101011011
1000011111000	43	.0000824	−.0016317	.000565570●	.000381600	.000366338	.000245902	.000163512	.000030022	.000002726	.00000003	1110010111010
1010000011010	43	.0000824	−.0017661	.000563719	.000483123	.000364799	.000245902	.000146129	.000022364	.000001750	.00000001	0110010111010
1010000101100	43	.0000824	−.0017795	.000563719	.000482225	.000363073	.000244057	.000144405	.000022732	.000001664	.00000001	1010011011010
0110010011010	43	.0000766	−.0017743	.000562711	.000480633	.000363073	.000242528	.000143238	.000022391	.000001618	.00000002	1001010110010
1010010001010	42	.0000721	−.0017239	.000562882	.000482838	.000367923	.000249129	.000149956	.000025258	.000002060	.00000001	1010101110010
1010011000011	42	.0000721	−.0017396	.000562569	.000481726	.000366664	.000248236	.000147551	.000024257	.000001901	.00000001	1011010001011
0010111001010	42	.0000721	−.0017651	.000562077	.000480049	.000363097	.000243270	.000144403	.000023121	.000001748	.00000001	1011010010011
1010100100011	42	.0000678	−.0017301	.000611964	.000481096●	.000365746	.000246998	.000148231	.000024747	.000001997	.00000002	0011101101010
0011100101010	42	.0000678	−.0017664	.000561255	.000478642	.000361499	.000241844	.000143349	.000022872	.000001724	.00000001	1010101100011
0111010101010	42	.0000674	−.0017279	.000561898	.000480987	.000365567	.000246733	.000147911	.000024540	.000001952	.00000001	0111100001101
0110101000101	41	.0000674	−.0017702	.000561095	.000478279	.000361011	.000242968	.000142968	.000022781	.000001716	.00000001	0110110110010
1011000010100	41	.0000581	−.0017823	.000560862	.000477486	.000359666	.000239766	.000141497	.000021497	.000001647	.00000001	0011010110010
1011000010011	42	.0000581	−.0017003	.000560691	.000479916●	.000365498	.000247731	.000149476	.000025471	.000002116	.00000002	1010011110010
1010101100010	42	.0000642	−.0017642	.000559462	.000475718	.000358329	.000239135	.000141423	.000022452	.000001685	.00000001	1010101110011
0110110011010	42	.0000577	−.0017449	.000559728●	.000476701	.000359966	.000240995	.000143047	.000022937	.000001736	.00000001	1010011011010
1010011101010	44	.0000577	−.0017465	.000559700	.000476624	.000359865	.000240911	.000143004	.000022956●	.000001743	.00000001	1011011101010
1001011101010	44	.0000577	−.0017702	.000559233	.000474993	.000357024	.000237448	.000139716	.000021702	.000001565	.00000001	0101010111100
0101010101100	41	.0000576	−.0017763	.000559143	.000474816	.000356932	.000237579●	.000140059	.000022018	.000001633	.00000001	0110101010101
1100010001101	43	.0000572	−.0017063	.000560373●	.000479058	.000364069	.000245950	.000141700	.000024672	.000001979	.00000001	1001101011100
0110110011100	43	.0000572	−.0017554	.000559431	.000475850	.000358608	.000239427	.000141617	.000022428	.000001667	.00000001	1101011011000
0110100010101	41	.0000534	−.0017538	.000558769	.000474903	.000357819	.000239045	.000141626●	.000022663	.000001722	.00000001	0110101110001
0101010100001	41	.0000534	−.0017784	.000558305	.000473353	.000355237	.000236035	.000138891	.000021726	.000001602	.00000001	1100100101101
1101000100010	43	.0000529	−.0017566	.000558609●	.000474441	.000357004	.000237988	.000140545	.000022168	.000001640	.00000001	1010010110101
0100001110010	44	.0000437	−.0017499	.000556997	.000471972	.000354520	.000234542	.000139314	.000021998	.000001636●	.00000001	1010001110110
1010000110110	44	.0000437	−.0017617	.000557768	.000471214	.000354242	.000234642	.000137929	.000021512	.000001573	.00000001	1010100111010
0100011110110	43	.0000432	−.0017157	.000557568●	.000474112	.000358305	.000240685	.000143720	.000023683	.000001877	.00000001	1011000111100
1100010101010	43	.0000432	−.0017249	.000573370	.000473368	.000356920	.000238893	.000141922	.000022906	.000001754	.00000001	0101011111100
0101001011100	43	.0000432	−.0017624	.000556663	.000470997	.000352954	.000234238	.000137663	.000021415	.000001558	.00000001	1100010110101
0110100100100	41	.0000387	−.0017868	.000555385	.000469226	.000349274	.000230588	.000134769	.000020695	.000001495	.00000001	0110110101001
0110100101001	41	.0000345	−.0017902	.000554524	.000468226	.000347457	.000228908	.000133485	.000020368	.000001461	.00000001	1101011000111
0011110010100	42	.0000329	−.0016699	.000556529●	.000474000	.000360159	.000244229	.000147867	.000025770	.000002239	.00000002	1101010000101
1001011001010	42	.0000329	−.0017818	.000554372	.000466608	.000347494	.000228984	.000133525	.000020320	.000001444	.00000001	0101110110010
1001001000110	42	.0000329	−.0017966	.000554091	.000465664	.000345914	.000227136	.000131841	.000019742	.000001371	.00000001	0101101110110
1001100100110	42	.0000287	−.0017859	.000553498	.000465025	.000345605	.000227222●	.000132167	.000019970	.000001407	.00000001	0110110100010
1011001001001	42	.0000287	−.0017989	.000553251	.000464197	.000344221	.000225604	.000130694	.000019466	.000001343	.00000001	0110011110010
0011101001001	41	.0000287	−.0017540	.000553369●	.000465884	.000347755	.000230281	.000137640	.000021201	.000001581	.00000001	0110111001001
1110000000111	41	.0000247	−.0017816	.000552852	.000464169	.000344965	.000226991	.000132279	.000020197	.000001454	.00000001	0101101101001
0111010000011	40	.0000144	−.0016767	.000552917●	.000467712	.000352854	.000237514	.000142713	.000022713	.000002077	.00000002	0011111010001
1011100001100	40	.0000144	−.0017128	.000552229	.000465380	.000348888	.000232762	.000138249	.000021555	.000001819	.00000001	0111010000011
0011011001001	42	.0000140	−.0017284	.000551824	.000464081	.000346642	.000229986	.000135549	.000021555	.000001642	.00000001	1101100000011
1001011001001	42	.0000140	−.0017706	.000551029	.000461419	.000339757	.000224778	.000130795	.000019899	.000001425	.00000001	1011010001011
0110011100101	43	.0000103	−.0017934	.000550596	.000459963	.000339757	.000219188	.000128181	.000018991	.000001308	.00000001	1100011111100
1100011110111	43	.0000103	−.0017033	.000551587●	.000464427	.000347826	.000231707	.000137296	.000021306	.000001738	.00000001	1000111110100
0011000111000	43	.0000103	−.0017278	.000551131	.000462915	.000344319	.000228784	.000134631	.000021306	.000001612	.00000001	1000011111000
0011101100001	43	.0000103	−.0017090	.000551499●	.000464266	.000347755	.000231852	.000137640	.000022561	.000001810	.00000001	0111101000011
1110000000111	42	.0000102	−.0017090	.000553062●	.000472794	.000364123	.000252711	.000158113	.000031214	.000003270	.00000006	0111111111000
0001110111000	42	.0000000	−.0016354	.000550944	.000465582	.000351716	.000237556	.000143489	.000025009	.000002187	.00000002	1110010000111

Table A

m = 7, n = 6   Rank Orders 851 thru 900 of 1716   C(7, 6) = 1716   1/C(7, 6) = .00058275

z	w	c_1	c_2	d = .2	.4	.6	.8	1.0	1.5	2.0	3.0	z^{tc}
1101000001011	42	.0000000	−.0016385	.000550887	.000465384	.000351370	.000237124	.000143061	.000024813	.000002150	.000000002	0010111110100
1100100100111	42	.0000000	−.0016878	.000549956	.000462247	.000346066	.000230802	.000137152	.000022574	.000001819	.000000001	0011011101100
1011100100101	42	.0000000	−.0016975	.000549772	.000461631	.000345032	.000229582	.000136026	.000022166	.000001762	.000000001	0011101011100
1011100010011	42	.0000000	−.0017245	.000549262	.000459916	.000342135	.000226138	.000132817	.000020964	.000001586	.000000001	0011101110010
0111000001110	42	.0000000	−.0017253	.000549247	.000459865	.000342054	.000226046	.000132737	.000020941	.000001588	.000000001	1000111110001
0011011101100	42	.0000000	−.0017267	.000549220	.000459775	.000341899	.000225858	.000132558	.000020869	.000001575	.000000001	1001011110011
0011011011100	42	.0000000	−.0017306	.000549148	.000459533	.000341497	.000225392	.000132137	.000020728	.000001557	.000000001	1100011100011
0011010111100	42	.0000000	−.0017503	.000548776	.000458290	.000339421	.000221915	.000129915	.000019950	.000001455	.000000001	1100110001110
1001010110001	42	.0000000	−.0017645	.000548509	.000457397	.000337934	.000221224	.000128334	.000019404	.000001384	.000000001	0111001001110
1101000010101	42	.0000000	−.0017785	.000548245	.000456514	.000336461	.000219502	.000126764	.000018860	.000001313	.000000001	0101011101010
1101100011001	42	.0000000	−.0017838	.000548144	.000456180	.000335910	.000218867	.000126195	.000018675	.000001291	.000000001	0110011100110
0110100001110	42	.0000000	−.0017849	.000548125	.000456114	.000335796	.000218728	.000126063	.000018622	.000001283	.000000001	1001011001101
1001011000110	42	.0000000	−.0017897	.000548034	.000455812	.000335295	.000218145	.000125535	.000018443	.000001283	.000000001	1011000101101
1010101010011	42	.0000000	−.0017907	.000548016	.000455754	.000335202	.000218045	.000125453	.000018425	.000001260	.000000001	0110101010010
1011010101001	42	.0000000	−.0017984	.000547871	.000455272	.000334408	.000217130	.000124635	.000018160	.000001229	.000000001	0111010101010
1001010101110	42	.0000000	−.0017998	.000547843	.000455178	.000334248	.000216938	.000124454	.000018090	.000001219	.000000001	1001101011001
0101010001110	42	.0000000	−.0018007	.000547827	.000455124	.000334156	.000216830	.000124354	.000018054	.000001214	.000000001	0101011001110
1010101010101	42	.0000000	−.0018180	.000547502	.000454042	.000332371	.000214772	.000122512	.000017455	.000001142	.000000000	1101010101010
1000110111001	44	−.0000102	−.0017090	.000547611	.000455124	.000339838	.000224368	.000131661	.000020733	.000001563	.000000001	1001101110010
1011100000011	41	−.0000103	−.0017033	.000547738	.000458095	.000340975	.000225938	.000133319	.000021524	.000001700	.000000001	0011111000010
0011110000110	41	−.0000103	−.0017278	.000547269	.000456498	.000338252	.000222673	.000130254	.000020361	.000001530	.000000001	1001111000011
0110111010100	42	−.0000140	−.0017284	.000546552	.000455252	.000336793	.000221309	.000129183	.000020051	.000001491	.000000001	1101100110011
1001011100010	42	−.0000140	−.0017706	.000545760	.000452612	.000332401	.000216188	.000124526	.000018447	.000001283	.000000000	0111010110010
1010101000110	44	−.0000140	−.0017934	.000545333	.000451195	.000330061	.000213480	.000122086	.000017629	.000001181	.000000000	0110010110101
1100100101110	44	−.0000144	−.0017767	.000547412	.000458186	.000341646	.000226907	.000134216	.000021747	.000001708	.000000001	1000101111010
1100010011100	44	−.0000144	−.0017128	.000546735	.000455931	.000337881	.000222484	.000130152	.000020292	.000001509	.000000001	1100001101110
0101010101110	42	−.0000247	−.0017540	.000544044	.000450229	.000330262	.000214734	.000123755	.000018416	.000001292	.000000001	1001101110010
1000110100110	43	−.0000247	−.0017816	.000543522	.000448480	.000327340	.000211318	.000120647	.000017351	.000001157	.000000000	1001110011100
1001010011010	43	−.0000287	−.0017859	.000542715	.000447083	.000325795	.000210022	.000119780	.000017228	.000001156	.000000000	1011001011010
0110100110011	42	−.0000287	−.0017989	.000542472	.000446279	.000324471	.000208496	.000118412	.000016777	.000001101	.000000000	1010011001010
0011010111000	42	−.0000329	−.0016699	.000544056	.000452766	.000335810	.000221945	.000130704	.000020993	.000001639	.000000001	1110000101011
1010001101001	42	−.0000329	−.0017818	.000541994	.000446012	.000324756	.000209246	.000119313	.000017177	.000001156	.000000000	0110100111010
1001011001010	43	−.0000345	−.0017966	.000541717	.000445093	.000323242	.000207499	.000117744	.000016657	.000001092	.000000000	1011010011001
1001010011100	43	−.0000387	−.0017902	.000540792	.000444883	.000323062	.000207336	.000117584	.000016566	.000001075	.000000000	1001100101110
0011110001100	41	−.0000432	−.0017868	.000541287	.000446826	.000327818	.000213768	.000123912	.000018955	.000001394	.000000000	1011010010110
1010100011001	41	−.0000432	−.0017249	.000541138	.000446406	.000327250	.000213258	.000123589	.000018970	.000001414	.000000000	0011110101010
0111011010001	41	−.0000432	−.0017624	.000540433	.000444053	.000323322	.000209908	.000119377	.000017481	.000001214	.000000000	0011110011010
0110110000101	40	−.0000437	−.0017499	.000540570	.000444649	.000324376	.000209908	.000120518	.000017866	.000001262	.000000000	0101110011001
0101011001001	40	−.0000437	−.0017617	.000540351	.000443930	.000323195	.000208549	.000119300	.000017462	.000001212	.000000000	0111001010001
1010110101010	41	−.0000529	−.0017566	.000538704	.000441331	.000320472	.000206302	.000117759	.000017157	.000001186	.000000001	1010010010011
1010010011001	43	−.0000534	−.0017538	.000538652	.000441237	.000320296	.000206062	.000117443	.000016984	.000001156	.000000000	1000111011010
1000110011010	43	−.0000534	−.0017784	.000538189	.000439697	.000317741	.000203067	.000114764	.000016082	.000001044	.000000000	0010101001001
0011100110011	41	−.0000572	−.0017054	.000538839	.000443076	.000324067	.000210883	.000122134	.000018772	.000001406	.000000000	1010101110010
1011100110001	43	−.0000576	−.0017763	.000537928	.000440074	.000319120	.000205158	.000116955	.000016989	.000001172	.000000000	1011001011010
1001101010010	43	−.0000577	−.0017463	.000537429	.000438494	.000319479	.000206194	.000114053	.000015945	.000001033	.000000000	0010101011110
0110110000011	40	−.0000577	−.0018029	.000538029	.000440425	.000319753	.000206196	.000117900	.000017313	.000001212	.000000000	0111011100001
0110110000101	40	−.0000577	−.0017465	.000537994	.000440425	.000319753	.000205906	.000117628	.000017207	.000001197	.000000000	0111100101001
0101011100001	40	−.0000577	−.0017702	.000537567	.000439067	.000317594	.000203502	.000115541	.000016568	.000001124	.000000000	0111010100010

164

z	w	c_1	c_2	d = .2	.4	.6	.8	1.0	1.5	2.0	3.0	z^{tc}
010011010100	42	−.0000581	−.0017642	.000537566	.000439113	.000317602	.000203391	.000115320	.000016388	.000001089	.000000000	110101001101
100011110010	43	−.0000674	−.0017279	.000536497	.000438414	.000317997	.000204731	.000117001	.000017160	.000001198	.000000001	101100001110
101000101010	43	−.0000674	−.0017702	.000535707	.000438800	.000336800	.000199737	.000112498	.000015646	.000001009	.000000000	100101011010
100100011110	43	−.0000674	−.0017823	.000535484	.000435073	.000312502	.000198404	.000111325	.000015280	.000000967	.000000000	101101010101
101010001001	43	−.0000678	−.0017301	.000536350	.000438041	.000317403	.000204023	.000116326	.000016894	.000001159	.000000001	110001101101
110010010101	42	−.0000678	−.0017664	.000535690	.000435914	.000313979	.000200162	.000112932	.000015822	.000001033	.000000000	101100111100
110001001100	42	−.0000721	−.0017239	.000535666	.000437086	.000315141	.000203441	.000116025	.000016893	.000001165	.000000001	101100110101
101001110100	42	−.0000721	−.0017396	.000535387	.000436201	.000312642	.000201883	.000114670	.000016472	.000001116	.000000000	011011011001
010101001001	42	−.0000721	−.0017651	.000534916	.000434664	.000310677	.000199037	.000112149	.000015663	.000001020	.000000001	011011011010
011010010001	40	−.0000766	−.0017743	.000533935	.000432909	.000305338	.000197398	.000111076	.000015541	.000001026	.000000000	110101011100
000111100010	41	−.0000824	−.0016317	.000535457	.000439607	.000322560	.000211590	.000124129	.000020131	.000001642	.000000001	110110000111
100110100101	41	−.0000824	−.0017661	.000532978	.000431465	.000309176	.000196116	.000110128	.000015281	.000000993	.000000000	011101101010
011110101001	41	−.0000824	−.0017795	.000532732	.000430671	.000307897	.000194672	.000108860	.000014882	.000000946	.000000000	100101001101
011110010001	40	−.0000863	−.0017621	.000532323	.000430567	.000308441	.000195738	.000110069	.000015410	.000001021	.000000000	011101011001
011010010101	40	−.0000906	−.0017502	.000531736	.000429901	.000308008	.000195595	.000110115	.000015487	.000001033	.000000000	011010111001
011101000101	40	−.0000906	−.0017631	.000531508	.000429186	.000306889	.000194369	.000109069	.000015180	.000000999	.000000001	011101101001
101000111010	43	−.0001003	−.0017408	.000530064	.000427294	.000313160	.000193236	.000108400	.000015060	.000000984	.000000000	110100111010
010001111010	42	−.0001008	−.0016546	.000531530	.000432160	.000313160	.000202230	.000116375	.000017654	.000001304	.000000000	111000111100
110000101001	42	−.0001008	−.0017121	.000530482	.000428762	.000307645	.000195938	.000110761	.000015788	.000001068	.000000000	011100101100
011000111100	42	−.0001008	−.0017374	.000530028	.000427314	.000305338	.000193360	.000108512	.000015087	.000000986	.000000000	110101011010
011110000011	39	−.0001008	−.0016172	.000532224	.000434526	.000317189	.000207061	.000120918	.000019405	.000001568	.000000001	001111100001
001111010000	39	−.0001008	−.0016681	.000531306	.000431569	.000301575	.000201575	.000115393	.000017689	.000001332	.000000000	011110100001
001111110011	41	−.0001110	−.0016063	.000530495	.000431919	.000314565	.000204973	.000113460	.000019147	.000001547	.000000001	111001000001
110001010100	41	−.0001110	−.0016661	.000529405	.000428369	.000308759	.000198272	.000110384	.000017008	.000001251	.000000000	001110001111
001011010100	41	−.0001110	−.0016972	.000528839	.000426534	.000305773	.000194850	.000108196	.000015953	.000001110	.000000000	111010100011
100011100001	41	−.0001110	−.0017204	.000528417	.000425172	.000303579	.000192369	.000106923	.000015251	.000001025	.000000000	011101001011
101000110001	41	−.0001110	−.0017342	.000528168	.000424370	.000302292	.000190920	.000104556	.000014849	.000000977	.000000000	011011001110
101101101001	41	−.0001110	−.0017596	.000527705	.000422880	.000299896	.000188222	.000104536	.000014104	.000000889	.000000000	010111001010
011001110001	42	−.0001110	−.0017598	.000527702	.000422870	.000299880	.000188203	.000103838	.000014094	.000000887	.000000000	011001100110
100100110001	41	−.0001110	−.0017677	.000527559	.000422411	.000299149	.000187392	.000103838	.000013890	.000000866	.000000000	110010110010
010011000110	41	−.0001110	−.0017692	.000527531	.000422322	.000299004	.000187225	.000103688	.000013838	.000000859	.000000000	100111001001
010010100110	41	−.0001110	−.0017878	.000527195	.000421243	.000297281	.000185304	.000102023	.000013336	.000000803	.000000000	101011001011
110010100001	41	−.0001153	−.0016706	.000528524	.000428771	.000306848	.000196484	.000112074	.000016639	.000001209	.000000000	011110101010
001110100100	41	−.0001153	−.0017184	.000527655	.000423959	.000302294	.000191297	.000107452	.000015100	.000001012	.000000000	100101010101
100100101010	41	−.0001153	−.0017626	.000526855	.000421390	.000298178	.000186676	.000103408	.000013836	.000000864	.000000000	011010110101
100011011001	41	−.0001153	−.0017691	.000526737	.000421011	.000297575	.000186006	.000102830	.000013665	.000000845	.000000001	011101100110
100101011010	41	−.0001153	−.0017703	.000526715	.000420939	.000297457	.000185870	.000102708	.000013622	.000000840	.000000000	100101001100
010110010100	41	−.0001153	−.0017873	.000526406	.000419951	.000295881	.000184114	.000101186	.000013164	.000000789	.000000000	101010100101
010100010001	42	−.0001250	−.0016369	.000527291	.000425616	.000306482	.000196903	.000112882	.000017091	.000001278	.000000000	001101110100
001101010100	41	−.0001250	−.0017123	.000525931	.000421252	.000299463	.000188960	.000105845	.000014777	.000000960	.000000000	110101010110
011000010101	41	−.0001250	−.0017253	.000525689	.000420448	.000298134	.000187420	.000104454	.000014306	.000000924	.000000000	010101111001
011100010110	41	−.0001250	−.0017289	.000525622	.000420230	.000297779	.000187014	.000104092	.000014188	.000000910	.000000000	100101011010
100101011010	43	−.0001250	−.0017552	.000525154	.000418755	.000295465	.000184476	.000101927	.000013558	.000000842	.000000000	011100101010
010101001010	41	−.0001255	−.0017776	.000524750	.000417466	.000293417	.000182198	.000099955	.000012962	.000000776	.000000000	110001010101
100011110110	43	−.0001255	−.0016744	.000526501	.000423120	.000302386	.000192130	.000108515	.000015529	.000001063	.000000001	100101011011
100011010101	43	−.0001255	−.0016891	.000525828	.000420959	.000301200	.000190863	.000107460	.000015240	.000001034	.000000000	110001010110
100111001010	43	−.0001250	−.0017116	.000525247	.000422343	.000300130	.000188915	.000105068	.000014427	.000000930	.000000000	101010110100
110000011011	42	−.0001357	−.0017401	.000523396	.000416189	.000292956	.000182521	.000100641	.000013310	.000000819	.000000000	100111100100
110000110110	43	−.0001395	−.0016702	.000523939	.000418946	.000297848	.000188230	.000105720	.000014905	.000001003	.000000000	100101111100
100100111100	43	−.0001395	−.0016935	.000523529	.000417658	.000295828	.000186006	.000103808	.000014329	.000000938	.000000000	110001110110

165

z	w	c₁	c₂	d = .2	.4	.6	.8	1.0	1.5	2.0	3.0	z^{tc}
1110000001011	41	−.0001439	−.0015460	.000525332	.000424730	.000307913	.000200132	.000116614	.000018742	.000001525	.000000001	0010111111000
0010110111000	41	−.0001439	−.0016460	.000523541	.000418993	.000298665	.000189600	.000107187	.000015510	.000001088	.000000001	1110000100101
1010100011001	41	−.0001439	−.0017461	.000521764	.000413390	.000289839	.000179840	.000098762	.000012947	.000000794	.000000000	0110011101010
0110100101010	41	−.0001439	−.0017578	.000521556	.000412725	.000288783	.000178668	.000097747	.000012641	.000000759	.000000000	1010001101001
1010010101001	41	−.0001439	−.0017595	.000521525	.000412631	.000288638	.000178514	.000097623	.000012612	.000000757	.000000000	1010101011010
0101010101010	41	−.0001439	−.0017739	.000521266	.000411805	.000287330	.000177065	.000096375	.000012239	.000000716	.000000000	1001010111001
1001100100110	42	−.0001497	−.0017459	.000520662	.000411483	.000287605	.000177753	.000097131	.000012499	.000000743	.000000001	1001011100110
0101010100011	42	−.0001497	−.0017606	.000520369	.000410652	.000286298	.000176320	.000095909	.000012148	.000000706	.000000000	0011110011010
1010110000011	40	−.0001542	−.0016631	.000521333	.000414989	.000294002	.000185431	.000104154	.000014872	.000001040	.000000001	0011101011010
0011110001010	40	−.0001542	−.0016994	.000520674	.000412860	.000290560	.000181519	.000100678	.000013724	.000000894	.000000000	1010110000110
0010011111000	41	−.0001579	−.0016405	.000521004	.000414893	.000294239	.000185818	.000104489	.000014911	.000001031	.000000000	1100110001101
1010001110010	41	−.0001579	−.0017240	.000519530	.000410269	.000286989	.000177837	.000097627	.000012841	.000000795	.000000000	0111000111010
0110001100010	41	−.0001579	−.0017416	.000519215	.000285412	.000285412	.000176091	.000096121	.000012388	.000000744	.000000000	0011101011001
1011001000011	40	−.0001682	−.0016551	.000518842	.000411044	.000289831	.000181942	.000101718	.000014357	.000000992	.000000000	0111011010010
0011011101010	40	−.0001682	−.0016847	.000518301	.000409283	.000286967	.000178667	.000098790	.000013377	.000000867	.000000000	1011100010011
0011101010010	40	−.0001682	−.0017058	.000517936	.000408165	.000285255	.000176829	.000097250	.000012944	.000000821	.000000000	1001110111010
0011100001110	42	−.0001686	−.0017031	.000517880	.000408059	.000285066	.000176560	.000096955	.000012802	.000000798	.000000000	1001011010110
1000110101010	42	−.0001686	−.0017172	.000517172	.000408487	.000281487	.000172585	.000093520	.000011768	.000000683	.000000000	0100111110010
1010101001101	42	−.0001687	−.0017424	.000518314	.000409478	.000286298	.000179175	.000099270	.000013547	.000000888	.000000000	0111011010010
0101110010010	39	−.0001687	−.0016789	.000518314	.000409478	.000287369	.000179175	.000099270	.000013547	.000000888	.000000001	0011101011010
0101100010101	39	−.0001687	−.0016994	.000517882	.000408129	.000285267	.000176879	.000097312	.000012970	.000000824	.000000000	1110100011100
0100111001100	41	−.0001691	−.0017121	.000517625	.000407388	.000284073	.000175492	.000096046	.000012528	.000000767	.000000000	1100110010101
1010101100000	39	−.0001729	−.0016964	.000517208	.000407186	.000284395	.000176247	.000096935	.000012920	.000000822	.000000000	0111001100110
1010100011010	41	−.0001784	−.0017331	.000515506	.000403279	.000278947	.000170575	.000092196	.000011532	.000000665	.000000000	1100110110010
1000101111000	42	−.0001784	−.0017449	.000515297	.000402625	.000277928	.000169464	.000091257	.000011267	.000000638	.000000000	0111011000110
1010001011100	42	−.0001826	−.0017096	.000515125	.000402290	.000279532	.000171546	.000091822	.000011882	.000000706	.000000000	1001101101010
0101001101010	42	−.0001826	−.0017271	.000514818	.000402304	.000279043	.000169927	.000091822	.000011493	.000000665	.000000001	0100110111010
1010101010010	42	−.0001831	−.0017404	.000514581	.000401563	.000278886	.000168666	.000090754	.000011190	.000000634	.000000001	0100111110010
1010010001101	42	−.0001831	−.0016244	.000516545	.000407886	.000282603	.000179752	.000100312	.000014070	.000000959	.000000001	0110111010000
1010010001101	41	−.0001831	−.0016761	.000515625	.000404975	.000282293	.000174623	.000095851	.000012681	.000000795	.000000000	0111011011100
1100010101001	41	−.0001831	−.0017035	.000515431	.000404360	.000281325	.000173554	.000094932	.000012406	.000000764	.000000000	1010100010011
0101010110100	41	−.0001831	−.0017011	.000515190	.000403630	.000280220	.000172381	.000093960	.000012139	.000000737	.000000000	1100110100101
0101100110100	41	−.0001831	−.0017086	.000515056	.000403206	.000279549	.000171638	.000093319	.000011948	.000000716	.000000000	1100111100101
0110100011100	42	−.0001876	−.0017271	.000514729	.000402179	.000277946	.000169889	.000091834	.000011522	.000000670	.000000000	0111011000011
1001110000001	39	−.0001876	−.0017006	.000514378	.000402390	.000279031	.000171547	.000093529	.000012159	.000000751	.000000001	1100101000001
1001100110001	40	−.0001934	−.0016890	.000513477	.000401113	.000277779	.000170535	.000092814	.000011977	.000000729	.000000000	1001011100010
0101010110100	40	−.0001934	−.0017020	.000513247	.000400396	.000276661	.000169312	.000091771	.000011672	.000000696	.000000000	0101001101001
0010100110010	41	−.0001971	−.0016957	.000512657	.000399564	.000275879	.000168731	.000091408	.000011612	.000000691	.000000000	1010100111100
1100110001100	41	−.0002016	−.0016767	.000512167	.000399318	.000276220	.000169546	.000092363	.000012022	.000000747	.000000000	0110111000001
0101100111100	39	−.0002016	−.0016883	.000511967	.000398704	.000275284	.000168545	.000091527	.000011790	.000000722	.000000001	0111011010100
0110101010010	39	−.0002016	−.0017015	.000511740	.000398018	.000274247	.000167448	.000090623	.000011546	.000000698	.000000000	1010110011100
1010001011010	42	−.0002113	−.0017082	.000509769	.000394435	.000269905	.000163309	.000087359	.000010657	.000000601	.000000000	1010110011010
0110010011001	41	−.0002118	−.0016885	.000510449	.000396680	.000273451	.000167192	.000090659	.000011607	.000000702	.000000000	0110011011100
0110010011100	41	−.0002118	−.0015671	.000510026	.000395382	.000271464	.000165060	.000088875	.000011107	.000000649	.000000001	0111011010001
0111110000001	38	−.0002119	−.0015671	.000512151	.000402065	.000271968	.000176668	.000098912	.000014205	.000001017	.000000000	0111001100001
0111001100001	39	−.0002156	−.0016706	.000509647	.000395298	.000271981	.000165964	.000089903	.000011524	.000000704	.000000001	1010100110001
1001011110100	41	−.0002160	−.0016228	.000510363	.000397535	.000275329	.000169513	.000092755	.000012250	.000000772	.000000000	1010100111100
1100110110100	41	−.0002160	−.0016566	.000509647	.000395764	.000272630	.000166629	.000090352	.000011585	.000000703	.000000001	0110100111000
0110110111000	41	−.0002160	−.0016845	.000509299	.000394282	.000270359	.000164189	.000088306	.000011009	.000000642	.000000000	0110100111001
0110101011000	40	−.0002263	−.0015137	.000510369	.000400465	.000270359	.000181519	.000100204	.000014920	.000001131	.000000001	1110101101100

166

m = 7, n = 6   Rank Orders 1001 thru 1050 of 1716   C(7, 6) = 1716   1/C(7, 6) = .00058275

z	w	c_1	c_2	d = .2	.4	.6	.8	1.0	1.5	2.0	3.0	z^{tc}
0010111100100	40	−.0002263	−.0016034	.000508793	.000395514	.000273741	.000168686	.000092556	.00012428	.000000810	.000000000	1101100001011
1000110001001	40	−.0002263	−.0016377	.000508190	.000393635	.000270810	.000165477	.000089810	.00011608	.000000717	.000000000	0111100001110
0011011100010	40	−.0002263	−.0016465	.000508048	.000393229	.000270231	.000164900	.000089365	.00011510	.000000710	.000000000	1101010100101
0100011100010	40	−.0002263	−.0016648	.000507718	.000392169	.000288536	.000162999	.000087702	.00010989	.000000648	.000000000	1011100001101
1010010010001	40	−.0002263	−.0016963	.000507297	.000390900	.000266628	.000160993	.000086059	.00010559	.000000593	.000000000	0110111010010
1001101001001	40	−.0002263	−.0016976	.000507177	.000390529	.000266058	.000160382	.000085551	.00010420	.000000589	.000000000	1001110101001
0110110001010	40	−.0002263	−.0017150	.000507154	.000390459	.000265951	.000160264	.000085449	.00010388	.000000552	.000000000	0110110100101
0101101001010	40	−.0002263	−.0016368	.000506385	.000389520	.000264504	.000158706	.000084145	.00010027	.000000690	.000000000	1101010101001
1001011001010	40	−.0002360	−.0016789	.000505650	.000390708	.000267685	.000162870	.000088015	.00011260	.000000534	.000000000	0110110100011
1000101101001	40	−.0002360	−.0016477	.000505285	.000388430	.000264164	.000159057	.000084796	.00010339	.000000609	.000000000	1001100100110
0101010010010	40	−.0002360	−.0017018	.000505253	.000387207	.000262290	.000157044	.000083113	.000009872	.000000543	.000000000	1101010010101
1100010001110	42	−.0002365	−.0016154	.000506644	.000391547	.000268918	.000164106	.000088967	.000011465	.000000704	.000000000	1000111001110
1011101000011	40	−.0002365	−.0016515	.000506016	.000389602	.000265911	.000160844	.000086205	.000010665	.000000617	.000000000	1100111001011
0011110010100	40	−.0002403	−.0015717	.000506719	.000392880	.000271642	.000167510	.000092145	.000012570	.000000846	.000000000	0111011001010
1011010001100	40	−.0002403	−.0016323	.000505666	.000389627	.000266603	.000162016	.000087455	.000010471	.000000682	.000000000	1101011100011
0111010001010	40	−.0002403	−.0016612	.000505156	.000388029	.000264108	.000159282	.000085119	.000010355	.000000593	.000000000	1001011100001
1001010010110	40	−.0002403	−.0016662	.000505068	.000387753	.000263677	.000158812	.000084719	.000010388	.000000578	.000000000	1000110110001
1001100110010	40	−.0002403	−.0016778	.000504873	.000387174	.000262821	.000157928	.000084010	.000010027	.000000578	.000000000	0110110100110
1001100110100	40	−.0002403	−.0016989	.000504508	.000386052	.000261102	.000156085	.000082471	.000009755	.000000534	.000000000	0111011010010
1000101010100	42	−.0002407	−.0016477	.000505285	.000388487	.000264780	.000159941	.000085606	.000010557	.000000609	.000000000	1001100101110
1000011100110	42	−.0002505	−.0015954	.000504373	.000388317	.000265912	.000161947	.000087716	.000011339	.000000705	.000000000	1101000011110
1100000101110	42	−.0002505	−.0016222	.000503888	.000386759	.000259145	.000159145	.000085263	.000010571	.000000615	.000000000	1100010011100
1001010011100	42	−.0002505	−.0016463	.000503479	.000385525	.000261556	.000157180	.000083645	.000010139	.000000572	.000000000	1100101011010
0011011011010	40	−.0002550	−.0015636	.000504082	.000388611	.000266978	.000163437	.000089150	.000011828	.000000764	.000000000	0111011001011
1011100100100	40	−.0002550	−.0016755	.000501907	.000387981	.000258090	.000153945	.000081237	.000009621	.000000531	.000000000	1010101011001
0111011001010	40	−.0002592	−.0016903	.000501365	.000381981	.000256884	.000152656	.000080164	.000009328	.000000502	.000000000	1011010101010
1010100101010	40	−.0002592	−.0016757	.000501140	.000381452	.000255602	.000151527	.000080382	.000009448	.000000517	.000000000	1010110101010
1001101010010	40	−.0002608	−.0016887	.000500998	.000380761	.000255093	.000151600	.000079446	.000009193	.000000491	.000000000	1010101010110
1001100110110	40	−.0002608	−.0016732	.000500998	.000380998	.000255093	.000151151	.000079890	.000009288	.000000498	.000000000	1001101100110
0011011001010	41	−.0002650	−.0016723	.000500322	.000379741	.000254753	.000151034	.000079121	.000009138	.000000486	.000000000	0110101100110
1110000010011	40	−.0002689	−.0014838	.000502812	.000388380	.000268492	.000166096	.000091939	.000012887	.000000903	.000000001	0011011111000
1010101101000	40	−.0002689	−.0015710	.000501322	.000383832	.000261515	.000158541	.000085518	.000010964	.000000678	.000000000	1110001010010
0011001011010	40	−.0002689	−.0016323	.000500279	.000380201	.000253474	.000153474	.000081315	.000009804	.000000557	.000000000	0111000110011
1010001011010	40	−.0002689	−.0016411	.000500125	.000380577	.000256008	.000152679	.000080645	.000009614	.000000537	.000000000	1010001110010
1011000101010	40	−.0002689	−.0016557	.000499586	.000377479	.000254945	.000151584	.000079770	.000009399	.000000518	.000000000	1011100101001
0010101010011	39	−.0002792	−.0016728	.000499882	.000377577	.000253575	.000150129	.000078565	.000009074	.000000485	.000000000	0011011110010
0011110000011	39	−.0002792	−.0015477	.000499821	.000382023	.000256252	.000158066	.000085615	.000011229	.000000727	.000000001	1110111101001
0001111001010	41	−.0002797	−.0015968	.000498973	.000377416	.000251775	.000153718	.000081928	.000011229	.000000603	.000000000	1011011001010
1000011101010	41	−.0002834	−.0016480	.000497990	.000376490	.000256252	.000148915	.000077890	.000008995	.000000481	.000000000	0111110100010
1100000101010	42	−.0002834	−.0015717	.000498588	.000377240	.000256383	.000151034	.000082205	.000009138	.000000602	.000000000	1010110100110
1010000111100	42	−.0002834	−.0015832	.000498398	.000378683	.000255566	.000153179	.000081529	.000010018	.000000586	.000000000	1100011011010
0011101100010	39	−.0002835	−.0015898	.000498295	.000378453	.000255326	.000153034	.000081500	.000010072	.000000598	.000000000	1011100100101
0101111000101	38	−.0002840	−.0015482	.000498895	.000380361	.000258226	.000156100	.000084025	.000010755	.000000668	.000000000	0101011101010
1011000100010	41	−.0002840	−.0015876	.000498239	.000378412	.000252603	.000153074	.000081551	.000010093	.000000600	.000000000	1000110100101
1010011010010	41	−.0002937	−.0015775	.000496574	.000375825	.000255330	.000150769	.000079906	.000009717	.000000563	.000000000	1111011100010
1010001011001	41	−.0002937	−.0016297	.000495681	.000373116	.000248502	.000146419	.000076310	.000008742	.000000464	.000000000	1101011101010
1010110001010	41	−.0002937	−.0016343	.000495603	.000372881	.000248151	.000146050	.000076009	.000008664	.000000457	.000000000	1011011101010
1001100010110	41	−.0002937	−.0016449	.000495422	.000372334	.000247329	.000145186	.000075304	.000008481	.000000439	.000000000	1011101101010
1010010100110	41	−.0002937	−.0016479	.000495372	.000372184	.000247105	.000144954	.000075117	.000008435	.000000435	.000000000	1001101011010
1001010101010	41	−.0002937	−.0016608	.000495150	.000371515	.000246100	.000143901	.000074260	.000008216	.000000415	.000000000	1010101011010

167

Table A

m = 7, n = 6   Rank Orders 1051 thru 1100 of 1716   C(7, 6) = 1716   1/C(7, 6) = .00058275

z	w	c_1	c_2	d = .2	.4	.6	.8	1.0	1.5	2.0	3.0	z^{tc}
1100011000101	40	-.002942	-.0015877	.000496303•	.000375124	.000251593	.000149716	.000079040	.000009478	.000000538	.000000000	0101110011101
0101101001100	40	-.002942	-.0016034	.000496045	.000374370	.000250492	.000148587	.000078136	.000009250	.000000516	.000000000	1101010001101
0101101010100	40	-.002942	-.0016289	.000495607	.000373038	.000248475	.000146451	.000076375	.000008779	.000000469	.000000000	1100101011010
0110110001010	40	-.002984	-.0015889	.000495488	.000373756•	.000250084•	.000148416	.000078116	.000009284	.000000521	.000000000	0101101110100
1001001110010	40	-.002984	-.0016252	.000494876	.000371923	.000247346	.000145552	.000075781	.000008674	.000000462	.000000000	1100110011010
1100100101010	41	-.003076	-.0016212	.000493205	.000369237	.000244471	.000143124	.000074078	.000008320	.000000431	.000000000	1011001010110
0101011010100	40	-.003081	-.0015477	.000494357•	.000372826	.000249918	.000148884	.000078814	.000009580	.000000556	.000000000	0101011110100
0110101011000	40	-.003081	-.0016116	.000493282	.000369619	.000245139	.000143890	.000074744	.000008514	.000000452	.000000000	0110101010101
0110110001001	38	-.003126	-.0015690	.000493126	.000369507•	.000247117	.000146348	.000076972	.000009202	.000000527	.000000000	1101001010001
0110110010001	38	-.003126	-.0015819	.000492970	.000369885	.000246205	.000145412	.000076223	.000009014	.000000509	.000000000	0111011001001
0110101100001	38	-.003169	-.0015720	.000492339	.000369059	.000245494	.000144937	.000075962	.000008988	.000000509	.000000000	0111110010001
1010010011010	41	-.003223	-.0016178	.000490520	.000369922	.000239990	.000139338	.000071490	.000007828	.000000394•	.000000000•	1010100111010
1010100101010	41	-.003266	-.0016105	.000489849	.000363991	.000239049	.000138733	.000071138	.000007785	.000000392•	.000000000•	0110101011010
0111011000101	38	-.003266	-.0015631	.000490666•	.000366516	.000242926	.000142894	.000074613	.000008745	.000000490	.000000000	0100111110001
1100110001001	40	-.003271	-.0014425	.000492567•	.000372206	.000251348	.000151637	.000081708	.000010611	.000000678	.000000000	1110011010101
1110010100001	40	-.003271	-.0015310	.000491085	.000367781	.000244716	.000144636	.000075919	.000009007	.000000508	.000000000	0110011101100
1001011011000	40	-.003271	-.0015542	.000490704	.000364109	.000243109	.000143002	.000074624	.000008692	.000000479	.000000000	0110100111100
1100101011001	40	-.003271	-.0015664	.000490499	.000366053	.000242177	.000142021	.000073819	.000008479	.000000458	.000000000	1100011111001
0101010101001	40	-.003271	-.0015762	.000490340	.000365598	.000241524	.000141363	.000073301	.000008354	.000000447	.000000000	1100101110001
0110110101000	40	-.003271	-.0015935	.000490047	.000364720	.000242021	.000139996	.000072191	.000008067	.000000419	.000000000	1110000000111
1100110000011	39	-.003373	-.0014674	.000490258•	.000367978	.000246443	.000147293	.000078579	.000009958	.000000625	.000000000	0011111001100
0011100011100	39	-.003373	-.0015152	.000489449	.000365539	.000242755	.000143362	.000075295	.000009020	.000000521	.000000000	1101110010011
1010110010101	39	-.003373	-.0015593	.000488705	.000363314	.000240426	.000139987	.000072430	.000008252	.000000443	.000000000	0111001001010
1001110001010	39	-.003373	-.0015669	.000488595	.000362985	.000238938	.000139359	.000072019	.000008148	.000000434	.000000000	0111000100110
0101110010010	39	-.003373	-.0015671	.000488575	.000362925	.000238849	.000137926	.000071941	.000008126	.000000431	.000000000	1010011000101
0110110000110	39	-.003373	-.0015841	.000488288	.000362065	.000237566	.000141601	.000070852	.000007844	.000000404	.000000000	1110001010101
0101101011001	40	-.003410	-.0015182	.000488677•	.000364094	.000240952	.000141601	.000073879	.000008621	.000000477	.000000000	1110001111100
0110101001011	40	-.003410	-.0015213	.000488088	.000362391	.000240858	.000141552	.000073880•	.000008649	.000000482	.000000000	0111101011101
0110110100100	40	-.003410	-.0015544	.000480088	.000362391	.000238494	.000139113	.000071916	.000008148	.000000434	.000000000	1100011100001
0011111111000	39	-.003513	-.0013380	.000489780•	.000370012	.000251222	.000153283	.000084049	.000011735	.000000844	.000000000	1110000000111
1101101001100	39	-.003513	-.0014521	.000487890	.000364408	.000244408	.000144420	.000076661	.000009594	.000000595	.000000000	0011110100100
0011101001100	39	-.003513	-.0014948	.000487168	.000362228	.000239547	.000140902	.000073721	.000008752	.000000502	.000000000	1101010010011
0011101011100	39	-.003513	-.0015147	.000486844	.000361288	.000238185	.000139516	.000072619	.000008479	.000000432	.000000000	0111001011010
1010101110100	39	-.003513	-.0015344	.000486501	.000360231	.000236558	.000137762	.000071145	.000008064	.000000424	.000000000	1001110001101
1101011001000	39	-.003513	-.0015421	.000486378	.000359879	.000236058	.000137264	.000070758	.000007973	.000000415	.000000000	1011100001101
1100110100101	39	-.003513	-.0015475	.000486286	.000359603	.000235645	.000136831	.000070405	.000007880	.000000403	.000000000	1011100010101
0101011011001	39	-.003513	-.0015586	.000486096	.000359030	.000234780	.000135917	.000069653	.000007680	.000000395	.000000000	1011101010101
0101011100101	39	-.003513	-.0015595	.000486092	.000359047•	.000234850•	.000136037	.000069789	.000007739	.000000372	.000000000	0111101010101
1101101001100	39	-.003513	-.0015733	.000485733	.000357983	.000233277	.000134409	.000068476	.000007406	.000000372	.000000000	1011101101001
1100100001110	41	-.003518	-.0014947	.000487068•	.000361994	.000239182	.000140474	.000073325	.000008613	.000000484	.000000000	0100111101100
1000110101100	41	-.003518	-.0015455	.000486215	.000359453	.000235401	.000136527	.000070110	.000007771	.000000401	.000000000	1011110111100
1100110010110	41	-.003615	-.0015172	.000484856	.000357811	.000234124	.000135782	.000069777	.000007768	.000000404•	.000000000	1011011011100
0101101011100	41	-.003615	-.0015405	.000484474	.000356702	.000232516	.000134150	.000068487	.000007457	.000000376	.000000000	1100011101110
1001101110100	41	-.003658	-.0015059	.000484259	.000357118	.000233651	.000135594	.000069782	.000007829	.000000413	.000000000	1001100111100
1001100110100	41	-.003658	-.0015079	.000484219	.000356975	.000233408	.000135312	.000069529	.000007751	.000000405	.000000000	1011110001101
1001011011000	41	-.003658	-.0015343	.000483786	.000355715	.000231579	.000133453	.000068057	.000007394	.000000373	.000000000	1110100001011
0010111110100	41	-.003658	-.0014152	.000484936•	.000360247	.000241354	.000141581	.000074001	.000009191	.000000556	.000000000	1110010101010
1010101110000	39	-.003702	-.0014192	.000482723	.000353762	.000229369	.000131581	.000066791	.000007185	.000000361	.000000000	1011011101010
0101101010100	39	-.003702	-.0015492	.000482498	.000353099	.000228395	.000130577	.000065987	.000006984	.000000342	.000000000	1011011101010

168

z	w	c_1	c_2	d = .2	.4	.6	.8	1.0	1.5	2.0	3.0	z^{tc}
0011011011000	39	−.0003800	−.0014425	.000482662	.000355920	.000233649	.000136537	.000071025	.000008334	.000000474	.000000000	1110011010011
1011001010001	39	−.0003800	−.0015260	.000481292	.000351926	.000227833	.000130593	.000065281	.000007148	.000000361	.000000000	0111011001010
0111001001000	39	−.0003800	−.0015436	.000480999	.000351067	.000226575	.000129303	.000065250	.000006893	.000000338	.000000000	1011010011001
1110000100011	39	−.0003842	−.0013629	.000483175	.000358452	.000237891	.000141220	.000074985	.000009462	.000000597	.000000000	0011101111000
1011000101001	39	−.0003842	−.0014353	.000481986	.000354961	.000232742	.000135859	.000070602	.000008268	.000000469	.000000000	1110110010011
1011000010010	39	−.0003842	−.0015137	.000480698	.000351203	.000227259	.000130243	.000066110	.000007137	.000000361	.000000000	0111011001010
1010010101000	39	−.0003842	−.0015238	.000480536	.000350743	.000226607	.000129598	.000066513	.000007025	.000000352	.000000000	1010110110001
0111000101010	39	−.0003842	−.0015240	.000480527	.000350692	.000226502	.000129458	.000066475	.000006975	.000000346	.000000000	0110110110001
1010100101010	39	−.0003842	−.0015396	.000480275	.000349975	.000225485	.000128449	.000064697	.000006799	.000000331	.000000000	1110110110010
1000001111000	41	−.0003944	−.0013904	.000480802	.000353909	.000232324	.000136009	.000070978	.000008442	.000000489	.000000000	1110000111110
1100001011010	41	−.0003944	−.0014707	.000479475	.000350005	.000226593	.000130058	.000066163	.000007190	.000000365	.000000000	1010101111100
1010010101100	41	−.0003944	−.0014826	.000479285	.000349471	.000225842	.000129317	.000065593	.000007061	.000000355	.000000000	1100101011010
0011100110010	38	−.0003748	−.0013748	.000481070	.000354792	.000233769	.000137622	.000072401	.000008900	.000000545	.000000000	0011111011010
0011110100010	40	−.0003945	−.0014387	.000480022	.000351720	.000229250	.000132932	.000068583	.000007873	.000000437	.000000000	1011010100001
0101111000001	37	−.0003950	−.0014183	.000480251	.000352473	.000230362	.000134059	.000069467	.000008085	.000000457	.000000000	0111110000011
1010110000010	40	−.0004047	−.0014897	.000477264	.000346073	.000222146	.000126213	.000063462	.000006662	.000000325	.000000000	1001101011110
1011011000010	40	−.0004047	−.0015071	.000476977	.000345239	.000220937	.000124988	.000062496	.000006432	.000000305	.000000000	1001101011010
0110110010010	40	−.0004047	−.0015204	.000476758	.000344599	.000220209	.000124048	.000061757	.000006258	.000000290	.000000000	1010011010100
0110100100100	40	−.0004279	−.0013827	.000476220	.000344055	.000219716	.000124005	.000061843	.000006317	.000000297	.000000000	1001011010010
1010101010010	40	−.0004089	−.0015168	.000476027	.000343490	.000218898	.000123177	.000061193	.000006165	.000000284	.000000000	1010101100110
1010011100100	39	−.0004094	−.0014383	.000477222	.000347086	.000224168	.000128552	.000065451	.000007191	.000000375	.000000000	1010011101010
1100101001010	39	−.0004094	−.0014475	.000477071	.000346647	.000223532	.000127908	.000064943	.000007070	.000000365	.000000000	0101011101100
0101101001010	39	−.0004094	−.0014850	.000476462	.000344890	.000221001	.000125354	.000062933	.000006588	.000000322	.000000000	1100101010001
1000101110010	40	−.0004187	−.0014564	.000475193	.000343368	.000219808	.000124630	.000062576	.000006559	.000000320	.000000000	1110010101010
1010011010010	39	−.0004192	−.0014957	.000474554	.000341531	.000217180	.000122000	.000060529	.000006088	.000000281	.000000000	0111010010111
1010100100101	39	−.0004234	−.0014678	.000474925	.000342746	.000219018	.000123911	.000062064	.000006465	.000000314	.000000000	1001011011100
0111000101010	39	−.0004234	−.0014125	.000475028	.000344022	.000220737	.000126450	.000064161	.000006997	.000000363	.000000000	0111100110001
0111001011010	39	−.0004234	−.0014616	.000474235	.000341750	.000218064	.000123191	.000061613	.000006395	.000000310	.000000000	1010101100110
0110101001010	37	−.0004279	−.0013827	.000474685	.000344107	.000221980	.000127427	.000065095	.000007293	.000000395	.000000000	0110110110010
0110110010010	37	−.0004279	−.0014103	.000474252	.000342897	.000220290	.000125766	.000063820	.000007002	.000000370	.000000000	1110010101001
1010100011010	40	−.0004376	−.0014633	.000471561	.000337356	.000213331	.000119222	.000058874	.000005866	.000000269	.000000000	0101001110010
1101010101010	40	−.0004376	−.0014780	.000471321	.000336665	.000212342	.000118233	.000058107	.000005692	.000000255	.000000000	1010011101100
0111010010010	37	−.0004376	−.0013978	.000472637	.000340526	.000217973	.000123993	.000062677	.000006810	.000000356	.000000000	0110110110001
0100111011000	39	−.0004381	−.0013803	.000472808	.000341041	.000218647	.000124569	.000063056	.000006847	.000000354	.000000000	1110010001101
1100010110010	39	−.0004381	−.0014141	.000472266	.000339500	.000216460	.000122391	.000061366	.000006456	.000000321	.000000000	0110110011101
0110010110001	39	−.0004381	−.0014420	.000471818	.000338226	.000214650	.000120590	.000059967	.000006131	.000000292	.000000000	1001011001001
0110100101010	39	−.0004419	−.0013855	.000472000	.000339793	.000217382	.000123613	.000062490	.000006796	.000000356	.000000000	0111100110001
1011010010010	37	−.0004423	−.0014126	.000471500	.000338289	.000215198	.000121365	.000060676	.000006329	.000000311	.000000000	0110101011110
0110101000010	39	−.0004423	−.0014371	.000471107	.000337172	.000213615	.000119790	.000059456	.000006048	.000000287	.000000000	1001011010110
1010010111000	38	−.0004516	−.0014307	.000469482	.000334641	.000210997	.000117659	.000058017	.000005778	.000000266	.000000000	1011001100111
1100100011010	39	−.0004521	−.0013052	.000471391	.000340164	.000218877	.000125526	.000064141	.000007211	.000000392	.000000000	0101011111000
1101010101001	39	−.0004521	−.0013653	.000470446	.000337516	.000215160	.000121857	.000061308	.000006555	.000000334	.000000000	0110011101100
0111000011100	39	−.0004521	−.0013808	.000470185	.000336742	.000214012	.000120664	.000060342	.000006306	.000000310	.000000000	1100011101001
1100010110010	39	−.0004521	−.0013851	.000470126	.000336598	.000214054	.000120745	.000060435	.000006345	.000000315	.000000000	0111101011100
0101010110100	39	−.0004521	−.0014174	.000469612	.000335144	.000211793	.000118505	.000058706	.000005950	.000000282	.000000000	0011111010100
1101010010100	39	−.0004623	−.0012768	.000469975	.000338602	.000217956	.000125366	.000064448	.000007486	.000000434	.000000000	1101001010100
0011110100100	38	−.0004623	−.0013375	.000468996	.000335779	.000213872	.000121201	.000061116	.000006629	.000000348	.000000000	1101111010010
1011010000010	38	−.0004623	−.0013664	.000468522	.000334400	.000211865	.000119148	.000059475	.000006216	.000000309	.000000000	0101011010010
0111010000110	38	−.0004623	−.0013714	.000468440	.000334164	.000211523	.000118801	.000059201	.000006149	.000000302	.000000000	1001111010001

Table A

m = 7, n = 6   Rank Orders 1151 thru 1200 of 1716   C(7, 6) = 1716   1/C(7, 6) = .00058275

z	w	c_1	c_2	d = .2	.4	.6	.8	1.0	1.5	2.0	3.0	z^{tc}
1001110010001	38	−.0004623	−.0013830	.000468258	.000333657	.000210820	.000118119	.000058686	.000006039	.000000294	.000000000	0111011000110
0101110010010	38	−.0004623	−.0014040	.000467919	.000332684	.000209428	.000116723	.000057594	.000005780	.000000271	.000000000	1011011000101
1000111001100	40	−.0004628	−.0013832	.000468152	.000333409	.000210446	.000117699	.000058316	.000005924	.000000281	.000000000	1100100001110
1100111001100	41	−.0004665	−.0012999	.000468789	.000336058	.000214604	.000122044	.000061793	.000006772	.000000358	.000000000	1100001111100
0111101100001	38	−.0004666	−.0013287	.000467660	.000334887	.000213058	.000120611	.000060760	.000006576	.000000345	.000000000	1010100111100
1001101100010	38	−.0004666	−.0013708	.000467293	.000332917	.000210218	.000117739	.000058491	.000006023	.000000294	.000000000	0111011000110
1010100011110	40	−.0004768	−.0013937	.000466977	.000331867	.000208722	.000116243	.000057324	.000005748	.000000269	.000000000	1000111100100
1010010010110	40	−.0004768	−.0012940	.000466344	.000333261	.000211760	.000119780	.000060303	.000006511	.000000339	.000000000	1000111101100
1100011011010	40	−.0004768	−.0013485	.000466099	.000330738	.000208137	.000116128	.000057428	.000005816	.000000276	.000000000	1001111101100
1000110110100	40	−.0004768	−.0013577	.000465957	.000330354	.000207620	.000115640	.000057070	.000005744	.000000271	.000000000	1011011011010
1100010100110	40	−.0004768	−.0013608	.000465899	.000330168	.000207324	.000115316	.000056797	.000005670	.000000263	.000000000	1001010011100
1001100100110	40	−.0004768	−.0013693	.000465769	.000329814	.000206846	.000114865	.000056464	.000005603	.000000258	.000000000	1100110011000
1010010101100	40	−.0004768	−.0013865	.000465494	.000329032	.000205739	.000113768	.000055620	.000005414	.000000243	.000000000	0101110010010
1010011001001	38	−.0004813	−.0013662	.000465008	.000328731	.000205879	.000114228	.000056148	.000005601	.000000263	.000000000	0110110110110
0111001100001	38	−.0004813	−.0013792	.000464799	.000328134	.000205028	.000113380	.000055489	.000005448	.000000249	.000000000	1010110001001
0111000111100	39	−.0004850	−.0013069	.000465243	.000330180	.000208246	.000116695	.000058088	.000006042	.000000299	.000000000	1110000111010
1001110000110	39	−.0004871	−.0013554	.000464086	.000327385	.000204529	.000113134	.000055396	.000005445	.000000249	.000000000	1001111000110
1001011110100	40	−.0004908	−.0013400	.000463636	.000326937	.000204276	.000113062	.000055418	.000005470	.000000252	.000000000	1011110101001
0010111110100	40	−.0004952	−.0011789	.000465356	.000332802	.000213059	.000122059	.000062555	.000007215	.000000413	.000000000	1110010101011
1110010000011	38	−.0004952	−.0013577	.000465199	.000332414	.000212591	.000121676	.000062327	.000007209	.000000419	.000000000	0011011011000
0011011101100	38	−.0004952	−.0012461	.000464308	.000329881	.000208963	.000118005	.000059405	.000006461	.000000343	.000000000	1110100010011
0011101100001	38	−.0004952	−.0012648	.000464016	.000329068	.000207829	.000116892	.000058551	.000006265	.000000326	.000000000	1110001011010
1010011110000	38	−.0004952	−.0013258	.000463046	.000326326	.000203952	.000113041	.000055561	.000005565	.000000263	.000000000	0111100011010
0111001100010	38	−.0004952	−.0013416	.000462799	.000325640	.000203005	.000112123	.000054868	.000005415	.000000251	.000000000	1110000100100
0110011001001	38	−.0004952	−.0013448	.000462743	.000325468	.000202741	.000111842	.000054635	.000005352	.000000245	.000000000	1010001111001
1010010100110	38	−.0004952	−.0013521	.000462637	.000325196	.000202397	.000111543	.000054436	.000005325	.000000244	.000000000	0111010101010
1110010100100	38	−.0004952	−.0013523	.000462627	.000325148	.000202302	.000111421	.000054321	.000005287	.000000240	.000000000	1011010101001
0110110110100	38	−.0004952	−.0013683	.000462378	.000324463	.000201365	.000110863	.000054066	.000005148	.000000238	.000000000	1010101010010
0110100110110	40	−.0005054	−.0013124	.000461649	.000323737	.000201224	.000110523	.000054066	.000005264	.000000238	.000000000	1001101100100
1010001010110	40	−.0005054	−.0013239	.000461164	.000323258	.000200646	.000110250	.000053613	.000005173	.000000232	.000000000	1100110110010
0011110010001	37	−.0005055	−.0012279	.000462683	.000327587	.000208800	.000116390	.000058400	.000006303	.000000332	.000000000	1011110000011
1011000110001	38	−.0005092	−.0013094	.000460710	.000322880	.000200587	.000110459	.000053919	.000005299	.000000246	.000000000	0111100110010
1011001110010	38	−.0005092	−.0013224	.000460502	.000322289	.000199750	.000109628	.000053276	.000005151	.000000233	.000000000	1011100010010
1000101111000	40	−.0005097	−.0012452	.000461608	.000325418	.000204075	.000113798	.000056405	.000005814	.000000285	.000000000	1110000111110
1010011101100	40	−.0005097	−.0013019	.000460719	.000322938	.000200618	.000110419	.000053827	.000005242	.000000230	.000000000	1100100111010
1010010110100	40	−.0005097	−.0013150	.000460518	.000322392	.000199878	.000109717	.000053308	.000005137	.000000230	.000000000	1000111101110
0110111100001	38	−.0005200	−.0012799	.000459168	.000320891	.000199851	.000109241	.000053191	.000005185	.000000237	.000000000	1011110000110
1010101010010	39	−.0005200	−.0013222	.000458509	.000319069	.000196342	.000106082	.000051377	.000004802	.000000207	.000000000	1011110101010
1010010100101	38	−.0005200	−.0013343	.000458316	.000318527	.000195585	.000106083	.000050816	.000004681	.000000198	.000000000	1010101010010
1100101000101	39	−.0005205	−.0012495	.000459559	.000322111	.000200632	.000111035	.000054594	.000005513	.000000266	.000000000	0101110010100
0101011001100	38	−.0005205	−.0012858	.000458990	.000320522	.000198417	.000108869	.000052940	.000005144	.000000235	.000000000	1100110010101
1010101101010	39	−.0005297	−.0013096	.000456896	.000316705	.000194042	.000105063	.000050261	.000004623	.000000195	.000000000	1010110110010
1010110100010	39	−.0005339	−.0012790	.000456588	.000316766	.000194543	.000105744	.000050858	.000004771	.000000208	.000000000	1001011100010
1010110011010	39	−.0005339	−.0013036	.000456205	.000315712	.000193098	.000104358	.000049826	.000004558	.000000191	.000000000	0101110110100
1010100100110	38	−.0005344	−.0012251	.000457347	.000319004	.000197726	.000108892	.000053279	.000005315	.000000253	.000000000	0101011010010
1011001001100	38	−.0005344	−.0012547	.000456558	.000317708	.000195921	.000107126	.000051929	.000005013	.000000228	.000000000	1101010100101
0101011101100	38	−.0005344	−.0012758	.000456558	.000316825	.000194721	.000105985	.000051084	.000004840	.000000214	.000000000	1101011010001
0011011001001	38	−.0005389	−.0011876	.000457108	.000319334	.000199669	.000110070	.000051929	.000005598	.000000281	.000000000	0111110011001
1010110011010	39	−.0005486	−.0012842	.000453790	.000312211	.000189744	.000101841	.000048265	.000004324	.000000177	.000000000	1010110011010

Table A

m = 7, n = 6   Rank Orders 1201 thru 1250 of 1716   C(7, 6) = 1716   1/C(7, 6) = .00058275

z	w	c_2	c_1	d = .2	.4	.6	.8	1.0	1.5	2.0	3.0	z^{tc}
0111100010001	36	−.0011634	−.0005529	.00454891	.000316209	.000195733	.000107890	.000052960	.000005388	.000000267	.000000000	0111011100001
0111010100001	36	−.0011781	−.0005529	.00454671	.000315627	.000194961	.000107168	.000052432	.000005281	.000000258	.000000000	0111101001001
0101011001001	38	−.0011607	−.0005534	.00454831	.000316095	.000195553	.000107666	.000052744	.000005744	.000000257	.000000000	1110101001001
1100101001001	38	−.0012182	−.0005534	.00453946	.000313672	.000192240	.000104488	.000050364	.000004803	.000000216	.000000000	0110110101100
0110101001010	38	−.0012435	−.0005626	.00453534	.000312602	.000190776	.000103085	.000049315	.000004433	.000000199	.000000000	0110011110010
1011000101010	39	−.0012246	−.0005626	.00452125	.000310610	.000188905	.000101680	.000048428	.000004581	.000000188	.000000000	1010011110010
0101011011001	38	−.0012522	−.0005631	.00451699	.000309449	.000187327	.000100182	.000047322	.000004210	.000000172	.000000000	1010011101010
0101010101001	38	−.0011841	−.0005631	.00452660	.000312186	.000191112	.000103823	.000050042	.000004777	.000000215	.000000000	1100101001001
0110100001001	38	−.0011872	−.0005631	.00452620	.000312101	.000191028	.000103775	.000050032	.000004789	.000000217	.000000000	1101101001001
0110100110001	38	−.0012202	−.0005631	.00452112	.000310715	.000189141	.000101972	.000048688	.000004507	.000000195	.000000000	0111010011001
1110000100101	38	−.0011143	−.0005673	.00452941	.000313833	.000193758	.000106549	.000052170	.000005258	.000000255	.000000000	1010101111000
0110110011000	38	−.0011752	−.0005673	.00452011	.000311303	.000190311	.000103248	.000049699	.000004728	.000000212	.000000000	1110011011010
0110100110001	38	−.0011755	−.0005673	.00452013	.000311331	.000190380	.000103344	.000049793	.000004761	.000000216	.000000000	0110101101100
1101000110001	38	−.0011832	−.0005673	.00451896	.000311012	.000189948	.000102935	.000049491	.000004700	.000000211	.000000000	0111011011100
0101001001100	38	−.0011949	−.0005673	.00451713	.000310504	.000189242	.000102246	.000048966	.000004583	.000000202	.000000000	1010101101100
0110100010100	38	−.0012129	−.0005673	.00451441	.000309773	.000188264	.000101329	.000048297	.000004450	.000000192	.000000000	1100010111100
1100001011100	40	−.0011311	−.0005775	.00450788	.000310070	.000189555	.000102946	.000049649	.000004759	.000000216	.000000000	0011111100001
0011110000011	37	−.0010340	−.0005776	.00452309	.000314377	.000189661	.000104696	.000054424	.000005918	.000000325	.000000000	1101101100011
0111110000101	37	−.0011094	−.0005776	.00451147	.000311173	.000191221	.000104696	.000051077	.000005130	.000000252	.000000000	1101011000011
0111100000110	37	−.0011224	−.0005776	.00450927	.000310512	.000190234	.000103669	.000050244	.000004915	.000000231	.000000000	0111111000010
1001110100001	37	−.0011260	−.0005776	.00450869	.000310348	.000189001	.000103436	.000050062	.000004872	.000000227	.000000000	1001111100001
0101110110010	37	−.0011523	−.0005776	.00450476	.000309306	.000188620	.000102152	.000048689	.000004689	.000000213	.000000000	0111011010101
0101110010010	37	−.0011747	−.0005878	.00450128	.000308345	.000187295	.000100868	.000048159	.000004477	.000000196	.000000000	1010111011000
1100011100010	39	−.0011498	−.0005878	.00448616	.000306337	.000185502	.000099592	.000047396	.000004365	.000000189	.000000000	1001111011100
1001101100100	39	−.0011517	−.0005878	.00448578	.000306208	.000185295	.000099365	.000047206	.000004315	.000000184	.000000000	1001110110110
1001001100110	39	−.0011781	−.0005878	.00448175	.000305121	.000183831	.000097985	.000046193	.000004112	.000000169	.000000000	1010110110110
1001100101010	39	−.0011479	−.0005921	.00447497	.000305355	.000184220	.000098536	.000046679	.000004231	.000000179	.000000000	1100101111000
1011001011000	39	−.0011712	−.0005921	.00448576	.000307909	.000182934	.000097325	.000045791	.000004054	.000000166	.000000000	1001110011010
0011001011100	39	−.0010526	−.0005960	.00447909	.000304163	.000188385	.000102604	.000049862	.000004928	.000000235	.000000000	0110110011011
0110101011000	38	−.0011165	−.0005960	.00447608	.000305295	.000184850	.000099332	.000047362	.000004401	.000000193	.000000000	1110011001001
1101000010110	39	−.0010953	−.0006018	.00446854	.000304433	.000184202	.000098951	.000047177	.000004385	.000000192	.000000000	1001011110010
0101011110000	39	−.0011462	−.0006018	.00446080	.000302344	.000191388	.000096295	.000045224	.000003992	.000000163	.000000000	1100101110010
1110000011011	37	−.0009617	−.0006063	.00448090	.000303807	.000190716	.000105485	.000052342	.000005614	.000000305	.000000000	0011111011010
0011110010001	37	−.0010341	−.0006063	.00446984	.000305779	.000186546	.000101485	.000049232	.000004890	.000000238	.000000000	0111010000111
1011011001001	37	−.0011126	−.0006063	.00445786	.000302525	.000181590	.000096760	.000046064	.000004211	.000000182	.000000000	1011011011000
1001100100010	37	−.0011226	−.0006063	.00445636	.000302125	.000181512	.000096663	.000045620	.000004142	.000000177	.000000000	0110011101010
0111011010010	37	−.0011228	−.0006063	.00445627	.000302085	.000181460	.000095893	.000045460	.000004114	.000000174	.000000000	1010011101001
0111011010001	37	−.0011384	−.0006105	.00445392	.000301460	.000180663	.000100952	.000045918	.000004007	.000000167	.000000000	1111010001001
1100010110001	37	−.0011338	−.0006105	.00444349	.000304933	.000185792	.000100952	.000048918	.000004846	.000000235	.000000000	1101010010011
1010010101100	38	−.0011078	−.0006105	.00445076	.000301480	.000178363	.000099469	.000045580	.000004136	.000000177	.000000000	0111011001010
0110101100010	37	−.0011254	−.0006105	.00444806	.000300744	.000180105	.000095527	.000044873	.000003988	.000000166	.000000000	1011100110000
0011011110000	37	−.0009573	−.0006203	.00445543	.000304703	.000186324	.000101840	.000049740	.000005063	.000000254	.000000000	1111000010011
1010110010010	37	−.0010917	−.0006203	.00443525	.000299300	.000179074	.000094995	.000044680	.000004006	.000000170	.000000000	0111010110010
0111000100110	37	−.0011051	−.0006203	.00443318	.000298733	.000178523	.000094255	.000044129	.000003891	.000000161	.000000000	1000111111000
1001000100111	39	−.0009682	−.0006207	.00445284	.000304068	.000185464	.000100997	.000049085	.000004904	.000000239	.000000000	1010011100001
1100100100011	39	−.0010437	−.0006207	.00444140	.000300971	.000181261	.000096980	.000046078	.000004256	.000000181	.000000000	1110100110001
1010011010010	39	−.0010879	−.0006207	.00443473	.000299189	.000178880	.000094749	.000044448	.000003931	.000000162	.000000000	1010011100011
0111011010001	39	−.0010982	−.0006207	.00443321	.000298792	.000178363	.000094277	.000044112	.000003870	.000000158	.000000000	1111010001001
1100010100011	39	−.0011015	−.0006207	.00443265	.000298620	.000178106	.000094041	.000043902	.000003820	.000000154	.000000000	1101010010011
1010011011100	39	−.0011142	−.0006207	.00443077	.000298129	.000177466	.000093429	.000043489	.000003745	.000000149	.000000000	1101011011010

171

Table A
m = 7, n = 6 Rank Orders 1251 thru 1300 of 1716 C(7, 6) = 1716 1/C(7, 6) = .00058275

z	w	c_1	c_2	.2	.4	.6	.8	1.0	1.5	2.0	3.0	z^{tc}
1010110001010110	38	−.006310	−.0010819	.000441680	.000296486	.000176226	.000092727	.000043185	.000003743	•000000151	.000000000	1001111001010010
1001110001001000	38	−.006310	−.0010936	.000441499	.000295995	.000175561	.000092098	.000042722	.000003650	.000000144	.000000000	1010111001010100
1001000111100000	39	−.006347	−.0010187	.000441933	.000297874	.000178361	.000094836	.000044757	.000004055	.000000173	.000000000	1110000110010110
1100001110010000	39	−.006347	−.0010446	.000441548	.000296859	.000177025	.000093601	.000043869	.000003885	.000000161	.000000000	1011000110010110
1010000111010000	39	−.006347	−.0010601	.000441321	.000296267	.000176255	.000092900	.000043371	.000003793	.000000155	.000000000	1101000011101010
1001000111001000	38	−.006450	−.0010501	.000439583	.000293701	.000173769	.000091021	.000042202	.000003618	.000000144	.000000000	1001100010101010
1001101100001000	38	−.006450	−.0010517	.000439552	.000293605	.000173623	.000090868	.000042079	.000003589	.000000142	.000000000	1011000010101010
1101001010010000	38	−.006450	−.0010754	.000439522	.000293689	.000172431	.000089783	.000041311	.000003449	.000000132	.000000000	1011001010010100
1101101000010100	37	−.006455	−.0009814	.000440530	.000296351	.000177375	.000094443	.000044729	.000004134	.000000184	.000000000	0101110110010100
0101110000010000	37	−.006455	−.0010306	.000439792	.000294380	.000174744	.000091976	.000042923	.000003770	.000000156	.000000000	1101110001010101
0101101001000100	37	−.006497	−.0010186	.000439190	.000293623	.000174111	.000091560	.000042695	.000003745	.000000155	.000000000	0101110110001100
1010100101001000	37	−.006639	−.0010436	.000436200	.000288492	.000168552	.000086975	.000039630	.000003219	.000000120	.000000000	0111100011001100
0111010001000001	35	−.006639	−.0009094	.000438213	.000293882	.000175757	.000093732	.000044577	.000004215	.000000195	.000000000	1110100010010100
0101101100001100	37	−.006644	−.0009669	.000437261	.000291412	.000172469	.000090635	.000042288	.000003736	.000000157	.000000000	0110110110100100
0111100010100001	35	−.006644	−.0009444	.000436895	.000290446	.000171217	.000089448	.000041428	.000003567	.000000144	.000000000	0111101010101100
1011010101010010	38	−.006682	−.0008996	.000437578	.000293035	.000175003	.000093201	.000044265	.000004172	.000000193	.000000000	1110101011010100
1011001010001010	38	−.006736	−.0010128	.000434869	.000286896	.000167293	.000086204	.000039245	.000003188	.000000139	.000000000	1100110110011001
0111010101001010	38	−.006779	−.0009928	.000434385	.000286443	.000167052	.000086139	.000039262	.000003205	.000000132	.000000000	1010101011001001
0101010101010010	37	−.006779	−.0010056	.000434197	.000285959	.000166430	.000085580	.000038873	.000003137	.000000116	.000000000	1010110110001010
0100111110000000	37	−.006784	−.0008469	.000436450	.000291973	.000174371	.000092922	.000044165	.000004160	.000000190	.000000000	1100011011011010
1110010000101001	37	−.006784	−.0008735	.000436061	.000290957	.000173040	.000091695	.000043280	.000003987	.000000155	.000000000	0101110111110000
0101001110100001	37	−.006784	−.0009157	.000435445	.000289354	.000170952	.000089781	.000041909	.000003723	.000000158	.000000000	0111110001101100
0110101001001001	37	−.006784	−.0009164	.000435429	.000289297	.000170855	.000089669	.000041810	.000003694	.000000150	.000000000	1110100010100101
1011010101010001	37	−.006784	−.0009351	.000435161	.000288615	.000169990	.000088902	.000041280	.000003604	.000000146	.000000000	0110110100100101
1011010101010001	37	−.006784	−.0009363	.000435136	.000288531	.000169853	.000088570	.000041153	.000003554	.000000146	.000000000	0111010101010100
1101010101000100	37	−.006784	−.0009432	.000435042	.000288304	.000169584	•000088531	.000041016	.000003488	•000000146	• .000000000	0111010110010101
0111001100100100	37	−.006784	−.0009514	.000434918	.000287971	.000169136	.000088106	.000040701	.000003488	.000000141	.000000000	1100101010011001
0111001010001010	37	−.006784	−.0009553	.000434860	.000287817	.000168833	.000087919	.000040567	.000003463	.000000139	.000000000	1010101010001001
1010101010010100	37	−.006784	−.0009737	.000434591	.000287121	.000168034	.000087105	.000039993	.000003359	.000000132	.000000000	1110110101010101
1100101011001100	39	−.006886	−.0009044	.000433717	.000286741	.000168316•	.000087684	.000040522	.000003475	.000000190	.000000000	1100011011010101
0011011100000100	36	−.006886	−.0008307	.000434825	.000289758	.000172414	.000091597	.000043444	.000004098	.000000191	.000000000	1101111000000011
1001111100000001	36	−.006886	−.0008728	.000434189	.000288044	.000170099	.000089394	.000041803	.000003748	.000000162	.000000000	0111110000110110
0101011101000010	36	−.006886	−.0008957	.000433847	.000287131	.000168880	.000088249	.000040960	.000003575	.000000148	•000000000	0111001110100101
0111000010001000	37	−.006924	−.0009182	.000432839	.000285229	.000166728	.000086439	.000039742	.000003366	.000000134	.000000000	1011001110001001
0100011010010000	39	−.006928	−.0008922	.000433119	•000285997	.000167703	.000087289	.000040312	.000003455	.000000139	.000000000	1100100111110000
1000111100100100	38	−.007031	−.0008700	.000431573	.000283961	.000165944	.000086106	.000039662	.000003386	.000000137	.000000000	1110110101110110
1001010010010100	38	−.007031	−.0008968	.000431174	.000282907	.000164555	.000084823	.000038738	.000003209	.000000115	.000000000	1110101011010110
1001101001011000	38	−.007031	−.0009208	.000430821	.000281990	.000163365	.000083743	.000037973	.000003070	.000000115	.000000000	1100101011010110
0110011011011000	37	−.007070	−.0008664	.000430902	.000282975	.000164997	.000085401	.000039230	.000003325	.000000133	.000000000	1100110101001001
1110001010010010	37	−.007113	−.0008072	.000430983	.000283975	.000166630	.000087039	.000040460	.000003573	.000000152	.000000000	0110101011111000
1001100011000010	36	−.007113	−.0008564	.000430271	.000282139	.000164259	.000084886	.000038931	.000003286	.000000131	.000000000	0111100010010010
1010101011010010	38	−.007128	−.0008922	.000429455	.000280298	.000161977	.000082850	.000037499	.000003022	.000000124	.000000000	1110100100101110
1010010100010010	36	−.007171	−.0008468	.000429340	•000280801	.000161217	.000082319	.000083901	.000003182	.000000161	.000000000	1011010010100110
1001100100100100	38	−.007171	−.0008829	.000428814	.000279447	.000161217	.000082319	.000037192	.000002982	.000000111	.000000000	0011110100001000
1010100000000011	36	−.007171	−.0006644	.000431215	.000286634	.000166685	.000087576	.000041084	.000004428	.000000227	.000000000	1110100100011010
0011101010101010	36	−.007215	−.0007516	.000429939	.000283238	.000163469	.000084566	.000038879	.000003774	.000000171	.000000000	1110101000011011
0011100101001001	36	−.007215	−.0007575	.000429027	.000282907	.000163365	.000082985	.000038542	.000003323	.000000130	.000000000	1011011110000110
0111110001010010	36	−.007215	−.0008129	.000429027	.000282975	.000163469	.000082985	.000038542	.000003323	.000000130	.000000000	0111011110000110
0111111000010101	36	−.007215	−.0008217	.000428893	.000282975	.000162985	.000082985	.000038542	.000003323	.000000130	.000000000	1101011100001001

172

Table A

m = 7, n = 6   Rank Orders 1301 thru 1350 of 1716   C(7, 6) = 1716   1/C(7, 6) = .00058275

z	w	c_1	c_2	d = .2	.4	.6	.8	1.0	1.5	2.0	3.0	z^{tc}
011011010010	36	−.0007215	−.0008535	.000428440	.000279360	.000161530	.000082816	.000037646	.000003097	.000000120	.000000000	101101010010
011101010001	36	−.0007313	−.0008166	.000427198	.000277997	.000160579	.000082331	.000037468	.000003108	.000000123	.000000000	011101010001
011101010010	36	−.0007313	−.0008296	.000427004	.000277486	.000161871	.000081707	.000037017	.000003020	.000000116	.000000000	101101010001
100011011000	38	−.0007317	−.0007827	.000427585	.000279026	.000159087	.000080922	.000038220	.000003226	.000000129	.000000000	111010011000
101001100110	38	−.0007317	−.0008394	.000426757	.000276877	.000158497	.000080403	.000036428	.000002896	.000000107	.000000000	101001100110
101001101100	38	−.0007355	−.0008525	.000426571	.000276408	.000161678	.000080273	.000036073	.000002837	.000000103	.000000000	111100011000
100111011000	36	−.0007355	−.0006874	.000426874	.000281613	.000165605	.000087072	.000040924	.000003791	.000000174	.000000000	111011001100
101001101010	36	−.0007355	−.0007993	.000426670	.000277426	.000161678	.000082117	.000037379	.000003106	.000000123	.000000000	011110011010
110010010110	36	−.0007360	−.0008141	.000426451	.000276849	.000159421	.000081419	.000036876	.000003009	.000000116	.000000000	101110011010
111000010110	38	−.0007360	−.0008329	.000426075	.000275915	.000158193	.000080273	.000036039	.000002843	.000000104	.000000000	101101010010
100101011000	38	−.0007457	−.0008444	.000425911	.000275503	.000157676	.000079819	.000035729	.000002792	.000000101	.000000000	110010110100
101101000110	38	−.0007457	−.0007089	.000426080	.000277725	.000161276	.000083386	.000038366	.000003296	.000000135	.000000000	100101110100
100110011000	38	−.0007457	−.0007715	.000425178	.000275412	.000158308	.000080711	.000036482	.000002952	.000000112	.000000000	110010110001
101010011000	38	−.0007457	−.0007742	.000425147	.000275354	.000158258	.000080688	.000036481	.000002958	.000000113	.000000000	100111011000
100101011010	38	−.0007457	−.0007821	.000425034	.000275067	.000157896	.000080367	.000036260	.000002920	.000000110	.000000000	110011001100
110011001010	38	−.0007457	−.0007982	.000424796	.000274442	.000157079	.000079619	.000035727	.000002822	.000000103	.000000000	101010011010
101101001010	37	−.0007560	−.0008133	.000424583	.000273907	.000155883	.000079034	.000035329	.000002756	.000000099	.000000000	100101011010
110110010010	37	−.0007560	−.0007642	.000423421	.000272846	.000154709	.000078941	.000034691	.000002683	.000000104	.000000000	100110101010
101101010000	37	−.0007560	−.0007888	.000423064	.000271929	.000154709	.000077888	.000034691	.000002683	.000000096	.000000000	010111011010
110101010000	37	−.0007602	−.0007735	.000422508	.000271287	.000154219	.000077595	.000034547	.000002672	.000000096	.000000000	101110100110
011110001010	36	−.0007607	−.0006674	.000423953	.000275101	.000159170	.000082092	.000037732	.000003260	.000000137	.000000000	010111110010
010111001000	37	−.0007607	−.0007313	.000423037	.000272760	.000156174	.000079394	.000035831	.000002910	.000000112	.000000000	110101100101
101011001000	37	−.0007749	−.0007500	.000420169	.000268020	.000151214	.000077889	.000033270	.000002502	.000000087	.000000000	101011001001
011111100000	38	−.0007786	−.0006686	.000420649	.000269895	.000157699	.000081462	.000035004	.000002808	.000000106	.000000000	110000111010
110101010010	37	−.0007792	−.0005889	.000421714	.000272791	.000156410	.000074512	.000037598	.000003324	.000000145	.000000000	011110011010
101001110000	37	−.0007889	−.0006992	.000422791	.000265927	.000149636	.000074065	.000032825	.000002470	.000000086	.000000000	111100011010
101010101000	37	−.0007889	−.0007109	.000418344	.000265516	.000149123	.000073566	.000032523	.000002421	.000000083	.000000000	101101010100
110101010010	37	−.0007889	−.0007241	.000418180	.000265051	.000148547	.000079468	.000032187	.000002367	.000000080	.000000000	011011010010
110101010000	37	−.0007894	−.0005769	.000417993	.000270236	.000155159	.000079468	.000036295	.000003095	.000000128	.000000000	110110101010
010111001000	36	−.0007894	−.0006378	.000420002	.000268041	.000152369	.000076972	.000034546	.000002776	.000000106	.000000000	010111011000
101101011000	36	−.0007894	−.0006381	.000419136	.000268066	.000152426	.000077045	.000034614	.000002795	.000000108	.000000000	011011010100
110100010010	36	−.0007894	−.0006458	.000419139	.000267789	.000152076	.000076735	.000034400	.000002758	.000000106	.000000000	011101010100
110010110000	37	−.0007894	−.0006575	.000419029	.000267352	.000151511	.000076222	.000034035	.000002690	.000000101	.000000000	110011010001
011011010100	37	−.0007936	−.0006755	.000418860	.000266716	.000150715	.000075523	.000033557	.000002609	.000000096	.000000000	101010010001
100111001010	36	−.0007936	−.0006246	.000418550	.000267322	.000151782	.000076594	.000034344	.000002754	.000000106	.000000000	101010101010
101101010100	36	−.0007936	−.0006277	.000418151	.000267238	.000151694	.000076534	.000034317	.000002756	.000000106	.000000000	011011011000
110101010100	37	−.0008029	−.0006608	.000418041	.000266051	.000150195	.000075204	.000033394	.000002594	.000000095	.000000000	110110010100
100111010010	36	−.0008034	−.0006664	.000416265	.000263199	.000152894	.000072894	.000031912	.000002361	.000000081	.000000000	110100101010
101011001010	36	−.0008034	−.0005502	.000417816	.000267179	.000152297	.000077344	.000034979	.000002889	.000000115	.000000000	011101010010
110101000110	36	−.0008034	−.0006077	.000417023	.000265232	.000149906	.000075281	.000033589	.000002662	.000000101	.000000000	101101011000
011011010100	36	−.0008034	−.0006330	.000416661	.000264316	.000148744	.000074247	.000032870	.000002534	.000000092	.000000000	110101011000
110001011100	38	−.0008038	−.0006088	.000416903	.000264956	.000149522	.000074893	.000033282	.000002590	.000000095	.000000000	111011011100
110010101100	37	−.0008038	−.0006235	.000416689	.000264407	.000148816	.000074258	.000032837	.000002510	.000000089	.000000000	100101110100
100111010010	37	−.0008141	−.0005886	.000415323	.000261995	.000147624	.000146583	.000032523	.000002496	.000000090	.000000000	100111000110
100101110010	37	−.0008141	−.0006119	.000414993	.000262786	.000146583	.000147840	.000031893	.000002389	.000000083	.000000000	101000111010
111001011010	36	−.0008178	−.0005598	.000415049	.000263206	.000147840	.000147840	.000032773	.000002547	.000000093	.000000000	101000101010
111001001001	36	−.0008223	−.0005135	.000414894	.000262185	.000148725	.000074826	.000033510	.000002699	.000000105	.000000000	010101011010
011011010100	36	−.0008223	−.0005431	.000414482	.000262474	.000147455	.000073713	.000032746	.000002566	.000000096	.000000000	111100110010
101101000110	36	−.0008223	−.0005642	.000414186	.000261446	.000146533	.000072296	.000032195	.000002474	.000000090	.000000000	111100010010
110100010110	37	−.0008281	−.0005496	.000413330	.000260304	.000145505	.000072167	.000031749	.000002408	.000000086	.000000000	100111010100

173

Table A

m = 7, n = 6   Rank Orders 1351 thru 1400 of 1716 $C(7,6) = 1716$ $1/C(7,6) = .00058275$

Column headings for the density columns: $d = .2, .4, .6, .8, 1.0, 1.5, 2.0, 3.0$

z	w	c_1	c_2	.2	.4	.6	.8	1.0	1.5	2.0	3.0	z^{tc}
1001011100100	37	−.0008281	−.0005643	.000413120	.000259775	.000144838	.000071578	.000031344	.000002338	.000000081	.000000000	1011100010110
1001101010100	37	−.0008281	−.0005868	.000412806	.000259000	.000143884	.000070756	.000030791	.000002250	.000000076	.000000000	1101010010010
0011111001000	35	−.0008326	−.0004199	.000414360	.000263735	.000150226	.000076663	.000034908	.000003009	.000000129	.000000000	1110110000010
1010111001000	35	−.0008326	−.0005034	.000413175	.000260730	.000146401	.000073132	.000032494	.000002560	.000000097	.000000000	0111110010010
0110111000010	35	−.0008326	−.0005211	.000412923	.000260092	.000145590	.000072406	.000031987	.000002468	.000000091	.000000000	1011110010010
1110001110001	36	−.0008363	−.0004648	.000413032	.000261003	.000146989	.000073746	.000032952	.000002645	.000000102	.000000000	1110001100001
0111000111000	36	−.0008363	−.0005010	.000412527	.000259750	.000145429	.000072379	.000032014	.000002482	.000000092	.000000000	0111001110010
1111000000011	35	−.0008466	−.0002687	.000413919	.000265078	.000152966	.000079501	.000037200	.000003514	.000000172	.000000000	0011110011000
0111011100001	35	−.0008466	−.0003687	.000412524	.000261574	.000148516	.000075487	.000034339	.000002947	.000000126	.000000000	1110101000010
0111000100110	35	−.0008466	−.0004688	.000411119	.000258053	.000144083	.000071549	.000031596	.000002449	.000000091	.000000000	0111011100010
0111100100010	35	−.0008466	−.0004804	.000410952	.000257624	.000143532	.000071052	.000031245	.000002384	.000000087	.000000000	1011011000010
0111010100001	35	−.0008466	−.0004821	.000410939	.000257621	.000143565	.000071112	.000031308	.000002404	.000000088	.000000000	0111110100010
1011011001000	35	−.0008470	−.0004966	.000410734	.000257100	.000142903	.000070522	.000030896	.000002330	.000000083	.000000000	1011010101010
1001110101000	37	−.0008470	−.0004484	.000411308	.000258570	.000144717	.000072080	.000031937	.000002496	.000000093	.000000000	1011010101010
1010101010010	37	−.0008470	−.0005288	.000410181	.000255762	.000141221	.000069027	.000029857	.000002148	.000000072	.000000000	1010101010010
1001010110001	37	−.0008568	−.0005407	.000410019	.000255367	.000140743	.000068623	.000029591	.000002107	.000000069	.000000000	1001110010010
1010011001100	37	−.0008568	−.0004646	.000408568	.000255270	.000141296	.000069382	.000030212	.000002229	.000000077	.000000000	1011010010110
1010011010010	37	−.0008568	−.0004905	.000408948	.000254392	.000140222	.000068462	.000029597	.000002132	.000000071	.000000000	1011010011010
1110010011010	37	−.0008610	−.0005060	.000408737	.000253882	.000139607	.000067943	.000029257	.000002080	.000000071	.000000000	1010011011000
1101000010110	37	−.0008610	−.0004046	.000409363	.000256124	.000142635	.000070661	.000031122	.000002388	.000000087	.000000000	1001011111000
1001100111000	37	−.0008610	−.0004525	.000408704	.000254509	.000140656	.000068957	.000029976	.000002202	.000000076	.000000000	1110001100110
1101000101010	37	−.0008610	−.0004711	.000408449	.000253886	.000139897	.000068309	.000029545	.000002134	.000000072	.000000000	1010011101010
1100100101100	37	−.0008610	−.0004802	.000408323	.000253580	.000139525	.000067992	.000029336	.000002101	.000000070	.000000000	1011100101010
1001001110010	37	−.0008610	−.0004816	.000408303	.000253529	.000139460	.000067934	.000029295	.000002095	.000000069	.000000000	1011010010010
1011101001100	36	−.0008713	−.0004955	.000408114	.000253073	.000138911	.000067472	.000028992	.000002049	.000000067	.000000000	0111110110010
1001110110010	37	−.0008713	−.0004054	.000407500	.000253310	.000139915	.000068641	.000029898	.000002220	.000000078	.000000000	1011010101100
0011011010010	37	−.0008718	−.0004447	.000406960	.000251989	.000138303	.000067261	.000028977	.000002073	.000000069	.000000000	0111110100010
1110000011010	37	−.0008718	−.0003843	.000407711	.000253938	.000141296	.000069407	.000030440	.000002318	.000000084	.000000000	1110011000010
1010001111000	35	−.0008897	−.0003176	.000405368	.000251164	.000138523	.000067964	.000029648	.000002221	.000000079	.000000000	1010010111000
1110001110010	37	−.0008897	−.0003685	.000404674	.000249476	.000136471	.000066212	.000028479	.000002035	.000000068	.000000000	1001011111000
0111110000001	33	−.0008902	−.0002204	.000406627	.000254419	.000142668	.000071672	.000032246	.000002698	.000000113	.000000000	0111111000001
1011011000010	36	−.0008999	−.0003716	.000402784	.000246612	.000133713	.000064183	.000027267	.000001877	.000000060	.000000000	1011010101010
1010110101010	36	−.0008999	−.0003845	.000402609	.000246190	.000133207	.000063759	.000026990	.000001836	.000000058	.000000000	1010110101010
1011000011010	36	−.0009042	−.0003665	.000402088	.000245629	.000132809	.000063539	.000026891	.000001829	.000000058	.000000000	1011101000010
1010001110010	35	−.0009047	−.0002084	.000404167	.000250790	.000139156	.000069013	.000030578	.000002435	.000000095	.000000000	1011110010010
1110100000101	35	−.0009047	−.0002685	.000403350	.000248804	.000136733	.000066928	.000029172	.000002200	.000000080	.000000000	1110010001010
1101010010010	35	−.0009047	−.0002840	.000403136	.000248283	.000136092	.000066373	.000028795	.000002136	.000000076	.000000000	1101010010010
0101011010100	35	−.0009047	−.0002852	.000403115	.000248220	.000136004	.000066288	.000028732	.000002124	.000000075	.000000000	1010101101000
0111001100100	37	−.0009047	−.0002883	.000403083	.000248166	.000135968	.000066283	.000028747	.000002134	.000000076	.000000000	0111010110100
0110101010100	35	−.0009047	−.0003206	.000402641	.000247087	.000134649	.000065151	.000027986	.000002010	.000000068	.000000000	0110101011100
1010100101010	36	−.0009139	−.0003389	.000400708	.000243903	.000131380	.000062611	.000026391	.000001777	.000000055	.000000000	1011010110010
1101010010010	35	−.0009144	−.0002647	.000401647	.000246297	.000134356	.000065190	.000028132	.000002060	.000000073	.000000000	0111011010010
0111011001010	35	−.0009144	−.0002892	.000401309	.000245470	.000133342	.000064317	.000027544	.000001963	.000000066	.000000000	1010110101010
1100011001100	37	−.0009149	−.0002935	.000401146	.000245096	.000132855	.000063860	.000027209	.000001898	.000000062	.000000000	1110010010010
0101100101000	35	−.0009149	−.0002103	.000401602	.000246843	.000135247	.000066026	.000028707	.000002150	.000000078	.000000000	0111110010100
0111010011100	35	−.0009187	−.0002441	.000401159	.000245810	.000134044	.000065040	.000028077	.000002061	.000000073	.000000000	1011010011100
0111001100100	35	−.0009187	−.0002720	.000400777	.000244878	.000132906	.000064064	.000027422	.000001954	.000000066	.000000000	0111011001100
1100101010100	35	−.0009191	−.0002838	.000400513	.000244264	.000132130	.000063366	.000026930	.000001864	.000000066	.000000000	1100101101100
1100100101100	37	−.0009191	−.0002175	.000399660	.000243815	.000132246	.000063748	.000027289	.000001946	.000000066	.000000000	1100101011100
1101000101100	37	−.0009289	−.0002175	.000399993	.000248891	.000131120	.000062785	.000026655	.000001844	.000000060	.000000000	1101010111100

174

z	w	c_1	c_2	z	.2	.4	.6	.8	1.0	1.5	2.0	3.0	z^{tc}
1110010001001	35	−.0009333	−.0001635	1111000001010	.000399594 •	.000244412	.000133297	.000064797	.000028067	.000002089	.000000075	.000000000	0110111011000
0110110011000	35	−.0009333	−.0002127	0101110010000	.000398932	.000242821	.000131380	.000063172	.000026989	.000001918	.000000065	.000000000	1110110011001
0110101011000	35	−.0009376	−.0001968	1010101000010	.000398382	.000242188	.000130896	.000062842	.000026842	.000001904	.000000064	.000000000	0110100101001
1010101000100	36	−.0009391	−.0001957	0111100010010	.000398109	.000241765	.000130464	.000062619	.000026619	.000001867	.000000062	.000000000	1010111010100
1001101100110	36	−.0009391	−.0002319	0110111001100	.000397619	.000240585	.000129044	.000061339	.000025830	.000001747	.000000055	.000000000	1101011000110
1001101100100	36	−.0009434	−.0002149	1001010010010	.000397084	.000239989	.000128603	.000061083	.000025707	.000001737	.000000055	.000000000	1110110000100
0110100101000	35	−.0009473	−.0001162	1100101001100	.000397702 •	.000242108	.000131401	.000063551	.000027377	.000002005	.000000071	.000000000	0110110011001
1110010100001	35	−.0009473	−.0001254	1100100101100	.000397586	.000241850	.000131116	.000063332	.000027247	.000001991	.000000070 •	.000000000	1110001011001
0111010101000	35	−.0009473	−.0001628	1010010101100	.000397084	.000240648	.000129675	.000062115	.000026443	.000001864	.000000063	.000000000	1110110111000
0011101011000	34	−.0009576	−.0000086	1100100110100	.000397551 •	.000224365	.000133790	.000065947	.000029148	.000002346	.000000095	.000000000	1110100000011
1010110100001	34	−.0009576	−.0001033	0111110010010	.000396041	.000239820	.000129362	.000062146	.000026585	.000001914	.000000067	.000000000	0111110010010
0110110000010	34	−.0009576	−.0001057	1011011001100	.000395838	.000239321	.000128749	.000061616	.000026226	.000001854	.000000063	.000000000	1011011001100
1010110001100	36	−.0009580	−.0001672	1010111001100	.000395082	.000237558	.000126634	.000059822	.000025035	.000001666	.000000052	.000000000	1010111001100
1000110011010	36	−.0009580	−.0001788	1011100111010	.000394930	.000237201	.000126216	.000059480	.000024817	.000001635	.000000049	.000000000	1011100111010
1011100100010	34	−.0009618	−.0000789	1011000111010	.000395595 •	.000239367	.000129020	.000061912	.000026436	.000001884	.000000064 •	.000000000	1011000111010
1011101000010	34	−.0009618	−.0000903	1101000011010	.000395453	.000239089	.000128768	.000061761	.000026377	.000001891	.000000066	.000000000	1101000011010
1001010101010	36	−.0009618	−.0001033	1010110101010	.000394671	.000238653 •	.000128222	.000061287	.000026055	.000001837	.000000062	.000000000	1010110101010
1001110100010	36	−.0009720	−.0001461	1011010101010	.000394078	.000238802 •	.000129001	.000062176	.000026722	.000001953	.000000069	.000000000	1011010101010
1001111100000	35	−.0009720	−.0000555	1011001101010	.000391397	.000237385	.000127300	.000060738	.000025771	.000001803	.000000060	.000000000	1011001101010
1100001000110	36	−.0009720	−.0000822	1011001110010	.000391245	.000236511	.000126268	.000059882	.000025215	.000001720	.000000055	.000000000	1011001110010
1001101011000	36	−.0009720	−.0001035	1110010011010	.000393423	.000235836	.000125469	.000059218	.000024785	.000001657	.000000052	.000000000	1110010011010
1100110001010	36	−.0009720	−.0001057	1010101011100	.000393392	.000235763	.000125382	.000059145	.000024737	.000001649	.000000051	.000000000	1010101011100
1010010000110	36	−.0009720	−.0001209	1010111001010	.000393191	.000235292	.000124830	.000058661	.000024447	.000001608	.000000049	.000000000	1010111001010
1010011001010	36	−.0009720	−.0001224	1011010011010	.000393174	.000235254	.000124791	.000058663	.000024431	.000001606	.000000049	.000000000	1011010011010
0110011010010	36	−.0009720	−.0001318	1010110101010	.000393046	.000234996	.000124484	.000058408	.000024266	.000001582	.000000048	.000000000	1010110101010
1010101010100	36	−.0009720	−.0001461	1011011010110	.000392859	.000234947	.000124423	.000058355	.000024230	.000001577	.000000048	.000000000	1011011010110
1010101010010	36	−.0009823	−.0000555	1010101101010	.000392218	.000234511	.000123977	.000057944	.000023987	.000001541	.000000046	.000000000	1010101101010
1001010101100	35	−.0009860	−.0000099	0110111100010	.000391397	.000234603 •	.000124648	.000058801	.000024617	.000001652	.000000052	.000000000	0110111100010
1001110100010	36	−.0009860	−.0000673	1011001101010	.000391397	.000233251	.000123289	.000057776	.000023996	.000001569	.000000048	.000000000	1011001101010
1100001010010	36	−.0009860	−.0000788	1011001110010	.000391245	.000232898	.000122878	.000057441	.000023783	.000001540	.000000048	.000000000	1011001110010
1001101011000	36	−.0010007	−.0000082	0110110011010	.000389548	.000231183	.000121748	.000056871	.000023554	.000001532	.000000047 •	.000000000	0110110011010
1010010011010	36	−.0010049	−.0000412	1010101111000	.000389430	.000231545 •	.000122408	.000057502	.000023988	.000001599	.000000050	.000000000	1010101111000
1010101010010	36	−.0010049	−.0000050	1010111010010	.000388962	.000230463	.000121160	.000056492	.000023251	.000001510	.000000046	.000000000	1010111010010
1010110000100	35	−.0010152	−.0000428	1011010101010	.000387623	.000228916	.000119972	.000055774	.000022990	.000001478	.000000044	.000000000	1011010101010
1011011100010	34	−.0010152	.0000152	1011001010110	.000387267	.000228099	.000119037	.000055024	.000022522	.000001415	.000000041	.000000000	1011001010110
0101111001000	34	−.0010157	.0001159	0110111100010	.000388509 •	.000231099	.000122599	.000057981	.000024437	.000001699	.000000057	.000000000	0110111100010
0110110001010	34	−.0010157	−.0001129	1011011001010	.000388472	.000231023	.000122522	.000057930	.000024413	.000001699	.000000057	.000000000	1011011001010
0110111001100	34	−.0010157	−.0000798	1011001101100	.000380035	.000229993	.000121304	.000056916	.000023751	.000001598	.000000051	.000000000	1011001101100
1001010100010	35	−.0010249	−.0001305	1101011010010	.000385935	.000226491	.000117749	.000054214	.000022100	.000001374	.000000040	.000000000	1101011010010
1100001010010	36	−.0010292	−.0000668	1011001110100	.000385484	.000225993	.000117419	.000054044	.000022029	.000001371	.000000040	.000000000	1011001110100
1111000001010	34	−.0010297	.0002648	0101111110000	.000387942 •	.000231896	.000124360	.000059784	.000025746	.000001929	.000000072	.000000000	0101111110000
0101110010000	34	−.0010297	.0001763	1110110000101	.000386792	.000229218	.000121219	.000057179	.000024046	.000001665	.000000056	.000000000	1110110000101
1010101000010	34	−.0010297	.0001531	0110110110100	.000386499	.000228560	.000120480	.000056596	.000023688	.000001619	.000000054	.000000000	0110110110100
0111100010010	34	−.0010297	.0001410	1101110101001	.000386345	.000228209	.000120080	.000055900	.000023487	.000001591	.000000052	.000000000	1101110101001
0110111001100	34	−.0010297	.0001312	1101011010100	.000386207	.000227865	.000119651	.000055900	.000023231	.000001548	.000000049	.000000000	1101011010100
1001010010010	36	−.0010297	.0001138	1110101010001	.000385986	.000227366	.000119085	.000055276	.000022950	.000001510	.000000047	.000000000	1110101010001
1100101001100	36	−.0010301	.0000883	1101001011100	.000385549	.000226367	.000117891	.000055438	.000022278	.000001404	.000000041	.000000000	1101001011100
1100100101100	34	−.0010399	.0001305	1101010101100	.000384362	.000225091	.000116986	.000053936	.000022050	.000001389	.000000041	.000000000	1101010101100
1010010101100	35	−.0010441	.0001556	1101010101100	.000383931	.000224731	.000116815	.000053895	.000022061	.000001397	.000000042	.000000000	1101010101100
1100100110100	36	−.0010441	.0001427	1101001101100	.000383764	.000224345	.000116370	.000053536	.000021835	.000001366	.000000040	.000000000	1101001101100

175

Table A

m = 7, n = 6 Rank Orders 1451 thru 1500 of 1716 C(7, 6) = 1716 1/C(7, 6) = .00058275

z	w	c_1	c_2	d = .2	.4	.6	.8	1.0	1.5	2.0	3.0	z^{tc}
1110100001001	34	-.0010486	.0002607	.000384512●	.000226787	.000119505	.000056237	.000023624	.00001641	.00000056	.00000000	0110111001001
0110110101000	34	-.0010486	.0001968	.000383687	.000224883	.000117301	.000054438	.000022473	.00001473	.00000046	.00000000	1110101001001
1011010000110	35	-.0010544	.0002330	.000383110	.000224389	.000117029	.000054323	.000022432	.00001470	.00000046	.00000000	1001111100100
1110110100100	35	-.0010544	.0001822	.000382452	.000222874	.000115283	.000052910	.000021539	.00001345	.00000039	.00000000	1101101011000
1110100100001	34	-.0010583	.0002706	.000382904●	.000224533	.000117460	.000054800	.000022794	.00001537	.00000050	.00000000	0111010011000
0111010010000	34	-.0010583	.0002343	.000382434	.000224107	.000117206	.000053772	.000022136	.00001441	.00000051	.00000000	1111001110000
1110100110000	34	-.0010626	.0002929	.000382447●	.000224107	.000116602	.000054689	.000022759	.00001539	.00000047	.00000000	0111011001000
0111010001100	34	-.0010626	.0002782	.000382237	.000223604	.000117206	.000054178	.000022420	.00001485	.00000045	.00000000	1110101001001
0111000011100	34	-.0010626	.0002528	.000381915	.000222884	.000115795	.000053540	.000022026	.00001432	.00000045	.00000000	1111010011001
1100000111100	36	-.0010728	.0003285	.000381055	.000222370	.000115727	.000053674●	.000022172	.00001460	.00000046	.00000000	1000111111000
1100010111000	36	-.0010728	.0002891	.000380547	.000221201	.000114377	.000052578	.000021475	.00001360	.00000041	.00000000	1110001011100
0011111100001	33	-.0010729	.0004650	.000382826●	.000226533	.000120654	.000057801	.000024893	.00001895	.00000074	.00000000	1111111100000
0111101000010	33	-.0010729	.0003306	.000381095	.000222528	.000115979	.000053937	.000022376	.00001503	.00000049	.00000000	0111110100010
0111101000010	33	-.0010729	.0003172	.000380916	.000222104	.000115473	.000053511	.000022095	.00001458	.00000047	.00000000	1001110100010
1001110011000	35	-.0010830	.0003527	.000379544	.000220429	.000114088	.000052593	.000021583	.00001395	.00000043	.00000000	1011101000110
1010100001010	35	-.0010830	.0003048	.000378930	.000219028	.000112485	.000051304	.000020772	.00001284	.00000037	.00000000	1010111000110
1010101001010	35	-.0010830	.0002863	.000378693	.000218489	.000111874	.000050818	.000020470	.00001243	.00000035	.00000000	1011101001010
1010011001010	35	-.0010830	.0002771	.000378576	.000218225	.000111575	.000050580	.000020323	.00001224	.00000034	.00000000	1100111010010
1010101010100	35	-.0010830	.0002758	.000378557	.000218179	.000111519	.000050533	.000020293	.00001220	.00000034	.00000000	1001111010010
1010011001000	35	-.0010830	.0002619	.000378381	.000217783	.000111075	.000050184	.000020079	.00001192	.00000033	.00000000	1101101001010
1001011101000	35	-.0010873	.0003230	.000378409●	.000218460	.000112075	.000051070	.000020662	.00001275	.00000037	.00000000	1101010010110
1100101001000	35	-.0010873	.0002970	.000378075	.000217698	.000111205	.000050374	.000020227	.00001217	.00000034	.00000000	1011001001100
1010101010010	35	-.0010873	.0002816	.000377880	.000217261	.000110716	.000049991	.000019993	.00001187	.00000033	.00000000	1011000010010
1010111000010	35	-.0010970	.0004089	.000377766	.000218395●	.000112517	.000051623	.000021079	.00001343	.00000041	.00000000	1111000010010
1001011001010	35	-.0010970	.0003285	.000376750	.000216109	.000109943	.000049588	.000019821	.00001178	.00000033	.00000000	1010101010010
1010101010010	35	-.0010970	.0003166	.000376600	.000215771	.000109563	.000049290	.000019639	.00001155	.00000032	.00000000	1010101101010
1001001000010	35	-.0011160	.0004447	.000374865	.000214546	.000109180	.000049372●	.000019829	.00001202	.00000034	.00000000	1110101101010
1010010011010	35	-.0011160	.0004299	.000374685	.000214158	.000108760	.000049053	.000019639	.00001180	.00000033	.00000000	1110101011010
1010101011000	35	-.0011160	.0004074	.000374399	.000213509	.000108027	.000048472	.000019280	.00001133	.00000033	.00000000	1010101101010
1010111000010	34	-.0011262	.0004578	.000373220	.000212321	.000107240	.000048069	.000019113	.00001125	.00000031	.00000000	1011101001010
1100000110010	35	-.0011299	.0005056	.000373159	.000212697●	.000107845	.000048612	.000019469	.00001174	.00000033	.00000000	1011000111000
1110001110000	35	-.0011299	.0004823	.000372871	.000212063	.000107146	.000048071	.000019142	.00001133	.00000032	.00000000	1110001110010
1011010100010	34	-.0011402	.0005089	.000371392	.000210203	.000105610	.000047081	.000018616	.00001080	.00000029	.00000000	1011011100010
0101011100010	34	-.0011402	.0004941	.000371211	.000209811	.000105185	.000046757	.000018424	.00001057	.00000028	.00000000	1011010110010
0101011101000	33	-.0011407	.0006219	.000372735●	.000213306	.000109150	.000049912	.000020386	.00001322	.00000042	.00000000	1011010000101
1011011010100	35	-.0011407	.0005881	.000372321	.000212403	.000108160	.000049149	.000019925	.00001264	.00000039	.00000000	0111110010001
0111011011000	33	-.0011407	.0005602	.000371965	.000211595	.000107239	.000048408	.000019457	.00001198	.00000036	.00000000	1101110010001
1100110010100	33	-.0011412	.0005181	.000371339	.000210216	.000105671	.000047146	.000018661	.00001087	.00000030	.00000000	0111101010001
1101110100100	35	-.0011449	.0006028	.000371756●	.000211736	.000107637	.000048825	.000019758	.00001248	.00000039	.00000000	1110111101000
0111100100100	33	-.0011449	.0005783	.000371442	.000211020	.000106816	.000048162	.000019338	.00001188	.00000035	.00000000	0111101100001
1100111000100	35	-.0011551	.0005853	.000369712	.000208544	.000104536	.000046544	.000018398	.00001071	.00000029	.00000000	1101100011000
1010100100100	35	-.0011551	.0005736	.000369566	.000208221	.000104177	.000046263	.000018228	.00001049	.00000028	.00000000	1010101010100
1100101010100	35	-.0011551	.0005604	.000369400	.000207851	.000103765	.000045942	.000018033	.00001025	.00000027	.00000000	1010110001010
0101011101000	33	-.0011596	.0006501	.000369747●	.000209280	.000105627	.000047522	.000019052	.00001169	.00000035	.00000000	1011011000100
1001110010100	34	-.0011654	.0006394	.000368576	.000206847	.000103852	.000046218	.000018277	.00001069	.00000030	.00000000	1010101110100
1100100110100	35	-.0011691	.0006345	.000367860	.000207450	.000102845	.000046373	.000017871	.00001022	.00000027	.00000000	0101111110000
1110100001001	33	-.0011736	.0007945	.000369063●	.000209672	.000106767	.000048693	.000019826	.00001298	.00000042	.00000000	0111011101000
1110100011000	33	-.0011736	.0007529	.000368435	.000208306	.000105269	.000047351	.000019168	.00001205	.00000037	.00000000	0111111011000

176

Table A

m = 7, n = 6   Rank Orders 1501 thru 1550 of 1716   C(7, 6) = 1716   1/C(7, 6) = .00058275

d = .2

z	w	c_1	c_2	.2	.4	.6	.8	1.0	1.5	2.0	3.0	z^{tc}
0111100011000	33	−.0011736	.0007104	.000368028	.000207392	.000104239	.000046712	.000018657	.000001134	.000000033	.000000000	1110011100001
1101010101000	33	−.0011736	.0006918	.000367803	.000206906	.000103712	.000046309	.000018415	.000001105	.000000032	.000000000	1101101010010
1011001011000	33	−.0011838	.0007177	.000366305	.000204942	.000101995	.000045128	.000017741	.000001022	.000000028	.000000000	1110010010010
0111111000001	32	−.0011839	.0008089	.000367436	.000207487	.000104866	.000047409	.000019163	.000001217	.000000038	.000000000	0111110001100
0111001110000	32	−.0011876	.0007959	.000367269	.000207102	.000104419	.000047043	.000018928	.000001181	.000000036	.000000000	1011110010001
1110000101100	35	−.0011881	.0007800	.000366423	.000205752	.000103120	.000046108	.000018385	.000001115	.000000033	.000000000	1011100010001
1100011011000	35	−.0011881	.0007572	.000366042	.000204931	.000102186	.000045350	.000017901	.000001046	.000000029	.000000000	1100101111000
1110010000110	34	−.0011983	.0007326	.000365742	.000204278	.000101474	.000044805	.000017575	.000001006	.000000027	.000000000	1110001101100
1100100110100	34	−.0011983	.0008360	.000365271	.000204530 •	.000102186	.000045643	.000018165	.000001096	.000000032	.000000000	1001111101010
1010111010100	34	−.0012270	.0007734	.000364447	.000202844	.000100427	.000044208	.000017295	.000000985	.000000027	.000000000	1101111010000
1010100010010	34	−.0011983	.0007707	.000364410	.000202752	.000100316	.000044114	.000017234	.000000976	.000000026	.000000000	1010110110000
1010110101000	34	−.0011983	.0007628	.000364311	.000202535	.000100077	.000043930	.000017122	.000000963	.000000026	.000000000	1100111100010
1010100100100	34	−.0011983	.0007467	.000364116	.000202115	.000099623	.000043586	.000016919	.000000939	.000000024	.000000000	0111101001100
1010100010010	34	−.0012081	.0007316	.000363933	.000201717	.000099193	.000043259	.000016726	.000000916	.000000023	.000000000	1010110100100
1001101110000	34	−.0012081	.0007817	.000362838	.000200634	.000098487	.000042901	.000016577	.000000909	.000000023	.000000000	1010101010100
1001101011000	34	−.0012123	.0007702	.000362697	.000200328	.000098155	.000042649	.000016428	.000000891	.000000023	.000000000	1011101010010
0101110110100	34	−.0012123	.0008621	.000363071 •	.000201694	.000098833	.000044000	.000017256	.000000994	.000000023	.000000000	1011100100010
1100110100100	34	−.0012123	.0008054	.000362384	.000200209	.000098227	.000042779	.000017023	.000000907	.000000023	.000000000	0111111010100
1010011001100	34	−.0012123	.0007923	.000362225	.000199864	.000097855	.000042497	.000016365	.000000888	.000000029	.000000000	0110110101100
1110010100010	34	−.0012270	.0009048	.000361020	.000199181	.000097803	.000042715 •	.000016584	.000000928	.000000026	.000000000	1101111011000
1010110011000	34	−.0012270	.0008687	.000360584	.000198243	.000096794	.000041953	.000016134	.000000875	.000000022	.000000000	1100111001010
1010011101000	34	−.0012312	.0008895	.000360096	.000197745	.000096458	.000041775	.000016056	.000000870	.000000022	.000000000	1110100011010
1110001011000	34	−.0012410	.0009815	.000359502	.000197728	.000096887	.000042266	.000016065	.000000920	.000000025	.000000000	1101011001100
1110010011000	34	−.0012410	.0009547	.000359177	.000197028	.000096132	.000041693	.000016405	.000000879	.000000023	.000000000	1101010011000
1011010011000	34	−.0012410	.0009307	.000357728	.000196420	.000095486	.000041212	.000015785	.000000848	.000000021	.000000000	1110101101000
1110010010000	33	−.0012512	.0009827	.000357205	.000195260	.000094721	.000040819	.000015619	.000000839	.000000021	.000000000	1110110100010
0101111110000	34	−.0012555	.0010007	.000358914 •	.000198439	.000098398	.000040579	.000015505	.000000830	.000000021	.000000000	1011101100010
1110110000100	32	−.0012560	.0010920	.000358241	.000197029	.000096910	.000042612	.000017394	.000001066	.000000032	.000000000	0111011010100
0111110000010	32	−.0012560	.0010667	.000357929	.000196342	.000096149	.000042016	.000016379	.000000941	.000000026	.000000000	1101011011000
1110010010100	34	−.0012662	.0010491	.000355921	.000193315	.000093326	.000040029	.000015246	.000000810	.000000020	.000000000	1100111010100
1100010110100	34	−.0012662	.0010362	.000355765	.000192979	.000092965	.000039756	.000015086	.000000791	.000000019	.000000000	1010101001100
1100101010010	34	−.0012704	.0010592	.000355301	.000192532	.000092680	.000039614	.000015028	.000000788	.000000019	.000000000	1101100111000
1001101000001	32	−.0012802	.0010982	.000354072	.000191158	.000091661	.000039019	.000014740	.000000765	.000000019	.000000000	1110100111000
0101111100010	32	−.0012846	.0012324	.000354916 •	.000193572	.000094496	.000041270	.000016120	.000000941	.000000027	.000000000	1011110100010
0101101100100	32	−.0012846	.0012167	.000354721	.000193140	.000094012	.000040887	.000015881	.000000908	.000000025	.000000000	1110111000100
1110010100001	32	−.0012846	.0011911	.000354422	.000192512	.000093352	.000040398	.000015596	.000000875	.000000024	.000000000	0111110001001
1110110000010	32	−.0012889	.0012482	.000354367	.000192943	.000094017	.000040983	.000015976	.000000928	.000000027	.000000000	0111011101000
0111100100100	34	−.0012889	.0012120	.000353930	.000192003	.000092998	.000040202	.000015506	.000000868	.000000024	.000000000	1101011011000
1110110000100	33	−.0012986	.0013291	.000353619	.000192558 •	.000094020	.000041134	.000016115	.000000952	.000000026	.000000000	1001111110000
0111101010100	32	−.0012986	.0012653	.000352865	.000190965	.000092322	.000039854	.000015356	.000000859	.000000023	.000000000	1100101010001
1110001101100	34	−.0012991	.0012287	.000352330	.000189840	.000091093	.000038901	.000014775	.000000784	.000000020	.000000000	1100101110000
1100111010100	34	−.0012991	.0012270	.000352315	.000189820	.000091084	.000038902 •	.000014781	.000000786	.000000019	.000000000	1110100111100
1001111001000	33	−.0012991	.0012033	.000352031	.000189213	.000090436	.000038415	.000014495	.000000753	.000000018	.000000000	1011101110000
1001110011000	33	−.0013094	.0012835	.000351204	.000188765	.000089717	.000038588 •	.000014660	.000000782	.000000020	.000000000	1011101010110
1010110010100	33	−.0013094	.0012576	.000350893	.000188101	.000088101	.000038053	.000014345	.000000745	.000000018	.000000000	1011111001000
1110000110100	34	−.0013094	.0012421	.000350765	.000187721	.000089320 •	.000037761	.000014177	.000000727	.000000018	.000000000	1110110001010
1110000101100	33	−.0013131	.0012968	.000350708	.000188165 •	.000089940	.000038273	.000014490	.000000764	.000000018	.000000000	1101110100100
1010101110100	34	−.0013131	.0012850	.000350573	.000187888 •	.000089655	.000038068	.000014374	.000000751	.000000019	.000000000	1010101110100
1111000011000	33	−.0013233	.0014339	.000350541	.000189093 •	.000091377	.000039519	.000015282	.000000867	.000000024	.000000000	1001111110000

177

Table A

$m = 7$, $n = 6$ Rank Orders 1551 thru 1600 of 1716 $C(7, 6) = 1716$ $1/C(7, 6) = .00058275$

z	w	c_1	c_2	$d = .2$	.4	.6	.8	1.0	1.5	2.0	3.0	z^{tc}
1001111010010	33	−.0013233	.0013585	.000349658	.000187239	.000089422	.000038062	.00014430	.000000767	.000000020	.000000000	1111010001110
1101100010100	33	−.0013233	.0013142	.000349138	.000186151	.000088279	.000037218	.00013943	.000000712	.000000017	.000000000	1110110110010
1011010010010	33	−.0013233	.0013039	.000349016	.000185894	.000088008	.000037011	.00013827	.000000699	.000000017	.000000000	1101011100100
1101010100010	33	−.0013233	.0013007	.000348981	.000185829	.000087947	.000036978	.00013807	.000000698	.000000017	.000000000	1011101010100
1010100100100	33	−.0013233	.0012880	.000348832	.000185517	.000087622	.000037047	.00013944	.000000727	.000000016	.000000000	1100111101000
1110100001010	33	−.0013423	.0014437	.000347378	.000184783	.000087649	.000037047	.00013671	.000000683	.000000016	.000000000	1010111011000
1010110101000	33	−.0013423	.0013929	.000346788	.000183563	.000086384	.000036125	.00013944	.000000670	.000000018	.000000000	1101101011000
1100110011000	33	−.0013520	.0013923	.000345893	.000182878	.000086074	.000036042	.00013419	.000000677	.000000016	.000000000	1110110110100
1110101011000	33	−.0013520	.0014376	.000345625	.000182329	.000085510	.000035636	.00013420	.000000677	.000000016	.000000000	1011010110010
1010101110000	33	−.0013562	.0014893	.000345488	.000182543	.000085896	.000035973	.00013191	.000000653	.000000015	.000000000	1110110010010
1100100100110	33	−.0013562	.0014873	.000345467	.000182504	.000085861	.000035950	.00013388	.000000676	.000000016	.000000000	1011101101000
1010110010100	33	−.0013562	.0014609	.000345165	.000181886	.000085228	.000035495	.00013133	.000000650	.000000016	.000000000	1110110010001
1011101000010	32	−.0013665	.0015314	.000344211	.000181146	.000084886	.000035400	.00013131	.000000657	.000000016	.000000000	1101010010000
0111111000000	31	−.0013670	.0016382	.000345381	.000183672	.000087584	.000037584	.00014326	.000000800	.000000022	.000000000	0111111000010
1011010001100	33	−.0013670	.0016137	.000345089	.000183048	.000086912	.000036915	.00014018	.000000761	.000000017	.000000000	1101011000001
1100101010100	33	−.0013815	.0016047	.000342483	.000179364	.000083658	.000034251	.00012826	.000000635	.000000015	.000000000	1101011001100
1001101101000	33	−.0013815	.0015772	.000342166	.000178715	.000082993	.000034251	.00012556	.000000607	.000000014	.000000000	1101100101100
1010100010100	33	−.0013912	.0016196	.000340980	.000177435	.000082075	.000033733	.00012313	.000000588	.000000016	.000000000	1011100101010
1010011010100	33	−.0013954	.0016451	.000340543	.000177039	.000081838	.000033623	.00012272	.000000587	.000000013	.000000000	1100011101000
0101011110000	31	−.0013999	.0018000	.000341555	.000179651	.000084723	.000035796	.00013539	.000000731	.000000020	.000000000	1010101010000
1110101010001	31	−.0013999	.0017908	.000341468	.000179510	.000084617	.000035747	.00013528	.000000735	.000000020	.000000000	0111110101000
0111011010000	31	−.0013999	.0017533	.000341034	.000178608	.000083671	.000035045	.00013118	.000000686	.000000018	.000000000	1110110010001
0111011010100	31	−.0014097	.0018102	.000340006	.000177639	.000083067	.000034740	.00012991	.000000679	.000000017	.000000000	1101100010001
1110010001100	33	−.0014101	.0017571	.000339901	.000176210	.000081567	.000033632	.00012345	.000000603	.000000014	.000000000	1010110001010
1101000101100	33	−.0014101	.0017325	.000339021	.000175643	.000080991	.000033221	.00012115	.000000580	.000000013	.000000000	1011011001000
1111000000001	33	−.0014139	.0018826	.000340101	.000178329	.000083943	.000035448	.00013421	.000000732	.000000017	.000000000	1011101100001
0111011000010	31	−.0014139	.0018335	.000339541	.000177182	.000082758	.000034580	.00012920	.000000655	.000000016	.000000000	1110010001000
1001011001000	31	−.0014144	.0017566	.000338569	.000175216	.000080724	.000033089	.00012061	.000000577	.000000014	.000000000	1110100111000
1100011110000	33	−.0014241	.0018561	.000338027	.000175235	.000081120	.000033506	.00012339	.000000612	.000000015	.000000000	1011100111000
1110001010010	33	−.0014241	.0018138	.000337546	.000174259	.000080126	.000032793	.00011939	.000000570	.000000013	.000000000	1010101101000
1101010011000	31	−.0014241	.0018017	.000337412	.000173994	.000079863	.000032610	.00011839	.000000560	.000000013	.000000000	1111010101000
1001111001000	32	−.0014344	.0019143	.000336931	.000174204	.000080481	.000033199	.00012219	.000000607	.000000015	.000000000	1011011000010
1011011000010	32	−.0014344	.0018576	.000336290	.000172910	.000079169	.000032262	.00011695	.000000553	.000000013	.000000000	1101100110010
1010110001000	32	−.0014344	.0018445	.000336142	.000172611	.000078868	.000032048	.00011576	.000000541	.000000013	.000000000	1101101010010
1101010101000	33	−.0014386	.0018773	.000335787	.000172382	.000078801	.000032059	.00011602	.000000546	.000000012	.000000000	1011101000100
1010100001000	32	−.0014386	.0018657	.000335656	.000172117	.000078532	.000031868	.00011496	.000000535	.000000012	.000000000	1100110101010
1010111001000	33	−.0014533	.0019564	.000334169	.000170789	.000077750	.000031510	.00011362	.000000530	.000000012	.000000000	1110011001010
1010011010010	32	−.0014570	.0020175	.000332003	.000171291	.000078386	.000032003	.00011649	.000000574	.000000013	.000000000	1110101111000
1110000110010	33	−.0014673	.0021023	.000333407	.000170812	.000078280	.000032059	.00011722	.000000612	.000000014	.000000000	1010111110000
1101000010010	32	−.0014673	.0020479	.000332807	.000169629	.000077108	.000031239	.00011272	.000000529	.000000012	.000000000	1011011101000
1010110100100	32	−.0014673	.0020386	.000332702	.000169418	.000076894	.000031087	.00011188	.000000521	.000000012	.000000000	1111010010010
1001101010100	32	−.0014673	.0020356	.000332670	.000169358	.000076838	.000031050	.00011169	.000000519	.000000011	.000000000	1011011000010
1011100011000	32	−.0014673	.0020270	.000332574	.000169166	.000076646	.000030915	.00011094	.000000512	.000000011	.000000000	1101100100010
1010110101000	32	−.0014673	.0020099	.000332386	.000168798	.000076285	.000030666	.00010960	.000000499	.000000011	.000000000	1101110100100
1011101010100	31	−.0014775	.0021199	.000331856	.000168866	.000076717	.000031092	.00011233	.000000531	.000000011	.000000000	1011110010100
1011001111000	32	−.0014812	.0021131	.000331150	.000167879	.000075873	.000030550	.00010952	.000000505	.000000011	.000000000	1110110001100
1100011001000	32	−.0014925	.0021560	.000329792	.000166285	.000074704	.000029878	.00010633	.000000480	.000000010	.000000000	1110110001010
1010101101000	32	−.0015065	.0022234	.000328092	.000164602	.000073544	.000029248	.00010348	.000000450	.000000010	.000000000	1011010110010
1100110100100	32	−.0015065	.0022086	.000327931	.000164291	.000073242	.000029123	.00010238	.000000440	.000000009	.000000000	1101110101100

178

Rank Orders 1601 thru 1650 of 1716

$m = 7, \quad n = 6$ $C(7, 6) = 1716$ $1/C(7, 6) = .00058275$

z	w	c_1	c_2	$d = .2$	.4	.6	.8	1.0	1.5	2.0	3.0	z^{tc}
0111110010000000	30	−.0015109	.0023537	.000328784	.00166486	.000075607	.000030773	.000011218	.000000554	.000000014	.000000000	1110111010000001
0111010000000001	30	−.0015249	.0024712	.000327705	.00165883	.000075523	.000030901	.000011355	.000000579	.000000015	.000000000	0111110110000001
0111011010010000	30	−.0015249	.0024416	.000327371	.00165207	.000074830	.000030398	.000011066	.000000532	.000000013	.000000000	1111100010001000
0111000001100000	32	−.0015254	.0024205	.000326465	.00164783	.000074423	.000030121	.000010918	.000000546	.000000013	.000000000	1111101010000100
0111010010010000	32	−.0015254	.0023672	.000326036	.00163448	.000073105	.000029153	.000010371	.000000472	.000000010	.000000000	1101101001001000
1101001001010000	32	−.0015351	.0023279	.000325046	.000162611	.000072257	.000028589	.000010067	.000000443	.000000009	.000000000	1101010011001000
1101010010100000	32	−.0015351	.0023883	.000324920	.00161713	.000071712	.000028328	.000009962	.000000438	.000000009	.000000000	1111010110011000
1101010111000000	32	−.0015351	.0023765	.000324823	.00161473	.000071483	.000028173	.000009880	.000000430	.000000009	.000000000	1111000010010000
1101011000010000	32	−.0015394	.0024339	.000324823	.00161725	.000071866	.000028481	.000010059	.000000449	.000000010	.000000000	1110101101000000
1101001100000000	32	−.0015394	.0024164	.000324635	.00161363	.000071517	.000028243	.000009932	.000000437	.000000008	.000000000	1101101110011000
1101010110010000	32	−.0015394	.0024031	.000322493	.00161094	.000071259	.000028070	.000009842	.000000429	.000000009	.000000000	1101010010010000
1001111010000000	31	−.0015496	.0025566	.000324423	.00162035	.000072527	.000029063	.000010416	.000000489	.000000011	.000000000	1111110000010010
1011001100010000	31	−.0015496	.0024763	.000323557	.00160364	.000070904	.000027949	.000009817	.000000432	.000000009	.000000000	1011110100001000
1101000010100000	32	−.0015496	.0024644	.000323427	.00160109	.000070655	.000027778	.000009724	.000000423	.000000009	.000000000	1101110100000010
1101000101100000	31	−.0015680	.0025875	.000321653	.00158593	.000069796	.000027396	.000009584	.000000417	.000000009	.000000000	1110010111000000
1011101011000000	31	−.0015783	.0026478	.000320576	.00157592	.000069176	.000027094	.000009462	.000000411	.000000009	•	1110010100001000
1101011001010000	31	−.0015783	.0026458	.000320557	.00157556	.000069144	.000027074	.000009452	.000000410	.000000009	.000000000	1110111000010000
1101010100100000	31	−.0015783	.0026194	.000320275	.00157020	.000068631	.000026728	.000009270	.000000394	.000000008	•	1111100010100001
1101010011010000	31	−.0015825	.0026666	.000320068	.00157059	.000068807	.000026894	.000009372	.000000394	.000000008	.000000000	1101011110100010
1101101100000000	31	−.0015825	.0026434	.000319819	.00156582	.000068349	.000026585	.000009209	.000000390	.000000008	.000000000	1101101101000010
1111100010000010	31	−.0015923	.0027589	.000319410	.00156781	.000068845	.000027019	.000009469	.000000417	.000000009	.000000000	1011011110000000
1011010110000000	31	−.0015923	.0027081	.000318874	.00155768	.000067883	.000026375	.000009130	.000000387	.000000008	.000000000	1101100110010000
1011010110100000	31	−.0016175	.0027433	.000315986	.00152834	.000065899	.000025316	.000008595	.000000356	.000000008	.000000000	1101010101001000
1101100110000000	31	−.0016217	.0028564	.000315520	.00152372	.000065603	.000025164	.000008661	.000000352	.000000007	.000000000	1101010100100010
1111110010000000	29	−.0016359	.0031160	.000315904	.00154572	.000068197	.000027105	.000009697	.000000466	.000000011	.000000000	0111110010000100
0111110000010000	29	−.0016359	.0030669	.000315382	.00153572	.000067225	.000026434	.000009330	.000000429	.000000010	.000000000	1101111000001000
1100011010100000	31	−.0016402	.0029597	.000314158	.00151268	.000065010	.000024921	.000008516	.000000350	.000000007	.000000000	1111000010010010
0111010110000000	29	−.0016402	.0030959	.000314976	.00153224	.000067024	.000026341	.000009294	.000000427	.000000010	•	1110001010100000
1111000101101000	31	−.0016504	.0031033	.000313325	.00150613	.000064547	.000025217	.000008713	.000000372	.000000008	.000000000	1110011110100000
1101000110010000	31	−.0016504	.0030511	.000312783	.00150061	.000064301	.000024591	.000008387	.000000344	.000000007	•	1110111100001100
1101010010010000	31	−.0016504	.0030465	.000312735	.00149974	.000064221	.000024538	.000008360	.000000342	.000000007	.000000000	1101011110100100
1101000101100000	31	−.0016504	.0030359	.000312626	.00149771	.000064032	.000024415	.000008297	.000000336	.000000007	.000000000	1110011100110000
1101010101010000	31	−.0016504	.0030330	.000312594	.00149710	.000063974	.000024375	.000008276	.000000334	.000000006	.000000000	1101010100110000
1101011011000000	30	−.0016607	.0030200	.000312461	.00149460	.000063748	.000024229	.000008202	.000000328	.000000006	.000000000	1101011010101000
1111101100000000	30	−.0016607	.0031327	.000311924	.00149229	.000064051	.000024381	.000008388	.000000348	.000000007	.000000000	1101111000100100
1101001101001000	31	−.0016644	.0031211	.000311307	.00148668	.000063831	.000024149	.000008310	.000000341	.000000007	.000000000	1101111000010010
1111100011000000	30	−.0016790	.0031328	.000311302	.00147722	.000063420	.000024147	.000008202	.000000333	.000000007	.000000000	1110001010110000
0111110010010000	30	−.0016833	.0032153	.000309726	.00147147	.000062441	.000023649	.000007989	.000000320	.000000006	.000000000	1110011010111000
1010011001010000	31	−.0016833	.0032446	.000309325	.00146813	.000062256	.000023569	.000007961	.000000319	.000000006	.000000000	1101001110000010
1011011010000000	30	−.0016936	.0033458	.000308663	.00146560	.0000623318	.000023705	.000008059	.000000330	.000000007	.000000000	1101100110000100
1101010100100000	30	−.0016936	.0033190	.000308395	.00146082	.000061888	.000023432	.000007923	.000000320	.000000006	.000000000	1011011101010000
1011011010100000	30	−.0016936	.0032950	.000308148	.00145625	.000061464	.000023154	.000007780	.000000308	.000000006	.000000000	1011010010100000
1101011010000000	30	−.0017033	.0033632	.000307235	.00144866	.000061043	.000022971	.000007714	.000000305	.000000006	.000000000	1101110100000010
1011011010010000	30	−.0017076	.0034258	.000307171	.00144417	.000061413	.000023248	.000007867	.000000319	.000000006	.000000000	1011110010000010
1011100110010000	30	−.0017076	.0033896	.000306804	.00144147	.000060802	.000022853	.000007666	.000000302	.000000006	.000000000	1101010100100100
1101001000110000	30	−.0017328	.0035233	.000304010	.00141722	.000058999	.000021920	.000007263	.000000277	.000000005	.000000000	0111110111000100
1111100001000000	28	−.0017512	.0038402	.000304275	.00143950	.000061615	.000023831	.000008318	.000000380	.000000009	.000000000	0111111000100000
0111110100000000	28	−.0017512	.0037844	.000303631	.00142771	.000060519	.000023102	.000007935	.000000344	.000000008	.000000000	1110101000000001
1100110001000000	30	−.0017614	.0037286	.000301368	.00139561	.000056519	.000021419	.000007087	.000000271	.000000005	.000000000	1110110010000010
1110011000100000	30	−.0017614	.0037112	.000301193	.00139246	.000057539	.000021236	.000006995	.000000264	.000000005	.000000000	1101110011000000

179

Table A

m = 7, n = 6   Rank Orders 1651 thru 1700 of 1716   C(7, 6) = 1716   1/C(7, 6) = .00058275

z	w	c_1	c_2	d = .2	.4	.6	.8	1.0	1.5	2.0	3.0	z^{tc}
1101011100100	30	−.017614	.0036978	.00301061	.00139011	.00057330	.00021104	.00006930	.00000259	.00000005	.000000000	1111011001010100
1110101100100	30	−.017657	.0037353	.00300740	.00138816	.00057267	.00021100	.00006939	.00000261	.00000005	.000000000	1110101011000
1110100101100	30	−.017657	.0037235	.00300623	.00138608	.00057081	.00020983	.00006881	.00000256	.00000005	.000000000	1110101111000
1111001010100	30	−.017754	.0038350	.00300136	.00138597	.00057323	.00021214	.00007018	.00000269	.00000005	.000000000	1110101110000
1101010110000	30	−.017754	.0037957	.00299748	.00137910	.00056712	.00020475	.00006828	.00000254	.00000005	.000000000	1101011101000
1110100011000	30	−.017943	.0039292	.00297984	.00136445	.00055900	.00020327	.00006699	.00000249	.00000004	.000000000	1110011101000
1110010101000	29	−.017943	.0039144	.00297837	.00136184	.00055667	.00020501	.00006626	.00000244	.00000005	.000000000	1101011011000
1101011001000	29	−.018046	.0040283	.00297291	.00136121	.00055883	.00020583	.00006764	.00000257	.00000005	.000000000	1011011001100
1011011001000	29	−.018046	.0040051	.00297059	.00135705	.00055508	.00020312	.00006644	.00000248	.00000005 •	.000000000	1101111001000
1110100110000	30	−.018083	.0040349	.00296745	.00135472	.00055393	.00020268	.00006630	.00000247	.00000005	.000000000	1110011101100
1011011000010	29	−.018186	.0041241	.00295954	.00134972	.00055218	.00020244	.00006645 •	.00000251 •	.00000005	.000000000	1011011010010
1011011010000	29	−.018186	.0041093	.00295805	.00134703	.00054973	.00020086	.00006565	.00000245	.00000004	.000000000	1110101001010
0111011010000	29	−.018186	.0040869	.00295589	.00134331	.00054651	.00019888	.00006469	.00000238	.00000004	.000000000	1101011100100
0111111000000	27	−.018438	.0042480	.00293066	.00132062	.00053277	.00019225	.00006201	.00000223	.00000006	.000000000	1111110000001
1101011010000	29	−.018623	.0045325	.00292856	.00133304 •	.00054879	.00020405	.00006844	.00000282	.00000004	.000000000	1110110100000
1110011100000	29	−.018767	.0044978	.00290157	.00129773	.00052077	.00018731	.00006032	.00000218	.00000003	.000000000	1101101000010
1101101001000	29	−.018767	.0044555	.00289755	.00129084	.00051483	.00018368	.00005858	.00000205	.00000005	.000000000	1101101010100
1011011000100	29	−.018767	.0044434	.00289638	.00128885	.00051310	.00018263	.00005808	.00000201	.00000005	.000000000	1110110010010 •
1011010101000	29	−.018864	.0045191	.00288794	.00128234	.00050977	.00018129	.00005763	.00000200	.00000005	.000000000	1101101010010
1110000010100	30	−.018907	.0045719	.00288612	.00128264	.00051099	.00018231	.00005820	.00000205	.00000004	.000000000	1101011100000
1011100110000	29	−.018907	.0045473	.00288379	.00127871	.00050763	.00018028	.00005724	.00000198	.00000003	.000000000	1011011010010
1101101001000	29	−.019054	.0046456	.00286957	.00126636	.00050044	.00017695	.00005594	.00000192	.00000003	.000000000	1110100010100
1110100100000	29	−.019096	.0046724	.00286531	.00126255	.00049815	.00017586	.00005551	.00000190	.00000003	.000000000	1110100110000
1111000011000	29	−.019193	.0047783	.00285969	.00126078	.00049884 •	.00017693	.00005620	.00000189	.00000003	.000000000	1101011011000
1101011010000	29	−.019193	.0047508	.00285709	.00125639	.00049509	.00017467	.00005513	.00000202	.00000004	.000000000	1110110010000
1011011000010	29	−.019296	.0048791	.00285280	.00125741 •	.00049838	.00017742	.00005667	.00000191	.00000003	.000000000	1111100100100
1011010110000	28	−.019296	.0048430	.00284938	.00125158	.00049336	.00017436	.00005520	.00000191	.00000003	.000000000	1110110100010
1011011010000	28	−.019338	.0048770	.00284578	.00124886	.00049198	.00017381	.00005501	.00000165	.00000003	.000000000	1011111000100 •
1011011000100	28	−.019877	.0052332	.00279308	.00120260	.00046470	.00016099	.00004996	.00000200	.00000003	.000000000	1101011100010
1011110000000	28	−.019877	.0052214	.00279199	.00120079	.00046317	.00016008	.00004954	.00000162	.00000004	.000000000	1101111100000
1111000000100	28	−.020017	.0053381	.00278055	.00119251	.00045921	.00015861	.00004910	.00000161	.00000003 •	.000000000	1111011010110
1101011100000	28	−.020017	.0053365	.00278038	.00119221	.00045894	.00015844	.00004901	.00000160	.00000003	.000000000	1111000010100
1101010010000	28	−.020017	.0053128	.00277824	.00118873	.00045609	.00015679	.00004825	.00000156	.00000003	.000000000	1110110010010
1110010110000	28	−.020206	.0054456	.00276048	.00117377	.00044762	.00015297	.00004681	.00000149	.00000002	.000000000	1110100111000
1100110100000	28	−.020304	.0055275	.00275255	.00116803	.00044486	.00015193	.00004649	.00000148	.00000004	.000000000	1101011111000
1111000010000	28	−.020346	.0055697	.00274966	.00116643	.00044438	.00015189	.00004653 •	.00000149	.00000002	.000000000	1110110100000
1101011000000	28	−.020346	.0055568	.00274852	.00116460	.00044289	.00015104	.00004615	.00000146	.00000002	.000000000	1101111101000
1011010110000	28	−.020449	.0057261	.00274766	.00117079 •	.00045011	.00015589	.00004857	.00000164	.00000003	.000000000	1011111100100
1011011100000	27	−.020449	.0056752	.00274308	.00116334	.00043397	.00015229	.00004690	.00000153	.00000002	.000000000	1101101110000
1110100100000	27	−.021127	.0061614	.00268014	.00111075	.00041431	.00013892	.00004183	.00000129	.00000002	.000000000	1111010110100
1101100100100	27	−.021127	.0061368	.00267797	.00110731	.00041155	.00013735	.00004112	.00000125	.00000002	.000000000	1111011010100
1101011010000	27	−.021170	.0061742	.00267464	.00110498	.00041046	.00013695	.00004100	.00000124	.00000002	.000000000	1111100110010
1011011010000	27	−.021316	.0062766	.00266083	.00109334	.00040388	.00013399	.00003989	.00000119	.00000002	.000000000	1111100111000
1110010110000	27	−.021456	.0064006	.00264993	.00108578	.00040039	.00013273	.00003951	.00000118	.00000002	.000000000	1111100110000
1101011010000	27	−.021456	.0063889	.00264893	.00108421	.00039915	.00013203	.00003920	.00000117	.00000002	.000000000	1110101101000
1011011010000	27	−.021456	.0063757	.00264779	.00108245	.00039777	.00013127	.00003887	.00000115	.00000002	.000000000	1110110010100
1011111000000	26	−.021559	.0065336	.00264559	.00108598	.00040242	.00013441	.00004041	.00000125	.00000002	.000000000	1111111010000
1111000110000	27	−.021596	.0065065	.00263748	.00107579	.00039498	.00013039	.00003866	.00000115	.00000002	.000000000	1101111000010
1111010010000	26	−.022280	.0070796	.00258124	.00103420	.00037407	.00012200	.00003582	.00000105	.00000002	.000000000	1101111101000

180

Table A

m = 7, n = 6   Rank Orders 1701 thru 1716 of 1716   C(7, 6) = 1716   1/C(7, 6) = .00058275

z	w	c_1	c_2	d = .2	.4	.6	.8	1.0	1.5	2.0	3.0	z^{tc}
1111010001000	26	−.0022567	.0072655	.000255314	.000100969	.000035999	.000011557	.000003337	.000000093	.000000001	.000000000	1110111010000
1110110010000	26	−.0022567	.0072527	.000255207	.000100809	.000035877	.000011491	.000003309	.000000092	.000000001	.000000000	1111011001000
1110101011000	26	−.0022609	.0072927	.000254895	.000100604	.000035787	.000011460	.000003300	.000000092	.000000001	.000000000	1111100101000
1111001010000	26	−.0022706	.0073714	.000254069	.000099972	.000035460	.000011325	.000003253	.000000090	.000000001	.000000000	1111010101000
1111110000000	25	−.0023390	.0079669	.000248622	.000096049	.000033546	.000010576	.000003006	.000000081	.000000001	.000000000	1111110000100
1110110010000	25	−.0023719	.0082394	.000245885	.000093987	.000032493	.000010144	.000002856	.000000076	.000000001	.000000000	1111011110000
1110101010000	25	−.0023719	.0082118	.000245667	.000093674	.000032265	.000010026	.000002807	.000000073	.000000001	.000000000	1111101001000
1111010010000	25	−.0023817	.0082940	.000244870	.000093084	.000031969	.000009907	.000002767	.000000072	.000000001	.000000000	1111101110000
1111001100000	25	−.0023859	.0083365	.000244575	.000092901	.000031894	.000009883	.000002760	.000000072	.000000001	.000000000	1111100110000
1110110110000	24	−.0024830	.0091918	.000236914	.000087415	.000029225	.000008839	.000002415	.000000060	.000000001	.000000000	1111110001000
1111100010000	24	−.0024969	.0093160	.000235817	.000086634	.000028846	.000008692	.000002366	.000000058	.000000001	.000000000	1111011100000
1111010100000	24	−.0024969	.0093012	.000235706	.000086483	.000028741	.000008639	.000002346	.000000057	.000000001	.000000000	1111101010000
1111011000000	23	−.0026080	.0103272	.000227297	.000080690	.000026024	.000007612	.000002016	.000000047	.000000001	.000000000	1111101100000
1111100100000	23	−.0026122	.0103673	.000226981	.000080476	.000025926	.000007576	.000002005	.000000046	.000000001	.000000000	1111101101000
1111101000000	22	−.0027232	.0114353	.000218871	.000075072	.000023466	.000006671	.000001722	.000000038	.000000001	.000000000	1111110100000
1111110000000	21	−.0028343	.0125612	.000211153	.000070155	.000021318	.000005910	.000001492	.000000031	.000000000	.000000000	1111111000000

181

Table A

m = 7, n = 7   Rank Orders 1 thru 50 of 1780   C(7, 7) = 3432   1/C(7, 7) = .00029138

z	w	c_1	c_2	d = .2	.4	.6	.8	1.0	1.5	2.0	3.0	z^{tc}
0000000111111111	77	.015400	.0074775	.000801716	.002202742	.004681248	.010001998	.01982265	.079638358	.216172856	.645528042	0000000111111111x
0000001011111111	76	.014886	.0069118	.000772180	.001865311	.004118123	.008331869	.015491892	.051445189	.107543475	.142135451	0000001011111111x
0000001101111111	75	.014364	.0063624	.000743669	.001723554	.003635702	.006996915	.012315000	.034835421	.059753730	.046822530	0000001101111111x
0000010011111111	74	.013842	.0058310	.000716168	.001592118	.003207433	.005866903	.009763296	.024016776	.032442281	.013932074	0000010011111111x
0000010101111111	74	.013815	.0058112	.000715153	.001589144	.003203933	.005874068•	.009816392	.024016776	.034843345	.018333586	0000011001111111x
0000011001111111	73	.013328	.0053345	.000690512	.001476396	.002853226	.004993319	.007928831	.016684110	.020017592	.006166708	0000010110111111x
0000100011111111	73	.013294	.0052988	.000688607	.001467614	.002824823	.004919389	.007765203	.016008058	.018612527	.005114297	0000011010111111x
0000100101111111	73	.013215	.0052370	.000685460	.001456470	.002801135	.004887086	.007753029	.016482979	.020491275	.007656606	0000010101111111x
0000101001111111	72	.012780	.0048208	.000663980	.001360575	.002511175	.004181302	.006291548	.011348642	.011310127	.002147399	0000011001011111x
0000100110111111	72	.012745	.0047867	.000662201	.001352582	.002486640	.004120916	.006165670	.010909262	.010561321	.001807683	0000010011011111x
0100000011111111	72	.012693	.0047455	.000660047	.001344812	.002468415	.004088618	.006121997	.010928263●	.010812163	.002042607	0000010111011111
0100001011111111	72	.012517	.0046080	.000652835	.001318789	.002407149	.003979026	.005970964	.010973410	.011699114	.003164349	0000011100011111x
0001000011111111	71	.012258	.0043608	.000640408	.001262518	.002239445	.003578271	.005159937	.008317407	.007359488	.001091943	0000011001101111
0001001001111111	71	.012231	.0043272	.000638401	.001253593	.002209020	.003497815	.004983530	.007767243	.006309673	.000711432	0000010101101111x
0001000101111111	71	.012179	.0042880	.000636329	.001246437	.002172000	.003422235	.004854639	.007705924	.006649379	.000826257	0000010011101111
0001001101111111	71	.012144	.0042555	.000634606	.001239197	.002172000	.003422235	.004854294	.007419410	.006085855	.000704458	0000100011101111x
0001010001111111	71	.011995	.0041410	.000628606	.001217459	.002120246	.003325906	.004707483	.007243111	.006111038●	.000816325	0000100101101111
0000110101111111	70	.011709	.0038862	.000615417	.001172994	.001968935	.002990251	.004079468	.005599638	.004061807	.000349483	0000101001101111
0001010101111111	70	.011657	.0038486	.000613410	.001156278	.001954416	.002966726	.004050780	.005614507	.004174301	.000405473	0000100110101111x
0000111001111111	70	.011630	.0038165	.000611810	.001148233	.001928369	.002901413	.003915752	.005192178	.003589349	.000264935	0000100011011111x
0100000111111111	71	.011623	.0038604	.000613694●	.001165332	.001994288	.003091050	.004346227	.006771421	.006105493	.001182641	0000011110111011x
0001100011111111	70	.011543	.0037486	.000608230	.001135152	.001896509	.002840127	.003817615	.005031482	.003486715	.000270411	0000110001110111x
0001100101111111	70	.011482	.0037074	.000604400	.001121674	.001882755	.002820891	.003800376	.005084379	.003635847	.000320997	0000110011011011x
0001101001111111	69	.011187	.0034631	.000593229	.001078688	.001753947	.002553595	.003333469	.004064979	.003180012	.000167476	0000111000011011x
0001010011111111	69	.011108	.0034138	.000590499	.001070676	.001739537	.002536508	.003325127	.004155701	.002590722	.000221050	0000110110011011
0001100110111111	69	.011108	.0033964	.000589758	.001065048	.001718006	.002478247	.003200760	.003772119	.002783851	.000127221	0000101101011011x
0001011001111111	69	.011101	.0033961	.000589745	.001065048	.001717585	.002477057	.003198112	.003763165	.002292464	.000124815	0000110101011011
0100001101111111	70	.011030	.0034249	.000590693●	.001074675	.001755819	.002581482	.003421601	.004451847	.003162628	.000296942	0000111001011011x
0001010110111111	68	.010600	.0033301	.000586307	.001051573	.001682770	.002405047	.003073192	.003500072	.002028387	.000096676	0000110011101011x
0010100011111111	69	.010960	.0032921	.000584126	.001046282	.001676828	.002408061●	.003103572	.003686029	.002313829	.000153254	0000111101011011x
0010000111111111	69	.010933	.0032617	.000582617	.001039111	.001654883	.002356089	.003002228	.003414232	.001995047	.000100558	0000110110110111
0001110001111111	68	.010846	.0031982	.000579219	.001027362	.001627902	.002307357	.002929325	.003316740	.001949613	.000105112●	0000111100011011x
0010000101111111	68	.010674	.0030728	.000572488	.001003969	.001573582	.002207376	.002775331	.003078375	.001784935	.000096396	0000110011101101x
0010001001111111	69	.010588	.0030221	.000569603	.000995639	.001558368	.002187437	.002757963	.003112410	.001866314	.000114108	0000111001101011x
0010010011111111	68	.010586	.0029940	.000568457	.000987556	.001529329	.002113442	.002609314	.002720502	.001441509	.000058163	0000110101101011
0001110101111111	68	.010560	.0029815	.000567694	.000985924	.001528062	.002114759	.002621475	.002774759	.001508486	.000066414	0000110011011011
0010001101111111	69	.010553	.0029938	.000568119●	.000989978	.001543863	.002157801	.002709065	.003004892	.001759517	.000099295	0000111001011101x
0001111001111111	68	.010508	.0029307	.000565137	.000975110	.001498139	.002051465	.002506210	.002526497	.001277531	.000044349	0000110100111011
0010010101111111	68	.010411	.0028825	.000562226	.000968501	.001491991	.002055861	.002541564	.002712295	.001525638	.000081158	0000110011010111x
0010011001111111	68	.010411	.0028660	.000561594	.000963668	.001473757	.002010932	.002450986	.002472651	.001266349	.000047513	0000110011011101
0010100101111111	68	.010411	.0028655	.000561570	.000963484	.001473064	.002009085	.002447128	.002461631	.001253775	.000045882	0000111001010111
0010101001111111	70	.010332	.0028036	.000558313	.000951511	.001443658	.001951850	.002353911	.002299734	.001121727	.000036189	0000111000111011x
1000000011111111	70	.010180	.0027751	.000556565	.000958785●	.001494423	.002111509	.002711009	.003389109	.002475941	.000324097	0010100011111110x
0010000110111111	68	.010148	.0026808	.000551575	.000929699	.001396958	.001873595	.001825096	.002182310	.001086473	.000040581	0010010011110110x
0010011101111111	67	.010073	.0026413	.000548554	.000918948	.001371343	.001825086	.002168880	.002055167	.000984790	.000032703	0000111100110111x
0101000011111111	68	.010066	.0026381	.000549021●	.000923145	.001387089	.001865236	.002248210	.002246285	.001177387	.000055259	0000111001101101
0001111101111111	68	.010039	.0026096	.000547617	.000916915	.001369265	.001825769	.002176279	.002038846	.001017855	.000035112	0000110101011011
0011000011111111	67	.010038	.0025983	.000547176	.000914097	.001359937	.001803849	.002135767	.001998755	.000946999	.000030401	0000111001100111
0101010011111111	67	.009986	.0025492	.000544701	.000903987	.001332975	.001747869	.002039889	.001814122	.000795530	.000019609	0000111010010111x

182

Table A

m = 7, n = 7 **Rank Orders 51 thru 100 of 1780** **C(7, 7) = 3432** **1/C(7, 7) = .00029138**

z	w	c_1	c_2	d = .2	.4	.6	.8	1.0	1.5	2.0	3.0	z^{tc}
0001100110111	67	.0009959	.025381	.000543986	.000902615	.001332335	.001751270●	.002051976	.001857515	.000840350	.000232246	0001001100111
0100001110111	68	.0009952	.025517	.000544433●	.000906625	.001347232	.001788779	.002125043	.002028929	.000999948	.000374336	0000100111101
0010010100111	67	.0009889	.024877	.000541277	.000893315	.001311063	.001712314	.001993988	.001773542	.000787133	.000209956	0010110101101
0010110010111	67	.0009862	.024760	.000540829	.000891694	.001309482	.001713747	.002001035	.001803999	.000819465	.000243623	0000101011011
0110000110111	67	.0009810	.024653	.000539544	.000891845●	.001317954	.001744165	.002000198	.001999273	.000819465	.000245603	0001010111011
0010100110111	67	.0009810	.024286	.000538132	.000882172	.001284733	.001663602	.001917163	.001651658	.000701769	.000016293	0001010100111
1000000110111	67	.0009810	.024283	.000538121	.000882090	.001284422	.001662283	.001915731	.001648213	.000698541	.000016035	0001000111011
0100010110111	67	.0009658	.024107	.000535660	.000884780●	.001315099	.001761836	.002131081	.002219347	.001272240	.000079466	0000100110111
0010100010111	69	.0009635	.023101	.000531640	.000860868	.001238178	.001583238	.001801479	.001502758	.000617970	.000013423	0000101011011
0110010010111	66	.0009524	.022470	.000527988	.000850397	.001218804	.001556234	.001772514	.001501768	.000642608●	.000016834	0010011101001x
0101000001111	67	.0009517	.022610	.000528430●	.000854238	.001232673	.001590217	.001836989	.001644069	.000768566	.000027525	0001011101101
0100100001111	67	.0009517	.022454	.000527836	.000850203	.001218930	.001557198	.001774662	.001505002	.000642735	.000016385	0000110011101
0010101000111	67	.0009517	.022445	.000527802	.000849963	.001218086	.001555107	.001770600	.001495385	.000633702	.000015608	0000110101101
0100101000111	66	.0009472	.021998	.000526602	.000840998	.001194637	.001507877	.001692724	.001362029	.000538556	.000010734	0001011001011x
0010110000111	67	.0009438	.021880	.000524757	.000839509	.001194139	.001511683●	.001704797	.001398120	.000569914	.000010058	0010010101110
1000010010111	66	.0009437	.021756	.000524289	.000836316	.001184880	.001490751	.001667679	.001328658	.000519266	.000019266	0001010011011
0010011001111	66	.0009375	.021418	.000522300	.000831088	.001175074	.001477849	.001655085	.001332828●	.000534174	.000011586	0001011001101
0100011001111	66	.0009340	.021172	.000520971	.000826586	.001164910	.001464707	.001641013	.001294754	.000510525	.000010564	0001010110111
0010110010111	66	.0009288	.020713	.000518640	.000817609	.001142354	.001415643	.001557176	.001179223	.000431700	.000006924	0010101010111
0110010010111	66	.0009262	.020809	.000518669●	.000820987	.001156838	.001452787	.001627981	.001325646	.000545200	.000013038	0010101110011
0100100110111	66	.0009262	.020619	.000520619	.000816471	.001142170	.001419305	.001568162	.001211486	.000459610	.000008454	0011000110111
0101000010111	66	.0009262	.020610	.000520610	.000816235	.001141374	.001417413	.001564651	.001204188	.000453744	.000008115	0010011101101
0100010010111	68	.0009255	.020767	.000520767	.000820347	.001155809	.001491413	.001628114	.001332833	.000555725	.000014381	0010110101101
0100100010111	66	.0009144	.019593	.000516332	.000797921	.001166578	.001347766	.001464201	.001545502	.000744970	.000029924	0000110111110
0010011010111	68	.0009113	.020334	.000518435●	.000819520	.001155911	.001471667	.001684123	.001075177	.000382116	.000005828	0010100111110
1000000111011	68	.0009109	.019109	.000512392	.000792392	.001095443	.001350539	.001485801	.001494140	.000704336	.000026274	0010101101111
0011000101111	65	.0009002	.019160	.000514995●	.000791832	.001083699	.001324815	.001441013	.001156771	.000456691	.000010361	0001000111101
0100100101111	66	.0008995	.018983	.000509440	.000787929	.001081988	.001324633	.001444538	.001074186	.000395538	.000007020	0000110110111
0101001001111	66	.0008969	.018876	.000508709	.000787502	.001081988	.001324633●	.001441013	.001089287	.000409871	.000006819	0010110110111
0010101010111	65	.0008923	.018448	.000505878	.000778136	.001061339	.001284788	.001381649	.000993695	.000348912	.000005408	0001011100111
0101000011011	66	.0008916	.018791	.000507022●	.000786271	.001088275	.001346531	.001491907	.001203670	.000508288	.000015042	0000011110101
0100010011011	66	.0008916	.018444	.000505769	.000778202	.001088266	.001287505	.001386851	.001092954	.000354553	.000005551	0010100101011
0010100011011	66	.0008916	.018439	.000505747	.000778059	.001061790	.001286394	.001384825	.000998923	.000351454	.000005389	0001011011011
0011010100111	65	.0008889	.018222	.000504622	.000774126	.001052829	.001270552	.001361829	.000970446	.000337212	.000005097	0001100100111x
0101010010111	65	.0008774	.017899	.000502679	.000768828	.001043513	.001258200	.001349164	.000969188	.000343874●	.000005734●	0001110101011
1000010011011	66	.0008774	.017458	.000500431	.000760485	.001023313	.001220188	.001290296	.000882152	.000290178	.000003718	0001001011101
0011000110111	65	.0008741	.017369	.000499673	.000760909	.001023905	.001225618	.001303726	.000913612	.000313001	.000004608	0011000101111
0101010000111	65	.0008740	.017431	.000499881●	.000760909	.001028700	.001236504	.001322938	.000948648	.000337781	.000005752	0001101011011
0010110011011	65	.0008740	.017235	.000499188	.000756561	.001019094	.001220641	.001271748	.000861481	.000280405	.000003506	0010110101101
0101010001011	65	.0008740	.017229	.000499165	.000756412	.001014613	.001205551	.001269827	.000858014	.000278036	.000003415	0010101010111
0110100110111	65	.0008661	.016902	.000496976	.000751410	.001008006	.001200874	.001271748●	.000882849	.000300568	.000004432	0001110101011
1000100110111	67	.0008622	.017247	.000497640●	.000759634	.001037684	.001270229	.001395348	.001110441	.000466076	.000013688	0010110111011
0100100011011	67	.0008595	.016969	.000496382	.000754597	.001037684	.001243941	.001351698	.001031798	.000404025	.000009070	0001010101110
0101010100111	65	.0008591	.016264	.000493812	.000739394	.000978708	.001145940	.001187621	.000764759	.000233179	.000002413	0010010101110
1000010101111	65	.0008564	.016177	.000493158	.000738244	.000978181	.001148107	.001194666	.000783592	.000247042	.000002915	0010010110111
1000100011011	67	.0008508	.016503	.000493507	.000746201	.001008377	.001219308	.001320934	.001000690	.000398950	.000009843	0001001111110
0101001010111	65	.0008481	.015832	.000490851	.000732901	.000970860	.001142146	.001194549	.000804807	.000266927	.000003831	0001110011011
0010110100111	65	.0008447	.015603	.000489586	.000728830	.000962143	.001117421	.001173826	.000779700	.000253686	.000003435	0011011001011
0011100100111	64	.0008401	.015347	.000488086	.000724419	.000955361	.001114102	.001156680	.000762977	.000246461	.000003274	0011011000111
0101010010111	65	.0008394	.015183	.000487419	.000721059	.000943953	.001094317	.001124249	.000712442	.000215831	.000002285	0010101101101

183

Table A

m = 7, n = 7   Rank Orders 101 thru 150 of 1780 $C(7, 7) = 3432$ $1/C(7, 7) = .00029138$

z	w	c_1	c_2	d = .2	.4	.6	.8	1.0	1.5	2.0	3.0	z^{tc}
0001101010111	64	.0008375	.0015099	.000486874	.000719851	.000942539	.001093902	.001126397x	.000722851	.000224861	.000002687	0011010100011x
0101000010111	65	.0008368	.0015278	.000487388	.000723688	.000954869	.001120815	.001171725	.000795263	.000269163	.000004185	0001011110101
0011000110111	65	.0008368	.0015109	.000486806	.000720146	.000944096	.001097823	.001133387	.000734181	.000231380	.000002859	0011000110101
0100100110111	65	.0008368	.0015090	.000486741	.000719731	.000942791	.001094943	.001128426	.000725679	.000225794	.000002649	0011001101101
0100110000111	66	.0008361	.0015270	.000487273	.000723779	.000956055	.001124548	.001179483	.000812909	.000283624	.000005078	0100001110101x
0011110000111	64	.0008305	.0014830	.000484990	.000715705	.000937361	.001090759	.001129033	.000743529	.000242530	.000003464	0000111100101x
0101110000111	64	.0008226	.0014362	.000482305	.000707598	.000921074	.001064925	.001094873	.000707969	.000226290	.000003086	0000111011001
1000011000111	64	.0008226	.0014166	.000481635	.000703539	.000908817	.001038993	.001052063	.000641871	.000187032	.000001846	0001101101011
0110101001011	64	.0008226	.0014160	.000481613	.000703399	.000908381	.001038038	.001050439	.000639198	.000185374	.000001794	0010101101011
0110010010111	65	.0008219	.0014772	.000483597	.000716123	.000947594	.001122220	.001191618	.000869205	.000334627	.000008453	0000101111011
0100110011011	65	.0008219	.0014175	.000481563	.000703800	.000910240	.001042564	.001058362	.000651694	.000192404	.000001970	0101010011101
0101000110111	65	.0008219	.0014172	.000481553	.000703735	.000910038	.001042122	.001057606	.000650433	.000191604	.000001943	0100001101101
0010100110111	64	.0008191	.0013952	.000480424	.000699816	.000901226	.001026815	.001035840	.000624998	.000179585	.000001705	0010101101011
0110001100111	64	.0008139	.0013731	.000478942	.000696215	.000895685	.001020592	.001030821	.000627090	.000183054	.000001835	0010011001011x
1001000110111	64	.0008077	.0013248	.000476452	.000687595	.000876298	.000986835	.000982629	.000569967	.000155575	.000001269	0010100110101
0100100011111	66	.0008073	.0014007	.000478933	.000702674	.000921399	.001081329	.001137464	.000808388	.000301313	.000006920	0000111101110
1000001010111	66	.0008073	.0013862	.000478431	.000699568	.000911751	.001060171	.001100964	.000743199	.000253909	.000004189	0001100110101
0101001001011	66	.0008073	.0013850	.000478388	.000699298	.000910895	.001058257	.001097604	.000736984	.000249315	.000003940	0100101101011
0010100100111	64	.0008042	.0013037	.000475283	.000683952	.000868967	.000975247	.000967447	.000554962	.000149382	.000001172	0010100101011
1000010010111	66	.0007995	.0013368	.000475645	.000690798	.000893292	.001029357	.001057899	.000690823	.000225389	.000003204	0010100110110
0011000011111	64	.0007963	.0012752	.000473184	.000679530	.000863641	.000972220	.000970251	.000573440	.000163113	.000001581	0010010111001
0100100101011	64	.0007933	.0012638	.000472373	.000677769	.000861842	.000970968	.000971428x	.000581739	.000169818	.000001835	0010111011011
0010110010111	64	.0007881	.0012237	.000470284	.000670550	.000845204	.000942448	.000930335	.000531278	.000144215	.000001209	0010011010011
0011001001011	63	.0007853	.0012188	.000469729	.000670032	.000845224	.000947448	.000941844	.000553459	.000158006	.000001609	0011011000111
0100010110111	64	.0007847	.0012181	.000469716	.000669905	.000846605	.000948366	.000942870	.000554623	.000158142	.000001575	0100001011110x
0010010001111	64	.0007846	.0012215	.000469714	.000670614	.000848750	.000953208	.000950936	.000567174	.000165606	.000001146	0010011011011
0100101001011	64	.0007846	.0012007	.000469172	.000667129	.000838528	.000932216	.000917331	.000519346	.000139630	.000001146	0011001101011
1000010100111	64	.0007846	.0012024	.000469079	.000666867	.000837740	.000930559	.000914622	.000515327	.000137397	.000001091	0010101011011
1000000110111	64	.0007811	.0012440	.000469100	.000675106	.000868466	.000989165	.001011307	.000660443	.000221175	.000003766	0010100101110
1000100110111	64	.0007767	.0011735	.000466988	.000662269	.000831756	.000925908	.000914413	.000528245	.000147599	.000001401	0010100110110
0011010000111	63	.0007704	.0011499	.000465305	.000658546	.000826926	.000922316	.000914862	.000541346	.000158703	.000001843	0001111010011
0010011000111	63	.0007704	.0011309	.000464677	.000654866	.000816186	.000900351	.000879810	.000491577	.000131463	.000001105	0010111001011
0101000101011	63	.0007704	.0011300	.000464646	.000654679	.000815627	.000899179	.000877896	.000488733	.000129868	.000001064	0010111001101
0100101010111	64	.0007697	.0011157	.000464072	.000652004	.000808508	.000885388	.000856748	.000461136	.000116145	.000000789	0010101010111
0110001010111	63	.0007670	.0011069	.000463504	.000650616	.000806417	.000883223	.000855430	.000463641	.000118745	.000000876	0010101111001
0100100010111	64	.0007670	.0011517	.000464841	.000659029	.000831152	.000933539	.000934913	.000572789	.000176292	.000002320	0010100101110
0100011010111	64	.0007670	.0011091	.000463471	.000651072	.000808325	.000887609	.000862747	.000473949	.000123914	.000000977	0001011101001
1000010000111	64	.0007670	.0011081	.000463439	.000650885	.000807771	.000887609	.000860901	.000471302	.000122487	.000000943	0010011101011
0011001010111	63	.0007625	.0010866	.000462078	.000647289	.000801530	.000878034	.000851574	.000465685	.000121336	.000000951●	0011001100111
0100110010111	63	.0007590	.0010675	.000460944	.000644030	.000795325	.000868759	.000840848	.000455792	.000117772	.000000908	0001111011110
1000010000111	65	.0007551	.0010898	.000461061	.000648141	.000800852	.000900852	.000892203	.000527882	.000154787	.000001749	0001101010110
0010101010111	63	.0007528	.0010208	.000458519	.000635887	.000777604	.000838964	.000799089	.000411821	.000098812	.000000602	0100011010011
1000010010111	63	.0007525	.0010811	.000460385	.000643354	.000777120	.000900058	.000893340	.000534139	.000159597	.000001926	0010100101101
0010100110111	63	.0007494	.0010012	.000457327	.000632529	.000771219	.000882354	.000787120	.000401470	.000095111	.000000560	0010100101101x
1000100001111	65	.0007473	.0010764	.000459471	.000646438	.000812539	.000913788	.000920907	.000587780	.000196892	.000003686	0000111011110
1001000010111	65	.0007473	.0010443	.000458382	.000640233	.000794309	.000876040	.000859565	.000494179	.000137703	.000001405	0001010011101
1000010010111	65	.0007473	.0010434	.000458382	.000640049	.000793758	.000874882	.000857656	.000491218	.000137703	.000001347	0010010011110
0011010010111	63	.0007441	.0009823	.000455985	.000629444	.000766892	.000825192	.000784702	.000444802	.000119003	.000000624	0010011110101
0011100000111	63	.0007411	.0009881	.000455724	.000630787	.000770563	.000840929	.000811309	.000444706	.000110780	.000001093	0001110010101
0100110010111	63	.0007332	.0009454	.000453176	.000623504	.000759593	.000820224	.000785888	.000421797	.000110192	.000000953	0010100110110

184

Table A

m = 7, n = 7   Rank Orders 151 thru 200 of 1780   C(7, 7) = 3432   1/C(7, 7) = .0029138

z	w	c_1	c_2	d = .2	.4	.6	.8	1.0	1.5	2.0	3.0	z^{tc}
01000101001011	63	.0007332	.0009276	.000452605	.000620252	.000750393	.000802019	.000757839	.000385681	.000092552	.000000594	00011101001101
01010100100111	63	.0007332	.0009264	.000452563	.000620006	.000749679	.000800571	.000755554	.000382597	.000091001	.000000563	00011011010101
00011100100111	62	.0007331	.0009456	.000453164	.000623533	.000759762	.000820625	.000786545	.000422612	.000110519	.000000955	00011110100111 x
01001001100111	64	.0007295	.0009299	.000452567	.000623655	.000752065	.000805775	.000745934	.000393634	.000098217	.000000653	01000100101101
10000010010111	63	.0007297	.0009079	.000451452	.000616884	.000743878	.000792121	.000745373	.000374509	.000088356	.000000539	00110001001101
10100010010011	65	.0007245	.0009543	.000452895	.000624842	.000765734	.000834158	.000808408	.000450686	.000123433	.000001174	01010011001101
01011001000111	63	.0007245	.0008878	.000450024	.000613514	.000738762	.000786274	.000740187	.000373869	.000089211	.000000566	00101001110110
01110001001011	63	.0007183	.0008451	.000447736	.000606208	.000723577	.000761820	.000707911	.000342561	.000077012	.000000408	00101011101101
01010110001011	63	.0007155	.0008769	.000447319	.000611917	.000741735	.000799402	.000766977	.000419511	.000114466	.000001177	00110001111011
01001011000111	62	.0007155	.0008402	.000447164	.000605489	.000723907	.000764759	.000714463	.000354010	.000082998	.000000517	00011100010111
01011001000111	62	.0007155	.0008399	.000447155	.000605438	.000723761	.000764470	.000714020	.000353454	.000082737	.000000513	00101011010011
01110001010101	63	.0007148	.0008698	.000447980	.000610599	.000738511	.000793409	.000757944	.000407493	.000107972	.000000992	00010101010101
01001010101011	63	.0007148	.0008263	.000446616	.000603033	.000717619	.000753059	.000697250	.000333867	.000074085	.000000379	00101010010111
00110010001011	62	.0007103	.0008257	.000446595	.000602913	.000717284	.000752401	.000696244	.000332623	.000073522	.000000371	00101010110111
01100100001011	62	.0007077	.0008213	.000445772	.000598628	.000711319	.000759833	.000710606	.000354846	.000077389	.000000557	00101010011011
01100010010011	63	.0007077	.0007996	.000444688	.000603022	.000711290	.000744621	.000729011	.000336987	.000096337	.000000460	00110100110110
01010010010111	63	.0007069	.0008258	.000445385	.000599221	.000723785	.000770803	.000700181	.000379745	.000090736	.000000770	00110010011110
01010000110111	63	.0007069	.0008031	.000444406	.000598845	.000713487	.000755129	.000697172	.000346301	.000087544	.000000645	00110001111010
01010001100111	63	.0007038	.0008010	.000444619	.000595566	.000715216	.000749223	.000732165	.000342412	.000079787	.000000495	01000001111011
10000100100111	64	.0007038	.0008249	.000444852	.000592746	.000725362	.000765980	.000738626	.000394339	.000103921	.000000944	00101110011110
01011000010111	62	.0007006	.0007592	.000442348	.000591747	.000689741	.000726233	.000667090	.000314037	.000068875	.000000352	00101110110111
00101010010111	62	.0007003	.0008049	.000443689	.000593303	.000718584	.000765452	.000725006	.000380975	.000098221	.000000834	00010101010111 x
01010101010011	62	.0006980	.0007368	.000441243	.000587965	.000689740	.000712695	.000648994	.000296563	.000062332	.000000280	01010101010111 x
10010010010011	62	.0006951	.0007688	.000441750	.000593050	.000705364	.000743667	.000695400	.000349283	.000084078	.000000563	01010101011011
00110100100111	64	.0006928	.0007789	.000439907	.000585089	.000712643	.000709266	.000647251	.000299405	.000064421	.000000316	01010101010111
10010001001011	64	.0006924	.0007645	.000441209	.000592382	.000712643	.000760023	.000722337	.000387021	.000103518	.000001003	00110101011110
10001000101011	64	.0006924	.0007616	.000441115	.000591848	.000667336	.000746545	.000701841	.000361053	.000090736	.000000719	00110001111010
10010000110011	64	.0006917	.0007824	.000441667	.000595578	.000706294	.000746845?	.000697172	.000354835	.000087544	.000001326	00110001111010
10100001100011	65	.0006917	.0007018	.000438813	.000582061	.000680396	.000701316	.000637811	.000292163	.000062124	.000000295	01000001110011
00011110000111	61	.0006817	.0007025	.000437653	.000582097	.000686227	.000717782	.000666871	.000332897	.000081407	.000000628	00111110000111 x
01010110000111	62	.0006810	.0006903	.000437178	.000580140	.000681570	.000709661	.000655612	.000321291	.000076732	.000000560	00011101110101
01010001100111	62	.0006810	.0006734	.000436652	.000577257	.000673673	.000694564	.000633138	.000294723	.000063684	.000000353	00110001110101
01010010011011	62	.0006803	.0006716	.000436593	.000576923	.000672760	.000692789	.000630454	.000291461	.000063322	.000000329	00110101011110
00101010011011	62	.0006803	.0006596	.000436112	.000574879	.000667737	.000683767	.000614617	.000277599	.000057680	.000000256	00101010101101
10010100010011	63	.0006783	.0006504	.000435532	.000573399	.000665308	.000680754	.000614612	.000276835	.000058014 ●	.000000272	00111101011101
10010001001011	63	.0006776	.0006646	.000435429	.000573964	.000667336	.000685018	.000621268	.000284599	.000061261	.000000314	01000010011101 x
10100000110011	64	.0006776	.0006407	.000438156 ●	.000588306	.000706294	.000723229	.000732229	.000240408	.000127563	.000002007	00110001111010
10010001010011	64	.0006775	.0006860	.000436460	.000578949	.000680440	.000709024	.000655457	.000320549	.000075468	.000000496	00100001110110
10001010010011	64	.0006775	.0006854	.000436440	.000578836	.000680121	.000708397	.000655491	.000319297	.000074847	.000000483	00011110110110
01010100100111	62	.0006731	.0006323	.000434157	.000570285	.000660788	.000678045	.000610590	.000276624	.000058703	.000000287	00110101010101
01010010100111	62	.0006696	.0006155	.000433098	.000567447	.000655767	.000668887	.000602621	.000277072	.000057123	.000000277	00110111010101
01100010100111	62	.0006635	.0006177	.000432188	.000577600 ●	.000660659	.000668214	.000625668	.000302264	.000071520	.000000518	00101010101101
01010100010111	62	.0006635	.0005742	.000430870	.000560541	.000641569	.000647272	.000571116	.000246912	.000048685	.000000191	00110111010111
00101010010111	61	.0006633	.0005911	.000431369 ●	.000563328	.000649404	.000646697	.000574270	.000245963	.000048297	.000000187	00101010110111
10100100010011	61	.0006600	.0005563	.000429778	.000557530	.000636388	.000661385	.000595585	.000269103	.000057542	.000000300	00110110010101
00110101010011	61	.0006555	.0005704	.000429483	.000559578 ●	.000636388	.000639454	.000595906	.000240038	.000046589	.000000175	00100101010111
01110010010011	61	.0006555	.0005539	.000428994	.000557017	.000636388	.000658149	.000595120	.000274289	.000066648	.000000351	00101011001011
00101010010111	61	.0006555	.0005533	.000428976	.000556923	.000638053	.000645589	.000577367	.000255994	.000053582	.000000265	00110110100111

185

z	w	c_1	c_2	d = .2	.4	.6	.8	1.0	1.5	2.0	3.0	z^{tc}
0110010101011	62	.0006548	.0005645	.000429188	.000558540	.000642623	.000654114	.000589463	.000268058	.000057937	.000000307	00110101011110
0100101110011	62	.0006548	.0005413	.000428496	.000554906	.000633153	.000636887	.000565122	.000243123	.000048456	.000000201	00110001011101
0101001011011	62	.0006548	.0005394	.000428441	.000554607	.000633354	.000658651	.000569167	.000240813	.000047548	.000000190	00101011101101
1000110010011	63	.0006489	.0005585	.000428066	.000557196	.000643089	.000658651	.000599167	.000283351	.000065376	.000000440	00011101101110
0010110010101	61	.0006458	.0004943	.000425662	.000547046	.000618803	.000616428	.000541079	.000225429	.000043185	.000000161	00101110011110
1000011000101	61	.0006437	.0005244	.000426198	.000551389	.000631268	.000639907	.000574676	.000258669	.000055874	.000000294	00101100111110
0011010001011	61	.0006406	.0004789	.000424367	.000544341	.000615321	.000613352	.000539490	.000227509	.000044623	.000000182	00011011100111
1000010100111	64	.0006403	.0005236	.000425638	.000551104	.000632902	.000645173	.000584258	.000273220	.000062227	.000000400	01100010111110
0011000100111	63	.0006402	.0005237	.000425625	.000551160	.000633163	.000645843	.000585454	.000275124	.000063289	.000000426	00001101110110
1000010100101	63	.0006402	.0005081	.000425152	.000548611	.000626374	.000633128	.000566884	.000264091	.000054198	.000000280	10000110100111
1000100010011	63	.0006402	.0005060	.000425089	.000548273	.000625462	.000631402	.000564340	.000251168	.000052935	.000000261	00110101101110
0011010010111	61	.0006379	.0004862	.000423317	.000540914	.000602134	.000602139	.000525138	.000215074	.000040445	.000000145	00101101010101
0010010001111	62	.0006372	.0004862	.000424039	.000545213	.000619799	.000623444	.000555341	.000245786	.000051944	.000000271	00010011110011
0110100001111	63	.0006323	.0004837	.000423160	.000544318	.000620551	.000627467	.000563145	.000256443	.000056497	.000000332	00110110011011x
1001100110011	61	.0006292	.0004268	.000420994	.000535527	.000600099	.000592749	.000516424	.000212255	.000038003	.000000135	01100011001100111x
0010011000111	61	.0006289	.0004198	.000420740	.000534395	.000597318	.000587839	.000509621	.000205532	.000038003	.000000130	00011111100010
0100010010111	61	.0006261	.0004498	.000421185	.000538896	.000610926	.000614322	.000548565	.000247891	.000054915	.000000352	00111110001101
1000110000111	62	.0006261	.0004152	.000420161	.000533544	.000597021	.000589065	.000512858	.000210875	.000040375	.000000159	00111011001101
0110010001111	61	.0006261	.0004146	.000420144	.000533453	.000596783	.000588630	.000512242	.000210243	.000040137	.000000156	00111011001101
0010010101101	62	.0006254	.0004027	.000419673	.000531520	.000592202	.000580712	.000501418	.000199760	.000036346	.000000119	01100110110101
1000010101111	63	.0006253	.0004350	.000420578	.000536186	.000603980	.000631402	.000529511	.000225662	.000045096	.000000193	00101011011110
1010000101011	61	.0006227	.0004690	.000421156	.000541308	.000619243	.000631140	.000573716	.000266037	.000066686	.000000537	00101000111110
1000011101011	63	.0006227	.0004314	.000420046	.000535495	.000603284	.000603284	.000533818	.000232652	.000048437	.000000244	00101100111110
1001010100111	63	.0006227	.0004293	.000419985	.000535167	.000603178	.000601659	.000531461	.000230035	.000047338	.000000228	00101110101110
0011010011011	61	.0006209	.0003981	.000418812	.000530525	.000592653	.000585258	.000508670	.000209948	.000040574	.000000165	00101011110101
1010010001101	61	.0006183	.0003785	.000417804	.000527336	.000586217	.000574328	.000496231	.000199657	.000037246	.000000137	00110101010101
0101001010101	62	.0006176	.0003832	.000417830	.000528011	.000586917	.000578690	.000502651	.000206428	.000040164	.000000164	00011111101011
0101001100011	62	.0006120	.0003779	.000416585	.000525792	.000582853	.000578038	.000504326	.000211344	.000042151	.000000195	00111100010011
0110010000111	61	.0006113	.0003867	.000416906	.000528006	.000592504	.000589743	.000521258	.000229587	.000049567	.000000301	00110011001101
0010010101101	62	.0006113	.0003441	.000415658	.000521550	.000575908	.000559941	.000479632	.000187853	.000033795	.000000111	00110110011110
0011100001111	63	.0006113	.0003432	.000415630	.000521402	.000575522	.000559239	.000478642	.000186856	.000033430	.000000107	00101011001101
1010010101011	61	.0006086	.0003228	.000414602	.000518120	.000568883	.000549019	.000465923	.000176664	.000030308	.000000086	00101010101101
0101000101011	61	.0006034	.0003329	.000414037	.000519087	.000574553	.000561821	.000485678	.000197798	.000037939	.000000154	00110011001101
0110100100111	61	.0006034	.0003096	.000413368	.000515700	.000566049	.000546940	.000465476	.000179195	.000031640	.000000100	00101100101101
0101001010011	61	.0006034	.0003078	.000413314	.000515421	.000565328	.000545643	.000463671	.000177441	.000031021	.000000094	00111101010101
0011010100011	60	.0006033	.0003255	.000413815	.000518101	.000572287	.000558159	.000481090	.000194435	.000037107	.000000154	00111011010101
0111000100111	60	.0006006	.0003225	.000413284	.000517330	.000571916	.000556837	.000483052	.000197433	.000038307	.000000166	01100011001101
0110000101011	60	.0005999	.0003179	.000413032	.000516532	.000570259	.000556166	.000479518	.000194002	.000036981	.000000149	00110011001101
0101001001011	61	.0005999	.0002934	.000412327	.000512960	.000561289	.000540467	.000458204	.000174373	.000030333	.000000092	00100011001101
0101001101001	61	.0005999	.0002922	.000412292	.000512784	.000560841	.000539672	.000457113	.000173353	.000029988	.000000089	00110011001101
1000110000111	62	.0005967	.0003335	.000412952	.000518435	.000577017	.000569853	.000499839	.000215859	.000045542	.000000259	00111011001110
0010010100011	60	.0005936	.0002697	.000410216	.000508869	.000552711	.000527117	.000450505	.000170837	.000029757	.000000092	00101101001101x
1010010001111	62	.0005888	.0002978	.000410618	.000512319	.000566367	.000555235	.000483142	.000193916	.000041850	.000000221	00101101101110
1000110100011	61	.0005888	.0002822	.000410160	.000509950	.000560282	.000542716	.000467673	.000180098	.000035720	.000000143	00101101011110
0010010100111	63	.0005888	.0002801	.000410100	.000509634	.000559436	.000542716	.000465533	.000185864	.000034848	.000001090	00101110101111
1100000001111	63	.0005881	.0003778	.000412788	.000523885	.000596367	.000609747	.000560774	.000289326	.000088088	.000001090	01000010011110
0010010011101	63	.0005881	.0002865	.000410491	.000511493	.000562050	.000547488	.000472428	.000193337	.000037651	.000000164	01000101011110
1010010100011	63	.0005881	.0002861	.000410157	.000511040	.000561912	.000547488	.000472428	.000192955	.000037497	.000000162	01100110110110
0101001011101	60	.0005857	.0002537	.000408847	.000505655	.000530978	.000539672	.000451421	.000175509	.000031914	.000000106	00110111100011

186

Table A

m = 7, n = 7 Rank Orders 251 thru 300 of 1780 C(7, 7) = 3432 1/C(7, 7) = .00029138

z	w	c_1	c_2	d = .2	.4	.6	.8	1.0	1.5	2.0	3.0	z^{tc}
0011010010011	60	.0005857	.0002360	.000409100	.000503145	.000545358	.000520273	.000437117	.000162886	.000027836	.000000082	0010101101011
1000100110011	62	.0005853	.0002653	.000507101	.000551192	.000557138	.000528360	.000459466	.000182134	.000033908	.000000128	0001011011110
0110000101011	61	.0005850	.0002491	.000408593	.000504855	.000549923	.000532396	.000448016	.000172382	.000030791	.000000103	0010010110101
1001000011001	60	.0005778	.0002491	.000407769	.000504148	.000550923	.000536179	.000455227	.000180877	.000033913	.000000131	0010101110110
0011010011001	61	.0005740	.0002033	.000406107	.000497566	.000536179	.000508463	.000424576	.000156692	.000026159	.000000075	0010011110011x
0101001010011	60	.0005740	.0001814	.000404857	.000493969	.000529441	.000498697	.000412994	.000147202	.000023715	.000000059	0100011101101
1000101010011	62	.0005739	.0001973	.000405295	.000496271	.000535275	.000508917	.000426823	.000159670	.000027786	.000000090	0001110100101
1001001001011	61	.0005706	.0002144	.000405751	.000498417	.000540269	.000516992	.000436920	.000167111	.000029733	.000000099	0011001001101
0101000110011	60	.0005705	.0001659	.000405837	.000491344	.000524991	.000492799	.000406550	.000143255	.000022742	.000000054	0100110110101x
1010000101011	62	.0005705	.0002382	.000405850	.000501432	.000535949	.000535949	.000464477	.000176557	.000040563	.000000225	0001110111010
1000010101011	62	.0005705	.0001994	.000404742	.000495836	.000535898	.000511181	.000430520	.000163027	.000028654	.000000092	0010110010110
1000100101011	61	.0005705	.0001979	.000404701	.000495629	.000535372	.000510249	.000429239	.000161805	.000028225	.000000088	0010111001110
0110100010011	60	.0005661	.0001797	.000403476	.000492733	.000531037	.000505433	.000425159	.000161554	.000028887	.000000102	0011100110101
0101011000011	60	.0005661	.0001642	.000403042	.000490602	.000525850	.000496638	.000413592	.000151717	.000025809	.000000079	0011001111010
0101010100011	61	.0005654	.0001529	.000403017	.000490476	.000525535	.000490168	.000412850	.000151044	.000025586	.000000077	0010111011010
1000011010011	62	.0005626	.0002021	.000403503	.000488876	.000521939	.000520784	.000446895	.000182529	.000023431	.000000061	0011101011101
0101010010011	62	.0005626	.0001844	.000403008	.000495233	.000539063	.000510383	.000432920	.000169743	.000031246	.000000176	0001001111010
1010001000011	60	.0005626	.0001798	.000402878	.000492125	.000531372	.000507465	.000428924	.000165911	.000030436	.000000131	0011000111110
0011001110011	62	.0005705	.0001714	.000401626	.000490329	.000530764	.000509972	.000435450	.000176557	.000035288	.000000188	0011110101001
0100101010011	60	.0005564	.0001117	.000399946	.000401986	.000510189	.000474548	.000388041	.000133957	.000020902	.000000049	0010110010110
1000110010011	60	.0005564	.0001114	.000399938	.000481945	.000510088	.000474376	.000387815	.000133767	.000020484	.000000048	0010111001110
0110010010011	59	.0005519	.0001268	.000399613	.000483487	.000516452	.000487459	.000406910	.000152327	.000027070	.000000099	0011100110101
1010001010011	60	.0005512	.0001229	.000399380	.000479709	.000507593	.000472670	.000404059	.000149679	.000026091	.000000088	0011001111010
0101001100011	60	.0005512	.0001002	.000398750	.000479407	.000506839	.000477362	.000387581	.000135842	.000021822	.000000057	0010111011010
0101010001011	60	.0005512	.0000981	.000398690	.000479930	.000509560	.000477173	.000385825	.000134284	.000021321	.000000053	0011101011101
1000110001011	60	.0005485	.0001049	.000398428	.000476601	.000501506	.000463615	.000394378	.000142487	.000023989	.000000073	0001101100101
1001001100011	60	.0005485	.0000803	.000397747	.000476436	.000501101	.000462924	.000375789	.000127880	.000019598	.000000044	0011001010101
0110010001011	60	.0005485	.0000791	.000397714	.000476436	.000501101	.000462924	.000376702	.000127109	.000019365	.000000043	0011010010101
0011000011011	59	.0005484	.0001136	.000398660	.000481185	.000512769	.000482830	.000402086	.000149627	.000026432	.000000095	0011101101011
0110000111110	61	.0005478	.0001109	.000398479	.000480673	.000511789	.000481347	.000400221	.000147990	.000025854	.000000089	0100001111001
1000000111110	64	.0005473	.0001804	.000400316	.000490003	.000534901	.000521327	.000454197	.000198847	.000044672	.000000346	1000000111110x
0101011100011	59	.0005422	.0000744	.000396519	.000474870	.000507748	.000465202	.000380981	.000133754	.000021671	.000000059	0011011100101
0101100010011	60	.0005398	.0000536	.000395538	.000471731	.000499516	.000455897	.000369818	.000125702	.000019462	.000000046	0011000111101
1001010100011	61	.0005366	.0000934	.000396095	.000476560	.000502793	.000479879	.000402383	.000154751	.000024952	.000000128	0001110101110
1000110100011	61	.0005366	.0000790	.000395686	.000474516	.000500170	.000471002	.000390332	.000143436	.000024923	.000000085	0011011101110
1001010010011	61	.0005359	.0000761	.000395226	.000472704	.000498687	.000469161	.000387826	.000141087	.000024099	.000000076	0010011101110
1000101100011	62	.0005359	.0000669	.000394652	.000471251	.000496301	.000464213	.000383141	.000135638	.000022338	.000000062	0100110110101
0101100001011	61	.0005340	.0000580	.000394652	.000471251	.000496301	.000464213	.000378353	.000134174	.000022122	.000000050	0011011101010
0110011000011	60	.0005336	.0000414	.000394124	.000469095	.000491424	.000453376	.000368545	.000126663	.000019987	.000000050	0100110010011
0011010100011	59	.0005335	.0000323	.000393888	.000467954	.000488851	.000449306	.000363550	.000123354	.000019093	.000000046	0001011010011
1000010110011	62	.0005333	.0001393	.000396759	.000481965	.000522696	.000506510	.000438962	.000189536	.000041686	.000000289	0001011111110
1000011101011	62	.0005333	.0000644	.000394710	.000472018	.000498558	.000465371	.000384130	.000139491	.000023886	.000000078	0100001110110
0011100011011	62	.0005333	.0000633	.000394679	.000471864	.000498180	.000464718	.000383255	.000138697	.000023616	.000000076	0100001110110
0011100100111	59	.0005309	.0000310	.000393394	.000467312	.000486675	.000450104	.000365379	.000125354	.000019821	.000000051	0010101110101
1001000110011	60	.0005302	.0000277	.000393370	.000466521	.000485585	.000448517	.000363466	.000123853	.000019353	.000000048	0010011101110
1000110010111	61	.0005288	.0000437	.000391727	.000462747	.000486880	.000457200	.000374969	.000133360	.000022168	.000000066	0011010110110
0011010100111	59	.0005256	.0000024	.000392417	.000466220	.000488858	.000452625	.000353031	.000117622	.000017924	.000000042	0110011001011
1001001010111	61	.0005253	.0000305	.000391727	.000466220	.000488858	.000452625	.000370250	.000130840	.000021623	.000000064	0011001011010
0101011100111	59	.0005226	.0000090	.000391387	.000463118	.000483055	.000444456	.000361008	.000125078	.000020263	.000000058	0001110001101

Table A

m = 7, n = 7   Rank Orders 301 thru 350 of 1780   C(7, 7) = 3432   1/C(7, 7) = .00029138

z	w	c_1	c_2	d = .2	.4	.6	.8	1.0	1.5	2.0	3.0	z^{tc}
0101100011101	60	.005219	−.0000175	.000390541	.000459497	.000474767	.000430974	.000343870	.000111564	.000016344	.000000034	0100011100101
0101101011101	60	.005192	−.0000368	.000389558	.000456520	.000469076	.000422725	.000334228	.000105037	.000014682	.000000026	0101010101010
0101011001011	61	.005191	−.0000362	.000391521●	.000466016	.000491739	.000460401	.000382836	.000144639	.000026687	.000000116	0000111011101 01x
1000100011011	61	.005191	−.0000026	.000390451	.000460798	.000479094	.000438980	.000354607	.000120175	.000018718	.000000046	0010101001110
1000100110011	61	.005191	−.0000040	.000390412	.000460604	.000478618	.000438168	.000348537	.000119246	.000018425	.000000041	0010100110110
0101001101101	60	.005156	−.0000176	.000389447	.000458251	.000474825	.000433379	.000353531	.000116498	.000017810	.000000029	0101001011010
0101101001101	59	.005140	−.0000479	.000388368	.000454242	.000466395	.000420557	.000333189	.000105995	.000015180	.000000044	0011101010010
0101010100111	59	.005139	−.0000331	.000388758●	.000456213	.000471189	.000428610	.000343625	.000114415	.000017629	.000000048	0100110101010
0110100100111	59	.005112	−.0000350	.000388245	.000455465	.000470726	.000428879●	.000344715	.000115952	.000018138	.000000027	0011101101001
0101000100111	60	.005105	−.0000618	.000387395	.000455185	.000462539	.000428129	.000343262	.000100262	.000014588		0101011001101
1010000101011	61	.005104	−.0000081	.000388811●	.000458612	.000478394	.000441562	.000360836	.000128374	.000021642	.000000069	0001010111010
1000010111011	61	.005104	−.0000267	.000388308	.000456207	.000472687	.000432111	.000348677	.000118532	.000018693	.000000048	0110100101110
0101010010011	61	.005104	−.0000306	.000388200	.000455683	.000471429	.000430004	.000345938	.000116267	.000018006	.000000027	0010110101010
0010101000111	59	.005042	−.0000817	.000385780	.000448239	.000457240	.000409673	.000322545	.000101108	.000014292	.000000061	0011110100011
0111010000111	58	.004970	−.0000665	.000384959	.000448947●	.000462539	.000421358	.000339708	.000116957	.000019213	.000000122	0001111011010
0101010001011	59	.004963	−.0000314	.000385773●	.000452240	.000473002	.000438927	.000362583	.000136302	.000025501	.000000054	0100100111001
0101100000111	59	.004963	−.0000703	.000384728	.000448268	.000461215	.000419338	.000337173	.000114789	.000016570	.000000042	0011100111010
0110010001011	59	.004963	−.0000850	.000384337	.000446432	.000456937	.000412389	.000328412	.000108071	.000016570	.000000042	0111001110011
0110010100011	59	.004963	−.0000862	.000384306	.000446282	.000456581	.000411800	.000327656	.000107465	.000016392	.000000041	0011011101001
0100011100011	59	.004963	−.0000918	.000384154	.000444559	.000454871	.000408983	.000324057	.000104617	.000015562	.000000035	0100011001110
0101100001011	59	.004963	−.0000945	.000384079	.000445196	.000454003	.000407539	.000322194	.000103115	.000015122	.000000033	0101011100101
1001000100111	59	.004963	−.0001101	.000383663	.000443251	.000449496	.000400312	.000313216	.000096590	.000013419	.000000025	0011100101010
0010101010011	59	.004963	−.0001101	.000383663	.000443251●	.000449508	.000400317	.000313224	.000096601	.000013423	.000000025	0011100101001
0101011000011	59	.004963	−.0001116	.000383639	.000443251	.000449230	.000399855	.000312631	.000096136	.000013293	.000000024	0101011001010
0010101100011	63	.004960	−.0000283	.000383623	.000443060	.000449051	.000399564	.000312264	.000095860	.000013219	.000000024	0010101010110
1000010011110	60	.004956	−.0000951	.000385765●	.000453313	.000473069	.000438712	.000361799	.000134422	.000024439	.000000101	1000010111110 0x
0100000101111	61	.004928	−.0000688	.000383940	.000444972	.000445794	.000407427	.000322197	.000103203	.000015140	.000000033	0100010111101
0110000010111	61	.004884	−.0001390	.000384152	.000447698	.000441478	.000420993	.000340138	.000117902	.000019477	.000000061	0011001110010
1000001100111	59	.004884	−.0001228	.000381533	.000438201	.000441622	.000390705	.000303573	.000091830	.000012478	.000000022	0110100101110
1000011001101	61	.004846	−.0001228	.000381274	.000439319●	.000445875	.000398734	.000314306	.000100105	.000014632	.000000031	0100011001110
0011011000101	58	.004821	−.0001424	.000380353	.000436605	.000440957	.000392071	.000307086	.000096156	.000013844	.000000029	0101011001101
1001000000111	60	.004818	−.0000940	.000381571●	.000442527	.000454900	.000414864	.000336007	.000118888	.000020572	.000000079	0011100111010
1001010001011	60	.004818	−.0001262	.000380707	.000438421	.000445191	.000398814	.000315355	.000101976	.000015337	.000000037	0011100100110
1000100011011	60	.004818	−.0001271	.000380683	.000438304	.000444917	.000398365	.000314783	.000101526	.000015206	.000000036	0001101001110
0100010001011	59	.004814	−.0001442	.000380181	.000436219	.000440359	.000391306	.000306246	.000095565	.000013665	.000000028	0010110101011
1100000010011	61	.004811	−.0000600	.000382324●	.000446392	.000464063	.000429849	.000355035	.000134079	.000025258	.000000120	0100100101111
1000101010101	61	.004811	−.0001362	.000380314	.000436983	.000442120	.000394002	.000309374	.000097385	.000014019	.000000028	0100010110111
0011000110111	61	.004811	−.0001369	.000380293	.000436886	.000441892	.000393629	.000308900	.000097017	.000013915	.000000029	0101010110011
0110010101001	59	.004788	−.0001615	.000379259	.000433556	.000435459	.000384430	.000298429	.000090548	.000012430	.000000022	0101011001110
0011100010011	58	.004787	−.0001537	.000379451●	.000434550	.000437869	.000388439	.000303561	.000094538	.000013542	.000000029	0011101101110
0011001000101	60	.004766	−.0001397	.000379441	.000435738●	.000441469	.000394815	.000311850	.000100829	.000015217	.000000037	0010111101010
1001010010101	60	.004734	−.0001563	.000378539	.000433168	.000436784	.000388272	.000304423	.000096018	.000014000	.000000031	0010101011010
1011010010101	58	.004734	−.0001807	.000377835	.000430272	.000430536	.000378753	.000293124	.000088664	.000012205	.000000023	0010101011110 11x
1000000110011	61	.004732	−.0001473	.000380117●	.000430890●	.000434896	.000417749	.000341676	.000125124	.000022599	.000000094	0011001111110
1000011110101	61	.004732	−.0001498	.000378658	.000434171	.000439474	.000392944	.000310558	.000101021	.000015477	.000000038	0101000111100
0011110001101	58	.004708	−.0001806	.000378592	.000433861	.000438754	.000391772	.000309076	.000099863	.000015139	.000000025	0110011110001
0011001100101	61	.004701	−.0001826	.000377372	.000429732	.000430500	.000379623	.000294851	.000090360	.000012724	.000000025	0110011101001
0110100101001	59	.004701	−.0002167	.000377191	.000429307	.000429816	.000377728	.000293860	.000093860	.000012531	.000000014	0100101101101
0101100100101	59	.004670	−.0002440	.000376745●	.000424578	.000420442	.000365008	.000277900	.000079329	.000010049	.000000066	0001111001110
0101100100111	59	.004669	−.0002440	.000377652●	.000433389	.000440736	.000397553	.000318392	.000109424	.000018363		0101101001010

Table A

z	w	c_1	c_2	d = .2	.4	.6	.8	1.0	1.5	2.0	3.0	z^{tc}
10000111001011	60	.0004669	−.0001816	.000376650	.000428672	.000429696	.000379501	.000295434	.000091231	.000012946	.000000026	0010011100011110
10001101011011	60	.0004669	−.0001837	.000376595	.000428413	.000429091	.000378511	.000294179	.000090261	.000012672	.000000025	0010011101011110
01010011001011	60	.0004642	−.0002010	.000375674	.000425764	.000424241	.000377745	.000286535	.000089439	.000011509	.000000020	0010101010111010
01010100100111	58	.0004618	−.0002009	.000374612	.000425653	.000425036	.000374033	.000290291	.000089389	.000012796	.000000028	0010111001011010
01010101011101	59	.0004591	−.0002262	.000373679	.000422435	.000417990	.000363065	.000276959	.000080018	.000010386	.000000016	0100101110010101
10101100100101	59	.0004590	−.0002440	.000373492	.000419740	.000413069	.000356250	.000269346	.000075422	.000009346	.000000013	0100100110010101
10110010011011	60	.0004590	−.0002123	.000375065	.000426120	.000421881	.000370285	.000286607	.000087629	.000012448	.000000027	0001011011010010
01110010010111	60	.0004590	−.0001896	.000374579	.000423881	.000378895	.000378895	.000286844	.000094370	.000014069	.000000034	0001011011010010
01110010011011	60	.0004590	−.0002082	.000374579	.000423881	.000422379	.000370748	.000286784	.000087073	.000012131	.000000024	0101001101110110
10010100001011	60	.0004590	−.0002121	.000374474	.000423391	.000421243	.000368916	.000284492	.000085361	.000011667	.000000021	0010010110101110
10100011001011	60	.0004555	−.0002009	.000374161	.000424052	.000424370	.000375195	.000293226	.000092679	.000013751	.000000033	0011001110011010
10001101011011	60	.0004555	−.0002215	.000373621	.000421562	.000418671	.000366108	.000281986	.000085488	.000011563	.000000021	0011001110011010
10011100001011	58	.0004555	−.0002240	.000373554	.000421253	.000417965	.000364984	.000280599	.000083492	.000011304	.000000020	0010100111011010
01001100110111	58	.0004528	−.0002462	.000372524	.000418213	.000412448	.000357489	.000272450	.000078952	.000010360	.000000014	0011101001010101
00111110000111	57	.0004504	−.0002652	.000371611	.000415486	.000407396	.000350462	.000264612	.000074290	.000009324	.000000014	0101001101110101
01100010001011	59	.0004442	−.0002235	.000371706	.000419218	.000418318	.000369682	.000289452	.000093220	.000014442	.000000042	0001111100000011
01100100110111	59	.0004442	−.0002488	.000370350	.000413371	.000405328	.000349295	.000264660	.000075555	.000009757	.000000016	0100101110010101
01010101010111	58	.0004441	−.0002581	.000370694	.000415043	.000409228	.000355575	.000274467	.000081241	.000011255	.000000023	0011101110001001
10010100010011	58	.0004441	−.0002826	.000370058	.000412147	.000402701	.000345348	.000260066	.000072722	.000009134	.000000012	0011011101010101
10101100100111	58	.0004441	−.0002838	.000370027	.000412003	.000402369	.000344819	.000259413	.000072260	.000009018	.000000013	0011011101010101
10000110011110	62	.0004438	−.0002147	.000371718	.000419786	.000419825	.000371992	.000292177	.000094644	.000014643	.000000040	1000001011011110
01010001000111	58	.0004415	−.0002593	.000370192	.000414306	.000408705	.000355620	.000273126	.000082211	.000011554	.000000025	0001101110101001
01101100110011	58	.0004415	−.0002820	.000369604	.000411630	.000402669	.000346145	.000261606	.000074208	.000009527	.000000015	0011101110100101
01100001001011	59	.0004408	−.0002841	.000369549	.000411376	.000402090	.000345229	.000260484	.000073423	.000009331	.000000015	0010101110100011
10100001011011	58	.0004406	−.0002548	.000370142	.000411221	.000414476	.000354331	.000260735	.000073671	.000009391	.000000015	0101010001101110
01100100001111	58	.0004362	−.0002488	.000369530	.000414306	.000409259	.000356472	.000274004	.000082376	.000011455	.000000023	0010100101101101
01010100110111	58	.0004362	−.0002581	.000368008	.000407470	.000410898	.000360721	.000280505	.000084432	.000013298	.000000035	0011101111010001
01010100010111	58	.0004362	−.0003085	.000365929	.000403801	.000391600	.000336996	.000252134	.000069408	.000008668	.000000012	0011011101010101
01011100010011	58	.0004297	−.0003088	.000368810	.000407434	.000395679	.000345599	.000251978	.000069076	.000008477	.000000012	0011011011110001
01000100000111	60	.0004297	−.0003051	.000368810	.000414789	.000414766	.000369000	.000292317	.000098868	.000016554	.000000061	0001011011111100
10000110101011	60	.0004297	−.0003073	.000366867	.000406022	.000395067	.000338051	.000254453	.000071524	.000009091	.000000014	0100101011011110
10001100111011	60	.0004297	−.0003081	.000366848	.000405931	.000394861	.000337726	.000254055	.000071246	.000009020	.000000014	0100100101011110
10001010000111	59	.0004296	−.0002933	.000367213	.000407683	.000398906	.000344181	.000262012	.000076923	.000010484	.000000021	0011101101010111
00111010000111	57	.0004273	−.0003099	.000366392	.000405357	.000394796	.000337701	.000256121	.000073654	.000009781	.000000018	0010111010100001
01110010000111	58	.0004266	−.0003116	.000365824	.000404974	.000394205	.000337954	.000255316	.000073130	.000009637	.000000017	0010101110010111
01100100110011	58	.0004266	−.0003272	.000365801	.000403091	.000390253	.000331861	.000248048	.000068349	.000008503	.000000013	0011100110111001
01001010010111	60	.0004262	−.0003199	.000365929	.000403801	.000390028	.000331512	.000247630	.000068072	.000008438	.000000012	0010101011101101
01011001010111	57	.0004221	−.0003350	.000364834	.000401369	.000391600	.000333823	.000250200	.000069408	.000008668	.000000015	0011101101010010
10011011001011	59	.0004217	−.0003051	.000365515	.000404558	.000388187	.000330251	.000247303	.000068979	.000008809	.000000015	0011011011010011x
10010110011011	59	.0004217	−.0003195	.000365149	.000402923	.000395270	.000341095	.000260122	.000077248	.000010753	.000000023	0010111101101110
10010100000111	59	.0004217	−.0003207	.000365117	.000402780	.000391652	.000335517	.000253454	.000072788	.000009660	.000000018	0011100110110110
11000100110011	60	.0004210	−.0002698	.000365010	.000408223	.000393330	.000335010	.000252835	.000072353	.000009549	.000000017	0011101011110000
11001010111011	60	.0004210	−.0003269	.000364831	.000401626	.000393556	.000335975	.000275601	.000087796	.000013432	.000000036	0010101011111100
01101010100111	60	.0004210	−.0003291	.000364776	.000401626	.000389005	.000332402	.000249768	.000070264	.000009012	.000000015	0101001101010100
10010100110111	58	.0004187	−.0003513	.000363818	.000398716	.000383674	.000331557	.000248754	.000088847	.000008989	.000000014	0100010011010110
01110010010011	57	.0004186	−.0003446	.000363974	.000399502	.000385516	.000327312	.000240913	.000065319	.000007989	.000000011	0011101011011001
01010101000111	58	.0004148	−.0003787	.000362441	.000394786	.000376666	.000327264	.000244548	.000067872	.000008628	.000000014	0011010110001001
01010100000111	59	.0004120	−.0003063	.000363778	.000398716	.000372264	.000314891	.000230738	.000059715	.000006836	.000000008	0100110011010010x
10101001010111	59	.0004120	−.0003063	.000363760	.000402022	.000394502	.000343307	.000230738	.000064416	.000012969	.000000041	0010111101011010
10010110011011	59	.0004120	−.0003610	.000362379	.000396002	.000380324	.000321101	.000238387	.000064634	.000007905	.000000011	0010110011010110

189

Table A

m = 7, n = 7 Rank Orders 401 thru 450 of 1780 C(7, 7) = 3432 1/C(7, 7) = .00029138

z	w	c_1	c_2	d = .2	.4	.6	.8	1.0	1.5	2.0	3.0	z^{tc}
10011001010011	59	.004120	−.0003616	.000362363	.000395931	.000380167	.000322860	.000238102	.000064451	.000007863	.000000011	00110101100110
01011010100011	57	.004076	−.0003617	.000361604	.000395118	.000380389	.000322887	.000241841	.000068297	.000009007	.000000017	00111011100101
01101100011101	58	.004069	−.0003856	.000360871	.000392239	.000374307	.000313743	.000231097	.000061337	.000007344	.000000010	01000111100101
01010100010101	58	.004069	−.0004030	.000360433	.000390306	.000370024	.000307393	.000232702	.000056812	.000006365	.000000007	01001010110101
01010010010101	58	.004069	−.0004033	.000360426	.000390272	.000370279	.000307279	.000223569	.000056730	.000006347	.000000007	01001011101010x
10000111010011	59	.004068	−.0003489	.000361763	.000396213	.000382982	.000326891	.000246519	.000071118	.000009602	.000000019	00101101101010
10100100010011	58	.004068	−.0003665	.000361316	.000394230	.000378614	.000320188	.000238544	.000065888	.000008335	.000000013	00110110001101
10100010010011	59	.004068	−.0003711	.000361199	.000393700	.000377430	.000318345	.000236322	.000064347	.000007962	.000000005	00110111101010
10100110010011	59	.004041	−.0003638	.000360911	.000393913	.000377430	.000321590	.000240780	.000067803	.000008851	.000000016	00100110101110
10010101010011	59	.004041	−.0003844	.000360390	.000391594	.000373886	.000313755	.000231480	.000061737	.000007419	.000000010	00110110101110
10010010101011	59	.004041	−.0003870	.000360325	.000391305	.000373249	.000312779	.000230323	.000060989	.000007247	.000000009	00101101101010
10010001010011	60	.004034	−.0003197	.000361878	.000398529	.000389335	.000337566	.000229896	.000080719	.000012116	.000000033	01100001111100
10000011111101	60	.004034	−.0003510	.000361110	.000395200	.000382150	.000326712	.000247123	.000072338	.000010038	.000000022	01100000111110
10100101010011	58	.004034	−.0003560	.000360985	.000394652	.000380952	.000324880	.000244942	.000070871	.000009667	.000000020	01001010100101
10100011001011	59	.003991	−.0003560	.000358435	.000385873	.000363639	.000300063	.000216869	.000054045	.000005926	.000000006	01010101110101
01010001001011	59	.003954	−.0004036	.000358362	.000387477	.000369004	.000308166	.000226831	.000060436	.000007314	.000000010	00111000110110
10010010011011	57	.003928	−.0004020	.000357037	.000387174	.000363094	.000301127	.000229948	.000059032	.000008142	.000000015	00111110001101
10010101010011	57	.003928	−.0004266	.000357325	.000384465	.000363094	.000300660	.000219413	.000056636	.000006890	.000000009	00110111001001
10010110010011	57	.003928	−.0004277	.000357295	.000384329	.000362792	.000300660	.000218854	.000056269	.000006504	.000000009	01011101010101
01100100010011	58	.003920	−.0004260	.000357211	.000384338	.000364003	.000301272	.000219686	.000056862	.000006640	.000000008	01001111010011
10000101011110	61	.003916	−.0003832	.000358160	.000388637	.000372330	.000315008	.000235413	.000066181	.000008631	.000000015	10000101011110x
01110010100101	58	.003894	−.0004423	.000356330	.000381921	.000358871	.000295715	.000213754	.000053647	.000005983	.000000007	01010101011001
01101100010111	57	.003893	−.0004122	.000357064	.000385237	.000360595	.000296848	.000216871	.000061998	.000007900	.000000014	01110100100101
10010110001011	57	.003893	−.0004355	.000356482	.000382676	.000360145	.000298405	.000216971	.000055729	.000006456	.000000008	00111101101001
10100010011011	59	.003893	−.0004373	.000356437	.000382472	.000365529	.000297717	.000216158	.000055211	.000006338	.000000008	00101101110010
10010001011011	61	.003889	−.0004124	.000365529	.000388002	.000372080	.000305583	.000225044	.000060294	.000007379	.000000011	10000011101110
10010001010011	59	.003858	−.0003838	.000357677	.000380002	.000362732	.000315559	.000236799	.000067885	.000009188	.000000019	10000011101101
10010101010011	57	.003858	−.0004245	.000356165	.000378778	.000354265	.000302732	.000222053	.000059032	.000007164	.000000006	00101101010011
10010110010011	57	.003840	−.0004604	.000354936	.000378778	.000354265	.000294265	.000208974	.000051859	.000005739	.000000006	01010101110101
01100100010111	57	.003814	−.0004157	.000355556	.000382864	.000364003	.000305743	.000227030	.000063216	.000008290	.000000016	00110111010011
01010011100011	57	.003814	−.0004583	.000354511	.000378348	.000354339	.000291343	.000210403	.000053046	.000006016	.000000007	00111001110101
01010010100111	57	.003807	−.0004592	.000354364	.000378246	.000354118	.000291017	.000210017	.000052810	.000005965	.000000007	00110011100101
10000111000111	58	.003782	−.0004590	.000354364	.000380079	.000354008	.000291013	.000210119	.000052923	.000005988	.000000013	01010010110101
10010011001111	59	.003775	−.0004315	.000354593	.000385664	.000359830	.000318367	.000242809	.000060112	.000007615	.000000035	00011100011110
10010011001011	59	.003775	−.0003802	.000355725	.000377622	.000371565	.000291949	.000211645	.000074606	.000011343	.000000008	00100111001011
10010011010011	59	.003775	−.0004551	.000353879	.000377501	.000354134	.000291553	.000211183	.000054087	.000006242	.000000012	01001100011101
00111001001011	57	.003751	−.0004552	.000353851	.000378455	.000356925	.000296924	.000218011	.000053798	.000006177	.000000009	00101111000011
00111001011011	59	.003748	−.0004425	.000353765	.000375097	.000349473	.000286087	.000205384	.000058569	.000007321	.000000011	00101010011110
01110000110011	59	.003744	−.0004758	.000352971	.000375097	.000349362	.000285918	.000205526	.000050708	.000005558	.000000006	01011101010101
01100101010011	57	.003744	−.0004758	.000352824	.000374779	.000349362	.000285918	.000205526	.000051206	.000005735	.000000006	00110111010011
01010010010011	57	.003717	−.0004754	.000352355	.000374157	.000349017	.000286114	.000206220	.000051953	.000005921	.000000007	00101011101001
11000011100111	59	.003696	−.0004205	.000353297	.000379233	.000350474	.000303522	.000226476	.000054386	.000008696	.000000018	01010101111100
10100101101011	59	.003696	−.0004776	.000351914	.000373318	.000347920	.000284936	.000205119	.000051412	.000005802	.000000007	01010010110101
10010101010011	59	.003696	−.0004797	.000351861	.000373089	.000347428	.000284202	.000204273	.000050901	.000005691	.000000007	01000110110110
10010100100111	58	.003695	−.0004660	.000352183	.000374576	.000350722	.000289236	.000210206	.000054669	.000006549	.000000009	00111011010011
00110110100101	56	.003672	−.0004803	.000351430	.000373899	.000347440	.000285135	.000206077	.000052733	.000006202	.000000011	00110111100110
01110000011011	57	.003668	−.0004660	.000351703	.000374248	.000350248	.000289234	.000210672	.000055294	.000006717	.000000011	00101011111010
01110010010011	57	.003665	−.0004618	.000351744	.000374779	.000351106	.000295078	.000212252	.000056248	.000006991	.000000006	00110100011101
01100110010011	57	.003665	−.0004964	.000350908	.000370698	.000343643	.000279668	.000199906	.000049110	.000005425	.000000006	01011101011010
01100100010011	57	.003665	−.0004969	.000350894	.000370637	.000343512	.000279472	.000199680	.000048976	.000005396	.000000006	00110011001001

190

Table A

m = 7, n = 7   Rank Orders 451 thru 500 of 1780   C(7, 7) = 3432   1/C(7, 7) = .00029138

d = .2

z	w	c_1	c_2	.2	.4	.6	.8	1.0	1.5	2.0	3.0	z^{tc}
11001001101100	59	.003661	.0004295	.000352450	.000377413	.000357883	.000300626	.000223796	.000063285	.000008513	.000000017	0011001111100
10001001110101	59	.003661	.0004886	.000351019	.000371298	.000344915	.000281444	.000201776	.000049943	.000005543	.000000006	0101000110110
00110001101011	59	.003661	.0004899	.000350985	.000371153	.000344608	.000280993	.000203851	.000049645	.000005481	.000000009	0101001101110
01001110001001	56	.003637	.0004884	.000350607	.000370879	.000345063	.000282619	.000194670	.000051923	.000006082	.000000009	0111001100011x
01001111000011	57	.003634	.0005123	.000349971	.000368272	.000339623	.000274573	.000197813	.000046503	.000004931	.000000006	0010111000111x
00111100000111	58	.003598	.0005038	.000349520	.000368175	.000344645	.000276902	.000205368	.000048596	.000005368	.000000006	0011111000101
01011100000101	58	.003554	.0004875	.000349135	.000368869	.000343845	.000282984	.000205774	.000054266	.000006728	.000000012	0101011100010
01011100100101	56	.003547	.0005443	.000347631	.000362825	.000331366	.000264896	.000185410	.000042583	.000004295	.000000004	0101011100101
01001011100101	57	.003521	.0005426	.000347191	.000362315	.000331229	.000265352	.000186340	.000043369	.000004465	.000000004	0100101110010
11000000101111	58	.003520	.0004515	.000349376	.000371711	.000351202	.000294951	.000220405	.000064301	.000009286	.000000026	0011111110010
10101010000111	58	.003520	.0004890	.000348459	.000367753	.000342681	.000282092	.000205269	.000054282	.000006743	.000000011	0011111010010
10110010000111	58	.003520	.0005026	.000348131	.000366354	.000339720	.000277729	.000200280	.000051294	.000006086	.000000009	0011110110101
11001000100011	57	.003520	.0005041	.000348095	.000366194	.000339377	.000277214	.000199681	.000050920	.000005929	.000000008	0011011010110
10010001110011	58	.003520	.0005047	.000348078	.000366113	.000339185	.000276903	.000199293	.000050640	.000005604	.000000007	0011011101110
10011010001011	58	.003520	.0005107	.000347931	.000365479	.000337825	.000274872	.000196942	.000049192	.000005045	.000000006	0010011100110
10010110100011	58	.003520	.0005238	.000347613	.000364129	.000334988	.000270729	.000192265	.000046524	.000005045	.000000005	0011001100110
10011101010011	58	.003520	.0005241	.000347605	.000364094	.000334915	.000270622	.000192144	.000046454	.000004952	.000000005	0011010110101
10011100100101	58	.003520	.0005261	.000347558	.000363890	.000334479	.000269975	.000191402	.000046013	.000004910	.000000005	0010101101101
10101000101011	57	.003520	.0005270	.000347535	.000363792	.000334272	.000269669	.000191053	.000045810	.000004952?	.000000005	0101010111011
11000000101011	59	.003513	.0004742	.000348666	.000368898	.000345155	.000285645	.000209147	.000056261	.000007093	.000000012	1100001111100
10000010111001	59	.003513	.0005062	.000347910	.000365739	.000338608	.000276176	.000198517	.000050167	.000005812	.000000008	0110000101110
10100010111001	59	.003513	.0005109	.000347798	.000365268	.000337623	.000274742	.000196898	.000049230	.000005614	.000000007	0100110110110
01010010111001	57	.003469	.0005652	.000345708	.000358721	.000326617	.000258630	.000179803	.000040538	.000004003	.000000003	0011010110110
01010010111010	58	.003441	.0005461	.000345654	.000358570?	.000326843	.000263843	.000185907	.000043980	.000004654	.000000005	0110010110110
10000110011110	60	.003402	.0005227	.000344761	.000359853	.000332734	.000270333	.000193706	.000048685	.000005621	.000000007	1000011001110x
10101101010011	56	.003379	.0005376	.000344761	.000359295	.000330029	.000267389	.000191237	.000048215	.000005691	.000000009	0011110101010
10011101010010	56	.003379	.0005609	.000344198	.000356900	.000324984	.000259994	.000182840	.000043296	.000004640	.000000005	0011110000101
10011100100110	56	.003379	.0005627	.000344154	.000356709	.000324675	.000259387	.000182141	.000042884	.000004552	.000000005	0101101010101
01101000110011	57	.003372	.0005603	.000344083	.000356762	.000324924	.000260059	.000183002	.000043434	.000004668	.000000005	0101010111001
01101100100101	57	.003372	.0005768	.000343686	.000355085	.000321430	.000255008	.000177366	.000040327	.000004060	.000000004	0101011001101
01101000101101	57	.003372	.0005772	.000343672	.000355026	.000321308	.000254832	.000177170	.000040221	.000004040	.000000004	0101010110101
10100110100011	58	.003371	.0005473	.000344361	.000357947	.000327338	.000263447	.000186665	.000045299	.000005006	.000000006	0101100100111
10010001111000	60	.003367	.0005332	.000344615	.000359134	.000329809	.000266940	.000190452	.000047227	.000005354	.000000007	1000010100110
10100101101010	58	.003327	.0005615	.000343536	.000355803	.000323779	.000258933	.000182025	.000044564	.000004983	.000000004	0110010110110
01010101101001	56	.003327	.0005838	.000342707	.000353268	.000319280	.000253111	.000176086	.000040273	.000004120	.000000003	0110110110101
01110001101001	57	.003292	.0005968	.000341784	.000351064	.000315845	.000248961	.000171985	.000038380	.000003783	.000000003	0101010110101
01110010010011	56	.003292	.0005476	.000342934	.000355946	.000326011	.000263657	.000188414	.000047586	.000005651	.000000006	0011110010001
01010101010011	56	.003292	.0005911	.000341903	.000351644	.000317123	.000250881	.000174190	.000039661	.000004020	.000000004	0011101001010
10101011010010	56	.003292	.0005918	.000341888	.000351579	.000316989	.000250688	.000173976	.000039545	.000004020	.000000004	0010101101101
10010100010011	60	.003288	.0005378	.000343086	.000356706	.000327601	.000265893	.000190821	.000048799	.000005874	.000000009	1000011101110
10100001100111	58	.003257	.0005737	.000341671	.000352349	.000319659	.000255210	.000179378	.000042666	.000004624	.000000005	0011001111010
10011001100101	56	.003230	.0005959	.000340648	.000349545	.000314878	.000249136	.000173216	.000039788	.000004122	.000000004	0101110110110
11001100100111	58	.003226	.0005100	.000342631	.000357918	.000332388	.000277647	.000202104	.000056938	.000007974	.000000021	0001111101100
00011100100111	58	.003226	.0006013	.000340459	.000348791	.000313332	.000246855	.000170586	.000038206	.000003788	.000000003	0001011001011
10010010100011	58	.003195	.0006042	.000339840	.000348758	.000313264	.000246760	.000170482	.000038152	.000003778	.000000003	0101001100110
01011010010011	57	.003181	.0005838	.000339840	.000347815	.000312499	.000252207	.000170961?	.000038980	.000004006	.000000005	0011110010101
10011001000101	56	.003174	.0006063	.000339413	.000349413	.000316225	.000252207	.000177365	.000042600	.000004733	.000000006	0011110010110
01010110010011	58	.003174	.0005486	.000340748	.000352538	.000322784	.000251645	.000177843	.000044393	.000005909	.000000010	0010101110101
01010100110011	58	.003174	.0006047	.000339434	.000351689	.000316709	.000245816	.000170323	.000038723	.000004020	.000000004	0101010110101
10010100100111	58	.003174	.0006072	.000333375	.000346870	.000311177	.000245079	.000169504	.000038274	.000003866	.000000004	0100011101110

191

Table A

m = 7, n = 7				Rank Orders 501 thru 550 of 1780				C(7, 7) = 3432		1/C(7, 7) = .00029138		
z	w	c_1	c_2	d = .2	.4	.6	.8	1.0	1.5	2.0	3.0	z^{tc}
0011110001010011	55	.0003150	−.0005920	.000339305	.000347810	.000314004	.000249817	.000175254	.000041852	.00004627	.00000006	0011011100001011
1100010101010011	58	.0003148	−.0005615	.000339954	.000350525	.000319488	.000257488	.000183561	.000046182	.000065455	.00000008	0011010101111100
1001010101010101	58	.0003148	−.0006207	.000338572	.000344831	.000307865	.000240970	.000166377	.000036344	.000003526	.00000003	0101010101010110
1001100100000111	58	.0003148	−.0006220	.000338539	.000344696	.000307589	.000240580	.000164950	.000036121	.000003485	.00000006	0100010101010110
1001001010010011	57	.0003146	−.0005923	.000339229	.000347655	.000313777	.000249536	.000174958	.000041678	.000004588	.00000006	0001110110110011
1100000001110010	56	.0003141	−.0006241	.000338424	.000344516	.000307514	.000240775	.000166442	.000036688	.000003639	.00000003	0011010101010001
1100000001110011	59	.0003141	−.0005524	.000340045	.000351274	.000321328	.000262042	.000187057	.000048469	.000006003	.00000011	0111010101010010
0011101010100011	55	.0003123	−.0006056	.000338507	.000345124	.000310868	.000246042	.000171575	.000040239	.000004354	.00000005	0011101010100011x
0110001110000111	56	.0003116	−.0006034	.000338415	.000345721	.000310760	.000245867	.000171285	.000039880	.000004235	.00000005	0010101110001001
0110001110000111	56	.0003116	−.0006203	.000338027	.000344162	.000307657	.000241578	.000165703	.000038390	.000003829	.00000004	0011100110001001
0101000001010011	56	.0003116	−.0006222	.000337984	.000343984	.000307295	.000241066	.000166144	.000037306	.000003774	.00000004	0011010001001110
0110000111110011	57	.0003109	−.0006198	.000337909	.000344005	.000307548	.000241558	.000166755	.000037659	.000003839	.00000004	0110000111000110x
1001010101010101	57	.0003085	−.0006180	.000337492	.000343443	.000307137	.000241444	.000166894	.000037853	.000003874	.00000004	0101000110001110
1001001110000101	58	.0003061	−.0006343	.000336673	.000341234	.000303434	.000236774	.000162169	.000035650	.000003492	.00000003	0101001110010101
0101010100100101	56	.0003034	−.0006574	.000335654	.000338391	.000298645	.000230795	.000156232	.000033067	.000003082	.00000002	0100111100101101
1100000111001011	56	.0002999	−.0006647	.000334852	.000336774	.000296503	.000228586	.000154361	.000032471	.000003007	.00000002	0101001010100101
1000101010010011	58	.0002999	−.0006009	.000336307	.000342628	.000308149	.000244691	.000171584	.000041077	.000004550	.00000006	0001001011011100
1000110111000011	58	.0002999	−.0006329	.000335576	.000340045	.000302289	.000236552	.000162826	.000036600	.000003721	.00000005	0110000111010110
0110100110101101	58	.0002999	−.0006376	.000335468	.000339249	.000301402	.000235309	.000161477	.000035899	.000003590	.00000003	0101010101010101x
0101000101000111	57	.0002998	−.0006249	.000335751	.000340500	.000304054	.000239188	.000165853	.000038390	.000004100	.00000004	0011010011001010
1000101010010011	57	.0002998	−.0006456	.000335264	.000338484	.000299921	.000233298	.000159360	.000034886	.000003419	.00000003	0011010100001110
1001010100100011	57	.0002981	−.0006481	.000335204	.000338232	.000299401	.000232553	.000158537	.000034441	.000003333	.00000003	0010011100110110
1010010000110011	57	.0002971	−.0006242	.000335281	.000338103	.000303549	.000239059	.000166086	.000038764	.000004193	.00000005	0011011100001110
1001011100010011	57	.0002971	−.0006419	.000334863	.000339835	.000299989	.000232682	.000159039	.000035679	.000003582	.00000003	0011010111000110
1100001010010101	58	.0002964	−.0006464	.000334756	.000338103	.000305829	.000242195	.000159363	.000034927	.000003438	.00000003	0011001110010100
1000101011001011	58	.0002964	−.0006087	.000335490	.000340935	.000299715	.000233706	.000160230	.000040250	.000004424	.00000005	0100010011100110
1001011010101011	58	.0002964	−.0006421	.000334547	.000337869	.000301569	.000233869	.000159161	.000035584	.000003561	.00000003	0110000110110110
0101010101010101	56	.0002947	−.0006460	.000334638	.000337508	.000298996	.000232709	.000159161	.000035047	.000003464	.00000002	0101001011101010x
0101100010010101	57	.0002947	−.0006856	.000333415	.000333407	.000291422	.000222709	.000148847	.000030295	.000002686	.00000003	0101010011011010
0101100100100011	57	.0002920	−.0006825	.000333003	.000332978	.000291385	.000223215	.000149730	.000030929	.000002806	.00000003	0101011100100101
0110000010111111	57	.0002919	−.0006098	.000334653	.000339728	.000304975	.000242222	.000170303	.000041579	.000004807	.00000007	0010111110010010
1001011010010111	57	.0002919	−.0006645	.000333386	.000334557	.000294518	.000227496	.000154232	.000033049	.000003157	.00000003	0101001101010110
1001001010000011	57	.0002919	−.0006652	.000333371	.000334495	.000294392	.000227318	.000154036	.000032946	.000003138	.00000003	0011001011001110
0101001110000011	55	.0002857	−.0006393	.000332854	.000335354	.000293187	.000234162	.000162526	.000038219	.000004239	.00000006	0011111110001001
0101011100000111	55	.0002857	−.0006620	.000332323	.000333165	.000293713	.000227793	.000154564	.000034362	.000003461	.00000003	0011110011001001
1011010101000011	59	.0002853	−.0006641	.000332274	.000332959	.000293288	.000227185	.000154564	.000033992	.000003388	.00000005	1000010111010110
0110101010101011	56	.0002850	−.0006542	.000332413	.000333614	.000294559	.000228821	.000156411	.000034564	.000003449	.00000003	0101101010101110
0110101010101011	56	.0002850	−.0006940	.000331460	.000330008	.000287640	.000219548	.000146790	.000030125	.000002723	.00000003	0101001101010110
0101001010101011	56	.0002823	−.0006916	.000331028	.000329498	.000287436	.000219815	.000147416	.000030637	.000002825	.00000004	0011011100101110
1001001011011011	57	.0002822	−.0006617	.000331686	.000332190	.000292798	.000225867	.000155284	.000033500	.000003394	.00000003	0010011100110101
1010001100011011	57	.0002822	−.0006761	.000331323	.000330824	.000290048	.000223361	.000151121	.000032359	.000003107	.00000003	0010101101011010
1010101010101011	57	.0002822	−.0006773	.000331323	.000330709	.000289814	.000223031	.000150762	.000032173	.000003073	.00000003	0011010101011110
0111010100101011	59	.0002818	−.0006632	.000331568	.000331810	.000292018	.000226020	.000155863	.000033576	.000003295	.00000006	1000100101101110x
0110110100100011	55	.0002778	−.0006525	.000331102	.000331918	.000229604	.000218409	.000158668	.000036936	.000004059	.00000006	0011110110001011
0101101010100101	55	.0002778	−.0006960	.000330105	.000327893	.000285676	.000218239	.000146599	.000030722	.000002889	.00000002	0111010010100101
1110000111000011	59	.0002771	−.0006967	.000330090	.000327832	.000285554	.000218239	.000146417	.000030629	.000002873	.00000003	0110111110001001
0110100101000101	56	.0002771	−.0007107	.000330472	.000329657	.000289329	.000222723	.000152037	.000032455	.000003343	.00000002	0100011011110001
0110010010101011	56	.0002771	−.0007107	.000329635	.000326312	.000282723	.000214454	.000142455	.000028727	.000002546	.00000003	0010111010011001
0110110010110011	56	.0002766	−.0007110	.000329629	.000326282	.000282672	.000214385	.000142385	.000028693	.000002541	.00000002	0101010010110001
1001100101011110	59	.0002766	−.0006663	.000330543	.000330105	.000290252	.000224730	.000153247	.000033814	.000003394	.00000003	1000100101110010

192

z	w	c_1	c_2	d = .2	.4	.6	.8	1.0	1.5	2.0	3.0	z^{tc}
0101100110011	55	.0002743	-.0007024	.000329321	.000332344	.000283662	.000216366	.00144892	.000030188	.000002822	.000000002	0011100110101
1010010110011	57	.0002743	-.0006923	.000329538	.000327166	.000285187	.000218320	.00146816	.000030945	.000002923	.000000002	0011100111010
1001101001101	57	.0002705	-.0007125	.000328364	.000324229	.000280535	.000212727	.00141393	.000028646	.000002558	.000000002	0100110100110
1001011001011	55	.0002682	-.0007056	.000327538	.000324328	.000281524	.000214690	.00147447	.000032395	.000002866	.000000004	0100101010011
0110101010011	56	.0002633	-.0006915	.000326749	.000321256	.000277037	.000217418	.00139260	.000028428	.000002603	.000000002	0011110100110
1011001001011	55	.0002629	-.0007236	.000329033	.000325649	.000295669	.000209635	.00167505	.000043551	.000005679	.000000013	0011011001100
0110100100111	57	.0002626	-.0006198	.000328193	.000327124	.000288642	.000235649	.00156171	.000036823	.000004129	.000000005	0001111101100
1010010010111	57	.0002626	-.0006562	.000327904	.000325954	.000286290	.000225556	.00152563	.000034880	.000003741	.000000005	0001110111100
1100101000111	57	.0002626	-.0006706	.000327863	.000325786	.000285950	.000221769	.00152031	.000034588	.000003681	.000000004	0001101011100
1000011110101	57	.0002626	-.0007104	.000326971	.000322259	.000279030	.000212310	.00142005	.000029634	.000002788	.000000002	0101100011110
0110101011001	57	.0002626	-.0007136	.000326897	.000321963	.000278442	.000211500	.00141143	.000029206	.000002712	.000000002	0100011100110
1010011001010	57	.0002626	-.0007288	.000326549	.000320565	.000275674	.000207700	.00137118	.000027268	.000002384	.000000002	0101010010110
0110010101011	57	.0002626	-.0007289	.000326548	.000320562	.000275670	.000207696	.00137117	.000027270•	.000002385	.000000002	0100100110110
1010011001011	57	.0002626	-.0007299	.000326524	.000320467	.000275481	.000207438	.00136843	.000027139	.000002363	.000000002	0101001010110
0101101001011	57	.0002626	-.0007306	.000326510	.000320408	.000275365	.000207277	.00136673	.000027057	.000002349	.000000002	0101001001110
1011010001011	58	.0002619	-.0006758	.000327610•	.000325068	.000284677	.000220071	.00150208	.000033584	.000003470	.000000004	0001110011010
0011110001011	54	.0002601	-.0006983	.000326813	.000322770	.000280852	.000215436	.00145781	.000031879	.000003239	.000000004	0011011011001x
0110010111001	56	.0002596	-.0007337	.000325897	.000319363	.000274270	.000206483	.00136311	.000027224	.000002410	.000000002	0110101011001
0110010101011	55	.0002594	-.0007113	.000326381•	.000321330	.000278150	.000211771	.00141870	.000029864	.000002557	.000000003	0101011011000
0110011001011	55	.0002594	-.0007290	.000325987	.000319796	.000275198	.000207829	.00137803	.000028016	.000002557	.000000002	0111001110101
0110101010101	55	.0002594	-.0007303	.000325958	.000319683	.000274978	.000207532	.00137494	.000027871	.000002533	.000000002	0100001110110
1010010101001	59	.0002591	-.0007459	.000324718	.000321121	.000284372	.000220400	.00135130	.000034588	.000003744	.000000005	0101010101110
1010010111001	57	.0002547	-.0007459	.000324602	.000316841	.000270699	.000202525	.00132698	.000025832	.000002202	.000000001	0101010011110
0110101010011	55	.0002516	-.0007259	.000324298	.000317803•	.000273540	.000207062	.00137903	.000025149	.000002149	.000000001	0010111010010
0111010001011	55	.0002485	-.0007592	.000324182•	.000314029	.000266792	.000199001	.00129910	.000025149	.000002149	.000000001	0101110110001
0111101001011	56	.0002484	-.0007187	.000323711	.000315570	.000274040	.000208423	.00139794	.000029848	.000002951	.000000003	0011110011010
0011101001011	56	.0002484	-.0007394	.000323653	.000315685	.000270301	.000202100	.00134288	.000027100	.000002456	.000000003	0011011011010
0110011011001	56	.0002484	-.0007419	.000323685	.000315448	.000269828	.000202609	.00133585	.000026747	.000002393	.000000003	0101011011010
1101010100011	57	.0002477	-.0007046	.000324348	.000318456	.000275819	.000210772	.00142139	.000030778	.000003073	.000000003	0101011011100
0110011101001	57	.0002477	-.0007360	.000323657	.000315776	.000270671	.000203889	.00135012	.000027479	.000002522	.000000002	0110000110110
1100010101011	57	.0002477	-.0007410	.000323545	.000315334	.000269811	.000202725	.00133793	.000026901	.000002424	.000000002	0100011100110
1010010101011	57	.0002450	-.0007168	.000323579	.000316577	.000272874	.000207230	.00138673	.000029231	.000002805	.000000002	0101010111100
1000010101011	57	.0002450	-.0007502	.000322843	.000313724	.000267402	.000199936	.00131156	.000025815	.000002252	.000000001	0110010111010
0101000110011	56	.0002449	-.0007541	.000322756	.000313387	.000266755	.000199073	.00130268	.000025416	.000002189	.000000001	0011101101110
0111010001011	56	.0002449	-.0007260	.000323377	.000315933	.000271844	.000206116	.00137792	.000029151	.000002852	.000000003	0011110011010
1011010001011	56	.0002449	-.0007446	.000322863	.000314235	.000268477	.000201475	.00133842	.000026687	.000002409	.000000002	0011100110010
0111001001011	56	.0002449	-.0007486	.000322863	.000313878	.000267773	.000200511	.00131821	.000026192	.000002324	.000000001	0101011010100
0101011001011	55	.0002433	-.0007782	.000321912	.000310892	.000262527	.000193882	.00125261	.000023455	.000001917	.000000001	0101011100101x
0101100101011	55	.0002398	-.0007838	.000321147	.000309423	.000260677	.000192070	.00123806	.000023052	.000001873	.000000001	0101011011100
1010101001011	56	.0002397	-.0007656	.000321522•	.000310913	.000263525	.000195819	.00127599	.000024667	.000002114	.000000001	0011010101010
1011000100111	56	.0002370	-.0007229	.000321985	.000313893	.000260063	.000205155	.00137672	.000029634	.000002979	.000000003	0011101110010
1011001001011	56	.0002370	-.0007605	.000321146	.000310601	.000263663	.000196494	.00135294	.000025294	.000002227	.000000002	0011100110110
1011000101011	56	.0002370	-.0007625	.000321101	.000310423	.000263318	.000196030	.00128114	.000025071	.000002190	.000000001	0010111001011
1010001001011	57	.0002363	-.0007235	.000321828•	.000313529	.000269450	.000204319	.00136733	.000029057	.000002850	.000000003	0011100111010
1010001110011	57	.0002363	-.0007594	.000321038	.000310475	.000263601	.000196523	.00128687	.000025361	.000002237	.000000003	0101010010110
0100101110101	57	.0002363	-.0007620	.000320983	.000310264	.000263202	.000195997	.00128152	.000025127	.000002201	.000000001	0101011010100
0110110011011	55	.0002336	-.0007831	.000320025	.000307730	.000259336	.000191155	.00131561	.000023342	.000001947	.000000001	0101011100101x
1000110011110	58	.0002331	-.0007513	.000320622	.000310195	.000263880	.000197403	.00129880	.000026040	.000002353	.000000001	1000011001110
1000101010110	58	.0002305	-.0007656	.000319809	.000308155	.000260647	.000193517	.00126112	.000024441	.000002098	.000000001	1000101010111x0

Table A

m = 7, n = 7 Rank Orders 601 thru 650 of 1780   C(7, 7) = 3432   1/C(7, 7) = .00029138

z	w	c₁	c₂	d = .2	.4	.6	.8	1.0	1.5	2.0	3.0	z^tc
011010010011 0101	55	.0002301	−.0007891	.000319251	.000306223	.000257210	.000189239	.000121995	.000022885	.000001895	.000000001	010101101101 01
101010001011 0101	56	.0002300	−.0007740	.000319554	.000305489	.000259489	.000192213	.000124978	.000024133	.000002078	.000000001	010101101011 010
101010001010 1011	56	.0002274	−.0007716	.000319116	.000306861	.000259133	.000192216	.000125273	.000024433	.000002137	.000000001	001011011011 010
011110000001 1011	54	.0002256	−.0007336	.000319649	.000309735	.000265263	.000200922	.000134699	.000029218	.000003014	.000000004	000111101000 101
010101110000 0101	54	.0002256	−.0007762	.000318703	.000306038	.000258096	.000191240	.000124555	.000024331	.000002148	.000000002	010111100001 0101
100100010011 1110	58	.0002252	−.0007771	.000318683	.000305957	.000257939	.000191029	.000124336	.000024229	.000002131	.000000001	100001101101 0101
010101010110 1101?	55	.0002249	−.0007668	.000318822	.000306582	.000259097	.000192464	.000125659	.000024649	.000002169	.000000002	010101101011 1010
010010010110 1011	55	.0002230	−.0008069	.000317896	.000303214	.000252898	.000184493	.000117752	.000021394	.000001697	.000000001	001101011001 1001
010110100000 0111	54	.0002230	−.0007888	.000317940	.000304239	.000255414	.000188208	.000121789	.000023320	.000002008	.000000001	011010101010 0101
110000100110 1101	55	.0002223	−.0007831	.000317922	.000304422	.000255864	.000188811	.000122361	.000023483	.000002016	.000000001	010101111000 0001
011000111000 0101	55	.0002223	−.0008021	.000317511	.000302857	.000252931	.000185000	.000118543	.000021885	.000001780	.000000001	010110000111 1001
010100010101 0101	56	.0002223	−.0008031	.000317491	.000302782	.000252791	.000184819	.000118363	.000021810	.000001770	.000000002	001011111000 0010
010100010110 0011	56	.0002221	−.0007541	.000318535	.000306838	.000260511	.000195002	.000128735	.000026379	.000002488	.000000002	001001101101 1010
010100110100 1011	56	.0002221	−.0007863	.000317830	.000304121	.000255326	.000188126	.000121682	.000023202	.000001975	.000000002	001010010101 1010
100100110011 1110	58	.0002221	−.0007872	.000317810	.000304040	.000255169	.000187914	.000121461	.000023099	.000001958	.000000001	001011010010 1010
010010010001 1011	56	.0002218	−.0007742	.000318017	.000304943	.000256899	.000190150	.000123667	.000023972	.000002077	.000000001	100100110110 1110
011011000001 0111	54	.0002191	−.0007829	.000317050	.000302391	.000252832	.000185345	.000119148	.000022252	.000001837	.000000003	010011100111 0001
001101100011 0011	54	.0002160	−.0007666	.000316770	.000302670	.000254269	.000187888	.000122173	.000023886	.000002125	.000000002	001011111000 0001
011100110110 0011	55	.0002111	−.0007666	.000316223	.000302617	.000255479	.000190371	.000125279	.000025664	.000002456	.000000002	011010001110 0001
110000100111 1101	57	.0002102	−.0007717	.000315982	.000301857	.000254032	.000188348	.000123047	.000024428	.000002209	.000000001	010010111001 0010
110010010000 0111	56	.0002104	−.0007600	.000316225	.000302884	.000256107	.000191244	.000126171	.000026032	.000002508	.000000002	001110101001 1100
100110101001 0101	56	.0002104	−.0008191	.000314936	.000297939	.000246720	.000177853	.000113517	.000020386	.000001601	.000000001	010101001001 0110
100010010001 1011	54	.0002081	−.0008205	.000314905	.000297821	.000246495	.000178556	.000113214	.000020255	.000001581	.000000001	100001001111 1010
011100110001 0011	54	.0002081	−.0007922	.000315105	.000299579	.000250482	.000184258	.000119301	.000023097	.000002037	.000000002	001011011010 1001
010101010001 0011	54	.0002081	−.0008100	.000314724	.000298037	.000247810	.000180805	.000115853	.000021651	.000001803	.000000001	001011011010 1001
010101010011 1011	53	.0002081	−.0008113	.000314696	.000298037	.000247610	.000180543	.000115588	.000021537	.000001787	.000000001	001110101010 0101
001110010001 1011	56	.0002080	−.0007730	.000315508	.000301198	.000253636	.000188516	.000123749	.000025211	.000002404	.000000003	011111010000 011x
010110100011 1011	56	.0002077	−.0007582	.000315771	.000302251	.000255606	.000191046	.000126256	.000026248	.000002560	.000000002	001101000101 1010
010110110001 0101	54	.0002077	−.0008144	.000314547	.000297559	.000246693	.000179262	.000114191	.000020810	.000001672	.000000001	011010001011 0110
110010001011 1101	55	.0002077	−.0008169	.000314493	.000297352	.000246302	.000178751	.000113674	.000020589	.000001639	.000000001	001010111100 0110
011010001011 1001	55	.0002074	−.0008235	.000314295	.000296777	.000245379	.000177696	.000112739	.000020308	.000001612	.000000001	011000101001 1001
110000001011 1101	57	.0002070	−.0007795	.000315204	.000300324	.000252035	.000186308	.000121324	.000023868	.000002135	.000000002	010100110111 1010
101000010111 1110	58	.0002069	−.0008152	.000315148	.000300187	.000251849	.000186140	.000121221	.000023891	.000002149	.000000002	100001011100 1001
110100000001 0111	54	.0002046	−.0007418	.000313967	.000296700	.000245982	.000179030	.000129314	.000021242	.000001773	.000000004	001101111101 1100
010100000011 0011	56	.0002025	−.0008331	.000315157	.000302079	.000256675	.000193416	.000109892	.000028100	.000002928	.000000002	001011011010 1110
001101101000 0101	56	.0002025	−.0008335	.000313174	.000294493	.000242296	.000174433	.000109821	.000019325	.000001480	.000000001	010100101101 1010
100010010101 0101	56	.0002025	−.0008335	.000313167	.000294465	.000242242	.000174363	.000109825?	.000019295	.000001476	.000000001	010100110111 0001
010111100000 0111	54	.0001994	−.0008131	.000313040	.000295305	.000244688	.000178149	.000114016	.000021320	.000001801	.000000001	010111100001 1010
010110100000 1101	54	.0001963	−.0008291	.000312125	.000293077	.000241274	.000174193	.000110321	.000019894	.000001594	.000000002	010101110000 0001
100101010000 0111	55	.0001935	−.0008013	.000312217	.000294596	.000244896	.000179459	.000115997	.000022603	.000002044	.000000001	011101110100 1010
100011110000 0011	55	.0001935	−.0008199	.000311808	.000293009	.000241853	.000175401	.000111808	.000020679	.000001724	.000000001	001110001000 1110
101010001011 0011	55	.0001935	−.0008238	.000311721	.000292674	.000241213	.000174552	.000110937	.000020287	.000001661	.000000002	010101111111 000
111000000011 1011	56	.0001928	−.0007027	.000314201	.000304201	.000265982	.000199711	.000137190	.000033093	.000004056	.000000009	000111111101 1100
110100000011 0011	56	.0001928	−.0007865	.000312387	.000295457	.000244564	.000181591	.000118059	.000023365	.000002138	.000000002	001011100111 1100
100010100011 0011	56	.0001928	−.0008060	.000312090	.000294310	.000244369	.000178664	.000115035	.000021969	.000001904	.000000002	001101011001 0101
110001010000 0111	56	.0001928	−.0008015	.000312057	.000294184	.000244128	.000178344	.000114705	.000021818	.000001880	.000000001	011010110110 1100
100001111010 0101	54	.0001928	−.0008164	.000311751	.000292977	.000242154	.000175902	.000112375	.000020954	.000001768	.000000001	011010000101 1010
101010001101 0101	56	.0001928	−.0008222	.000311626	.000292605	.000241268	.000175405?	.000111207	.000020448	.000001689	.000000001	010100111010 0010
100110110001 0101	56	.0001928	−.0008340	.000311366	.000291601	.000239357	.000172224	.000108641	.000019329	.000001517	.000000001	001101100111 0110

194

m = 7, n = 7   Rank Orders 651 thru 700 of 1780   C(7, 7) = 3432   1/C(7, 7) = .00029138

z	w	c_1	c_2	d = .2	.4	.6	.8	1.0	1.5	2.0	3.0	z^{tc}
10100011001101	56	.0001928	−.0008376	.000311290	.000291314	.000238826	.000171542	.000107965	.000019053	.000001478	.000000001	0101100111010
10101001010101	56	.0001928	−.0008385	.000311269	.000291237	.000238970	.000171350	.000107771	.000018970	.000001466	.000000001	0100110110010
01010101001010	54	.0001884	−.0008579	.000310051	.000288468	.000234721	.000167073	.000104059	.000017799	.000001331	.000000001	0101011001001
10010110010011	55	.0001883	−.0008390	.000310427	.000288913	.000233390	.000170470	.000104470	.000019103	.000001510	.000000001	0101101100101
01010010010001	54	.0001849	−.0008620	.000309318	.000287122	.000233097	.000165550	.000102886	.000017506	.000001302	.000000001	0101100100101
11000010110011	54	.0001849	−.0008112	.000310386	.000291066	.000240292	.000174655	.000111781	.000022018	.000001793	.000000001	0101100100101x
10100100101001	56	.0001849	−.0008471	.000309623	.000288225	.000235058	.000167958	.000105161	.000018318	.000001378	.000000001	0011001111100
10110001000111	56	.0001849	−.0008496	.000309570	.000288028	.000234699	.000167504	.000104718	.000018145	.000001402	.000000001	0110100110110
10110010000111	55	.0001848	−.0008038	.000310541	.000291770	.000241772	.000176777	.000114114	.000022241	.000002020	.000000002	0101101011010
10111001100011	55	.0001848	−.0008426	.000309704	.000285598	.000235820	.000169009	.000106267	.000018829	.000001484	.000000001	0011110100110
10011001010011	55	.0001848	−.0008440	.000309674	.000288485	.000235611	.000168741	.000106001	.000018720	.000001468	.000000001	0011011010110
01101010100111	54	.0001788	−.0008598	.000308221	.000285490	.000231613	.000164631	.000102560	.000017683	.000001348	.000000001	0100111011001
10100110010101	55	.0001786	−.0008439	.000308529	.000286665	.000233765	.000167346	.000105194	.000018698	.000001486	.000000001	0101101101001
10010110011110	57	.0001783	−.0008437	.000308451	.000286466	.000233389	.000166798	.000104567	.000018343	.000001420	.000000001	1000101001110
10101010010101	55	.0001752	−.0008493	.000307767	.000285203	.000231921	.000165532	.000103726	.000018278	.000001438	.000000001•	0101011001110
01110010011110	54	.0001735	−.0008756	.000306914	.000282688	.000227733	.000160502	.000099989	.000016526	.000001207	.000000001	0110110011001
01011011000111	57	.0001731	−.0008433	.000307495	.000284987	.000231972	.000165847	.000104159	.000018497	.000001467	.000000001	0101011011110
10000111100011	53	.0001708	−.0007903	.000308208	.000288625	.000239405	.000175985	.000114719	.000022377	.000002287	.000000003	1000011110010
01011011100011	55	.0001708	−.0008520	.000306946	.000283913	.000230632	.000164694	.000103379	.000018434	.000001389	.000000001	0011110011010
10111001100011	53	.0001708	−.0008503	.000306939	.000283890	.000230630	.000164640	.000103326	.000018413	.000001357	.000000001	0011100100101
10101010101110	57	.0001704	−.0008561	.000306723	.000283141	.000229181	.000162646	.000101198	.000017376	.000001311	.000000001	1000101101110
01110010110101	54	.0001701	−.0008598	.000306599	.000282873	.000228879	.000162449	.000101166	.000017506•	.000001347	.000000001	0100110101001
01010101010101	54	.0001701	−.0008794	.000306189	.000281367	.000226154	.000159034	.000097869	.000016216	.000001181	.000000001	0011101011001
10101010101010	54	.0001699	−.0008600	.000306176	.000281319	.000226069	.000158929	.000097769	.000017253	.000001176	.000000001	0101010101010
11000011110011	58	.0001697	−.0007836	.000306502	.000282560	.000238333	.000161771	.000100511	.000017253	.000001313	.000000001	1100011110010
11000001110011	55	.0001673	−.0008420	.000308136	.000288741	.000239810	.000176594	.000115352	.000023632	.000002322	.000000003	1000011111100
01011001100011	55	.0001673	−.0008564	.000306452	.000283401	.000230483	.000164874	.000103760	.000018644	.000001513	.000000001	0011111100110
01011011100011	53	.0001673	−.0008489	.000306152	.000282514	.000228514	.000162411	.000101382	.000017730	.000001389	.000000002	0011110011010
01101011001001	55	.0001673	−.0008594	.000306090	.000282075	.000228083	.000161859	.000100835	.000017507	.000001350	.000000001	0011100111010
10110001111001	56	.0001666	−.0008541	.000306068	.000282268•	.000228599	.000162615	.000101633	.000017853	.000001408	.000000001	0011100111001
01110000101101	54	.0001622	−.0008703	.000304911	.000279705	.000224965	.000158677	.000098175	.000016692	.000001259	.000000001	0101011011001
11000110010010	57	.0001617	−.0008608	.000305009•	.000280213	.000225933	.000159883	.000099307	.000017078	.000001304	.000000001	1001001110001
10011001010101	55	.0001590	−.0008230	.000305315•	.000282439	.000230742	.000166544	.000106250	.000020213	.000001800	.000000002	0011110011010x
10010100100101	55	.0001590	−.0008821	.000304069	.000277817	.000222256	.000155705	.000095536	.000020202	.000001144	.000000001	0011110001110
10101010101010	56	.0001583	−.0008835	.000304039	.000277707	.000222052	.000155444	.000095278	.000015789	.000001129	.000000000	0101011001110
11000100111001	55	.0001583	−.0008439	.000304734•	.000280516	.000227294	.000162128	.000101818	.000015685	.000001485	.000000001	0101010101100
01111110000011	52	.0001566	−.0008211	.000304908•	.000281887	.000230331	.000166421	.000106384	.000018255	.000001852	.000000002	0011111000011x
01110100100011	53	.0001559	−.0008658	.000304193	.000279513	.000226130	.000161132	.000101171	.000020424	.000001505	.000000001	0111010101010
01100110010011	53	.0001559	−.0008232	.000303842	.000278232	.000223817	.000158231	.000157405?	.000017120	.000001350	.000000000	0111010001001
01011001010101	53	.0001503	−.0009028	.000302017	.000278090	.000223560	.000157906	.000090038	.000016994	.000001332	.000000001	0011101010101
11000101010101	56	.0001556	−.0008550	.000303998•	.000278809	.000224764	.000159260	.000099183	.000017262	.000001342	.000000001	0101011011100
11000101001001	55	.0001555	−.0008286	.000304548•	.000280951	.000228834	.000164627	.000104657	.000019702	.000001736	.000000001	0011101100110
10011010010101	55	.0001555	−.0008857	.000303347	.000276506	.000220691	.000154251	.000094426	.000015516	.000001118	.000000000	0100111001110
10010100101101	57	.0001555	−.0008861	.000303302	.000276339	.000220388	.000153868	.000094053	.000015372	.000001098	.000000000	0101011011010
10100010011110	54	.0001552	−.0008953	.000303960•	.000277741	.000224715	.000159275	.000099266	.000017360	.000001365	.000000001	1000101011001x
01101010010011	53	.0001532	−.0008897	.000303090	.000275726	.000219416	.000152772	.000093095	.000015088	.000001070	.000000000	0110100101001
01101010100101	54	.0001525	−.0008615	.000303103	.000276552	.000221412	.000155620	.000096071	.000016369	.000001256	.000000001	0110011011001
10010100011110	57	.0001520	−.0008615	.000302712	.000275364	.000219379	.000153116	.000093651	.000015412	.000001120	.000000000	0110100111010
10010101010101	54	.0001503	−.0006871	.000303195•	.000277266	.000222837	.000157405	.000097731	.000016899	.000001310	.000000001	1000100111010
10010101010101	55	.0001503	−.0009028	.000302017	.000273635	.000216713	.000150033	.000090805	.000014379	.000000985	.000000001	0101011001110

Table A

m = 7, n = 7   Rank Orders 701 thru 750 of 1780   C(7, 7) = 3432   1/C(7, 7) = .00029138

z	w	c_1	c_2	d = .2	.4	.6	.8	1.0	1.5	2.0	3.0	z^{tc}
0111001001011	53	.0001480	−.0008736	.000322100•	.000275260	.000220248	.000154877	.000095754	.000016440	.000001278	.000000001	0011011011000001
1101000010101	55	.0001476	−.0008216	.000303217•	.000279070	.000227224	.000163729	.000104448	.000019989	.000001808	.000000002	0011011110100
1001001100101	55	.0001476	−.0008965	.000301653	.000273319	.000216753	.000150463	.000091433	.000014716	.000001034	.000000000	0101000011010
1001000111010	55	.0001476	−.0008976	.000301630	.000273236	.000216605	.000150279	.000091258	.000014652	.000001026	.000000001	0101001110110
1100000111011	56	.0001469	−.0008573	.000302333•	.000276052	.000221810	.000156852	.000097627	.000017090	.000001354	.000000001	0101000111100
1010011000011	53	.0001445	−.0008758	.000301514	.000274040	.000218840	.000153610	.000094817	.000016226	.000001258	.000000001	0011111001100
1000111100001	54	.0001413	−.0008522	.000301415	.000274895•	.000221142	.000157005	.000098434	.000017860	.000001514	.000000000	0011111001010
1000111100011	54	.0001413	−.0008699	.000301037	.000273478	.000218505	.000153586	.000095002	.000016382	.000001282	.000000001	0011110001110
1001110001011	54	.0001413	−.0008745	.000300940	.000273118	.000217841	.000152736	.000094159	.000016032	.000001230	.000000001	0011101011000
1001010001011	55	.0001406	−.0008667	.000300962•	.000273431	.000218513	.000153614	.000095007	.000016336	.000001267	.000000001	0011101010100
1000110101101	55	.0001406	−.0009001	.000300276	.000270958	.000214107	.000148166	.000089808	.000014400	.000001013	.000000000	0110010100110
1101010001101	55	.0001406	−.0009040	.000300195	.000270664	.000213580	.000147512	.000089182	.000014167	.000000982	.000000000	0100110110100
1100100010101	55	.0001380	−.0008633	.000300538•	.000272892	.000218149	.000153641	.000095161	.000016516	.000001300	.000000001	0101011011010
1000101110101	55	.0001380	−.0008946	.000299895	.000270575	.000214020	.000148431	.000090274	.000014678	.000001055	.000000000	0100101110110
1010100010101	55	.0001380	−.0008997	.000299791	.000270199	.000213346	.000147594	.000089472	.000014377	.000001015	.000000000	0101011101010
0101100010011	53	.0001362	−.0009069	.000299324	.000269178	.000211948	.000146152	.000088281	.000014049	.000000983	.000000001	0101011101000101x
0101101010001	54	.0001336	−.0009179	.000298615	.000267620	.000209789	.000143884	.000086364	.000013479	.000000921	.000000001	0101101011010
1001100100011	54	.0001334	−.0008586	.000299810•	.000272015	.000217719	.000153811	.000095967	.000017210	.000001443	.000000001	0011101110010
1010110010011	54	.0001334	−.0008974	.000299001	.000269049	.000212335	.000147013	.000089321	.000014544	.000001055	.000000001	0011101011010
1010101010011	54	.0001334	−.0008972	.000298972	.000268943	.000212146	.000146776	.000089093	.000014457	.000001044	.000000001	0011101010100
1110010000111	55	.0001327	−.0007963	.000300942•	.000276344	.000225633	.000163840	.000105830	.000021315	.000002084	.000000002	0010111111000
1101000011011	55	.0001327	−.0008437	.000299970	.000272803	.000219197	.000155640	.000097681	.000017799	.000001511	.000000001	0011011110100
1000011110001	55	.0001327	−.0008711	.000299409	.000270778	.000215571	.000151114	.000093301	.000016076	.000001264	.000000000	0110001111010
1100010110011	55	.0001327	−.0008737	.000299347	.000270524	.000215063	.000150417	.000092566	.000015731	.000001208	.000000000	0011010011010
1100010011011	55	.0001327	−.0008749	.000299322	.000270431	.000214893	.000150200	.000092353	.000015644	.000001195	.000000001	0100001111100
1110000011101	55	.0001327	−.0008782	.000299262	.000270243	.000214609	.000149910	.000092133	.000015614	.000001198•	.000000001	0100011110001
1010100011001	55	.0001327	−.0009102	.000298599	.000267834	.000210284	.000144525	.000086959	.000013657	.000000937	.000000000	0110010110010
1010100011010	55	.0001327	−.0009103	.000298598	.000267829	.000210275	.000144513	.000086947	.000013652	.000000936	.000000000	0010101011010
1010010011010	54	.0001334	−.0009125	.000298553	.000267671	.000209999	.000144178	.000086633	.000013542	.000000923	.000000001	0101010010110
1001001011010	54	.0001327	−.0009131	.000298540	.000267623	.000209912	.000144070	.000086530	.000013503	.000000918	.000000001	0011010101010
1001101100011	54	.0001300	−.0009009	.000298280	.000267740•	.000210770	.000145554	.000088201	.000014262	.000001027	.000000000	0011100100110
0110110010011	53	.0001266	−.0009056	.000297553	.000266385	.000209119	.000143992	.000086990	.000013954	.000000996	.000000000	0101011100010
1000110100110	56	.0001261	−.0009039	.000297481	.000266245	.000208876	.000143635	.000086574	.000013718	.000000955	.000000000	1000110101110x
1010110001011	54	.0001238	−.0009007	.000297132	.000265888	.000208832	.000141230	.000087195•	.000014122	.000001023	.000000000	0100111010110
0110010000111	53	.0001187	−.0009295	.000295993	.000263674	.000206021	.000138239	.000084964	.000013499	.000000956	.000000000	0100101010010
0110101001011	53	.0001187	−.0009295	.000295596	.000262265	.000203554	.000138146	.000082169	.000012519	.000000835	.000000000	0101011010101
0101101001011	53	.0001187	−.0009301	.000295583	.000262220	.000203477	.000138146	.000082084	.000012491	.000000832	.000000000	0011010100010
0101110001011	52	.0001186	−.0008935	.000296311•	.000264875	.000208208	.000143976	.000087623	.000014528	.000001097	.000000001	0011011101010
1010101001011	54	.0001186	−.0009126	.000295619	.000263428	.000205606	.000140729	.000084493	.000013329	.000000934	.000000001	0010111010010
1001100000111	56	.0001182	−.0008959	.000296179•	.000264485	.000207504	.000143047	.000086660	.000014086	.000001027	.000000000	1000011101000
1000110010110	56	.0001182	−.0009133	.000295817	.000263167	.000205132	.000140086	.000083809	.000013004	.000000882	.000000000	1000110010100
1000100101110	56	.0001182	−.0009136	.000295811	.000263144	.000205089	.000140031	.000083756	.000012983	.000000880	.000000000	1000110101110
1100100001110	57	.0001175	−.0008563	.000296847•	.000267120	.000212306	.000149043	.000092425	.000016282	.000001336	.000000001	1000001111100
0110101000111	53	.0001152	−.0009324	.000294889	.000261010	.000202093	.000136915	.000081185	.000012294	.000000815	.000000000	0101001101001
0101100100111	55	.0001152	−.0009079	.000295172•	.000261987	.000203751	.000138859	.000082937	.000012850	.000000876	.000000001	0011011011010
1010010000111	54	.0001151	−.0008988	.000295541	.000263358	.000206208	.000141897	.000085632	.000013918	.000001015	.000000001	0011011110010
1010010110011	54	.0001151	−.0009144	.000295227	.000262099	.000204270	.000139553	.000083643	.000013148	.000000905	.000000001	0110100101010
1010101000011	54	.0001151	−.0009164	.000295185	.000262099	.000204009	.000139233	.000083342	.000013039	.000000905	.000000001	0011001101010
1001001000111	56	.0001103	−.0008923	.000294162	.000260122	.000201451	.000136620	.000081129	.000012325	.000000816	.000000000	1001001010110x

Table A

m = 7, n = 7   Rank Orders 751 thru 800 of 1780 $C(7, 7) = 3432$ $1/C(7, 7) = .00029138$

z	w	c_1	c_2	$d = .2$	.4	.6	.8	1.0	1.5	2.0	3.0	z^{tc}
1011000110011	54	.0001072	−.0009054	.000293934	.000260479	.000202812	.000138770	.000083459	.000013338	.000000958	.000000000	0011001110010
1100101000011	54	.0001041	−.0008730	.000294025	.000261931	.000206073	.000143245	.000088019	.000015248	.000001240	.000000000	0011011101100
1001110010101	54	.0001041	−.0009301	.000292865	.000257778	.000198717	.000134179	.000079370	.000011985	.000000795	.000000000	0011011001100
1001011001001	54	.0001041	−.0009322	.000292821	.000257622	.000198441	.000133842	.000079053	.000011871	.000000781	.000000000	0110011010010
0110010011001	53	.0001038	−.0009400	.000292607	.000257015	.000197504	.000132813	.000078176	.000011628	.000000758	.000000001	0110110010110
1100100011101	55	.0001034	−.0008926	.000293486	.000260222	.000203124	.000139605	.000084497	.000013831	.000001029	.000000000	0110011011100
1100010010101	55	.0001034	−.0009070	.000293187	.000259134	.000201167	.000137159	.000082133	.000012918	.000000904	.000000001	0110010111100
1100100010011	55	.0001034	−.0009082	.000293162	.000259045	.000201008	.000136962	.000081945	.000012848	.000000894	.000000001	0101011011100
1010010100101	56	.0001033	−.0009084	.000293143	.000259046	.000201082	.000137123	.000082158	.000012977	.000000918	.000000001	1000110111010
1011100001011	52	.0001010	−.0008798	.000293296	.000260429	.000204026	.000141060	.000086093	.000014554	.000001138	.000000001	0010111100001
0110101100011	52	.0001010	−.0009144	.000292611	.000258039	.000199901	.000136104	.000081482	.000012918	.000000928	.000000000	0011110010001
0101100100011	52	.0001010	−.0009149	.000292599	.000258000	.000199833	.000136022	.000081407	.000012891	.000000925	.000000000	0011110110001
1010010101110	56	.0001006	−.0009196	.000292414	.000257393	.000198704	.000134523	.000079864	.000012208	.000000821	.000000000	1001000111010
0101101101001	53	.0001003	−.0009425	.000291907	.000255784	.000196083	.000131538	.000077237	.000011418	.000000739	.000000000	0110101101001
0111011001001	54	.0000989	−.0009453	.000291582	.000255103	.000195134	.000130506	.000076319	.000011101	.000000699	.000000000	0101011010110
1100010110011	52	.0000955	−.0009093	.000291733	.000256782	.000198764	.000135353	.000081123	.000012944	.000000941	.000000000	0111011001110
1100010100101	54	.0000955	−.0009141	.000291570	.000256219	.000197697	.000133947	.000079679	.000012307	.000000844	.000000000	0111011010001
1010100001011	54	.0000954	−.0008711	.000292429	.000259134	.000201394	.000141000	.000086497	.000014973	.000001222	.000000001	0101110111100
1001011010101	54	.0000954	−.0009472	.000291415	.000253914	.000193785	.000129318	.000075461	.000010919	.000000685	.000000000	0101100010110
1001100011011	54	.0000954	−.0009475	.000290878	.000253861	.000193696	.000129212	.000075365	.000010888	.000000681	.000000000	0101010100110
0111010100011	52	.0000931	−.0009178	.000291067	.000255384	.000196889	.000133433	.000079528	.000012477	.000000888	.000000000	0011110110001
0111000110011	53	.0000924	−.0009287	.000290710	.000254349	.000195186	.000131407	.000077633	.000011782	.000000796	.000000000	0111000111000
1010000110110	56	.0000919	−.0009199	.000290786	.000254751	.000195925	.000132287	.000078421	.000012019	.000000820	.000000000	1001000111010
1100011001001	54	.0000892	−.0009058	.000290565	.000254924	.000196780	.000133657	.000079893	.000012633	.000000903	.000000000	0101011101110
1001011000101	54	.0000892	−.0009392	.000289902	.000252614	.000192801	.000128902	.000075503	.000011128	.000000721	.000000000	0110011001110
1000101100011	54	.0000892	−.0009431	.000289824	.000252338	.000192323	.000128327	.000074971	.000010945	.000000699	.000000000	0110011101010
1001010100101	54	.0000895	−.0009095	.000289839	.000253595	.000195177	.000132149	.000078724	.000012331	.000000872	.000000000	0101101010010
1000110101001	54	.0000858	−.0009415	.000289205	.000251392	.000191394	.000127639	.000074573	.000010919	.000000702	.000000000	0101001101110
1011100110001	54	.0000858	−.0009462	.000289112	.000251065	.000190830	.000126964	.000073950	.000010707	.000000677	.000000001	0101011010010
0111100100101	52	.0000814	−.0009475	.000288271	.000249682	.000189365	.000125787	.000073199	.000010618	.000000678	.000000000	0101011000101
1011010000011	53	.0000812	−.0008889	.000289420	.000253769	.000196511	.000134461	.000081343	.000013568	.000001065	.000000001	0011110100010
1001011100011	53	.0000812	−.0009265	.000288662	.000251075	.000191767	.000128646	.000075819	.000011496	.000000796	.000000000	0111000101110
1011100010001	53	.0000812	−.0009285	.000288621	.000250934	.000191523	.000128353	.000075547	.000011401	.000000770	.000000000	0011011011110
1100011010001	53	.0000805	−.0009195	.000288661	.000251288	.000192229	.000129229	.000076356	.000011663	.000000799	.000000000	0111010101100
1010101010001	54	.0000805	−.0009554	.000287950	.000248825	.000188014	.000123886	.000071182	.000010038	.000000621	.000000000	0110010101010
1001101010001	54	.0000786	−.0009580	.000287901	.000248653	.000187722	.000123886	.000071472	.000010038	.000000610	.000000000	0101010101010
1010010100011	54	.0000786	−.0009371	.000287957	.000249551	.000189690	.000126504	.000074039	.000010985	.000000727	.000000000	1010010011010
1100000111011	54	.0000779	−.0008504	.000289531	.000255263	.000199688	.000138624	.000085415	.000015138	.000001284	.000000001	0011011111000
1010000101011	55	.0000779	−.0008984	.000288581	.000251932	.000193877	.000131526	.000078661	.000012553	.000000915	.000000001	0011011111100
1010110001011	53	.0000779	−.0009107	.000288339	.000251092	.000192430	.000129800	.000077075	.000012006	.000000848	.000000000	0011011111100
1011010000111	54	.0000779	−.0009146	.000288262	.000250822	.000191962	.000129236	.000076547	.000011817	.000000824	.000000000	0011001110010
1001011110001	54	.0000779	−.0009272	.000288012	.000249952	.000190463	.000127438	.000074881	.000011236	.000000751	.000000000	0111000101110
0110110001101	54	.0000779	−.0009336	.000287886	.000249514	.000189701	.000126542	.000074056	.000010954	.000000717	.000000000	0011010111110
1001101010101	54	.0000779	−.0009485	.000287593	.000248503	.000187994	.000124522	.000072225	.000010361	.000000650	.000000000	0110011001110
1010000110101	54	.0000779	−.0009487	.000287561	.000248489	.000187969	.000124489	.000072197	.000010351	.000000649	.000000000	0101010101010
1010010101101	54	.0000779	−.0009501	.000287561	.000248394	.000187808	.000124303	.000072026	.000010296	.000000643	.000000000	0111001100110
1010100100101	54	.0000772	−.0009071	.000288278	.000248255	.000187571	.000124022	.000071770	.000010212	.000000633	.000000000	0101001101010
1000111100001	55	.0000747	−.0009118	.000288278	.000251118	.000192609	.000130092	.000077386	.000012132	.000000864	.000000000	0011010111100
1011100000110	55	.0000747	−.0009372	.000287215	.000248262	.000188164	.000125047	.000072865	.000010628	.000000683	.000000000	1000111000110
1010101100011	53	.0000716	−.0009272	.000286840	.000248068	.000188507	.000125895	.000073913	.000011130	.000000754	.000000000	0010111100010

197

m = 7, n = 7 Rank Orders 801 thru 850 of 1780 C(7, 7) = 3432 1/C(7, 7) = .00029138

d = .2

z	w	c_1	c_2	.2	.4	.6	.8	1.0	1.5	2.0	3.0	z^{tc}
01011110000101	51	.0000672	−.0009107	.000286342	.000247849	.000189001•	.000127012	.000075246	.000011742	.000000842	.000000000	0011111000101
01101110001001	52	.0000665	−.0009351	.000285732	.000245975	.000185928	.000123429	.000071983	.000010628	.000000703	.000000000	0100011110101
01101110000101	52	.0000665	−.0009541	.000285359	.000244692	.000183749	.000120865	.000069655	.000009866	.000000615	.000000000	0101011100101
01011011000101	52	.0000660	−.0009550	.000285341	.000244632	.000183649	.000120749	.000069552	.000009834	.000000612	.000000000	1000011011100
11000100011110	56	.0000661	−.0009020	.000286295•	.000247991	.000189353	.000127438	.000075598	.000011800	.000000839	.000000000•	1000110101010
01011010001110	55	.0000638	−.0009528	.000285278	.000244526	.000183472	.000120493	.000069257	.000009679	.000000588	.000000000	0101011101010
01101010010101	55	.0000639	−.0009639	.000284672	.000243238	.000181810	.000118911	.000068071	.000009442	.000000574	.000000001	0011101010101
10110010000101	53	.0000637	−.0009296	.000285315•	.000245477	.000185605	.000123353	.000072077	.000010730	.000000719	.000000000	1001110101010
01101010010011	53	.0000637	−.0009452	.000285011	.000244438	.000183852	.000121302	.000070225	.000010130	.000000651	.000000000	0011101110101
10101010010011	53	.0000637	−.0009473	.000284971	.000244299	.000183615	.000121021	.000069968	.000010044	.000000641	.000000000	0011011110101
10011000101110	55	.0000633	−.0009461	.000284913	.000244172	.000183385	.000120693	.000069603	.000009861	.000000612	.000000000	0001111001100
11100000011011	55	.0000630	−.0008652	.000286432	.000249536	.000192643	.000131786	.000077891	.000013483	.000001073	.000000001	0101111000110
01100101001001	54	.0000630	−.0009566	.000284651	.000243414	.000182211	.000119408	.000068511	.000009558	.000000583	.000000000	0100100111010
01011001010101	54	.0000630	−.0009569	.000284644	.000243390	.000182169	.000119357	.000068464	.000009541	.000000581	.000000000	0110100111010
10000001101110	56	.0000626	−.0009052	.000285579•	.000244692	.000187794	.000125970	.000074455	.000011496	.000000806	.000000000	1000100111100
10101000101101	53	.0000602	−.0009475	.000284315	.000243219	.000182447	.000120039	.000069291	.000009917	.000000631	.000000000	0011110101010
01110100100101	52	.0000586	−.0009560	.000283847	.000242152	.000180943	.000118451	.000067948	.000009519	.000000588	.000000000	0101011011010
10010000101101	52	.0000581	−.0009588	.000283685	.000241708	.000180200	.000117535	.000067064	.000009181	.000000544	.000000000	1000011101010
01100011010101	52	.0000551	−.0009523	.000283184	.000241047	.000179733	.000117419	.000067225•	.000009375	.000000577	.000000000	1001001110101
10100010010011	53	.0000550	−.0009415	.000283451	.000241993	.000181340	.000119297	.000068913	.000009913	.000000637	.000000000	0011011110101
10011100000111	53	.0000519	−.0008928	.000283836•	.000244365	.000186004	.000125230	.000074612	.000012023	.000000916	.000000001	0001111001100
10011110000101	53	.0000519	−.0009489	.000282732	.000240540	.000179438	.000117380	.000067341	.000009463	.000000588	.000000000	0101111000110
10011000100101	53	.0000519	−.0009423	.000282683	.000240372	.000179153	.000117044	.000067034	.000009362	.000000576	.000000000	0101011011010
11000010100101	54	.0000512	−.0009687	.000282721	.000240710	.000179811	.000117841	.000067752	.000009577	.000000598	.000000000	0110100111100
01101001001001	52	.0000489	−.0009687	.000281795	.000238375	.000176416	.000114256	.000068776	.000010154	.000000521	.000000000	0011110101010
01110010000111	51	.0000488	−.0009338	.000282457	.000240684	.000180347	.000118884	.000068976	.000009705	.000000682	.000000000	1000011111010
10010100010101	54	.0000486	−.0009368	.000282331	.000240263	.000179562	.000117848	.000067922	.000009847	.000000616	.000000000	0011100111010
10100001010101	55	.0000485	−.0009358	.000282334•	.000240339	.000179753	.000118133	.000068231	.000009181	.000000638	.000000000	1000100111100
10100010100011	52	.0000485	−.0009523	.000282006	.000239198	.000177789	.000115791	.000066075	.000009115	.000000550	.000000000	0011011110101
10100010010011	53	.0000485	−.0009415	.000281995	.000239157	.000177719	.000115706	.000065997	.000009088	.000000547	.000000000	0101011110110
01010010101001	53	.0000455	−.0009697	.000281125	.000237247	.000175170	.000113185	.000064019	.000008622	.000000509	.000000001	0101001011001x
10100111000011	53	.0000441	−.0008950	.000282312•	.000241760	.000183053	.000122602	.000072670	.000011563	.000000870	.000000000	0001110110110
10011011000101	53	.0000441	−.0009711	.000280827	.000236668	.000174408	.000112393	.000063342	.000008406	.000000484	.000000000	0101011011110
10011010001001	53	.0000441	−.0009719	.000280813	.000236619	.000174326	.000112300	.000063260	.000008382	.000000481	.000000000	0110100111010
10100010011001	54	.0000434	−.0009133	.000281812•	.000240242	.000180533	.000119601	.000069864	.000010521	.000000726	.000000000	0100100111100
11000010001101	54	.0000434	−.0009455	.000281183	.000238076	.000176841	.000115229	.000065862	.000009169	.000000563	.000000000	0011010111100
10010001010101	54	.0000434	−.0009464	.000281165	.000238014	.000176737	.000115107	.000065750	.000009132	.000000559	.000000000	0101011011010
01011000100011	51	.0000409	−.0009180	.000281273•	.000239218	.000179283	.000118433	.000066639	.000000321	.000000710	.000000000	1000101111000
11001100000011	51	.0000409	−.0009336	.000280981	.000239254	.000177706	.000116639	.000067400	.000009832	.000000655	.000000000	0011110110101
01110011000011	51	.0000409	−.0009345	.000280964	.000238200	.000177617	.000116537	.000067310	.000009804	.000000652	.000000000	0101011011010
10011001100101	53	.0000406	−.0009723	.000280154	.000235529	.000173145	.000111305	.000062572	.000008252	.000000472	.000000000	0101100110010
10100101011010	55	.0000406	−.0009574	.000280431	.000236439	.000174613	.000112942	.000063976	.000008644	.000000509	.000000000	0110101111001
01011001010110	55	.0000402	−.0009605	.000280322	.000236226	.000174396	.000112817	.000063963	.000008709•	.000000523	.000000000	0011011100001
11001011011001	53	.0000344	−.0009300	.000279811	.000236379	.000175680	.000114884	.000066113	.000009508	.000000619	.000000000	0101101001001
10011110110001	53	.0000344	−.0009620	.000279199	.000234323	.000171455	.000110951	.000062614	.000008414	.000000497	.000000000	0101011001110
01011011011010	51	.0000344	−.0009666	.000279108	.000234017	.000171755	.000110358	.000062084	.000008247	.000000479	.000000000	0101111010100
01011100100101	51	.0000292	−.0009595	.000278675	.000232903	.000170862	.000109893	.000061979	.000008348•	.000000498	.000000000	0101110101101x
11001100001001	51	.0000292	−.0009380	.000278988	.000234233	.000173032	.000112349	.000064127	.000008993	.000000566	.000000000	0011011001110
10011001100101	53	.0000292	−.0009740	.000277988	.000231932	.000169225	.000107984	.000060266	.000007808	.000000439	.000000000	0101011011010
10010110010101	53	.0000292	−.0009765	.000277940	.000231771	.000168961	.000107683	.000060002	.000007730	.000000431	.000000000	0101100110110

Table A

m = 7, n = 7 Rank Orders 851 thru 900 of 1780 C(7, 7) = 3432 1/C(7, 7) = .00029138

z	w	c_1	c_2	d = .2	.4	.6	.8	1.0	1.5	2.0	3.0	z^{tc}
1011100000111	52	.0000264	−.0008925	.000279037•	.000232396	.000177228	.000117629	.000069150	.000010815	.000000803	.000000001	000111111100010
1001101100011	52	.0000264	−.0009472	.000277987	.000232840	.000177252	.000110629	.000062787	.000008666	.000000537	.000000000	001111011001100
1100110010011	52	.0000264	−.0009478	.000277975	.000232800	.000177188	.000110555	.000062722	.000008646	.000000535	.000000000	001110011000110
1110001001111	52	.0000258	−.0009400	.000278002•	.000233031	.000171600	.000111002	.000070680	.000008707	.000000535	.000000000	011000010111100
1110000100111	53	.0000257	−.0008754	.000279220•	.000237203•	.000178655	.000119317	.000065009	.000011321	.000000611	.000000001	001101011101000
1101000100101	53	.0000257	−.0009237	.000278033	.000234079	.000173394	.000113124	.000063501	.000009352	.000000558	.000000000	001101101110100
1100010100011	53	.0000257	−.0009375	.000277982	.000233195	.000171643	.000111432	.000063210	.000008876	.000000548	.000000000	001101101101100
1100101001001	53	.0000257	−.0009401	.000277735	.000233024	.000170267	.000111105	.000061798	.000008784	.000000498	.000000000	001101101101100
1100110110001	53	.0000257	−.0009531	.000277619	.000232194	.000169520	.000109520	.000061798	.000008341	.000000498	.000000000	001101110111100
1100100100111	53	.0000257	−.0009592	.000277619	.000231808	.000169628	.000108786	.000061148	.000008140	.000000476	.000000000	001101110111100
0111010101001	53	.0000257	−.0009747	.000277324	.000230823	.000168012	.000106952	.000059546	.000007669	.000000429	.000000000	011010010110110
1001010101001	53	.0000257	−.0009748	.000277321	.000230814	.000167997	.000106935	.000059531	.000007664	.000000428	.000000000	011001101100110
1010101100011	53	.0000257	−.0009766	.000277288	.000230702	.000167813	.000106726	.000059348	.000007610	.000000422	.000000000	011011011001010
1010101001011	55	.0000222	−.0009780	.000277261	.000230615	.000167670	.000106563	.000059206	.000007568	.000000418	.000000000	010101010101010
1010000111010	53	.0000222	−.0009459	.000277218	.000231578•	.000169863	.000109410	.000061890	.000008448	.000000514	.000000000	101000011101010x
1010000110011	53	.0000178	−.0009253	.000276785	.000231529•	.000170553•	.000110648	.000063228	.000008968	.000000578	.000000000	001100011011010
1001001110001	53	.0000178	−.0009567	.000276189	.000229535	.000167255	.000106860	.000059865	.000007918	.000000462	.000000000	001101011011100
1001100110001	53	.0000178	−.0009617	.000276094	.000229222	.000166745	.000106281	.000059360	.000007768	.000000446	.000000000	010100111100010
1010011000110	54	.0000146	−.0009656	.000275420	.000227964	.000165203	.000104813	.000058213	.000007475	.000000418	.000000000	010101011101010
1100001000110	55	.0000139	−.0009232	.000276091	.000230417	.000169355	.000109620	.000062486	.000008796	.000000561	.000000000	100001111011100
0111100001101	51	.0000116	−.0009328	.000275487	.000229152	.000167707	.000108002	.000061210	.000008475	.000000531	.000000000	010101111100001
0110101100101	51	.0000116	−.0009695	.000274801	.000226901	.000164059	.000103898	.000057647	.000007425	.000000421	.000000000	011001110100110
0110100101011	51	.0000116	−.0009698	.000274796	.000226885	.000164032	.000103869	.000057622	.000007418	.000000421	.000000000	011001011001001
1010101010001	52	.0000115	−.0009355	.000275417•	.000228975	.000167462	.000107761	.000060090	.000008437	.000000528	.000000000	010101110010010
0110110001001	52	.0000115	−.0009499	.000275145	.000228075	.000165990	.000106090	.000059561	.000007994	.000000481	.000000000	001111001100101
1010110010001	55	.0000115	−.0009529	.000275089	.000227889	.000165685	.000105740	.000059252	.000007899	.000000470	.000000000	101010101010010
1100010101001	55	.0000113	−.0009323	.000275411•	.000228967	.000167385	.000107577	.000060771	.000008282	.000000503	.000000000	101010101111100
1001101010111	53	.0000111	−.0009663	.000274756	.000226855	.000163987	.000103774	.000057485	.000007332	.000000407	.000000000	100110110100110
1010101011010	53	.0000108	−.0009776	.000274484	.000226083	.000162823	.000102535	.000056462	.000007067	.000000383	.000000000	010100111100110
1010101100101	52	.0000088	−.0009598	.000274463•	.000226654•	.000164131	.000104250	.000058097	.000007617	.000000445	.000000000	001110101010010
1110000010101	53	.0000081	−.0008941	.000275549•	.000230416	.000170336	.000111306	.000064285	.000009493	.000000648	.000000000	010101011111000
1010011101001	53	.0000081	−.0009690	.000274150	.000225817	.000162849	.000102829	.000056858	.000007239	.000000403	.000000000	011000111011010
0110100011001	53	.0000091	−.0009702	.000274129	.000225747	.000162734	.000102697	.000056742	.000007204	.000000400	.000000000	010100111010100
1001010101010	54	.0000059	−.0009602	.000273996	.000225861•	.000163105	.000101018	.000057508	.000007477	.000000431	.000000000	100101010111010x
0111010100111	51	.0000037	−.0009774	.000273566	.000224548•	.000161630	.000101907	.000056306	.000007184	.000000404	.000000000	010101010101010
0110010010011	52	.0000037	−.0009677	.000273359	.000225519	.000163146	.000103619	.000057793	.000007619	.000000449	.000000000	010101011101100
1001010100101	54	.0000036	−.0009519	.000273630•	.000225518	.000163146	.000101243	.000055679	.000006958	.000000377	.000000000	100110011100110
1001100110110	54	.0000026	−.0009692	.000273882•	.000224250	.000161077	.000105824	.000055706	.000008166	.000000503	.000000000	100110011100110
1001100010011	53	.0000001	−.0009270	.000273004	.000226635	.000165077	.000102128	.000057253	.000007528	.000000443	.000000000	001110011011100
1011001000101	53	−.0000001	−.0009505	.000272942	.000224374	.000161856	.000102451	.000056902	.000007384	.000000424	.000000000	010011001110000
0110111001001	50	−.0000026	−.0009270	.000272943•	.000225175	.000163652	.000104813	.000059179	.000008191	.000000521	.000000000	001111001001001
0111010011001	51	−.0000033	−.0009692	.000272201	.000225175	.000159176	.000099814	.000054840	.000006875	.000000382	.000000000	011001110011001
1010010100101	53	−.0000036	−.0009519	.000272263•	.000223216	.000160557	.000101317	.000056088	.000007213	.000000410	.000000000	010101101011000
1010010101110	54	−.0000037	−.0009677	.000271943	.000222201	.000158930	.000099497	.000054516	.000006759	.000000364	.000000000	100110101011010
0110101010011	51	−.0000064	−.0009774	.000271372	.000221052	.000158604	.000099608	.000053616	.000006604	.000000357	.000000000	010101010101001x
1010101000110	54	−.0000064	−.0008941	.000271586•	.000221849	.000158847	.000099608	.000054743	.000006875	.000000378	.000000000	100101010101001
1001010000111	52	−.0000081	−.0009270	.000272513•	.000225498	.000165147	.000107097	.000061528	.000009096	.000000640	.000000000	001111101110110
1100110000111	52	−.0000081	−.0008941	.000271100	.000220805	.000157430	.000098258	.000056528	.000006622	.000000356	.000000000	010101110010110
1001110000101	52	−.0000081	−.0009702	.000271080	.000220739	.000157325	.000098143	.000053589	.000006595	.000000354	.000000000	010101110011100

Table A

m = 7, n = 7   Rank Orders 901 thru 950 of 1780   C(7, 7) = 3432   1/C(7, 7) = .00029138

z	w	c_1	c_2	d = .2	.4	.6	.8	1.0	1.5	2.0	3.0	z^{tc}
1100010101010101	53	−.0000088	−.0009598	.000271135•	.000221110	.000158002	.000098923	.000054260	.000006775	.000000370	.000000000	0101010101111110
1011010100101	52	−.0000108	−.0009776	.000270445•	.000219485	.000155756	.000096655	.000052453	.000006334	.000000332	.000000000	0101101011001100
0111010100001	51	−.0000111	−.0009663	.000270599•	.000220108	.000156842	.000097926	.000056677	.000006677	.000000367	.000000000	0111011101110100
1101010100001	50	−.0000113	−.0009323	.000271214•	.000222180	.000160247	.000101800	.000056987	.000007710	.000000479	.000000000	0011101010010001
1000011100101	53	−.0000115	−.0009355	.000271091	.000221784	.000159537	.000100900	.000056105	.000006962	.000000433	.000000000	0101011101111100
1001001001101	53	−.0000115	−.0009499	.000270822	.000220903	.000158114	.000099302	.000054722	.000006883	.000000391	.000000000	0011100111101100
1011000011110	53	−.0000116	−.0009529	.000270767	.000220726	.000157828	.000098983	.000054447	.000007365	.000000384	.000000000	0101010110110100
1010001101010	54	−.0000116	−.0009328	.000271123•	.000221945	.000159849	.000101298	.000054491	.000006883	.000000451	.000000000	1000011101111100
1010001101010	54	−.0000116	−.0009695	.000269667	.000219667	.000156137	.000097102	.000052830	.000006491	.000000338	.000000000	1001010011011010
1010010110110	54	−.0000116	−.0009698	.000270427	.000219649	.000156108	.000097068	.000052801	.000006414	.000000337	.000000000	1001010101011010
0111100010000011	50	−.0000139	−.0009232	.000270880•	.000221875	.000160161	.000101917	.000057192	.000007805	.000000491	.000000000	0111101110000001
0111000110100011	51	−.0000146	−.0009656	.000269959•	.000219077	.000155752	.000097028	.000052977	.000006566	.000000359	.000000000	0110100010100001
1001111001001	52	−.0000178	−.0009253	.000270104•	.000220491	.000158512	.000100367	.000055981	.000007478	.000000454	.000000000	0010111100111100
1001110001101	52	−.0000178	−.0009567	.000269524	.000218606	.000155481	.000096982	.000053059	.000006626	.000000366	.000000000	0110110101001010
0101101100111	50	−.0000222	−.0009617	.000269430	.000218299	.000154986	.000096426	.000052578	.000006486	.000000352	.000000000	0100111100101010x
1001101000011	52	−.0000257	−.0009459	.000269548•	.000217893	.000155038•	.000103688	.000053166	.000006731	.000000380	.000000000	0001111011011100
1001101000011	52	−.0000257	−.0008754	.000268338•	.000221068	.000160764	.000103688	.000059330	.000008728	.000000614	.000000000	0001111011011100
1001010000111	52	−.0000257	−.0009237	.000268655	.000218148	.000156010	.000096275	.000054537	.000007195	.000000432	.000000000	0011101110011010
1100011010011	53	−.0000264	−.0009375	.000268401	.000217322	.000154684	.000096797	.000053263	.000006825	.000000394	.000000000	0011011001111010
1010010010011	52	−.0000264	−.0009401	.000268351	.000217161	.000154426	.000096509	.000053014	.000006752	.000000386	.000000000	0011011010011100
1000111011000	52	−.0000257	−.0009531	.000268113	.000216389	.000153191	.000095136	.000051835	.000006414	.000000352	.000000000	0111001000111110
1011001001001	52	−.0000257	−.0009592	.000268001	.000216027	.000152611	.000094492	.000051283	.000006256	.000000336	.000000000	0110101001011010
1001011010101	52	−.0000257	−.0009747	.000267715	.000215106	.000151150	.000092888	.000049927	.000005889	.000000302	.000000000	0110101011001010
1001101001001	52	−.0000257	−.0009748	.000267713	.000215098	.000151137	.000092873	.000049914	.000005885	.000000301	.000000000	0010110110101010
1010001010101	52	−.0000257	−.0009780	.000267655	.000214911	.000150840	.000092690	.000049759	.000005843	.000000297	.000000000	0101010101010010
1011010001011	51	−.0000258	−.0009400	.000268338•	.000217171	.000154911	.000096617	.000049639	.000005811	.000000294	.000000000	0011101010101100
1100010000111	53	−.0000264	−.0009237	.000269087	.000219730	.000158628	.000096617	.000053137	.000006807	.000000394	.000000000	0110011110111000
0110101100101	53	−.0000264	−.0009472	.000268083	.000216479	.000153409	.000101233	.000052057	.000007108	.000000515	.000000000	0100110111000100
1000100011001	52	−.0000264	−.0009478	.000268071	.000216440	.000153346	.000095398	.000052057	.000007961	.000000355	.000000000	0101100011011100
1100100110001	52	−.0000292	−.0009380	.000267737	.000216212	.000153462•	.000095746	.000052518	.000006671	.000000381	.000000000	0011100101101100
1011001010001	52	−.0000292	−.0009740	.000267077	.000214083	.000150073	.000092007	.000049336	.000005785	.000000294	.000000000	0101010100110100
1011001101001	52	−.0000292	−.0009765	.000267030	.000213933	.000149836	.000091747	.000049116	.000005725	.000000289	.000000000	1010000101101010x
1010010011010	54	−.0000292	−.0009595	.000267334•	.000214888	.000151315	.000095063	.000050418	.000006053	.000000317	.000000000	1010101011011100
1101000010101	52	−.0000344	−.0009300	.000266904	.000213170	.000152444	.000091724	.000052158	.000006651	.000000383	.000000000	0111010001101010
1001010110001	52	−.0000344	−.0009620	.000266317	.000212899	.000149424	.000091029	.000048919	.000005848	.000000303	.000000000	0101010101101100
1011001011100	53	−.0000402	−.0009666	.000266232	.000211445	.000148999	.000091029	.000048256	.000005646	.000000293	.000000000	1000111010010110
0110101010101	50	−.0000406	−.0009605	.000265247	.000211607	.000147548	.000090194	.000048256	.000005646	.000000288	.000000000	1001101101001001
1010010011001	52	−.0000406	−.0009574	.000265255•	.000211607	.000147957	.000090693	.000048748	.000005822	.000000308	.000000000	0111010001001010
1010010011001	52	−.0000406	−.0009723	.000264975	.000210696	.000146492	.000089062	.000047350	.000005429	.000000270	.000000000	0111001100110010
1101000011110	54	−.0000409	−.0009180	.000265891•	.000213716	.000151298	.000094339	.000051812	.000006653	.000000388	.000000000	1001011101011100
1100011001110	54	−.0000409	−.0009336	.000265599	.000212750	.000149719	.000092545	.000050236	.000006173	.000000336	.000000000	1000111001011100
1011000101110	54	−.0000409	−.0009345	.000265583	.000212697	.000149633	.000092447	.000050151	.000006148	.000000333	.000000000	1000101011001100
1011100000101	51	−.0000434	−.0009133	.000265527	.000213270•	.000150984	.000094220	.000051835	.000006712	.000000397	.000000000	0011011001100010
1010101000011	51	−.0000434	−.0009455	.000264950	.000211445	.000148129	.000091117	.000049225	.000005998	.000000327	.000000000	0011100110010010
1100011010011	51	−.0000434	−.0009464	.000264934	.000211393	.000148048	.000091029	.000049152	.000005998	.000000325	.000000000	0011011101011100
1100010001011	52	−.0000441	−.0008950	.000265719•	.000214057	.000152297	.000096687	.000053085	.000007061	.000000432	.000000000	1001011011111000
0101100100101	53	−.0000441	−.0009711	.000264345	.000209700	.000145458	.000088229	.000046799	.000005336	.000000264	.000000000	1001011011001010
1010100101101	52	−.0000441	−.0009719	.000264332	.000209655	.000145389	.000088153	.000046736	.000005319	.000000262	.000000000	0110010110101010
1001101011011	53	−.0000455	−.0009697	.000264100	.000209296	.000144995	.000087798	.000046459	.000005243	.000000254	.000000000	0110101001001010

z	w	c_1	c_2	d = .2	.4	.6	.8	1.0	1.5	2.0	3.0	z^{tc}
011110000101	50	-.0000485	-.0009358	.000264163	.000210385	.000147195	.000090476	.000048863	.000005952	.000000324	.000000000	0101011110001
011110001001	50	-.0000485	-.0009523	.000263871	.000209478	.000145803	.000088994	.000047646	.000005641	.000000296	.000000000	0101101110001
101101001001	50	-.0000485	-.0009529	.000263861	.000209447	.000145756	.000088946	.000047606	.000005631	.000000295	.000000000	0110101010001
110100101101	51	-.0000486	-.0009368	.000264126	.000210323	.000147145	.000090463	.000048946	.000005977	.000000329	.000000000	0011011011100
101101010011	54	-.0000488	-.0009338	.000264117	.000210304	.000147055	.000090279	.000048883	.000005856	.000000311	.000000000	1001001011100
101100101101	53	-.0000489	-.0009687	.000263467	.000208290	.000143947	.000086987	.000045895	.000005147	.000000248	.000000000	1001010110010
101101010101	51	-.0000512	-.0009423	.000263531	.000209201	.000145792	.000089216	.000047951	.000005769	.000000311	.000000000	0011101101010
111010001001	52	-.0000519	-.0008928	.000264278	.000211732	.000149813	.000093605	.000051641	.000006677	.000000409	.000000000	0110110111000
101100110011	52	-.0000519	-.0009489	.000263272	.000208557	.000144851	.000088208	.000047100	.000005525	.000000286	.000000000	0110110111010
101101001101	52	-.0000519	-.0009514	.000263272	.000208416	.000144632	.000087972	.000046902	.000005473	.000000281	.000000000	0110001110010
110100100101	52	-.0000550	-.0009415	.000262831	.000208023	.000144454	.000088008	.000047042	.000005545	.000000289	.000000000	0100111010101
101011001001	53	-.0000551	-.0009566	.000262534	.000207114	.000143046	.000086486	.000045757	.000005207	.000000258	.000000000	1000101101010
010101100101	50	-.0000581	-.0009588	.000261948	.000206161	.000142076	.000085756	.000045164	.000005179	.000000261	.000000000	0110101110001
101010010011	53	-.0000586	-.0009560	.000261893	.000206081	.000141954	.000085587	.000045098	.000005207	.000000250	.000000000	0101011110101
011101100011	52	-.0000602	-.0009475	.000261743	.000206062	.000142158	.000085930	.000045509	.000006210	.000000261	.000000000	1000101101010
011111000011	49	-.0000626	-.0009052	.000262067	.000207828	.000145361	.000089698	.000048851	.000007232	.000000366	.000000000	0011111001100
100110100011	51	-.0000630	-.0008652	.000262718	.000210002	.000143604	.000093604	.000052235	.000007232	.000000481	.000000000	0011110100011
100011001011	51	-.0000630	-.0009569	.000261066	.000204739	.000140541	.000084444	.000044414	.000004979	.000000243	.000000000	1000101110100
011100110011	51	-.0000630	-.0009569	.000261060	.000204721	.000140513	.000084415	.000044391	.000004973	.000000243	.000000000	0011101011010
011101001001	50	-.0000633	-.0009461	.000261198	.000205270	.000141448	.000085481	.000045315	.000005236	.000000268	.000000000	0110001110010
110101001001	52	-.0000637	-.0009296	.000261413	.000206014	.000142600	.000086689	.000046283	.000005462	.000000286	.000000000	0100110111010
110001011001	52	-.0000637	-.0009452	.000261132	.000205123	.000141208	.000085183	.000045025	.000005129	.000000256	.000000000	0110100111010
110010010101	52	-.0000637	-.0009473	.000261096	.000205010	.000141035	.000084998	.000044874	.000005091	.000000252	.000000000	0101101011010
101010101001	53	-.0000638	-.0009639	.000260771	.000204022	.000139515	.000083371	.000043531	.000004747	.000000222	.000000000	1001010101010
011101010101	50	-.0000638	-.0009528	.000261469	.000204085	.000140005	.000084147	.000044321	.000005018	.000000251	.000000000	0011010101100
011110010001	49	-.0000661	-.0009020	.000260581	.000206921	.000144451	.000088981	.000048383	.000006129	.000000360	.000000000	0011110010010
101100010011	53	-.0000665	-.0009351	.000260792	.000204839	.000141173	.000085361	.000045280	.000005231	.000000266	.000000000	1000101110110
101010011010	51	-.0000665	-.0009541	.000260452	.000203774	.000139446	.000083093	.000043827	.000004860	.000000234	.000000000	1001001110110
101010001101	52	-.0000666	-.0009336	.000260436	.000203722	.000139446	.000083512	.000043755	.000004841	.000000232	.000000000	0110101101010
110000111001	54	-.0000672	-.0009107	.000261097	.000205999	.000143086	.000087492	.000047095	.000005733	.000000315	.000000000	1000011101100
110100000101	52	-.0000716	-.0009272	.000259980	.000203723	.000140188	.000084706	.000044939	.000005214	.000000268	.000000000	0101000110101
100111110001	50	-.0000747	-.0009372	.000259228	.000202277	.000138453	.000083144	.000043814	.000004995	.000000253	.000000000	0111000110011
111010010001	51	-.0000772	-.0009071	.000259297	.000203171	.000140177	.000085172	.000045582	.000005490	.000000301	.000000000	0011111000011
111001000011	51	-.0000779	-.0008504	.000260178	.000206156	.000144979	.000090513	.000050183	.000006853	.000000451	.000000000	0011111000001x
110001110001	51	-.0000779	-.0008984	.000259320	.000203440	.000140696	.000085787	.000046123	.000005645	.000000316	.000000000	0011111000110
110011100001	51	-.0000779	-.0009107	.000259100	.000202747	.000139616	.000084617	.000045142	.000005378	.000000290	.000000000	0011011011100
110011001001	51	-.0000779	-.0009146	.000259031	.000202532	.000139285	.000084260	.000044845	.000005298	.000000283	.000000000	0011011001100
100011110001	51	-.0000779	-.0009272	.000258807	.000201831	.000138200	.000083093	.000043874	.000005040	.000000258	.000000000	0111011001110
100101110010	51	-.0000779	-.0009336	.000258893	.000201474	.000137646	.000082497	.000043378	.000004908	.000000246	.000000000	0110110111010
110011001010	51	-.0000779	-.0009485	.000258428	.000200645	.000136371	.000081139	.000042264	.000004627	.000000221	.000000000	0110011101100
100110000101	51	-.0000779	-.0009487	.000258424	.000200634	.000136354	.000081121	.000042249	.000004623	.000000221	.000000000	1100110110001
101001110001	51	-.0000779	-.0009501	.000258399	.000200554	.000136230	.000080988	.000042139	.000004595	.000000218	.000000000	1010110110001
111001001001	51	-.0000786	-.0009521	.000258363	.000200444	.000136062	.000080810	.000041995	.000004559	.000000215	.000000000	0011111001001
110010101001	52	-.0000786	-.0009371	.000258492	.000201020	.000137015	.000081844	.000042841	.000004765	.000000232	.000000000	0110101011010
110101010010	51	-.0000805	-.0009195	.000258447	.000201451	.000138003	.000083097	.000043989	.000005113	.000000267	.000000000	0101010111100
101011100001	51	-.0000805	-.0009554	.000257809	.000199463	.000134945	.000079839	.000041310	.000004429	.000000206	.000000000	1010101101010
101010101010	51	-.0000812	-.0009580	.000257764	.000199323	.000134731	.000079610	.000041124	.000004382	.000000202	.000000000	1010101010110
110101001010	51	-.0000812	-.0008889	.000258841	.000202821	.000139610	.000085360	.000045825	.000005563	.000000306	.000000000	0101011111100
110001011010	52	-.0000812	-.0009265	.000258195	.000200805	.000137072	.000081934	.000043174	.000004889	.000000245	.000000000	0110101111010
110010010110	52	-.0000812	-.0009285	.000258149	.000200693	.000136900	.000081934	.000043023	.000004850	.000000242	.000000000	0110011101100

Table A

m = 7, n = 7   Rank Orders 1001 thru 1050 of 1780   C(7, 7) = 3432   1/C(7, 7) = .00029138

z	w	c_1	c_2	d = .2	.4	.6	.8	1.0	1.5	2.0	3.0	z^{tc}
1010100111010	53	−.0000814	−.0009475	.000257787	.000199597	.000135229	.000080159	.000041567	.000004481	.000000209	.000000000	1010100111010
1101011001011	51	−.0000858	−.0009095	.000257645	.000200387●	.000137096	.000082515	.000043697	.000005101	.000000269	.000000000	0011001101100
1001011000011	51	−.0000858	−.0009415	.000257077	.000198619	.000134374	.000079207	.000041299	.000004400	.000000212	.000000000	0111001001010
1011001100011	51	−.0000892	−.0009462	.000256995	.000198365	.000133988	.000081880	.000040969	.000005043	.000000206	.000000000	0101011001100
1011011100011	51	−.0000892	−.0009058	.000257059	.000199522	.000136257	.000078533	.000043303	.000004338	.000000266	.000000000	0011011001010
1011100110001	51	−.0000892	−.0009392	.000256467	.000197684	.000133432	.000078868	.000040823	.000004405	.000000208	.000000000	0111001101100
0110111000101	49	−.0000919	−.0009431	.000256399	.000197473	.000133113	.000077901	.000040551	.000004685	.000000202	.000000000	0101100111010
1001110000110	52	−.0000924	−.0009199	.000256308	.000197935●	.000134205	.000078533	.000041777	.000004444	.000000235	.000000000	1000111100110
1100101010110	53	−.0000931	−.0009287	.000256053	.000197255	.000133195	.000078826	.000040882	.000004563	.000000212	.000000000	1000110101110
			−.0009178	.000256106	.000197588	.000133769	.000079451	.000041390		.000000221	.000000000	
1110100001011	51	−.0000954	−.0008711	.000256502	.000199456●	.000136980	.000083077	.000044494	.000005423	.000000304	.000000000	0011011011000
1010101010101	51	−.0000954	−.0009472	.000255175	.000195385	.000130805	.000076570	.000039194	.000004087	.000000185	.000000000	0110101001010
1011010011011	51	−.0000954	−.0009479	.000255161	.000195343	.000130742	.000076503	.000039139	.000004074	.000000184	.000000000	0110101011010
1011101001001	50	−.0000955	−.0009141	.000255739●	.000197171	.000133561	.000079506	.000041604	.000004698	.000000239	.000000000	0011110011010
1100110100011	53	−.0000958	−.0009093	.000255760●	.000197247	.000133624	.000079500	.000041532	.000004635	.000000229	.000000000	1000101101100
1001100101101	51	−.0000989	−.0009453	.000254558	.000194431	.000129837	.000075806	.000038699	.000004007	.000000174	.000000000	1001010100110
1001110100110	52	−.0001010	−.0009425	.000254834	.000196648	.000129490	.000075503	.000042565	.000003951	.000000270	.000000000	0011100100110x
0111010100110	53	−.0001010	−.0009196	.000254689●	.000195314	.000131430	.000077621	.000040242	.000005054	.000000186	.000000000	0110110010110
1100110011101	53	−.0001010	−.0008798	.000255307●	.000197283	.000131435	.000080784	.000042811	.000004273	.000000183	.000000000	0101001110010
			−.0009144	.000254691	.000195357	.000131452	.000077573	.000040138	.000004343	.000000206	.000000000	0100111111100
1001001100110	53	−.0001010	−.0009149	.000254681	.000195325	.000131404	.000077522	.000040096	.000004333	.000000205	.000000000	1001001011100
0111110000101	49	−.0001033	−.0009084	.000254382	.000195068	.000131393	.000077747●	.000040418	.000004475	.000000246	.000000000	0011011100001
0111011000011	50	−.0001034	−.0008926	.000254633●	.000195871	.000132628	.000079056	.000041488	.000004747	.000000242	.000000000	0011101111000
1011010010011	50	−.0001034	−.0009070	.000254387	.000195135	.000131537	.000077934	.000040596	.000004535	.000000228	.000000000	0011110010010
1001011001100	52	−.0001038	−.0009082	.000254366	.000195072	.000131444	.000077837	.000040519	.000004516	.000000227	.000000000	1001100100110x
1001110011001	51	−.0001041	−.0009400	.000253731	.000193180	.000128571	.000074791	.000038020	.000003881	.000000170	.000000000	0011001100110
1010101011001	51	−.0001041	−.0008730	.000254834	.000196648	.000133864	.000080376	.000042565	.000005015	.000000270	.000000000	0110110010110
0110101011001	51	−.0001041	−.0009301	.000253846	.000193640	.000129335	.000075635	.000038728	.000004064	.000000186	.000000000	0011011100110
1100010011101	51	−.0001072	−.0009322	.000253809	.000193529	.000129168	.000075463	.000038591	.000004032	.000000183	.000000000	0110110111010
			−.0009054	.000253703	.000194011●	.000130292	.000076840	.000039795	.000004351	.000000211	.000000000	0100111111100
1010100010011	52	−.0001100	−.0009263	.000252818	.000192047	.000127705	.000074332	.000037869	.000003916	.000000176	.000000000	1000111010010
0101010100101	49	−.0001103	−.0009223	.000252838●	.000192236	.000128090	.000074814	.000038314	.000004053	.000000190	.000000000	0110110100100x
1100011010010	51	−.0001151	−.0008988	.000252345	.000191956	.000128245●	.000075249	.000038776	.000004189	.000000201	.000000000	0110101011100
1100010100011	51	−.0001151	−.0009144	.000252038	.000191122	.000126987	.000073932	.000037714	.000003930	.000000179	.000000000	0011010011100
1001110100101	50	−.0001152	−.0009179	.000251994	.000191017	.000126830	.000073769	.000037584	.000003900	.000000177	.000000000	0011110011010
1001110100101	51	−.0001152	−.0009324	.000251736	.000190930	.000127254	.000073700	.000037549	.000003902	.000000178	.000000000	0011110100110
1010100100110	52	−.0001175	−.0008563	.000252638●	.000193520●	.000130129	.000072407	.000036497	.000003645	.000000156	.000000000	1000110011010
0111110001001	48	−.0001182	−.0008959	.000251816	.000191183	.000130983	.000078356	.000041440	.000004929	.000000274	.000000000	0011111001000
0111100001001	49	−.0001182	−.0009133	.000251522	.000190308	.000127516	.000074726	.000038478	.000004162	.000000202	.000000000	0110111001100
						.000126235	.000073423	.000037455	.000003929	.000000183	.000000000	0101110110010
0111010010011	49	−.0001182	−.0009136	.000251517	.000190294	.000126215	.000073403	.000037440	.000003926	.000000183	.000000000	0110101101001
1100101100011	51	−.0001186	−.0009126	.000251453	.000190157	.000126000	.000073146	.000037200	.000003845	.000000174	.000000000	0101010010110
1100001011010	53	−.0001187	−.0008935	.000251775●	.000191116	.000127406	.000074574	.000038318	.000004096	.000000193	.000000000	1010001011100
0111010110010	52	−.0001187	−.0009295	.000251475	.000189251	.000126148	.000073299	.000037319	.000003870	.000000176	.000000000	1001101110010
1010101100110	52	−.0001187	−.0009301	.000251136	.000189192	.000124619	.000071721	.000036067	.000003571	.000000153	.000000000	1001010110100
1010100110110	52	−.0001238	−.0009007	.000251125	.000189187	.000126192	.000071671	.000036999	.000003571	.000000152	.000000000	1010101110100
0110011110010	52	−.0001261	−.0009039	.000250663	.000189187	.000125225●	.000072690	.000036026	.000003848	.000000176	.000000000	0101010111010
0111010011010	49	−.0001261	−.0009211	.000250211	.000188399	.000125225●	.000072055	.000037690	.000003615	.000000176	.000000000	0110110011001x
1011000110010	52	−.0001266	−.0009056	.000250077	.000188080	.000123930	.000072125	.000036615	.000003808●	.000000163●	.000000000	1001011110001x

202

d = .2   C(7,7) = 3432

z	w	c_1	c_2	.2	.4	.6	.8	1.0	1.5	2.0	3.0	z^{tc}
1101000000111	50	-.001327	-.0007963	.000250809•	.000191942	.000130613	.000079044	.000042554	.000005438	.000000338	.000000000	0001111101000
1101000000011	50	-.001327	-.0008437	.000249991	.000189436	.000126781	.000074938	.000039123	.000004479	.000000237	.000000000	0010111011000
1000111110001	50	-.001327	-.0008711	.000249519	.000188002	.000124621	.000072672	.000037281	.000004013	.000000194	.000000000	0111000001110
1100101100001	50	-.001327	-.0008737	.000249481	.000187908	.000124512	.000072592	.000037244	.000004021•	.000000196	.000000000	0011011001100
1100110010011	50	-.001327	-.0008749	.000249396	.000187848	.000124424	.000072503	.000037174	.000004004	.000000195	.000000000	0011001101100
1011100000110	50	-.001327	-.0008782	.000248854	.000187631	.000124067	.000072096	.000036817	.000003898	.000000184	.000000000	0110110011010
1011100000101	50	-.001327	-.0009102	.000248853	.000186021	.000121705	.000069694	.000034934	.000003472	.000000150	.000000000	0110111001010
1011100000011	50	-.001327	-.0009103	.000248854	.000186018	.000121700	.000069690	.000034931	.000003471	.000000150	.000000000	0110111001010
1010101000101	50	-.001327	-.0009125	.000248816	.000185904	.000121530	.000069513	.000034790	.000003438	.000000148	.000000000	0101101100110
1010100110101	50	-.001327	-.0009131	.000248805	.000185873	.000121485	.000069469	.000034756	.000003431	.000000147	.000000000	0111011001010
1110000001101	51	-.001334	-.0008586	.000249592•	.000188373	.000125222	.000073298	.000037771	.000004118	.000000202	.000000000	0100111011000
1100101101001	51	-.001334	-.0008974	.000248937	.000186430	.000122370	.000070393	.000035485	.000003593	.000000159	.000000000	0110110011100
1100111011001	51	-.001334	-.0008988	.000248913	.000186359	.000122266	.000070287	.000035402	.000003574	.000000158	.000000000	0110101111000
1010101011010	52	-.001336	-.0009175	.000248568	.000185352	.000120788	.000068777	.000034213	.000003303	.000000137	.000000000	1010101011010x
1010100101001	52	-.001362	-.0009069	.000248255	.000185101	.000120761	.000068929•	.000034418	.000003376	.000000143	.000000000	1010010101110x
1010100011001	50	-.001380	-.0008633	.000248682•	.000186846	.000123594	.000071982	.000036918	.000003992	.000000196	.000000000	0010101110100
1010011010001	50	-.001380	-.0008946	.000248144	.000185225	.000121178	.000069483	.000034919	.000003512	.000000155	.000000000	0011011101000
1010011000101	50	-.001380	-.0008997	.000248059	.000184971	.000120804	.000069101	.000034618	.000003444	.000000150	.000000000	0111011010010
1010011000011	50	-.001406	-.0008667	.000248127•	.000185865	.000122479	.000071011	.000036229	.000003856	.000000186	.000000000	0111011001010
1010010110011	50	-.001406	-.0009001	.000247556	.000184149	.000119933	.000068389	.000034145	.000003364	.000000145	.000000000	1000111001010
1101000101001	51	-.001406	-.0009040	.000247490	.000183952	.000119644	.000068095	.000033915	.000003313	.000000141	.000000000	0101101011010
1100101011010	51	-.001413	-.0008522	.000248225•	.000186285	.000123113	.000071630	.000036682	.000003933	.000000189	.000000000	0101000111100
1100111001001	51	-.001413	-.0008699	.000247933	.000185438	.000121896	.000070416	.000035745	.000003728	.000000173	.000000000	0110001111000
1100110001001	51	-.001413	-.0008745	.000247857	.000185213	.000121570	.000070087	.000035488	.000003670	.000000169	.000000000	0110011111100
1100011000101	52	-.001445	-.0008758	.000247234	.000184140	.000120355	.000069018	.000034718	.000003509	.000000156	.000000000	1000111011010
1010101100011	49	-.001469	-.0008573	.000247112•	.000184429•	.000121144•	.000070044	.000035654	.000003781	.000000182	.000000000	0011011101010
1010011010011	50	-.001476	-.0008216	.000247576•	.000185970	.000123482	.000072470	.000037592	.000004246	.000000222	.000000000	0101011101010
1010101001001	50	-.001476	-.0008965	.000246313	.000182223	.000117977	.000066847	.000033149	.000003205	.000000135	.000000000	1010011010010
1010101000011	50	-.001476	-.0008976	.000246294	.000182167	.000117896	.000066764	.000033084	.000003191	.000000134	.000000000	0110110110010
1010100110011	52	-.001480	-.0008736	.000246623•	.000183201	.000119405	.000068269	.000034233	.000003430	.000000151	.000000000	1000110110100
1010101010011	50	-.001503	-.0009028	.000245714	.000181119	.000116700	.000065729	.000032364	.000003059	.000000125	.000000000	0101101011010
0110100110011	48	-.001520	-.0008615	.000246088•	.000182664	.000119172	.000068343	.000034458	.000003545	.000000164	.000000000	0101011011000
1100101010110	51	-.001525	-.0008897	.000245511	.000181045	.000116819	.000065943	.000032560	.000003102	.000000128	.000000000	1001011000010
1100101010011	52	-.001532	-.0008776	.000245578•	.000181401	.000117397	.000066544	.000033026	.000003200	.000000134	.000000000	1001101010010
1010010101110	51	-.001552	-.0008953	.000244924	.000179981	.000115602	.000064891	.000031828	.000002970	.000000119	.000000000	1001101000100
0111010000101	48	-.001555	-.0008561	.000245530•	.000181877	.000118439	.000067811	.000034138	.000003501	.000000161	.000000000	0111011010010
1010011010011	50	-.001555	-.0008286	.000245982•	.000183196	.000120344	.000069725	.000035628	.000003839	.000000190	.000000000	0111011001010
1010011001011	50	-.001555	-.0008857	.000245028	.000180389	.000116258	.000065594	.000032399	.000003105	.000000130	.000000000	0111001101010
0110111001001	50	-.001555	-.0008878	.000244992	.000180284	.000116107	.000065443	.000032323	.000003080	.000000128	.000000000	0110111001010
1010101001110	52	-.001556	-.0008550	.000245529•	.000181921	.000118543	.000067947	.000034268	.000003543	.000000166	.000000000	0011101101100
1101000011110	52	-.001559	-.0008489	.000245570•	.000182053	.000118690	.000068035	.000034282	.000003512	.000000160	.000000000	1000101111010
1100001100110	52	-.001559	-.0008658	.000245284	.000181201	.000117437	.000066757	.000033276	.000003282	.000000142	.000000000	1001001110110
1100001110110	52	-.001559	-.0008676	.000245253	.000181110	.000117305	.000066623	.000033172	.000003259	.000000140	.000000000	1001001111100
1100000111110	53	-.001566	-.0008211	.000245907•	.000183228	.000120507	.000069943	.000035821	.000003888	.000000193	.000000000	1100000111110x
1011100010011	49	-.001583	-.0008439	.000245215	.000181642	.000118438	.000067987	.000034362	.000003581	.000000169	.000000000	0011011001110
1010001100101	49	-.001590	-.0008230	.000245426•	.000182415	.000119620	.000069201	.000035316	.000003798	.000000187	.000000000	0111001101110
1010011100001	50	-.001590	-.0008821	.000244439	.000179518	.000115411	.000064952	.000032000	.000003047	.000000126	.000000000	0111001101010
0110110110011	50	-.001590	-.0008835	.000244417	.000179452	.000115318	.000064860	.000031930	.000003032	.000000125	.000000000	0110110110010
0110111000101	48	-.001617	-.0008608	.000244300•	.000179785•	.000116142	.000065870	.000032804	.000003256	.000000144	.000000000	0110111000101•
1010111000110	51	-.001622	-.0008703	.000244041	.000179119	.000115195	.000064907	.000032040	.000003074	.000000129	.000000000	1000111001010

Table A

m = 7, n = 7   **Rank Orders 1101 thru 1150 of 1780**   **1/C(7,7) = .00029138**

C(7,7) = 3432

z	w	c_1	c_2	d = .2	.4	.6	.8	1.0	1.5	2.0	3.0	z^{tc}
1001111000010101	49	−0001666	−0008541	.000243493	.000178590	.000114950	.000064923	.000032174	.000003139	.000000135	.000000000	0101111000000110
1101010000010011	50	−0001673	−0008420	.000243567	.000178986	.000115616	.000065645	.000032761	.000003279	.000000147	.000000000	0101110110100110
1100011100010001	50	−0001673	−0008564	.000243324	.000178262	.000114555	.000064566	.000031915	.000003086	.000000131	.000000000	0101110011001100
1100110001010011	50	−0001673	−0008594	.000243275	.000178121	.000114353	.000064365	.000031761	.000003053	.000000129	.000000000	0011110011001100
0111110000000011	47	−0001697	−0007836	.000244091	.000181130	.000119057	.000069319	.000035756	.000004027	.000000214	.000000000	0011111100001100
1101010100100110	50	−0001699	−0008632	.000242716	.000177139	.000113254	.000063428	.000031115	.000002937	.000000121	.000000000	0111010101001100
1010010010010110	51	−0001701	−0008598	.000242749	.000177262	.000113440	.000063615	.000031258	.000002966	.000000123	.000000000	1001010101010100
1010010100100110	51	−0001701	−0008794	.000242422	.000176303	.000112057	.000062235	.000030198	.000002741	.000000107	.000000000	0101010100110010
0111010001001001	48	−0001704	−0008561	.000242761	.000177414	.000113753	.000064002	.000031607	.000003067	.000000133	.000000000	0101101010110001
1110000011011110	52	−0001708	−0007903	.000243771	.000180419	.000118096	.000068352	.000034975	.000003820	.000000193	.000000000	1000011111111000
1100100101011010	52	−0001708	−0008500	.000242775	.000177490	.000113825	.000064014	.000031563	.000003027	.000000127	.000000000	1010010100111100
1100100011011010	52	−0001708	−0008503	.000242727	.000177476	.000113804	.000063994	.000031548	.000003023	.000000137	.000000000	1010010011011100
0111000101001001	48	−0001731	−0008433	.000242474	.000177209	.000113748	.000064133	.000031766	.000003116	.000000104	.000000000	0110101011000001
1010100010011010	51	−0001735	−0008756	.000241837	.000175444	.000111230	.000061617	.000029820	.000002688	.000000123	.000000000	1001101101010010
0101100100100110	50	−0001752	−0008493	.000241975	.000176241	.000112566	.000062820	.000030954	.000002941	.000000128	.000000000	0111011101010001
0111010001100001	48	−0001783	−0008437	.000241499	.000175642	.000112114	.000062522	.000030905	.000002975 •	.000000128	.000000000	0110010101010001
1010010100101010	50	−0001786	−0008439	.000241417	.000175457	.000111843	.000062522	.000030648	.000002902	.000000121	.000000000	0101010101011010
1010001010011010	51	−0001788	−0008598	.000241127	.000174637	.000110668	.000061349	.000029745	.000002706	.000000106	.000000000	1001010110011010
1111000010011101	50	−0001848	−0008038	.000240923	.000175489 •	.000112576	.000063586	.000031616	.000003153	.000000141	.000000000	0101110110111000
1100110101011001	50	−0001848	−0008426	.000240290	.000173673	.000110001	.000061050	.000029688	.000002746	.000000111	.000000000	0111100011001100
1010010100110011	50	−0001848	−0008440	.000240267	.000173607	.000109906	.000060957	.000029617	.000002731	.000000143	.000000000	0110010110101010
1100101100000011	49	−0001849	−0008112	.000240794 •	.000175186	.000112215	.000063293	.000031441	.000003142	.000000109	.000000000	0011101101000100
1011101010010001	49	−0001849	−0008471	.000240198	.000173449	.000109714	.000060792	.000029508	.000002716	.000000109	.000000000	0111011001000110
1010110100100101	49	−0001849	−0008496	.000240156	.000173327	.000109537	.000060615	.000029372	.000002686	.000000107	.000000000	1011011001001110
1001110100110110	51	−0001849	−0008620	.000239944	.000172699	.000108619	.000059687	.000028651	.000002529	.000000096	.000000000	1010100110011010x
1011010010101010	50	−0001883	−0008390	.000239702	.000172805 •	.000109159	.000060414	.000029293	.000002688	.000000108	.000000000	1010100110101010
1100100101101001	49	−0001884	−0008579	.000239365	.000171857	.000107816	.000059093	.000028291	.000002480	.000000093	.000000000	1010101101011010
1110000010011011	49	−0001928	−0007027	.000241101 •	.000177960	.000117151	.000068777	.000036010	.000004334	.000000257	.000000000	0011111111100000
1101010001010011	49	−0001928	−0007865	.000239727	.000173961	.000111337	.000062846	.000031290	.000003162	.000000146	.000000000	0011011110100100
1101010101000011	49	−0001928	−0008000	.000239510	.000173344	.000110471	.000062000	.000030649	.000003027	.000000136	.000000000	0011110011001100
1001011111000001	49	−0001928	−0008015	.000239486	.000173276	.000110374	.000061906	.000030577	.000003012	.000000135	.000000000	0011011011010010
1001110110001011	49	−0001928	−0008164	.000239231	.000172510	.000109239	.000060736	.000029647	.000002792	.000000116	.000000000	0111011110000010
1011110010010001	49	−0001928	−0008222	.000239136	.000172239	.000108854	.000060357	.000029358	.000002730	.000000112	.000000000	0111011001000110
1010101101010110	49	−0001928	−0008340	.000238947	.000171712	.000108127	.000059663	.000028849	.000002633	.000000105	.000000000	1011010010010110
1011010010010110	49	−0001928	−0008343	.000238944	.000171702	.000108114	.000059651	.000028840	.000002631	.000000103	.000000000	0111011010010110
1101010010101010	49	−0001928	−0008376	.000238876	.000171542	.000107886	.000059426	.000028668	.000002595	.000000102	.000000000	1011010110101010
1100010001100101	49	−0001928	−0008385	.000238874	.000171501	.000107829	.000059371	.000028628	.000002587	.000000124	.000000000	1010101101011010
1100010011010011	50	−0001935	−0008013	.000239340 •	.000172968	.000109937	.000061435	.000030175	.000002897	.000000257	.000000000	0110010111111000
1100010011010011	50	−0001935	−0008199	.000239043	.000172135	.000108782	.000060323	.000029349	.000002733	.000000000	.000000000	0110111100111100
1101010011001001	50	−0001935	−0008238	.000238980	.000171957	.000108534	.000060084	.000029171	.000002697	.000000110	.000000000	0110101110110100
1010110001011010	51	−0001963	−0008291	.000238374	.000170869	.000107297	.000059016	.000028428	.000002561	.000000106	.000000000	1010001110010010
1100100010010011	49	−0001994	−0008131	.000238067	.000170695	.000107379 •	.000059252	.000028675	.000002629	.000000166	.000000000	0111011100001010
1110101000010011	49	−0002025	−0007418	.000238653 •	.000173125	.000111212	.000063252	.000031848	.000003362	.000000091	.000000000	0010110111101000
1010101010010110	49	−0002025	−0008331	.000237170	.000168889	.000105214	.000057344	.000027344	.000002394	.000000091	.000000000	1011010101010110
1100110110010110	49	−0002025	−0008363	.000237165	.000168887	.000105193	.000057323	.000027329	.000002391	.000000094	.000000000	0111011001010110
1100101001010110	51	−0002046	−0008152	.000237059	.000169020	.000105575	.000057755	.000027663	.000002452	.000000122	.000000000	1001010110011100
1011110010010110	47	−0002069	−0007777	.000237226	.000170066	.000107352	.000059569	.000029205	.000002808		.000000000	0111011010010010

204

Table A

m = 7, n = 7   Rank Orders 1151 thru 1200 of 1780   C(7, 7) = 3432   1/C(7, 7) = .00029138

z	w	c_1	c_2	d = .2	.4	.6	.8	1.0	1.5	2.0	3.0	z^{tc}
1110010001011	49	−.0002077	−.0007582	.000237413	.000170782	.000108445	.000060782	.000030069	.000002999	.000000137	.000000000	0011011011000
1010001110001	49	−.0002077	−.0008144	.000236505	.000168200	.000104807	.000057370	.000027370	.000002430	.000000095	.000000000	0111010011010
1011000010101	49	−.0002080	−.0008169	.000236465	.000168086	.000104649	.000057065	.000027255	.000002407	.000000093	.000000000	0110011101010
1110000101110	52	−.0002081	−.0007730	.000237116	.000169958	.000107255	.000059570	.000029109	.000002770	.000000118	.000000000	1100010111100x
1100100101010	51	−.0002081	−.0007922	.000236788	.000169061	.000106021	.000058385	.000028228	.000002593	.000000105	.000000000	1001011011100
1100010101010	51	−.0002081	−.0008100	.000236497	.000168225	.000104835	.000057219	.000027345	.000002410	.000000092	.000000000	1001100101100
1110100010010	51	−.0002104	−.0008113	.000236476	.000168167	.000104755	.000057142	.000027287	.000002399	.000000091	.000000000	1010011011100
1010101100001	49	−.0002104	−.0007600	.000236888	.000169900	.000107486	.000059978	.000029520	.000002899	.000000130	.000000000	0011101011000
1011010010001	49	−.0002104	−.0008191	.000235935	.000167195	.000103690	.000056278	.000026732	.000002321	.000000087	.000000000	0011010110010
1010011010010	49	−.0002104	−.0008205	.000235913	.000167134	.000103606	.000056197	.000026673	.000002309	.000000087	.000000000	0111010110010
1010110000011	48	−.0002105	−.0007717	.000236686	.000169370	.000106778	.000059315	.000029037	.000002805	.000000124	.000000000	0011110100010
1110010010001	50	−.0002111	−.0007666	.000236644	.000169343	.000106741	.000059248	.000028953	.000002766	.000000119	.000000000	0110011111000
1010100110010	51	−.0002160	−.0007829	.000236343	.000167138	.000104165	.000056986	.000027360	.000002465	.000000098	.000000000	1010001110010
1100001110010	49	−.0002191	−.0007960	.000235474	.000165668	.000102480	.000055551	.000026387	.000002303	.000000088	.000000000	0110110010010
0110110001001	51	−.0002218	−.0007742	.000234694	.000165895	.000103105	.000056316	.000027038	.000002459	.000000100	.000000000	0111001000001
1010011001010	47	−.0002221	−.0007541	.000234552	.000166644	.000104152	.000057315	.000027768	.000002597	.000000109	.000000000	0110111000010
1100110001001	49	−.0002221	−.0007863	.000234797	.000165191	.000102127	.000055355	.000026303	.000002301	.000000088	.000000000	0101111100010
1011001010001	49	−.0002221	−.0007872	.000234282	.000165152	.000102074	.000055305	.000026267	.000002294	.000000088	.000000000	0101101100100
1011010001010	50	−.0002223	−.0007831	.000234268	.000165292	.000102278	.000055504	.000026414	.000002322	.000000090	.000000000	0111011001010
1010011100010	50	−.0002223	−.0008021	.000234001	.000164420	.000101059	.000054324	.000025535	.000002148	.000000078	.000000000	1001110011010
1010110001010	50	−.0002223	−.0008031	.000233986	.000164378	.000101001	.000054269	.000025494	.000002140	.000000077	.000000000	1001011001010
1010101010010	51	−.0002230	−.0007888	.000234079	.000164784	.000101616	.000054874	.000025942	.000002224	.000000083	.000000000	1010101011010
1010101010001	50	−.0002249	−.0008069	.000233432	.000163427	.000099968	.000053419	.000024933	.000002051	.000000072	.000000000	1001110101010
0110111000010	47	−.0002252	−.0007668	.000234023	.000165188	.000102482	.000055888	.000026794	.000002431	.000000099	.000000000	1011011010010
1100011001001	51	−.0002256	−.0007336	.000234467	.000166470	.000104234	.000057534	.000027985	.000002650	.000000113	.000000000	1010111001010
1011000101010	51	−.0002256	−.0007762	.000233776	.000164539	.000101619	.000055021	.000026117	.000002277	.000000087	.000000000	1010001101100
1100101001010	49	−.0002274	−.0007771	.000233548	.000164305	.000101442	.000054968	.000026079	.000002270	.000000086	.000000000	1010100111100
1011001100010	49	−.0002300	−.0007716	.000233018	.000164305	.000100486	.000054958	.000026125•	.000002296	.000000089	.000000000	0101000111100
1011010001001	49	−.0002300	−.0007740	.000233018	.000163415	.000100486	.000054175	.000025605	.000002211	.000000084	.000000000	0111011001010
1010011100010	50	−.0002301	−.0007891	.000232680	.000162680	.000099470	.000053196	.000024877	.000002067	.000000074	.000000000	1001011011010
0111010100001	47	−.0002305	−.0007656	.000233078•	.000163717	.000099996	.000054729	.000026057	.000002318	.000000092	.000000000	0111010010001x
0111110001001	47	−.0002331	−.0007513	.000232808	.000163544	.000110014•	.000054863•	.000026205	.000002358	.000000096	.000000000	0111011011100
1100111000011	50	−.0002336	−.0007831	.000232201	.000161925	.000098788	.000052717	.000024601	.000002033	.000000072	.000000000	1001100110010
1001110110001	48	−.0002363	−.0007235	.000232661	.000163838	.000101748	.000055729	.000026261	.000002525	.000000109	.000000000	0011110001010
1010111001001	48	−.0002363	−.0007594	.000232085	.000162211	.000099474	.000053519	.000025143	.000002180	.000000083	.000000000	0101111001010
1100110110001	48	−.0002363	−.0007620	.000232044	.000162096	.000099896	.000053362	.000026685	.000002157	.000000081	.000000000	0101110001100
1101100010001	49	−.0002370	−.0007229	.000232530•	.000163599	.000101447	.000055432	.000025091	.000002463	.000000103	.000000000	0101111000110
1011001001001	49	−.0002370	−.0007605	.000231937	.000161951	.000099184	.000053272	.000025007	.000002150	.000000081	.000000000	0101110100110
1011010001001	49	−.0002370	−.0007625	.000231905	.000161863	.000099063	.000053157	.000024472	.000002134	.000000080	.000000000	0111011100110
1001011011000	49	−.0002397	−.0007656	.000231363	.000160945	.000098075	.000052348	.000024472	.000002048	.000000075	.000000000	1010110101100
1010100110110	50	−.0002398	−.0007838	.000231048	.000160090	.000096907	.000051238	.000023659	.000001893	.000000065	.000000000	1010011010010
1010110101001	50	−.0002433	−.0007782	.000230494	.000159319	.000096204	.000050740	.000023371	.000001858	.000000063	.000000000	1010100110010x
1110011010001	49	−.0002449	−.0007260	.000231015•	.000161104	.000098747	.000053247	.000025222	.000002210	.000000086	.000000000	0101011001010
1010100100101	49	−.0002449	−.0007486	.000230729	.000160327	.000097746	.000052279	.000024528	.000002084	.000000078	.000000000	0011100100110
1010111011001	49	−.0002450	−.0007168	.000230667	.000160160	.000097521	.000052003	.000024377	.000002056	.000000076	.000000000	0101111010100
1101110101001	48	−.0002450	−.0007502	.000231156•	.000161571	.000099511	.000054011	.000025840	.000002359	.000000098	.000000000	0011101010010
1100101001001	48	−.0002450	−.0007541	.000230623	.000160074	.000097430	.000052003	.000024340	.000002055	.000000076	.000000000	0011101101100
1101100101001	48	−.0002477	−.0007541	.000230561	.000159901	.000097192	.000051775	.000024173	.000002022	.000000074	.000000000	0101101100110
1101000110001	48	−.0002477	−.0007046	.000230853•	.000161303	.000099397	.000054020	.000025895	.000002382	.000000100	.000000000	0011101011100
1001011110001	48	−.0002477	−.0007360	.000230352	.000159892	.000097431	.000052114	.000024466	.000002088	.000000079	.000000000	1001110010110

Table A

m = 7, n = 7   Rank Orders 1201 thru 1250 of 1780   C(7, 7) = 3432   1/C(7, 7) = .00029138

z	w	c₁	c₂	d = .2	.4	.6	.8	1.0	1.5	2.0	3.0	z^tc
1011100010101	48	-.0002477	-.0007410	.000230273	.000159673	.000097132	.000051830	.000024258	.000002048	.000000076	.000000000	0101101110010
1100001100011	49	-.0002484	-.0007187	.000230486	.000160396	.000096164	.000052819	.000024980	.000002182	.000000085	.000000000	0101100111000
1100100110001	49	-.0002484	-.0007394	.000230167	.000159533	.000097009	.000051745	.000024209	.000002042	.000000076	.000000000	0110001100100
1010100110001	49	-.0002485	-.0007419	.000230128	.000159429	.000096869	.000051088	.000024118	.000002026	.000000065	.000000000	0110010111000
1011000011110	50	-.0002516	-.0007592	.000229829	.000158617	.000095760	.000050561	.000023343	.000001877	.000000065	.000000000	1010010111010
1001110000110	50	-.0002547	-.0007259	.000228789	.000159171	.000096817	.000051701	.000024234	.000002059	.000000077	.000000000	1000111000100
0111101010001	48	-.0002591	-.0007459	.000228907	.000159155	.000094800	.000049975	.000023065	.000001866	.000000066	.000000000	0111011011000
1010100001110	46	-.0002591	-.0006754	.000229192	.000159192	.000097546	.000052758	.000025178	.000002290	.000000095	.000000000	1001011100000
1101100001110	50	-.0002594	-.0007113	.000228561	.000157491	.000095303	.000050303	.000023619	.000001983	.000000074	.000000000	1000110001100
1100011010110	50	-.0002594	-.0007290	.000228280	.000156709	.000094231	.000049618	.000022873	.000001841	.000000064	.000000000	1001101001100
1100101001010	50	-.0002594	-.0007303	.000228260	.000156655	.000094158	.000049550	.000022824	.000001832	.000000064	.000000000	1001011010110x
1000100111001	49	-.0002596	-.0007337	.000228189	.000156499	.000093974	.000049398	.000022726	.000001819	.000000063	.000000000	1001110011110x
1011010000011	51	-.0002601	-.0006983	.000228627	.000157805	.000095772	.000051088	.000023944	.000002040	.000000077	.000000000	1100001101110
1011100000011	47	-.0002619	-.0006758	.000228680	.000158369	.000096780	.000052191	.000024841	.000002250	.000000094	.000000000	0011111010000
1110000010011	48	-.0002626	-.0006198	.000229408	.000160495	.000099724	.000055001	.000026931	.000002675	.000000126	.000000000	0011111011000
1110100010011	48	-.0002626	-.0006562	.000228848	.000158965	.000097640	.000053022	.000025454	.000002370	.000000102	.000000000	0010111110010
1110001000001	48	-.0002653	-.0006689	.000228653	.000158433	.000096920	.000052343	.000024956	.000002272	.000000095	.000000000	0011011010010
1100010100001	48	-.0002626	-.0006706	.000228626	.000158626	.000096821	.000052250	.000024889	.000002259	.000000094	.000000000	0111011011000
1100111110001	48	-.0002626	-.0007104	.000228002	.000156634	.000094456	.000049994	.000023222	.000001928	.000000071	.000000000	0111101011000
1011001101001	48	-.0002626	-.0007136	.000227952	.000156498	.000094272	.000049822	.000023099	.000001905	.000000069	.000000000	0111001001100
1010101010001	48	-.0002626	-.0007288	.000227718	.000155870	.000093440	.000049058	.000022558	.000001811	.000000064	.000000000	0111010010010
1010101100001	48	-.0002626	-.0007289	.000227717	.000155867	.000093436	.000049054	.000022554	.000001810	.000000064	.000000000	1000110010010
1011010100001	48	-.0002626	-.0007299	.000227701	.000155824	.000093379	.000049001	.000022517	.000001804	.000000063	.000000000	0110111010010
1011010010001	48	-.0002629	-.0007306	.000227692	.000155798	.000093344	.000048969	.000022495	.000001800	.000000063	.000000000	0110101110010
1110000100110	50	-.0002633	-.0007236	.000227722	.000155928	.000093512	.000049103	.000022570	.000001803	.000000062	.000000000	1001100110010
1110000011110	49	-.0002682	-.0006915	.000228160	.000157199	.000095264	.000050763	.000023778	.000002029	.000000077	.000000000	0101101111000
1010001010110	50	-.0002682	-.0007056	.000227038	.000155168	.000092986	.000048843	.000022482	.000001810	.000000063	.000000000	1001010111010
1010101010010	48	-.0002705	-.0007125	.000226518	.000154267	.000092034	.000048096	.000022017	.000001749	.000000061	.000000000	1001011010100
1001101001001	48	-.0002743	-.0006923	.000226115	.000153982	.000092007	.000048232	.000022181	.000001793	.000000064	.000000000	1011110101100
1110011001010	50	-.0002743	-.0007024	.000225948	.000153509	.000091345	.000047591	.000021703	.000001699	.000000058	.000000000	0101110011100
0111110001001	46	-.0002766	-.0006663	.000226098	.000154435	.000092854	.000049146	.000022900	.000001946	.000000075	.000000000	0101110100001
1011001110010	49	-.0002771	-.0006740	.000225879	.000153914	.000092161	.000048482	.000022403	.000001842	.000000067	.000000000	1000111011000
1010101000110	49	-.0002771	-.0007107	.000225314	.000152385	.000090121	.000046597	.000021060	.000001604	.000000053	.000000000	0110111001100
1110000100101	49	-.0002771	-.0007110	.000225309	.000152373	.000090090	.000046583	.000021050	.000001602	.000000052	.000000000	1000110111000
1110000110110	50	-.0002778	-.0006525	.000226073	.000154567	.000093079	.000049347	.000023023	.000001953	.000000074	.000000000	1010101111000
1001011010010	50	-.0002778	-.0006960	.000225406	.000152768	.000090679	.000047126	.000021437	.000001667	.000000056	.000000000	1010010101100
1001011011010	50	-.0002782	-.0006967	.000225396	.000152742	.000090646	.000047095	.000021415	.000001664	.000000056	.000000000	1010110101100
0111011010001	46	-.0002818	-.0006632	.000225186	.000153061	.000091508	.000048127	.000022270	.000001855	.000000070	.000000000	0110100100001x
1101000001001	48	-.0002822	-.0006617	.000225131	.000152955	.000091355	.000047958	.000022125	.000001816	.000000066	.000000000	0101111010110
1001001100001	48	-.0002822	-.0006761	.000224911	.000152369	.000090584	.000047253	.000021627	.000001729	.000000061	.000000000	0111001011100
1101010010101	48	-.0002822	-.0006773	.000224893	.000152321	.000090521	.000047196	.000021587	.000001722	.000000061	.000000000	0101101110010
1011011010110	49	-.0002823	-.0006916	.000224647	.000151669	.000089649	.000046382	.000021000	.000001614	.000000054	.000000000	1001011110100
1011001010110	49	-.0002850	-.0006940	.000224120	.000150805	.000088748	.000045669	.000020805	.000001548	.000000050	.000000000	1001101011010
0111100100111	46	-.0002853	-.0006542	.000224681	.000152416	.000090965	.000047769	.000022075	.000001835	.000000069	.000000000	0111010100010
1110001110010	50	-.0002857	-.0006393	.000224821	.000152814	.000091456	.000048171	.000022320	.000001857	.000000069	.000000000	1001100111100
1010010011101	50	-.0002857	-.0006620	.000224479	.000151909	.000090270	.000047091	.000021560	.000001725	.000000061	.000000000	1010000111100
1101010001010	48	-.0002919	-.0006641	.000224448	.000151828	.000090165	.000046997	.000021495	.000001715	.000000060	.000000000	0100111110100
1001011001101	48	-.0002919	-.0006098	.000224114	.000152279	.000091340	.000048357	.000022590	.000001947	.000000076	.000000000	0101101111100
1001011001001	48	-.0002919	-.0006645	.000223311	.000150056	.000088396	.000045646	.000020661	.000001600	.000000054	.000000000	0110111011010

206

Table A

m = 7, n = 7 Rank Orders 1251 thru 1300 of 1780 $C(7,7) = 3432$ $1/C(7,7) = .00029138$

z	w	c_1	c_2	d = .2	.4	.6	.8	1.0	1.5	2.0	3.0	z^{tc}
1010110001010	49	−.002920	−.0006825	.000223010	.000149266	.000087352	.000044685	.000019978	.000001480	.000000047	.000000000	1010011001010 10x
1010101010101	49	−.002947	−.0006856	.000222473	.000149378	.000086425	.000043955	.000019516	.000001414	.000000044	.000000000	1010101010101010x
1101011000001	47	−.002964	−.0006087	.000223339•	.000151079•	.000090195	.000047526	.000022106	.000001892	.000000075	.000000000	0011110000010
1001111000010	47	−.002964	−.0006421	.000222823	.000151021	.000088305	.000045753	.000020818	.000001647	.000000058	.000000000	0111100000110
1101111000010	47	−.002964	−.0006460	.000222763	.000149676	.000088988	.000045551	.000020673	.000001620	.000000056	.000000000	0111110000010
1110001100011	48	−.002971	−.0006242	.000222959•	.000149514	.000088988	.000046387	.000021265	.000001722	.000000062	.000000000	0101010110010
1011001101100	48	−.002971	−.0006419	.000222695	.000150163	.000088092	.000045581	.000020705	.000001628	.000000057	.000000000	0110110110010
1101001010010	48	−.002971	−.0006464	.000222626	.000149472	.000087856	.000045368	.000020555	.000001602	.000000055	.000000000	0110110101010
1110011000110	48	−.002998	−.0006249	.000222457	.000149291	.000088164	.000045738	.000020850	.000001659	.000000059	.000000000	0110111000110x
1011010101100	48	−.002998	−.0006456	.000222150	.000149362•	.000087122	.000044803	.000020202	.000001551	.000000052	.000000000	0111110100100
1101001010100	48	−.002998	−.0006481	.000222112	.000148458	.000086996	.000044691	.000020125	.000001539	.000000052	.000000000	0101010110010
1101110110000	47	−.002999	−.0006009	.000222815•	.000150383	.000089584	.000047104	.000021865	.000001863	.000000073	.000000000	0011110110010
1001111001101	47	−.002999	−.0006329	.000222321	.000149040	.000087774	.000045407	.000020632	.000001628	.000000057	.000000000	0111101100010
1011001000101	47	−.002999	−.0006376	.000222250	.000148849	.000087521	.000045174	.000020466	.000001598	.000000055	.000000000	1010010110010
1011010011010	49	−.002999	−.0006574	.000221833	.000147729	.000086035	.000043808	.000019499	.000001429	.000000045	.000000000	1001101100010
1010110101000	47	−.003034	−.0005923	.000221304	.000147025	.000085421	.000043391	.000019268	.000001404	.000000044	.000000000	1010010110010
1010100100010	48	−.003061	−.0006207	.000221165	.000147213•	.000085920	.000043967	.000019725	.000001495	.000000050	.000000000	0011101111000
1010101010010	48	−.003085	−.0006180	.000220968	.000147167	.000086066•	.000044191	.000019919	.000001536	.000000052	.000000000	0011110101010
1101011001110	48	−.003109	−.0006198	.000219755	.000146355	.000085198	.000043483	.000019452	.000001460	.000000048	.000000000	0111010011010
1011011001110	49	−.003116	−.0006034	.000220480	.000146825	.000085887	.000044146	.000019936	.000001549	.000000053	.000000000	0111110100100
1100111000110	49	−.003116	−.0006203	.000220342	.000146127	.000084958	.000043290	.000019326	.000001441	.000000047	.000000000	1001110011100
1100011001110	50	−.003116	−.0006222	.000220315	.000146054	.000084864	.000043205	.000019268	.000001431	.000000046	.000000000	0111110001010
1010111001110	46	−.003123	−.0006056	.000220430•	.000146486	.000085471	.000043767	.000019661	.000001496	.000000050	.000000000	1100010101010x
1001010010010	49	−.003141	−.0005524	.000219929	.000148193	.000087921	.000046131	.000021404	.000001834	.000000073	.000000000	0111011001010
1001011001011	48	−.003143	−.0006241	.000219798	.000145221	.000084011	.000042541	.000018851	.000001372	.000000043	.000000000	1001101010010
1110010100010	47	−.003146	−.0005923	.000220214•	.000146389	.000085566	.000043964	.000019852	.000001544	.000000053	.000000000	0110101101010
1010101000100	47	−.003148	−.0005615	.000220666•	.000147643	.000086920	.000045565	.000021018	.000001767	.000000069	.000000000	0011101111000
1011010100010	47	−.003148	−.0006013	.000219776	.000145281	.000084164	.000042730	.000019012	.000001410	.000000046	.000000000	0111010101010
1010101000110	47	−.003150	−.0006220	.000219755	.000145227	.000084094	.000042667	.000018968	.000001403	.000000045	.000000000	1001110001010
1101011001110	50	−.003150	−.0005920	.000220150•	.000146283	.000085448	.000043862	.000019779	.000001529	.000000052	.000000000	1100001110100
1110101001000	47	−.003174	−.0005486	.000220369•	.000147380	.000087146	.000045555	.000021052	.000001783	.000000070	.000000000	0011101101010
1010101110000	47	−.003174	−.0006047	.000219523	.000145131	.000084185	.000042840	.000019124	.000001435	.000000047	.000000000	0111000101010
1011000101001	47	−.003174	−.0006072	.000219486	.000145036	.000084062	.000042730	.000019048	.000001423	.000000047	.000000000	0111101110010
1011010010110	48	−.003181	−.0005838	.000219703•	.000145729	.000083545	.000043590	.000019647	.000001522	.000000052	.000000000	0110100101010
1010100010010	49	−.003195	−.0006042	.000219139	.000144518	.000083545	.000042323	.000018783	.000001379	.000000044	.000000000	1001101010010
1011001101000	47	−.003226	−.0005100	.000219977•	.000147350	.000087532	.000046094	.000021510	.000001883	.000000077	.000000000	0011101111000
1010101010100	47	−.003226	−.0006013	.000218620	.000143796	.000082909	.000041902	.000018563	.000001363	.000000044	.000000000	0111100111000
1010100100110	47	−.003226	−.0006017	.000218615	.000143782	.000082891	.000041887	.000018553	.000001361	.000000043	.000000000	0111101110100
1010100100110	47	−.003230	−.0005959	.000218626•	.000143855	.000082981	.000041950	.000018580	.000001358	.000000043	.000000000	1001110101010
1001110010110	47	−.003257	−.0005737	.000218468	.000143982•	.000083390	.000042436	.000018971	.000001437	.000000048	.000000000	0101101110010
0111110000011	45	−.003288	−.0005378	.000218433	.000144506•	.000084347	.000043432	.000019728	.000001584	.000000058	.000000000	0110111100000
1100010000110	49	−.003292	−.0005476	.000218209	.000143965	.000083639	.000042772	.000019246	.000001491	.000000052	.000000000	1000111011000
1001010010100	48	−.003292	−.0005911	.000217564	.000142283	.000080834	.000040807	.000017909	.000001267	.000000039	.000000000	1010101011100
1010100101101	49	−.003292	−.0005918	.000217555	.000142259	.000081442	.000040605	.000017891	.000001250	.000000038	.000000000	1001101010010
1001010011010	49	−.003293	−.0005838	.000217462	.000142051	.000081197	.000040419	.000017762	.000001245	.000000038	.000000000	1001101101010
1010101010010	45	−.003327	−.0005615	.000217036•	.000141581	.000080860	.000041136	.000017678	.000001245	.000000044	.000000000	0101110101010
0111010010010	45	−.003344	−.0005332	.000217059•	.000141998	.000081566	.000041980	.000018213	.000001344	.000000044	.000000000	0111110101010
0111001000010	45	−.003367	−.0005473	.000217062•	.000142469	.000082388	.000041980	.000018213	.000001466	.000000052	.000000000	0111101010010
1011100100110	47	−.003371	−.0005473	.000216780	.000141778	.000081494	.000041162	.000018267	.000001359	.000000052	.000000000	1001101010001
1011100000110	48	−.003372	−.0005603	.000216558	.000141210	.000080756	.000040492	.000017795	.000001278	.000000040	.000000000	1001011100010

Table A

m = 7, n = 7   Rank Orders 1301 thru 1350 of 1780   C(7, 7) = 3432   1/C(7, 7) = .00029138

z	w	c_1	c_2	d = .2	.4	.6	.8	1.0	1.5	2.0	3.0	z^{tc}
1011001100110	48	−0003372	−0005768	.000216317	.000140595	.000079981	.000039815	.000017340	.000001208	.000000036	.000000000	1001110110010
1011010010010	48	−0003372	−0005774	.000216309	.000140573	.000079954	.000039792	.000017325	.000001206	.000000036	.000000000	1001101010010
1110000101010	49	−0003379	−0005376	.000216758●	.000141849	.000081614	.000041264	.000018324	.000001361	.000000045	.000000000	1001010111000
1100010111010	49	−0003379	−0005609	.000216420	.000140984	.000080522	.000040307	.000017678	.000001260	.000000039	.000000000	1001000110010
0111100110010	49	−0003379	−0005627	.000216394	.000140919	.000080442	.000040238	.000017632	.000001254	.000000039	.000000000	1010010110100
1011001100001	45	−0003402	−0005227	.000216579●	.000141876	.000081908	.000041675	.000018690	.000001450	.000000051	.000000000	0111001100001x
1100101001010	47	−0003441	−0005461	.000215518	.000139853	.000079609	.000039749	.000017407	.000001244	.000000039	.000000000	0110110100100
1011010010001	48	−0003469	−0005652	.000214724	.000138322	.000077873	.000038307	.000016462	.000001102	.000000032	.000000000	1010101001010
1101010000001	46	−0003513	−0004742	.000215268●	.000140592	.000081166	.000041422	.000018688	.000001491	.000000055	.000000000	0011110010010
1011110100001	46	−0003513	−0005062	.000214791	.000139334	.000079522	.000039925	.000017631	.000001302	.000000043	.000000000	0110110000110
0111101000101	46	−0003513	−0005109	.000214722●	.000139155	.000079292	.000039718	.000017487	.000001278	.000000042	.000000000	0101110100010
1111010000101	47	−0003520	−0004515	.000215453●	.000141155	.000081886	.000042039	.000019091	.000001547	.000000058	.000000000	0100111110000
1010000010101	47	−0003520	−0004890	.000214908	.000139773	.000080097	.000040444	.000017986	.000001359	.000000046	.000000000	0011110011000
1110001000101	47	−0003520	−0005026	.000214710	.000139247	.000079453	.000039876	.000017598	.000001296	.000000043	.000000000	0101011011000
1110010100101	47	−0003520	−0005041	.000214689	.000139194	.000079386	.000039818	.000017559	.000001290	.000000042	.000000000	0101110111000
1100111100000	47	−0003520	−0005047	.000214682	.000139179	.000079373	.000039810	.000017557	.000001290●	.000000042	.000000000	0111100011000
1011011001001	47	−0003520	−0005107	.000214596	.000138961	.000079099	.000039572	.000017396	.000001265	.000000041	.000000000	0111001100100
1001011010001	47	−0003520	−0005238	.000214405	.000138474	.000078486	.000039035	.000017034	.000001209	.000000038	.000000000	0110100100110
1010101001001	47	−0003520	−0005241	.000214400	.000138462	.000078470	.000039021	.000017025	.000001207	.000000038	.000000000	0110110010100
1010011001001	47	−0003520	−0005261	.000214373	.000138393	.000078384	.000038947	.000016976	.000001200	.000000037	.000000000	0110110101000
1101010101001	47	−0003520	−0005270	.000214360	.000138359	.000078343	.000038911	.000016952	.000001196	.000000037	.000000000	0101011010010
1010100011010	48	−0003521	−0005426	.000214102	.000137710	.000077514	.000038174	.000016445	.000001114	.000000033	.000000000	1000110011000
1010101001010	48	−0003547	−0005443	.000213593	.000136903	.000076706	.000037561	.000016071	.000001065	.000000030	.000000000	1010010111000
1100000011010	49	−0003554	−0004875	.000214288●	.000138802	.000079147	.000039712	.000017526	.000001291	.000000042	.000000000	0111001111000
1010010011010	47	−0003598	−0005037	.000213260	.000137015	.000077261	.000038233	.000016603	.000001164	.000000036	.000000000	0111001011000
1010001110010	48	−0003634	−0005123	.000212473	.000135665	.000075827	.000037180	.000015875	.000001059	.000000031	.000000000	1011000110010x
1100010111010	49	−0003637	−0004884	.000212763●	.000136438	.000076802	.000037933	.000016432	.000001141	.000000035	.000000000	1100011001110x
1110001001010	46	−0003661	−0004295	.000213194●	.000138033	.000079072	.000040073	.000017967	.000001416	.000000052	.000000000	0011110010010
1010111001001	46	−0003661	−0004886	.000212333	.000135821	.000076252	.000037569	.000016244	.000001127	.000000035	.000000000	0110110010010
1010011001001	47	−0003661	−0004899	.000212313	.000135770	.000076188	.000037514	.000016207	.000001122	.000000034	.000000000	0110110101000
1101100000110	48	−0003665	−0004618	.000212650●	.000136682	.000077347	.000038525	.000016885	.000001225	.000000040	.000000000	1000111110100
1001011001001	48	−0003665	−0004964	.000212146	.000135391	.000075713	.000037089	.000015911	.000001072	.000000031	.000000000	1001011001000
1001100100110	48	−0003665	−0004969	.000212138	.000135370	.000075688	.000037068	.000015898	.000001070	.000000031	.000000000	1001011001100
1100100101110	47	−0003668	−0004660	.000212526●	.000136428	.000077053	.000038278	.000016722	.000001201	.000000038	.000000000	0110010111010
1100100101100	49	−0003672	−0004803	.000212246	.000135764	.000076223	.000037545	.000016219	.000001117	.000000034	.000000000	1100010110010
1010100001001	47	−0003695	−0004660	.000212040	.000135678	.000076311●	.000037717	.000016378	.000001154	.000000036	.000000000	0111101011000
1101010001001	47	−0003696	−0004205	.000212688●	.000137384	.000078519	.000039704	.000017762	.000001392	.000000051	.000000000	0011110011000
1010110100001	46	−0003696	−0004776	.000211857	.000135250	.000075801	.000037292	.000016103	.000001115	.000000034	.000000000	0111100010010
1010111001001	46	−0003696	−0004797	.000211827	.000135173	.000075706	.000037209	.000016048	.000001107	.000000034	.000000000	0110110110010
1110100001010	48	−0003717	−0004754	.000211499	.000134708	.000075261	.000036872	.000015836	.000001074	.000000032	.000000000	1001011101000
1101010100110	48	−0003744	−0004758	.000211008	.000133952	.000074518	.000036317	.000015501	.000001031	.000000030	.000000000	1001101010010
1010100111100	46	−0003748	−0004725	.000210983	.000133996	.000074640	.000036465	.000015624	.000001058	.000000032	.000000000	0111101010100
1010100111100	49	−0003751	−0004425	.000211360●	.000134968	.000075841	.000037488	.000016294	.000001154	.000000036	.000000000	1100001111100
1111000000011	46	−0003775	−0003802	.000211820●	.000136595	.000078104	.000039585	.000017778	.000001415	.000000053	.000000000	0011101110000
1011001110001	46	−0003775	−0004562	.000210745	.000133872	.000074676	.000036575	.000015727	.000001079	.000000033	.000000000	0111110011000
1110000110001	46	−0003775	−0004315	.000210730	.000133833	.000074629	.000036534	.000015700	.000001075	.000000033	.000000000	0111001100010
1110011000001	47	−0003782	−0004590	.000210951●	.000134507	.000075508	.000037306	.000016218	.000001153	.000000037	.000000000	0011110010010
1010111000110	47	−0003807	−0004315	.000210106	.000132820	.000073586	.000035720	.000015187	.000001001	.000000029	.000000000	1001111000110
1011011000110	48	−0003814	−0004157	.000210599	.000134188	.000075343	.000037263	.000016228	.000001162	.000000037	.000000000	1001011101110

m = 7, n = 7 Rank Orders 1351 thru 1400 of 1780 C(7, 7) = 3432 1/C(7, 7) = .00029138

z	w	c_1	c_2	d = .2	.4	.6	.8	1.0	1.5	2.0	3.0	z^{tc}
1100110001101 0	48	−.0003814	−.0004592	.000209974	.000132613	.000073378	.000035561	.000015090	.000000988	.000000028	.000000000	1010011100110 0
1100101010010	48	−.0003840	−.0004604	.000209474	.000131835	.000072613	.000034945	.000014747	.000000945	.000000026	.000000000	1010101010100
1010110000101	46	−.0003858	−.0004225	.000209708 •	.000132752	.000073900	.000037082	.000015563	.000001075	.000000033	.000000000	1010111001010 0
0111110000101	46	−.0003889	−.0003838	.000209696	.000133291 •	.000074818 •	.000037082	.000016227	.000001192	.000000040	.000000000	0111011100100
1100100100001	44	−.0003893	−.0004124	.000209219	.000132147	.000073407	.000035856	.000015398	.000001059	.000000032	.000000000	1011010110100
1100010010110	48	−.0003893	−.0004122	.000209208	.000132091	.000073301	.000035733	.000015295	.000001035	.000000031	.000000000	1001100101100
1100001001101 0	48	−.0003893	−.0004355	.000208881	.000131283	.000072315	.000034899	.000014751	.000000958	.000000027	.000000000	1011001001100
1010110001101 0	48	−.0003893	−.0004373	.000208856	.000131223	.000072243	.000034839	.000014713	.000000952	.000000027	.000000000	1001101100010
1001101001010	47	−.0003894	−.0004423	.000208767	.000131030	.000072025	.000034665	.000014606	.000000939	.000000026	.000000000	1001110101010
0111101010001	44	−.0003916	−.0003832	.000209225 •	.000132592	.000074156	.000036602	.000015944	.000001157	.000000039	.000000000	0111110101000 1x
1011100100110	47	−.0003920	−.0004260	.000208512	.000130860	.000072000	.000034720	.000014669	.000000953	.000000027	.000000000	1001101110010
1100001101100	48	−.0003928	−.0004020	.000208721 •	.000131491	.000072814	.000035425	.000015135	.000001020	.000000030	.000000000	1011000111100
1100100110010	48	−.0003928	−.0004266	.000208376	.000130642	.000071781	.000034552	.000014566	.000000939	.000000026	.000000000	1101110011000
1100010011010	48	−.0003928	−.0004277	.000208360	.000130604	.000071736	.000034515	.000014543	.000000936	.000000026	.000000000	0111111001100
1010110010010	46	−.0003954	−.0004036	.000208220	.000130748 •	.000072127 •	.000034950	.000014874	.000000995	.000000030	.000000000	0011110101001 0x
1011011000011	47	−.0003991	−.0004269	.000207233	.000128922	.000070137	.000033361	.000013871	.000000857	.000000023	.000000000	0111110000110
1001111000011	45	−.0004034	−.0003197	.000207961 •	.000131373	.000073706	.000036587	.000016087	.000001213	.000000043	.000000000	0111110000110
1001111100011	45	−.0004034	−.0003510	.000207508	.000130373	.000072234	.000035281	.000015186	.000001060	.000000033	.000000000	0101111000110
0111110101010	48	−.0004034	−.0003560	.000207437	.000130194	.000072010	.000035086	.000015055	.000001039	.000000032	.000000000	0101110111000
1110000010101	46	−.0004041	−.0003638	.000207272	.000129742	.000071512	.000034688	.000014804	.000001005	.000000031	.000000000	0101110111000
1011010100110	46	−.0004041	−.0003844	.000206915	.000129039	.000070661	.000033972	.000014339	.000000939	.000000027	.000000000	0111010110100
1010101010010	46	−.0004041	−.0003870	.000208880	.000128953	.000070557	.000033885	.000014283	.000000931	.000000027	.000000000	0111010101010
1101101000101	46	−.0004068	−.0003489	.000206929 •	.000129525	.000071432	.000034697	.000014840	.000001016	.000000031	.000000000	1011001010010 0x
1101011000101	46	−.0004068	−.0003665	.000206683	.000128919	.000070695	.000034074	.000014432	.000000957	.000000028	.000000000	1100010101100
1011000111010	46	−.0004068	−.0003711	.000206619	.000128766	.000070511	.000033921	.000014334	.000000943	.000000028	.000000000	1001110101100
1011011001010	47	−.0004069	−.0003856	.000206144	.000128198	.000069806	.000033311	.000013926	.000000880	.000000024	.000000000	0011111001010 0
1011010101010	47	−.0004069	−.0004030	.000206140	.000127604	.000069093	.000032716	.000013539	.000000829	.000000022	.000000000	0101110010010
1110000101010	48	−.0004076	−.0003617	.000206591 •	.000128811	.000070589	.000033982	.000014364	.000000942	.000000027	.000000000	0101110111000
1110000010101	46	−.0004120	−.0003063	.000206571	.000129529 •	.000071788	.000035144	.000015190	.000001078	.000000035	.000000000	0101110011000
1010110101010	46	−.0004120	−.0003610	.000205817	.000127696	.000069579	.000033292	.000013987	.000000906	.000000026	.000000000	0111010101100
1101010110010	46	−.0004148	−.0003616	.000205809	.000127675	.000069553	.000033270	.000013973	.000000904	.000000026	.000000000	0111011010100
1011001001110	47	−.0004186	−.0003787	.000205060	.000126300	.000068062	.000032083	.000013226	.000000801	.000000021	.000000000	1011001010010 0x
1001101001100	48	−.0004187	−.0003446	.000204853	.000126424 •	.000068457 •	.000032510	.000013535	.000000849	.000000023	.000000000	1100010101100
1001101010100	47	−.0004210	−.0003513	.000204742	.000126182	.000068184	.000032294	.000013403	.000000833	.000000023	.000000000	0011110101100
1010111000011	45	−.0004210	−.0003698	.000205476 •	.000126418	.000071118	.000034896	.000015170	.000001113	.000000038	.000000000	0011110101100
1010110100110	45	−.0004210	−.0003269	.000204673	.000126420	.000068649	.000032768	.000013747	.000000890	.000000026	.000000000	0101110011000
1110010001100	45	−.0004210	−.0003291	.000204643	.000126347	.000068562	.000032695	.000013699	.000000884	.000000025	.000000000	1100101011000
1111000010001	46	−.0004217	−.0003051	.000204446 •	.000126955	.000069336	.000033339	.000014134	.000000946	.000000028	.000000000	0111011011000
1011001010101		−.0004217	−.0003195	.000204644	.000126462	.000068738	.000032854	.000013805	.000000898	.000000026	.000000000	0111010011000
1110010010100	46	−.0004217	−.0003207	.000204628	.000126422	.000068691	.000032816	.000013781	.000000895	.000000026	.000000000	0101011011010 0
1100110110001	48	−.0004221	−.0003350	.000204358	.000125806	.000067950	.000032186	.000013365	.000000833 •	.000000023 •	.000000000	1110010011010 0x
1011011000011	45	−.0004262	−.0003199	.000203828	.000125249	.000067598	.000032041	.000013337	.000000848	.000000024	.000000000	0111011100010
1110001100110	47	−.0004266	−.0003116	.000203869 •	.000125385	.000067757	.000032159	.000013401	.000000804	.000000022	.000000000	1001110111000
1011010011010	47	−.0004266	−.0003272	.000203654	.000124865	.000067136	.000031644	.000013072	.000000801	.000000021	.000000000	1101101110010
1010110011010	48	−.0004273	−.0003281	.000203642	.000124837	.000067102	.000031617	.000013055	.000000841	.000000023	.000000000	1100010111000
1001110011100	48	−.0004273	−.0003099	.000203763 •	.000125242	.000067763	.000032066	.000013346	.000001346	.000000025	.000000000	0011100111000
1111000010001	47	−.0004296	−.0002933	.000203584	.000125206	.000067757 •	.000032256	.000013504	.000001103	.000000038	.000000000	1100100111000
1110001010001	46	−.0004297	−.0002312	.000204429 •	.000127313	.000070352	.000034484	.000014990	.000001103	.000000038	.000000000	0111011100010
1011010100001	45	−.0004297	−.0003073	.000203374	.000124729	.000067206	.000031810	.000013225	.000000836	.000000023	.000000000	0111101011010

209

Table A

m = 7, n = 7   Rank Orders 1401 thru 1450 of 1780   C(7, 7) = 3432   1/C(7, 7) = .00029138

z	w	c_1	c_2	z	d = .2	.4	.6	.8	1.0	1.5	2.0	3.0	z^{tc}
1011100010001	45	−.0004297	−.0003081	1011100010001	.000203364	.000124704	.000067177	.000031786	.000013210	.000000834	.000000023	.000000000	0111010110010
1110100000110	47	−.0004362	−.0002488	1101010011110	.000202994	.000124886•	.000067816	.000032490	.000013728	.000000917	.000000028	.000000000	1000111110100
1100101100110	47	−.0004362	−.0003085	1101010101010	.000202170	.000122882	.000065407	.000030478	.000012429	.000000735	.000000019	.000000000	1010110010100
1101100100101	47	−.0004362	−.0003088	1011100100110	.000202124	.000122873	.000065397	.000030470	.000012424	.000000735	.000000019	.000000000	1010111010010
1011011000101	45	−.0004408	−.0002840	1110000111100	.000201696	.000123520•	.000066481	.000031507	.000013146	.000000846	.000000024	.000000000	0101111010010
1110100010110	46	−.0004415	−.0002593	1010111000010	.000201903•	.000123099	.000066004	.000031108	.000012477	.000000751	.000000020	.000000000	1001111010010
1110010100110	47	−.0004415	−.0002820	1110001100010	.000201594	.000122362	.000065134	.000031108	.000012882	.000000805	.000000022	.000000000	1000101110010
1100011101010	47	−.0004415	−.0002841	1101010010101	.000201566•	.000122297	.000065058	.000030334	.000012393	.000000740	.000000019	.000000000	1010100101100
0111110100001	43	−.0004438	−.0002147	1011011001001	.000202116•	.000124032	.000067322	.000032315	.000013716	.000000940	.000000030	.000000000	0111011000001
1110001010010	47	−.0004441	−.0002581	1110001010010	.000201439	.000122418	.000065364	.000030648	.000012614	.000000773	.000000021	.000000000	1001101011000
1101010110010	47	−.0004441	−.0002826	1101010110010	.000201106	.000121626	.000064432	.000029888	.000012136	.000000711	.000000018	.000000000	1010101001100
1101010101010	47	−.0004441	−.0002838	1101010101010	.000201091	.000121590	.000064391	.000029856	.000012117	.000000709	.000000018	.000000000	1010101100010
1011100100110	46	−.0004442	−.0002719	1011100100110	.000201234•	.000121957	.000064837	.000030229	.000012356	.000000741	.000000019	.000000000	1001110010100
1110000111100	48	−.0004448	−.0002235	1110000111100	.000201787•	.000123378	.000066561	.000031667	.000013280	.000000868	.000000026	.000000000	1100001111000
1010111000010	46	−.0004504	−.0002652	1010111000010	.000200214	.000120523	.000063539	.000029327	.000011848	.000000685	.000000017	.000000000	1010110100010x
1110001100010	47	−.0004528	−.0002462	1110001100010	.000200041	.000120500	.000063665•	.000029492	.000011975	.000000705	.000000018	.000000000	1010011001100
1101010010101	45	−.0004555	−.0002009	1101010010101	.000200177•	.000121275	.000064770	.000030484	.000012640	.000000803	.000000023	.000000000	0111101001100
1011011001001	45	−.0004555	−.0002215	1011011001001	.000199900	.000120617	.000063996	.000029852	.000012241	.000000749	.000000019	.000000000	0111011001100
1101100010001	45	−.0004555	−.0002240	1101100010001	.000199865	.000120536	.000063901	.000029775	.000012192	.000000743	.000000020	.000000000	0111100110100
1110100100101	45	−.0004590	−.0001896	1110100100101	.000199705•	.000120711•	.000064324	.000030209	.000012500	.000000790	.000000023	.000000000	0101111011000
1101011001001	45	−.0004590	−.0002082	1101011001001	.000199454	.000120116	.000063624	.000029637	.000012138	.000000742	.000000020	.000000000	1011001000100
1101010110010	46	−.0004590	−.0002121	1101010110010	.000199402	.000119993	.000063482	.000029522	.000012067	.000000732	.000000019	.000000000	1001111001100
1110000011010	47	−.0004590	−.0002123	1110000011010	.000199385	.000119927	.000063371	.000029404	.000011974	.000000714	.000000018	.000000000	0011111011000
1011010010010	46	−.0004591	−.0002440	1011010010010	.000198941	.000118906	.000062198	.000028467	.000011397	.000000643	.000000016	.000000000	1010011011000
1011100011010	46	−.0004618	−.0002262	1011100011010	.000198704	.000118764	.000062192	.000028524•	.000011455	.000000653	.000000016	.000000000	1010111010010
1110000110010	47	−.0004625	−.0002009	1110000110010	.000198915•	.000119367	.000062933	.000029135	.000011839	.000000702	.000000016	.000000000	0101010101010
1101010010101	45	−.0004642	−.0002010	1101010010101	.000198616•	.000118968•	.000062610	.000028948	.000011758	.000000701	.000000024•	.000000000	0111010100110
1101010010101	45	−.0004669	−.0001440	1101010010101	.000198895•	.000120034	.000064004	.000030137	.000012524	.000000805	.000000019	.000000000	0111011001100
1101001100001	45	−.0004669	−.0001816	1101001100001	.000198398	.000118870	.000062655	.000029047	.000011842	.000000716	.000000019	.000000000	1011100110100
1101100110001	45	−.0004669	−.0001837	1101100110001	.000198371	.000118808	.000062584	.000028990	.000011807	.000000711	.000000019	.000000000	0111001100100
1010100100101	46	−.0004670	−.0002167	1010100100101	.000197897	.000117695	.000061276	.000027921	.000011132	.000000622	.000000015	.000000000	1011001010100
1100111000110	46	−.0004701	−.0001826	1100111000110	.000197808	.000117963	.000061772	.000028393	.000011451	.000000666	.000000017	.000000000	1001111001100
1011100011001	47	−.0004708	−.0001806	1011100011001	.000197708	.000117835	.000061660	.000028316	.000011408	.000000661	.000000017	.000000000	1000111000100
1011011100001	44	−.0004732	−.0000911	1011011100001	.000198491	.000120120	.000064556	.000030803	.000013047	.000000904	.000000030	.000000000	0011111010010
1010111100100	44	−.0004732	−.0001473	1010111100100	.000197726	.000118273	.000062338	.000028941	.000011831	.000000724	.000000020	.000000000	0111111000010
1011100100001	44	−.0004732	−.0001498	1011100100001	.000197692	.000118195	.000062247	.000028867	.000011785	.000000717	.000000020	.000000000	0101111100000
1110101010100	47	−.0004734	−.0001807	1110101010100	.000197229	.000117125	.000060995	.000027843	.000011137	.000000631	.000000016	.000000000	1100101010110x
1101010001001	45	−.0004739	−.0001563	1101010001001	.000197481•	.000117811	.000061846	.000028563•	.000011603	.000000695	.000000019	.000000000	0101111101000
1101010010001	45	−.0004766	−.0001397	1101010010001	.000197226	.000117627	.000061791	.000028582•	.000011638	.000000704	.000000019	.000000000	0101101111000
1101001011100	47	−.0004787	−.0001537	1101001011100	.000196656	.000116603	.000060710	.000027745	.000011125	.000000637	.000000016	.000000000	1100110110100
1010101001010	46	−.0004788	−.0001615	1010101001010	.000196535	.000116345	.000060428	.000027529	.000010997	.000000623	.000000015	.000000000	0011111010100
1110110101010	44	−.0004811	−.0000600	1110110101010	.000197488•	.000118985	.000063699	.000030292	.000012793	.000000881	.000000029	.000000000	0011110010100
1011110100001	44	−.0004811	−.0001362	1011110100001	.000196468	.000116565	.000060841	.000027933	.000011279	.000000666	.000000018	.000000000	0111110010010
1011110010001	46	−.0004814	−.0001369	1011110010001	.000196458	.000116541	.000060815	.000027912	.000011266	.000000664	.000000016	.000000000	1001011100010
1011001001110	44	−.0004818	−.0001442	1011001001110	.000196288	.000116179	.000060392	.000027560	.000011038	.000000631	.000000016	.000000000	1001110010010
1111010011001	46	−.0004818	−.0000940	1111010011001	.000196894	.000117654	.000062129	.000028977	.000011930	.000000750	.000000021	.000000000	0110010100100
1110011010001	45	−.0004818	−.0001262	1110011010001	.000196472	.000116675	.000061003	.000028073	.000011369	.000000677	.000000018	.000000000	0111001011000
1101001100011	45	−.0004818	−.0001271	1101001100011	.000196460	.000116647	.000060971	.000028048	.000011353	.000000675	.000000018	.000000000	1100101110000
1101010110001	45	−.0004818	—	1101010110001	.000196269	.000116269	.000063901	.000027474	.000010988	.000000626	.000000016	.000000000	1101011110000

210

Table 11

m = 7, n = 7   Rank Orders 1451 thru 1500 of 1780   C(7,7) = 3432   1/C(7,7) = .00029138

z	w	c_1	c_2	d = .2	.4	.6	.8	1.0	1.5	2.0	3.0	z^{tc}
1011001100001	44	−.0004846	−.0001228	.000196023	.000116064•	.000060468	.000027716	.000011175	.000000658	.000000017	.000000000	0111011001100010
1100110100010	46	−.0004884	−.0001390	.000195106	.000114507	.000058883	.000026521	.000010460	.000000570	.000000013	.000000000	1010110100101100
1101110000101	44	−.0004928	−.0000688	.000195255•	.000115543	.000060351	.000027813	.000010306	.000000686	.000000015	.000000000	1001011110011000
1011101000110	45	−.0004956	−.0000951	.000194408	.000113991	.000058715	.000026558	.000010545	.000000591	.000000015	.000000000	0111101010000001x
0111110000110	42	−.0004960	−.0000314	.000195239•	.000116005	.000061098	.000028515	.000011790	.000000762	.000000023	.000000000	1000111101010010
1111000001110	46	−.0004963	−.0000703	.000195119	.000115749	.000060782	.000028234	.000011594	.000000728	.000000021	.000000000	1001011111101000
1010100001010	46	−.0004963	−.0000850	.000194605	.000114550	.000059391	.000027109	.000010887	.000000634	.000000016	.000000000	1010011111101000
1100100100010	46	−.0004963	−.0000862	.000194411	.000114099	.000058871	.000026692	.000010630	.000000602	.000000015	.000000000	1001110111101000
1110101000010	46	−.0004963	−.0000862	.000194397	.000114065	.000058833	.000026662	.000010612	.000000599	.000000015	.000000000	1011011011011000
1100111100010	46	−.0004963	−.0000918	.000194323	.000113897	.000058643	.000026513	.000010521	.000000588	.000000014	.000000000	1011100001111100
1101100011010	46	−.0004963	−.0000945	.000194288	.000113818	.000058555	.000026444	.000010480	.000000584	.000000014	.000000000	0110011111100100
1100100011010	46	−.0005112	−.0001101	.000194082	.000113341	.000058010	.000026012	.000010216	.000000552	.000000013	.000000000	1010110101011000
1001011010010	46	−.0005139	−.0001101	.000194082	.000113341	.000058009	.000026012	.000010216	.000000552	.000000013	.000000000	1011011001011000
1010110010010	46	−.0005140	−.0001116	.000194071	.000113316	.000057982	.000025991	.000010204	.000000550	.000000013	.000000000	1010110101010010
1010101010010	46	−.0005156	−.0000665	.000194527•	.000114469	.000057963	.000025976	.000010195	.000000559	.000000013	.000000000	1010101010110010
1010110001100	47	−.0004970	−.0000817	.000193052	.000112161	.000057121	.000025491	.000009965	.000000532	.000000012	.000000000	1100010111111000
1010101100100	46	−.0005042	−.0000081	.000192924	.000112812•	.000058236	.000026530	.000010663	.000000630	.000000017	.000000000	0110111101010100
1101010001001	44	−.0005104	−.0000081	.000192682	.000112255	.000058316	.000026026	.000010353	.000000591	.000000014	.000000000	0101110110101000
1100111010001	44	−.0005104	−.0000267	.000192682	.000112255	.000057601	.000026026	.000010353	.000000591	.000000014	.000000000	0111101101011000
1011001110010	44	−.0005104	−.0000306	.000192631	.000112139	.000057471	.000025924	.000010291	.000000584	.000000014	.000000000	0110110110101000
1010110010010	45	−.0005105	−.0000618	.000192194	.000111140	.000056325	.000025009	.000009726	.000000513	.000000012	.000000000	1010110010010010
1110001010010	46	−.0005112	−.0000350	.000192416	.000111746	.000057043	.000025583	.000010075	.000000554	.000000013	.000000000	1010110011011000
1100110101010	46	−.0005139	−.0000331	.000191966	.000111110	.000056468	.000025186	.000009854	.000000530	.000000012	.000000000	1010110101100010
1011100100010	45	−.0005140	−.0000479	.000191757	.000110658	.000055976	.000024812	.000009635	.000000506	.000000012	.000000000	0110111011100010
1010110010010	44	−.0005156	−.0000176	.000191872•	.000111183	.000056682	.000025420	.000010027	.000000559	.000000014	.000000000	0111101101010010
1010100100101	44	−.0005191	−.0000362	.000191944•	.000111849	.000057619	.000026226	.000010542	.000000625	.000000017	.000000000	0111110110101000
1011000100011	44	−.0005191	−.0000026	.000191447	.000110726	.000056359	.000025242	.000009946	.000000553	.000000014	.000000000	1011011010100100
1001001010001	44	−.0005191	−.0000040	.000191430	.000110687	.000056316	.000025209	.000009926	.000000551	.000000014	.000000000	0111101010100100
1110001001001	45	−.0005192	−.0000368	.000190975	.000109654	.000055142	.000024280	.000009358	.000000481	.000000011	.000000000	1011010101010010
1011100110010	45	−.0005219	−.0000175	.000190752	.000109535	.000055150•	.000024339	.000009410	.000000490	.000000011	.000000000	1100101110100010
1110000111010	46	−.0005226	.0000090	.000190967	.000110121	.000055840	.000024886	.000009743	.000000529	.000000013	.000000000	1011100111100010
1100110010010	44	−.0005253	.0000305	.000190777	.000110104	.000055994•	.000025085	.000009897	.000000554	.000000013	.000000000	1101110110101000
1011010010100	44	−.0005256	.0000024	.000190341	.000109148	.000054917	.000024234	.000009375	.000000490	.000000011	.000000000	1100010111001100
1010100011010	45	−.0005288	.0000437	.000190329	.000109597•	.000055614	.000024863	.000009790	.000000546	.000000014	.000000000	0110101101010010
1010101010010	44	−.0005302	.0000277	.000189869	.000108748	.000054725	.000024187	.000009386	.000000497	.000000012	.000000000	1001011110100010
1010101110000	46	−.0005309	.0000310	.000189786	.000108656	.000054655	.000024144	.000009363	.000000495	.000000012	.000000000	1100110110101000
1101011000011	44	−.0005333	.0001393	.000190764•	.000111275	.000057800	.000026723	.000010993	.000000715	.000000022	.000000000	0011101110100010
1010111000001	44	−.0005333	.0000644	.000189792	.000109037	.000055197	.000024658	.000009700	.000000541	.000000014	.000000000	0111101110100010
1011111010010	43	−.0005333	.0000633	.000189778	.000109006	.000055197	.000024632	.000009685	.000000539	.000000014	.000000000	0111011110100010
1011010010010	46	−.0005335	.0000323	.000189331	.000108011	.000054076	.000023748	.000009145	.000000472	.000000011	.000000000	1100101011010010
1101100100110	45	−.0005336	.0000414	.000189429	.000108259	.000054367	.000023982	.000009289	.000000490	.000000011	.000000000	1001011101100110
1100101010010	45	−.0005340	.0000580	.000189587•	.000108677	.000054868	.000024392	.000009546	.000000523	.000000013	.000000000	0111011010100100
1011101010001	43	−.0005359	.0000669	.000189357	.000108446	.000054720	.000024321	.000009521	.000000523	.000000013	.000000000	0111010100100010
1110011010001	44	−.0005366	.0000934	.000189563•	.000108995	.000055354	.000024816	.000009816	.000000557	.000000014	.000000000	0111101100100010
1100111000001	44	−.0005366	.0000790	.000189383	.000108598	.000054919	.000024485	.000009620	.000000535	.000000013	.000000000	0111100111100010
1010111010010	45	−.0005398	.0000761	.000189347	.000108518	.000054832	.000024418	.000009581	.000000530	.000000013	.000000000	1010111010100010
1011110110010	46	−.0005422	.0000536	.000188490	.000107030	.000053324	.000023301	.000008927	.000000455	.000000010	.000000000	0111011100100010
0111111000010	41	−.0005473	.0000744	.000188331	.000107019	.000053441	.000023441	.000009028	.000000469	.000000010	.000000000	1101000111010010x
1011110000010	46	−.0005473	.0001804	.000188784•	.000108793	.000055727	.000025358	.000010246	.000000631	.000000018	.000000000	0111111100000010x
1011110000110	44	−.0005478	.0001109	.000187804	.000106633	.000053305	.000023447	.000009071	.000000480	.000000011	.000000000	1001111110000010

Table A

z	w	c_1	c_2	d = .2	.4	.6	.8	1.0	1.5	2.0	3.0	z^{tc}
1110010011100	46	−.0005484	.0001136	.000187734	.000106556	.000053244	.000023408	.000009050	.000000478	.000000011	.000000000	1100011011111000
1110010100110	45	−.0005485	.0001049	.000187605	.000106291	.000052966	.000023204	.000008932	.000000465	.000000011	.000000000	1001101011011000
1101101010010	45	−.0005485	.0000803	.000187294	.000105599	.000052205	.000022622	.000008589	.000000427	.000000009	.000000000	1011010100011000
1101010100110	45	−.0005485	.0000791	.000187280	.000105568	.000052171	.000022597	.000008575	.000000426	.000000009	.000000000	1010101011001100
1001011100010	45	−.0005512	.0001229	.000187363	.000106127	.000052922	.000023221	.000008961	.000000471	.000000011	.000000000	1001101110100110
1011011100100	45	−.0005512	.0001002	.000187075	.000105484	.000052211	.000022676	.000008638	.000000435	.000000009	.000000000	1010101100110100
1011010100110	45	−.0005512	.0000981	.000187050	.000105428	.000052152	.000022632	.000008613	.000000432	.000000011	.000000000	1010101111001000
1011010101010	46	−.0005519	.0001268	.000187285	.000106046	.000052862	.000023184	.000008941	.000000469	.000000009	.000000000	1100101011100110
1001011110010	45	−.0005519	.0001714	.000187050	.000106156	.000053217	.000023546	.000009189	.000000502	.000000012	.000000000	1001011001100010
1011010110010	45	−.0005564	.0001117	.000186302	.000104504	.000051410	.000022172	.000008381	.000000413	.000000009	.000000000	1011010110010010
1101010010010	45	−.0005564	.0001114	.000186299	.000104497	.000051402	.000022166	.000008377	.000000412	.000000009	.000000000	1011001010010100
1101011000010	43	−.0005626	.0002021	.000186359	.000105507	.000052845	.000023405	.000009162	.000000511	.000000012	.000000000	1011111000001100
1100111000010	43	−.0005626	.0001844	.000186135	.000105007	.000052291	.000022976	.000008905	.000000480	.000000011	.000000000	0111110000100100
1011010001010	43	−.0005626	.0001798	.000186078	.000104883	.000052155	.000022873	.000008844	.000000473	.000000011	.000000000	0110110111000100
1011010001010	45	−.0005654	.0001529	.000185244	.000103411	.000050670	.000021784	.000008214	.000000403	.000000009	.000000000	1010110111010000
1101000010010	45	−.0005661	.0001797	.000185455	.000103972	.000051314	.000022283	.000008509	.000000436	.000000010	.000000000	1001101110010000
1110001001010	45	−.0005661	.0001642	.000185259	.000103534	.000050833	.000021915	.000008292	.000000412	.000000009	.000000000	1010111100101000
1100101001010	45	−.0005661	.0001633	.000185248	.000103512	.000050809	.000021897	.000008282	.000000410	.000000009	.000000000	0110111010010100
1111000010010	45	−.0005705	.0002382	.000185419	.000104513	.000052148	.000023020	.000008986	.000000498	.000000013	.000000000	0111010010010100
1011011010001	43	−.0005705	.0001994	.000184938	.000103461	.000051004	.000022151	.000008474	.000000440	.000000010	.000000000	0111010010110000
1011010100001	43	−.0005705	.0001979	.000184921	.000103424	.000050964	.000022122	.000008457	.000000438	.000000010	.000000000	0111011010100100
1011011100010	44	−.0005706	.0001659	.000184490	.000102475	.000049917	.000021317	.000007979	.000000383	.000000008	.000000000	1100110010010010x
1011010001001	43	−.0005739	.0002144	.000184514	.000103005	.000050679	.000021971	.000008391	.000000434	.000000010	.000000000	1011011001001000
1100010110010	45	−.0005740	.0001973	.000184287	.000102483	.000050082	.000021496	.000008100	.000000398	.000000010	.000000000	1011010111000000
1011010101010	45	−.0005741	.0001814	.000184072	.000102035	.000049612	.000021153	.000007906	.000000379	.000000008	.000000000	1010101010000100
1101010001100	45	−.0005778	.0002033	.000183684	.000101694	.000049423	.000021076	.000007884	.000000379	.000000008	.000000000	1100110100101000x
1101010000110	43	−.0005801	.0002491	.000183858	.000102407	.000050331	.000021820	.000008342	.000000434	.000000010	.000000000	0110111010010000
1100101000110	44	−.0005850	.0002491	.000182996	.000101171	.000049216	.000021057	.000007921	.000000390	.000000009	.000000000	1001110110100000
1100110100001	45	−.0005853	.0002653	.000183142	.000101550	.000049657	.000021408	.000008135	.000000416	.000000010	.000000000	0110110100110100
1101010100001	43	−.0005857	.0002537	.000182927	.000101112	.000049180	.000021040	.000007913	.000000390	.000000009	.000000000	1100011110010000
1101010100110	45	−.0005857	.0002363	.000182711	.000100639	.000048671	.000020658	.000007692	.000000366	.000000008	.000000000	1100110011010100
1101010100110	45	−.0005857	.0002360	.000182708	.000100632	.000048662	.000020651	.000007688	.000000366	.000000008	.000000000	1100101110100100
1111000000011	42	−.0005881	.0003778	.000184049	.000103930	.000052420	.000023598	.000009475	.000000585	.000000018	.000000000	0011111110000000
1011110000001	42	−.0005881	.0002865	.000182915	.000101428	.000049659	.000021459	.000008182	.000000424	.000000010	.000000000	0111110110000010
1111100100001	42	−.0005881	.0002861	.000182911	.000101419	.000049650	.000021453	.000008178	.000000424	.000000010	.000000000	0111010111011000
1111001011100	43	−.0005888	.0002978	.000182928	.000101540	.000049805	.000021576	.000008251	.000000431	.000000010	.000000000	1100010111010000
1100101001010	43	−.0005888	.0002822	.000182739	.000101137	.000049379	.000021261	.000008071	.000000412	.000000009	.000000000	0111011001110000
1101010101010	43	−.0005888	.0002801	.000182715	.000101087	.000049326	.000021222	.000008049	.000000410	.000000009	.000000000	0111010110010100
1010010110100	45	−.0005936	.0002697	.000181743	.000099595	.000047929	.000020247	.000007504	.000000353	.000000007	.000000000	1101010010100100x
1110000100001	43	−.0005967	.0003335	.000181985	.000100544	.000049108	.000021193	.000008077	.000000419	.000000010	.000000000	0111001110000000
1100110000110	44	−.0005999	.0003179	.000181235	.000099335	.000047951	.000020376	.000007618	.000000371	.000000008	.000000000	1001110110010100
1001110001010	44	−.0005999	.0002934	.000180934	.000098686	.000047259	.000019862	.000007324	.000000340	.000000007	.000000000	1010110001000100
1011010001010	44	−.0005999	.0002922	.000180920	.000098657	.000047228	.000019840	.000007311	.000000339	.000000007	.000000000	1010110110001100
1110010010100	45	−.0006006	.0003225	.000181165	.000099273	.000047913	.000020356	.000007609	.000000370	.000000008	.000000000	1100110110110000
1110010011010	45	−.0006033	.0003255	.000180735	.000098692	.000047413	.000020027	.000007434	.000000353	.000000007	.000000000	1100101001010000
1010100010010	45	−.0006033	.0003329	.000180810	.000098878	.000047627	.000020196	.000007536	.000000365	.000000007	.000000000	1011110011010100
1001101101000	44	−.0006034	.0003096	.000180503	.000098263	.000046970	.000019709	.000007256	.000000336	.000000007	.000000000	1110010100010100
1011010010100	44	−.0006034	.0003078	.000180524	.000098218	.000046924	.000019676	.000007238	.000000334	.000000007	.000000000	1010110100010010
1101010100001	44	−.0006086	.0003228	.000179775	.000097344	.000046242	.000019265	.000007037	.000000318	.000000007	.000000000	0111001110010000

212

m = 7, n = 7 Rank Orders 1551 thru 1600 of 1780 C(7, 7) = 3432 1/C(7, 7) = .00029138

z	w	c_1	c_2	d = .2	.4	.6	.8	1.0	1.5	2.0	3.0	z^{tc}
1101001110010	44	−.0006113	.0003441	.000179570	.000097248	.000046259	.000019320	.000007082	.000000325	.000000007	.000000000	1011000110100
1110100111010	44	−.0006113	.0003432	.000179559	.000097225	.000046236	.000019304	.000007073	.000000324	.000000007	.000000000	1011000110100
1110100011010	45	−.0006120	.0003719	.000179784 •	.000097792 •	.000046864	.000019774	.000007342	.000000351	.000000007	.000000000	1101000111000
1011100101010	43	−.0006176	.0003832	.000178945	.000096723	.000045987	.000019222	.000007060	.000000327	.000000006	.000000000	1010101101000
1011010010010	44	−.0006183	.0003785	.000178766	.000096435	.000045718	.000019039 •	.000006961 •	.000000318	.000000007	.000000000	1011011101000
1011010010010	44	−.0006209	.0003981	.000178540	.000096296	.000045690	.000019061 •	.000006988 •	.000000322	.000000007	.000000000	0101101101000
1011011000001	42	−.0006227	.0004690	.000179099 •	.000097726 •	.000047302	.000020296	.000007714	.000000403	.000000010	.000000000	0111110010100
1101011000001	42	−.0006227	.0004314	.000178648	.000096767	.000046288	.000019546	.000007283	.000000357	.000000008	.000000000	0111110010100
1101001010001	42	−.0006227	.0004293	.000178624	.000096718	.000046237	.000019509	.000007263	.000000355	.000000008	.000000000	0111101010100
1101010010001	42	−.0006253	.0004350	.000178231	.000096228	.000045845	.000019268	.000007142	.000000345	.000000008	.000000000	0111101010100
1011010101010	43	−.0006254	.0004027	.000177812	.000095336	.000044893	.000018559	.000006733	.000000301	.000000006	.000000000	1011010101010
1111010001010	44	−.0006261	.0004498	.000178250 •	.000096343	.000045969	.000019347	.000007178	.000000345	.000000007	.000000000	1011011110000
1011010001010	44	−.0006261	.0004152	.000177837	.000095473	.000045061	.000018688	.000006807	.000000308	.000000006	.000000000	1011010110000
1011010010010	44	−.0006261	.0004146	.000177831	.000095460	.000045047	.000018678	.000006802	.000000308	.000000006	.000000000	1100100110010
1011010010010	43	−.0006289	.0004198	.000177412	.000094932	.000044624	.000018419	.000006674	.000000297	.000000006	.000000000	1011011011000
1001110001100	44	−.0006292	.0004268	.000177447 •	.000095038	.000044742	.000018506	.000006722	.000000302	.000000008	.000000000	1100111000010x
1011110001001	42	−.0006323	.0004837	.000177595	.000095762	.000045658	.000019240	.000007162	.000000351	.000000007	.000000000	1001111001000
1011110001010	43	−.0006372	.0004862	.000176770	.000094622	.000044662	.000018578	.000006806	.000000316	.000000006	.000000000	1001111001000
1101010001100	44	−.0006379	.0004579	.000176311	.000093744	.000043779	.000017950	.000006459	.000000283	.000000005	.000000000	1001101010100
1011010001001	42	−.0006402	.0005237	.000176701 •	.000094864 •	.000045059	.000018925	.000007026	.000000343	.000000008	.000000000	0101101010100
1110100110001	42	−.0006402	.0005081	.000176518	.000094486	.000044671	.000018647	.000006871	.000000327	.000000007	.000000000	0111101010100
1101010100001	42	−.0006402	.0005060	.000176495	.000094438	.000044623	.000018613	.000006852	.000000326	.000000007	.000000000	0111101010100
1011101010001	41	−.0006403	.0005236	.000176681 •	.000094840	.000045043	.000018918	.000007023	.000000343	.000000008	.000000000	1100111011000
1011100100001	44	−.0006406	.0004789	.000176098	.000093630	.000043773	.000017985	.000006492	.000000287	.000000006	.000000000	1011001101010
1011010100001	44	−.0006437	.0005244	.000176108 •	.000094060	.000044378	.000018489	.000006800	.000000323	.000000007	.000000000	0111101101000
1010100110100	42	−.0006458	.0004943	.000175378	.000092776	.000043116	.000017596	.000006303	.000000324	.000000008	.000000000	1001101101000
1011010100010	42	−.0006489	.0005585	.000175604 •	.000093845	.000044163	.000018410	.000006780	.000000273	.000000006	.000000000	1011110010100
1101010100010	43	−.0006548	.0005645	.000174661	.000092389	.000043100	.000017723	.000006421	.000000290	.000000005	.000000000	1011110001100
1011010100010	44	−.0006548	.0005413	.000174385	.000091813	.000042503	.000017293	.000006181	.000000267	.000000005	.000000000	1011101010100
1011010010010	43	−.0006548	.0005394	.000174365	.000091771	.000042461	.000017264	.000006165	.000000266	.000000005	.000000000	1011110101000
1011000011100	44	−.0006555	.0005704	.000174606 •	.000092356 •	.000043090	.000017723	.000006422	.000000291	.000000006	.000000000	1100011101000
1110011001100	44	−.0006555	.0005539	.000174410	.000091943	.000042661	.000017412	.000006248	.000000273	.000000005	.000000000	1100101011000
1101011001010	44	−.0006555	.0005533	.000174403	.000091930	.000042844	.000017403	.000006243	.000000273	.000000005	.000000000	1100101101000
1101101010010	43	−.0006600	.0005563	.000173662	.000090955	.000041844	.000016903	.000005994	.000000253	.000000005	.000000000	1101001101000
1101000100010	44	−.0006633	.0005911	.000173487	.000090992 •	.000042014	.000017068	.000006097	.000000264	.000000005	.000000000	1001101011000
1011010110010	43	−.0006635	.0006177	.000173277	.000091608	.000042651	.000017526	.000006351	.000000289	.000000006	.000000000	1011100101010
1101011001010	43	−.0006635	.0005742	.000173272	.000090569	.000041593	.000016776	.000005940	.000000250	.000000005	.000000000	1011100100100
1101100100010	43	−.0006696	.0005736	.000173264	.000090555	.000041579	.000016767	.000005935	.000000250	.000000005	.000000000	1101010100100
1101001001010	43	−.0006731	.0006155	.000172688	.000090108	.000041373	.000016707	.000005930	.000000253 •	.000000005	.000000000	1011011010100
1101100111000	43	−	.0006323	.000172285	.000089698	.000041098	.000016563	.000005868	.000000249	.000000005	.000000000	1011101010100
1111000011101	41	−.0006775	.0007407	.000172789 •	.000091264 •	.000042888	.000017914	.000006644	.000000329	.000000008	.000000000	0101111001000
1101101001001	41	−.0006775	.0006860	.000172162	.000089989	.000041593	.000016992	.000006134	.000000279	.000000006	.000000000	0111110100100
1101101001001	41	−.0006775	.0006854	.000172155	.000089975	.000041579	.000016983	.000006129	.000000278	.000000006	.000000000	1011011011000
1011100101010	42	−.0006776	.0006540	.000171759	.000089158	.000040731	.000016368	.000005783	.000000243	.000000004	.000000000	1011010110010
1011010101010	43	−.0006783	.0006504	.000171598	.000088913	.000040513	.000016225	.000005709	.000000237	.000000004	.000000000	1011101010100
1011100010010	42	−.0006810	.0006596	.000171370	.000088683	.000040363	.000016150	.000005678	.000000236	.000000004	.000000000 •	1011101010010x
1111000010010	42	−.0006810	.0006903	.000171598	.000089219	.000040919	.000016542	.000005891	.000000255	.000000005	.000000000	1010101110000
1011011001010	43	−.0006734	.0006734	.000171406	.000088834	.000040537	.000016278	.000005749	.000000242	.000000004	.000000000	1011011101000
1101010010010	43	−.0006810	.0006716	.000171385	.000088794	.000040498	.000016251	.000005735	.000000241	.000000004	.000000000	1011010101000
1110000111000	44	−.0006817	.0007025	.000171620 •	.000089352 •	.000041088	.000016675	.000005968	.000000263	.000000005	.000000000	1110000111000x

Table A

m = 7, n = 7   Rank Orders 1601 thru 1650 of 1780   C(7, 7) = 3432   1/C(7, 7) = .00029138

z	w	c_1	c_2	d = .2	.4	.6	.8	1.0	1.5	2.0	3.0	z^{tc}
1101011000100	43	−.0006893	.0007018	.00170316	.00087593	.00039617	.00015751	.000005502	.00000225	.00000004	.000000000	1100111001100
1011111000001	40	−.0006917	.0007824	.00170830•	.00088926•	.00041067	.00016812	.000006097	.00000283	.00000006	.000000000	0111111001000
1110100001001	41	−.0006924	.0007789	.00170672	.00088691	.00040859	.00016677	.000006027	.00000277	.00000006	.000000000	0101111011000
1110110001001	41	−.0006924	.0007645	.00170508	.00088362	.00040530	.00016447	.000005879	.00000265	.00000005	.000000000	0111011011000
1101001010010	41	−.0006924	.0007616	.00170475	.00088298	.00040467	.00016403	.000005879	.00000263	.00000005	.000000000	1100110110000
1111010001001	43	−.0006928	.0007198	.00169926	.00087207	.00039366	.00015623	.000005448	.00000222	.00000004	.000000000	1100110011000
1110110010001	41	−.0006951	.0007368	.00170104	.00087858•	.00040130	.00016205	.000005783	.00000256	.00000005	.000000000	0111010101000
1101010101001	43	−.0006980	.0007368	.00169228	.00086406	.00038769	.00015280	.000005288	.00000210	.00000004	.000000000	1101010101010x
1111000010010	41	−.0007003	.0008049	.00169619•	.00087472•	.00039939	.00016137	.000005767	.00000257	.00000005	.000000000	1101010101000
1101100010010	43	−.0007006	.0007592	.00169029	.00086310	.00038773	.00015317	.000005318	.00000214	.00000004	.000000000	1101001100100
1110001100001	41	−.0007038	.0008249	.00169249•	.00087126	.00039724	.00016032	.000005725	.00000255	.00000005	.000000000	0111100111000
1001111000010	42	−.0007069	.0008258	.00168710	.00086392	.00039097	.00015626	.000005512	.00000235	.00000005	.000000000	1001111001000
1100111100010	42	−.0007069	.0008031	.00168450	.00085863	.00038564	.00015252	.000005309	.00000216	.00000004	.000000000	1011110001100
1011100001010	42	−.0007069	.0008010	.00168426	.00085818	.00038520	.00015223	.000005294	.00000215	.00000004	.000000000	1010111000100
1110100100001	43	−.0007077	.0007996	.00168291	.00085626	.00038356	.00015118	.000005241	.00000211	.00000004	.000000000	1100111011000
1110100001100	42	−.0007103	.0008213	.00168083	.00085511	.00038342	.00015142	.000005264	.00000214	.00000004	.000000000	1100101101000
1101101001001	42	−.0007148	.0008698	.00167861	.00085582•	.00038584	.00015371	.000005407	.00000229	.00000004	.000000000	1110011001000
1011010101010	43	−.0007148	.0008263	.00167371	.00084608	.00037623	.00014710	.000005055	.00000198	.00000003	.000000000	1101101011000x
1101010101010	42	−.0007148	.0008257	.00167364	.00084595	.00037610	.00014701	.000005050	.00000198	.00000003	.000000000	1101011011000
1110000011000	43	−.0007155	.0008769	.00167819•	.00085571	.00038584	.00015383	.000005414	.00000230	.00000004	.000000000	1101101101000
1110001101010	43	−.0007155	.0008402	.00167406	.00084750	.00037784	.00014825	.000005117	.00000204	.00000004	.000000000	1101010111000
1100100110010	43	−.0007155	.0008399	.00167402	.00084744	.00037778	.00014822	.000005115	.00000204	.00000004	.000000000	1101011101000
1011010010010	42	−.0007245	.0008451	.00166990	.00084241	.00037389	.00014594	.000005007	.00000196	.00000003	.000000000	1010110011000
1011101000010	42	−.0007245	.0008878	.00166419	.00083796	.00037176	.00014524	.000004994	.00000197•	.00000003	.000000000	1010111011000
1110010100001	40	−.0007285	.0009543	.00166285	.00084127•	.00037710	.00014969	.000005260	.00000225	.00000005	.000000000	0111101001000
1101100101010	42	−.0007297	.0009079	.00165756	.00083064	.00036648	.00014228	.000004859	.00000188	.00000003	.000000000	0111101101000
1011100100010	41	−.0007297	.0009299	.00165529	.00082923	.00036609	.00014236	.000004873	.00000191	.00000003	.000000000	1110011000010
1101010010010	43	−.0007325	.0009456	.00165602	.00083125	.00036822	.00014383	.000004951	.00000197	.00000003	.000000000	1101101011000x
1101000110010	42	−.0007331	.0009454	.00165580	.00083094	.00036795	.00014367	.000004943	.00000196	.00000003	.000000000	1101011011000
1110000011100	42	−.0007332	.0009276	.00165385	.00082715	.00036432	.00014123	.000004817	.00000186	.00000003	.000000000	0111001011000
1110100101010	42	−.0007332	.0009264	.00165371	.00082690	.00036408	.00014108	.000004809	.00000186	.00000003	.000000000	1010101101000
1101001001010	42	−.0007441	.0009881	.00164717	.00082259	.00036261	.00014098	.000004831	.00000190	.00000003	.000000000	1101010101000
1011010001010	41	−.0007441	.0009823	.00164140	.00081469	.00035609	.00013699	.000004637	.00000176	.00000003	.000000000	1100110101000
1110010000110	40	−.0007473	.0010764	.00164653•	.00082816	.00036399	.00014713	.000005188	.00000226	.00000005	.000000000	0111110101000
1011011000010	40	−.0007473	.0010443	.00164306	.00082150	.00036382	.00014285	.000004963	.00000207	.00000004	.000000000	0111011001000
1101100100110	42	−.0007473	.0010434	.00164296	.00082132	.00036382	.00014274	.000004958	.00000206	.00000004	.000000000	0111011011000
1010110010010	40	−.0007494	.0010012	.00163466	.00080721	.00035069	.00013398	.000004500	.00000167	.00000003	.000000000	1110011001000
1101000101010	40	−.0007525	.0010811	.00163825•	.00081768	.00035609	.00014213	.000004944	.00000207	.00000004	.000000000	1101010010100x
1110000100110	40	−.0007528	.0010208	.00163094	.00080370	.00034851	.00013292	.000004457	.00000165	.00000003	.000000000	0111011011000
1100100010110	40	−.0007551	.0010898	.00163472•	.00081367	.00035909	.00014044	.000004865	.00000201	.00000004	.000000000	0111101011000
1101100001100	42	−.0007590	.0010675	.00162562	.00079995	.00034698	.00013258	.000004461	.00000167	.00000003	.000000000	1100111001100
1111001001000	42	−.0007625	.0010866	.00162185	.00079634	.00034471	.00013146	.000004415	.00000165	.00000003	.000000000	1001111101000
1101010001010	41	−.0007670	.0011517	.00162133	.00080012•	.00034986	.00013544	.000004638	.00000185	.00000003	.000000000	0011111001000
1010111000010	41	−.0007670	.0011091	.00161668	.00079114	.00034126	.00012967	.000004338	.00000160	.00000003	.000000000	1011100010100
1010101010010	41	−.0007670	.0011081	.00161658	.00079096	.00034109	.00012957	.000004333	.00000160	.00000003	.000000000	1101010101000
1100101010010	42	−.0007677	.0011069	.00161526	.00078915	.00033958	.00012864	.000004288	.00000157	.00000003	.000000000	1101010101000
1110000101010	41	−.0007697	.0011157	.00161296	.00078682	.00033806	.00012787	.000004256	.00000155	.00000003	.000000000	1101011101000
1101000011100	42	−.0007697	.0011543	.00161543•	.00079219•	.00034322	.00013137	.000004435	.00000169	.00000003	.000000000	1100101111000
1110000110010	41	−.0007704	.0011543	.00161340	.00078834	.00033970	.00012900	.000004315	.00000160	.00000003	.000000000	1101011011000
1100100011100	42	−.0007704	.0011309	.00161309	—	—	—	—	—	—	—	—

Table A

				Rank Orders 1651 thru 1700 of 1780					C(7, 7) = 3432			1/C(7, 7) = .00029138

m = 7, n = 7

z	w	c₁	c₂	d = .2	.4	.6	.8	1.0	1.5	2.0	3.0	zᵗᶜ
1110110001010	41	−.0007767	.0011735	.00160743	.00078346	.00033713	.00012799	.000004284	.000000159●	.000000003	.000000000	1010111001000
1101110000010	39	−.0007811	.0012440	.00160771●	.00078869	.00034370	.00013295	.000004562	.000000186	.000000004	.000000000	0111110000100
1110010001010	41	−.0007846	.0012215	.00159935	.00077614	.00033273	.00012591	.000004203	.000000156	.000000003	.000000000	1010111011000
1110010101010	41	−.0007846	.0012037	.00159746	.00077259	.00032943	.00012377	.000004095	.000000147	.000000003	.000000000	1011011101000
1110010101010	41	−.0007846	.0012024	.00159733	.00077236	.00032921	.00012363	.000004089	.000000147	.000000003	.000000000	1101011110000
1011101001010	40	−.0007847	.0012181	.00159880●	.00077524	.00033194	.00012542	.000004179	.000000152	.000000002	.000000000	1011110100010x
1110010011100	42	−.0007853	.0012188	.00159788	.00077406	.00033099	.00012484	.000004151	.000000152	.000000003	.000000000	1110001101000
1110100011010	41	−.0007887	.0012237	.00159378	.00076916	.00032731	.00012275	.000004054	.000000145	.000000002	.000000000	1011001101000
1111000001010	41	−.0007933	.0012638	.00158931	.00076593	.00032589	.00012233	.000004048	.000000146●	.000000002	.000000000	1011111010000
1110110001100	41	−.0007963	.0012752	.00158544	.00076175	.00032295	.00012074	.000003978	.000000142	.000000002	.000000000	1101110000100
1110101000001	39	−.0007995	.0013368	.00158689●	.00076783	.00032987	.00012576	.000004252	.000000167	.000000003	.000000000	0111011001000
1101110010100	41	−.0008042	.0013037	.00157533	.00075068	.00031512	.00011648	.000003789	.000000130	.000000003	.000000000	1101010100100
1110010010001	39	−.0008073	.0014007	.00158043●	.00076343	.00032810	.00012534	.000004253	.000000169	.000000003	.000000000	0111011011000
1101110001001	39	−.0008073	.0013862	.00157894	.00076071	.00032561	.00012375	.000004173	.000000163	.000000003	.000000000	0111011101000
1110010011001	39	−.0008073	.0013850	.00157882	.00076050	.00032542	.00012362	.000004167	.000000162	.000000003	.000000000	0111101011000
1011101001100	41	−.0008077	.0013248	.00157176	.00074745	.00031320	.00011558	.000003754	.000000129	.000000002	.000000000	1011100110010x
1110110010100	41	−.0008139	.0013731	.00156656	.00074382	.00031171	.00011521	.000003754	.000000131●	.000000002	.000000000	1010111001000
1110011010100	41	−.0008191	.0013952	.00156020	.00073711	.00030708	.00011274	.000003646	.000000124●	.000000002	.000000000	1010111101000
1111010010010	40	−.0008219	.0014772	.00156415●	.00073711	.00031712	.00011948	.000003992	.000000151	.000000003	.000000000	1001110110010
1110110010010	40	−.0008219	.0014175	.00155793	.00073561	.00030654	.00011266	.000003650	.000000125	.000000002	.000000000	1011100011000
1110110100010	39	−.0008219	.0014172	.00155790	.00073556	.00030649	.00011264	.000003649	.000000125	.000000002	.000000000	1011101001000
1101110100100	41	−.0008226	.0014362	.00155869●	.00073764	.00030856	.00011399	.000003717	.000000130	.000000002	.000000000	1100110111000
1110010100001	39	−.0008226	.0014166	.00155666	.00073394	.00030519	.00011186	.000003612	.000000122	.000000002	.000000000	0111010101000
1011110010001	39	−.0008305	.0014160	.00155660	.00073383	.00030510	.00011181	.000003609	.000000122	.000000002	.000000000	1011010111000
1110010110001	41	−.0008305	.0014830	.00155049	.00073009	.00030395	.00011179	.000003629	.000000126●	.000000002	.000000000	1011100111000
1011101010001	41	−.0008361	.0015270	.00154578	.00072676	.00030251	.00011138	.000003624	.000000127●	.000000002	.000000000	1011110001010x
1011011010010	40	−.0008368	.0015278	.00154471	.00072551	.00030161	.00011088	.000003602	.000000125	.000000002	.000000000	1010111010000
1110011100100	40	−.0008368	.0015109	.00154297	.00072233	.00029873	.00010906	.000003512	.000000119	.000000002	.000000000	1010111100000
1110011001000	40	−.0008368	.0015090	.00154279	.00072201	.00029845	.00010889	.000003504	.000000118	.000000002	.000000000	0110111011000
1110101001000	41	−.0008375	.0015099	.00154170	.00072026	.00029739	.00010827	.000003475	.000000116	.000000002	.000000000	1011001011000x
1110101010001	40	−.0008394	.0015183	.00153936	.00071828	.00029584	.00010749	.000003444	.000000115	.000000002	.000000000	1101010101000
1101110011000	41	−.0008401	.0015347	.00153987●	.00071994	.00029742	.00010853	.000003495	.000000116	.000000002	.000000000	1100110011000
1110010011000	40	−.0008447	.0015603	.00153505	.00071529	.00029458	.00010714	.000003439	.000000116	.000000002	.000000000	0111010011000
1110011000100	41	−.0008481	.0015832	.00153166	.00071236	.00029290	.00010639	.000003411	.000000115	.000000002	.000000000	1011011001000
1011110100010	40	−.0008508	.0016503	.00153421●	.00071974	.00030056	.00011164	.000003686	.000000137	.000000002	.000000000	0111110001000
1101110100010	40	−.0008564	.0016177	.00152157	.00070179	.00028571	.00010260	.000003248	.000000105	.000000001	.000000000	1010111000100
1110101100010	40	−.0008591	.0016264	.00151810	.00069801	.00028308	.00010121	.000003188	.000000102	.000000001	.000000000	1010110110010
1111000010010	38	−.0008595	.0016993	.00152488●	.00071088	.00029498	.00010887	.000003571	.000000131	.000000002	.000000000	1010110110100x
1110101000100	40	−.0008622	.0017247	.00152304	.00070995	.00029483	.00010898	.000003582	.000000132	.000000002	.000000000	0111011010100
1110011010010	40	−.0008661	.0016902	.00151307	.00069535	.00028258	.00010148	.000003218	.000000105	.000000001	.000000000	1100101010100
1110101100001	40	−.0008740	.0017431	.00150548	.00068888	.00027892	.00009984	.000003158	.000000103	.000000001	.000000000	1100111001000
1101110110000	40	−.0008740	.0017235	.00150352	.00068541	.00027586	.00009797	.000003068	.000000097	.000000001	.000000000	1010110010000
1110100100010	40	−.0008740	.0017229	.00150346	.00068530	.00027578	.00009792	.000003066	.000000097	.000000001	.000000000	1011100010100
1011100110010	39	−.0008741	.0017369	.00150468●	.00068757	.00027781	.00009918	.000003127	.000000101	.000000001	.000000000	1011110000100
1101001100010	40	−.0008774	.0017458	.00150007	.00068239	.00027412	.00009691	.000003038	.000000096	.000000001	.000000000	1010101101000
1011011100010	40	−.0008827	.0017899	.00149593	.00067965	.00027303	.00009691	.000003037	.000000096	.000000001	.000000000	1010110010100
1110110001010	40	−.0008889	.0018222	.00148909	.00067303	.00026883	.00009483	.000002952	.000000092	.000000001	.000000000	1010111100010x
1110111000100	40	−.0008916	.0018791	.00149019	.00067736	.00027331	.00009778	.000003098	.000000102	.000000002	.000000000	1010101111000
1110101100010	39	−.0008916	.0018444	.00148679	.00067145	.00026818	.00009466	.000002951	.000000093	.000000001	.000000000	1011100101000
1101100100010	39	−.0008916	.0018439	.00148674	.00067136	.00026810	.00009462	.000002949	.000000093	.000000001	.000000000	1011110010000

Table A

m = 7, n = 7   Rank Orders 1701 thru 1750 of 1780   C(7, 7) = 3432   1/C(7, 7) = .00029138

z	w	c_1	c_2	d = .2	.4	.6	.8	1.0	1.5	2.0	3.0	z^{tc}
1110100011000	40	−.0008923	.0018448	.000148567	.000067006	.00026713	.00009406	.00002924	.00000091	.000000001	.000000000	1110101011010000
1110100100010	39	−.0008969	.0018876	.000148255	.000066854	.00026694	.00009430	.00002945	.00000093	.000000001	.000000000	1011011101010000
1110010100010	39	−.0008995	.0018983	.000147929	.000066513	.00026465	.00009311	.00002895	.00000090	.000000001	.000000000	1011101011010000
1110000111000	40	−.0009002	.0019160	.000147988	.000066673	.00026619	.00009409	.00002942	.00000108	.000000002	.000000000	1110011100010000
1110110000001	37	−.0009109	.0020334	.000147419	.000066624	.00026868	.00009660	.00003093	.00000107	.000000002	.000000001	0111101001000000
1101110100000	37	−.0009113	.0019593	.000146622	.000065245	.00025657	.00008910	.00002731	.00000082	.000000002	.000000001	0111101100100000
1111110000001	37	−.0009144	.0020580	.000147094	.000063353	.00026717	.00008803	.00003068	.00000107	.000000002	.000000001	0111101100100000x
1110110100010	38	−.0009255	.0020767	.000145471	.000064447	.00025310	.00008789	.00002709	.00000083	.000000001	.000000000	1100111011010000
1110111000010	39	−.0009262	.0020809	.000145400	.000064385	.00025276	.00008789	.00002704	.00000083	.000000001	.000000000	1100111011010000
1110011000100	39	−.0009262	.0020619	.000145216	.000064068	.00025004	.00008626	.00002629	.00000078	.000000001	.000000000	1011101011010000
1110110010100	39	−.0009262	.0020610	.000145207	.000064055	.00024993	.00008620	.00002626	.00000078	.000000001	.000000000	1101110111001000
1110101010100	39	−.0009288	.0020713	.000144879	.000063714	.00024767	.00008504	.00002578	.00000075	.000000001	.000000000	0111110101010000x
1110010101010	39	−.0009340	.0021172	.000144480	.000063461	.00024671	.00008482	.00002577	.00000075	.000000001	.000000000	1100111011010000
1110000110100	39	−.0009375	.0021418	.000144156	.000063194	.00024525	.00008419	.00002555	.00000072	.000000001	.000000000	1011101011010000
1110101011010	39	−.0009437	.0021756	.000143489	.000062565	.00024137	.00008328	.00002481	.00000074	.000000001	.000000000	1011110111001000
1110101101010	38	−.0009438	.0021880	.000143590	.000062748	.00024295	.00008328	.00002525	.00000071	.000000001	.000000000	1011100101010000x
1110110010010	39	−.0009472	.0021998	.000143162	.000062293	.00023987	.00008168	.00002458	.00000076	.000000001	.000000000	1011011011001000
1110011011000	38	−.0009517	.0022610	.000143018	.000062409	.00024183	.00008312	.00002531	.00000073	.000000001	.000000000	1011101011001000
1110110001100	38	−.0009517	.0022454	.000142872	.000062168	.00023984	.00008191	.00002479	.00000073	.000000001	.000000000	1011101011010000
1110100011010	38	−.0009517	.0022445	.000142864	.000062156	.00023974	.00008191	.00002477	.00000073	.000000001	.000000000	1011101011010000
1110100111000	39	−.0009524	.0022470	.000142773	.000062054	.00023900	.00008150	.00002459	.00000071	.000000001	.000000000	1100101011001000
1111010001000	38	−.0009635	.0023101	.000141606	.000060976	.00023246	.00007840	.00002338	.00000066	.000000001	.000000000	0111110110001000x
1111000000100	36	−.0009658	.0024107	.000142199	.000061177	.00024328	.00008507	.00002656	.00000088	.000000001	.000000000	1100111111001000
1110110000110	38	−.0009810	.0024653	.000140263	.000060106	.00022899	.00007747	.00002325	.00000067	.000000001	.000000000	1011101011010000
1110101100100	38	−.0009810	.0024286	.000139924	.000059548	.00022441	.00007484	.00002207	.00000061	.000000001	.000000000	1011110111001000
1110011010010	38	−.0009810	.0024283	.000139921	.000059544	.00022438	.00007482	.00002206	.00000061	.000000001	.000000000	1011101011010000x
1110100010010	38	−.0009862	.0024760	.000139534	.000059304	.00022350	.00007462	.00002205	.00000059	.000000001	.000000000	1011101011010000
1110101000110	38	−.0009889	.0024877	.000139222	.000058994	.00022151	.00007364	.00002166	.00000059	.000000001	.000000000	1101101011001000
1110111000100	37	−.0009952	.0025517	.000138816	.000058805	.00022127	.00007388	.00002187	.00000058	.000000001	.000000000	1011101111001000
1110110001010	38	−.0009959	.0025381	.000138580	.000058472	.00021869	.00007244	.00002124	.00000058	.000000001	.000000000	1011100011010000
1110101010010	38	−.0009986	.0025492	.000138262	.000058153	.00021664	.00007144	.00002083	.00000056	.000000001	.000000000	1101101010100100x
1110011000110	38	−.0010038	.0025983	.000137887	.000057931	.00021588	.00007129	.00002085	.00000056	.000000001	.000000000	1100111011001000
1110100100110	37	−.0010039	.0026096	.000137974	.000058085	.00021719	.00007206	.00002120	.00000058	.000000001	.000000000	1011101011010000
1110101000010	37	−.0010066	.0026381	.000137812	.000058016	.00021713	.00007217	.00002128	.00000059	.000000001	.000000000	1011110111001000
1110001101100	38	−.0010073	.0026241	.000137574	.000057682	.00021456	.00007074	.00002066	.00000056	.000000001	.000000000	1011101011010000
1111011000001	37	−.0010148	.0026808	.000136898	.000057144	.00021171	.00006955	.00002024	.00000054	.000000001	.000000000	1011111111001000
1111100000001	35	−.0010180	.0027951	.000137458	.000058307	.00022201	.00007577	.00002314	.00000073	.000000001	.000000000	1011110111001000100x
1110110100100	37	−.0010332	.0028036	.000135129	.000055640	.00020324	.00006581	.00001887	.00000049	.000000001	.000000000	1011101011001000
1110011001100	37	−.0010411	.0028825	.000134640	.000055356	.00020243	.00006575	.00001895	.00000050	.000000001	.000000000	1011101111001000
1110100011100	38	−.0010411	.0028660	.000134455	.000055128	.00020065	.00006477	.00001853	.00000048	.000000001	.000000000	1011100011010000
1110101001000	37	−.0010411	.0028655	.000134450	.000055121	.00020060	.00006475	.00001852	.00000047	.000000001	.000000000	1101101011010000
1110011010100	37	−.0010508	.0029307	.000133525	.000054339	.00019622	.00006282	.00001782	.00000045	.000000001	.000000000	1011110111001000
1110101100010	36	−.0010553	.0029938	.000133381	.000054428	.00019774	.00006389	.00001834	.00000048	.000000001	.000000000	1011100101010000
1110100101010	37	−.0010560	.0029815	.000133162	.000054130	.00019552	.00006269	.00001783	.00000045	.000000001	.000000000	1100111011001000
1110101010010	37	−.0010586	.0029940	.000132859	.000053839	.00019372	.00006183	.00001749	.00000044	.000000001	.000000000	1011101011010000
1111101000001	36	−.0010588	.0030221	.000133089	.000054209	.00019665	.00006346	.00001820	.00000047	.000000001	.000000000	1011110111001000100x
1110110110000	37	−.0010674	.0030728	.000132201	.000053406	.00019188	.00006125	.00001735	.00000044	.000000001	.000000000	1011101111001000
1110111010000	36	−.0010846	.0031982	.000130631	.000052136	.00018504	.00005834	.00001633	.00000040	.000000000	.000000000	1101101011001000
1110011100100	36	−.0010933	.0032617	.000129841	.00051500	.00018165	.00005692	.00001583	.00000038	.000000001	.000000000	1011101011010000

Table A

m = 7, n = 7   Rank Orders 1751 thru 1780 of 1780   C(7, 7) = 3432   1/C(7, 7) = .00029138

z	w	c_1	c_2	d = .2	.4	.6	.8	1.0	1.5	2.0	3.0	z^{tc}
1110101001000	36	−.0011030	.0033301	.000128945	.000050767	.000017766	.000005521	.000001523	.000000036	.000000000	.000000000	1110110100100x
1111010000010	35	−.0011101	.0034249	.000128653	.000050787	.000017899	.000005624	.000001573	.000000039	.000000000	.000000000	1011110100000
1111000011000	36	−.0011108	.0034138	.000128450	.000050520	.000017705	.000005522	.000001531	.000000037	.000000000	.000000000	1110011100000
1110011001000	36	−.0011108	.0033964	.000128303	.000050301	.000017541	.000005435	.000001495	.000000035	.000000000	.000000000	1110110100100x
1111010010000	36	−.0011187	.0033961	.000128301	.000050298	.000017538	.000005434	.000001495	.000000035	.000000000	.000000000	1110101010000
1110010110000	36	−.0011187	.0034631	.000127664	.000049840	.000017318	.000005350	.000001468	.000000034	.000000000	.000000000	1110101011000x
1110110000100	35	−.0011447	.0036767	.000125513	.000048250	.000016533	.000005044	.000001369	.000000031	.000000000	.000000000	1101110010000
1101001000100	35	−.0011482	.0037074	.000125240	.000048058	.000016443	.000005010	.000001358	.000000031	.000000000	.000000000	1011011010100
1101110001000	35	−.0011543	.0037486	.000124648	.000047562	.000016170	.000004893	.000001316	.000000029	.000000000	.000000000	1101110001000
1111100000010	34	−.0011623	.0038604	.000124359	.000047618	.000016329	.000005006	.000001370	.000000032	.000000000	.000000000	1011111000100x
1110101010000	35	−.0011630	.0038165	.000123895	.000046985	.000015875	.000004774	.000001276	.000000028	.000000000	.000000000	1110101010000
1111000101000	35	−.0011657	.0038486	.000123756	.000046938	.000015877	.000004784	.000001283	.000000028	.000000000	.000000000	1101011010000
1110100110000	35	−.0011709	.0038862	.000123280	.000046555	.000015675	.000004700	.000001254	.000000027	.000000000	.000000000	1110001110000
1111010010001	34	−.0011995	.0041410	.000121060	.000045017	.000014961	.000004438	.000001173	.000000025	.000000000	.000000000	1110101110000
1110110100000	34	−.0012144	.0042555	.000119761	.000044012	.000014444	.000004228	.000001103	.000000023	.000000000	.000000000	1101101100000
1111010100000	34	−.0012179	.0042880	.000119502	.000043839	.000014367	.000004201	.000001095	.000000023	.000000000	.000000000	1110101100000
1111010010010	34	−.0012231	.0043272	.000119040	.000043480	.000014182	.000004126	.000001070	.000000022	.000000000	.000000000	1101101110000
1111001100010	34	−.0012258	.0043608	.000118909	.000043440	.000014187	.000004136	.000001075	.000000022	.000000000	.000000000	1110101110000
1110111000100	34	−.0012517	.0046080	.000117015	.000042203	.000013645	.000003948	.000001021	.000000021	.000000000	.000000000	1101101110000
1111010001000	33	−.0012693	.0047455	.000115509	.000041059	.000013068	.000003719	.000000945	.000000019	.000000000	.000000000	1101110110000
1111011000100	33	−.0012745	.0047867	.000115062	.000040722	.000012899	.000003652	.000000923	.000000018	.000000000	.000000000	1111011010010100x
1111001100100	32	−.0012780	.0048208	.000114815	.000040564	.000012832	.000003629	.000000917	.000000018	.000000000	.000000000	1110101101000x
1111100000100	32	−.0013215	.0052370	.000111645	.000037985	.000011915	.000003307	.000000822	.000000015	.000000000	.000000000	1111101001000
1111000110000	32	−.0013294	.0052988	.000110973	.000037985	.000011668	.000003211	.000000791	.000000014	.000000000	.000000000	1111010010000x
1111001110000	32	−.0013328	.0053345	.000110736	.000037840	.000011607	.000003191	.000000785	.000000014	.000000000	.000000000	1111001100000
1111000100010	31	−.0013815	.0058112	.000107258	.000035602	.000010636	.000002855	.000000687	.000000012	.000000000	.000000000	1111010010100x
1111101000000	31	−.0013842	.0058310	.000107026	.000035428	.000010551	.000002822	.000000677	.000000012	.000000000	.000000000	1111010100100x
1111100100000	30	−.0014364	.0063624	.000103438	.000033200	.000009615	.000002508	.000000588	.000000010	.000000000	.000000000	1111011000000x
1111110000000	29	−.0014886	.0069118	.000099967	.000031108	.000008760	.000002227	.000000510	.000000008	.000000000	.000000000	1111101000000x
1111111000000	28	−.0015400	.0074775	.000096705	.000029223	.000008018	.000001992	.000000447	.000000007	.000000000	.000000000	1111110000000x

217

Tables B-1 through B-10.   Power tables

These tables contain exact values of the power of four nonparametric two-sample tests for location against the normal shift alternative, d:

    Tables B-1 and B-2: Wilcoxon test (Chapter 2, Section 3);
    Tables B-3 and B-4: normal scores test (Chapter 2, Section 4);
    Tables B-5 and B-6: median test (Chapter 2, Section 5);
    Tables B-7 and B-8: Kolmogorov-Smirnov test (Chapter 2, Section 6).

In addition, the power of the two-sample Student t test is given for comparison:

    Tables B-9 and B-10: Student t test (Chapter 2, Section 3).

Odd-numbered tables are for one-sided tests and even-numbered tables are for two-sided tests.

Values are given to 8 decimal places for all sample sizes $2 \leq n \leq m \leq 7$ that yield nontrivial results. If the sample of size m (or n) is from a normal population with mean μ_1 (or μ_2), $\mu_2 > \mu_1$, and variance σ^2, the shift alternative is $d = (\mu_2 - \mu_1)/\sigma$. Values are tabulated for $d = .2(.2)1.0,1.5,2.0,3.0$. Entries in the table are ordered according to increasing values of $m + n$, from $2 \leq m + n \leq 14$.

In Tables B-1 through B-8, the nominal levels of significance α are $\alpha = .25, .10, .05, .025, .01, .005$. The α's appearing in the tables are the attainable levels of significance nearest to but less than the nominal α's. Below each power value, on the line labeled t, is the power of the corresponding Student t test (same d, m, n, and attainable α).

In Tables B-9 and B-10, the power values are given for exactly the nominal α's.

For additional description of tests and tables, see Chapter 2.

Table B-1 **Power of one-sided two-sample Wilcoxon test vs. normal shift alternative d.**

m	n	α	d = .2	.4	.6	.8	1.0	1.5	2.0	3.0
3	1	.25000000	.30408113	.36269560	.42457845	.48823116	.55203144	.70186270	.82279295	.95637437
		t	.30449150	.36361864	.42609511	.49038893	.55483521	.70599835	.82740912	.95939012
2	2	.16666667	.21449982	.26918351	.32975727	.39480253	.46254547	.63069050	.77491573	.94286569
		t	.21464351	.26953440	.33038014	.39575540	.46387086	.63296654	.77780527	.94517173
4	1	.20000000	.24963115	.30500135	.36508824	.42852600	.49369886	.65286476	.78783898	.94531133
		t	.25032310	.30659396	.36776176	.43240605	.49883389	.66074251	.79690073	.95143652
3	2	.20000000	.26093064	.32992304	.40493202	.48327052	.56190350	.74083046	.87065102	.98101432
		t	.26248246	.33355963	.41106493	.49209458	.57333095	.75653193	.88562196	.98660576
3	2	.10000000	.13715522	.18246273	.23573967	.29619213	.36242771	.54013749	.70716641	.92051393
		t	.13764968	.18373401	.23810279	.29995788	.36785606	.55011471	.72034528	.93109575
5	1	.16666667	.21227386	.26431674	.32202364	.38421340	.44936488	.61355475	.75845157	.93530472
		t	.21313871	.26634481	.32548849	.38932530	.45623488	.62445464	.77133214	.94431785
4	2	.13333333	.18352688	.24379506	.31304421	.38922832	.46952386	.66669899	.82396356	.97163148
		t	.18473301	.24680609	.31843609	.39743953	.48074387	.68409691	.84235014	.97966390
4	2	.06666667	.05962001	.13348103	.17966062	.23432814	.29662075	.47430205	.65376746	.90079395
		t	.09669406	.13544603	.18346000	.24060300	.30596193	.49263260	.67909779	.92186603
3	3	.20000000	.26923889	.34847726	.43461849	.52362304	.61107987	.79802119	.91608846	.99264607
		t	.27168253	.35418233	.44409868	.53691720	.62767933	.81789939	.93158974	.99590951
3	3	.10000000	.14354009	.19803289	.26305310	.33706822	.41750902	.62412482	.79725992	.96662593
		t	.14434951	.20015750	.26703942	.34340772	.42652614	.63938177	.81455945	.97496723
3	3	.05000000	.07478896	.10770305	.14955634	.20057409	.26024891	.43720951	.62357041	.88989100
		t	.07524361	.10898187	.15213899	.20501563	.26711407	.45183760	.64520209	.90979992
6	1	.14285714	.18497832	.23393612	.28918808	.34974515	.41421617	.58099353	.73315697	.92614917
		t	.18594645	.23624214	.29318643	.35572692	.42236119	.59429777	.74925955	.93779281
5	2	.19047619	.25574643	.33085021	.41318427	.49921510	.58493750	.77379569	.89987551	.98959357
		t	.25900555	.33840256	.42568103	.51672004	.60083718	.80048835	.92143065	.99468927
5	2	.09523810	.13673482	.18888197	.25173848	.32311870	.41162190	.60648935	.78270364	.96219527
		t	.13783289	.19174898	.25679668	.33161108	.41367238	.62681649	.80575143	.97335340
5	2	.04761905	.07124868	.10272598	.14292084	.19216342	.25009137	.42386889	.61011394	.88311041
		t	.07209714	.10508679	.14763924	.20019622	.26238445	.44941709	.64690883	.91492479
4	3	.20000000	.27422919	.35975222	.45270407	.54804689	.64038613	.82955401	.93803876	.99635492
		t	.27786098	.36807910	.46619654	.56636393	.66236668	.85263006	.95309947	.99829787
4	3	.05714286	.08818756	.13005634	.18362191	.24868709	.32379285	.53572249	.73427219	.95187804
		t	.08879698	.13178295	.18710139	.25460181	.33274364	.55306579	.75622038	.96420451
4	3	.02857143	.04572295	.07006039	.10296429	.14539961	.19766839	.36566437	.55944910	.86310011
		t	.04618485	.07144500	.10592862	.15077637	.20639037	.38620864	.59213559	.89557396

219

Table B-1 **Power of one-sided two-sample Wilcoxon test vs. normal shift alternative d.**

m	n	α / t	d = .2	.4	.6	.8	1.0	1.5	2.0	3.0
7	1	.25000000	.31011727	.37573900	.44511493	.51615713	.58862730	.74667863	.86610908	.97686117
		t	.31169979	.37918801	.45057135	.52358377	.59579993	.75826070	.87672539	.98097672
6	2	.21428571	.28673146	.36683270	.45712940	.54728378	.63472019	.81695205	.92721879	.99424338
		t	.29006014	.37643573	.46947970	.56419881	.65532908	.84006531	.94399541	.99725518
		.07142857	.10616315	.15146150	.20774847	.27443110	.34979948	.55669415	.74615836	.95289806
		t	.10721653	.15431017	.21324982	.28342682	.36294126	.58034163	.77444105	.96766953
		.03571429	.05518162	.08197661	.11725866	.16174143	.21549021	.38378931	.57348251	.86705914
		t	.05607058	.08452682	.12250198	.17090392	.22985133	.41517524	.62039669	.90918495
5	3	.19642857	.27453128	.36523898	.46395425	.56446709	.66104273	.85142671	.95183378	.99796816
		t	.27791252	.37305271	.47662644	.58176676	.68129893	.87141610	.96360501	.99907719
		.07142857	.11105070	.16406632	.23084455	.31009222	.39877476	.63097574	.81981105	.97913442
		t	.11216717	.16715780	.23688505	.31996903	.41303760	.65472009	.84440135	.98721534
		.03571429	.05842612	.09093765	.13491776	.19119547	.25938639	.46718908	.68038068	.93731675
		t	.05897716	.09258893	.13842347	.19744954	.26928178	.48828947	.70916279	.95510447
		.01785714	.03019552	.04867975	.07495852	.11045937	.15609400	.31300578	.50829462	.83925962
		t	.03064967	.05011262	.07817364	.11654647	.16635979	.33921357	.55248834	.88576964
4	4	.24285714	.33267618	.43314353	.53795002	.63995538	.73262187	.89858204	.97260671	.99929383
		t	.33718101	.44303983	.55312697	.65921712	.75396457	.91596606	.98074656	.99971977
		.10000000	.15271391	.22088778	.30341154	.39698108	.49643738	.73102208	.89091904	.99239520
		t	.15439095	.22530359	.31156541	.40950194	.51331109	.75400771	.90963661	.99578362
		.02857143	.04819218	.07711919	.11731554	.17001998	.23530126	.44098141	.65973631	.93201156
		t	.04859831	.07837925	.12007854	.17509902	.24356358	.45969667	.68654734	.94978613
		.01428571	.02488926	.04116277	.06490506	.09769854	.14069827	.29308250	.48883124	.83037372
		t	.02520858	.04226833	.06746078	.10267197	.14930078	.31627733	.52970003	.87617395
7	2	.25000000	.32991054	.41822945	.51062497	.60215621	.68806047	.85640843	.94842739	.99677817
		t	.33389279	.42703639	.52444164	.62038702	.70940740	.87786682	.96220755	.99865560
		.05555556	.08502411	.12467199	.17542080	.23725963	.30904394	.51481376	.71359633	.94382950
		t	.08604505	.12751402	.18106134	.24672314	.32320651	.54169722	.74719558	.96249691
		.02777778	.04411074	.06722768	.09847644	.13885665	.18878582	.35105400	.54213099	.85235078
		t	.04499754	.06983994	.10398124	.14869790	.20453783	.38704822	.59776710	.90414290
6	3	.19047619	.27028476	.36376768	.46590574	.57004943	.66920395	.86136067	.95797368	.99854325
		t	.27386573	.37210562	.47946385	.58830394	.69067646	.88175320	.96920298	.99939865
		.08333333	.12952728	.19072434	.26668257	.35507490	.45158599	.69051352	.86565780	.98879063
		t	.13154874	.19612024	.27679853	.37086918	.47325838	.72186106	.89250855	.99443303
		.04761905	.07799793	.12094916	.17794527	.24899681	.33228424	.56716663	.77645437	.97131844
		t	.07891281	.12362845	.18346516	.25848499	.34664615	.59373266	.80652146	.98261021
		.02380952	.04089251	.06658693	.10299657	.15165565	.21305609	.41276282	.63394360	.92316984
		t	.04141183	.06821598	.10660509	.15835077	.22403881	.43805869	.67055892	.94748541

220

Table B-1 **Power of one-sided two-sample Wilcoxon test vs. normal shift alternative d.**

m	n	α	d = .2	.4	.6	.8	1.0	1.5	2.0	3.0
5	4	.20634921	.29460859	.39695801	.50678911	.61589523	.71627571	.89644120	.97418223	.99950932
		t	.29865287	.40613682	.52121401	.63450345	.73706343	.91311754	.98148928	.99979572
		.09523810	.14966185	.22142122	.30931844	.40939965	.51540073	.75899369	.91302545	.99587441
		t	.15173010	.22685288	.31924827	.42438287	.53509120	.78333615	.93014179	.99786446
		.03174603	.05514336	.09018877	.13921484	.20332793	.28181484	.51803344	.74378822	.96634848
		t	.05569493	.09191168	.14298077	.21017329	.29271763	.54053202	.77153443	.97798247
		.01587302	.02879235	.04926907	.07970342	.12217678	.17790925	.37037032	.59763631	.91265994
		t	.02911559	.05034455	.08221909	.12708498	.18634239	.39183280	.63127646	.93760353
		.00793651	.01481297	.02613929	.04370374	.06939151	.10488232	.24152600	.43371037	.80145386
		t	.01508419	.02708390	.04602534	.07416971	.11357945	.26761238	.48345073	.86228265
7	3	.19166667	.27428010	.37122224	.47690135	.58397261	.68484805	.87500917	.96502488	.99903613
		t	.27795849	.37974885	.49064899	.60225160	.70599524	.89393699	.97460641	.99961435
		.09166667	.14319513	.21112498	.29460052	.39028463	.49266884	.73433519	.89608812	.99343131
		t	.14535360	.21686075	.30524813	.40665835	.51468105	.76364504	.91890591	.99704212
		.03333333	.05786314	.09222076	.14095990	.20429388	.28150302	.51350084	.73685994	.96319345
		t	.05786939	.09465401	.14619153	.21366302	.29623551	.54316288	.77288361	.97818942
		.01666667	.02984193	.05051348	.08099153	.12326818	.17850577	.36861873	.59356225	.90952884
		t	.03030044	.05210744	.08465399	.13029752	.19040212	.39790592	.63817817	.94101480
		.00833333	.01536117	.02682575	.04446886	.07012065	.10541139	.24074999	.43111242	.79821359
		t	.01575883	.02818704	.04776204	.07680054	.11740674	.27565017	.49583120	.87306591
6	4	.23809524	.33703129	.44864551	.56433867	.67456170	.77112166	.92848375	.98542176	.99983629
		t	.34177805	.45904623	.57999957	.69377253	.79137541	.94209002	.99019207	.99994209
		.08571429	.13901283	.21105286	.30093341	.40451852	.51487664	.76712816	.92112750	.99696024
		t	.14104385	.21652809	.31113976	.42011855	.53550827	.79232754	.93797138	.99855466
		.03333333	.05923856	.09854468	.15385045	.22609089	.31378694	.56894160	.79390370	.98060438
		t	.05997624	.10084940	.15885449	.23540494	.32774361	.59532544	.82238373	.98866508
		.01904762	.03529900	.06129982	.09999153	.15360989	.22289381	.45031898	.69102388	.95516483
		t	.03577284	.06273137	.10331847	.16000554	.23363108	.47517955	.72467537	.97121522
		.00952381	.01836972	.03327097	.05677160	.09119533	.13870872	.31622054	.54535289	.89406082
		t	.01864467	.03424018	.05910298	.09609583	.14751573	.34087059	.58688015	.92758450
		.00476190	.00943043	.01757558	.03089671	.05135630	.08092233	.20326640	.38936920	.77555698
		t	.00965437	.01840061	.03303185	.05596335	.08967610	.23198250	.44784432	.85222784
5	5	.21031746	.30567405	.41628112	.53379902	.64821630	.75034588	.92074514	.98378819	.99982470
		t	.31008617	.42623597	.54918480	.66752743	.77111266	.93520809	.98894623	.99993663
		.07539683	.12504985	.19372362	.28121100	.38393943	.49519941	.75521794	.91717294	.99700714
		t	.12703510	.19914392	.29142496	.39968995	.51617204	.78102073	.93427580	.99849735
		.04761905	.08300919	.13501380	.20550792	.29373914	.39580473	.66442990	.86526622	.99234008
		t	.08399998	.13796836	.21159683	.30402494	.41083808	.68822842	.88563171	.99559460
		.01587302	.03017531	.05363124	.08933519	.13984391	.20634066	.43096201	.67609887	.95235319
		t	.03051620	.05479927	.09212021	.14532638	.21574878	.45380214	.70820665	.96846979

221

Table B-1 **Power of one-sided two-sample Wilcoxon test vs. normal shift alternative d.**

m	n	α	d = .2	.4	.6	.8	1.0	1.5	2.0	3.0
5	5	.00793651	.01568267	.02903507	.05048395	.08264280	.12771946	.30027435	.53027082	.88890352
		t	.01590300	.02983218	.05249851	.08686369	.13546172	.32332209	.56979941	.92244607
5	5	.00396825	.00804477	.01531453	.02744193	.04640347	.07424634	.19235802	.37663746	.76825953
		t	.00822603	.01599887	.02925343	.05039410	.08197445	.21876320	.43217128	.84443135
7	4	.20606061	.30168351	.41317139	.53199201	.64779841	.75106529	.92228858	.98451098	.99984330
		t	.30603357	.42303109	.54727228	.66699436	.77168964	.93651215	.98948556	.99994535
7	4	.08181818	.13560309	.20940956	.30238446	.40997628	.52444013	.78184617	.93148242	.99792959
		t	.13787549	.21506161	.31298026	.42616712	.54572098	.80679054	.94691712	.99906354
7	4	.03636364	.06536731	.10952257	.17151451	.25185103	.34804566	.61697084	.83494848	.98865413
		t	.06644435	.11282623	.17852036	.26403331	.36637981	.64798669	.86386510	.99419889
7	4	.02121212	.03995571	.07015888	.11514031	.17712817	.25627142	.50611832	.74920000	.97349847
		t	.04053365	.07207845	.11955614	.18547477	.26995660	.53511870	.78366367	.98478590
7	4	.00606061	.01232711	.02345277	.04183349	.07013570	.11081863	.27367020	.50082539	.87629020
		t	.01256535	.02433565	.04411503	.07501447	.11993510	.30137661	.55049113	.91933637
7	4	.00303030	.00631809	.01234863	.02267924	.03923919	.06413002	.17392025	.35287009	.75212534
		t	.00650540	.01307328	.02464107	.04365068	.07283319	.20479686	.41945331	.84476636
6	5	.21428571	.31547367	.43270582	.55606876	.67401839	.77652908	.93716474	.98919672	.99992979
		t	.32018249	.44322479	.57202517	.69350262	.79673465	.94952654	.99286062	.99997638
6	5	.08874459	.14805742	.22910653	.33012644	.44504817	.56445859	.81831049	.95033531	.99902346
		t	.15039354	.23538007	.34163172	.46213342	.58613665	.84088843	.96220908	.99956508
6	5	.04112554	.07463766	.12558836	.19654234	.28713089	.39326633	.67382254	.87736862	.99457520
		t	.07576214	.12900363	.20366307	.29921794	.41089602	.70052559	.89662083	.99710795
6	5	.01515152	.02997005	.05501653	.09398428	.14986406	.22382439	.47044267	.72468903	.97045840
		t	.03034534	.05633070	.09715916	.15614275	.23455667	.49530753	.75636225	.98174748
6	5	.00865801	.01776314	.03387517	.06020282	.09998828	.15567732	.36332036	.61687038	.93780041
		t	.01800602	.03477212	.06249583	.10480357	.16444921	.38755074	.65460376	.95949199
6	5	.00432900	.00920239	.01822469	.03370973	.05839195	.09499628	.24886767	.47470307	.86627547
		t	.00936646	.01886130	.03542600	.06220781	.10238641	.27312009	.52090709	.90982890
7	5	.21590909	.32166728	.44419029	.57217576	.69276472	.79529605	.94793086	.99224141	.99996661
		t	.32640120	.45473341	.58799871	.71173502	.81446495	.95866540	.99497008	.99998838
7	5	.07449495	.12944987	.20735616	.30745740	.42411271	.54750714	.81371034	.95117465	.99920140
		t	.13155338	.21320897	.31850524	.44088820	.56912989	.83648100	.96280932	.99964720
7	5	.03661616	.06887500	.11937244	.19128670	.28454518	.39478045	.68558158	.88949294	.99617181
		t	.06994125	.12269951	.19836206	.29670726	.41261608	.71212945	.90937100	.99802621
7	5	.02398990	.04692894	.08464305	.14119704	.21862845	.31553697	.60030042	.83565379	.99125537
		t	.04768591	.08714265	.14684024	.22896056	.33173140	.62922489	.86229074	.99534905
7	5	.00883838	.01872235	.03662503	.06633782	.11161469	.17503904	.40695176	.67300383	.96081125
		t	.01899943	.03766244	.06901587	.11723921	.18520730	.43359314	.71052083	.97622647
7	5	.00252525	.00572074	.01201575	.02346116	.04270472	.07268491	.20930213	.42824947	.84482325
		t	.00584860	.01254283	.02498385	.04622184	.07982414	.23517367	.48137010	.89944786

Table B-1 **Power of one-sided two-sample Wilcoxon test vs. normal shift alternative d.**

m	n	α	d = .2	.4	.6	.8	1.0	1.5	2.0	3.0
6	6	.24242424	.35560395	.48315083	.61232022	.72991495	.82619564	.96022199	.99480350	.99998411
6	6	t	.36076561	.49428054	.62843166	.74847714	.84415240	.96900914	.99674530	.99999527
6	6	.08982684	.15353472	.24145554	.35100989	.47434366	.59987572	.85134898	.96594852	.99961014
6	6	t	.15603811	.24819840	.36327926	.49222659	.62190685	.87168368	.97481368	.99983917
6	6	.04653680	.08605209	.14629707	.22953200	.33382941	.45246361	.74217628	.92087507	.99813806
6	6	t	.08743428	.15045185	.23800967	.34775412	.47189359	.76722616	.93665079	.99907818
6	6	.02056277	.04125129	.07609811	.12950550	.20407494	.29904454	.58530018	.82803185	.99099026
6	6	t	.04181485	.07802262	.13398535	.21250805	.31259827	.61079332	.85239070	.99483430
6	6	.00757576	.01639925	.03271839	.06033367	.10315599	.16410449	.39254253	.66151209	.95901729
6	6	t	.01662835	.03359762	.06264767	.10811211	.17322765	.41739305	.69759615	.97443739
6	6	.00432900	.00968650	.02001186	.03827689	.06798442	.11249187	.29776131	.55335276	.91997413
6	6	t	.00984292	.02064081	.04001686	.07191505	.12015059	.32220829	.59600291	.94839097
7	6	.22261072	.33630299	.46729815	.60181998	.72499000	.82558822	.96261924	.99562756	.99999063
7	6	t	.34136227	.47841482	.61807965	.74375730	.84362342	.97098698	.99729273	.99999734
7	6	.09032634	.15746594	.25087741	.36722097	.49702071	.62682739	.87423976	.97516584	.99981856
7	6	t	.16015915	.25814530	.38035139	.51585125	.64944829	.89308780	.98218310	.99993139
7	6	.03671329	.07150328	.12711383	.20707977	.31061270	.43141886	.73408264	.93218699	.99846485
7	6	t	.07269212	.13085271	.21500129	.32402126	.45055630	.75935886	.93696989	.99924449
7	6	.01748252	.03666343	.07022712	.12327775	.19910612	.27924054	.59556844	.84260861	.99365112
7	6	t	.03723962	.07224244	.12804881	.20817483	.31185095	.62233444	.86637266	.99645205
7	6	.00699301	.01580685	.03268486	.06201506	.10834051	.17495032	.42354750	.70334315	.97334156
7	6	t	.01603675	.03359130	.06444522	.11359810	.18464438	.44910827	.73746461	.98399874
7	6	.00407925	.00954764	.02047635	.04035779	.07333808	.12331609	.33084897	.60501009	.94676154
7	6	t	.00970406	.02112477	.04219141	.07753639	.13153740	.35646251	.64642380	.96723729
7	7	.22785548	.34815346	.48604729	.62551305	.75001670	.84816507	.97200042	.99735271	.99999693
7	7	t	.35339937	.49743661	.64179925	.76820080	.86489108	.97864176	.99841195	.99999919
7	7	.08245921	.14898521	.24401186	.36438549	.49969022	.63473838	.88561942	.98011228	.99990586
7	7	t	.15163043	.25129202	.37766860	.51874094	.65740136	.90326938	.98583584	.99996519
7	7	.04865967	.09431764	.16551245	.26424088	.38609466	.52006602	.81463611	.95835906	.99962949
7	7	t	.09598096	.17055888	.27442769	.40231932	.54158407	.83702850	.96841801	.99984280
7	7	.01893939	.04077051	.07947720	.14083514	.22789092	.33856982	.65578645	.88656723	.99734690
7	7	t	.04145551	.08187521	.14645747	.23836266	.35492489	.68222821	.90591276	.99859973
7	7	.00874126	.02020751	.04234255	.08069845	.14044307	.22422071	.51383256	.79239947	.99008539
7	7	t	.02051210	.04353172	.08381961	.14697804	.23574698	.53966089	.81956248	.99423665
7	7	.00349650	.00864648	.01942814	.03979122	.07454900	.12826653	.35382988	.64357796	.96361431
7	7	t	.00878688	.02003421	.04155967	.07868821	.13648155	.37926227	.68225181	.97810358

223

Table B-2 **Power of two-sided two-sample Wilcoxon test vs. normal shift alternative d.**

m	n	α	d = .2	.4	.6	.8	1.0	1.5	2.0	3.0
3	2	.20000000	.20779141	.23074858	.26765048	.31656128	.37497491	.54331879	.70780192	.92052591
		t	.20801832	.23164801	.26964058	.32000708	.38015588	.55320219	.72095621	.93110707
4	2	.13333333	.14066165	.16237316	.19764379	.24509596	.30281746	.47560365	.65397722	.90079632
		t	.14103443	.16386536	.20099587	.25101045	.31189799	.49385204	.67929014	.92186812
3	3	.20000000	.21039553	.24086475	.28931537	.35246136	.42612477	.62576519	.79748973	.96662766
		t	.21079887	.24245507	.29280011	.35840281	.43486038	.64094087	.81477418	.97496879
3	3	.10000000	.10692590	.12753479	.16129173	.20722544	.26385577	.43784911	.62365475	.88989158
		t	.10717943	.12856370	.16365200	.21149914	.27060768	.45244777	.64528144	.90980045
5	2	.19047619	.20046793	.22980646	.27662459	.33795904	.40998603	.60812258	.78294124	.96219728
		t	.20100744	.23193240	.28127901	.34588721	.42163219	.62832819	.80596529	.97335512
5	2	.09523810	.10187743	.12166246	.15416186	.19856368	.25358325	.42450228	.61020039	.88311107
		t	.10234166	.12353964	.15844392	.20626224	.26564714	.44998809	.64698409	.91492533
4	3	.22857143	.24141232	.27876939	.33728940	.41185152	.49622287	.70859580	.86692529	.98638295
		t	.24216254	.28168217	.34351116	.42209059	.51061395	.73004192	.88674886	.99180737
4	3	.05714286	.06279054	.07979248	.10825393	.14813656	.19901486	.36584636	.55946664	.86310017
		t	.06307547	.08097417	.11105263	.15339933	.20766715	.38637685	.59215143	.89557401
7	1	.25000000	.25696759	.27751287	.31058916	.35453670	.40720397	.56038335	.71288467	.91777544
		t	.25735194	.27902173	.31387794	.36012266	.41541969	.57516021	.73154068	.93179788
6	2	.14285714	.15222937	.17990295	.22453956	.28390028	.35489565	.55757194	.74626940	.95288874
		t	.15278618	.18211903	.22946631	.29245507	.36773604	.58113970	.77453856	.96767009
6	2	.07142857	.07737682	.09520518	.12481150	.16586803	.21764565	.38413636	.57352394	.86705937
		t	.07788961	.09729926	.12965928	.17473995	.23181615	.41547631	.62043092	.90918512
5	3	.25000000	.26444287	.30624039	.37102850	.45226821	.54223672	.75726257	.90209567	.99297174
		t	.26567503	.31095090	.38083919	.46786779	.56334352	.78448435	.92308628	.99656829
5	3	.07142857	.07916102	.10235340	.14086924	.19412968	.26075279	.46734550	.68039294	.93731678
		t	.07950576	.10377355	.14419201	.20026257	.27057738	.48843379	.70917381	.95510449
5	3	.03571429	.04024663	.05405067	.07768673	.11177122	.15669090	.31307074	.50829950	.83925963
		t	.04054574	.05532182	.08077487	.11777713	.16691079	.33927123	.55249253	.88576964
4	4	.20000000	.21459018	.25699244	.32324223	.40721779	.50139672	.73163108	.89096817	.99239529
		t	.21547798	.26043124	.33054961	.41913060	.51788864	.75454152	.90967732	.99578369
4	4	.05714286	.06423327	.08563356	.12158151	.17203468	.23619696	.44107133	.65974235	.93201156
		t	.06449867	.08864257	.12453029	.17704166	.24441932	.45978068	.68655287	.94978614
4	4	.02857143	.03265717	.04518500	.06686995	.09860505	.14109268	.29312029	.48883369	.83037373
		t	.03288872	.04617928	.06934446	.10352891	.14966852	.31631143	.52970217	.87617396

Table B-2 **Power of two-sided two-sample Wilcoxon test vs. normal shift alternative d.**

m	n	α	d = .2	.4	.6	.8	1.0	1.5	2.0	3.0
7	2	.22222222	.23390769	.26801924	.32182087	.39107846	.47053144	.67720615	.84128788	.98002665
		t	.23501093	.27229793	.33094888	.40609048	.49164565	.70904130	.87161454	.98936018
7	2	.05555556	.06087670	.07691493	.10382841	.14166094	.19020801	.35126127	.54215313	.85235088
		t	.06140956	.07911150	.10898474	.15127518	.20580394	.38722148	.59778448	.90414297
6	3	.16666667	.18021161	.21981420	.28241310	.36307861	.45541315	.69097181	.86569436	.98879070
		t	.18130284	.22406836	.29154643	.37817031	.47664552	.72195237	.89253516	.99443307
6	3	.09523810	.10546743	.13589559	.18560413	.25268763	.33395482	.56734042	.77646633	.97131846
		t	.10603109	.13818051	.19081327	.26197225	.34819989	.59388787	.80653171	.98261022
6	3	.04761905	.05401281	.07341850	.10635268	.15320910	.21373273	.41282728	.63394770	.92316985
		t	.05435649	.07486149	.10982244	.15981793	.22466830	.43811644	.67056247	.94748541
6	3	.02380952	.02746025	.03871538	.05839047	.08755373	.12715713	.27267087	.46631517	.81777278
		t	.02775722	.03999727	.06160380	.09399770	.13846465	.30369138	.52132688	.87861247
5	4	.19047619	.20657737	.25329580	.32601644	.41756968	.51912946	.75937810	.91305064	.99587443
		t	.20768798	.25756370	.33496309	.43186866	.53840627	.78364833	.93016003	.99786448
5	4	.06349206	.07231583	.09890051	.14335028	.20516511	.28257631	.51809526	.74379140	.96634848
		t	.07268401	.10043328	.14698093	.21192814	.29343617	.54058861	.77153725	.97798247
5	4	.03174603	.03703121	.05328802	.08154276	.12296537	.17822550	.37039417	.59763746	.91265994
		t	.03725864	.05426917	.08399341	.12783639	.18664008	.39185459	.63127748	.93760353
5	4	.01587302	.01882511	.02804939	.04455860	.06975056	.10502363	.24153624	.43371085	.80145386
		t	.01902220	.02892360	.04683324	.07450273	.11370812	.26762130	.48345113	.86228265
7	3	.18333333	.19844926	.24242224	.31122993	.39856067	.49652165	.73475766	.89611803	.99343135
		t	.19962155	.24696169	.32086543	.41422687	.51810297	.76398994	.91892800	.99704215
7	3	.06666667	.07540999	.10171048	.14557009	.20639379	.28239867	.51358009	.73686449	.96319345
		t	.07592784	.10385011	.15058529	.21562793	.29705835	.54323233	.77288741	.97818942
7	3	.03333333	.03862594	.05487504	.08302875	.12416205	.17887372	.36864876	.59356386	.90952884
		t	.03896304	.05631612	.08658286	.13112755	.19073712	.39793193	.63817950	.94101480
7	3	.01666667	.01963615	.02889600	.04541375	.07052650	.10557520	.24076280	.43111308	.79821359
		t	.01992073	.03014742	.04863180	.07716356	.11754912	.27566055	.49583170	.87306591
6	4	.17142857	.18835147	.23750360	.31411415	.41061331	.51748801	.76735127	.92113908	.99696025
		t	.18950796	.24196449	.32350827	.42568132	.53781908	.79250773	.93797975	.99855466
6	4	.06666667	.07675521	.10712350	.15775895	.22774470	.31443589	.56898622	.79390556	.98060438
		t	.07725312	.10918591	.16259263	.23660485	.32834348	.59536489	.82238531	.98866508
6	4	.03809524	.04490985	.06582548	.10197687	.15441995	.22320078	.45033845	.69102464	.95516483
		t	.04521556	.06713832	.10522472	.16077226	.23391745	.47519707	.72467603	.97121522
6	4	.01904762	.02299661	.03537331	.05760342	.09154794	.13883847	.31622825	.54535317	.89406082
		t	.02319994	.03627690	.05995290	.09642636	.14763536	.34087741	.58688039	.92758450
6	4	.00952381	.01168840	.01827896	.03131376	.05151815	.08098088	.20326975	.38936932	.77555698
		t	.01185841	.01935624	.03341938	.05611009	.08972792	.23198531	.44784441	.85222784

225

Table B-2 **Power of two-sided two-sample Wilcoxon test vs. normal shift alternative d.**

m	n	α	d = .2	.4	.6	.8	1.0	1.5	2.0	3.0
5	5	.22222222	.24158665	.29717054	.38183919	.48501857	.59458959	.82866919	.95144888	.99888407
5	5	t	.24297750	.30241117	.39247194	.50125735	.61524639	.84996548	.96281598	.99947681
5	5	.09523810	.10907464	.14761603	.21129701	.30633105	.39677024	.66469513	.86528883	.99234008
5	5	t		.15020720	.21712416		.41172571	.68828531	.88563386	.99559460
5	5	.03174603	.03796420	.05718938	.09084562	.14043872	.20655762	.43097436	.67609929	.95235319
5	5	t	.03821271	.05827180	.09357565	.14589220	.21595252	.45381338	.70820703	.96846979
5	5	.01587302	.01943766	.03069268	.05116540	.08290324	.12781184	.30067930	.53027098	.88890352
5	5	t	.01960384	.03144175	.05315034	.08710910	.13554749	.32332653	.56979955	.92244607
5	5	.00793651	.00987867	.01610688	.02776134	.04652338	.07428819	.19236018	.37663753	.76825953
5	5	t	.01001862	.01675578	.02955164	.05050354	.08201181	.21876503	.43217134	.84443135
7	4	.23030303	.25037106	.30730106	.39486204	.50099368	.61067867	.84143611	.95725061	.99913210
7	4	t	.25188553	.31351720	.40635287	.51749619	.63257696	.86317621	.96826589	.99963307
7	4	.07272727	.08417698	.11855073	.17552813	.25350114	.34867217	.61700984	.83494992	.98865413
7	4	t	.08489629	.12149035	.18227848	.26553689	.36693395	.64801835	.86386616	.99419889
7	4	.04242424	.05042917	.07495838	.11717778	.17792810	.25656442	.50613437	.74920053	.97349847
7	4	t	.05084267	.07672281	.12149269	.18622092	.27022195	.53513268	.78366411	.98478592
7	4	.02424242	.02950848	.04596657	.07532869	.11962286	.18019000	.39531264	.64400985	.94371248
7	4	t	.02977465	.04714277	.07835347	.12579411	.19104575	.42296754	.68437386	.96499739
7	4	.00606061	.00767746	.01291776	.02290118	.03931966	.06415710	.17392152	.35287013	.75212534
7	4	t	.00782434	.01360875	.02484351	.04372184	.07285641	.20479787	.41945334	.84476636
6	5	.24675325	.26865221	.33094188	.42411462	.53460223	.64774671	.87029132	.96972761	.99959721
6	5	t	.27028664	.33700958	.43612513	.55232325	.66931379	.88937862	.97797406	.99983637
6	5	.08225108	.09559028	.13543606	.20080364	.28882571	.39388499	.67385649	.87736969	.99457520
6	5	t	.09634175	.13847592	.20766556	.30076797	.41144536	.70055324	.89862161	.99710796
6	5	.03030303	.03706159	.05808267	.09520649	.15031242	.22397551	.47044948	.72468920	.97045840
6	5	t	.03734199	.05931296	.09833044	.15656695	.23469712	.49531362	.75636241	.98174748
6	5	.01731602	.02167659	.03551204	.06083513	.10021346	.15575112	.36332349	.61687046	.93780041
6	5	t	.02186373	.03636186	.06310068	.10501569	.16451768	.38755353	.65460382	.95949199
6	5	.00865801	.01109520	.01899225	.03399781	.05849184	.09502822	.24886895	.47470310	.86627547
6	5	t	.01122480	.01960084	.03569834	.06230046	.10241548	.27312120	.52090711	.90982890
6	5	.00432900	.00563711	.00994101	.01834530	.03256154	.05467304	.15692874	.33155579	.73871176
6	5	t	.00574187	.01044725	.01981768	.03601746	.06173307	.18384981	.39298261	.83104742
7	5	.20202020	.22433704	.28816914	.38460362	.50038604	.62046818	.86044986	.96821762	.99962765
7	5	t	.22592787	.29413385	.39657897	.51833730	.64264670	.88053936	.97683735	.99985210
7	5	.07323232	.08675945	.12737804	.19455393	.28576987	.39519758	.68560025	.88949340	.99617181
7	5	t	.08749784	.13039091	.20143456	.29782449	.41298532	.71214458	.90937133	.99802621
7	5	.04797980	.05820596	.08950640	.14311739	.21932151	.31576525	.60030987	.83565401	.99125537
7	5	t	.05875319	.09181436	.14864121	.22959356	.33193395	.62923260	.86229090	.99534905
7	5	.01767677	.02256567	.03815904	.06690082	.11180360	.17509696	.40695377	.67300387	.96081125
7	5	t	.02278252	.03914940	.06955160	.11741584	.18526051	.43359491	.71052086	.97622647

Table B-2 Power of two-sided two-sample Wilcoxon test vs. normal shift alternative d.

m	n	α	d = .2	.4	.6	.8	1.0	1.5	2.0	3.0
7	5	.00505051	.00675168	.01240412	.02359586	.04274765	.07269746	.20930252	.42824947	.84482325
		t	.00685609	.01291351	.02508939	.04626091	.07983529	.23517400	.48137011	.89944786
7	5	.00252525	.00342925	.00647847	.01267696	.02364888	.04149597	.13048108	.29506986	.71212563
		t	.00351025	.00688533	.01391830	.02670625	.04802994	.15788153	.36214760	.82097846
6	6	.24025974	.26437075	.33265597	.43386168	.55209259	.67056121	.89112084	.97846333	.99982860
		t	.26614365	.33918811	.44661739	.57052921	.69236461	.90837449	.98472785	.99993471
6	6	.09307359	.10911876	.15674899	.23385204	.33545525	.45301992	.74220121	.92087568	.99813806
		t	.11004820	.16045837	.24203279	.34922104	.47237789	.76724564	.93665120	.99907818
6	6	.04112554	.05065040	.08002802	.13100526	.20459633	.29920937	.58530623	.82803197	.99099026
		t	.05106937	.08182607	.13540999	.21299355	.31274857	.61079853	.85239080	.99483430
6	6	.01515152	.01961651	.03397152	.06078039	.10330146	.16414768	.39254391	.66151212	.95901729
		t	.01979829	.03481333	.06307418	.10824877	.17326757	.41739426	.69759617	.97443739
6	6	.00865801	.01146830	.02068572	.03851055	.06805856	.11251336	.29776195	.55335277	.91997413
		t	.01159502	.02129206	.04023874	.07198423	.12017029	.32220886	.59600292	.94839097
6	6	.00432900	.00586718	.01101860	.02134786	.03927323	.06768905	.20020308	.41749467	.84001667
		t	.00595598	.01145775	.02265719	.04240575	.07415569	.22446450	.46869486	.89471040
7	6	.23426573	.26022368	.33349050	.44128596	.56569201	.68812591	.90590304	.98381838	.99991648
		t	.26211294	.34041022	.45465342	.58469300	.71007395	.92172819	.98878583	.99997009
7	6	.07342657	.08865308	.13438235	.20988842	.31157937	.43172115	.73409309	.92138717	.99846485
		t	.08949427	.13780183	.21759264	.32489085	.45081847	.75936697	.93697002	.99924449
7	6	.03496503	.04425546	.07322219	.12434879	.19945265	.29734181	.59557141	.84260865	.99365112
		t	.04469363	.07512205	.12905467	.20849195	.31194104	.62233687	.86637269	.99645205
7	6	.02214452	.02878942	.04995528	.08873741	.14846533	.23056377	.50842220	.77817765	.98632263
		t	.02906620	.05120848	.09203233	.15524022	.24233194	.53469245	.80721733	.99210110
7	6	.00815851	.01114038	.02104327	.04054134	.07339202	.12333045	.33084930	.60501009	.94676154
		t	.01126956	.02167208	.04236637	.07758658	.13155051	.35646280	.64642381	.96723729
7	6	.00466200	.00651003	.01277065	.02552159	.04792529	.08362646	.24749589	.49897042	.90211892
		t	.00660183	.01323186	.02697731	.05129310	.09058012	.27259454	.54714106	.93832198
7	7	.20862471	.23603260	.31348838	.42758450	.55914488	.68795570	.91178175	.98646153	.99995238
		t	.23794566	.32051085	.44118836	.57841293	.71004061	.92688243	.99071508	.99998364
7	7	.09731935	.11695550	.17497891	.26778968	.38728475	.52042241	.81464668	.95835921	.99962949
		t	.11811523	.17956318	.27769478	.40337429	.54188668	.83703637	.96841810	.99984279
7	7	.03787879	.04871216	.08247501	.14185151	.22819981	.33865384	.65578843	.88656725	.99734690
		t	.04923841	.08474441	.14740425	.23864176	.35499826	.68222977	.90591277	.99859973
7	7	.01748252	.02362453	.04354641	.08107980	.14055145	.22424830	.51383311	.79239947	.99008539
		t	.02387191	.04469358	.08418030	.14707842	.23577196	.53966136	.81956249	.99423665
7	7	.00699301	.00992754	.01985229	.03991782	.07458300	.12827473	.35383003	.64357796	.96361431
		t	.01004601	.02044361	.04167961	.07872081	.13648902	.37926240	.68225181	.97810358
7	7	.00407925	.00593280	.01233694	.02575391	.04996050	.08930542	.27216377	.54535276	.93190580
		t	.00601694	.01277160	.02710867	.05330758	.09632289	.29748001	.59132330	.95840062

Table B-3 **Power of one-sided two-sample normal scores test vs. normal shift alternative d.**

m	n	α	d = .2	.4	.6	.8	1.0	1.5	2.0	3.0
3	1	.25000000	.30408113	.36269560	.42457845	.48823116	.55203144	.70186270	.82279295	.95637437
		t	.30449150	.36361864	.42609511	.49038893	.55483521	.70599835	.82740912	.95939012
2	2	.16666667	.21449982	.26918351	.32975727	.39480253	.46254547	.63069050	.77491573	.94286569
		t	.21464351	.26953440	.33038014	.39575540	.46387086	.63296654	.77780527	.94517173
4	1	.20000000	.24963115	.30500135	.36508824	.42852600	.49369886	.65286476	.78783898	.94531133
		t	.25032310	.30659396	.36776176	.43240605	.49883389	.66074251	.79690073	.95143652
3	2	.20000000	.26093064	.32992304	.40493202	.48327052	.56190350	.74083046	.87065102	.98101432
		t	.26248246	.33355963	.41106493	.49209458	.57333095	.75653193	.88562196	.98660576
3	2	.10000000	.13715522	.18246273	.23573967	.29619213	.36242771	.54013749	.70716641	.92051393
		t	.13764968	.18373401	.23810279	.29995788	.36785606	.55011471	.72034528	.93109575
5	1	.16666667	.21227386	.26431674	.32202364	.38421340	.44936488	.61355475	.75845157	.93530472
		t	.21313871	.26634481	.32548849	.38932530	.45623488	.62445464	.77133214	.94431785
4	2	.20000000	.26418818	.33711145	.41632503	.49864904	.58055973	.76250361	.88808730	.98574325
		t	.26726595	.34429426	.42833009	.51568573	.60221910	.79036365	.91230560	.99278551
4	2	.06666667	.09596201	.13348103	.17966062	.23432814	.29662075	.47430205	.65376746	.90079395
		t	.09669406	.13544603	.18346000	.24060300	.30596193	.49263260	.67909779	.92186603
3	3	.25000000	.32666919	.41100762	.49918048	.58685909	.66985969	.83707671	.93498265	.99440704
		t	.33114999	.42109142	.51532481	.60866268	.69607119	.86561145	.95523862	.99801083
3	3	.10000000	.14354009	.19803289	.26305310	.33706822	.41750902	.62412482	.79725992	.96662593
		t	.14434951	.20015750	.26703942	.34340772	.42652614	.63938177	.81455945	.97496723
3	3	.05000000	.07478896	.10770305	.14955634	.20057409	.26024891	.43720951	.62357041	.88989100
		t	.07524361	.10898187	.15213899	.20501563	.26711407	.45183760	.64520209	.90979992
6	1	.14285714	.18497832	.23393612	.28918808	.34974515	.41421617	.58099353	.73315697	.92614917
		t	.18594645	.23624214	.29318643	.35572692	.42236119	.59429777	.74925955	.93779281
5	2	.23809524	.31248781	.39523441	.48279148	.57095812	.65547682	.82906602	.93266662	.99489828
		t	.31574711	.40249876	.49432614	.58642594	.67395670	.84891294	.94654149	.99721078
5	2	.09523810	.13673482	.18888197	.25143848	.32311870	.40162190	.60648935	.78270364	.96219527
		t	.13783269	.19174898	.25679568	.33161108	.41367238	.62681649	.80575143	.97335340
5	2	.04761905	.07124868	.10272598	.14292084	.19216342	.25009137	.42386889	.61011394	.88311041
		t	.07209714	.10508679	.14763924	.20019622	.26238445	.44941709	.64690883	.91492479
4	3	.22857143	.30809637	.39758223	.49254490	.58762857	.67750452	.85421214	.94962780	.99730740
		t	.31237841	.40718740	.50775554	.60778951	.70110330	.87741136	.96375607	.99890013
4	3	.08571429	.12768593	.18176380	.24777409	.32418472	.40814978	.62502135	.80420989	.97061997
		t	.12933929	.18620115	.25622813	.33774454	.42747429	.65698419	.83837373	.98401739
4	3	.02857143	.04572295	.07006039	.10296429	.14539961	.19766839	.36566437	.55944910	.86310011
		t	.04618485	.07144500	.10592862	.15077637	.20639037	.38620864	.59213559	.89557396

228

Table B-3 **Power of one-sided two-sample normal scores test vs. normal shift alternative d.**

m	n	α	d = .2	.4	.6	.8	1.0	1.5	2.0	3.0
7	1	.25000000	.31011727	.37573900	.44511493	.51615713	.58662730	.74667863	.86610908	.97686117
		t	.31169979	.37918801	.45057135	.52358377	.59579993	.75826070	.87672539	.98097672
6	2	.25000000	.32881559	.41592294	.50719288	.59788399	.68337346	.82258181	.94665837	.99678993
		t	.33203507	.42302560	.51831871	.61255331	.70054819	.86986689	.95775845	.99826838
6	2	.07142857	.10616315	.15146150	.20774847	.27443110	.34979948	.55669415	.74615836	.95289806
		t	.10721653	.15431017	.21324982	.28342682	.36294126	.58034163	.77444105	.96766953
6	2	.03571429	.05518162	.08197661	.11725866	.16174143	.21549021	.38378931	.57348251	.86705914
		t	.05607058	.08452682	.12250198	.17090392	.22985133	.41517524	.62039669	.90918495
5	3	.23214286	.31733818	.41322538	.51428577	.61408413	.70648273	.87917660	.96322050	.99857418
		t	.32125271	.42204491	.52821880	.63238441	.72757033	.89856425	.97386295	.99947050
5	3	.08928571	.13588961	.19658449	.27094788	.35672704	.44996174	.68152726	.85552398	.98626274
		t	.13752068	.20092981	.27910487	.36952153	.46765787	.70778278	.87948506	.99215255
5	3	.03571429	.05842612	.09093765	.13491776	.19119547	.25938639	.46718908	.68038068	.93731675
		t	.05897716	.09258893	.13842347	.19744954	.26928178	.48828947	.70916279	.95510447
5	3	.01785714	.03019552	.04867975	.07495852	.11045937	.15609400	.31300578	.50829462	.83925962
		t	.03064967	.05011262	.07817364	.11654647	.16635979	.33921357	.55248834	.88576964
4	4	.24285714	.33267618	.43314353	.53795002	.63995538	.73262187	.89858204	.97260671	.99929383
		t	.33718101	.44303983	.55312697	.65921712	.75396457	.91596606	.98074656	.99971977
4	4	.10000000	.15271391	.22088778	.30341154	.39698108	.49643738	.73102208	.89091904	.99239520
		t	.15439095	.22530359	.31156541	.40950194	.51331109	.75400771	.90963661	.99578362
4	4	.02857143	.04819218	.07711919	.11731554	.17001998	.23530126	.44098141	.65973631	.93201156
		t	.04859831	.07837925	.12007854	.17509902	.24356358	.45969667	.68654734	.94978613
4	4	.01428571	.02486926	.04116277	.06490506	.09769854	.14069827	.29308250	.48883124	.83037372
		t	.02520858	.04226833	.06746078	.10267197	.14930078	.31627733	.52970003	.87617395
7	2	.25000000	.33034159	.41918880	.51215450	.60422799	.69057630	.85930129	.95067378	.99726922
		t	.33389279	.42703639	.52444164	.62038702	.70940740	.87786682	.96220755	.99865560
7	2	.08333333	.12351607	.17535880	.23888688	.31287296	.39482141	.61018382	.79307769	.96854551
		t	.12502192	.17937016	.24649215	.32504072	.41215498	.63904102	.82435365	.98121115
7	2	.02777778	.04411074	.06722768	.09847644	.13885665	.18878582	.35105400	.54213099	.85235078
		t	.04499754	.06983994	.10398124	.14869790	.20453783	.38704822	.59776710	.90414290
6	3	.22619048	.31386966	.41330825	.51837726	.62186342	.71693178	.89014812	.96937906	.99907699
		t	.31759811	.42173243	.53165694	.63918132	.73664741	.90736406	.97809753	.99965971
6	3	.09523810	.14615247	.21234791	.29289662	.38471684	.48287239	.71712195	.88058089	.99016285
		t	.14837957	.21829883	.30404527	.40208369	.50661398	.75066536	.90901238	.99592019
6	3	.04761905	.07799793	.12099490	.17794527	.24899681	.33228424	.56716663	.77645437	.97131844
		t	.07891281	.12362845	.18346516	.25848499	.34664615	.59373266	.80652146	.98261021
6	3	.02380952	.04089251	.06658693	.10299657	.15165565	.21305609	.41276282	.63394360	.92316984
		t	.04141183	.06821598	.10660509	.15835077	.22403881	.43805869	.67055892	.94748541

Table B-3 Power of one-sided two-sample normal scores test vs. normal shift alternative d.

m	n	α	d = .2	.4	.6	.8	1.0	1.5	2.0	3.0
5	4	.23809524	.33324793	.44037142	.55183828	.65909182	.75457239	.91663808	.98080657	.99969067
		t	.33730756	.44936647	.56562169	.67640807	.77338993	.93066232	.98648379	.99987898
		.09523810	.14976646	.22158022	.30939937	.40920431	.51471985	.75674336	.91023985	.99506754
		t	.15173010	.22685288	.31924827	.42438287	.53509120	.78333615	.93014179	.99786446
		.04761905	.07998673	.12643057	.18854009	.26603404	.35634917	.60429411	.81128358	.98048072
		t	.08096752	.12935527	.19463414	.27655867	.37225267	.63281032	.84148241	.98960707
		.02380952	.04199882	.06982756	.10967862	.16316375	.23053711	.44573257	.67285833	.94082945
		t	.04260370	.07177144	.11406175	.17139051	.24410692	.47663707	.71566723	.96489533
		.00793651	.01481297	.02613929	.04370374	.06939151	.10488232	.24152600	.43371037	.80145386
		t	.01508419	.02708390	.04602534	.07416971	.11357945	.26761238	.48345073	.86228265
7	3	.23333333	.32485530	.43718408	.53645875	.64181108	.73704783	.90443949	.97567203	.99942330
		t	.32889201	.43719609	.55043672	.65967110	.75688595	.92043543	.98297188	.99980095
		.10000000	.15500252	.22670002	.31375060	.41226076	.51624082	.75546146	.90899737	.99519033
		t	.15709613	.23220350	.32384013	.42755685	.53647584	.78110666	.92770000	.99760267
		.05000000	.08293376	.12973872	.19187571	.26900258	.35859697	.60422158	.81011327	.98024187
		t	.08400050	.13287239	.19832102	.28001059	.37507421	.63325403	.84052293	.98929687
		.02500000	.04358386	.07176827	.11185932	.16540668	.23264218	.44692252	.67334755	.94138980
		t	.04431020	.07406102	.11694806	.17481505	.24795561	.48081265	.71916538	.96599238
		.00833333	.01536117	.02682575	.04446886	.07012065	.10541139	.24074999	.43111242	.79821359
		t	.01575883	.02818704	.04776204	.07680054	.11740674	.27565017	.49583120	.87306591
6	4	.24761905	.34880987	.46190602	.57794687	.68730549	.78202254	.93347154	.98673491	.99985336
		t	.35317631	.47147597	.59235047	.70495572	.80060285	.94589145	.99106907	.99995031
		.09523810	.15273564	.22921787	.32304177	.42928032	.54042125	.78651730	.93012489	.99743902
		t	.15487261	.23495558	.33368054	.44543487	.56162210	.81180605	.94655377	.99890034
		.04761905	.08207663	.13234752	.20021974	.28510662	.38355778	.64638266	.84970067	.98955023
		t	.08310163	.13541214	.20657313	.29593920	.39959157	.67275081	.87413151	.99435933
		.02380952	.04336429	.07399197	.11857292	.17895145	.25513567	.49420104	.73111298	.96656238
		t	.04396282	.07593208	.12295170	.18711471	.26846665	.52238290	.76569656	.97970686
		.00952381	.01836972	.03327097	.05671160	.09119533	.13870872	.31622054	.54535289	.89406082
		t	.01864467	.03424018	.05910298	.09609583	.14751573	.34087059	.58688015	.92758450
		.00476190	.00943043	.01757558	.03089671	.05135630	.08092233	.20326640	.38936920	.77555698
		t	.00965437	.01840061	.03303185	.05596335	.08967610	.23198250	.44784432	.85222784
5	5	.24603175	.34920607	.46477213	.58318770	.69423729	.78960296	.93875365	.98862516	.99989947
		t	.35362085	.47442866	.59763621	.71176981	.80780751	.95030828	.99235774	.99996556
		.09523810	.15406572	.23254942	.32877540	.43731214	.55007151	.79528349	.93381573	.99720536
		t	.15631729	.23868454	.34029233	.44598182	.57346507	.82372166	.95276438	.99918987
		.04761905	.08300919	.13501380	.20550792	.29373914	.39580473	.66462990	.86528622	.99234008
		t	.08399998	.13796836	.21159683	.30402494	.41083808	.68822842	.88563171	.99559460
		.01984127	.03704921	.06464770	.10568701	.16234702	.23508519	.46951573	.70941893	.95957792
		t	.03756892	.06639740	.10978233	.17025345	.24837916	.50007192	.74991014	.97757521

Table B-3 Power of one-sided two-sample normal scores test vs. normal shift alternative d.

m	n	α	d = .2	.4	.6	.8	1.0	1.5	2.0	3.0
5	5	.00793651	.01568267	.02903507	.05048395	.08264280	.12771946	.30067435	.53027082	.88890352
		t	.01590300	.02983218	.05249851	.08686369	.13546172	.32332209	.56979941	.92244607
		.00396825	.00804477	.01531453	.02744193	.04640347	.07424634	.19235802	.37663746	.76825953
		t	.00822603	.01599887	.02925343	.05039410	.08197445	.21876320	.43217128	.84443135
7	4	.24848485	.35358309	.47110418	.59103163	.70276468	.79784833	.94351905	.99012340	.99992804
		t	.35780113	.48027831	.60464808	.71911450	.81460475	.95369515	.99321336	.99997337
		.10000000	.16209615	.24474163	.34554230	.45829709	.57407008	.81843770	.94754659	.99863396
		t	.16430445	.25062940	.35631581	.47434157	.59459461	.84080225	.96035008	.99945213
		.04848485	.08505786	.13890355	.21182437	.30275496	.40727901	.67793507	.87364181	.99271623
		t	.08623874	.14246719	.21923844	.31536899	.42580641	.70713798	.89873530	.99678524
		.02424242	.04517250	.07845943	.12736366	.19381159	.27740941	.53400701	.77323145	.97917520
		t	.04584226	.08064066	.13227844	.20290017	.29197061	.56295617	.80518139	.98782658
		.00909091	.01806138	.03352626	.05828541	.09515321	.14629571	.33695626	.57662565	.91347478
		t	.01841076	.03478560	.06144486	.10169944	.15812436	.36971644	.62945456	.94928558
		.00303030	.00631808	.01234863	.02267924	.03923919	.06413002	.17392025	.35287009	.75212534
		t	.00650540	.01307328	.02464107	.04365068	.07283319	.20479686	.41945331	.84476636
6	5	.24891775	.35805363	.48011389	.60389626	.71776607	.81283488	.95211906	.99260923	.99996208
		t	.36245414	.48964791	.61789817	.73429148	.82937175	.96131234	.99506459	.99998712
		.09740260	.16089413	.24634565	.35111892	.46825046	.58775694	.83387144	.95614949	.99917193
		t	.16317002	.25245727	.36231104	.48482543	.60870656	.85519677	.96728116	.99966623
		.04978355	.08879019	.14666932	.22518496	.32266262	.43354762	.71120293	.89745535	.99566628
		t	.08999219	.15030502	.23271806	.33534579	.45186632	.73813987	.91823387	.99811172
		.02380952	.04547515	.08051706	.13254724	.20355446	.29273975	.56144613	.79947838	.98419597
		t	.04604908	.08243406	.13695044	.21182676	.30612745	.58820547	.82845688	.99146521
		.00865801	.01776314	.03387517	.06020282	.09998828	.15567732	.36332036	.61687038	.93780041
		t	.01800602	.03477217	.06249583	.10480357	.16444921	.38755074	.65460376	.95949199
		.00432900	.00920239	.01822049	.03370973	.05839195	.09499628	.24886767	.47470307	.86627547
		t	.00936646	.01886130	.03542600	.06220781	.10238641	.27312009	.52090700	.90982890
7	5	.25000000	.36376227	.49091364	.61882039	.73467340	.82920643	.96062480	.99475581	.99998227
		t	.36554786	.50020821	.63241272	.75040544	.84452959	.96833151	.99663497	.99999424
		.09848485	.16782352	.25644147	.36782406	.49130603	.61527052	.85894223	.96785895	.99962146
		t	.16995746	.26252485	.37882406	.50727177	.63488416	.87705102	.97584441	.99984046
		.04924242	.09120567	.15129298	.23515386	.33931818	.45698315	.74246023	.91900323	.99775811
		t	.09205743	.15507465	.24294241	.35225216	.47526449	.76704680	.93546722	.99899131
		.02398990	.04703743	.08496498	.14184692	.21968056	.31699699	.60208588	.83655498	.99108916
		t	.04768591	.08714265	.14684024	.22896056	.33173140	.62922489	.86229074	.99534905
		.00883838	.01872235	.03662308	.06633783	.11161469	.17503904	.40695176	.67300383	.96081125
		t	.01899943	.03766244	.06901587	.11723921	.18520730	.43359314	.71052083	.97622647
		.00378788	.00840013	.01724941	.03288763	.05838925	.09683984	.26066508	.49875093	.88606441
		t	.00858419	.01798970	.03494319	.06306619	.10605070	.29134442	.55592334	.93292515

Table B-3 Power of one-sided two-sample normal scores test vs. normal shift alternative d.

m	n	α	d = .2	.4	.6	.8	1.0	1.5	2.0	3.0
6	6	.25000000	.36553653	.49468533	.62422128	.74085127	.83516398	.96357647	.99543988	.99988735
6	6	t	.36991110	.50410071	.63781111	.75644602	.85017514	.97080194	.99700231	.99999586
6	6	.09523810	.16188555	.25296886	.36521433	.49004726	.61545124	.86090163	.96897534	.99965154
6	6	t	.16415920	.25911790	.37643633	.50643615	.63566513	.87956526	.97710599	.99986489
6	6	.04545455	.08441348	.14403076	.22666061	.33048679	.44887778	.73913577	.91925663	.99802245
6	6	t	.08561787	.14770567	.23426821	.34315848	.46680238	.76302804	.93481846	.99902512
6	6	.02380952	.04713981	.08579262	.14401001	.22382018	.32353264	.61345497	.84669603	.99274520
6	6	t	.04779414	.08799560	.14906043	.23317320	.33830495	.63992771	.87064094	.99609004
6	6	.00865801	.01854790	.03661156	.06678061	.11292735	.17768071	.41388110	.68171477	.96346945
6	6	t	.01883084	.03768437	.06956825	.11881832	.18837308	.44192171	.72078297	.97878798
6	6	.00432900	.00968650	.02001186	.03827689	.06798442	.11249187	.29776131	.55335276	.91997413
6	6	t	.00984292	.02064081	.04001686	.07191505	.12015059	.32220829	.59600291	.94839097
7	6	.24883450	.36920733	.50393665	.63805051	.75683216	.85048133	.97072670	.99692561	.99999494
7	6	t	.37353939	.51322100	.65127100	.77166265	.86430850	.97662883	.99799381	.99999837
7	6	.09848485	.17012752	.26832676	.38860549	.52035655	.64953149	.88717707	.97886715	.99986077
7	6	t	.17251203	.27474336	.40014691	.53681098	.66915450	.90314488	.98463649	.99994753
7	6	.04778555	.09056005	.15648818	.24767126	.36095142	.48750256	.78040607	.94147920	.99905910
7	6	t	.09186276	.16049591	.25595495	.37459578	.50640777	.80338486	.95439739	.99959982
7	6	.02389277	.04874984	.09073123	.15461144	.24229032	.35098156	.65545318	.87882554	.99589123
7	6	t	.04943389	.09306901	.16000460	.25225456	.36654934	.68164757	.90005168	.99807415
7	6	.00932401	.02062225	.04168419	.07725064	.13173880	.20759503	.47335388	.74650022	.97996009
7	6	t	.02092783	.04286306	.08033908	.13825958	.21931346	.50218174	.78220246	.98960094
7	6	.00466200	.01080618	.02294372	.04475594	.08047801	.13389045	.34996526	.62529492	.95201890
7	6	t	.01099459	.02371646	.04691662	.08536667	.14334433	.37843300	.66963260	.97228033
7	7	.24883450	.37485878	.51586205	.65473550	.77518624	.86723417	.97747271	.99807404	.99999826
7	7	t	.37917985	.52506265	.66761498	.78923878	.87984289	.98214137	.99876050	.99999946
7	7	.09848485	.17401836	.27850732	.40635731	.54482529	.67767766	.90794605	.98563843	.99994521
7	7	t	.17644565	.28503999	.41798647	.56106251	.69644360	.92143242	.98967327	.99998022
7	7	.04982517	.09661261	.16938466	.26989654	.39329206	.52808238	.82060577	.96038560	.99965044
7	7	t	.09799132	.17361858	.27853804	.40719497	.54669257	.84037184	.96943331	.99985235
7	7	.02476690	.05200369	.09875083	.17030547	.26809780	.38757684	.70529973	.91204426	.99832642
7	7	t	.05275675	.10133534	.17622966	.27885664	.40392020	.72970793	.92835666	.99920379
7	7	.00990676	.02266139	.04694890	.08841331	.15197775	.23962026	.53310237	.80383940	.98975989
7	7	t	.02300515	.04829484	.09195494	.15941058	.25276160	.56279539	.83563967	.99532579
7	7	.00495338	.01195670	.02619150	.05223704	.09520976	.15925929	.40929295	.69838386	.97380201
7	7	t	.01214692	.02699192	.05451005	.10038068	.16921408	.43769394	.73791943	.98629527

Table B-4 **Power of two-sided two-sample normal scores test vs. normal shift alternative d.**

m	n	α	d = .2	.4	.6	.8	1.0	1.5	2.0	3.0
3	2	.20000000	.20779141	.23074858	.26765048	.31656128	.37497491	.54331879	.70780192	.92052591
		t	.20801832	.23164801	.26964058	.32000708	.38015588	.55320219	.72095621	.93110707
4	2	.13333333	.14066165	.16237316	.19764379	.24509596	.30281746	.47560365	.65397722	.90079632
		t	.14103443	.16386536	.20099587	.25101045	.31189799	.49385204	.67929014	.92186812
3	3	.20000000	.21039553	.24086475	.28931537	.35246136	.42612477	.62576519	.79748973	.96662766
		t	.21079887	.24245507	.29280011	.35840281	.43486038	.64094087	.81477418	.97496879
3	3	.10000000	.10692590	.12753479	.16129173	.20722544	.26385577	.43784911	.62365475	.88989158
		t	.10717943	.12856370	.16365200	.21149914	.27060768	.45244777	.64528144	.90980045
5	2	.19047619	.20046793	.22980646	.27662459	.33795904	.40998603	.60812258	.78294124	.96219728
		t	.20100744	.23193240	.28127901	.34588721	.42163219	.62832819	.80596529	.97335512
5	2	.09523810	.10187743	.12166246	.15416186	.19856368	.25358325	.42450228	.61020039	.88311107
		t	.10234166	.12353964	.15844392	.20626224	.26564714	.44998809	.64698409	.91492533
4	3	.22857143	.24141232	.27876939	.33728940	.41185152	.49622287	.70859580	.86692529	.98638295
		t	.24216254	.28168217	.34351116	.42209059	.51061395	.73004192	.88674886	.99180737
4	3	.05714286	.06279054	.07979248	.10825393	.14813656	.19901486	.36584636	.55946664	.86310017
		t	.06307547	.08097417	.11105263	.15339933	.20766715	.38637685	.59215143	.89557401
7	1	.25000000	.25696759	.27751287	.31058916	.35453670	.40720397	.56038335	.71288467	.91777544
		t	.25735194	.27902173	.31387794	.36012266	.41541969	.57516021	.73154068	.93179788
6	2	.21428571	.22538572	.25784013	.30919505	.37563546	.45238488	.65554795	.82269886	.97448058
		t	.22623120	.26114417	.31633080	.38756574	.46950578	.68305098	.85102914	.98503538
6	2	.07142857	.07737682	.09520518	.12481150	.16586803	.21764565	.38413636	.57352394	.86705937
		t	.07788961	.09729926	.12965928	.17473995	.23181615	.41547631	.62043092	.90918512
5	3	.25000000	.26444287	.30624039	.37102850	.45226821	.54223672	.75726257	.90209567	.99297174
		t	.26567503	.31095090	.38083919	.46786779	.56324352	.78448435	.92308628	.99656829
5	3	.07142857	.07916102	.10235340	.14086924	.19412968	.26075279	.46734550	.68039294	.93731678
		t	.07950576	.10377355	.14419201	.20026257	.27057738	.48843379	.70917382	.95510449
5	3	.03571429	.04024663	.05405567	.07768673	.11177122	.15669090	.31307074	.50829950	.83925963
		t	.04054574	.05532182	.08077487	.11777713	.16691079	.33927123	.55249253	.88576964
4	4	.22857143	.24372595	.28758249	.35554702	.44070255	.53483872	.75823064	.90568721	.99392207
		t	.24482723	.29181371	.36442409	.45493607	.55416213	.78358398	.92511929	.99701927
4	4	.05714286	.06423327	.08563356	.12158151	.17203468	.23619696	.44107133	.65974235	.93201156
		t	.06449867	.08674257	.12432029	.17704166	.24441932	.45978068	.68655287	.94978614
4	4	.02857143	.03265717	.04518500	.06686995	.09860505	.14109268	.29312029	.48883369	.83037373
		t	.03288872	.04617928	.06934446	.10352891	.14966852	.31631143	.52970217	.87617396

233

Table B-4 Power of two-sided two-sample normal scores test vs. normal shift alternative d.

m	n	α	d = .2	.4	.6	.8	1.0	1.5	2.0	3.0
7	2	.22222222	.23390769	.26801924	.32182087	.39107846	.47053144	.67720615	.84128788	.98002665
		t	.23501093	.27229794	.33094888	.40609048	.49164565	.70904130	.87161454	.98936018
7	2	.05555556	.06087670	.07691493	.10382841	.14168094	.19020801	.35126127	.54215313	.85235088
		t	.06140956	.07911150	.10898474	.15127518	.20580394	.38722148	.59778448	.90414297
6	3	.23809524	.25376225	.29902026	.36890220	.45597801	.55151939	.77418728	.91593532	.99530007
		t	.25486113	.30322296	.37776425	.46987148	.57015007	.79767635	.93302631	.99766442
6	3	.09523810	.10546743	.13589559	.18560413	.25268763	.33395482	.56734042	.77646633	.97131846
		t	.10603109	.13818051	.19081327	.26197225	.34819989	.59388787	.80653171	.98261022
6	3	.04761905	.05401281	.07341850	.10635268	.15320910	.21373273	.41282728	.63394770	.92316985
		t	.05435649	.07486149	.10982244	.15981793	.22466830	.43811644	.67056247	.94748541
6	3	.02380952	.02746025	.03871538	.05839047	.08755373	.12715713	.27267087	.46631517	.81777278
		t	.02775722	.03999727	.06160380	.09399770	.13846465	.30369138	.52132688	.87861247
5	4	.22222222	.23915886	.28801674	.36322346	.46642960	.55784281	.78857504	.92742719	.99685971
		t	.24043630	.29289118	.37332763	.47234351	.57892624	.81377544	.94416978	.99858603
5	4	.09523810	.10660303	.14036980	.19536773	.26915684	.35768097	.60440964	.81128989	.98048072
		t	.10722815	.14291043	.20117285	.27950211	.37348762	.63291306	.84148779	.98960708
5	4	.04761905	.05469451	.07618268	.11265986	.16447229	.23107372	.44577513	.67286048	.94082945
		t	.05510830	.07793292	.11690580	.17261859	.24460236	.47667482	.71566907	.96489533
5	4	.01587302	.01882511	.02804939	.04455860	.06975056	.10502363	.24153624	.43371085	.80145386
		t	.01902220	.02892360	.04683324	.07450273	.11370812	.26762130	.48345113	.86228265
7	3	.23333333	.24995039	.29787475	.37161888	.46299252	.56243035	.78915397	.92663801	.99666765
		t	.25106065	.30211056	.38040033	.47683519	.58080770	.81139870	.94179506	.99837726
7	3	.10000000	.11130466	.14486163	.19943611	.27254230	.36014702	.60436730	.81012209	.98024188
		t	.11197012	.14755239	.20553993	.28333245	.37650302	.63338226	.84053033	.98929688
7	3	.05000000	.05710304	.07864858	.11514963	.16688313	.23326304	.44697570	.67335051	.94138980
		t	.05759014	.08069300	.12005569	.17618240	.24851883	.48085843	.71916780	.96599238
7	3	.01666667	.01963615	.02889600	.04541375	.07052650	.10557520	.24076280	.43111308	.79921359
		t	.01992073	.03014742	.04863180	.07716356	.11754912	.27566055	.49583170	.87306591
6	4	.24761905	.26678097	.32161211	.40465881	.50512025	.61096968	.83469925	.95171534	.99859535
		t	.26817892	.32688613	.41538613	.52157335	.63203733	.85709315	.96439964	.99947482
6	4	.09523810	.10785863	.14534554	.20630986	.28775404	.38462391	.64646052	.84978411	.98955023
		t	.10852224	.14802973	.21238631	.29842149	.40057270	.67281902	.87413437	.99435933
6	4	.04761905	.05558428	.07984283	.12138151	.18003267	.25555167	.49422838	.73111408	.96656238
		t	.05600113	.08160446	.12543864	.18812779	.26878959	.52238696	.76569749	.97970686
6	4	.01904762	.02299661	.03537331	.05760342	.09154794	.13883847	.31622825	.54535317	.89406082
		t	.02319994	.03627660	.05995290	.09642636	.14763536	.34087741	.58688039	.92758450
6	4	.00952381	.01168840	.01857896	.03131376	.05151815	.08098088	.20326975	.38936932	.77555698
		t	.01185841	.01935624	.03341938	.05611009	.08972792	.23198530	.44784441	.85222784

234

Table B-4 **Power of two-sided two-sample normal scores test vs. normal shift alternative d.**

m	n	α	d = .2	.4	.6	.8	1.0	1.5	2.0	3.0
5	5	.23809524	.25787030	.31444381	.40007268	.50349983	.61215982	.83958508	.95538158	.99896825
		t	.25927901	.31975014	.41082273	.51991915	.63302258	.86101514	.96678853	.99957331
5	5	.09523810	.10843155	.14761603	.21129701	.29619917	.39677024	.66469513	.86528883	.99234008
		t	.10907464	.15020720	.21712416	.30633105	.41172571	.68828531	.88563386	.99559460
5	5	.03968254	.04695548	.06922407	.10767142	.16314050	.23537881	.46953300	.70941953	.95957792
		t	.04732970	.07086193	.11167781	.17099931	.24865080	.50008729	.74991067	.97757521
5	5	.01587302	.01943766	.03069268	.05116540	.08290324	.12781184	.30067930	.53027098	.88890352
		t	.01960384	.03144175	.05315034	.08710910	.13554749	.32332653	.56979955	.92244607
5	5	.00793651	.00987867	.01610688	.02776134	.04652338	.07428819	.19236018	.37663753	.76825953
		t	.01001862	.01675578	.02955164	.05050354	.08201181	.21876503	.43217134	.84443135
7	4	.24848485	.26903556	.32763422	.41575007	.52115820	.63051166	.85316136	.96102316	.99911425
		t	.27050739	.33316338	.42691129	.53808139	.65183951	.87448465	.97203247	.99970692
7	4	.09696970	.11071460	.15147754	.21751981	.30513500	.40819513	.67799334	.87364397	.99271623
		t	.11149389	.15462961	.22464393	.31758270	.42664064	.70718815	.89873706	.99678524
7	4	.04848485	.05774290	.08406213	.12976679	.19476465	.27775836	.53402679	.77323211	.97917520
		t	.05777076	.08605376	.13455612	.20378544	.29228802	.56297321	.80518193	.98782658
7	4	.02424242	.02950848	.04596657	.07532869	.11962286	.18019000	.39531264	.64400985	.94371248
		t	.02977465	.04714277	.07835347	.12579411	.19104575	.42296754	.68437386	.96499739
7	4	.00606061	.00767746	.01291776	.02290118	.03931966	.06415710	.17392152	.35287013	.75212534
		t	.00782434	.01360875	.02484351	.04372184	.07285641	.20479787	.41945333	.84476636
6	5	.24675325	.26880904	.33150689	.42518072	.53607874	.64941147	.87144979	.97008926	.99959860
		t	.27028664	.33700958	.43612513	.55232325	.66931379	.88937862	.97797406	.99983637
6	5	.09956710	.11456456	.15895900	.23056972	.32482640	.43434367	.71124693	.89745671	.99566628
		t	.11536743	.16219444	.23782866	.33735767	.45259060	.73817765	.91823497	.99811172
6	5	.04761905	.05700152	.08566466	.13466368	.20435412	.29301696	.56145946	.79947875	.98419597
		t	.05701706	.08743593	.13897862	.21257931	.30638400	.58821729	.82845719	.99146521
6	5	.01731602	.02167659	.03551204	.06083513	.10021346	.15575112	.36332349	.61687046	.93780041
		t	.02186373	.03636186	.06310068	.10501569	.16451768	.38755353	.65460382	.95949199
6	5	.00865801	.01109520	.01899225	.03399781	.05849184	.09502822	.24886895	.47470310	.86627547
		t	.01122480	.01960084	.03569834	.06230046	.10241548	.27312120	.52090711	.90982890
6	5	.00432900	.00563711	.00994101	.01834530	.03256154	.05467304	.15692874	.33155579	.73871176
		t	.00574187	.01044725	.01981768	.03601746	.06173307	.18384981	.39298261	.83104742
7	5	.25000000	.27390741	.34154189	.44159313	.55821820	.67485141	.89179439	.97820312	.99980987
		t	.27541683	.34711570	.45251961	.57410017	.69377613	.90719828	.98404201	.99992394
7	5	.09848485	.11469253	.16266293	.23992533	.34114255	.45761758	.74248983	.91900397	.99775811
		t	.11553685	.16604916	.24745137	.35393658	.47583574	.76707166	.93546779	.99899131
7	5	.04797980	.05827764	.08979129	.14374193	.22036476	.31721872	.60209478	.83655518	.99108916
		t	.05875319	.09181436	.14864121	.22959356	.33193395	.62932260	.86229090	.99534905
7	5	.01767677	.02256567	.03815904	.06690082	.11180360	.17509696	.40695377	.67300387	.96081125
		t	.02278252	.03914940	.06955160	.11741584	.18526051	.43359491	.71052086	.97622647

Table B-4 **Power of two-sided two-sample normal scores test vs. normal shift alternative d.**

m	n	α	d = .2	.4	.6	.8	1.0	1.5	2.0	3.0
7	5	.00757576	.00997782	.01785497	.03310137	.05845848	.09686039	.26066575	.49875094	.88606441
		t	.01012659	.01856813	.03514260	.06312928	.10606899	.29134498	.55592335	.93292515
7	5	.00252525	.00342925	.00647847	.01267696	.02364888	.04149597	.13048108	.29506986	.71212563
		t	.00351025	.00688533	.01391830	.02670625	.04802994	.15788153	.36214760	.82097846
6	6	.24242424	.2680340	.33575746	.43769738	.55636602	.67476252	.89342772	.97903665	.99982915
		t	.26836596	.34153796	.44905503	.57290743	.69449149	.90940010	.98497810	.99993659
6	6	.09090909	.10682880	.15412643	.23080386	.33203316	.44940179	.73915852	.91925716	.99802245
		t	.10765118	.15743187	.23816955	.34457783	.46726998	.76304675	.93481885	.99902512
6	6	.04761905	.05816216	.09045765	.14581105	.22445320	.32373484	.61346257	.84669618	.99274520
		t	.05864432	.09250836	.15077008	.23376216	.33848915	.63393425	.87064107	.99609004
6	6	.02164502	.02753815	.04620373	.08016088	.13225834	.20406543	.45408781	.71823608	.97045949
		t	.02780339	.04740982	.08336557	.13897364	.21606610	.48415224	.75811946	.98474150
6	6	.00865801	.01146830	.02068572	.03851055	.06805856	.11251336	.29776195	.55335277	.91997413
		t	.01159502	.02129206	.04023874	.07198423	.12017028	.32220886	.59600292	.94839097
6	6	.00432900	.00586718	.01101860	.02134786	.03927323	.06768905	.20020308	.41749467	.84001667
		t	.00595598	.01145775	.02265719	.04240576	.07415569	.22446451	.46869486	.89471040
7	6	.24941725	.27605333	.35090578	.46011413	.58467358	.70549402	.91449455	.98591624	.99993172
		t	.27768654	.35688112	.47163534	.60100528	.72429070	.92787958	.99005368	.99997580
7	6	.09557110	.11346553	.16643097	.25156943	.36232873	.48794028	.78042159	.94147948	.99905910
		t	.11437357	.17007131	.25962394	.37585959	.50679845	.80339767	.95439760	.99959982
7	6	.04778555	.05939735	.09503374	.15618416	.24280928	.35113586	.65545785	.87882561	.99589123
		t	.05991126	.09722717	.16149494	.25273608	.36668935	.68165156	.90005174	.99807415
7	6	.02331002	.03021504	.05215599	.09218883	.15351814	.23732374	.51771807	.78541974	.98722476
		t	.03050821	.05347851	.09564828	.16058930	.24952756	.54447704	.81440864	.99274290
7	6	.00932401	.01264350	.02360354	.04497138	.08054183	.13390757	.34996566	.62529492	.95201890
		t	.01279845	.02435198	.04712009	.08542576	.14335987	.37843335	.66963260	.97228033
7	6	.00466200	.00651003	.01277065	.02552159	.04792529	.08362646	.24749589	.49897042	.90211892
		t	.00660183	.01323186	.02691731	.05129310	.09058012	.27259454	.54714106	.93832198
7	7	.25000000	.27907449	.36028966	.47730756	.60815474	.73156804	.93140325	.99066020	.99997410
		t	.28075620	.36638034	.48884518	.62408846	.74927156	.94245979	.99347392	.99999122
7	7	.09965035	.11974882	.17902432	.27349003	.39448777	.52843681	.82061588	.96038573	.99965044
		t	.12072159	.18289159	.28191192	.40828733	.54700672	.84038006	.96943340	.99985235
7	7	.04953380	.06263312	.10284952	.17172180	.26853544	.38769753	.70530263	.91204429	.99832642
		t	.06320619	.10528661	.17756560	.27925971	.40402853	.72971036	.92835668	.99920379
7	7	.02447552	.03242372	.05776590	.10412340	.17492635	.27056459	.57571585	.83623559	.99375839
		t	.03273703	.05918892	.10785985	.18253517	.28352385	.60201472	.86100188	.99677166
7	7	.00990676	.01381347	.02681961	.05242834	.09526210	.15927213	.40929319	.69838386	.97380201
		t	.01397256	.02759865	.05469149	.10042941	.16922582	.43769414	.73791943	.98629527
7	7	.00466200	.00673405	.01385117	.02861874	.05495701	.09723504	.28884789	.56537036	.93807251
		t	.00683321	.01435966	.03018850	.05879424	.10518927	.31666392	.61416433	.96419890

Table B-5 Power of one-sided two-sample median test vs. normal shift alternative d.

m	n	α	d = .2	.4	.6	.8	1.0	1.5	2.0	3.0
2	2	.16666667	.21449982	.26918351	.32975727	.39480253	.46254547	.63069050	.77491573	.94286569
		t	.21464351	.26953440	.33038014	.39575540	.46387086	.63296654	.77780527	.94517173
3	2	.10000000	.13715522	.18246273	.23573967	.29619213	.36242771	.54013749	.70716641	.92051393
		t	.13764968	.18373401	.23810279	.29995788	.36785606	.55011471	.72034528	.93109575
4	2	.20000000	.26073485	.32940785	.40397682	.48178412	.55984858	.73764381	.86736328	.97967387
		t	.26726595	.34429426	.42833009	.51568573	.60221910	.79036365	.91230560	.99278551
3	3	.05000000	.07478896	.10770305	.14955634	.20057409	.26024891	.43720951	.62357041	.88989100
		t	.07524361	.10898187	.15213899	.20501563	.26711407	.45183760	.64520209	.90979992
5	2	.14285714	.19481534	.25650122	.32661975	.40298700	.48273825	.67603465	.82838155	.97152811
		t	.19985414	.26863084	.34752107	.43355487	.52277255	.73143043	.88013121	.98893495
4	3	.02857143	.04572295	.07006039	.10296429	.14539961	.19766839	.36566437	.55944910	.86310011
		t	.04618485	.07144500	.10592862	.15077637	.20639037	.38620864	.59213559	.89557396
6	2	.21428571	.28127808	.35655801	.43739585	.52041425	.60020039	.77957032	.89860253	.98785082
		t	.29006014	.37643573	.46947970	.56419881	.65532908	.84006531	.94399541	.99725518
5	3	.07142857	.10783265	.15558299	.21498741	.28516059	.36396596	.57629874	.76406702	.95846206
		t	.11216717	.16715780	.23688505	.31996903	.41303760	.65472009	.84440135	.98721534
4	4	.24285714	.32309597	.41226420	.50590396	.59888162	.68622564	.85714958	.94982934	.99721688
		t	.33718101	.44303983	.55312697	.65921712	.75396457	.91596606	.98074656	.99971977
4	4	.01428571	.02486926	.04116277	.06490506	.09769854	.14069827	.29308250	.48883124	.83037372
		t	.02520858	.04226833	.06746078	.10267197	.14930078	.31627733	.52970003	.87617395
7	2	.16666667	.22652652	.29647951	.37443091	.45733151	.54154872	.73443091	.87246968	.98351048
		t	.23380143	.31367855	.40337632	.49844929	.59357115	.79894632	.92483230	.99582301
6	3	.04761905	.07574444	.11466040	.16554318	.22849121	.30227253	.51481631	.71820016	.94669384
		t	.07891281	.12362845	.18346516	.25848499	.34464615	.59373266	.80652146	.98261021
5	4	.16666667	.23604725	.31856186	.41081865	.50786229	.60397765	.80726958	.92832392	.99568874
		t	.24846139	.34774965	.45873837	.57300568	.68136875	.88290524	.97172889	.99957890
5	4	.00793651	.01481297	.02613929	.04370374	.06939151	.10488232	.24152600	.43371037	.80145386
		t	.01508419	.02708390	.04602534	.07416971	.11357945	.26761238	.48345073	.86228265

Table B-5 Power of one-sided two-sample median test vs. normal shift alternative d.

m	n	α	d = .2	.4	.6	.8	1.0	1.5	2.0	3.0
7	3	.08333333	.12652266	.18273437	.25170625	.33160154	.41910466	.64184264	.81990055	.97560872
		t	.13345324	.20109336	.28586112	.38453225	.49123686	.74404255	.90855572	.99630606
6	4	.02380952	.04172567	.06895781	.10773892	.15956757	.22468227	.43282405	.65541623	.93093828
		t	.04396282	.07593208	.12295170	.18711471	.26840665	.52236290	.76569656	.97970686
5	5	.10317460	.15792918	.22853029	.31354768	.40925879	.51010546	.74375049	.89856553	.99334114
		t	.16774698	.25372566	.35838366	.47487524	.59346836	.83745594	.95818409	.99935242
5	5	.00396825	.00804477	.01531453	.02744193	.04640347	.07424634	.19235802	.37663746	.76825953
		t	.00822603	.01599887	.02925343	.05039410	.08197445	.21876320	.43217128	.84443135
7	4	.01515152	.02810442	.04893728	.08020154	.12405892	.18167615	.37934329	.60836384	.91614680
		t	.02967632	.05414366	.09222017	.14700338	.21992942	.46655184	.72546850	.97441928
7	4	.19696970	.27875429	.37391009	.47707785	.58138859	.67982593	.86779279	.96046059	.99868395
		t	.29462187	.41004134	.53386365	.65444195	.76103217	.93189289	.98839674	.99993559
6	5	.06709957	.10977026	.16869328	.24427775	.33445852	.43466783	.68609126	.86889759	.99069776
		t	.11748952	.19000388	.28484725	.39762910	.51960291	.79365640	.94373639	.99908308
6	5	.00216450	.00471081	.00957265	.01820948	.03251520	.05465844	.15692817	.33155577	.73871176
		t	.00484102	.01009894	.01969280	.03597602	.06172036	.18384935	.39298260	.83104742
7	5	.12121212	.18720891	.27130308	.37032535	.47820100	.58706435	.81564576	.94130618	.99789252
		t	.20071965	.30526496	.42867829	.55949051	.68400797	.90343708	.98311685	.99991601
7	5	.00757576	.01543183	.02925926	.05178724	.08583326	.13366788	.31561175	.55047127	.89222853
		t	.01645600	.03297943	.06112737	.10507068	.16800209	.40548266	.68333814	.97067349
6	6	.04004329	.07085812	.11691136	.18043752	.26144007	.35707201	.61990468	.83189277	.98703912
		t	.07644132	.13367947	.21493626	.31912158	.43982856	.73998665	.92434997	.99869130
6	6	.00108225	.00255582	.00559789	.01140386	.02167308	.03854840	.12442825	.28865162	.70609606
		t	.00263849	.00595956	.01250022	.02440747	.04447775	.15011032	.35114827	.81449595
7	6	.20862471	.30312187	.41271125	.52921572	.64284931	.74463204	.91635887	.98192354	.99976151
		t	.32377743	.45891359	.59900369	.72729721	.83100429	.96732654	.99680379	.99999651
7	6	.02505828	.04750039	.08343631	.13626046	.20766536	.29658861	.56175087	.79625087	.98307155
		t	.05160525	.09669595	.16543951	.25956500	.37538105	.69053655	.90468105	.99825408
7	6	.00058275	.00147877	.00346023	.00748907	.01504079	.02812584	.10076897	.25087434	.67690208
		t	.00153464	.00372221	.00833544	.01727840	.03324262	.12554045	.31846529	.80166764
7	7	.14306527	.22350799	.32442983	.43975280	.55991446	.67422177	.88426423	.97322130	.99961252
		t	.24152873	.36834500	.51149286	.65312196	.77595767	.95256914	.99507912	.99999463
7	7	.01456876	.02988045	.05635417	.09823220	.15862679	.23846246	.49937153	.75470803	.97794168
		t	.03269591	.06625698	.12166665	.20330631	.31073092	.63429746	.88016885	.99765968
7	7	.00029138	.00080172	.00202074	.00468125	.01000200	.01978227	.07963836	.21617286	.64552804
		t	.00083608	.00219496	.00528607	.01170993	.02392958	.10241463	.28448399	.78622782

Table B-6 Power of two-sided two-sample median test vs. normal shift alternative d.

m	n	α	d = .2	.4	.6	.8	1.0	1.5	2.0	3.0
2	2	.33333333	.34064370	.36209376	.39629321	.44109262	.49380843	.64093656	.77766711	.94297162
		t	.34069974	.36231417	.39677505	.44191438	.49502325	.64315905	.78053801	.94527667
3	2	.20000000	.20779141	.23074858	.26765048	.31656128	.37497491	.54331879	.70780192	.92052591
		t	.20801832	.23164801	.26964058	.32000708	.38015588	.55320219	.72095621	.93110707
3	3	.10000000	.10692590	.12753479	.16129173	.20722544	.26385577	.43784911	.62365475	.88989158
		t	.10717943	.12856370	.16365200	.21149914	.27060768	.45244777	.64528144	.90980045
4	3	.05714286	.06279054	.07979248	.10825393	.14813656	.19901486	.36584636	.55946664	.86310017
		t	.06307547	.08097417	.11105263	.15339933	.20766715	.38637685	.59215143	.89557401
5	3	.14285714	.15296617	.18273972	.23052273	.29359792	.36831073	.57694884	.76413520	.95846231
		t	.15525668	.19174362	.25013488	.32670525	.41626481	.65511300	.84443374	.98721541
4	4	.02857143	.03265717	.04518500	.06686995	.09860505	.14109268	.29312029	.48883369	.83037373
		t	.03288872	.04617928	.06934446	.10352891	.14966852	.31631143	.52970217	.87617396
6	3	.09523810	.10417853	.13075709	.17416802	.23285864	.30435971	.51507008	.71822114	.94669389
		t	.10603109	.13818051	.19081327	.26197225	.34819989	.59388787	.80653171	.98261022
5	4	.01587302	.01882511	.02804939	.04455860	.06975056	.10502363	.24153624	.43371085	.80145386
		t	.01902220	.02892360	.04683324	.07450273	.11370812	.26762130	.48345113	.86228265
7	3	.16666667	.17861282	.21357061	.26896104	.34071237	.42363720	.64244713	.81995469	.97560885
		t	.18230813	.22793335	.29965693	.39115757	.49420611	.74433572	.90857416	.99630608
6	4	.04761905	.05450864	.07540150	.11078280	.16091263	.22523739	.43286868	.65541851	.93093828
		t	.05600113	.08160446	.12543864	.18812779	.26878959	.52238696	.76569749	.97970686

239

Table B-6 **Power of two-sided two-sample median test vs. normal shift alternative d.**

m	n	α	d = .2	.4	.6	.8	1.0	1.5	2.0	3.0
5	5	.20634921	.22148362	.26535979	.33358618	.41947785	.51498290	.74432049	.89860857	.99334120
		t	.22659771	.28477534	.37350339	.48165822	.59626746	.83766669	.95819328	.99935242
5	5	.00793651	.00987867	.01610688	.02776134	.04652338	.07428819	.19236018	.37663753	.76825953
		t	.01001862	.01675578	.02955164	.05050354	.08201181	.21876503	.43217134	.84443135
7	4	.03030303	.03575407	.05254616	.08178923	.12470907	.18192354	.37935916	.60836447	.91614680
		t	.03686804	.05731159	.09351315	.14749164	.22009979	.46656044	.72546876	.97441928
6	5	.13419913	.14819735	.18926245	.25454764	.33923277	.43673112	.68627383	.86890758	.99069777
		t	.15277493	.20704583	.29239062	.40068368	.52073273	.79371914	.94373832	.99908308
6	5	.00432900	.00563711	.00994101	.01834530	.03256174	.05467304	.15692874	.33155579	.73871176
		t	.00574187	.01044725	.01981768	.03801746	.06173307	.18384981	.39298261	.83104742
7	5	.24242424	.26060754	.31275378	.39210442	.48882605	.59186904	.81611565	.94133408	.99789254
		t	.26765607	.33894864	.44408064	.56587561	.68640318	.90356913	.98312065	.99991601
7	5	.01515152	.01888427	.03071622	.05235531	.08603747	.13373543	.31561468	.55047134	.89822853
		t	.01966671	.03422967	.06157386	.10521668	.16804573	.40548409	.68333816	.97067349
6	6	.08008658	.09189792	.12716396	.18506073	.26336549	.35781132	.61995158	.83189451	.98703912
		t	.09559986	.14203033	.21824534	.32031146	.44021620	.74000177	.92435026	.99869130
6	6	.00216450	.00297970	.00575106	.01145480	.02168864	.03855275	.12442837	.28665163	.70609606
		t	.00304885	.00610310	.01254644	.02442115	.04448146	.15011043	.35114828	.81449595
7	6	.05011655	.05971369	.08892164	.13852536	.20852335	.29688626	.56176508	.79625125	.98307155
		t	.06264319	.10109556	.16702294	.26007861	.37553095	.69054086	.90468111	.99825408
7	6	.00116550	.00168992	.00353038	.00751039	.01504670	.02812733	.10076900	.25087434	.67690208
		t	.00173798	.00378726	.00835448	.01728349	.03324386	.12554047	.31846529	.80166764
7	7	.02913753	.03639529	.05904865	.09922638	.15896446	.23856665	.49937511	.75470809	.97794168
		t	.03854013	.06836222	.12234616	.20350246	.31078147	.63429848	.88016886	.99765968
7	7	.00058275	.00089842	.00204996	.00468927	.01000399	.01978271	.07963836	.21617286	.64552804
		t	.00092875	.00222179	.00529313	.01171161	.02392994	.10241463	.28448399	.78622782

Table B-7 Power of one-sided two-sample Kolmogorov-Smirnov test vs. normal shift alternative d.

m	n	α	d = .2	.4	.6	.8	1.0	1.5	2.0	3.0
3	1	.25000000	.30408113	.36269560	.42457845	.48823116	.55203144	.70186270	.82279295	.95637437
		t	.30449150	.36361884	.42609511	.49038893	.55483521	.70599835	.82740912	.95939012
2	2	.16666667	.21449982	.26918351	.32975727	.39480253	.46254547	.63069050	.77491573	.94286569
		t	.21464351	.26953440	.33038014	.39575540	.46387086	.63296654	.77780527	.94517173
4	1	.20000000	.24963115	.30500135	.36508824	.42852600	.49369886	.65286476	.78783898	.94531133
		t	.25032310	.30659396	.36776176	.43240605	.49883389	.66074251	.79690073	.95143652
3	2	.10000000	.13715522	.18246273	.23573967	.29619213	.36242771	.54013749	.70716641	.92051393
		t	.13764968	.18373401	.23810279	.29995788	.36785606	.55011471	.72034528	.93109575
5	1	.16666667	.21227386	.26431674	.32202364	.38421340	.44936488	.61355475	.75845157	.93530472
		t	.21313871	.26634481	.32548849	.38932530	.45623488	.62445464	.77133214	.94431785
4	2	.20000000	.26073485	.32940784	.40397682	.48178412	.55984858	.73764381	.86736328	.97967387
		t	.26726595	.34429426	.42833009	.51568573	.60221910	.79036365	.91230560	.99278551
4	2	.06666667	.09596201	.13348103	.17966062	.23432814	.29662075	.47430205	.65376746	.90079395
		t	.09669406	.13544603	.18346000	.24060300	.30596193	.49263260	.67909779	.92186603
3	3	.05000000	.07478896	.10770305	.14955634	.20057409	.26024891	.43720951	.62357041	.88989100
		t	.07524361	.10898187	.15213899	.20501563	.26711407	.45183760	.64520209	.90079992
6	1	.14285714	.18497832	.23393612	.28918808	.34974515	.41421617	.58099353	.73315697	.92614917
		t	.18594645	.23624214	.29318643	.35572692	.42236119	.59429777	.74925955	.93779281
5	2	.14285714	.19481534	.25650122	.32661975	.40298700	.48273825	.67603465	.82838155	.97152811
		t	.19985414	.26863084	.34752107	.43355487	.52277255	.73143043	.88013121	.98893495
5	2	.04761905	.07124868	.10272598	.14292084	.19216342	.25009137	.42386889	.61011394	.88311041
		t	.07209714	.10508679	.14763924	.20019622	.26238445	.44941709	.64690883	.91492479
4	3	.20000000	.27023957	.35089695	.43876386	.52959561	.61872309	.80769980	.92406633	.99456725
		t	.27786098	.36807910	.46619654	.56636393	.66236668	.85263006	.95309947	.99829787
4	3	.02857143	.04572295	.07006039	.10296429	.14539961	.19766839	.36556437	.55944910	.86310011
		t	.04618485	.07144500	.10592862	.15077637	.20639037	.38620864	.59213559	.89557396
7	1	.25000000	.31011727	.37573900	.44511493	.51615713	.58662730	.74667863	.86610908	.97686117
		t	.31169979	.37918801	.45057135	.52358377	.59579993	.75826070	.87672539	.98097672

241

Table B-7 Power of one-sided two-sample Kolmogorov-Smirnov test vs. normal shift alternative d.

m	n	α	d = .2	.4	.6	.8	1.0	1.5	2.0	3.0
6	2	.21428571	.28127808	.35655801	.43739585	.52041425	.60202039	.77957032	.89960253	.98785082
		t	.29006014	.37643573	.46947970	.56419881	.65532908	.84006531	.94399541	.99725518
6	2	.03571429	.05518162	.08197661	.11725866	.16174143	.21549021	.38378931	.57348251	.86705914
		t	.05607058	.08452682	.12250198	.17090392	.22985133	.41517524	.62039669	.90918495
5	3	.23214286	.31163680	.40096110	.49570126	.59056676	.68026564	.85663829	.95148547	.99772830
		t	.32125271	.42204491	.52821880	.63238441	.72757033	.89856425	.97386295	.99947050
5	3	.07142857	.10783265	.15558299	.21498741	.28516059	.36396596	.57629874	.76406702	.95846206
		t	.11216717	.16715780	.23688505	.31996903	.41303760	.65472009	.84440135	.98721534
5	3	.01785714	.03019552	.04867975	.07495852	.11045937	.15609400	.31300578	.50829462	.83925962
		t	.03064967	.05011262	.07817364	.11654647	.16635979	.33921357	.55248834	.88576964
4	4	.11428571	.16837581	.23638735	.31696844	.40698564	.50185625	.72583861	.88286902	.99054697
		t	.17408304	.25060448	.34187298	.44335955	.54859442	.78291706	.92506721	.99701919
4	4	.01428571	.02486926	.04116277	.06490506	.09769854	.14069827	.29308250	.48883124	.83037372
		t	.02520858	.04226833	.06746078	.10267197	.14930078	.31627733	.52970003	.87617396
7	2	.16666667	.22652652	.29647951	.37443091	.45733151	.54154872	.73443091	.87246968	.98351048
		t	.23380143	.31367855	.40337632	.49844929	.59357115	.79894632	.92483230	.98582301
7	2	.08333333	.12160694	.17047015	.22995203	.29905011	.37571656	.58020118	.76159162	.95530926
		t	.12502192	.17937016	.24649215	.32504072	.41215498	.63904102	.82435365	.98121115
7	2	.02777778	.04411074	.06722768	.09847644	.13885665	.18878582	.35105400	.54213099	.85235078
		t	.04499754	.06983994	.10398124	.14869790	.20453783	.38704822	.59776710	.90414290
6	3	.16666667	.23577522	.31813945	.41048053	.50789856	.60464309	.80971111	.93114109	.99638754
		t	.24380367	.33689643	.44120749	.54973812	.65460249	.85973795	.96076981	.99909069
6	3	.04761905	.07574444	.11466040	.16554318	.22849121	.30227253	.51481631	.71820016	.94669384
		t	.07891281	.12362845	.18346516	.25848499	.34664615	.59373266	.80652146	.98261021
6	3	.01190476	.02108574	.03548362	.05684158	.08685300	.12685829	.27264368	.46631351	.81777278
		t	.02151588	.03690113	.06015241	.09335557	.13819683	.30366830	.52132553	.87861247
5	4	.21428571	.29653053	.39059881	.49131930	.59232348	.68724001	.86876841	.95969422	.99855863
		t	.30843486	.41721379	.53273900	.64552184	.74674361	.91795798	.98291853	.99982149
5	4	.07142857	.11264528	.16828875	.23872976	.32241285	.41578184	.65638326	.84372576	.98607013
		t	.11714742	.18036792	.26140440	.35772676	.46393581	.72324807	.89844356	.99566754
5	4	.03968254	.06509445	.10125670	.14971033	.21092669	.28396206	.49930847	.70931472	.94605317
		t	.06846655	.11104770	.16960278	.24493276	.33490792	.59123359	.81148569	.98522411
5	4	.00793651	.01481297	.02613929	.04370374	.06939151	.10488232	.24152600	.43371037	.80145386
		t	.01508419	.02708390	.04602534	.07416971	.11357945	.26761238	.48345073	.86228265
7	3	.20000000	.27761442	.36743579	.46503127	.56462190	.66009460	.84983360	.95109029	.99803876
		t	.28831930	.39163831	.50324262	.61456522	.71710980	.90000287	.97661740	.99966408
7	3	.08333333	.12652266	.18273437	.25170625	.33160154	.41910466	.64184264	.81990055	.97560872
		t	.13345324	.20109336	.28586112	.38453225	.49123686	.74404255	.90855572	.99630606

Table B-7 Power of one-sided two-sample Kolmogorov-Smirnov test vs. normal shift alternative d.

m	n	α	d = .2	.4	.6	.8	1.0	1.5	2.0	3.0
7	3	.03333333	.05543315	.08743081	.13107794	.18724707	.25553968	.46400578	.67752001	.93512789
		t	.05786314	.09465401	.14619376	.21366302	.29623551	.54316288	.77288361	.97818942
7	3	.00833333	.01536117	.02682575	.04446886	.07012065	.10541139	.24074999	.43111242	.79821359
		t	.01575883	.02818704	.04776204	.07680054	.11740674	.27565017	.49583120	.87306591
6	4	.14761905	.21722242	.30248163	.39983504	.50352281	.60660133	.82120309	.94087774	.99761747
		t	.22737369	.32689687	.44053119	.55926365	.67280696	.88365484	.97370688	.99969286
6	4	.09047619	.14101882	.20767271	.28972926	.38409293	.48553084	.80241730	.89263775	.99361279
		t	.14798668	.22581947	.32255603	.43300041	.54886052	.82758037	.94248086	.99874391
6	4	.04761905	.07940762	.12489455	.18570354	.26171534	.35063972	.59737902	.80720977	.98129763
		t	.08310163	.13541214	.20657313	.29593920	.39959157	.67275081	.87413151	.99435933
6	4	.02380952	.04172567	.06895781	.10773892	.15956757	.22468227	.43282405	.65541623	.93093828
		t	.04396282	.07593208	.12295170	.18711471	.26840665	.52236290	.76569656	.97970686
6	4	.00476190	.00943043	.01575558	.03089671	.05135630	.08092233	.20326640	.38936920	.77555698
		t	.00965437	.01840061	.03303185	.05596335	.08967610	.23198250	.44784432	.85222784
5	5	.17857143	.25947998	.35595965	.46260870	.57191418	.67580979	.87308575	.96585343	.99925206
		t	.26991487	.38002081	.50084172	.62154363	.73133531	.91692482	.98429455	.99988225
5	5	.03968254	.06808887	.10984155	.16701803	.24004453	.32713296	.57572157	.79373322	.97955883
		t	.07115659	.11885559	.18542910	.26220453	.37255954	.64934722	.86176695	.99355123
5	5	.00396825	.00804477	.01531453	.02744193	.04640347	.07424634	.19235802	.37663746	.76235953
		t	.00822603	.01599887	.02925343	.05039410	.08197445	.21876320	.43217128	.84443135
7	4	.20909091	.29779900	.40073361	.51124402	.62098528	.72176869	.90118480	.97658910	.99963995
		t	.30981136	.42730090	.55164773	.67105975	.77511474	.93796685	.98982106	.99994822
7	4	.06060606	.10020780	.15573997	.22813470	.31595100	.41515734	.67080891	.86226096	.99092053
		t	.10539424	.17017569	.25593812	.35989351	.47530376	.75124696	.92168131	.99804826
7	4	.03333333	.05826770	.09574161	.14820576	.21668824	.30011301	.54691233	.77337787	.97633492
		t	.06137838	.10505083	.16755879	.24983224	.34946473	.62979646	.85217669	.99314028
7	4	.01515152	.02810442	.04893728	.08020154	.12405892	.18167615	.37934329	.60836384	.91614680
		t	.02967632	.05414366	.09222017	.14700338	.21992942	.46655183	.72546850	.97441928
7	4	.00303030	.00631809	.01234863	.02267924	.03923919	.06413002	.17392025	.35287009	.75212534
		t	.00650540	.01307328	.02464107	.04365068	.07283319	.20479686	.41945331	.84476636
6	5	.23809524	.33480445	.44397851	.55760225	.66662523	.76307292	.92356008	.98385838	.99981158
		t	.34941707	.47552422	.60413718	.72223275	.81986745	.95801486	.99447425	.99998451
6	5	.08874459	.14220025	.21371920	.30226814	.40381925	.51180395	.75988634	.91505331	.99625362
		t	.15039354	.23538007	.34163172	.46213342	.58613665	.84088843	.96220908	.99565508
6	5	.02380952	.04382101	.07538656	.12154013	.18417357	.26316748	.50925887	.74796572	.97266080
		t	.04604908	.08243406	.13695414	.21182676	.30612745	.58820547	.82845688	.99146521
6	5	.00216450	.00477081	.00957265	.01820948	.03251520	.05465844	.15692817	.33155577	.73871176
		t	.00484102	.01009894	.01969280	.03597602	.06172036	.18384935	.39298260	.83104742

Table B-7 **Power of one-sided two-sample Kolmogorov-Smirnov test vs. normal shift alternative d.**

m	n	α	d = .2	.4	.6	.8	1.0	1.5	2.0	3.0
7	5	.21717172	.31343439	.42443694	.54176575	.65545633	.75647113	.92369897	.98470180	.99984797
		t	.32797516	.45648573	.58974113	.71328253	.81569261	.95898544	.99504112	.99998964
7	5	.08333333	.13715467	.21076483	.30334230	.41044754	.52448454	.78166817	.93178127	.99809912
		t	.14512164	.23195061	.34180412	.46696285	.59540650	.85344284	.96852709	.99974119
7	5	.03282828	.05941568	.10026321	.15820247	.23416285	.32633035	.59125849	.81630255	.98648287
		t	.06338235	.11241813	.18376484	.27792709	.39071072	.69110238	.89831880	.99751035
7	5	.01515152	.02960564	.05382049	.09126954	.14480910	.21567230	.45398723	.70629283	.96552947
		t	.03130014	.05952363	.10448524	.16982263	.25649929	.53740597	.79867017	.98951074
7	5	.00757576	.01543133	.02952943	.05178724	.08583326	.13366788	.31561175	.55047127	.89822853
		t	.01645600	.03297943	.06112737	.10507068	.16800209	.40648266	.68333814	.97067349
7	5	.00126263	.00292398	.00629155	.01261318	.02362885	.04149020	.13048091	.29506986	.71212563
		t	.00302168	.00671063	.01386067	.02668877	.04802505	.15788139	.36214759	.82097845
6	6	.23809524	.34161502	.45883731	.57987568	.69381345	.79157866	.94237492	.99029319	.99994266
		t	.35550432	.48859208	.62296009	.74379485	.84058747	.96792748	.99658707	.99999489
6	6	.07142857	.12090148	.19037751	.27983796	.38553008	.50018142	.76574545	.92548937	.99787161
		t	.12774090	.20915812	.31495052	.43853056	.56834630	.83841725	.96421785	.99969162
6	6	.01298701	.02596893	.04819228	.08323471	.13421163	.20273769	.43865270	.69465836	.96356386
		t	.02741988	.05319584	.09509001	.15711112	.24081395	.51957533	.78706224	.98854163
6	6	.00108225	.00255582	.00559789	.01140386	.02167308	.03854840	.12442825	.28665162	.70609606
		t	.00263849	.00595956	.01250022	.02440747	.04447775	.15011033	.35114827	.81449595
7	6	.21911422	.32157179	.44016984	.56473237	.68335842	.78574216	.94291305	.99106959	.99995989
		t	.33699464	.47360359	.61340570	.73975281	.84059744	.97011970	.99717924	.99999715
7	6	.07342657	.12526332	.19831886	.29235244	.40291866	.52171611	.78890227	.93899847	.99872708
		t	.13388081	.22183947	.33577281	.46719284	.60222217	.86682641	.97513457	.99987337
7	6	.04545455	.08255126	.13861765	.21593663	.31331255	.42540869	.70981729	.90079109	.99681317
		t	.08788663	.15444423	.24773333	.36462168	.49559855	.79529685	.95138223	.99954679
7	6	.01923077	.03754174	.06789703	.11410004	.17875629	.26206103	.52500580	.77354796	.98168901
		t	.04061095	.07808346	.13714712	.22092487	.32793644	.64057421	.87725421	.99703840
7	6	.00757576	.01626933	.03223742	.05911270	.10063392	.15962803	.38143146	.64689115	.95420988
		t	.01727206	.03595853	.06855013	.12004551	.19382570	.46366976	.75024023	.98575603
7	6	.00407925	.00901812	.01842385	.03489260	.06146684	.10108373	.26638393	.50131534	.88125993
		t	.00970406	.02112477	.04219141	.07753639	.13153740	.35646251	.64642380	.96723729
7	7	.10606061	.17722442	.27277231	.38852156	.51506776	.63996167	.87653420	.97464419	.99979475
		t	.18788956	.30020677	.43562488	.57906571	.71257385	.92837069	.99100002	.99998453
7	7	.02651515	.05232537	.09475485	.15800806	.24362099	.34895429	.64497892	.86910681	.99524605
		t	.05607578	.10692296	.18457590	.28992398	.41695552	.74143460	.93344576	.99931489
7	7	.00407925	.00950823	.02030821	.03987263	.07221317	.12110040	.32402178	.59445013	.94262911
		t	.01014308	.02286961	.04689127	.08771645	.15026021	.40464221	.70730185	.98210625

244

Table B-8 **Power of two-sided two-sample Kolmogorov-Smirnov test vs. normal shift alternative d.**

m	n	α	d = .2	.4	.6	.8	1.0	1.5	2.0	3.0
3	2	.20000000 t	.20779141 .20801832	.23074858 .23164801	.26765048 .26964058	.31656128 .32000708	.37497491 .38015588	.54331879 .55320219	.70780192 .72095621	.92052591 .93110707
4	2	.13333333 t	.14066165 .14103443	.16237316 .16386536	.19764379 .20099587	.24509596 .25101045	.30281746 .31189799	.47560365 .49385204	.65397722 .67929014	.90079632 .92186812
3	3	.10000000 t	.10692590 .10717943	.12753479 .12856370	.16129173 .16365200	.20722544 .21149914	.26385577 .27060768	.43784911 .45244777	.62365475 .64528144	.88989158 .90980045
5	2	.09523810 t	.10187743 .10234166	.12166246 .12353964	.15416186 .15844392	.19856368 .20626224	.25358325 .26564714	.42450228 .44998809	.61020039 .64698409	.88311107 .91492533
4	3	.22857143 t	.23933199 .24216254	.27076172 .28168217	.32040513 .34351116	.38449209 .42209059	.45837849 .51061395	.65385738 .73004192	.81621909 .88674886	.97026581 .99180737
4	3	.05714286 t	.06279054 .06307547	.07979248 .08097417	.10825393 .11105263	.14813656 .15339933	.19901486 .20766715	.36584636 .38637685	.55946664 .59215143	.86310017 .89557401
7	1	.25000000 t	.25696759 .25735194	.27751287 .27902173	.31058916 .31387794	.35453670 .36012266	.40720397 .41541969	.56038335 .57516021	.71288467 .73154068	.91777544 .93179788
6	2	.21428571 t	.22438045 .22623120	.25394885 .26114417	.30091363 .31633080	.36204195 .38756574	.43327124 .46950578	.62633183 .68305098	.79358263 .85102914	.96336956 .98503538
6	2	.07142857 t	.07737682 .07788961	.09520518 .09729926	.12481150 .12965928	.16586803 .17473995	.21764565 .23181615	.38413636 .41547631	.57352394 .62043092	.86705937 .90918512
5	3	.14285714 t	.15296617 .15525668	.18273972 .19174362	.23052273 .25013488	.29359792 .32670525	.36831073 .41626481	.57694884 .65511300	.76413520 .84443374	.95846231 .98721541
5	3	.03571429 t	.04024663 .04054574	.05405567 .05532182	.07768673 .08077487	.11177122 .11777713	.15669090 .16691079	.31307074 .33927123	.50829950 .55249253	.83925963 .88576964
4	4	.22857143 t	.24214801 .24482723	.28159622 .29181371	.34323026 .36442409	.42142746 .45493607	.50936588 .55416213	.72697784 .78358398	.88298868 .92511929	.99054740 .99701927
4	4	.02857143 t	.03265717 .03288872	.04518500 .04617928	.06686995 .06934446	.09860505 .10352891	.14109268 .14966852	.29312029 .31631143	.48883369 .52970217	.83037373 .87617396
7	2	.16666667 t	.17637753 .17806681	.20494673 .21157580	.25071005 .26512246	.31099059 .33529959	.38227149 .41752787	.58138695 .63989353	.76174884 .82445115	.95531031 .98121161
7	2	.05555556 t	.06087670 .06140956	.07691493 .07911150	.10382841 .10898474	.14168094 .15127518	.19020801 .20580394	.35126127 .38722148	.54215313 .59778448	.85235088 .90414297
6	3	.09523810 t	.10417853 .10603109	.13075709 .13818051	.17416802 .19081327	.23285864 .26197225	.30435971 .34819989	.51507008 .59388787	.71822114 .80653171	.94669389 .98261022
6	3	.02380952 t	.02746025 .02775722	.03871538 .03999727	.05839047 .06160380	.08755373 .09399770	.12715713 .13846465	.27267087 .30369138	.46631517 .52132688	.81777278 .87861247

Table B-8 Power of two-sided two-sample Kolmogorov-Smirnov test vs. normal shift alternative d.

m	n	α	d = .2	.4	.6	.8	1.0	1.5	2.0	3.0
5	4	.14285714	.15546539	.19251397	.25164369	.32889082	.41883608	.65673655	.84375302	.98607018
		t	.15794021	.20214571	.27225433	.36276389	.46611212	.72344166	.89845432	.99566755
5	4	.07936508	.08798542	.11372748	.15611534	.21402301	.28536888	.49945648	.70932502	.94605318
		t	.09004489	.12203717	.17492640	.24726025	.33587328	.59131185	.81148970	.98522411
5	4	.01587302	.01882511	.02804939	.04455860	.06975056	.10502363	.24153624	.43371085	.80145386
		t	.01902220	.02892360	.04683324	.07450273	.11370812	.26762130	.48345113	.86228265
7	3	.25000000	.26486759	.30787481	.37446916	.45781906	.54984777	.76772962	.91083740	.99484450
		t	.26806117	.31993537	.39908317	.49586477	.59925909	.82424977	.94747526	.99864799
7	3	.06666667	.07440482	.09763169	.13625082	.18971714	.25664872	.46411927	.67752772	.93512790
		t	.07592784	.10385011	.15058529	.21562793	.29705835	.54323233	.77288741	.97818942
7	3	.01666667	.01963615	.02889600	.04541375	.07052650	.10557520	.24076280	.43111308	.79821359
		t	.01992073	.03014742	.04863180	.07716356	.11754912	.27566055	.49583170	.87306591
6	4	.18095238	.19573947	.23880854	.30636759	.39243063	.48944377	.72802038	.89266991	.99361284
		t	.19944789	.25298645	.33584105	.43900824	.55136947	.80261537	.94249016	.99874391
6	4	.09523810	.10628781	.13915283	.19279998	.26502517	.35208483	.59751466	.80721808	.98129764
		t	.10852224	.14802973	.21238631	.29842149	.40057270	.67281902	.87413437	.99435933
6	4	.04761905	.05450864	.07540150	.11078280	.16091263	.22523739	.43286868	.65541851	.93093828
		t	.05600113	.08160446	.12543864	.18812779	.26878959	.52238696	.76569749	.97970686
6	4	.00952381	.01168840	.01857896	.03131376	.05151815	.08098088	.20326975	.38936932	.77555698
		t	.01185841	.01935624	.03341938	.05611009	.08972792	.23198530	.44784441	.85222784
5	5	.07936508	.08978407	.12094613	.17232985	.24241535	.32811891	.57580165	.79373730	.97955883
		t	.09172359	.12874267	.18982963	.27285651	.37324734	.64939001	.86176853	.99355124
5	5	.00793651	.00987867	.01610688	.02776134	.04652338	.07428819	.19236018	.37663753	.76825953
		t	.01001862	.01675578	.02955164	.05050354	.08201181	.21876503	.43217134	.84443135
7	4	.21212121	.22864412	.27641121	.35024840	.44231410	.54326240	.77677099	.92135205	.99648829
		t	.23316864	.29349036	.38502331	.49573863	.61183682	.85042167	.96380428	.99953532
7	4	.06666667	.07615807	.10473529	.15243473	.21854610	.30087485	.54697274	.77338094	.97633493
		t	.07816696	.11287751	.17093037	.25117234	.34995554	.62982411	.85217760	.99314028
7	4	.03030303	.03575407	.05254616	.08178923	.12470907	.18192354	.37935916	.60836447	.91614680
		t	.03686804	.05731159	.09351315	.14749164	.22009979	.46656044	.72546876	.97441928
7	4	.00606061	.00767746	.01291776	.02290118	.03931966	.06415710	.17392152	.35287013	.75212534
		t	.00782434	.01360875	.02484351	.04372184	.07285641	.20479787	.41945333	.84476636
6	5	.23809524	.25678686	.31043827	.39218267	.49189582	.59802478	.82666623	.94930672	.99873557
		t	.26141485	.32764846	.42637075	.54266086	.66034444	.88455261	.97654942	.99981704
6	5	.04761905	.05588483	.08107481	.12403140	.18518536	.26354797	.50928225	.74796659	.97266080
		t	.05741706	.08743593	.13897862	.21257931	.30638400	.58821729	.82845719	.99146521
6	5	.00432900	.00563711	.00994101	.01834530	.03256154	.05467304	.15692874	.33155579	.73871176
		t	.00574187	.01044725	.01981768	.03601746	.06173307	.18384981	.39298261	.83104742

Table B-8 **Power of two-sided two-sample Kolmogorov-Smirnov test vs. normal shift alternative d.**

m	n	α	d = .2	.4	.6	.8	1.0	1.5	2.0	3.0
7	5	.23737374	.25728099	.31428246	.40068256	.50518934	.61505887	.84426569	.95875062	.99925075
		t	.26247175	.33346499	.43837809	.56028085	.68135456	.90104279	.98246805	.99991017
		.06565657	.07628439	.10830636	.16175397	.23561470	.32687649	.59129169	.81630374	.98648287
		t	.07895954	.11917405	.18643767	.27889005	.39102617	.69111505	.89831907	.99751035
		.03030303	.03680287	.05698704	.09255752	.14529262	.21583958	.45399541	.70629306	.96552947
		t	.03803609	.06226822	.10550828	.17017091	.25660742	.53740982	.79867024	.98951074
		.01515152	.01888427	.03071622	.05235531	.08603747	.13373543	.31561468	.55047134	.89822853
		t	.01966671	.03422967	.06157386	.10521668	.16804573	.40548409	.68333816	.97067349
		.00252525	.00342925	.00647847	.01267696	.02364888	.04149597	.13048108	.29506986	.71212563
		t	.00351025	.00688533	.01391830	.02670625	.04802994	.15788153	.36214760	.82097846
6	6	.14285714	.16004769	.21023111	.28913717	.38954597	.50177818	.76585556	.92549382	.99787161
		t	.16417612	.22606202	.32206603	.44124260	.56928067	.83845872	.96421880	.99969162
		.02597403	.03198116	.05076285	.08424765	.13457880	.20285992	.43865803	.69465849	.96356386
		t	.03305842	.05543514	.09590196	.15737944	.24089464	.51957796	.78706229	.98854163
		.00216450	.00297970	.00575106	.01145480	.02168864	.03855275	.12442837	.28665163	.70609606
		t	.00304886	.00610310	.01254644	.02442115	.04448146	.15011043	.35114828	.81449595
7	6	.21212121	.23294184	.29260738	.38314430	.49270505	.60775132	.84565866	.96107227	.99939939
		t	.23919299	.31575316	.42869608	.55922260	.68740413	.91138646	.98653091	.99995857
		.09090909	.10562640	.14939188	.22055442	.31512624	.42606058	.70985131	.90079210	.99681317
		t	.10917858	.16345140	.25116623	.36579811	.49596043	.79530858	.95138242	.99954679
		.03846154	.04664421	.07186883	.11569436	.17934394	.26225959	.52501472	.77354818	.98168901
		t	.04887892	.08130308	.13828034	.22128474	.32803938	.64057703	.87725425	.99703840
		.01515152	.01951545	.03351424	.05957270	.10078541	.15967355	.38143295	.64689118	.95420988
		t	.02029883	.03705767	.06891219	.12015351	.19385482	.46367046	.75024024	.98575603
		.00815851	.01071839	.01907513	.03512130	.06154030	.10110527	.26638459	.50131536	.88125993
		t	.01126956	.02167208	.04236537	.07758658	.13155051	.35646280	.64642381	.96723729
		.00116550	.00168992	.00353038	.00751039	.01504670	.02812733	.10076900	.25087434	.67690208
		t	.00173798	.00378726	.00835448	.01728349	.03324386	.12554047	.31846529	.80166764
7	7	.21212121	.23547015	.30203837	.40194441	.52065731	.64209483	.87665935	.97464821	.99979475
		t	.24159211	.32449225	.44540001	.58255822	.71367904	.92840690	.99100056	.99998453
		.05303030	.06461796	.09995527	.16001120	.24432217	.34917697	.64498711	.86910697	.99524605
		t	.06734136	.11121172	.18603540	.29036707	.41707532	.74143732	.93344579	.99931489
		.00815851	.01110725	.02087937	.04005809	.07226779	.12111496	.32402212	.59445013	.94262911
		t	.01162690	.02335666	.04703523	.08775469	.15026933	.40464236	.70730185	.98210625
		.00058275	.00089842	.00204996	.00468927	.01000399	.01978271	.07963836	.21617286	.64552804
		t	.00092875	.00222180	.00529313	.01171162	.02392994	.10241464	.28448401	.78622784

Table B-9 **Power of one-sided two-sample Student t test vs. normal shift alternative d.**

m	n	α	d = .2	.4	.6	.8	1.0	1.5	2.0	3.0
3	1	.250	.30449150	.36361864	.42609511	.49038893	.55483521	.70599835	.82740912	.95939012
3	1	.100	.12706432	.15856884	.19448031	.23458490	.27848610	.40077923	.53083021	.76241507
3	1	.050	.06445001	.08167943	.10184644	.12503592	.15125001	.22924825	.32183641	.52593346
3	1	.025	.03245837	.04145974	.05213895	.06460417	.07893148	.12306826	.17861006	.31633641
3	1	.010	.01303987	.01673553	.02115603	.02636312	.03240914	.05143812	.07626550	.14258412
3	1	.005	.00652940	.00839327	.01062883	.01327026	.01634776	.02610417	.03899023	.07429849
2	2	.250	.31336429	.38265862	.45580736	.53035810	.60372738	.76688108	.88338517	.98268405
2	2	.100	.13164658	.16921541	.21259135	.26133585	.31469378	.46101039	.60918701	.84162897
2	2	.050	.06692893	.08759978	.11223072	.14090400	.17355047	.27072476	.38388882	.61722223
2	2	.025	.03374634	.04457902	.05769656	.07324694	.09131778	.14750204	.21800447	.38738935
2	2	.010	.01356691	.01802276	.02347142	.03000257	.03768794	.06223645	.09450439	.17995629
2	2	.005	.00679491	.00904359	.01180232	.01512146	.01904367	.03168707	.04856997	.09479882
4	1	.250	.30761059	.37036410	.43673434	.50489487	.57286867	.72963921	.85061078	.97072827
4	1	.100	.13011436	.16583699	.20717951	.25387296	.30535365	.44887491	.59804386	.83914658
4	1	.050	.06664444	.08712084	.11176665	.14080255	.17430144	.27648574	.39908921	.65597368
4	1	.025	.03382807	.04493773	.05864341	.07522873	.09492704	.15866383	.24273108	.45360022
4	1	.010	.01369001	.01841582	.02435881	.03170274	.04062668	.07089553	.11397206	.23944606
4	1	.005	.00688009	.00930646	.01238360	.01622142	.02093232	.03725152	.06128043	.13622969
3	2	.250	.32131451	.39985313	.48261384	.56606761	.64660433	.81591951	.92257651	.99314636
3	2	.100	.13764968	.18373401	.23810279	.29995788	.36785606	.55011471	.72034528	.93109575
3	2	.050	.07089904	.09766674	.13085377	.17071610	.21714206	.35755604	.51697445	.79739470
3	2	.025	.03611465	.05075921	.06949403	.09279039	.12097873	.21326580	.33263349	.60229273
3	2	.010	.01465552	.02092553	.02914534	.03964851	.05274689	.09846321	.16417560	.34763430
3	2	.005	.00737420	.01060261	.01488078	.02041355	.02740654	.05252595	.09038983	.20698952
5	1	.250	.30949902	.37444089	.44313724	.51356613	.58354883	.74316287	.86321511	.97599025
5	1	.100	.13197886	.17029999	.21498652	.26570377	.32173799	.47731161	.63561978	.87440887
5	1	.050	.06805671	.09066850	.11829764	.15124849	.18961643	.30757218	.44818347	.72674257
5	1	.025	.03476236	.04736221	.06326364	.08289507	.10661416	.18509891	.29002811	.54607437
5	1	.010	.01416660	.01968716	.02685403	.03597539	.04736248	.08762711	.14734766	.32474571
5	1	.005	.00714947	.01003478	.01383378	.01874317	.02497383	.04776442	.08341030	.20083753
4	2	.250	.32623480	.41049264	.49907374	.58766162	.67194448	.84236709	.94092035	.99628889
4	2	.100	.14150613	.19319553	.25483058	.32522651	.40226693	.60414737	.77997043	.96228059
4	2	.050	.07358204	.10465843	.14404261	.19207408	.24847871	.41824525	.60261933	.88006909
4	2	.025	.03780456	.05534746	.07854236	.10816414	.14473478	.26657592	.42223643	.73667604
4	2	.010	.01548561	.02325993	.03392840	.04811985	.06643952	.13341278	.23269225	.49943325
4	2	.005	.00783524	.01192167	.01763473	.02539193	.03563416	.07491074	.13778874	.33454573

Table B-9 Power of one-sided two-sample Student t test vs. normal shift alternative d.

m	n	α	d = .2	.4	.6	.8	1.0	1.5	2.0	3.0
3	3	.250	.33114999	.42109142	.51532481	.60866268	.69607119	.86561145	.95523862	.99801083
3	3	.100	.14434951	.20015750	.26703942	.34340177	.42652614	.63938177	.81455945	.97496723
3	3	.050	.07524361	.10898187	.15213899	.20501563	.26711407	.45183760	.64520209	.90979992
3	3	.025	.03872416	.05784303	.08343524	.11638624	.15723610	.29293454	.46260813	.78255418
3	3	.010	.01588613	.02438749	.03623037	.05216384	.07289484	.14909868	.26127378	.55056355
3	3	.005	.00804394	.01252010	.01888150	.02763183	.03929899	.08448593	.15685082	.37777657
6	1	.250	.31077573	.37719517	.44745315	.51938826	.59068019	.75200429	.87119826	.97901628
6	1	.100	.13323464	.17331420	.22026179	.27368320	.33274173	.49600160	.65939162	.89398541
6	1	.050	.06903144	.09313631	.12286495	.15857401	.20035808	.32912004	.48119557	.76884367
6	1	.025	.03542828	.04911380	.06663885	.08854390	.11527727	.20474884	.32463113	.60765153
6	1	.010	.01452181	.02065477	.02878911	.03934402	.05274855	.10133325	.17488034	.39213039
6	1	.005	.00735683	.01061018	.01500733	.02082922	.02838405	.05699335	.10331091	.25873056
5	2	.250	.32955287	.41775028	.51003832	.60215080	.68867662	.85876108	.95124187	.99760069
5	2	.100	.14418306	.19981704	.26656355	.34288307	.42607398	.63961672	.81560941	.97581877
5	2	.050	.07549511	.10972585	.15370063	.20777923	.27147159	.46128576	.65886862	.92064572
5	2	.025	.03904852	.05880996	.08550205	.12014521	.16337854	.30799027	.48834676	.81502157
5	2	.010	.01612365	.02511899	.03785428	.05524753	.07818424	.16408972	.29198322	.61293933
5	2	.005	.00820076	.01301283	.02000030	.02981052	.04314167	.09630857	.18364431	.44463411
4	3	.250	.33769604	.43521640	.53683432	.63606165	.72686559	.89279506	.96989108	.99919204
4	3	.100	.14899553	.21173024	.28753641	.37401264	.46717381	.69591494	.86537240	.98870778
4	3	.050	.07836549	.11734623	.16815713	.23101811	.30487533	.51926882	.72627838	.95475001
4	3	.025	.04066814	.06333487	.09457249	.13561861	.18708912	.35732132	.55937944	.87643915
4	3	.010	.01684522	.02723333	.04231845	.06330520	.09131092	.19664373	.34984918	.69855809
4	3	.005	.00858232	.01415983	.02249091	.03444596	.05095007	.11764173	.22628697	.53405534
7	1	.250	.31169979	.37918801	.45057135	.52358377	.59579993	.75826070	.87672539	.98097672
7	1	.100	.13413829	.17548744	.22406640	.27943016	.34064228	.50921359	.67574480	.90626853
7	1	.050	.06974246	.09494585	.12622527	.16397142	.20826827	.34482372	.50468022	.79612354
7	1	.025	.03592366	.05042886	.06919138	.09283843	.12188443	.21971101	.35055365	.65018794
7	1	.010	.01479389	.02140690	.03031272	.04202541	.05707401	.11248134	.19723433	.44407017
7	1	.005	.00751925	.01106939	.01595994	.02254838	.03123191	.06488854	.12051208	.30751334
6	2	.250	.33203507	.42302560	.51831871	.61255331	.70054819	.86986689	.95775845	.99826838
6	2	.100	.14614824	.20470379	.27523008	.35587758	.44345150	.66446413	.83885263	.98276658
6	2	.050	.07692258	.11354819	.16103069	.21971520	.28888480	.49288760	.69760599	.94274846
6	2	.025	.03999582	.06149172	.09099872	.12960857	.17814052	.34024790	.53731937	.86210315
6	2	.010	.01662395	.02661362	.04107593	.06119028	.08808607	.19011336	.34097373	.69310316
6	2	.005	.00849378	.01391486	.02200837	.03364534	.04977343	.11566452	.22485833	.54031970

249

Table B-9 **Power of one-sided two-sample Student t test vs. normal shift alternative d.**

m	n	α	d = .2	.4	.6	.8	1.0	1.5	2.0	3.0
5	3	.250	.34238845	.44532707	.55208881	.65516304	.74780385	.90956399	.97768791	.99959194
5	3	.100	.15241018	.22031845	.30276879	.39659979	.49677631	.73400778	.89542731	.99398377
5	3	.050	.08071251	.12375000	.18054261	.25115391	.33391339	.56833360	.77908660	.97446317
5	3	.025	.04216512	.06767807	.10354509	.15127383	.21143884	.40818699	.62951696	.92480244
5	3	.010	.01760757	.02958102	.04749131	.07298046	.10752933	.23853076	.42406722	.79456815
5	3	.005	.00902019	.01555395	.02567588	.04064138	.06178535	.14922739	.29093642	.65450733
4	4	.250	.34562934	.45229349	.56251388	.66803616	.76163873	.91984369	.98196148	.99974830
4	4	.100	.15439095	.22530359	.31156541	.40950194	.51331109	.75400771	.90963661	.99578362
4	4	.050	.08191784	.12704460	.18688996	.26137501	.34843051	.59136437	.80153433	.98059976
4	4	.025	.04285783	.06969334	.10769698	.15845889	.22246334	.42992730	.65687130	.93833574
4	4	.010	.01792291	.03055592	.04963680	.07696837	.11413966	.25478052	.45065807	.82133631
4	4	.005	.00918932	.01609503	.02691197	.04303441	.06593159	.16080596	.31303230	.68752688
7	2	.250	.33389279	.42703639	.52444164	.62038702	.70940740	.87786682	.96220755	.99865560
7	2	.100	.14765197	.20845694	.28188806	.36582867	.45666996	.68275897	.85503758	.98678695
7	2	.050	.07802674	.11652772	.16676747	.22905886	.30246662	.51689051	.72552358	.95594646
7	2	.025	.04073892	.06362124	.09532827	.13722051	.19001967	.36576357	.57436696	.89192848
7	2	.010	.01702466	.02783301	.04374311	.06616338	.09642816	.21200367	.38099800	.74961184
7	2	.005	.00873233	.01466634	.02371406	.03695396	.05556260	.13275549	.26069133	.61118768
6	3	.250	.34592426	.45293407	.56347988	.66923490	.76292983	.92079645	.98234783	.99976100
6	3	.100	.15502674	.22694162	.31451485	.41390210	.51904881	.76106037	.91463811	.99637331
6	3	.050	.08253924	.12879740	.19036283	.26710599	.35674210	.60503403	.81501944	.98410217
6	3	.025	.04335019	.07118211	.11087339	.16413036	.23140738	.44849135	.68083015	.95095181
6	3	.010	.01822522	.03153550	.05188789	.08132065	.12161384	.27449072	.48433046	.85529003
6	3	.005	.00938115	.01674185	.02846364	.04617663	.07160348	.17803193	.34781092	.74093488
5	4	.250	.35148898	.46487128	.58118068	.69073932	.78550695	.93613999	.98793521	.99989956
5	4	.100	.15848459	.23570699	.32997705	.43640339	.54754713	.79302674	.93479529	.99812078
5	4	.050	.08467041	.13470505	.20181785	.28552858	.38267261	.64389473	.84906900	.99058052
5	4	.025	.04459033	.07486582	.11856376	.17751303	.25189542	.48730099	.72543448	.96744906
5	4	.010	.01879855	.03336096	.05599491	.08906086	.13454435	.30576180	.53224226	.89162469
5	4	.005	.00969199	.01777225	.03088550	.05096602	.08001817	.20158029	.39074125	.79149608
7	3	.250	.34868715	.45886976	.57231361	.68002899	.77435179	.92878252	.98537429	.99984452
7	3	.100	.15709613	.23220350	.32384013	.42755685	.53647584	.78110666	.92770000	.99760267
7	3	.050	.08400050	.13287239	.19832102	.28001059	.37507421	.63325403	.84052293	.98929687
7	3	.025	.04431020	.07406102	.11694608	.17481505	.24795561	.48081265	.71916538	.96599238
7	3	.010	.01873441	.03318047	.05564470	.08851533	.13380979	.30512261	.53293735	.89421446
7	3	.005	.00968274	.01776005	.03090171	.05108879	.08039502	.20375074	.39658455	.80169571

Table B-9 **Power of one-sided two-sample Student t test vs. normal shift alternative d.**

m	n	α	d = .2	.4	.6	.8	1.0	1.5	2.0	3.0
6	4	.250	.35600875	.47454566	.59538137	.70768220	.80283750	.94679605	.99127390	.99995215
6	4	.100	.16170381	.24394365	.34452949	.45745088	.57383679	.82056331	.95024517	.99903849
6	4	.050	.08686865	.14090654	.21396413	.30511271	.41010157	.68318323	.88046859	.99490880
6	4	.025	.04599567	.07914739	.12766538	.19352384	.27649425	.53284293	.77433202	.98127092
6	4	.010	.01952344	.03575061	.06152325	.09971468	.15256964	.34956350	.59683287	.93163535
6	4	.005	.01011436	.01923126	.03443556	.05818568	.09297606	.23873020	.45731062	.85862502
5	5	.250	.35835774	.47956005	.60268099	.71627105	.81145273	.95171635	.99265475	.99996772
5	5	.100	.16319325	.24775807	.35123732	.46705316	.58564101	.83215184	.95612172	.99929269
5	5	.050	.08779955	.14353925	.21910476	.31333002	.42144826	.69848853	.89158281	.99603827
5	5	.025	.04654445	.08082541	.13122599	.19974213	.28592762	.54936418	.79054164	.98470172
5	5	.010	.01978118	.03660441	.06349776	.10349662	.15890370	.36421507	.61674497	.94110783
5	5	.005	.01025562	.01972202	.03563047	.06060528	.09727886	.25053071	.47693466	.87415152
7	4	.250	.35960718	.48222782	.60655379	.72080159	.81595739	.95419993	.99331483	.99997401
7	4	.100	.16430045	.25062940	.35631581	.47434157	.59459461	.84080225	.96035008	.99945213
7	4	.050	.08866554	.14602722	.22402344	.32126395	.43246550	.71333281	.90211406	.99699235
7	4	.025	.04715838	.08274465	.13537531	.20709951	.29721954	.56940053	.80999625	.98845272
7	4	.010	.02013287	.03780476	.06634777	.10908086	.16843568	.38695145	.64782767	.95496795
7	4	.005	.01047371	.02050683	.03760274	.06471218	.10476096	.27195861	.51321126	.90165210
6	5	.250	.36374951	.49104205	.61924736	.73546561	.83029061	.96162527	.99511952	.99998735
6	5	.100	.16696414	.25747474	.36831942	.49134206	.61510944	.85937906	.96862310	.99969079
6	5	.050	.09034345	.15082384	.23341572	.33620028	.45281999	.73897763	.91864913	.99813050
6	5	.025	.04815660	.08584721	.14201419	.21869824	.31468047	.59834716	.83553043	.99219170
6	5	.010	.02060713	.03941316	.07012603	.11637373	.18064644	.41419204	.68174146	.96654566
6	5	.005	.01073579	.02144380	.03993243	.06949226	.11330682	.29489495	.54869387	.92246954
7	5	.250	.36810457	.50028021	.63241272	.75040544	.84452959	.96833151	.99653497	.99999424
7	5	.100	.17005785	.26548422	.38234532	.51103643	.63850133	.87910503	.97644527	.99984757
7	5	.050	.09245539	.15693602	.24544652	.35529616	.47860924	.76977459	.93666015	.99902728
7	5	.025	.04951075	.09013448	.15129439	.23498563	.33914086	.63744742	.86738557	.99570132
7	5	.010	.02131143	.04186489	.07599841	.12785270	.19998047	.45669694	.73175598	.98005474
7	5	.005	.01114978	.02297121	.04382622	.07763022	.12802986	.33445074	.60750506	.95078591
6	6	.250	.36991110	.50410071	.63781111	.75644602	.85017514	.97080194	.99700231	.99999586
6	6	.100	.17123634	.26853731	.38766605	.51843071	.64714561	.88594161	.97889827	.99988348
6	6	.050	.09320694	.15911633	.24972499	.36202947	.48757612	.77986621	.94202858	.99922482
6	6	.025	.04996238	.09156966	.15439566	.24039008	.34715648	.64957442	.87641764	.99644610
6	6	.010	.02152867	.04262492	.07781806	.13139093	.20587657	.46903169	.74511648	.98280835
6	6	.005	.01127087	.02342060	.04497272	.08001605	.13230594	.34542778	.62261148	.95642705

Table B-9 **Power of one-sided two-sample Student t test vs. normal shift alternative d.**

m	n	α	d = .2	.4	.6	.8	1.0	1.5	2.0	3.0
7	6	.250	.37494651	.51471752	.65267431	.77282191	.86515244	.97684837	.99801974	.99999840
7	6	.100	.17478414	.27776578	.40371400	.54054685	.67262720	.90486386	.98504191	.99995000
7	6	.050	.09561294	.16616058	.26358977	.38378095	.51627518	.81059470	.95701458	.99964306
7	6	.025	.051149698	.09651565	.16517018	.25920393	.37494637	.69010327	.90445787	.99824563
7	6	.010	.022323316	.04546201	.08471116	.14490815	.22846092	.51536487	.79255728	.99068756
7	6	.005	.01173699	.02519433	.04958581	.08974540	.14987589	.39022425	.68165527	.97467293
7	7	.250	.38059281	.52655585	.66899058	.79034352	.88061635	.98231545	.99877713	.99999947
7	7	.100	.17874842	.28811119	.42158152	.56475673	.69977663	.92289274	.98995813	.99998119
7	7	.050	.09829222	.17407549	.27915037	.40791947	.54744967	.84086382	.96958152	.99985372
7	7	.025	.05320105	.10208647	.17735659	.28035794	.40569712	.73132786	.92907025	.99921992
7	7	.010	.02320322	.04867051	.09259156	.16037538	.25407030	.56453137	.83681206	.99539981
7	7	.005	.01225287	.02720783	.05490511	.10103033	.17017525	.43931670	.73936681	.98647538

Table B-10 **Power of two-sided two-sample Student t test vs. normal shift alternative d.**

m	n	α	d = .2	.4	.6	.8	1.0	1.5	2.0	3.0
3	1	.250	.25490576	.26943135	.29301411	.32475661	.36348327	.48150386	.61090513	.82868385
3	1	.100	.10256135	.11020174	.12279145	.14011835	.16189385	.23331022	.32318717	.52602965
3	1	.050	.05138836	.05554128	.06242244	.07197193	.08410706	.12502409	.17925539	.31638186
3	1	.025	.02572185	.02788417	.03147738	.03648557	.04288664	.06478419	.09460169	.17465883
3	1	.010	.01029548	.01118136	.01265607	.01471697	.01736038	.02648395	.03911481	.07430719
3	1	.005	.00514887	.00559534	.00633900	.00737919	.00871498	.01333919	.01977687	.03793973
2	2	.250	.25653387	.27579594	.30679678	.34798132	.39735807	.54153420	.68735349	.89527654
2	2	.100	.10341351	.11357656	.13025961	.15308972	.18156436	.27320478	.38452473	.61724513
2	2	.050	.05185070	.05738118	.06652705	.07918239	.09520176	.14869158	.21830708	.38740013
2	2	.025	.02596234	.02884366	.03362692	.04028394	.04877563	.07768151	.11668006	.21925393
2	2	.010	.01039395	.01157483	.01353984	.01628430	.01980166	.03191738	.04862821	.09480088
2	2	.005	.00519849	.00579370	.00678493	.00817098	.00995022	.01610336	.02465356	.04867549
4	1	.250	.25594455	.27351591	.30194388	.33999963	.38607425	.52360671	.66780314	.88454224
4	1	.100	.10341339	.11360873	.13044882	.15369662	.18300532	.27943339	.39994946	.65601946
4	1	.050	.05195514	.05782538	.06762227	.08135532	.09901895	.16001760	.24311860	.45362025
4	1	.025	.02605900	.02924969	.03461136	.04220391	.05209994	.08739219	.13828856	.28341399
4	1	.010	.01044963	.01180809	.01410355	.01738075	.02169797	.03749873	.06134981	.13623318
4	1	.005	.00523063	.00592829	.00711009	.00880352	.01104503	.01934478	.03213167	.07406227
3	2	.250	.25890035	.28501435	.32663697	.38110820	.44506731	.62190321	.78141082	.95646758
3	2	.100	.10511729	.12036781	.14543850	.17978174	.22258901	.35886261	.51722358	.79739907
3	2	.050	.05293302	.06174242	.07644917	.09705681	.12351135	.21385786	.33274410	.60229462
3	2	.025	.02658937	.03138792	.03948123	.05099345	.06606270	.11999004	.19702691	.40390938
3	2	.010	.01067504	.01272163	.01620200	.02121272	.02787497	.05263259	.09040935	.20698985
3	2	.005	.00534630	.00639818	.00819360	.01079250	.01427226	.02742410	.04809313	.11603784
5	1	.250	.25659008	.27604534	.30744137	.34930520	.39971338	.54801004	.69879854	.90945206
5	1	.100	.10399164	.11591998	.13563821	.16288018	.19723423	.30994702	.44881541	.72676981
5	1	.050	.05236936	.05949927	.07144829	.08829176	.11009182	.18614429	.29029775	.54608544
5	1	.025	.02632280	.03032426	.03709953	.04679422	.05958614	.10629632	.17543887	.37350731
5	1	.010	.01057977	.01234205	.01535438	.01972452	.02559369	.04794269	.08345461	.20083925
5	1	.005	.00530293	.00622603	.00781167	.01012847	.01328840	.02548601	.04561815	.11731190
4	2	.250	.26051880	.29127410	.33994634	.40294243	.47578028	.66925217	.82968472	.97646050
4	2	.100	.10638295	.12541069	.15669235	.19949629	.25266707	.41908553	.60274929	.88007046
4	2	.050	.05379276	.06522517	.08443162	.11154935	.14660901	.26693661	.42229035	.73667658
4	2	.025	.02711915	.03356020	.04455771	.06045107	.08161517	.15940451	.27212432	.55771231
4	2	.010	.01092947	.01377637	.01871026	.02599463	.03596016	.07497043	.13777732	.33454581
4	2	.005	.00548583	.00697984	.00958910	.01348403	.01888695	.04073547	.07786732	.20795476

Table B-10 Power of two-sided two-sample Student t test vs. normal shift alternative d.

m	n	α	d = .2	.4	.6	.8	1.0	1.5	2.0	3.0
3	3	.250	.26182417	.29628500	.35046645	.41989470	.49907711	.70231054	.85957495	.98509912
3	3	.100	.10717943	.12856370	.16365200	.21149914	.27060768	.45244777	.64528144	.90980045
3	3	.050	.05426753	.06713784	.08877466	.11932928	.15879087	.29319470	.46264081	.78255439
3	3	.025	.02738504	.03464554	.04707461	.06508813	.08912856	.17744889	.30392107	.60951815
3	3	.010	.01104635	.01425926	.01985194	.02815275	.03956763	.08452865	.15685597	.37777660
3	3	.005	.00554699	.00723412	.01019630	.01464706	.02086237	.04623937	.08957842	.23940707
6	1	.250	.25703103	.27776918	.31117351	.35558776	.40885484	.56393544	.71822695	.92318651
6	1	.100	.10440360	.11756550	.13932807	.16939538	.20729300	.33115562	.48170152	.76886229
6	1	.050	.05267695	.06074400	.07429894	.09347000	.11836906	.20561403	.32483734	.60765862
6	1	.025	.02657709	.03115952	.03904438	.05040597	.06551645	.12146051	.20522514	.44175642
6	1	.010	.01068593	.01278053	.01639202	.02169335	.02891127	.05713178	.10334215	.25873155
6	1	.005	.00536403	.00647980	.00841760	.01129179	.01525659	.03123706	.05873510	.16078176
5	2	.250	.26168534	.29576283	.34940657	.41826525	.49696864	.69988295	.85795245	.98494416
5	2	.100	.10733489	.12920029	.16513035	.21421043	.27493795	.46189527	.65594928	.92064630
5	2	.050	.05446501	.06795726	.09071925	.12300291	.16488047	.30823994	.48837828	.81502178
5	2	.025	.02754993	.03533987	.04876096	.06837168	.09477212	.19305365	.33468857	.66598357
5	2	.010	.01114664	.01468813	.02091890	.03029565	.04338817	.09634664	.18364884	.44463414
5	2	.005	.00560891	.00750122	.01086982	.01602334	.02336644	.05447842	.10974243	.30451498
4	3	.250	.26400264	.30460672	.36778991	.44746233	.53631958	.75177662	.89987542	.99346610
4	3	.100	.10879918	.13499376	.17789385	.23612360	.30741687	.51962061	.72631289	.95475012
4	3	.050	.05536005	.07157780	.09898206	.13786507	.18817778	.35746337	.55939271	.87643919
4	3	.025	.02806278	.03745149	.05371705	.07761882	.10991594	.22964691	.39695454	.74810600
4	3	.010	.01137802	.01565785	.02325902	.03482232	.05112605	.11766299	.22628883	.53405534
4	3	.005	.00573201	.00802229	.01214696	.01854385	.02777850	.06754660	.13833909	.37689263
7	1	.250	.25735194	.27902173	.31387794	.36012266	.41541969	.57516021	.73154068	.93179788
7	1	.100	.10471054	.11879078	.14207242	.17423169	.21473747	.34463798	.50510820	.79613748
7	1	.050	.05291205	.06169622	.07648200	.09743891	.12471433	.22046040	.35072159	.65019296
7	1	.025	.02668757	.03181737	.04058185	.05327321	.07024314	.13360978	.22893288	.49290424
7	1	.010	.01077237	.01313933	.01724743	.02333085	.03169635	.06500190	.12053574	.30751397
7	1	.005	.00541507	.00669323	.00893237	.01229227	.01698916	.03640601	.07077970	.20076099
6	2	.250	.26256372	.29912980	.35645623	.42957534	.51240968	.72115841	.87618371	.98925965
6	2	.100	.10807008	.13212432	.17162687	.22549116	.29189109	.49336535	.69766204	.94274876
6	2	.050	.05499710	.07012439	.09571847	.13212083	.17940883	.34043602	.53734017	.86210325
6	2	.025	.02789990	.03679363	.05221975	.07493677	.10573602	.22102127	.38500345	.73937315
6	2	.010	.01132917	.01546214	.02282007	.03405428	.04997110	.11569115	.22486105	.54031971
6	2	.005	.00571492	.00795618	.01200827	.01832770	.02751418	.06769459	.14076172	.39250753

Table B-10 Power of two-sided two-sample Student t test vs. normal shift alternative d.

m	n	α	d = .2	.4	.6	.8	1.0	1.5	2.0	3.0
5	3	.250	.26567503	.31095090	.38083919	.46786779	.56324352	.78448435	.92308628	.99655829
5	3	.100	.11008371	.14008649	.18913489	.25542217	.33591378	.56856531	.77910489	.97446320
5	3	.050	.05625009	.07520811	.10736015	.15310338	.21226838	.40827624	.62952358	.92480245
5	3	.025	.02863003	.03982044	.05938784	.08842714	.12796435	.27446979	.47222734	.83306784
5	3	.010	.01166529	.01688744	.02631561	.04093303	.06191154	.14923965	.29093726	.65450733
5	3	.005	.00589617	.00873516	.01395842	.02226665	.03456192	.08937075	.18840874	.50149291
4	4	.250	.26670952	.31485113	.38877831	.48009803	.57906481	.80236553	.93448312	.99766362
4	4	.100	.11075457	.14273398	.19493116	.26525226	.35018722	.59154799	.80154706	.98059977
4	4	.050	.05666808	.07690689	.11125466	.16011344	.22318803	.42999756	.65687587	.93893575
4	4	.025	.02887386	.04083606	.06180637	.09299702	.13549863	.29223025	.49968681	.85695516
4	4	.010	.01177767	.01736775	.02750547	.04329650	.06604102	.16081552	.31303287	.68752688
4	4	.005	.00595681	.00899833	.01462568	.02363123	.03702786	.09701484	.20490927	.53540042
7	2	.250	.26324825	.30174597	.36190621	.43825493	.52414313	.73673714	.88879310	.99174922
7	2	.100	.10865263	.13443933	.17676033	.23437233	.30515727	.51728523	.72555566	.95594664
7	2	.050	.05542586	.07187322	.09975849	.13949125	.19112998	.36591391	.57438190	.89192857
7	2	.025	.02818711	.03799163	.05508506	.08040133	.11488771	.24419313	.42531463	.78967264
7	2	.010	.01148273	.01611837	.02444908	.03731037	.05572802	.13277542	.26063312	.61118768
7	2	.005	.00580574	.00835012	.01300844	.02038397	.03126911	.07993837	.16946195	.46729421
6	3	.250	.26699486	.31593020	.39098460	.48351411	.58350533	.80741478	.93767218	.99794668
6	3	.100	.11111983	.14418924	.19816171	.27081931	.35840172	.60520100	.81503054	.98410218
6	3	.050	.05698157	.07819946	.11427961	.16558747	.23207700	.44855311	.68083396	.95095182
6	3	.025	.02910533	.04181867	.06421006	.09767826	.14345286	.31220973	.53179068	.88470749
6	3	.010	.01191216	.01795693	.02901609	.04641420	.07169998	.17803978	.34781136	.74093488
6	3	.005	.00603986	.00936917	.01560373	.02572080	.04097204	.11046739	.23603143	.60287272
5	4	.250	.26886177	.32292466	.40506804	.50486625	.61053789	.83557882	.95352408	.99898555
5	4	.100	.11235257	.14904789	.20876143	.28866499	.38399280	.64400538	.84907484	.99058052
5	4	.050	.05776031	.08137393	.12157600	.17881706	.25242276	.48734139	.72543646	.96744906
5	4	.025	.02956569	.04375048	.06885099	.10650696	.15803952	.34577864	.58007237	.91592190
5	4	.010	.01212784	.01889097	.03136910	.05116248	.08009306	.20158533	.39074148	.79149608
5	4	.005	.00615757	.00988891	.01695138	.02853871	.04615890	.12685525	.27051609	.66144199
7	3	.250	.26806117	.31993537	.39908317	.49586477	.59925909	.82424977	.94747526	.99864799
7	3	.100	.11197012	.14755239	.20553993	.28333245	.37650302	.63338226	.84053033	.98929688
7	3	.050	.05759014	.08069300	.12005569	.17618240	.24851883	.48085843	.71916780	.96599238
7	3	.025	.02950650	.04351420	.06832645	.10560846	.15673585	.34388228	.57891510	.91688987
7	3	.010	.01212459	.01888631	.03139165	.05128955	.08047242	.20375617	.39658480	.80169571
7	3	.005	.00616529	.00992998	.01708377	.02887863	.04690801	.13020490	.27947085	.68133406

Table B-10 **Power of two-sided two-sample Student t test vs. normal shift alternative d.**

m	n	α	d = .2	.4	.6	.8	1.0	1.5	2.0	3.0
6	4	.250	.27060941	.32943356	.41803845	.52422987	.63454052	.85863342	.96495919	.99948946
6	4	.100	.11367599	.15426806	.22014477	.30776204	.41115255	.68325691	.88047170	.99490880
6	4	.050	.05867932	.08514339	.13030235	.19460117	.27690259	.53286875	.77433301	.98127091
6	4	.025	.03015714	.04626112	.07497050	.11830479	.17771389	.39091592	.64210879	.94816984
6	4	.010	.01243346	.02023876	.03484486	.05834093	.09303096	.23873318	.45731072	.85862502
6	4	.005	.00633544	.01069237	.01909715	.03316161	.05489563	.15562536	.33120136	.75368921
5	5	.250	.27145707	.33257300	.42423503	.53335343	.64564231	.86858855	.96941135	.99963179
5	5	.100	.11424438	.15650188	.22498290	.31579207	.42239944	.69855002	.89158515	.99603827
5	5	.050	.05904263	.08663003	.13372627	.20073945	.28629549	.54938560	.79054238	.98470171
5	5	.025	.03037440	.04718212	.07720710	.12258536	.18476890	.40629451	.66154595	.95587845
5	5	.010	.01253671	.02069407	.03601672	.06074815	.09732798	.25053316	.47693473	.87415152
5	5	.005	.00639236	.01094969	.01978381	.03463498	.05765768	.16436772	.34843627	.77470470
7	4	.250	.27205466	.33478442	.42859202	.53974789	.65338370	.87536911	.97231948	.99971222
7	4	.100	.11478677	.15864273	.22964577	.32357257	.43333771	.71338557	.90211591	.99699235
7	4	.050	.05946006	.08835288	.13774021	.20802057	.29755043	.56941837	.80999682	.98845272
7	4	.025	.03066587	.04843396	.08030000	.12861115	.19486081	.42886974	.69013464	.96643206
7	4	.010	.01270066	.02143062	.03795908	.06483980	.10480333	.27196053	.51321131	.90165210
7	4	.005	.00649284	.01141429	.02106039	.03745830	.06310185	.18257307	.38540155	.81886225
6	5	.250	.27360765	.34049797	.43973911	.55588145	.67256385	.89111834	.97847941	.99984298
6	5	.100	.11584004	.16277510	.23855497	.33822422	.45354888	.73901568	.91865024	.99813049
6	5	.050	.06013936	.09113915	.14416364	.21949991	.31495463	.59835988	.83553077	.99219170
6	5	.025	.03107570	.05018284	.08457494	.13681564	.20833809	.45705356	.72261040	.97558552
6	5	.010	.01289764	.02230991	.04025338	.06960208	.11334147	.29489630	.54869391	.92246954
6	5	.005	.00660231	.01191731	.02242830	.04044066	.06874906	.20031854	.41846809	.84984569
7	5	.250	.27541683	.34711570	.45251961	.57410017	.69377613	.90719828	.98404201	.99992394
7	5	.100	.11720105	.16815513	.25004657	.35701713	.47919372	.76980010	.93666075	.99902728
7	5	.050	.06108514	.09503697	.15319049	.23565414	.33935540	.63745565	.86738574	.99570132
7	5	.025	.03168720	.05281733	.08410095	.14940831	.22908458	.49956647	.76865383	.98573035
7	5	.010	.01321692	.02375755	.04410095	.07771823	.12805567	.33445155	.60750507	.95078591
7	5	.005	.00679002	.01279719	.02487880	.04590103	.07926478	.23388572	.47955220	.89850179
6	6	.250	.27613161	.34971652	.45749676	.58110081	.70178143	.91287059	.98581354	.99994270
6	6	.100	.11769100	.17003148	.25414516	.36365237	.48811549	.77988823	.94202905	.99922482
6	6	.050	.06140376	.09634781	.15621293	.24101843	.34735370	.64958148	.87641777	.99644610
6	6	.025	.03188101	.05365185	.09313997	.15335893	.23551925	.51209553	.78112332	.98781637
6	6	.010	.01331106	.02418473	.04523492	.08009833	.13232952	.34542847	.62261149	.95642705
6	6	.005	.00684274	.01304474	.02556815	.04743125	.08218745	.24283605	.49471350	.90803074

Table B-10 **Power of two-sided two-sample Student t test vs. normal shift alternative d.**

m	n	α	d = .2	.4	.6	.8	1.0	1.5	2.0	3.0
7	6	.250	.27828387	.35750854	.47227676	.60161605	.72481843	.92810303	.99009868	.99997599
7	6	.100	.11928963	.17628307	.26748748	.38512993	.51669410	.81060859	.95701481	.99964305
7	6	.050	.06250683	.10090313	.16674892	.25971592	.37509578	.69010757	.90445793	.99824563
7	6	.025	.03259066	.05673158	.10078990	.16813642	.25959861	.55787240	.82400739	.99354849
7	6	.010	.01368015	.02588177	.04980676	.08980980	.14989288	.39022464	.68165527	.97467291
7	6	.005	.00705945	.01407981	.02850709	.05406556	.09501134	.28232127	.55960388	.94268891
7	7	.250	.28075620	.36638034	.48884518	.62408846	.74927156	.94245979	.99347392	.99999122
7	7	.100	.12111207	.18338897	.28254036	.40901749	.54776556	.84087209	.96958162	.99985372
7	7	.050	.06375890	.10608241	.17870883	.28076627	.40580695	.73133032	.92907027	.99921992
7	7	.025	.03339374	.06023846	.10953546	.18498683	.28674681	.60594941	.86341902	.99689301
7	7	.010	.01409688	.02782105	.05508859	.10107964	.17018713	.43931691	.73936681	.98647538
7	7	.005	.00730397	.01526690	.03193166	.06187009	.11010967	.32701157	.62608092	.96698300

Tables C-1 and C-2

If a combined sample of m X's and n Y's is ordered from smallest to largest, the Wilcoxon statistics are

$$w_X = \text{the sum of the ranks of the } X\text{'s, and}$$

$$w_Y = \text{the sum of the ranks of the } Y\text{'s.}$$

In terms of the notation of Chapter 1,

$$w_Y = \sum_{i=1}^{m+n} i z_i \quad \text{and} \quad w_X = \sum_{i=1}^{m+n} i(1 - z_i).$$

Note that $w_X + w_Y = (m + n)(m + n + 1)/2$.

Table C-1.   Distribution of Wilcoxon two-sample statistic under normal shift alternative d

Table C-1 gives values to 6 decimal places of the distribution of w_Y (and w_X) under normal shift alternative $d = .2(.2)1.0,1.5,2.0,3.0$ for $1 \le n \le m \le 7$, where the X's and Y's are from normal populations with means μ_X and μ_Y, respectively ($\mu_Y - \mu_X > 0$), and common variance σ^2; $d = (\mu_Y - \mu_X)/\sigma$. Entries in the table are ordered according to increasing values of $m + n$, from $2 \le m + n \le 14$.

Table C-2.   Log R(w | d), the common logarithm of the rank-sum likelihood ratio under normal shift alternative d

Values of $\log R(w \mid d)$ are given to 4 decimal places in the same format as Table C-1. As defined in (4.2), Chapter 3, the rank-sum likelihood ratio is

$$R(w \mid d) = \frac{\Pr\{w \mid d\}}{\Pr\{w \mid 0\}};$$

common logarithms (log to the base 10) are given here to facilitate use of the SRS (sequential, rank-sum) test.

For certain combinations of large w_X, d, m, and n, the values in Table C-1 are zero to 6 places. Since Table C-2 is based on Table C-1, these zero values result in certain columns in Table C-2 ending with a string of nonchanging values. In such a case the value given is an upper bound for the true one.

Table C-1 Distribution of Wilcoxon two-sample statistic under normal shift alternative d.

m	n	w_y	w_x	d = 0.	.2	.4	.6	.8	1.0	1.5	2.0	3.0
1	1	2	1	.500000	.556232	.611351	.664313	.714196	.760250	.855578	.921350	.983053
		1	2	.500000	.443769	.388649	.335687	.285804	.239750	.144422	.078650	.016947
2	1	3	3	.333333	.391392	.451875	.513387	.574469	.633702	.765812	.865767	.968796
		2	4	.333333	.329678	.318953	.301853	.279454	.253096	.179532	.111166	.028514
		1	5	.333333	.278930	.229172	.184760	.146077	.113202	.054656	.023066	.002690
3	1	4	6	.250000	.304081	.362696	.424579	.488231	.552031	.701863	.822793	.956374
		3	7	.250000	.261934	.267537	.266425	.258715	.245012	.191848	.128923	.037263
		2	8	.250000	.232584	.210892	.186355	.160466	.134632	.077450	.037827	.005508
		1	9	.250000	.201402	.158875	.122642	.092588	.068325	.028840	.010457	.000854
2	2	7	3	.166667	.214500	.269184	.329757	.394803	.462546	.630691	.774916	.942866
		6	4	.166667	.189787	.209134	.223110	.230547	.230869	.202646	.147942	.047053
		5	5	.333333	.327996	.312497	.288298	.257573	.222888	.135194	.067523	.009613
		4	6	.166667	.141573	.116275	.092299	.070788	.052434	.021223	.006869	.000363
		3	7	.166667	.126144	.092910	.066536	.046290	.031263	.010246	.002751	.000106
4	1	5	10	.200000	.249631	.305001	.365088	.428526	.493699	.652865	.787839	.945311
		4	11	.200000	.217800	.230777	.237961	.238821	.233330	.195992	.139816	.044252
		3	12	.200000	.197168	.188909	.175908	.159198	.140028	.089708	.048122	.008148
		2	13	.200000	.178666	.155250	.131202	.107823	.086157	.043461	.018355	.001912
		1	14	.200000	.156735	.120062	.089841	.065633	.046786	.017974	.005869	.000376
3	2	9	6	.100000	.137155	.182463	.235740	.296192	.362428	.540138	.707166	.920514
		8	7	.100000	.123775	.147460	.169192	.187078	.199476	.200693	.163485	.060500
		7	8	.200000	.220229	.233070	.237143	.232068	.218516	.159333	.092361	.016232
		6	9	.200000	.195974	.184383	.166582	.144540	.120473	.064209	.026832	.002350
		5	10	.200000	.174510	.146264	.117729	.090989	.067513	.026772	.008221	.000362
		4	11	.100000	.077719	.058075	.041702	.028764	.019048	.005674	.001300	.000030
		3	12	.100000	.070636	.048286	.031911	.020369	.012547	.003181	.000636	.000012

259

Table C-1 Distribution of Wilcoxon two-sample statistic under normal shift alternative d.

m	n	w_y	w_x	d = 0.	.2	.4	.6	.8	1.0	1.5	2.0	3.0
5	1	6	15	.166667	.212274	.264317	.322024	.384213	.449365	.613555	.758452	.935305
		5	16	.166667	.186786	.203423	.215323	.221563	.221670	.196550	.146937	.050033
		4	17	.166667	.170927	.170096	.164256	.153926	.139986	.096879	.055666	.010564
		3	18	.166667	.157686	.144752	.128924	.111405	.093394	.052634	.024537	.003016
		2	19	.166667	.144490	.121687	.099540	.079076	.060999	.028010	.010675	.000882
		1	20	.166667	.127837	.095725	.069933	.049817	.034586	.012373	.003734	.000200
4	2	11	10	.066667	.095962	.133481	.179661	.234328	.296621	.474302	.653768	.900794
		10	11	.066667	.087565	.110314	.133384	.154900	.172903	.192397	.170196	.070838
		9	12	.133333	.157869	.178929	.194213	.201977	.201361	.166749	.107524	.022154
		8	13	.133333	.144922	.150405	.149101	.141247	.127927	.082344	.040576	.004671
		7	14	.200000	.195485	.182548	.162877	.138877	.113185	.055822	.020947	.001362
		6	15	.133333	.117092	.098124	.078445	.059814	.043491	.015908	.004310	.000128
		5	16	.133333	.107754	.083297	.061572	.043510	.029385	.009024	.002081	.000047
		4	17	.066667	.048651	.034011	.022764	.014580	.008931	.002152	.000389	.000005
		3	18	.066667	.044700	.028892	.017983	.010768	.006197	.001302	.000210	.000002
3	3	15	6	.050000	.074789	.107703	.149556	.200574	.260249	.437210	.623570	.889891
		14	7	.050000	.068751	.090330	.113497	.136494	.157260	.186915	.173690	.076735
		13	8	.100000	.125699	.150444	.171565	.186555	.193571	.173896	.118829	.026020
		12	9	.150000	.168427	.179984	.183103	.177401	.163762	.108896	.054263	.006009
		11	10	.150000	.154398	.150814	.139822	.123064	.102855	.052494	.019547	.001088
		10	11	.150000	.138272	.120929	.100336	.078977	.058976	.022583	.006238	.000182
		9	12	.150000	.127107	.102464	.078568	.057302	.039752	.012807	.003027	.000068
		8	13	.100000	.075702	.054499	.037291	.024240	.014960	.003558	.000607	.000006
		7	14	.050000	.034719	.023000	.014527	.008742	.005009	.001001	.000146	.000001
		6	15	.050000	.032137	.019832	.011735	.006651	.003607	.000640	.000084	.000001
6	1	7	21	.142857	.184978	.233936	.289188	.349745	.414216	.580994	.733157	.926149
		6	22	.142857	.163773	.182284	.197013	.206810	.210892	.195367	.151768	.054933
		5	23	.142857	.150926	.154560	.153435	.147665	.137779	.101232	.061392	.012766
		4	24	.142857	.140619	.134113	.123932	.110964	.096266	.058783	.029475	.004108
		3	25	.142857	.131064	.116544	.100437	.083384	.067892	.034864	.014699	.001443
		2	26	.142857	.120962	.099406	.079274	.061338	.046042	.019666	.006931	.000481
		1	27	.142857	.107677	.079157	.056721	.039594	.026912	.009095	.002578	.000120

Table C-1 **Distribution of Wilcoxon two-sample statistic under normal shift alternative d.**

m	n	w_y	w_x	d = 0.	.2	.4	.6	.8	1.0	1.5	2.0	3.0
5	2	13	15	.047619	.071249	.102726	.142921	.192163	.250091	.423869	.610114	.883110
		12	16	.047619	.065486	.086156	.108518	.130955	.151531	.182621	.172590	.079085
		11	17	.095238	.119012	.141968	.161746	.176096	.183316	.167306	.117172	.027398
		10	18	.095238	.110759	.122633	.129326	.129967	.124534	.091251	.050425	.007022
		9	19	.142857	.152473	.154826	.149624	.137667	.120648	.070361	.030696	.002603
		8	20	.142857	.139326	.129253	.114072	.095791	.076560	.035313	.012099	.000623
		7	21	.142857	.127308	.107884	.086920	.066571	.048462	.017541	.004619	.000124
		6	22	.095238	.077934	.060669	.044916	.031615	.021152	.006193	.001315	.000022
		5	23	.095238	.072721	.052960	.036772	.024334	.015343	.003913	.000734	.000010
		4	24	.047619	.033104	.021988	.013945	.008440	.004872	.001000	.000151	.000001
		3	25	.047619	.030629	.018937	.011241	.006400	.003492	.000633	.000087	.000001
4	3	18	10	.028571	.045723	.070060	.102964	.145400	.197668	.365664	.559449	.863100
		17	11	.028571	.042465	.059996	.080658	.103288	.126125	.170058	.174823	.088778
		16	12	.057143	.078586	.102337	.126297	.147841	.164302	.171567	.132507	.034504
		15	13	.085714	.107456	.127359	.142786	.151519	.152291	.122264	.071260	.009973
		14	14	.114286	.131976	.143798	.147881	.143599	.131730	.083104	.037292	.002884
		13	15	.114286	.120729	.120026	.112320	.098958	.082104	.039667	.013268	.000508
		12	16	.142857	.138665	.126818	.109290	.088766	.067963	.026986	.007475	.000203
		11	17	.114286	.101802	.085321	.067279	.049913	.034839	.010867	.002322	.000035
		10	18	.114286	.093351	.071910	.052233	.035773	.023100	.005989	.001080	.000012
		9	19	.085714	.064608	.045999	.030922	.019621	.011750	.002527	.000377	.000003
		8	20	.057143	.039315	.025574	.015718	.009122	.004996	.000857	.000101	.000000
		7	21	.028571	.018257	.011069	.006363	.003465	.001786	.000267	.000028	.000000
		6	22	.028571	.017068	.009732	.005290	.002737	.001347	.000182	.000018	.000000
7	1	8	28	.125000	.164122	.210302	.263200	.322010	.385481	.553379	.710998	.917697
		7	29	.125000	.145995	.165437	.181915	.194147	.201146	.193299	.155111	.059164
		6	30	.125000	.135220	.141683	.143803	.141391	.134684	.103888	.065854	.014775
		5	31	.125000	.126794	.124502	.118345	.108900	.097011	.063061	.033491	.005162
		4	32	.125000	.119289	.110196	.098536	.085288	.071455	.039809	.018090	.002027
		3	33	.125000	.111917	.097044	.081490	.066264	.052176	.024924	.009725	.000805
		2	34	.125000	.103817	.083626	.065323	.049473	.036324	.014636	.004844	.000292
		1	35	.125000	.092846	.067211	.047389	.032527	.021723	.007004	.001886	.000078

Table C-1 Distribution of Wilcoxon two-sample statistic under normal shift alternative d.

m	n	w_y	w_x	d = 0.	.2	.4	.6	.8	1.0	1.5	2.0	3.0
6	2	15	21	.035714	.055182	.081977	.117259	.161741	.215490	.383789	.573483	.867059
		14	22	.035714	.050982	.069485	.090490	.112690	.134309	.172905	.172676	.085839
		13	23	.071429	.093165	.115758	.137088	.154831	.166876	.164753	.123426	.032050
		12	24	.071429	.087403	.101613	.112293	.118022	.118045	.095505	.057634	.009296
		11	25	.107143	.121476	.130734	.133605	.129716	.119707	.078756	.038329	.003937
		10	26	.107143	.113533	.114049	.108652	.098209	.084269	.045969	.018427	.001275
		9	27	.142857	.139118	.128481	.112544	.093519	.073739	.032442	.010399	.000443
		8	28	.107143	.095824	.081197	.065171	.049538	.035655	.012332	.003026	.000065
		7	29	.107143	.089673	.071197	.053611	.038279	.025911	.007729	.001648	.000027
		6	30	.071429	.055436	.040843	.028556	.018940	.011914	.002960	.000525	.000003
		5	31	.071429	.052143	.036225	.023941	.015046	.008989	.001983	.000317	.000000
		4	32	.035714	.023871	.015213	.009238	.005343	.002941	.000531	.000070	.000000
		3	33	.035714	.022195	.013229	.007553	.004127	.002155	.000347	.000041	.000000
5	3	21	15	.017857	.030196	.048680	.074959	.110459	.156094	.313006	.508295	.839260
		20	16	.017857	.028231	.042258	.059959	.080736	.103292	.154183	.172086	.098057
		19	17	.035714	.052625	.073129	.095927	.118897	.139388	.163787	.139430	.041818
		18	18	.053571	.072648	.092754	.111568	.126522	.135381	.125093	.082164	.013837
		17	19	.071429	.090832	.108419	.121542	.128053	.126887	.095358	.049859	.004997
		16	20	.089286	.104337	.114304	.117436	.113201	.102429	.060747	.024669	.001418
		15	21	.107143	.116388	.118376	.112754	.100611	.084131	.040609	.013210	.000448
		14	22	.107143	.107153	.100239	.087718	.071812	.055009	.021155	.005405	.000107
		13	23	.107143	.100212	.087681	.071775	.054982	.039423	.012883	.002814	.000042
		12	24	.107143	.092331	.074475	.056224	.039727	.026273	.007004	.001241	.000012
		11	25	.089286	.071609	.053815	.037890	.024993	.015444	.003488	.000527	.000004
		10	26	.071429	.052695	.036451	.023633	.014355	.008167	.001494	.000180	.000001
		9	27	.053571	.037170	.024254	.014876	.008573	.004640	.000759	.000083	.000000
		8	28	.035714	.022839	.013750	.007788	.004147	.002074	.000278	.000025	.000000
		7	29	.017857	.010684	.006040	.003223	.001622	.000770	.000092	.000007	.000000
		6	30	.017857	.010051	.005376	.002728	.001312	.000597	.000065	.000005	.000000

Table C-1 Distribution of Wilcoxon two-sample statistic under normal shift alternative d.

m	n	w_y	w_x	d = 0.	.2	.4	.6	.8	1.0	1.5	2.0	3.0
4	4	26	10	.014286	.024869	.041163	.064905	.097699	.140698	.293083	.488831	.830374
		25	11	.014286	.023323	.035956	.052411	.072321	.094603	.147899	.170905	.101638
		24	12	.028571	.043663	.062746	.084880	.108203	.130133	.161295	.142901	.044923
		23	13	.042857	.060859	.081022	.101216	.118758	.131004	.128746	.088282	.015461
		22	14	.071429	.093131	.113749	.130214	.139789	.140827	.109194	.058024	.005714
		21	15	.071429	.086832	.098506	.104324	.103185	.095357	.058366	.023663	.001184
		20	16	.100000	.112807	.118671	.116465	.106676	.091238	.045913	.015276	.000519
		19	17	.100000	.103950	.100668	.090835	.076382	.059871	.024000	.006254	.000121
		18	18	.114286	.110316	.099217	.083156	.064959	.047310	.015802	.003444	.000048
		17	19	.100000	.089617	.074815	.058187	.042164	.028472	.007849	.001403	.000013
		16	20	.100000	.082643	.063657	.045691	.030556	.019038	.004278	.000618	.000004
		15	21	.071429	.054818	.039239	.026192	.016302	.009460	.001786	.000218	.000001
		14	22	.071429	.051298	.034485	.021694	.012768	.007030	.001181	.000131	.000001
		13	23	.042857	.028271	.017456	.010081	.005442	.002744	.000366	.000032	.000000
		12	24	.028571	.017564	.010134	.005483	.002780	.001320	.000153	.000012	.000000
		11	25	.014286	.008253	.004492	.002301	.001108	.000501	.000052	.000004	.000000
		10	26	.014286	.007788	.004022	.001965	.000907	.000394	.000038	.000002	.000000
7	2	17	28	.027778	.044111	.067228	.098476	.138857	.188786	.351054	.542131	.852351
		16	29	.027778	.040913	.057444	.076944	.098403	.120258	.163760	.171465	.091479
		15	30	.055556	.075075	.096485	.117997	.137404	.152450	.160758	.127477	.036196
		14	31	.055556	.070812	.085617	.098247	.107060	.110855	.097180	.062996	.011452
		13	32	.083333	.099000	.111455	.118960	.120433	.115712	.083657	.044358	.005291
		12	33	.083333	.093503	.099283	.099808	.095041	.085780	.052762	.023658	.002009
		11	34	.111111	.116378	.115294	.108070	.095885	.080565	.041258	.015289	.000868
		10	35	.111111	.108057	.099394	.086482	.071194	.055467	.023424	.007104	.000263
		9	36	.111111	.100310	.085610	.069060	.052649	.037928	.013034	.003141	.000063
		8	37	.083333	.070258	.056015	.042221	.030079	.020249	.005871	.001191	.000017
		7	38	.083333	.066449	.050177	.035871	.024272	.015541	.004000	.000727	.000008
		6	39	.055556	.041327	.029136	.019461	.012310	.007373	.001608	.000248	.000002
		5	40	.055556	.039078	.026115	.016573	.009984	.005707	.001118	.000157	.000001
		4	41	.027778	.017965	.011060	.006478	.003607	.001908	.000310	.000036	.000000
		3	42	.027778	.016766	.009687	.005352	.002824	.001422	.000207	.000022	.000000

Table C-1 **Distribution of Wilcoxon two-sample statistic under normal shift alternative d.**

m	n	w_y	w_x	d = 0.	.2	.4	.6	.8	1.0	1.5	2.0	3.0
6	3	24	21	.011905	.021086	.035484	.056842	.086853	.126858	.272644	.466314	.817773
		23	22	.011905	.019807	.031103	.046155	.064803	.086198	.140119	.167630	.105397
		22	23	.023810	.037105	.054362	.074949	.097341	.119228	.154404	.142511	.048149
		21	24	.035714	.051529	.069775	.088737	.106078	.119302	.123347	.089203	.017472
		20	25	.047619	.064994	.082993	.099217	.111131	.116722	.100229	.058939	.007202
		19	26	.059524	.075763	.090050	.100006	.103844	.100896	.070618	.033377	.002551
		18	27	.083333	.098790	.109306	.112921	.108969	.098281	.056742	.021858	.001053
		17	28	.083333	.092090	.094753	.090795	.081046	.067415	.031322	.009457	.000247
		16	29	.095238	.098303	.094494	.084605	.070573	.054861	.021530	.005499	.000104
		15	30	.095238	.091869	.082463	.068885	.053558	.038766	.012662	.002667	.000033
		14	31	.095238	.085923	.072186	.056477	.041157	.027942	.007804	.001415	.000014
		13	32	.083333	.070200	.055045	.040172	.027287	.017251	.004009	.000597	.000004
		12	33	.083333	.065590	.048159	.032982	.021068	.012553	.002539	.000334	.000002
		11	34	.059524	.043646	.029855	.019043	.011322	.006272	.001050	.000112	.000000
		10	35	.047619	.032620	.020881	.012484	.006967	.003629	.000524	.000049	.000000
		9	36	.035714	.023215	.014143	.008072	.004313	.002157	.000285	.000025	.000000
		8	37	.023810	.014349	.008115	.004303	.002137	.000994	.000109	.000008	.000000
		7	38	.011905	.006746	.003600	.001807	.000853	.000378	.000037	.000002	.000000
		6	39	.011905	.006375	.003232	.001549	.000701	.000299	.000027	.000002	.000000
5	4	30	15	.007937	.014813	.026139	.043704	.069392	.104882	.241526	.433710	.801454
		29	16	.007937	.013979	.023130	.036000	.052785	.073027	.128844	.163926	.111206
		28	17	.015873	.026351	.040920	.059509	.081151	.103906	.147663	.146152	.053689
		27	18	.023810	.037058	.053780	.072846	.092197	.109160	.125409	.097812	.020659
		26	19	.039683	.057460	.077453	.097260	.113875	.124426	.115552	.071425	.008867
		25	20	.047619	.064363	.080780	.094191	.102094	.102937	.076963	.037330	.002608
		24	21	.063492	.080584	.094756	.103280	.104401	.097938	.060485	.023827	.001027
		23	22	.071429	.084314	.092052	.092986	.086941	.075273	.037637	.011790	.000300
		22	23	.087302	.096151	.097934	.092270	.080441	.064917	.027190	.007129	.000129
		21	24	.087302	.090064	.085814	.075532	.061433	.046186	.016094	.003472	.000041
		20	25	.095238	.091525	.081233	.066596	.050437	.035298	.010262	.001840	.000015

Table C-1 Distribution of Wilcoxon two-sample statistic under normal shift alternative d.

m	n	w_y	w_x	d = 0.	.2	.4	.6	.8	1.0	1.5	2.0	3.0
5	4	19	26	.087302	.078146	.064591	.049295	.034740	.022610	.005462	.000807	.000004
		18	27	.087302	.073294	.056894	.040833	.027099	.016633	.003495	.000455	.000002
		17	28	.071429	.055955	.040522	.027125	.016781	.009594	.001679	.000180	.000001
		16	29	.063492	.046328	.031294	.019563	.011316	.006055	.000902	.000083	.000000
		15	30	.047619	.032700	.020835	.012314	.006749	.003430	.000454	.000038	.000000
		14	31	.039683	.025494	.015227	.008451	.004356	.002084	.000238	.000017	.000000
		13	32	.023810	.014249	.007936	.004110	.001977	.000883	.000085	.000005	.000000
		12	33	.015873	.008934	.004693	.002298	.001049	.000445	.000038	.000002	.000000
		11	34	.007937	.004227	.002109	.000985	.000430	.000175	.000014	.000001	.000000
		10	35	.007937	.004012	.001910	.000855	.000359	.000141	.000010	.000001	.000000
7	3	27	28	.008333	.015361	.026826	.044469	.070121	.105411	.240750	.431112	.798214
		26	29	.008333	.014481	.023688	.036523	.053148	.073094	.127869	.162450	.111315
		25	30	.016667	.027228	.041707	.059968	.081026	.102997	.144882	.143298	.053665
		24	31	.025000	.037970	.053983	.071902	.089800	.105267	.119530	.093699	.020844
		23	32	.033333	.048156	.064921	.081739	.096190	.105899	.101305	.065530	.009394
		22	33	.041667	.056552	.071493	.084243	.092597	.095023	.075563	.040131	.003790
		21	34	.058333	.074533	.088058	.098058	.101091	.097156	.065111	.028806	.001815
		20	35	.066667	.079767	.088670	.091611	.088011	.078670	.043560	.015674	.000620
		19	36	.075000	.084490	.088306	.085653	.077129	.064505	.029964	.008886	.000215
		18	37	.083333	.087660	.085517	.077380	.064954	.050594	.019577	.004793	.000076
		17	38	.083333	.082941	.076525	.065463	.051936	.038227	.012838	.002733	.000034
		16	39	.083333	.077606	.066986	.053591	.039743	.027325	.007707	.001366	.000011
		15	40	.083333	.073467	.060066	.045548	.032041	.020915	.005204	.000821	.000006
		14	41	.075000	.061752	.047152	.033386	.021918	.013342	.002773	.000361	.000002
		13	42	.066667	.051749	.037300	.024960	.015506	.008942	.001633	.000189	.000001
		12	43	.058333	.042457	.028725	.018059	.010546	.005720	.000896	.000088	.000000
		11	44	.041667	.028577	.018235	.010819	.005967	.003057	.000416	.000035	.000000
		10	45	.033333	.021517	.012944	.007253	.003784	.001837	.000219	.000017	.000000
		9	46	.025000	.015397	.008863	.004766	.002393	.001121	.000124	.000009	.000000
		8	47	.016667	.009557	.005128	.002573	.001206	.000528	.000049	.000003	.000000
		7	48	.008333	.004509	.002291	.001092	.000488	.000204	.000017	.000001	.000000
		6	49	.008333	.004275	.002070	.000945	.000406	.000164	.000013	.000001	.000000

Table C-1 Distribution of Wilcoxon two-sample statistic under normal shift alternative d.

m	n	w_y	w_x	d = 0.	.2	.4	.6	.8	1.0	1.5	2.0	3.0
6	4	34	21	.004762	.009430	.017576	.030897	.051356	.080922	.203266	.389369	.775557
		33	22	.004762	.008939	.015695	.025815	.039839	.057786	.112954	.155984	.118504
		32	23	.009524	.016929	.028029	.043280	.062415	.084185	.134098	.145671	.061104
		31	24	.014286	.023940	.037245	.053859	.072481	.090893	.118623	.102880	.025440
		30	25	.023810	.037367	.054356	.073350	.091912	.107059	.114767	.080160	.012056
		29	26	.028571	.042407	.058152	.073733	.086516	.094031	.083420	.047064	.004300
		28	27	.042857	.059663	.076641	.090897	.099602	.100916	.074461	.034439	.002002
		27	28	.047619	.062335	.075099	.083307	.085133	.080196	.048580	.017980	.000599
		26	29	.061905	.076020	.085853	.089202	.085308	.075134	.038315	.011876	.000276
		25	30	.066667	.076974	.081638	.079553	.071252	.058681	.025112	.006452	.000098
		24	31	.076191	.082770	.082518	.075517	.063462	.048995	.017773	.003852	.000042
		23	32	.076191	.077654	.072581	.062223	.048940	.035326	.010771	.001942	.000014
		22	33	.085714	.082107	.072172	.058223	.043117	.029321	.007725	.001211	.000007
		21	34	.076191	.068548	.056549	.042776	.029673	.018879	.004183	.000544	.000002
		20	35	.076191	.064348	.049857	.035435	.023103	.013818	.002625	.000293	.000001
		19	36	.066667	.053027	.038732	.025978	.016001	.009052	.001505	.000148	.000000
		18	37	.061905	.046346	.031892	.020167	.011719	.006257	.000901	.000077	.000000
		17	38	.047619	.033466	.021629	.012853	.007020	.003524	.000435	.000032	.000000
		16	39	.042857	.028391	.017338	.009756	.005058	.002415	.000265	.000017	.000000
		15	40	.028571	.017772	.010200	.005398	.002632	.001183	.000111	.000006	.000000
		14	41	.023810	.014050	.007672	.003875	.001809	.000780	.000067	.000004	.000000
		13	42	.014286	.007906	.004053	.001923	.000844	.000342	.000025	.000001	.000000
		12	43	.009524	.004984	.002423	.001094	.000458	.000177	.000012	.000001	.000000
		11	44	.004762	.002369	.001099	.000475	.000191	.000071	.000004	.000000	.000000
		10	45	.004762	.002258	.001003	.000417	.000162	.000059	.000003	.000000	.000000

Table C-1 Distribution of Wilcoxon two-sample statistic under normal shift alternative d.

m	n	w_y	w_x	d = 0.	.2	.4	.6	.8	1.0	1.5	2.0	3.0
5	5	40	15	.003968	.008045	.015315	.027442	.046404	.074246	.192358	.376638	.768260
		39	16	.003968	.007638	.013721	.023042	.036239	.053473	.108316	.153633	.120644
		38	17	.007937	.014493	.024596	.038851	.057201	.078621	.130288	.145828	.063450
		37	18	.011905	.020557	.032879	.048766	.067167	.086023	.117323	.105145	.026970
		36	19	.019841	.032277	.048504	.067406	.086729	.103441	.116345	.084043	.013017
		35	20	.027778	.042041	.058710	.075703	.090200	.099395	.090588	.051887	.004667
		34	21	.035714	.051364	.067937	.082695	.092699	.095772	.073140	.034260	.001877
		33	22	.043651	.058799	.072698	.082538	.086102	.082578	.051738	.019499	.000644
		32	23	.055556	.070461	.081922	.087356	.085477	.076797	.040650	.012856	.000297
		31	24	.063492	.075428	.082066	.081799	.074722	.062586	.027598	.007183	.000107
		30	25	.071429	.079912	.081799	.076629	.065719	.051620	.019284	.004219	.000044
		29	26	.075397	.079052	.075754	.066361	.053154	.038941	.012142	.002192	.000015
		28	27	.079365	.078440	.070865	.058533	.044215	.030556	.008248	.001294	.000007
		27	28	.079365	.073394	.062032	.047918	.033835	.021843	.004950	.000646	.000002
		26	29	.075397	.065716	.052343	.038098	.025343	.015410	.003003	.000336	.000001
		25	30	.071429	.058407	.043688	.029893	.018712	.010719	.001809	.000177	.000000
		24	31	.063492	.048936	.034529	.022303	.013187	.007138	.001047	.000089	.000000
		23	32	.055556	.040138	.026564	.016100	.008934	.004538	.000567	.000041	.000000
		22	33	.043651	.029730	.018570	.010634	.005582	.002685	.000294	.000019	.000000
		21	34	.035714	.022826	.013402	.007226	.003577	.001625	.000155	.000009	.000000
		20	35	.027778	.016925	.009506	.004918	.002344	.001029	.000091	.000005	.000000
		19	36	.019841	.011256	.005887	.002836	.001257	.000512	.000038	.000002	.000000
		18	37	.011905	.006378	.003158	.001443	.000608	.000236	.000015	.000001	.000000
		17	38	.007937	.004034	.001901	.000829	.000334	.000125	.000007	.000000	.000000
		16	39	.003968	.001921	.000865	.000362	.000141	.000051	.000003	.000000	.000000
		15	40	.003968	.001834	.000792	.000319	.000120	.000042	.000002	.000000	.000000

Table C-1 **Distribution of Wilcoxon two-sample statistic under normal shift alternative d.**

m	n	w_y	w_x	d = 0.	.2	.4	.6	.8	1.0	1.5	2.0	3.0
7	4	38	28	.003030	.006318	.012349	.022679	.039239	.064130	.173920	.352870	.752125
		37	29	.003030	.006009	.011104	.019154	.030897	.046689	.099750	.147955	.124165
		36	30	.006061	.011419	.019968	.032451	.049091	.069232	.121635	.143184	.067422
		35	31	.009091	.016210	.026739	.040856	.057902	.076221	.110813	.105190	.029786
		34	32	.015152	.025412	.039364	.056374	.074723	.091774	.110853	.085749	.015156
		33	33	.018182	.029035	.042687	.057832	.072272	.083400	.084867	.054067	.006102
		32	34	.027273	.041201	.057200	.073038	.085853	.092995	.080008	.042467	.003173
		31	35	.033333	.047421	.061894	.074164	.081637	.082618	.059290	.025754	.001203
		30	36	.042424	.057048	.070265	.079312	.082089	.077960	.047271	.017069	.000510
		29	37	.048485	.061612	.071602	.076132	.074096	.066047	.033881	.010206	.000202
		28	38	.057576	.069042	.075658	.075792	.069437	.058209	.025506	.006528	.000093
		27	39	.060606	.068694	.071063	.067116	.057890	.045624	.017012	.003663	.000035
		26	40	.069697	.074723	.073109	.065296	.053257	.039686	.012869	.002414	.000018
		25	41	.069697	.070418	.064882	.054524	.041801	.029245	.008047	.001265	.000006
		24	42	.072727	.069447	.060470	.048019	.034783	.022989	.005484	.000747	.000003
		23	43	.069697	.062904	.051771	.038857	.026602	.016615	.003438	.000405	.000001
		22	44	.069697	.059315	.046053	.032620	.021079	.012429	.002226	.000227	.000000
		21	45	.060606	.048792	.035839	.024017	.014683	.008191	.001276	.000113	.000000
		20	46	.057576	.043867	.030530	.019408	.011269	.005977	.000826	.000065	.000000
		19	47	.048485	.034881	.022936	.013781	.007565	.003794	.000455	.000031	.000000
		18	48	.042424	.028886	.018001	.010265	.005354	.002555	.000272	.000017	.000000
		17	49	.033333	.021486	.012695	.006873	.003408	.001548	.000146	.000008	.000000
		16	50	.027273	.016578	.009247	.004730	.002218	.000953	.000078	.000004	.000000
		15	51	.018182	.010474	.005546	.002697	.001204	.000493	.000036	.000002	.000000
		14	52	.015152	.008336	.004229	.001976	.000850	.000337	.000023	.000001	.000000
		13	53	.009091	.004711	.002253	.000994	.000404	.000151	.000009	.000000	.000000
		12	54	.006061	.002981	.001357	.000571	.000222	.000080	.000004	.000000	.000000
		11	55	.003030	.001422	.000620	.000251	.000094	.000033	.000002	.000000	.000000
		10	56	.003030	.001359	.000569	.000222	.000081	.000027	.000001	.000000	.000000

Table C-1 Distribution of Wilcoxon two-sample statistic under normal shift alternative d.

m	n	w_y	w_x	d = 0.	.2	.4	.6	.8	1.0	1.5	2.0	3.0
6	5	45	21	.002165	.004711	.009573	.018210	.032515	.054658	.156928	.331556	.738712
		44	22	.002165	.004492	.008652	.015500	.025877	.040338	.091940	.143147	.127564
		43	23	.004329	.008561	.015651	.026493	.041596	.060681	.114453	.142167	.071525
		42	24	.006494	.012207	.021141	.033782	.049876	.068147	.107122	.107819	.032658
		41	25	.010823	.019288	.031581	.047567	.065995	.084459	.111017	.091259	.017112
		40	26	.015152	.025380	.038991	.054991	.071272	.084983	.092363	.061421	.007004
		39	27	.021645	.034390	.049997	.066565	.081230	.090942	.083325	.045735	.003286
		38	28	.025974	.039030	.053521	.067019	.076687	.080250	.061163	.027232	.001163
		37	29	.034632	.049238	.063781	.075322	.081150	.079821	.051724	.019382	.000574
		36	30	.041126	.055156	.067296	.074734	.075584	.069661	.038021	.011827	.000229
		35	31	.049784	.063023	.072523	.075886	.072236	.062588	.029110	.007653	.000104
		34	32	.054113	.064770	.070351	.069363	.062105	.050520	.019894	.004366	.000039
		33	33	.062771	.071001	.072856	.067841	.057347	.044030	.014991	.002844	.000019
		32	34	.064935	.069279	.066986	.058712	.046659	.033634	.009714	.001544	.000007
		31	35	.069264	.069758	.063658	.052645	.039462	.026821	.006672	.000910	.000003
		30	36	.069264	.065928	.056855	.044429	.031468	.020208	.004364	.000518	.000001
		29	37	.069264	.062313	.050795	.037522	.025121	.015249	.002858	.000294	.000001
		28	38	.064935	.055152	.042444	.029597	.018703	.010711	.001732	.000153	.000000
		27	39	.062771	.050333	.036601	.024136	.014434	.007830	.001109	.000086	.000000
		26	40	.054113	.041015	.028198	.017583	.009943	.005100	.000628	.000042	.000000
		25	41	.049784	.035737	.023308	.013811	.007435	.003637	.000401	.000025	.000000
		24	42	.041126	.027885	.017188	.009628	.004900	.002265	.000216	.000011	.000000
		23	43	.034632	.022182	.012932	.006860	.003310	.001452	.000122	.000006	.000000
		22	44	.025974	.015765	.008722	.004397	.002019	.000844	.000063	.000003	.000000
		21	45	.021645	.012458	.006552	.003148	.001381	.000553	.000038	.000001	.000000
		20	46	.015152	.008289	.004152	.001903	.000797	.000305	.000019	.000001	.000000
		19	47	.010823	.005573	.002630	.001137	.000449	.000162	.000009	.000001	.000000
		18	48	.006494	.003178	.001429	.000590	.000223	.000077	.000004	.000000	.000000
		17	49	.004329	.002021	.000869	.000344	.000125	.000042	.000002	.000000	.000000
		16	50	.002165	.000967	.000399	.000152	.000054	.000017	.000001	.000000	.000000
		15	51	.002165	.000926	.000368	.000136	.000046	.000015	.000001	.000000	.000000

269

Table C-1 Distribution of Wilcoxon two-sample statistic under normal shift alternative d.

m	n	w_y	w_x	d = 0.	.2	.4	.6	.8	1.0	1.5	2.0	3.0
7	5	50	28	.001263	.002924	.006292	.012613	.023629	.041490	.130481	.295070	.712126
		49	29	.001263	.002797	.005724	.010848	.019076	.031195	.078821	.133180	.132698
		48	30	.002525	.005348	.010423	.018729	.031083	.047737	.100711	.136950	.078253
		47	31	.003788	.007654	.014184	.024148	.037827	.054617	.096938	.107804	.037735
		46	32	.006313	.012143	.021361	.034420	.050875	.069084	.103575	.095027	.021017
		45	33	.008838	.016064	.026659	.040439	.056139	.071414	.089774	.067623	.009428
		44	34	.012626	.021946	.034729	.050090	.065917	.079244	.085281	.053839	.004916
		43	35	.016414	.026969	.040274	.054704	.067646	.076227	.068926	.035814	.002052
		42	36	.021465	.033606	.047710	.061467	.071922	.076500	.059203	.025868	.000978
		41	37	.026515	.039333	.052821	.064258	.070862	.070893	.046617	.017039	.000426
		40	38	.032828	.046223	.058845	.067771	.070650	.066716	.037710	.011729	.000207
		39	39	.037879	.050521	.060843	.066188	.065071	.057849	.027884	.007295	.000087
		38	40	.044192	.056141	.064325	.066502	.062069	.052332	.022009	.005003	.000045
		37	41	.049242	.059174	.064067	.062518	.055006	.043659	.015676	.003004	.000018
		36	42	.054293	.061944	.063635	.058878	.049083	.036883	.011509	.001907	.000009
		35	43	.058081	.062792	.061097	.053516	.042212	.029995	.008117	.001162	.000004
		34	44	.060606	.062262	.057537	.047838	.035795	.024115	.005694	.000707	.000002
		33	45	.061869	.060174	.052634	.041410	.029309	.018669	.003822	.000410	.000001
		32	46	.061869	.057205	.047569	.035579	.023941	.014498	.002618	.000248	.000000
		31	47	.060606	.053061	.041781	.029591	.018853	.010808	.001698	.000139	.000000
		30	48	.058081	.048345	.036212	.024409	.014809	.008088	.001129	.000083	.000000
		29	49	.054293	.042840	.030428	.019455	.011197	.005802	.000710	.000046	.000000
		28	50	.049242	.036912	.024919	.015149	.008293	.004089	.000442	.000025	.000000
		27	51	.044192	.031363	.020063	.011566	.006009	.002813	.000268	.000013	.000000
		26	52	.037879	.025635	.015656	.008628	.004290	.001925	.000166	.000008	.000000
		25	53	.032828	.021072	.012219	.006400	.003027	.001293	.000099	.000004	.000000
		24	54	.026515	.016173	.008922	.004449	.002005	.000816	.000055	.000002	.000000
		23	55	.021465	.012424	.006513	.003090	.001327	.000515	.000031	.000001	.000000
		22	56	.016414	.009073	.004552	.002072	.000855	.000320	.000018	.000001	.000000
		21	57	.012626	.006607	.003142	.001357	.000532	.000189	.000009	.000000	.000000
		20	58	.008838	.004436	.002029	.000845	.000320	.000110	.000005	.000000	.000000
		19	59	.006313	.002998	.001299	.000513	.000184	.000060	.000002	.000000	.000000
		18	60	.003788	.001717	.000712	.000269	.000093	.000029	.000001	.000000	.000000
		17	61	.002525	.001096	.000436	.000159	.000053	.000016	.000001	.000000	.000000
		16	62	.001263	.000526	.000201	.000071	.000023	.000007	.000000	.000000	.000000
		15	63	.001263	.000505	.000187	.000064	.000020	.000006	.000000	.000000	.000000

Table C-1 Distribution of Wilcoxon two-sample statistic under normal shift alternative d.

m	n	w_x	w_y	d = 0.	.2	.4	.6	.8	1.0	1.5	2.0	3.0
6	6	21	57	.001082	.002556	.005598	.011404	.021673	.038548	.124428	.286652	.706096
		22	56	.001082	.002447	.005103	.009837	.017567	.029131	.075775	.130843	.133921
		23	55	.002165	.004684	.009311	.017037	.028745	.044812	.097559	.135858	.079958
		24	54	.003247	.006713	.012707	.022057	.035172	.051613	.094781	.108159	.039043
		25	53	.005411	.010671	.019210	.031615	.047640	.065837	.102400	.096552	.022001
		26	52	.007576	.014181	.024169	.037557	.053279	.069103	.090358	.069968	.009972
		27	51	.011905	.021052	.033861	.049581	.066162	.080552	.088946	.057077	.005194
		28	50	.014069	.023749	.036338	.050446	.063593	.072868	.067931	.035766	.001954
		29	49	.019481	.031236	.045321	.059550	.070919	.076619	.061113	.027181	.001026
		30	48	.023810	.036247	.049837	.061928	.069595	.070793	.048060	.017893	.000446
		31	47	.030303	.043739	.056945	.066910	.071002	.068093	.039618	.012510	.000219
		32	46	.034632	.047407	.058450	.064937	.065041	.058766	.029188	.007750	.000090
		33	45	.042208	.054855	.064155	.067546	.064055	.054745	.023573	.005403	.000047
		34	44	.045455	.056068	.062146	.061917	.055473	.044716	.016495	.003193	.000019
		35	43	.051948	.060743	.063778	.060147	.050968	.038824	.012383	.002060	.000009
		36	42	.055195	.061234	.060968	.054491	.043736	.031537	.008749	.001263	.000004
		37	41	.059524	.062621	.059104	.050058	.038055	.025978	.006268	.000783	.000002
		38	40	.059524	.059438	.053216	.042727	.030771	.019885	.004168	.000450	.000001
		39	39	.062771	.059440	.050474	.038439	.026261	.016099	.002959	.000280	.000000
		40	38	.059524	.053442	.043018	.031046	.020093	.011664	.001866	.000153	.000000
		41	37	.059524	.050756	.038824	.026643	.016407	.009068	.001288	.000094	.000000
		42	36	.055195	.044641	.032395	.021092	.012322	.006460	.000802	.000051	.000000
		43	35	.051948	.039884	.027487	.017003	.009441	.004706	.000516	.000029	.000000
		44	34	.045455	.033104	.021655	.012722	.006712	.003181	.000308	.000015	.000000
		45	33	.042208	.029216	.018189	.010183	.005127	.002321	.000202	.000009	.000000
		46	32	.034632	.022780	.013488	.007188	.003447	.001488	.000115	.000005	.000000
		47	31	.030303	.018927	.010653	.005402	.002466	.001014	.000069	.000003	.000000
		48	30	.023810	.014117	.007551	.003642	.001583	.000620	.000038	.000001	.000000
		49	29	.019481	.010986	.005600	.002578	.001072	.000402	.000022	.000001	.000000
		50	28	.014069	.007550	.003667	.001612	.000640	.000230	.000011	.000000	.000000
		51	27	.011905	.006118	.002855	.001209	.000464	.000162	.000008	.000000	.000000
		52	26	.007576	.003681	.001624	.000650	.000236	.000078	.000003	.000000	.000000
		53	25	.005411	.002501	.001053	.000403	.000140	.000044	.000002	.000000	.000000
		54	24	.003247	.001436	.000579	.000213	.000071	.000022	.000001	.000000	.000000
		55	23	.002165	.000917	.000356	.000126	.000041	.000012	.000000	.000000	.000000
		56	22	.001082	.000441	.000165	.000057	.000018	.000005	.000000	.000000	.000000
		57	21	.001082	.000424	.000153	.000051	.000016	.000004	.000000	.000000	.000000

Table C-1 **Distribution of Wilcoxon two-sample statistic under normal shift alternative d.**

m	n	w_y	w_x	d = 0.	.2	.4	.6	.8	1.0	1.5	2.0	3.0
7	6	63	28	.000583	.001479	.003460	.007489	.015041	.028126	.100769	.250874	.676902
		62	29	.000583	.001420	.003174	.006525	.012361	.021642	.063224	.119597	.138391
		61	30	.001166	.002727	.005830	.011411	.020495	.033852	.083503	.128499	.086826
		60	31	.001748	.003922	.008012	.014933	.025441	.039697	.083353	.106040	.044643
		59	32	.002914	.006259	.012209	.021657	.035002	.051634	.092699	.098333	.026580
		58	33	.004079	.008360	.015517	.026119	.039937	.055561	.084873	.074835	.012981
		57	34	.006410	.012496	.022026	.035144	.050829	.066730	.087148	.064431	.007329
		56	35	.008159	.015162	.025416	.038468	.052626	.065152	.072388	.044423	.003169
		55	36	.011072	.019678	.031471	.045334	.058880	.069026	.066126	.034356	.001645
		54	37	.013986	.023685	.036023	.049245	.060563	.067072	.055120	.024140	.000772
		53	38	.018065	.029157	.042195	.054793	.063898	.066977	.047569	.017783	.000401
		52	39	.021562	.033121	.045546	.056103	.061947	.061359	.037468	.011857	.000181
		51	40	.026807	.039234	.051355	.060151	.063081	.059275	.031563	.008650	.000098
		50	41	.030303	.042269	.052641	.058560	.058222	.051767	.023791	.005544	.000043
		49	42	.035548	.047211	.055928	.059122	.055797	.047041	.018828	.003792	.000021
		48	43	.039627	.050124	.056497	.056767	.050869	.040677	.014197	.002476	.000010
		47	44	.044289	.053339	.057203	.054645	.046517	.035305	.010770	.001631	.000005
		46	45	.047203	.054111	.055195	.050111	.040506	.029165	.007764	.001019	.000002
		45	46	.051282	.055996	.054386	.046996	.036142	.024749	.005801	.000669	.000001
		44	47	.052448	.054530	.050397	.041413	.030265	.019679	.004038	.000405	.000001
		43	48	.054779	.054242	.047745	.037365	.026006	.016104	.002926	.000260	.000000
		42	49	.054779	.051641	.043268	.032224	.021338	.012567	.002010	.000157	.000000
		41	50	.054779	.049170	.039229	.027821	.017541	.009836	.001390	.000096	.000000
		40	51	.052448	.044837	.034068	.023009	.013815	.007376	.000922	.000056	.000000
		39	52	.051282	.041769	.030255	.019491	.011170	.005695	.000636	.000035	.000000
		38	53	.047203	.036634	.025293	.015535	.008490	.004129	.000410	.000020	.000000
		37	54	.044289	.032739	.021543	.012617	.006578	.003053	.000270	.000012	.000000
		36	55	.039627	.027909	.017507	.009780	.004865	.002156	.000170	.000007	.000000
		35	56	.035548	.023868	.014287	.007623	.003626	.001537	.000109	.000004	.000000
		34	57	.030303	.019389	.011069	.005637	.002560	.001037	.000066	.000002	.000000

Table C-1 Distribution of Wilcoxon two-sample statistic under normal shift alternative d.

m	n	w_y	w_x	d = 0.	.2	.4	.6	.8	1.0	1.5	2.0	3.0
7	6	33	58	.026807	.016374	.008938	.004359	.001899	.000739	.000043	.000001	.000000
		32	59	.021562	.012559	.006543	.003047	.001268	.000472	.000024	.000001	.000000
		31	60	.018065	.010030	.004987	.002219	.000884	.000315	.000015	.000000	.000000
		30	61	.013986	.007412	.003523	.001501	.000573	.000196	.000008	.000000	.000000
		29	62	.011072	.005602	.002546	.001039	.000380	.000125	.000005	.000000	.000000
		28	63	.008159	.003956	.001728	.000679	.000240	.000076	.000003	.000000	.000000
		27	64	.006410	.002970	.001241	.000468	.000159	.000048	.000002	.000000	.000000
		26	65	.004079	.001800	.000717	.000258	.000084	.000024	.000001	.000000	.000000
		25	66	.002914	.001230	.000470	.000162	.000051	.000014	.000000	.000000	.000000
		24	67	.001748	.000708	.000261	.000087	.000026	.000007	.000000	.000000	.000000
		23	68	.001166	.000454	.000161	.000052	.000015	.000004	.000000	.000000	.000000
		22	69	.000583	.000219	.000075	.000024	.000007	.000002	.000000	.000000	.000000
		21	70	.000583	.000211	.000070	.000021	.000006	.000002	.000000	.000000	.000000
7	7	77	28	.000291	.000802	.002021	.004681	.010002	.019782	.079638	.216173	.645528
		76	29	.000291	.000772	.001865	.004118	.008332	.015492	.051445	.107544	.142136
		75	30	.000583	.001487	.003447	.007271	.013994	.024631	.069671	.119508	.093645
		74	31	.000874	.002147	.004777	.009615	.017615	.029396	.071409	.102129	.050597
		73	32	.001457	.003439	.007325	.014105	.024606	.038965	.081666	.098225	.031709
		72	33	.002040	.004616	.009401	.017260	.028619	.042935	.077410	.078200	.016516
		71	34	.003205	.006945	.013514	.023647	.037275	.053020	.082593	.070621	.009955
		70	35	.004371	.009005	.016637	.027596	.041148	.055232	.073499	.052523	.004748
		69	36	.005828	.011558	.020498	.032541	.046299	.059117	.068455	.041645	.002514
		68	37	.007576	.014359	.024287	.036697	.049584	.059980	.059986	.030950	.001260
		67	38	.009907	.017981	.029068	.041894	.053885	.061917	.053935	.023894	.000689
		66	39	.012238	.021208	.032680	.044814	.054735	.059599	.044928	.016948	.000334
		65	40	.015443	.025617	.037733	.049388	.057488	.059564	.039386	.012908	.000188
		64	41	.018357	.029050	.040766	.050757	.056108	.055108	.031598	.008846	.000088
		63	42	.021853	.033096	.044385	.052739	.055556	.051920	.026112	.006348	.000047
		62	43	.025350	.036681	.046943	.053157	.053295	.047339	.020837	.004393	.000024
		61	44	.029138	.040309	.049266	.053221	.050842	.042979	.016622	.003052	.000012

Table C-1 Distribution of Wilcoxon two-sample statistic under normal shift alternative d.

m	n	w_y	w_x	d = 0.	.2	.4	.6	.8	1.0	1.5	2.0	3.0
7	7	60	45	.032634	.043089	.050216	.051675	.046975	.037745	.012789	.002040	.000006
		59	46	.036422	.045993	.051225	.050336	.043659	.033444	.010021	.001407	.000003
		58	47	.039627	.047744	.050698	.047462	.039186	.028548	.007511	.000919	.000002
		57	48	.042541	.049038	.049788	.044538	.035116	.024416	.005703	.000618	.000001
		56	49	.045163	.049690	.048133	.041059	.030855	.020435	.004214	.000401	.000000
		55	50	.047203	.049637	.045936	.037421	.026843	.016961	.003103	.000261	.000000
		54	51	.048368	.048553	.042881	.033326	.022797	.013732	.002224	.000165	.000000
		53	52	.049242	.047274	.039927	.029674	.019411	.011181	.001620	.000108	.000000
		52	53	.049242	.045124	.036377	.025803	.016108	.008853	.001140	.000067	.000000
		51	54	.048368	.042390	.032684	.022173	.013239	.006958	.000801	.000042	.000000
		50	55	.047203	.039500	.029085	.018847	.010749	.005398	.000554	.000026	.000000
		49	56	.045163	.036134	.025449	.015778	.008612	.004140	.000382	.000016	.000000
		48	57	.042541	.032499	.021862	.012950	.006755	.003104	.000256	.000010	.000000
		47	58	.039627	.028983	.018678	.010607	.005308	.002342	.000175	.000006	.000000
		46	59	.036422	.025432	.015655	.008496	.004064	.001714	.000114	.000004	.000000
		45	60	.032634	.021809	.012858	.006687	.003068	.001242	.000075	.000002	.000000
		44	61	.029138	.018601	.010485	.005217	.002291	.000888	.000048	.000001	.000000
		43	62	.025350	.015488	.008363	.003990	.001682	.000626	.000031	.000001	.000000
		42	63	.021853	.012772	.006605	.003022	.001223	.000438	.000020	.000001	.000000
		41	64	.018357	.010280	.005101	.002241	.000872	.000300	.000012	.000000	.000000
		40	65	.015443	.008261	.003919	.001648	.000614	.000203	.000007	.000000	.000000
		39	66	.012238	.006275	.002857	.001154	.000414	.000132	.000004	.000000	.000000
		38	67	.009907	.004858	.002118	.000821	.000283	.000086	.000003	.000000	.000000
		37	68	.007576	.003564	.001493	.000557	.000185	.000055	.000002	.000000	.000000
		36	69	.005828	.002624	.001055	.000378	.000121	.000034	.000001	.000000	.000000
		35	70	.004371	.001900	.000739	.000257	.000080	.000022	.000001	.000000	.000000
		34	71	.003205	.001326	.000491	.000163	.000048	.000013	.000001	.000000	.000000
		33	72	.002040	.000810	.000288	.000092	.000026	.000007	.000000	.000000	.000000
		32	73	.001457	.000556	.000191	.000059	.000016	.000004	.000000	.000000	.000000
		31	74	.000874	.000322	.000107	.000032	.000009	.000002	.000000	.000000	.000000
		30	75	.000583	.000207	.000066	.000019	.000005	.000001	.000000	.000000	.000000
		29	76	.000291	.000100	.000031	.000009	.000002	.000001	.000000	.000000	.000000
		28	77	.000291	.000097	.000029	.000008	.000002	.000000	.000000	.000000	.000000

Table C-2

Log R(w|d), the common logarithm of the rank-sum likelihood ratio under normal shift alternative d.

m	n	w_y	w_x	d = .2	.4	.6	.8	1.0	1.5	2.0	3.0
1	1	2	1	0.0462	0.0873	0.1234	0.1548	0.1819	0.2332	0.2654	0.2936
		1	2	−0.0518	−0.1094	−0.1730	−0.2429	−0.3192	−0.5393	−0.8032	−1.4698
2	1	3	3	0.0697	0.1321	0.1875	0.2363	0.2790	0.3612	0.4145	0.4633
		2	4	−0.0047	−0.0191	−0.0430	−0.0765	−0.1195	−0.2687	−0.4769	−1.0678
		1	5	−0.0773	−0.1627	−0.2562	−0.3582	−0.4690	−0.7852	−1.1598	−2.0930
3	1	4	6	0.0850	0.1616	0.2300	0.2906	0.3440	0.4483	0.5173	0.5826
		3	7	0.0202	0.0294	0.0276	0.0148	−0.0087	−0.1149	−0.2876	−0.8266
		2	8	−0.0313	−0.0738	−0.1275	−0.1925	−0.2687	−0.5089	−0.8201	−1.6569
		1	9	−0.0938	−0.1968	−0.3093	−0.4313	−0.5633	−0.9379	−1.3785	−2.4663
2	2	7	3	0.1095	0.2082	0.2963	0.3745	0.4433	0.5779	0.6674	0.7526
		6	4	0.0564	0.0985	0.1266	0.1409	0.1415	0.0848	−0.0517	−0.5492
		5	5	−0.0070	−0.0280	−0.0630	−0.1119	−0.1747	−0.3919	−0.6934	−1.5400
		4	6	−0.0708	−0.1563	−0.2566	−0.3718	−0.5022	−0.8950	−1.3849	−2.6624
		3	7	−0.1209	−0.2537	−0.3987	−0.5563	−0.7268	−1.2112	−1.7822	−3.1969
4	1	5	10	0.0962	0.1832	0.2613	0.3309	0.3924	0.5137	0.5954	0.6745
		4	11	0.0370	0.0621	0.0754	0.0770	0.0669	−0.0087	−0.1554	−0.6550
		3	12	−0.0061	−0.0247	−0.0557	−0.0990	−0.1548	−0.3481	−0.6186	−1.3899
		2	13	−0.0489	−0.1099	−0.1830	−0.2683	−0.3657	−0.6629	−1.0372	−2.0196
		1	14	−0.1058	−0.2216	−0.3475	−0.4839	−0.6309	−1.0463	−1.5324	−2.7253
3	2	9	6	0.1372	0.2611	0.3724	0.4715	0.5592	0.7325	0.8495	0.9640
		8	7	0.0926	0.1686	0.2283	0.2720	0.2998	0.3025	0.2134	−0.2182
		7	8	0.0418	0.0664	0.0739	0.0645	0.0384	−0.0987	−0.3355	−1.0906
		6	9	−0.0088	−0.0353	−0.0794	−0.1410	−0.2201	−0.4934	−0.8723	−1.9299
		5	10	−0.0592	−0.1358	−0.2301	−0.3420	−0.4716	−0.8733	−1.3861	−2.7425
		4	11	−0.1094	−0.2360	−0.3798	−0.5411	−0.7201	−1.2461	−1.8861	−3.5171
		3	12	−0.1509	−0.3161	−0.4960	−0.6910	−0.9014	−1.4973	−2.1968	−3.9208

Table C-2

Log R(w|d), the common logarithm of the rank-sum likelihood ratio under normal shift alternative d.

m	n	w_y	w_x	d = .2	.4	.6	.8	1.0	1.5	2.0	3.0
5	1	6	15	0.1050	0.2002	0.2860	0.3627	0.4307	0.5660	0.6580	0.7491
		5	16	0.0494	0.0865	0.1112	0.1236	0.1238	0.0716	-0.0547	-0.5225
		4	17	0.0109	0.0088	-0.0063	-0.0345	-0.0757	-0.2356	-0.4762	-1.1980
		3	18	-0.0240	-0.0612	-0.1115	-0.1749	-0.2515	-0.5005	-0.8320	-1.7423
		2	19	-0.0620	-0.1366	-0.2238	-0.3238	-0.4365	-0.7745	-1.1934	-2.2766
		1	20	-0.1151	-0.2408	-0.3771	-0.5244	-0.6829	-1.1293	-1.6497	-2.9206
4	2	11	10	0.1581	0.3015	0.4305	0.5459	0.6482	0.8521	0.9915	1.1307
		10	11	0.1184	0.2187	0.3011	0.3661	0.4138	0.4602	0.4070	0.0263
		9	12	0.0733	0.1277	0.1633	0.1803	0.1790	0.0971	-0.0934	-0.7794
		8	13	0.0361	0.0523	0.0485	0.0250	-0.0179	-0.2093	-0.5166	-1.4555
		7	14	-0.0099	-0.0396	-0.0891	-0.1583	-0.2472	-0.5542	-0.9799	-2.1669
		6	15	-0.0564	-0.1331	-0.2303	-0.3481	-0.4865	-0.9233	-1.4904	-3.0173
		5	16	-0.0925	-0.2043	-0.3355	-0.4863	-0.6568	-1.1695	-1.8067	-3.4556
		4	17	-0.1368	-0.2922	-0.4666	-0.6601	-0.8729	-1.4910	-2.2339	-4.1079
		3	18	-0.1736	-0.3631	-0.5690	-0.7917	-1.0317	-1.7094	-2.5021	-4.4436
3	3	15	6	0.1748	0.3332	0.4758	0.6033	0.7164	0.9417	1.0959	1.2503
		14	7	0.1383	0.2568	0.3560	0.4361	0.4976	0.5726	0.5408	0.1860
		13	8	0.0993	0.1773	0.2344	0.2708	0.2868	0.2402	0.0749	-0.5846
		12	9	0.0503	0.0791	0.0866	0.0728	0.0381	-0.1390	-0.4415	-1.3973
		11	10	0.0125	0.0023	-0.0305	-0.0859	-0.1638	-0.4559	-0.8850	-2.1396
		10	11	-0.0353	-0.0935	-0.1746	-0.2785	-0.4054	-0.8223	-1.3810	-2.9172
		9	12	-0.0719	-0.1655	-0.2808	-0.4179	-0.5767	-1.0686	-1.6950	-3.3423
		8	13	-0.1208	-0.2636	-0.4283	-0.6154	-0.8250	-1.4487	-2.2170	-4.1938
		7	14	-0.1584	-0.3372	-0.5367	-0.7573	-0.9992	-1.6986	-2.5361	-4.6197
		6	15	-0.1919	-0.4016	-0.6294	-0.8760	-1.1418	-1.8930	-2.7731	-4.9208
6	1	7	21	0.1122	0.2141	0.3062	0.3888	0.4623	0.6092	0.7102	0.8117
		6	22	0.0593	0.1058	0.1395	0.1606	0.1691	0.1359	0.0262	-0.4150
		5	23	0.0238	0.0341	0.0310	0.0143	-0.0157	-0.1495	-0.3667	-1.0488
		4	24	-0.0068	-0.0274	-0.0617	-0.1097	-0.1714	-0.3856	-0.6854	-1.5412
		3	25	-0.0374	-0.0884	-0.1530	-0.2312	-0.3230	-0.6125	-0.9876	-1.9955
		2	26	-0.0722	-0.1574	-0.2557	-0.3671	-0.4917	-0.8611	-1.3141	-2.4732
		1	27	-0.1227	-0.2564	-0.4011	-0.5572	-0.7249	-1.1961	-1.7435	-3.0757

276

Table C-2

Log R(w|d), the common logarithm of the rank-sum likelihood ratio under normal shift alternative d.

m	n	w_y	w_x	d = .2	.4	.6	.8	1.0	1.5	2.0	3.0
5	2	13	15	0.1749	0.3339	0.4773	0.6058	0.7203	0.9494	1.1076	1.2682
		12	16	0.1383	0.2575	0.3577	0.4393	0.5027	0.5837	0.5592	0.2203
		11	17	0.0967	0.1733	0.2300	0.2669	0.2843	0.2447	0.0900	−0.5410
		10	18	0.0655	0.1097	0.1328	0.1350	0.1164	−0.0185	−0.2761	−1.1323
		9	19	0.0282	0.0349	0.0201	−0.0160	−0.0733	−0.3075	−0.6678	−1.7394
		8	20	−0.0108	−0.0434	−0.0977	−0.1735	−0.2709	−0.6069	−1.0721	−2.3604
		7	21	−0.0500	−0.1219	−0.2157	−0.3316	−0.4695	−0.9108	−1.4903	−3.0621
		6	22	−0.0870	−0.1958	−0.3264	−0.4789	−0.6534	−1.1869	−1.8600	−3.6285
		5	23	−0.1171	−0.2548	−0.4132	−0.5925	−0.7929	−1.3863	−2.1133	−3.9744
		4	24	−0.1578	−0.3355	−0.5333	−0.7514	−0.9900	−1.6778	−2.4985	−4.5638
		3	25	−0.1916	−0.4004	−0.6269	−0.8715	−1.1347	−1.8761	−2.7407	−4.8326
4	3	18	10	0.2042	0.3895	0.5567	0.7066	0.8400	1.1071	1.2918	1.4801
		17	11	0.1720	0.3221	0.4507	0.5581	0.6448	0.7746	0.7866	0.4923
		16	12	0.1383	0.2530	0.3444	0.4128	0.4586	0.4774	0.3652	−0.2190
		15	13	0.0981	0.1719	0.2216	0.2474	0.2496	0.1542	−0.0802	−0.9342
		14	14	0.0625	0.0997	0.1119	0.0991	0.0616	−0.1383	−0.4863	−1.5979
		13	15	0.0238	0.0212	−0.0075	−0.0625	−0.1436	−0.4595	−0.9351	−2.3523
		12	16	−0.0129	−0.0517	−0.1163	−0.2066	−0.3226	−0.7237	−1.2812	−2.8476
		11	17	−0.0502	−0.1269	−0.2301	−0.3597	−0.5159	−1.0218	−1.6920	−3.5164
		10	18	−0.0878	−0.2012	−0.3400	−0.5044	−0.6943	−1.2806	−2.0247	−3.9716
		9	19	−0.1227	−0.2703	−0.4427	−0.6403	−0.8630	−1.5304	−2.3562	−4.4858
		8	20	−0.1624	−0.3491	−0.5605	−0.7968	−1.0583	−1.8238	−2.7517	−5.1549
		7	21	−0.1945	−0.4118	−0.6522	−0.9162	−1.2040	−2.0299	−3.0118	−5.4559
		6	22	−0.2237	−0.4677	−0.7325	−1.0186	−1.3267	−2.1958	−3.2128	−5.4559
7	1	8	28	0.1182	0.2259	0.3233	0.4109	0.4890	0.6461	0.7549	0.8657
		7	29	0.0674	0.1217	0.1629	0.1912	0.2066	0.1893	0.0937	−0.3248
		6	30	0.0341	0.0544	0.0608	0.0535	0.0324	−0.0803	−0.2783	−0.9273
		5	31	0.0061	−0.0017	−0.0237	−0.0598	−0.1100	−0.2971	−0.5719	−1.3840
		4	32	−0.0203	−0.0547	−0.1033	−0.1660	−0.2428	−0.4969	−0.8394	−1.7901
		3	33	−0.0480	−0.1099	−0.1858	−0.2756	−0.3794	−0.7002	−1.1090	−2.1911
		2	34	−0.0806	−0.1745	−0.2818	−0.4025	−0.5367	−0.9314	−1.4116	−2.6310
		1	35	−0.1291	−0.2694	−0.4212	−0.5846	−0.7599	−1.2515	−1.8212	−3.2031

277

Table C-2

Log R(w|d), the common logarithm of the rank-sum likelihood ratio under normal shift alternative d.

m	n	w_y	w_x	d = .2	.4	.6	.8	1.0	1.5	2.0	3.0
6	2	15	21	0.1889	0.3608	0.5163	0.6559	0.7805	1.0312	1.2056	1.3852
		14	22	0.1545	0.2890	0.4037	0.4990	0.5752	0.6849	0.6843	0.3808
		13	23	0.1153	0.2096	0.2831	0.3359	0.3685	0.3629	0.2375	-0.3480
		12	24	0.0876	0.1530	0.1964	0.2180	0.2181	0.1261	-0.0931	-0.8855
		11	25	0.0545	0.0864	0.0958	0.0830	0.0481	-0.1336	-0.4464	-1.4348
		10	26	0.0251	0.0271	0.0060	-0.0378	-0.1042	-0.3675	-0.7645	-1.9243
		9	27	-0.0115	-0.0460	-0.1035	-0.1840	-0.2872	-0.6437	-1.1379	-2.5089
		8	28	-0.0484	-0.1204	-0.2159	-0.3350	-0.4778	-0.9389	-1.5491	-3.2177
		7	29	-0.0773	-0.1775	-0.3007	-0.4470	-0.6164	-1.1418	-1.8129	-3.5922
		6	30	-0.1100	-0.2427	-0.3981	-0.5764	-0.7778	-1.3825	-2.1333	-4.0757
		5	31	-0.1366	-0.2948	-0.4747	-0.6764	-0.9001	-1.5566	-2.3534	-4.3767
		4	32	-0.1749	-0.3706	-0.5872	-0.8250	-1.0843	-1.8279	-2.7102	-4.9507
		3	33	-0.2065	-0.4313	-0.6747	-0.9372	-1.2193	-2.0123	-2.9358	-5.2518
5	3	21	15	0.2281	0.4355	0.6230	0.7913	0.9415	1.2437	1.4543	1.6720
		20	16	0.1989	0.3740	0.5260	0.6552	0.7622	0.9362	0.9839	0.7396
		19	17	0.1683	0.3112	0.4290	0.5223	0.5913	0.6614	0.5915	0.0685
		18	18	0.1322	0.2383	0.3186	0.3732	0.4026	0.3683	0.1857	-0.5878
		17	19	0.1043	0.1812	0.2308	0.2535	0.2495	0.1254	-0.1561	-1.1551
		16	20	0.0676	0.1072	0.1190	0.1030	0.0596	-0.1672	-0.5586	-1.7991
		15	21	0.0359	0.0433	0.0221	-0.0273	-0.1050	-0.4213	-0.9090	-2.3782
		14	22	0.0000	-0.0289	-0.0868	-0.1737	-0.2895	-0.7045	-1.2971	-3.0009
		13	23	-0.0290	-0.0870	-0.1739	-0.2897	-0.4342	-0.9199	-1.5807	-3.4056
		12	24	-0.0646	-0.1579	-0.2800	-0.4308	-0.6104	-1.1846	-1.9362	-3.9580
		11	25	-0.0958	-0.2198	-0.3722	-0.5529	-0.7620	-1.4081	-2.2293	-4.3825
		10	26	-0.1321	-0.2921	-0.4803	-0.6968	-0.9418	-1.6795	-2.5976	-4.9507
		9	27	-0.1587	-0.3441	-0.5564	-0.7958	-1.0623	-1.8487	-2.8077	-5.2518
		8	28	-0.1941	-0.4145	-0.6613	-0.9351	-1.2359	-2.1084	-3.1566	-5.5528
		7	29	-0.2230	-0.4707	-0.7435	-1.0416	-1.3656	-2.2903	-3.3825	-5.5528
		6	30	-0.2495	-0.5213	-0.8159	-1.1339	-1.4759	-2.4388	-3.5616	-5.5528

Table C-2

Log R(w|d), the common logarithm of the rank-sum likelihood ratio under normal shift alternative d.

m	n	w_x	w_y	d = .2	.4	.6	.8	1.0	1.5	2.0	3.0
4	4	10	26	0.2407	0.4596	0.6573	0.8349	0.9933	1.3120	1.5342	1.7643
		11	25	0.2128	0.4008	0.5645	0.7043	0.8210	1.0150	1.0778	0.8521
		12	24	0.1841	0.3416	0.4728	0.5783	0.6584	0.7516	0.6991	0.1965
		13	23	0.1523	0.2765	0.3732	0.4426	0.4852	0.4777	0.3138	-0.4427
		14	22	0.1152	0.2020	0.2607	0.2916	0.2948	0.1843	-0.0902	-1.0969
		15	21	0.0848	0.1395	0.1645	0.1597	0.1254	-0.0877	-0.4797	-1.7804
		16	20	0.0523	0.0743	0.0661	0.0280	-0.0398	-0.3380	-0.8159	-2.2847
		17	19	0.0168	0.0028	-0.0417	-0.1170	-0.2227	-0.6197	-1.2038	-2.9157
		18	18	-0.0153	-0.0614	-0.1380	-0.2453	-0.3830	-0.8592	-1.5209	-3.3758
		19	17	-0.0476	-0.1260	-0.2351	-0.3750	-0.5455	-1.1051	-1.8529	-3.8961
		20	16	-0.0827	-0.1961	-0.3401	-0.5148	-0.7203	-1.3687	-2.2091	-4.4559
		21	15	-0.1149	-0.2601	-0.4357	-0.6416	-0.8779	-1.6019	-2.5152	-4.8996
		22	14	-0.1437	-0.3162	-0.5175	-0.7477	-1.0069	-1.7815	-2.7366	-5.1549
		23	13	-0.1806	-0.3900	-0.6285	-0.8962	-1.1936	-2.0683	-3.1337	-5.6320
		24	12	-0.2113	-0.4501	-0.7168	-1.0118	-1.3354	-2.2718	-3.3914	-5.6320
		25	11	-0.2382	-0.5024	-0.7929	-1.1102	-1.4548	-2.4380	-3.5985	-5.6320
		26	10	-0.2634	-0.5504	-0.8615	-1.1975	-1.5589	-2.5774	-3.7746	-5.6320
7	2	28	17	0.2008	0.3838	0.5496	0.6988	0.8322	1.1016	1.2904	1.4869
		29	16	0.1681	0.3155	0.4424	0.5493	0.6364	0.7705	0.7904	0.5176
		30	15	0.1307	0.2397	0.3271	0.3932	0.4384	0.4614	0.3607	-0.1860
		31	14	0.1053	0.1878	0.2475	0.2848	0.3000	0.2428	0.0545	-0.6858
		32	13	0.0748	0.1262	0.1545	0.1599	0.1425	0.0016	-0.2738	-1.1972
		33	12	0.0500	0.0760	0.0783	0.0570	0.0125	-0.1985	-0.5468	-1.6177
		34	11	0.0201	0.0160	-0.0120	-0.0640	-0.1396	-0.4302	-0.8613	-2.1074
		35	10	-0.0121	-0.0483	-0.1088	-0.1933	-0.3017	-0.6761	-1.1942	-2.6251
		36	9	-0.0444	-0.1132	-0.2065	-0.3243	-0.4668	-0.9306	-1.5487	-3.2464
		37	8	-0.0741	-0.1725	-0.2952	-0.4425	-0.6144	-1.1520	-1.8447	-3.6981
		38	7	-0.0983	-0.2203	-0.3660	-0.5357	-0.7293	-1.3187	-2.0593	-3.9965
		39	6	-0.1284	-0.2802	-0.4555	-0.6544	-0.8771	-1.5384	-2.3508	-4.4436
		40	5	-0.1527	-0.3278	-0.5253	-0.7454	-0.9883	-1.6962	-2.5499	-4.7033
		41	4	-0.1892	-0.3999	-0.6322	-0.8865	-1.1630	-1.9527	-2.8861	-5.1426
		42	3	-0.2192	-0.4575	-0.7151	-0.9927	-1.2907	-2.1270	-3.0993	-5.4436

Table C-2

Log R(w|d), the common logarithm of the rank-sum likelihood ratio under normal shift alternative d.

m	n	w_y	w_x	d = .2	.4	.6	.8	1.0	1.5	2.0	3.0
6	3	24	21	0.2482	0.4743	0.6789	0.8630	1.0275	1.3598	1.5929	1.8369
		23	22	0.2210	0.4170	0.5884	0.7358	0.8597	1.0707	1.1486	0.9471
		22	23	0.1926	0.3585	0.4980	0.6115	0.6996	0.8119	0.7770	0.3058
		21	24	0.1592	0.2908	0.3952	0.4727	0.5238	0.5382	0.3975	−0.3104
		20	25	0.1350	0.2412	0.3188	0.3680	0.3893	0.3232	0.0926	−0.8203
		19	26	0.1047	0.1797	0.2253	0.2416	0.2291	0.0742	−0.2512	−1.3679
		18	27	0.0738	0.1178	0.1319	0.1164	0.0716	−0.1669	−0.5812	−1.8983
		17	28	0.0433	0.0557	0.0372	−0.0120	−0.0920	−0.4249	−0.9450	−2.5282
		16	29	0.0137	−0.0034	−0.0514	−0.1301	−0.2395	−0.6457	−1.2385	−2.9605
		15	30	−0.0156	−0.0625	−0.1406	−0.2499	−0.3903	−0.8763	−1.5527	−3.4576
		14	31	−0.0447	−0.1203	−0.2269	−0.3643	−0.5325	−1.0864	−1.8280	−3.8484
		13	32	−0.0744	−0.1801	−0.3168	−0.4848	−0.6840	−1.3178	−2.1445	−4.3645
		12	33	−0.1039	−0.2381	−0.4025	−0.5971	−0.8220	−1.5161	−2.3965	−4.6903
		11	34	−0.1347	−0.2996	−0.4949	−0.7207	−0.9772	−1.7535	−2.7239	−5.2975
		10	35	−0.1642	−0.3580	−0.5814	−0.8347	−1.1180	−1.9588	−2.9902	−5.6777
		9	36	−0.1870	−0.4022	−0.6458	−0.9180	−1.2190	−2.0987	−3.1619	−5.5528
		8	37	−0.2199	−0.4674	−0.7429	−1.0468	−1.3794	−2.3381	−3.4846	−5.5528
		7	38	−0.2466	−0.5194	−0.8187	−1.1449	−1.4984	−2.5040	−3.6955	−5.5528
		6	39	−0.2712	−0.5662	−0.8856	−1.2301	−1.6003	−2.6411	−3.8452	−5.5528
5	4	30	15	0.2710	0.5176	0.7408	0.9416	1.1210	1.4833	1.7375	2.0042
		29	16	0.2458	0.4645	0.6566	0.8228	0.9638	1.2104	1.3150	1.1465
		28	17	0.2201	0.4112	0.5739	0.7086	0.8159	0.9686	0.9641	0.5292
		27	18	0.1921	0.3538	0.4856	0.5879	0.6613	0.7215	0.6136	−0.0616
		26	19	0.1607	0.2904	0.3893	0.4578	0.4963	0.4641	0.2552	−0.6508
		25	20	0.1308	0.2295	0.2962	0.3312	0.3347	0.2085	−0.1057	−1.2614
		24	21	0.1035	0.1738	0.2112	0.2159	0.1882	−0.0210	−0.4256	−1.7912
		23	22	0.0720	0.1101	0.1145	0.0853	0.0227	−0.2782	−0.7823	−2.3767
		22	23	0.0419	0.0499	0.0240	−0.0355	−0.1286	−0.5066	−1.0879	−2.8314
		21	24	0.0135	−0.0074	−0.0628	−0.1526	−0.2765	−0.7343	−1.4004	−3.3335
		20	25	−0.0172	−0.0690	−0.1553	−0.2760	−0.4310	−0.9675	−1.7140	−3.8144

Table C-2

Log R(w|d), the common logarithm of the rank-sum likelihood ratio under normal shift alternative d.

m	n	w_y	w_x	d = .2	.4	.6	.8	1.0	1.5	2.0	3.0
5	4	19	26	-0.0481	-0.1308	-0.2482	-0.4001	-0.5867	-1.2036	-2.0340	-4.3177
		18	27	-0.0759	-0.1859	-0.3300	-0.5080	-0.7200	-1.3975	-2.2832	-4.6622
		17	28	-0.1060	-0.2461	-0.4205	-0.6290	-0.8718	-1.6287	-2.5993	-5.1549
		16	29	-0.1368	-0.3072	-0.5112	-0.7490	-1.0206	-1.8477	-2.8862	-5.5016
		15	30	-0.1632	-0.3589	-0.5873	-0.8485	-1.1424	-2.0206	-3.1025	-5.6777
		14	31	-0.1921	-0.4159	-0.6717	-0.9595	-1.2796	-2.2220	-3.3681	-5.6777
		13	32	-0.2229	-0.4771	-0.7629	-1.0807	-1.4307	-2.4493	-3.6777	-5.6777
		12	33	-0.2496	-0.5292	-0.8392	-1.1800	-1.5521	-2.6208	-3.8996	-5.6777
		11	34	-0.2736	-0.5755	-0.9064	-1.2666	-1.6568	-2.7660	-4.0545	-5.6777
		10	35	-0.2962	-0.6185	-0.9677	-1.3444	-1.7494	-2.8910	-4.2006	-5.6777
7	3	27	28	0.2656	0.5077	0.7272	0.9250	1.1020	1.4607	1.7137	1.9813
		26	29	0.2399	0.4537	0.6417	0.8046	0.9430	1.1859	1.2899	1.1257
		25	30	0.2131	0.3983	0.5560	0.6867	0.7909	0.9391	0.9343	0.5078
		24	31	0.1815	0.3343	0.4587	0.5553	0.6243	0.6795	0.5737	-0.0789
		23	32	0.1597	0.2895	0.3895	0.4602	0.5020	0.4827	0.2935	-0.5500
		22	33	0.1326	0.2344	0.3057	0.3468	0.3580	0.2585	-0.0163	-1.0411
		21	34	0.1064	0.1815	0.2255	0.2387	0.2215	0.0477	-0.3064	-1.5071
		20	35	0.0779	0.1238	0.1380	0.1206	0.0719	-0.1848	-0.6287	-2.0318
		19	36	0.0517	0.0709	0.0576	0.0121	-0.0654	-0.3984	-0.9263	-2.5426
		18	37	0.0219	0.0112	-0.0321	-0.1082	-0.2167	-0.6290	-1.2401	-3.0377
		17	38	-0.0020	-0.0370	-0.1048	-0.2053	-0.3384	-0.8123	-1.4841	-3.3931
		16	39	-0.0309	-0.0948	-0.1917	-0.3215	-0.4842	-1.0339	-1.7854	-3.8716
		15	40	-0.0547	-0.1421	-0.2623	-0.4151	-0.6003	-1.2044	-2.0064	-4.1726
		14	41	-0.0844	-0.2015	-0.3515	-0.5342	-0.7498	-1.4320	-2.3176	-4.6709
		13	42	-0.1100	-0.2522	-0.4266	-0.6334	-0.8724	-1.6108	-2.5485	-4.9788
		12	43	-0.1379	-0.3076	-0.5092	-0.7428	-1.0085	-1.8137	-2.8209	-5.4648
		11	44	-0.1637	-0.3588	-0.5855	-0.8440	-1.1344	-2.0012	-3.0707	-5.6197
		10	45	-0.1900	-0.4107	-0.6623	-0.9449	-1.2588	-2.1818	-3.3027	-5.6197
		9	46	-0.2105	-0.4503	-0.7197	-1.0190	-1.3484	-2.3048	-3.4534	-5.6197
		8	47	-0.2415	-0.5118	-0.8114	-1.1405	-1.4994	-2.5298	-3.7594	-5.6197
		7	48	-0.2667	-0.5607	-0.8824	-1.2323	-1.6109	-2.6852	-3.9665	-5.6197
		6	49	-0.2898	-0.6048	-0.9454	-1.3125	-1.7065	-2.8136	-4.0757	-5.6197

281

Table C-2

Log R(w|d), the common logarithm of the rank-sum likelihood ratio under normal shift alternative d.

m	n	w_y	w_x	d = .2	.4	.6	.8	1.0	1.5	2.0	3.0
6	4	34	21	0.2967	0.5671	0.8121	1.0328	1.2302	1.6302	1.9125	2.2118
		33	22	0.2735	0.5179	0.7340	0.9225	1.0840	1.3751	1.5152	1.3959
		32	23	0.2498	0.4687	0.6574	0.8164	0.9464	1.1486	1.1845	0.8072
		31	24	0.2242	0.4161	0.5763	0.7053	0.8036	0.9192	0.8574	0.2506
		30	25	0.1957	0.3584	0.4886	0.5866	0.6528	0.6830	0.5272	-0.2955
		29	26	0.1715	0.3086	0.4117	0.4811	0.5173	0.4653	0.2167	-0.8224
		28	27	0.1436	0.2524	0.3265	0.3662	0.3719	0.2399	-0.0949	-1.3306
		27	28	0.1169	0.1978	0.2428	0.2523	0.2263	0.0086	-0.4230	-1.9006
		26	29	0.0892	0.1420	0.1586	0.1392	0.0841	-0.2083	-0.7170	-2.3509
		25	30	0.0624	0.0879	0.0767	0.0288	-0.0554	-0.4240	-1.0142	-2.8313
		24	31	0.0359	0.0346	-0.0038	-0.0793	-0.1917	-0.6321	-1.2962	-3.2628
		23	32	0.0082	-0.0210	-0.0879	-0.1922	-0.3338	-0.8496	-1.5936	-3.7420
		22	33	-0.0186	-0.0746	-0.1679	-0.2984	-0.4658	-1.0451	-1.8499	-4.1069
		21	34	-0.0459	-0.1294	-0.2507	-0.4095	-0.6059	-1.2604	-2.1467	-4.6031
		20	35	-0.0733	-0.1841	-0.3324	-0.5182	-0.7414	-1.4627	-2.4157	-4.9788
		19	36	-0.0994	-0.2358	-0.4093	-0.6197	-0.8671	-1.6464	-2.6527	-5.3467
		18	37	-0.1257	-0.2880	-0.4870	-0.7228	-0.9953	-1.8371	-2.9069	-5.7917
		17	38	-0.1531	-0.3427	-0.5687	-0.8314	-1.1307	-2.0396	-3.1780	-5.7917
		16	39	-0.1788	-0.3930	-0.6427	-0.9280	-1.2491	-2.2082	-3.3914	-5.7917
		15	40	-0.2061	-0.4473	-0.7237	-1.0355	-1.3830	-2.4090	-3.6635	-5.7917
		14	41	-0.2290	-0.4918	-0.7885	-1.1194	-1.4847	-2.5500	-3.8326	-5.7917
		13	42	-0.2569	-0.5471	-0.8708	-1.2286	-1.6208	-2.7535	-4.1135	-5.7917
		12	43	-0.2812	-0.5944	-0.9339	-1.3184	-1.7303	-2.9069	-4.2798	-5.7917
		11	44	-0.3032	-0.6367	-1.0012	-1.3972	-1.8253	-3.0343	-4.3767	-5.7917
		10	45	-0.3240	-0.6763	-1.0575	-1.4685	-1.9106	-3.1463	-4.6777	-5.7917

Table C-2

Log R(w|d), the common logarithm of the rank-sum likelihood ratio under normal shift alternative d.

m	n	w_y	w_x	d = .2	.4	.6	.8	1.0	1.5	2.0	3.0
5	5	40	15	0.3069	0.5864	0.8398	1.0679	1.2720	1.6855	1.9773	2.2869
		39	16	0.2843	0.5387	0.7639	0.9605	1.1295	1.4360	1.5878	1.4829
		38	17	0.2615	0.4912	0.6897	0.8577	0.9959	1.2152	1.2642	0.9028
		37	18	0.2372	0.4411	0.6123	0.7514	0.8588	0.9936	0.9460	0.3551
		36	19	0.2113	0.3882	0.5311	0.6405	0.7171	0.7681	0.6269	−0.1830
		35	20	0.1799	0.3250	0.4354	0.5115	0.5536	0.5133	0.2713	−0.7746
		34	21	0.1578	0.2792	0.3646	0.4142	0.4283	0.3113	−0.0180	−1.2794
		33	22	0.1293	0.2215	0.2766	0.2950	0.2768	0.0738	−0.3499	−1.8313
		32	23	0.1032	0.1686	0.1965	0.1871	0.1406	−0.1356	−0.6356	−2.2721
		31	24	0.0748	0.1114	0.1100	0.0707	−0.0062	−0.3618	−0.9464	−2.7733
		30	25	0.0487	0.0588	0.0305	−0.0361	−0.1410	−0.5686	−1.2286	−3.2153
		29	26	0.0205	0.0020	−0.0554	−0.1518	−0.2869	−0.7930	−1.5365	−3.7130
		28	27	−0.0050	−0.0492	−0.1322	−0.2540	−0.4145	−0.9832	−1.7875	−4.0735
		27	28	−0.0339	−0.1070	−0.2191	−0.3702	−0.5603	−1.2049	−2.0894	−4.5572
		26	29	−0.0596	−0.1584	−0.2964	−0.4734	−0.6895	−1.3997	−2.3504	−4.9742
		25	30	−0.0874	−0.2135	−0.3783	−0.5817	−0.8237	−1.5963	−2.6054	−5.2518
		24	31	−0.1130	−0.2645	−0.4543	−0.6825	−0.9491	−1.7827	−2.8533	−5.8027
		23	32	−0.1411	−0.3204	−0.5378	−0.7936	−1.0878	−1.9912	−3.1351	−5.8027
		22	33	−0.1668	−0.3711	−0.6132	−0.8932	−1.2110	−2.1716	−3.3704	−5.8027
		21	34	−0.1944	−0.4256	−0.6939	−0.9993	−1.3420	−2.3625	−3.6183	−5.8027
		20	35	−0.2151	−0.4657	−0.7518	−1.0737	−1.4314	−2.4832	−3.7624	−5.8027
		19	36	−0.2462	−0.5277	−0.8449	−1.1982	−1.5879	−2.7235	−4.0934	−5.8027
		18	37	−0.2710	−0.5763	−0.9164	−1.2917	−1.7026	−2.8882	−4.2975	−5.8027
		17	38	−0.2939	−0.6207	−0.9810	−1.3753	−1.8041	−3.0303	−4.4225	−5.8027
		16	39	−0.3150	−0.6614	−1.0398	−1.4509	−1.8953	−3.1514	−4.5986	−5.8027
		15	40	−0.3352	−0.6996	−1.0942	−1.5197	−1.9774	−3.2561	−4.5986	−5.8027

Table C-2

Log R(w|d), the common logarithm of the rank-sum likelihood ratio under normal shift alternative d.

m	n	w_y	w_x	d = .2	.4	.6	.8	1.0	1.5	2.0	3.0
7	4	38	28	0.3191	0.6101	0.8741	1.1122	1.3255	1.7588	2.0661	2.3948
		37	29	0.2973	0.5639	0.8007	1.0084	1.1877	1.5174	1.6886	1.6125
		36	30	0.2751	0.5178	0.7287	0.9084	1.0577	1.3025	1.3733	1.0462
		35	31	0.2511	0.4685	0.6526	0.8040	0.9234	1.0859	1.0633	0.5154
		34	32	0.2245	0.4146	0.5706	0.6929	0.7822	0.8642	0.7527	0.0001
		33	33	0.2032	0.3706	0.5025	0.5993	0.6615	0.6691	0.4732	−0.4741
		32	34	0.1791	0.3216	0.4278	0.4980	0.5327	0.4674	0.1923	−0.9342
		31	35	0.1530	0.2687	0.3473	0.3890	0.3941	0.2501	−0.1120	−1.4427
		30	36	0.1286	0.2191	0.2717	0.2866	0.2642	0.0469	−0.3954	−1.9204
		29	37	0.1040	0.1693	0.1959	0.1841	0.1342	−0.1556	−0.6767	−2.3806
		28	38	0.0788	0.1186	0.1193	0.0813	0.0047	−0.3535	−0.9454	−2.7931
		27	39	0.0544	0.0691	0.0443	−0.0199	−0.1233	−0.5517	−1.2187	−3.2372
		26	40	0.0302	0.0207	−0.0283	−0.1168	−0.2445	−0.7336	−1.4604	−3.5879
		25	41	0.0044	−0.0310	−0.1066	−0.2220	−0.3771	−0.9375	−1.7412	−4.0508
		24	42	−0.0200	−0.0801	−0.1802	−0.3203	−0.5001	−1.1225	−1.9882	−4.4145
		23	43	−0.0445	−0.1291	−0.2537	−0.4183	−0.6227	−1.3068	−2.2354	−4.8018
		22	44	−0.0700	−0.1799	−0.3297	−0.5193	−0.7487	−1.4957	−2.4881	−5.2411
		21	45	−0.0941	−0.2281	−0.4019	−0.6156	−0.8691	−1.6765	−2.7302	−5.4814
		20	46	−0.1181	−0.2755	−0.4722	−0.7083	−0.9837	−1.8433	−2.9453	−5.7602
		19	47	−0.1430	−0.3250	−0.5463	−0.8067	−1.1064	−2.0274	−3.1928	−5.7602
		18	48	−0.1669	−0.3723	−0.6162	−0.8989	−1.2202	−2.1930	−3.4075	−5.7602
		17	49	−0.1907	−0.4192	−0.6857	−0.9903	−1.3331	−2.3573	−3.6197	−5.7602
		16	50	−0.2162	−0.4697	−0.7608	−1.0898	−1.4568	−2.5425	−3.8675	−5.7602
		15	51	−0.2395	−0.5156	−0.8286	−1.1789	−1.5665	−2.7009	−4.0835	−5.7602
		14	52	−0.2594	−0.5542	−0.8846	−1.2509	−1.6534	−2.8206	−4.2262	−5.7602
		13	53	−0.2854	−0.6057	−0.9613	−1.3526	−1.7802	−3.0092	−4.4814	−5.7602
		12	54	−0.3081	−0.6498	−1.0255	−1.4359	−1.8816	−3.1490	−4.7825	−5.7602
		11	55	−0.3286	−0.6893	−1.0826	−1.5092	−1.9696	−3.2773	−4.7825	−5.7602
		10	56	−0.3481	−0.7262	−1.1353	−1.5756	−2.0485	−3.3675	−4.7825	−5.7602

Table C-2

Log R(w|d), the common logarithm of the rank-sum likelihood ratio under normal shift alternative d.

m	n	w_y	w_x	d = .2	.4	.6	.8	1.0	1.5	2.0	3.0
6	5	45	21	0.3377	0.6456	0.9249	1.1767	1.4022	1.8603	2.1851	2.5331
		44	22	0.3170	0.6017	0.8549	1.0775	1.2703	1.6281	1.8204	1.7703
		43	23	0.2961	0.5581	0.7867	0.9826	1.1466	1.4222	1.5164	1.2180
		42	24	0.2741	0.5126	0.7162	0.8854	1.0209	1.2174	1.2202	0.7015
		41	25	0.2509	0.4650	0.6429	0.7851	0.8923	1.0110	0.9259	0.1989
		40	26	0.2240	0.4105	0.5598	0.6724	0.7488	0.7850	0.6078	-0.3350
		39	27	0.2010	0.3635	0.4878	0.5743	0.6234	0.5854	0.3248	-0.8187
		38	28	0.1768	0.3139	0.4116	0.4701	0.4899	0.3719	0.0205	-1.3490
		37	29	0.1528	0.2652	0.3374	0.3698	0.3626	0.1742	-0.2520	-1.7807
		36	30	0.1274	0.2138	0.2594	0.2643	0.2288	-0.0340	-0.5412	-2.2544
		35	31	0.1024	0.1633	0.1830	0.1616	0.0994	-0.2330	-0.8132	-2.6813
		34	32	0.0780	0.1139	0.1078	0.0598	-0.0298	-0.4345	-1.0932	-3.1433
		33	33	0.0535	0.0647	0.0337	-0.0392	-0.1540	-0.6219	-1.3437	-3.5121
		32	34	0.0281	0.0135	-0.0437	-0.1435	-0.2856	-0.8250	-1.6239	-3.9736
		31	35	0.0030	-0.0366	-0.1191	-0.2443	-0.4120	-1.0162	-1.8814	-4.3781
		30	36	-0.0214	-0.0857	-0.1928	-0.3426	-0.5349	-1.2006	-2.1262	-4.7265
		29	37	-0.0459	-0.1346	-0.2662	-0.4404	-0.6572	-1.3844	-2.3726	-5.1415
		28	38	-0.0709	-0.1846	-0.3412	-0.5405	-0.7826	-1.5739	-2.6286	-5.5114
		27	39	-0.0959	-0.2342	-0.4150	-0.6383	-0.9040	-1.7528	-2.8627	-5.7977
		26	40	-0.1203	-0.2830	-0.4882	-0.7357	-1.0257	-1.9356	-3.1079	-5.7977
		25	41	-0.1439	-0.3295	-0.5568	-0.8257	-1.1363	-2.0940	-3.3079	-5.7977
		24	42	-0.1687	-0.3788	-0.6305	-0.9239	-1.2590	-2.2794	-3.5610	-5.7977
		23	43	-0.1934	-0.4278	-0.7031	-1.0197	-1.3775	-2.4531	-3.7912	-5.7977
		22	44	-0.2168	-0.4739	-0.7713	-1.1094	-1.4881	-2.6138	-3.9995	-5.7977
		21	45	-0.2399	-0.5189	-0.8373	-1.1952	-1.5928	-2.7613	-4.1892	-5.7977
		20	46	-0.2619	-0.5622	-0.9011	-1.2788	-1.6957	-2.9109	-4.4023	-5.7977
		19	47	-0.2882	-0.6143	-0.9787	-1.3818	-1.8240	-3.1049	-4.5572	-5.7977
		18	48	-0.3103	-0.6573	-1.0417	-1.4637	-1.9242	-3.2442	-4.8124	-5.7977
		17	49	-0.3309	-0.6972	-1.0995	-1.5384	-2.0141	-3.3811	-4.8124	-5.7977
		16	50	-0.3501	-0.7341	-1.1526	-1.6070	-2.0973	-3.4902	-4.8124	-5.7977
		15	51	-0.3686	-0.7690	-1.2024	-1.6697	-2.1710	-3.5572	-4.8124	-5.7977

Table C-2

Log R(w|d), the common logarithm of the rank-sum likelihood ratio under normal shift alternative d.

m	n	w_y	w_x	d = .2	.4	.6	.8	1.0	1.5	2.0	3.0
7	5	50	28	0.3647	0.6974	0.9995	1.2721	1.5166	2.0142	2.3686	2.7512
		49	29	0.3453	0.6564	0.9340	1.1792	1.3928	1.7953	2.0231	2.0215
		48	30	0.3258	0.6156	0.8702	1.0902	1.2765	1.6007	1.7342	1.4911
		47	31	0.3054	0.5734	0.8044	0.9994	1.1589	1.4080	1.4542	0.9983
		46	32	0.2840	0.5293	0.7365	0.9062	1.0391	1.2150	1.1776	0.5223
		45	33	0.2594	0.4794	0.6604	0.8028	0.9074	1.0067	0.8837	0.0280
		44	34	0.2400	0.4394	0.5984	0.7177	0.7976	0.8295	0.6298	-0.4096
		43	35	0.2156	0.3898	0.5228	0.6150	0.6668	0.6231	0.3388	-0.9030
		42	36	0.1946	0.3468	0.4569	0.5251	0.5519	0.4406	0.0810	-1.3415
		41	37	0.1712	0.2993	0.3844	0.4269	0.4271	0.2450	-0.1920	-1.7938
		40	38	0.1486	0.2534	0.3147	0.3328	0.3079	0.0602	-0.4469	-2.1994
		39	39	0.1250	0.2058	0.2423	0.2349	0.1838	-0.1330	-0.7153	-2.6413
		38	40	0.1039	0.1630	0.1774	0.1475	0.0734	-0.3027	-0.9460	-2.9921
		37	41	0.0797	0.1142	0.1036	0.0480	-0.0522	-0.4971	-1.2146	-3.4346
		36	42	0.0572	0.0689	0.0352	-0.0438	-0.1679	-0.6737	-1.4543	-3.8053
		35	43	0.0338	0.0219	-0.0355	-0.1385	-0.2869	-0.8546	-1.6989	-4.1842
		34	44	0.0117	-0.0225	-0.1027	-0.2286	-0.4002	-1.0271	-1.9329	-4.5520
		33	45	-0.0120	-0.0702	-0.1743	-0.3244	-0.5203	-1.2092	-2.1791	-4.9463
		32	46	-0.0340	-0.1141	-0.2402	-0.4123	-0.6301	-1.3735	-2.3971	-5.3143
		31	47	-0.0577	-0.1615	-0.3113	-0.5071	-0.7487	-1.5524	-2.6382	-5.7825
		30	48	-0.0796	-0.2051	-0.3764	-0.5935	-0.8561	-1.7112	-2.8465	-5.7640
		29	49	-0.1028	-0.2514	-0.4457	-0.6856	-0.9711	-1.8834	-3.0767	-5.7640
		28	50	-0.1251	-0.2958	-0.5119	-0.7736	-1.0807	-2.0471	-3.2961	-5.7640
		27	51	-0.1489	-0.3429	-0.5821	-0.8665	-1.1962	-2.2180	-3.5214	-5.7640
		26	52	-0.1695	-0.3837	-0.6424	-0.9459	-1.2940	-2.3590	-3.6975	-5.7640
		25	53	-0.1925	-0.4291	-0.7100	-1.0352	-1.4047	-2.5228	-3.9141	-5.7640
		24	54	-0.2146	-0.4730	-0.7752	-1.1213	-1.5116	-2.6807	-4.1224	-5.7640
		23	55	-0.2374	-0.5179	-0.8417	-1.2089	-1.6198	-2.8389	-4.3317	-5.7640
		22	56	-0.2574	-0.5569	-0.8988	-1.2832	-1.7102	-2.9647	-4.5162	-5.7640
		21	57	-0.2812	-0.6040	-0.9687	-1.3756	-1.8250	-3.1374	-4.8002	-5.7640
		20	58	-0.2994	-0.6391	-1.0197	-1.4413	-1.9041	-3.2474	-4.9463	-5.7640
		19	59	-0.3233	-0.6866	-1.0903	-1.5349	-2.0206	-3.4200	-4.8002	-5.7640
		18	60	-0.3436	-0.7261	-1.1480	-1.6099	-2.1115	-3.5370	-4.8002	-5.7640
		17	61	-0.3627	-0.7629	-1.2011	-1.6780	-2.1954	-3.7033	-4.8002	-5.7640
		16	62	-0.3805	-0.7972	-1.2506	-1.7414	-2.2687	-3.8002	-4.8002	-5.7640
		15	63	-0.3977	-0.8296	-1.2964	-1.8002	-2.3378	-3.8002	-4.8002	-5.7640

286

Table C-2

Log R(w|d), the common logarithm of the rank-sum likelihood ratio under normal shift alternative d.

m	n	w_y	w_x	d = .2	.4	.6	.8	1.0	1.5	2.0	3.0
6	6	57	21	0.3731	0.7136	1.0227	1.3015	1.5516	2.0605	2.4230	2.8145
		56	22	0.3542	0.6734	0.9584	1.2103	1.4300	1.8451	2.0824	2.0925
		55	23	0.3352	0.6336	0.8960	1.1231	1.3160	1.6539	1.7977	1.5675
		54	24	0.3154	0.5925	0.8320	1.0347	1.2013	1.4652	1.5226	1.0800
		53	25	0.2949	0.5502	0.7665	0.9446	1.0851	1.2769	1.2514	0.6091
		52	26	0.2722	0.5038	0.6952	0.8471	0.9600	1.0765	0.9654	0.1193
		51	27	0.2475	0.4539	0.6195	0.7448	0.8303	0.8734	0.6807	-0.3602
		50	28	0.2273	0.4120	0.5545	0.6551	0.7142	0.6837	0.4051	-0.8573
		49	29	0.2050	0.3667	0.4852	0.5611	0.5947	0.4965	0.1446	-1.2785
		48	30	0.1825	0.3208	0.4151	0.4658	0.4732	0.3050	-0.1240	-1.7271
		47	31	0.1593	0.2739	0.3440	0.3697	0.3516	0.1164	-0.3842	-2.1420
		46	32	0.1363	0.2273	0.2730	0.2737	0.2296	-0.0742	-0.6501	-2.5847
		45	33	0.1138	0.1818	0.2042	0.1811	0.1129	-0.2529	-0.8927	-2.9542
		44	34	0.0911	0.1358	0.1342	0.0865	-0.0071	-0.4402	-1.1534	-3.3904
		43	35	0.0679	0.0891	0.0636	-0.0082	-0.1264	-0.6227	-1.4017	-3.7810
		42	36	0.0450	0.0432	-0.0055	-0.1010	-0.2430	-0.7999	-1.6406	-4.1398
		41	37	0.0220	-0.0030	-0.0752	-0.1942	-0.3600	-0.9775	-1.8807	-4.5194
		40	38	-0.0006	-0.0486	-0.1439	-0.2865	-0.4761	-1.1547	-2.1219	-4.8716
		39	39	-0.0236	-0.0946	-0.2129	-0.3784	-0.5909	-1.3265	-2.3504	-5.1956
		38	40	-0.0468	-0.1410	-0.2826	-0.4716	-0.7078	-1.5037	-2.5902	-5.7746
		37	41	-0.0692	-0.1855	-0.3491	-0.5596	-0.8171	-1.6648	-2.7997	-5.7746
		36	42	-0.0921	-0.2314	-0.4177	-0.6512	-0.9316	-1.8375	-3.0334	-5.7746
		35	43	-0.1147	-0.2764	-0.4850	-0.7405	-1.0429	-2.0031	-3.2531	-5.7746
		34	44	-0.1376	-0.3220	-0.5530	-0.8307	-1.1550	-2.1693	-3.4728	-5.7746
		33	45	-0.1597	-0.3655	-0.6175	-0.9155	-1.2596	-2.3209	-3.6663	-5.7746
		32	46	-0.1819	-0.4095	-0.6829	-1.0020	-1.3669	-2.4791	-3.8767	-5.7746
		31	47	-0.2043	-0.4540	-0.7489	-1.0894	-1.4754	-2.6407	-4.0835	-5.7746
		30	48	-0.2270	-0.4987	-0.8154	-1.1771	-1.5841	-2.8004	-4.2975	-5.7746
		29	49	-0.2487	-0.5414	-0.8783	-1.2596	-1.6854	-2.9452	-4.5114	-5.7746
		28	50	-0.2703	-0.5839	-0.9410	-1.3419	-1.7867	-3.0913	-4.6711	-5.7746
		27	51	-0.2891	-0.6201	-0.9933	-1.4090	-1.8672	-3.2006	-4.7746	-5.7746
		26	52	-0.3135	-0.6687	-1.0662	-1.5065	-1.9895	-3.3880	-4.8794	-5.7746
		25	53	-0.3351	-0.7110	-1.1283	-1.5874	-2.0888	-3.5291	-4.8794	-5.7746
		24	54	-0.3544	-0.7485	-1.1828	-1.6583	-2.1749	-3.6663	-4.8794	-5.7746
		23	55	-0.3728	-0.7840	-1.2342	-1.7246	-2.2561	-3.7332	-4.8794	-5.7746
		22	56	-0.3903	-0.8173	-1.2822	-1.7863	-2.3267	-3.9332	-4.8794	-5.7746
		21	57	-0.4070	-0.8490	-1.3276	-1.8412	-2.3908	-4.0343	-4.8794	-5.7746

287

Table C-2

Log R(w|d), the common logarithm of the rank-sum likelihood ratio under normal shift alternative d.

m	n	w_y	w_x	d = .2	.4	.6	.8	1.0	1.5	2.0	3.0
7	6	63	28	0.4043	0.7735	1.1089	1.4117	1.6835	2.2378	2.6339	3.0650
		62	29	0.3867	0.7361	1.0490	1.3265	1.5697	2.0353	2.3122	2.3755
		61	30	0.3691	0.6991	0.9908	1.2451	1.4630	1.8551	2.0423	1.8721
		60	31	0.3508	0.6611	0.9315	1.1629	1.3561	1.6783	1.7828	1.4071
		59	32	0.3320	0.6222	0.8711	1.0796	1.2484	1.5026	1.5282	0.9600
		58	33	0.3116	0.5802	0.8063	0.9907	1.1341	1.3181	1.2635	0.5027
		57	34	0.2899	0.5360	0.7389	0.8992	1.0174	1.1333	1.0022	0.0581
		56	35	0.2691	0.4934	0.6734	0.8095	0.9023	0.9480	0.7359	-0.4106
		55	36	0.2497	0.4536	0.6121	0.7257	0.7947	0.7761	0.4917	-0.8281
		54	37	0.2287	0.4108	0.5466	0.6365	0.6808	0.5956	0.2370	-1.2579
		53	38	0.2078	0.3684	0.4818	0.5486	0.5690	0.4204	-0.0068	-1.6539
		52	39	0.1864	0.3247	0.4153	0.4583	0.4541	0.2399	-0.2597	-2.0769
		51	40	0.1654	0.2823	0.3510	0.3716	0.3446	0.0709	-0.4912	-2.4374
		50	41	0.1445	0.2398	0.2861	0.2835	0.2325	-0.1050	-0.7376	-2.8520
		49	42	0.1232	0.1968	0.2209	0.1958	0.1216	-0.2759	-0.9719	-3.2224
		48	43	0.1020	0.1540	0.1561	0.1084	0.0113	-0.4458	-1.2042	-3.5851
		47	44	0.0807	0.1111	0.0912	0.0213	-0.0984	-0.6141	-1.4338	-3.9473
		46	45	0.0593	0.0679	0.0259	-0.0664	-0.2091	-0.7838	-1.6660	-4.3122
		45	46	0.0381	0.0255	-0.0379	-0.1519	-0.3164	-0.9464	-1.8846	-4.6307
		44	47	0.0169	-0.0173	-0.1025	-0.2387	-0.4257	-1.1135	-2.1123	-5.0207
		43	48	-0.0042	-0.0596	-0.1661	-0.3235	-0.5316	-1.2723	-2.3234	-5.2614
		42	49	-0.0256	-0.1024	-0.2304	-0.4094	-0.6393	-1.4354	-2.5438	-5.7386
		41	50	-0.0469	-0.1450	-0.2942	-0.4945	-0.7457	-1.5955	-2.7586	-5.7386
		40	51	-0.0680	-0.1873	-0.3578	-0.5793	-0.8518	-1.7550	-2.9723	-5.7386
		39	52	-0.0891	-0.2291	-0.4201	-0.6619	-0.9544	-1.9063	-3.1696	-5.7386
		38	53	-0.1100	-0.2709	-0.4826	-0.7450	-1.0581	-2.0611	-3.3751	-5.7386
		37	54	-0.1312	-0.3130	-0.5453	-0.8282	-1.1615	-2.2150	-3.5818	-5.7386
		36	55	-0.1522	-0.3547	-0.6076	-0.9108	-1.2644	-2.3680	-3.7850	-5.7386
		35	56	-0.1730	-0.3958	-0.6686	-0.9914	-1.3641	-2.5137	-3.9710	-5.7386
		34	57	-0.1939	-0.4373	-0.7304	-1.0732	-1.4656	-2.6639	-4.1804	-5.7386

288

Table C-2

Log R(w|d), the common logarithm of the rank-sum likelihood ratio under normal shift alternative d.

m	n	w_y	w_x	d = .2	.4	.6	.8	1.0	1.5	2.0	3.0
7	6	33	58	-0.2140	-0.4770	-0.7888	-1.1496	-1.5594	-2.7978	-4.3490	-5.7386
		32	59	-0.2347	-0.5179	-0.8498	-1.2305	-1.6602	-2.9480	-4.5555	-5.7386
		31	60	-0.2555	-0.5590	-0.9106	-1.3106	-1.7590	-3.0924	-4.7797	-5.7386
		30	61	-0.2757	-0.5988	-0.9694	-1.3877	-1.8541	-3.2318	-4.8446	-5.7386
		29	62	-0.2959	-0.6383	-1.0276	-1.4639	-1.9476	-3.3629	-5.0442	-5.7386
		28	63	-0.3143	-0.6741	-1.0799	-1.5317	-2.0302	-3.4802	-5.0442	-5.7386
		27	64	-0.3341	-0.7130	-1.1370	-1.6065	-2.1220	-3.6027	-5.0442	-5.7386
		26	65	-0.3553	-0.7548	-1.1993	-1.6888	-2.2249	-3.7654	-5.0442	-5.7386
		25	66	-0.3746	-0.7926	-1.2544	-1.7611	-2.3121	-3.8624	-5.0442	-5.7386
		24	67	-0.3923	-0.8268	-1.3040	-1.8243	-2.3913	-3.9415	-5.0442	-5.7386
		23	68	-0.4091	-0.8591	-1.3505	-1.8846	-2.4644	-4.0665	-5.0442	-5.7386
		22	69	-0.4252	-0.8898	-1.3944	-1.9394	-2.5350	-4.0665	-5.0442	-5.7386
		21	70	-0.4408	-0.9191	-1.4371	-1.9946	-2.5894	-4.0665	-5.0442	-5.7386
7	7	77	28	0.4395	0.8410	1.2058	1.5355	1.8317	2.4366	2.8703	3.3454
		76	29	0.4232	0.8062	1.1502	1.4562	1.7256	2.2468	2.5670	2.6882
		75	30	0.4068	0.7719	1.0960	1.3804	1.6259	2.0775	2.3118	2.2059
		74	31	0.3901	0.7369	1.0414	1.3043	1.5267	1.9121	2.0675	1.7625
		73	32	0.3729	0.7013	0.9859	1.2276	1.4272	1.7486	1.8287	1.3377
		72	33	0.3547	0.6636	0.9274	1.1471	1.3232	1.5792	1.5836	0.9083
		71	34	0.3358	0.6249	0.8679	1.0655	1.2185	1.4111	1.3430	0.4921
		70	35	0.3139	0.5805	0.8003	0.9738	1.1016	1.2257	1.0798	0.0359
		69	36	0.2974	0.5462	0.7469	0.9000	1.0062	1.0699	0.8540	-0.3651
		68	37	0.2776	0.5059	0.6852	0.8159	0.8985	0.8986	0.6112	-0.7789
		67	38	0.2588	0.4674	0.6262	0.7355	0.7958	0.7359	0.3823	-1.1578
		66	39	0.2387	0.4265	0.5637	0.6505	0.6875	0.5648	0.1414	-1.5646
		65	40	0.2198	0.3879	0.5048	0.5708	0.5862	0.4066	-0.0778	-1.9145
		64	41	0.1993	0.3465	0.4417	0.4852	0.4774	0.2358	-0.3170	-2.3173
		63	42	0.1802	0.3077	0.3826	0.4052	0.3758	0.0773	-0.5368	-2.6720
		62	43	0.1604	0.2675	0.3215	0.3227	0.2712	-0.0851	-0.7612	-3.0329
		61	44	0.1409	0.2280	0.2616	0.2417	0.1688	-0.2437	-0.9798	-3.3852

289

Table C-2

Log R(w|d), the common logarithm of the rank-sum likelihood ratio under normal shift alternative d.

m	n	w_y	w_x	d = .2	.4	.6	.8	1.0	1.5	2.0	3.0
7	7	60	45	0.1206	0.1871	0.1996	0.1581	0.0631	-0.4068	-1.2039	-3.7428
		59	46	0.1013	0.1481	0.1405	0.0787	-0.0370	-0.5604	-1.4131	-4.0700
		58	47	0.0809	0.1070	0.0783	-0.0048	-0.1424	-0.7222	-1.6345	-4.4219
		57	48	0.0617	0.0683	0.0199	-0.0833	-0.2411	-0.8727	-1.8378	-4.7257
		56	49	0.0414	0.0276	-0.0413	-0.1654	-0.3444	-1.0300	-2.0514	-5.0527
		55	50	0.0218	-0.0118	-0.1008	-0.2451	-0.4445	-1.1822	-2.2574	-5.3729
		54	51	0.0016	-0.0522	-0.1617	-0.3266	-0.5468	-1.3373	-2.4668	-5.6845
		53	52	-0.0177	-0.0910	-0.2199	-0.4042	-0.6438	-1.4828	-2.6601	-5.6923
		52	53	-0.0379	-0.1315	-0.2806	-0.4853	-0.7452	-1.6355	-2.8656	-5.6923
		51	54	-0.0572	-0.1702	-0.3387	-0.5627	-0.8420	-1.7810	-3.0602	-5.6923
		50	55	-0.0773	-0.2102	-0.3987	-0.6425	-0.9417	-1.9305	-3.2589	-5.6923
		49	56	-0.0968	-0.2491	-0.4567	-0.7196	-1.0377	-2.0730	-3.4479	-5.6923
		48	57	-0.1169	-0.2891	-0.5165	-0.7991	-1.1368	-2.2212	-3.6465	-5.6923
		47	58	-0.1358	-0.3266	-0.5724	-0.8730	-1.2284	-2.3554	-3.8198	-5.6923
		46	59	-0.1559	-0.3666	-0.6321	-0.9523	-1.3272	-2.5029	-4.0172	-5.6923
		45	60	-0.1750	-0.4044	-0.6884	-1.0268	-1.4196	-2.6397	-4.1914	-5.6923
		44	61	-0.1949	-0.4439	-0.7470	-1.1043	-1.5158	-2.7814	-4.3852	-5.6923
		43	62	-0.2139	-0.4816	-0.8030	-1.1782	-1.6072	-2.9154	-4.5588	-5.6923
		42	63	-0.2332	-0.5196	-0.8592	-1.2521	-1.6983	-3.0472	-4.7374	-5.6923
		41	64	-0.2517	-0.5561	-0.9132	-1.3232	-1.7860	-3.1738	-4.9627	-5.6923
		40	65	-0.2716	-0.5955	-0.9717	-1.4004	-1.8818	-3.3194	-5.1887	-5.6923
		39	66	-0.2901	-0.6317	-1.0253	-1.4709	-1.9687	-3.4442	-5.0877	-5.6923
		38	67	-0.3095	-0.6699	-1.0816	-1.5449	-2.0599	-3.5809	-5.0877	-5.6923
		37	68	-0.3275	-0.7052	-1.1334	-1.6124	-2.1422	-3.7033	-5.0877	-5.6923
		36	69	-0.3464	-0.7423	-1.1881	-1.6841	-2.2301	-3.8112	-5.0877	-5.6923
		35	70	-0.3617	-0.7717	-1.2302	-1.7379	-2.2941	-3.9415	-5.0877	-5.6923
		34	71	-0.3832	-0.8144	-1.2941	-1.8236	-2.4020	-4.0287	-5.0877	-5.6923
		33	72	-0.4012	-0.8495	-1.3457	-1.8912	-2.4834	-4.3095	-5.0877	-5.6923
		32	73	-0.4183	-0.8828	-1.3940	-1.9539	-2.5613	-4.1634	-5.0877	-5.6923
		31	74	-0.4343	-0.9138	-1.4391	-2.0121	-2.6193	-4.1634	-5.0877	-5.6923
		30	75	-0.4497	-0.9433	-1.4822	-2.0665	-2.6863	-4.1634	-5.0877	-5.6923
		29	76	-0.4644	-0.9717	-1.5200	-2.1220	-2.7655	-4.1634	-5.0877	-5.6923
		28	77	-0.4790	-0.9991	-1.5614	-2.1634	-2.8624	-4.1634	-5.0877	-5.6923

Table D. $P(n'; r, m, n, d)$ A table for use in selection procedure

The probability that, in a random sample of $m + n$ observations (m X-observations from a standard normal distribution and n Y-observations from a normal distribution with mean d and variance 1, all $m + n$ observations mutually independent), at least n' of the Y-observations are among the r largest observations is given by

$$P(n'; r, m, n, d) = \sum_{\mathbf{z} \in \mathscr{Z}} P_{m,n}(\mathbf{z} \mid d)$$

for $n' \leq n, r$. The summation is over the set $\mathscr{Z}$ of $(m + n)$-element $\mathbf{z}$ vectors with at least n' ones among the r rightmost elements.

Values of $P(n'; r, m, n, d)$ are given in Table D to 6 decimal places for $1 \leq n \leq m \leq 7$ and $n = 1, m = 8(1)12; r = 1(1)[m/2]; n' = 1(1)n;$ and $d = .0(.2)1.0,1.5,2.0,3.0$. Entries in the table are ordered according to increasing values of $m + n$, from $2 \leq m + n \leq 14$.

See Chapter 4 for more details.

Table D

P(n';r,m,n,d)

m	n	r	n'	d = 0.	.2	.4	.6	.8	1.0	1.5	2.0	3.0
1	1	1	1	.500000	.556231	.611351	.664313	.714196	.760250	.855578	.921350	.983053
2	1	1	1	.333333	.391392	.451875	.513387	.574469	.633702	.765812	.865767	.968795
3	1	1	1	.250000	.304081	.362696	.424578	.488231	.552031	.701863	.822793	.956374
		2	1	.500000	.566015	.630233	.691003	.746946	.797043	.893710	.951716	.993638
2	2	1	1	.500000	.568285	.634566	.697016	.754136	.804859	.900933	.956619	.994725
		2	2	.166667	.214500	.269184	.329757	.394803	.462545	.630691	.774916	.942866
		2	1	.833333	.873856	.907090	.933464	.953710	.968737	.989754	.997249	.999894
4	1	1	1	.200000	.249631	.305001	.365088	.428526	.493699	.652865	.787839	.945311
		2	1	.400000	.467431	.535778	.603049	.667347	.727029	.848857	.927655	.989564
3	2	1	1	.400000	.471007	.542928	.613417	.680270	.741635	.863588	.938419	.992235
		2	2	.100000	.137155	.182463	.235740	.296192	.362428	.540137	.707166	.920514
		2	1	.700000	.762841	.817841	.864214	.901868	.931306	.975624	.993017	.999706
5	1	1	1	.166667	.212274	.264317	.322024	.384213	.449365	.613555	.758452	.935305
		2	1	.333333	.399060	.467740	.537347	.605776	.671035	.810105	.905389	.985338
		3	1	.500000	.569987	.637836	.701603	.759702	.811021	.906984	.961054	.995902
4	2	1	1	.333333	.403300	.476522	.550516	.622724	.690777	.831427	.921911	.989829
		2	2	.066667	.095962	.133481	.179661	.234328	.296621	.474302	.653767	.900794
		2	1	.600000	.674127	.742149	.802121	.852909	.894210	.960069	.987946	.999453
		3	2	.200000	.260735	.329408	.403977	.481784	.559849	.737644	.867363	.979674
		3	1	.800000	.851554	.893533	.926306	.950827	.968401	.991180	.998087	.999959
3	3	1	1	.500000	.575567	.648402	.716073	.776691	.829060	.922385	.970450	.997472
		2	2	.200000	.261888	.331982	.408106	.487428	.566785	.745993	.874358	.981760
		2	1	.800000	.852365	.894806	.927738	.952195	.969572	.991735	.998262	.999965
		3	3	.050000	.074789	.107703	.149556	.200574	.260249	.437210	.623570	.889891
		3	2	.500000	.583792	.663911	.737165	.801214	.854772	.943403	.982526	.999188
		3	1	.950000	.967863	.980168	.988265	.993349	.996393	.999360	.999916	.999999

Table D

P(n';r,m,n,d)

m	n	r	n'	d = 0.	.2	.4	.6	.8	1.0	1.5	2.0	3.0
6	1	1	1	.142857	.184978	.233936	.289188	.349745	.414216	.580994	.733157	.926149
		2	1	.285714	.348752	.416220	.486201	.556555	.625108	.776361	.884925	.981083
		3	1	.428571	.499678	.570780	.639637	.704220	.762888	.877593	.946317	.993848
5	2	1	1	.285714	.353299	.425907	.501126	.576263	.648638	.803241	.906789	.987499
		2	2	.047619	.071249	.102726	.142921	.192163	.250091	.423869	.610114	.883110
		2	1	.523810	.603305	.678978	.748073	.808566	.859331	.944175	.982396	.999147
		3	2	.142857	.194815	.256501	.326620	.402987	.482738	.676035	.828382	.971528
		3	1	.714286	.780360	.836905	.883193	.919424	.946527	.983911	.996273	.999912
4	3	1	1	.428571	.506730	.584621	.659247	.727993	.788903	.901353	.961664	.996652
		2	2	.142857	.196440	.260322	.333053	.412185	.494525	.691577	.842404	.976182
		2	1	.714286	.782076	.839743	.886549	.922785	.949532	.985483	.996809	.999933
		3	3	.028571	.045723	.070060	.102964	.145400	.197668	.365664	.559449	.863100
		3	2	.371429	.458231	.546962	.633265	.713157	.783565	.909242	.970221	.998494
		3	1	.885714	.922655	.949869	.968927	.981605	.989612	.997988	.999715	.999998
7	1	1	1	.125000	.164122	.210302	.263200	.322010	.385481	.553379	.710998	.917697
		2	1	.250000	.310117	.375739	.445115	.516157	.586627	.746679	.866109	.976861
		3	1	.375000	.445337	.517422	.588318	.657549	.721311	.850566	.931963	.991636
		4	1	.500000	.572131	.641924	.707262	.766448	.818323	.913628	.965455	.996798
6	2	1	1	.250000	.314775	.385896	.461117	.537749	.612942	.778198	.892831	.985239
		2	2	.035714	.055182	.081977	.117259	.161741	.215490	.383789	.573483	.867059
		2	1	.464286	.545919	.625967	.701171	.768836	.827120	.928455	.976578	.998798
		3	2	.107143	.151584	.206473	.271232	.344273	.423097	.624267	.793271	.963367
		3	1	.642857	.718077	.785001	.841878	.888025	.923755	.975615	.994031	.999846
		4	2	.214286	.281278	.356558	.437396	.520414	.602020	.779570	.898603	.987851
		4	1	.785714	.842643	.888808	.924509	.950823	.969300	.992206	.998514	.999978

Table D

P(n';r,m,n,d)

m	n	r	n'	d = 0.	.2	.4	.6	.8	1.0	1.5	2.0	3.0
5	3	1	1	.375000	.453271	.533452	.612267	.686609	.753915	.882063	.953308	.995843
		2	2	.107143	.153355	.210818	.278845	.355572	.438086	.645595	.813753	.970812
		2	1	.642857	.720568	.789299	.847168	.893529	.928856	.978509	.995088	.999891
		3	3	.017857	.030196	.048680	.074959	.110459	.156094	.313006	.508295	.839260
		3	2	.285714	.368781	.458338	.549884	.638640	.720283	.875506	.957011	.997660
		3	1	.821429	.874338	.915408	.945621	.966670	.980547	.995944	.999390	.999995
		4	3	.071429	.107833	.155583	.214987	.285161	.363966	.576299	.764067	.958462
		4	2	.500000	.592405	.679897	.758337	.824934	.878488	.959844	.990037	.999745
		4	1	.928571	.954866	.972843	.984465	.991563	.995655	.999350	.999932	1.000000
4	4	1	1	.500000	.580775	.658198	.729341	.792035	.845046	.935222	.977716	.998508
		2	2	.214286	.284596	.363891	.448965	.535868	.620472	.799745	.913506	.991084
		2	1	.785714	.844979	.892437	.928523	.954569	.972408	.993533	.998874	.999986
		3	3	.071429	.108284	.156753	.217142	.288503	.368579	.583410	.771303	.961279
		3	2	.500000	.593365	.681659	.760627	.827433	.880904	.961334	.990613	.999772
		3	1	.928571	.955093	.973138	.984736	.991772	.995797	.999385	.999937	1.000000
		4	4	.014286	.024869	.041163	.064905	.097699	.140698	.293083	.488831	.830374
		4	3	.242857	.323096	.412264	.505904	.598882	.686226	.857150	.949829	.997217
		4	2	.757143	.825340	.880062	.921501	.951106	.971056	.993794	.999050	.999992
		4	1	.985714	.992212	.995978	.998035	.999093	.999606	.999962	.999998	1.000000
8	1	1	1	.111111	.147643	.191343	.242045	.299112	.361434	.529532	.691318	.909839
		2	1	.222222	.279475	.343018	.411287	.482298	.553812	.720313	.848759	.972708
		3	1	.333333	.402043	.473903	.546599	.617735	.685074	.825777	.918160	.989322
		4	1	.444444	.517495	.589953	.659450	.723904	.781707	.891883	.954969	.995492
7	2	1	1	.222222	.284133	.353377	.427924	.505163	.582176	.755705	.879866	.983044
		2	2	.027778	.044111	.067228	.098476	.138857	.188786	.351054	.542131	.852351
		2	1	.416667	.498628	.581008	.660278	.733264	.797538	.913156	.970627	.998413
		3	2	.083333	.121508	.170470	.229952	.299050	.375717	.580201	.761592	.955309
		3	1	.583333	.664148	.738364	.803405	.857766	.901074	.966702	.991457	.999761
		4	2	.166667	.226527	.296480	.374431	.457332	.541549	.734431	.872470	.983510
		4	1	.722222	.789983	.847184	.893176	.928372	.953996	.987499	.997463	.999958

Table D

$$P(n';r,m,n,d)$$

m	n	r	n'	d = 0.	.2	.4	.6	.8	1.0	1.5	2.0	3.0
6	3	1	1	.333333	.410476	.491362	.572630	.650864	.723036	.864256	.945337	.995043
		2	2	.083333	.123373	.174963	.238093	.311518	.392754	.606081	.787821	.965632
		2	1	.583333	.667247	.743901	.810452	.865335	.908306	.971098	.993161	.999841
		3	3	.011905	.021086	.035484	.056842	.086853	.126858	.272644	.466314	.817773
		3	2	.226190	.303263	.390098	.482609	.575838	.664746	.843168	.943413	.996713
		3	1	.761905	.827208	.880093	.920599	.949916	.969954	.993331	.998943	.999991
		4	3	.047619	.075744	.114660	.165543	.228491	.302273	.514816	.718200	.946694
		4	2	.404762	.499816	.594818	.684436	.764244	.831356	.940184	.984206	.999555
		4	1	.880952	.921468	.950723	.970644	.983423	.991140	.998557	.999837	.999999
5	4	1	1	.444444	.527572	.609491	.686700	.756319	.816405	.921503	.972591	.998141
		2	2	.166667	.230367	.305336	.388968	.477480	.566443	.763745	.895458	.988947
		2	1	.722222	.793585	.853027	.899907	.934899	.959611	.990098	.998219	.999977
		3	3	.047619	.076343	.116301	.168723	.233663	.309729	.527445	.732047	.952677
		3	2	.404762	.501514	.598114	.688950	.769417	.836588	.943743	.985697	.999632
		3	1	.880952	.922070	.951552	.971448	.984074	.991604	.998686	.999858	.999999
		4	4	.007937	.014813	.026139	.043704	.069392	.104882	.241526	.433710	.801454
		4	3	.166667	.236047	.318562	.410819	.507862	.603978	.807270	.928324	.995689
		4	2	.642857	.731142	.806977	.868142	.914457	.947378	.987719	.997987	.999982
		4	1	.960317	.977109	.987529	.993595	.996904	.998593	.999852	.999990	1.000000
9	1	1	1	.100000	.134279	.175765	.224440	.279822	.340936	.508638	.673645	.902488
		2	1	.200000	.254552	.315970	.382882	.453430	.525417	.696685	.832700	.968639
		3	1	.300000	.366708	.437686	.510704	.583335	.653193	.803010	.904965	.986946
		4	1	.400000	.472713	.546338	.618388	.686536	.748834	.871310	.944550	.994074
		5	1	.500000	.573472	.644472	.710777	.770614	.822799	.917598	.967992	.997264
7	3	1	1	.300000	.375394	.456050	.538640	.619580	.695497	.847726	.937714	.994253
		2	2	.066667	.101610	.148032	.206492	.276329	.355535	.571662	.764168	.960625
		2	1	.533333	.620964	.703237	.776567	.838568	.888272	.963438	.991072	.999784
		3	3	.008333	.015361	.026826	.044469	.070121	.105411	.240750	.431112	.798214
		3	2	.183333	.253955	.336549	.427700	.522656	.616070	.812592	.929735	.995672
		3	1	.708333	.782955	.845562	.895167	.932253	.958392	.990249	.998381	.999985
		4	3	.033333	.055433	.087431	.131078	.187247	.255540	.464006	.677520	.935128
		4	2	.333333	.426534	.523970	.619880	.708791	.786437	.919607	.977608	.999314
		4	1	.833333	.886121	.926136	.954508	.973468	.985370	.997440	.999693	.999999
		5	3	.083333	.126523	.182734	.251706	.331602	.419105	.641843	.819901	.975609
		5	2	.500000	.597525	.689282	.770510	.838181	.891247	.967619	.993004	.999877
		5	1	.916667	.947910	.969164	.982745	.990889	.995467	.999396	.999946	1.000000

Table D

$$P(n';r,m,n,d)$$

m	n	r	n'	d = 0.	.2	.4	.6	.8	1.0	1.5	2.0	3.0
6	4	1	1	.400000	.483736	.568244	.649670	.724588	.790434	.908547	.967618	.997776
		2	2	.133333	.190695	.260717	.341509	.429693	.520842	.731386	.878495	.986844
		2	1	.666667	.746755	.815726	.871849	.914976	.946259	.986284	.997456	.999967
		3	3	.033333	.056052	.089208	.134676	.193343	.264666	.480776	.697146	.944420
		3	2	.333333	.428725	.528427	.626262	.716413	.794450	.925543	.980274	.999465
		3	1	.833333	.887247	.927629	.956023	.974746	.986317	.997726	.999743	.999999
		4	4	.004762	.009430	.017576	.030897	.051356	.080922	.203266	.389369	.775557
		4	3	.119048	.177801	.251768	.338956	.435262	.535043	.760793	.906552	.993961
		4	2	.547619	.647092	.737486	.814327	.875425	.920865	.980145	.996544	.999966
		4	1	.928571	.956893	.975475	.986873	.993401	.996890	.999645	.999974	1.000000
		5	4	.023810	.041726	.068958	.107739	.159568	.224682	.432824	.655416	.930938
		5	3	.261905	.352540	.452149	.554544	.653063	.741846	.900223	.971868	.999145
		5	2	.738095	.815191	.876468	.921957	.953488	.973891	.995293	.999429	.999997
		5	1	.976190	.987217	.993556	.996956	.998655	.999445	.999955	.999998	1.000000
5	5	1	1	.500000	.584818	.665739	.739418	.803472	.856685	.943871	.982150	.999006
		2	2	.222222	.298592	.384496	.475827	.567706	.655283	.832031	.934353	.994681
		2	1	.777778	.841888	.892672	.930609	.957328	.975067	.994820	.999226	.999994
		3	3	.083333	.128029	.186595	.258681	.342142	.433183	.661316	.837115	.980345
		3	2	.500000	.600374	.694450	.777100	.845184	.897789	.971207	.994188	.999913
		3	1	.916667	.948668	.970138	.983624	.991548	.995900	.999491	.999958	1.000000
		4	4	.023810	.041886	.069438	.108751	.161344	.227426	.438198	.662002	.934231
		4	3	.261905	.353224	.453609	.556725	.655768	.744786	.902548	.972962	.999210
		4	2	.738095	.815679	.877208	.922742	.954178	.974420	.995464	.999460	.999998
		4	1	.976190	.987275	.993619	.997003	.998684	.999461	.999957	.999998	1.000000
		5	5	.003968	.008045	.015315	.027442	.046403	.074246	.192358	.376637	.768260
		5	4	.103175	.157929	.228530	.313548	.409259	.510105	.743751	.898566	.993341
		5	3	.500000	.604336	.701631	.786242	.854875	.906815	.976102	.995776	.999958
		5	2	.896825	.936446	.963171	.979962	.989781	.995123	.999430	.999957	1.000000
		5	1	.996032	.998166	.999208	.999681	.999880	.999958	.999998	1.000000	1.000000
10	1	1	1	.090909	.123214	.162716	.209529	.263305	.323201	.490114	.657631	.895580
		2	1	.181818	.233864	.293203	.358645	.428469	.500546	.675346	.817779	.964666
		3	1	.272727	.337301	.407038	.479831	.553275	.624900	.782043	.892384	.984535
		4	1	.363636	.435325	.509197	.582740	.653475	.719210	.851932	.934319	.992573
		5	1	.454545	.528795	.602049	.671861	.736126	.793272	.900378	.959897	.996327

Table D

$$P(n';r,m,n,d)$$

m	n	r	n'	d = 0.	.2	.4	.6	.8	1.0	1.5	2.0	3.0
7	4	1	1	.363636	.446949	.532797	.617139	.696143	.766725	.896273	.962787	.997413
		2	2	.109091	.160729	.225806	.303145	.389892	.481814	.702085	.862497	.984772
		2	1	.618182	.704456	.780921	.844858	.895257	.932692	.982187	.996603	.999954
		3	3	.024242	.042490	.070257	.109838	.162765	.229255	.441239	.665839	.936478
		3	2	.278788	.370490	.470186	.571693	.668501	.755010	.907191	.974481	.999273
		3	1	.787879	.852503	.902739	.939328	.964272	.980174	.996525	.999590	.999998
		4	4	.003030	.006318	.012349	.022679	.039239	.064130	.173920	.352870	.752125
		4	3	.087879	.137419	.202911	.283707	.376812	.477130	.717993	.884989	.992071
		4	2	.469697	.574314	.674030	.762683	.836194	.893048	.971422	.994755	.999944
		4	1	.893939	.933572	.960816	.978282	.988713	.994509	.999328	.999947	1.000000
		5	4	.015152	.028104	.048937	.080202	.124059	.181676	.379343	.608364	.916184
		5	3	.196970	.278754	.373910	.477078	.581389	.679826	.867793	.960461	.998684
		5	2	.651515	.744129	.822095	.883187	.927734	.957954	.991775	.998930	.999995
		5	1	.954545	.974384	.986469	.993316	.996918	.998676	.999883	.999994	1.000000
6	5	1	1	.454545	.541491	.626445	.705496	.775575	.834808	.934108	.978805	.998809
		2	2	.181818	.252715	.335437	.426368	.520638	.612938	.806303	.922870	.993644
		2	1	.727273	.801449	.862210	.909032	.942959	.966070	.992686	.998879	.999990
		3	3	.060606	.097666	.148638	.214222	.293277	.382699	.619010	.811932	.976644
		3	2	.424242	.527978	.629792	.723121	.803044	.867012	.960674	.991767	.999871
		3	1	.878788	.922767	.953595	.973764	.986067	.993061	.999089	.999922	1.000000
		4	4	.015152	.028309	.049587	.081646	.126721	.185978	.388619	.620623	.922938
		4	3	.196970	.279846	.376381	.480073	.586468	.685607	.872846	.963034	.998855
		4	2	.651515	.745167	.823766	.885060	.929464	.959345	.992273	.999029	.999996
		4	1	.954545	.974569	.986681	.993485	.997029	.998739	.999893	.999994	1.000000
		5	5	.002165	.004711	.009573	.018209	.032515	.054658	.156928	.331556	.738712
		5	4	.067100	.109770	.168693	.244278	.334459	.434668	.686091	.868898	.990698
		5	3	.391775	.499980	.608067	.708229	.794368	.863145	.961953	.992817	.999921
		5	2	.824675	.886191	.930650	.960424	.978891	.989494	.998655	.999891	1.000000
		5	1	.987013	.993628	.997088	.998763	.999512	.999822	.999990	1.000000	
11	1	1	1	.083333	.113895	.151613	.196713	.248973	.307668	.473532	.643007	.889059
		2	1	.166667	.216406	.273750	.337686	.406629	.478536	.655940	.803863	.960790
		3	1	.250000	.312429	.380739	.452961	.526747	.599592	.762670	.880400	.982107
		4	1	.333333	.403626	.477169	.551484	.624015	.692388	.833704	.924342	.991008
		5	1	.416667	.490800	.565247	.637437	.705031	.766147	.883831	.951780	.995311
		6	1	.500000	.574388	.646212	.713170	.773440	.825821	.920234	.969638	.997547

Table D

$$P(n';r,m,n,d)$$

m	n	r	n'	d = 0.	.2	.4	.6	.8	1.0	1.5	2.0	3.0
7	5	1	1	.416667	.504447	.592041	.675143	.750122	.814501	.924736	.975525	.998613
		2	2	.151515	.216960	.295825	.385123	.480226	.575620	.782420	.911833	.992614
		2	1	.681818	.763758	.832874	.887613	.928294	.956655	.990334	.998481	.999987
		3	3	.045455	.076384	.120777	.180175	.254392	.341105	.581582	.788492	.973010
		3	2	.363636	.467245	.573107	.673836	.763110	.836843	.949599	.989089	.999823
		3	1	.840909	.895676	.935548	.962577	.979621	.989609	.998565	.999872	1.000000
		4	4	.010101	.019895	.036577	.062943	.101681	.154689	.347678	.584071	.912124
		4	3	.151515	.225356	.315804	.418480	.526587	.632262	.843579	.952570	.998448
		4	2	.575758	.679809	.771502	.846333	.902877	.942434	.988361	.998464	.999993
		4	1	.929293	.958888	.977658	.988679	.994662	.997663	.999787	.999988	1.000000
		5	5	.001263	.002924	.006292	.012613	.023629	.041490	.130481	.295070	.712126
		5	4	.045455	.078795	.127649	.193859	.276962	.373709	.634269	.839934	.987820
		5	3	.310606	.416072	.527821	.637207	.736170	.818968	.946013	.989192	.999871
		5	2	.752525	.832310	.893451	.936695	.964913	.981893	.997489	.999783	1.000000
		5	1	.973485	.986329	.993449	.997089	.998803	.999546	.999972	.999999	1.000000
		6	5	.007576	.015432	.029259	.051787	.085833	.133668	.315612	.550471	.898229
		6	4	.121212	.187209	.271303	.370325	.478201	.587064	.815646	.941306	.997893
		6	3	.500000	.611857	.715061	.802925	.871964	.922047	.983206	.997651	.999988
		6	2	.878788	.926601	.958549	.978221	.989375	.995195	.999530	.999972	1.000000
		6	1	.992424	.996548	.998543	.999432	.999796	.999932	.999997	1.000000	1.000000
6	6	1	1	.500000	.588113	.671844	.747483	.812483	.865677	.950140	.985119	.999283
		2	2	.227273	.308384	.399452	.495560	.591036	.680466	.853946	.947231	.996438
		2	1	.772727	.840555	.893765	.932901	.959881	.977316	.995744	.998440	.999997
		3	3	.090909	.141377	.207408	.287982	.379843	.477881	.711017	.874148	.988054
		3	2	.500000	.605918	.704436	.789687	.858346	.909841	.977397	.996071	.999960
		3	1	.909091	.945011	.968788	.983410	.991756	.996176	.999588	.999972	1.000000
		4	4	.030303	.053955	.089869	.140462	.206711	.287516	.527002	.749716	.965233
		4	3	.272727	.372098	.480504	.589988	.692440	.781354	.927227	.983160	.999694
		4	2	.727273	.811546	.877631	.925534	.957622	.977484	.996596	.999671	.999999
		4	1	.969697	.984123	.992258	.996494	.998529	.999429	.999962	.999998	1.000000
		5	5	.007576	.015486	.029446	.052238	.086726	.135209	.319428	.556077	.901790
		5	4	.121212	.187595	.272257	.371958	.480495	.589859	.818464	.942907	.998015
		5	3	.500000	.612408	.716035	.804112	.873148	.923066	.983628	.997745	.999989
		5	2	.878788	.926798	.958798	.978437	.989528	.995289	.999546	.999974	1.000000
		5	1	.992424	.996562	.998556	.999440	.999800	.999934	.999997	1.000000	1.000000

Table D

P(n';r,m,n,d)

m	n	r	n'	d = 0.	.2	.4	.6	.8	1.0	1.5	2.0	3.0
6	6	6	6	.001082	.002556	.005598	.011404	.021673	.038548	.124428	.286652	.706096
		6	5	.040043	.070858	.116911	.180438	.261440	.357072	.619905	.831893	.987039
		6	4	.283550	.387552	.500099	.612346	.715587	.803223	.940278	.987890	.999854
		6	3	.716450	.804976	.874354	.924399	.957618	.977905	.996873	.999726	1.000000
		6	2	.959957	.978960	.989747	.995377	.998075	.999261	.999953	.999998	1.000000
		6	1	.998918	.999576	.999847	.999949	.999984	.999996	1.000000	1.000000	1.000000
12	1	1	1	.076923	.105934	.142039	.185564	.236395	.293921	.458563	.629568	.882881
		2	1	.153846	.201466	.256920	.319335	.387328	.458885	.638187	.790839	.957012
		3	1	.230769	.291107	.357902	.429335	.503135	.576794	.744707	.868983	.979677
		4	1	.307692	.376395	.449247	.523839	.597584	.667988	.816557	.914652	.989397
		5	1	.384615	.458086	.533013	.606776	.676876	.741189	.868000	.943721	.994230
		6	1	.461538	.536600	.610375	.680362	.744448	.801089	.905993	.963062	.996823
7	6	1	1	.461538	.551587	.638973	.719434	.789767	.848191	.942779	.982767	.999165
		2	2	.192308	.268747	.357378	.453687	.551899	.646047	.834524	.939317	.995854
		2	1	.730769	.807269	.869070	.915776	.948782	.970590	.994307	.999235	.999995
		3	3	.069930	.113385	.172720	.247995	.336880	.434765	.678212	.856866	.986133
		3	2	.437063	.546204	.651895	.746801	.825856	.886977	.970472	.994713	.999944
		3	1	.877622	.923771	.955500	.975712	.987631	.994131	.999337	.999953	1.000000
		4	4	.020979	.039384	.068835	.112363	.171903	.247444	.484952	.720118	.959886
		4	3	.216783	.309329	.415533	.527904	.637619	.736573	.907851	.977839	.999580
		4	2	.657343	.755202	.835788	.896901	.939571	.966999	.994708	.999466	.999999
		4	1	.951049	.973331	.986505	.993672	.997257	.998902	.999922	.999997	1.000000
		5	5	.004662	.010150	.020448	.038234	.066569	.108312	.279041	.516047	.888243
		5	4	.086247	.141384	.216075	.309055	.415554	.527950	.779307	.927301	.997316
		5	3	.412587	.529023	.642930	.745198	.829491	.893305	.975668	.996461	.999981
		5	2	.820513	.886671	.933422	.963711	.981691	.991465	.999112	.999945	1.000000
		5	1	.983683	.992216	.996573	.998611	.999483	.999823	.999992	1.000000	1.000000
		6	6	.000583	.001479	.003460	.007489	.015041	.028126	.100769	.250874	.676902
		6	5	.025058	.047500	.083436	.136260	.207665	.296589	.561751	.796251	.983072
		6	4	.208625	.303122	.412711	.529216	.642849	.744632	.916359	.981924	.999762
		6	3	.616550	.723589	.813507	.882664	.931358	.962748	.994241	.999458	.999999
		6	2	.922494	.956894	.977829	.989481	.995406	.998157	.999871	.999995	1.000000
		6	1	.995921	.998300	.999349	.999771	.999927	.999978	.999999	1.000000	1.000000

299

Table D

P(n';r,m,n,d)

m	n	r	n'	d = 0.	.2	.4	.6	.8	1.0	1.5	2.0	3.0
7	7	1	1	.500000	.590889	.676956	.754168	.819852	.872908	.954917	.987238	.999454
		2	2	.230769	.315776	.411079	.511028	.609251	.699892	.869948	.955938	.997428
		2	1	.769231	.840063	.895148	.935106	.962137	.979194	.996425	.999578	.999998
		3	3	.096154	.151175	.223125	.310335	.408518	.511436	.745965	.897766	.991919
		3	2	.500000	.610504	.712605	.799795	.868648	.918971	.981601	.997178	.999979
		3	1	.903846	.942775	.968220	.983569	.992105	.996480	.999665	.999980	1.000000
		4	4	.034965	.062998	.105513	.164876	.241363	.332538	.587873	.802332	.978419
		4	3	.279720	.385454	.500118	.614317	.718876	.806994	.942652	.988551	.999857
		4	2	.720280	.809749	.879179	.928573	.960784	.980039	.997369	.999787	1.000000
		4	1	.965035	.982019	.991456	.996257	.998492	.999442	.999968	.999999	1.000000
		5	5	.010490	.021672	.041326	.072978	.119808	.183624	.407760	.658458	.945986
		5	4	.132867	.207828	.302754	.412687	.529277	.642679	.861450	.963558	.999210
		5	3	.500000	.618835	.727313	.817721	.886537	.934399	.988055	.998662	.999996
		5	2	.867133	.921205	.956799	.978162	.989845	.995664	.999647	.999984	1.000000
		5	1	.989510	.995329	.998092	.999287	.999756	.999924	.999997	1.000000	1.000000
		6	6	.002331	.005541	.012097	.024336	.045274	.078187	.227553	.459083	.865146
		6	5	.051282	.091550	.151067	.231331	.330262	.441904	.717699	.900108	.995896
		6	4	.296037	.409274	.530420	.648500	.753429	.838512	.959152	.993527	.999960
		6	3	.703963	.800335	.874981	.927585	.961306	.980969	.997775	.999849	1.000000
		6	2	.948718	.973541	.987463	.994558	.997841	.999219	.999959	.999999	1.000000
		6	1	.997669	.999102	.999684	.999899	.999971	.999992	1.000000	1.000000	1.000000
		7	7	.000291	.000802	.002021	.004681	.010002	.019782	.079638	.216173	.645528
		7	6	.014569	.029880	.056384	.098232	.158627	.238462	.499372	.754708	.977942
		7	5	.143065	.223508	.324430	.439753	.559914	.674222	.884264	.973221	.999613
		7	4	.500000	.621261	.731543	.822770	.891427	.938454	.989528	.998936	.999998
		7	3	.856935	.915277	.953746	.976788	.989318	.995501	.999652	.999985	1.000000
		7	2	.985431	.993485	.997336	.999006	.999662	.999896	.999996	1.000000	1.000000
		7	1	.999709	.999903	.999971	.999992	.999998	1.000000	1.000000	1.000000	1.000000

Index

Accuracy of computations, 13
  inequalities for checking, 15
  non-linear relationships for checking, 17
Algorithm, modified midpoint composite
  quadrature, 10

Back-recursive rule, 12, 13, 14
Barr, D. R., 58
Bauer, F. L., 11, 18
Bechhofer, R. E., 58, 59
Berk, R. H., 57
Bradley, R. A., 1, 2, 43, 45, 47, 48, 56, 57
Brown, G. W., 26

c_1, c_2, 2, 3, 23, 25, 71, 74
c_1 test, see Normal scores test
CDC 1604 computer, 7, 19
Chernoff, H., 42
Crossovers, 4, 71, 74

Davies, O. L., 56
Desu, M. M., 60
Dixon, W. J., 25
Dwass, M., 28

Efficiency, asymptotic relative, 29, 42
  Hodges-Lehmann, 29, 42, 71
  tables of, 29

Gaunt, J. A., 11
Ghosh, J. K., 57
Graybill, F. A., 26
Greenwood, J. A., 24
Gupta, S. S., 5, 58

Hall, W. J., 57
Hartley, H. O., 24

Henrici, P., 11
Hodges, J. L., Jr., 11, 29, 61
Hoeffding, W., 25

Integration, numerical, see Quadrature
Interpolation in tables, 19

Kiefer, J., 58
Klotz, J., 2, 5, 11, 14, 26
Kolmogorov, A. N., 27
Kolmogorov-Smirnov test, 21, 28
  efficiency of, 29, 42
  power of, 27, 218

Lehmann alternatives, 43, 47, 56, 57
Lehmann, E. L., 29, 61
Likelihood ratio, 44
  rank-sum, 46, 258

Mahamunulu, D. M., 58, 59
Mann, H. B., 23
Martin, D. C., 1, 56
Massey, F. J., 27
Median test, 21, 26, 28
  efficiency of, 29, 42
  power of, 26, 218
Merchant, S. D., 2, 57
Milton, R. C., 5
Mood, A. M., 26, 42
MPRT (most powerful rank test), 1, 21, 27
Multiple decision procedures, see Selection
  procedures

Noether, G. E., 29
Normal scores test, 21, 25, 28
  efficiency of, 29, 42
  power of, 25, 218

Order statistics, 1, 2, 25
expected value of, 2

Pitman, E. J. G., 29, 42
Power of two-sample tests, 1, 21
Kolmogorov-Smirnov test, 27, 218
median test, 26, 218
normal scores test, 25, 218
Student's *t* test, 24, 25, 26, 27, 218
Wilcoxon test, 23, 218
Probability ratio, *see* Likelihood ratio

Quadrature, 6
composite formulas, 6, 7
conditional integration, 6
extrapolation to the limit, 11
modified composite midpoint formula, 7
Romberg, 18, 24

Rank order, definition of, 1
complement of, 2, 71
graph of, 61, 64
total number of, 4, 71, 73
transpose of, 2, 71
transpose-complement of, 2, 3, 71
Ranking procedures, *see* Selection procedures
Rhodes, L. J., 2, 43, 45, 47, 48
Richardson, L. F., 11
Rizvi, M. H., 58
Romberg integration, 11, 18, 24
Rutishauser, H., 11

Savage, I. R., 2, 3, 4, 10, 12, 14, 15, 42, 43,
    57, 64
Savage, L. J., 57
Selection procedures, 1, 58
tables for, 291
Sequential rank tests, 1, 43, 44
ASN function, 46, 47, 48
OC function, 46, 47
SCR test (sequential, configural rank

test), 45, 46, 51
modified, 57
SRS test (sequential, rank-sum test), 45,
    46, 51, 56, 258
modified, 57
Type I and II errors, 44, 45
Smirnov, N. V., 27
Sobel, M., 2, 4, 15, 58, 59, 60, 64
Stiefel, E., 11
Student's *t* test, 21
power of, 24, 25, 26, 27, 218

Teichroew, D., 3, 5, 11, 13, 25, 59
Terry, M. E., 3, 25
Tsao, C. K., 25
Two-sample tests, admissibility of, 1
critical regions, 22
efficiency of, 29, 42
power of, 1, 21
comparison of tests by, 28
computation of, 23
tables of, 218
Type I and II errors, 22, 28, 44, 45

van der Vaart, H. R., 42

Wald, A., 44, 45, 46, 47, 48
Westenberg, J., 26
Whitney, D. R., 23
Wijsman, R. A., 57
Wilcoxon, F., 1, 3, 23, 45, 47, 48, 56, 57
Wilcoxon test, 21, 23, 28
efficiency of, 29, 42
power of, 23, 218
Wilcoxon two-sample statistic, 3, 23, 45, 46,
    71, 74
distribution of, 258
Wilf, H. S., 11, 18
Witting, H., 42
Woodworth, G., 2, 4, 15, 64